ENERGY PRIMER
SOLAR, WATER, WIND, AND BIOFUELS

UPDATED and REVISED EDITION

EDITED BY

RICHARD MERRILL
THOMAS GAGE

We dance round in a ring and suppose
But the Secret sits in the middle and knows.
— Robert Frost

Abstract: *The Energy Primer is a comprehensive, semi-technical book about renewable forms of energy — solar, water, wind and biofuels. The biofuels section covers biomass energy, agriculture, aquaculture, alcohol, methane and wood. Additional sections on energy conservation and solar architecture stress the need to conserve energy resources and utilize passive solar energy designs in building. The focus is on small-scale systems which can be applied to the needs of the individual, small group, or community. More than ¼ of the book is devoted to reviews of books and hardware sources. Hundreds of illustrations and a dozen original articles are used to describe the workings of solar water heaters, space heaters and dryers; waterwheels and water turbines; windmills and wind generators; wood burning heaters; alcohol stills; and methane digesters. The final section of the book describes the concepts and potential of integrating renewable energy systems together with descriptions of groups working in this area.*

FRONT COVER
Photo courtesy of the High Altitude Observatory and NASA. Sun's hot outer atmosphere, or corona, color-coded to distinguish levels of brightness, reaches outward for millions of miles. A coronography, one of Skylab's eight telescopes, masked the sun's disk, creating artificial eclipses. It permitted 8½ months of corony observation, compared to less than 80 hours from all natural eclipses since use of photography began in 1938.

PUBLISHED IN COOPERATION WITH PORTOLA INSTITUTE, INC.
A DELTA SPECIAL/SEYMOUR LAWRENCE

A DELTA SPECIAL / SEYMOUR LAWRENCE
Published by
Dell Publishing Co., Inc.
1 Dag Hammarskjold Plaza
New York, New York 10017

This is a revised and updated edition

Copyright © 1974, 1978 by Portola Institute

Delta ® TM 755118, Dell Publishing Co., Inc.

ISBN: 0-440-52311-7

Printed in the United States of America
First Edition published 1974
6 printings—82,000 copies
Second Delta Printing—July 1978

PORTOLA INSTITUTE

Portola Institute is a non-profit tax exempt corporation. Established in 1966 as a center for "educational risk taking" its purpose has been to promote research and social action projects, businesses and learning environments which stress learning in the world — "learning by doing." Portola Institute operates with a very small staff as a project initiating and facilitating foundation rather than a fund granting foundation; as such it has no money to give away.

Some of Portola Institute's projects are listed below:

WHOLE EARTH CATALOG
BIG ROCK CANDY MOUNTAIN
DESCHOOLING/DECONDITIONING
Whole Earth Truck Store
Briarpatch Network and BRIARPATCH REVIEW
Ortega Park Teachers Laboratory
Nature Explorations
San Andreas Health Council

Currently, Portola Institute is focusing attention on the field of appropriate technology. Research is being conducted in Arusha, Tanzania in cooperation with the Swedish Government and work is also being done in Menlo Park, California. Portola Institute is available to groups and individuals in need of assistance in developing new projects. Portola Institute welcomes substantial contributions to aid its activities.

PORTOLA INSTITUTE
558 Santa Cruz Avenue
Menlo Park, California 94025

HOW TO ORDER

A considerable amount of access information is provided in the Energy Primer - access to the information, tools, and hardware with which to build renewable energy systems.

—The reader will find addresses of publishers and suppliers under book and hardware reviews.

—If "or Whole Earth Truck Store" appears under the address of the supplier, the item may be ordered from:

Whole Earth Truck Store
558 Santa Cruz Avenue
Menlo Park, California 94025

—Usually, the price of the item is listed. **All items are postpaid by the supplier at the listed price unless otherwise noted.** *Shipping on items not postpaid, especially large items, can cost a good deal of money. Consult the supplier or your local post office for postage rates.*

—State sales tax must be added to the listed price or cost if the supplier and the customer are within the same state. Periodicals are exempt from sales tax. (California sales tax is 6%).

—Send a check or money order for the items you wish to purchase. Cash has been known to get lost in the mail. The Post Office charges extra for C.O.D.

—Many names and addresses are given for further information. The group may or may not have any product to sell. If you are requesting information, send a stamped, self-addressed envelope and perhaps a quarter or more to help defray the cost to the group. Small groups have gone broke trying to answer hundreds of inquiries. Besides, your quarter is likely to get your letter answered.

CONTENTS

CREDITS

EDITORS

Richard Merrill (R.M.)

Thomas Gage (T.G.)

ASSOCIATE EDITORS

Yedida Merrill . . . . . *Copy Editor, Water*
Michael Riordan . . . . *Energy Conservation,*
Solar Architecture
Berkeley Solar Group . . *Solar*
Chuck Missar (C.M.) . . . *Solar*
James Bukey (J.R.B.) . . *Wind*

CONTRIBUTORS

Solar
Dean Anson
Jim Augustyn
Chip Barnaby
David Burton
Harry Burris
Phillip Caesar
Doug Daniels
Fred Godfrey
Judith Jennings
Lee Johnson
Larry Lewis
Dian Missar
Marshall Merriam
Lynn Nelson
Carolyn Pesko
Judy Raak
RAIN Magazine Staff
Paul Shippee
Gary Smith
Priscilla Thomas
Dave Weiman
David Wentworth
E.N. Whitney
Wu Wicklein
Bruce Wilcox
Jerry Yudelson

Biofuels
Helen Bommarito
Evan Brown
Bill Campbell
Nancy Davenport
Len Dawson
Bill Day
Henry Esbenshade
Isao Fujimoto
Stuart Hill
Kurt Kline
John Jeavons
Max Kroeschl
Jim Lassoie
Larry Lewis
Perry McCarty
Darryl McLeod
Bill McLarney
R.R. Morrow
Steve Nelson
Mike Perelman
Ann Poole
Steve Serfling
Andrew Shapiro
Jay Shelton
Sim Van der Ryn

Energy Conservation
Larry Lewis
Michael Perelman
RAIN Magazine Staff
Lee Schipper

Solar Architecture
Bruce Anderson
Berkeley Solar Group
Alan Edgar
Larry Lewis

Water
Nancy Bellows
Bill Delp
Bruce Franklin
Byrd Helligas
Vic Marks

Wind
Skip Allen
Catherine Brown
Jim DeKorne
Richard Dickey
Lou Divone
Karen George
Lee Johnson
Don Kozak
Hans Meyer
Alvin Miller
Albert Nunez
Jack Park
Carolyn Pesko
Dick Schwinn
Carolyn Sheehan
Ken Smith
Alan Sondak
Ben Wolfe

Integrated Systems
John Jeavons
RAIN Magazine Staff

PRODUCTION

Layout and Design
Alan Edgar
Richard Gordon
Jeanne Campbell

Typesetting
Trudy Smith
Phyllis Grossman
Mary Lamprech
Evelyn Eldridge

Artwork
Beth Amine
Jim Phillips
Robin Saunders
Dennis Dahlin
Cayla Werner
Lynn Smith
Holly Bowers
David Baylon
Edward A. Wong
James Bukey
Connie Williams

Photography
Alan Edgar
John Gaylord
Lynn Houghton
Guy Bommarito
Buddy Metzger

Indexing
Larry Lewis
Yedida Merrill
Dian Missar

Conversion Factors
Kim Mitchell

Miscellaneous
Lottie Arko
Multi Fassett
Bob Gaylord
Becky Johnson
Elisabeth McAllister
Andre Merrill
Riana Merrill
Jacqueline Meyer
Ray Oszewski
Bob Parks
Richard Raymond
Wally Thompson
the Whole Earth
Truck Store Staff

PREFACE

This new edition of the *Energy Primer* was created to fulfill three goals. First, numerous advances in small scale renewable energy systems have been made since early 1974 when the first edition was published. Thus the information in that first edition was out–of–date and lacking in descriptions of state–of–the–art technical innovations and theories. This edition is an effort to make the *Energy Primer* current. To this end we have rewritten the wood article and the methane article and have vastly overhauled all the literature and hardware review sections. The solar hardware reviews, especially, have been put in a readable matrix format with representative manufacturers of all types of solar units listed. Second, this updated edition includes a meaty section on energy conservation. Both theory and practical applications are discussed in two articles that we hope our readers find to their liking. We have also removed a group of articles in the Architecture section of the first edition and replaced them with one article on passive applications of solar energy in building design. We have renamed that section Solar Architecture and moved it to a place of prominence in this edition. Third, we learned a great deal from our readers' and reviewers' comments and suggestions. We incorporated a lot of advice into this updated edition. We are particularly indebted to the people whose names are listed under the Credits, without whose long hours this book would not have been possible. We, of course, are alone responsible for any errors of fact or judgement.

One additional note. Even as we write this Preface, the Updated and Revised Edition of the *Energy Primer* becomes dated and obsolete. Changes are occurring too fast now, and we suspect that general access-information catalogues have done their work and will not be useful for much longer. The valuable books for the future should be those that focus on detailed how–to information for local resources and regional climates. Information, as well as energy production, needs to be decentralized.

Future Challenges

Probably the most important energy development since 1974 has been the growing acceptance of the potentials for energy conservation. Prior to 1974 it was widely believed that consuming energy by itself would lead to a better "standard of living". Indeed, the concepts that large centralized power generation promotes economic growth and that energy expansion produces jobs have become axioms among planners of public policy. Yet there is increasing evidence that these assumptions are no longer valid:

1) <u>Comparison with other countries.</u> A number of studies have shown that Americans use more than two times the energy per person as do Swedes, West Germans and the Swiss. Yet the living standards in these European countries are comparable or higher than this country's, and unemployment is much lower.

2) <u>The social advantages of saving energy.</u> Other recent studies have shown that by combining a variety of energy conservation strategies, the U.S. could reduce its energy use by 30–50 per cent without adverse social or economic effects. The important 1974 Ford Foundation Study, A TIME TO CHOOSE, gave a formal legitimacy to this conclusion by noting:

"... the U.S. can balance its energy budget, control pollution, and avoid reliance on insecure oil sources abroad by slowing its growth rate in energy consumption. In fact, our energy growth can be trimmed to about 2 per cent a year or, in time, to zero without adversely affecting the economy."

In fact the conclusion is inescapable: it is becoming cheaper to save energy than to exploit new energy resources . . . a new axiom for planners.

3) <u>The potential for creating more jobs.</u> The energy industry, utility and oil companies are the least labor intensive industries in the industrialized world. Once the well has been drilled or the power plant constructed the employees needed for maintenance are few. Furthermore, energy is a replacement for labor. In economic terms the choice is capital or labor and increasingly firms are choosing the capital (read energy) intensive alternative. But when we consider all the strategies for energy conservation of plugging leaks and designing and manufacturing new solar energy systems it seems obvious that a new broad based industry that will create more jobs and open up opportunities for small businesses is available at our footsteps.

4) <u>The emergence of new standards of living.</u> Our culture is accustomed to wasting cheap energy. Yet energy will never be cheap again. As a world we can only evolve successfully if we adopt a whole new attitude towards energy consumption; one that makes efficient use of energy and equitable distribution of energy resources first–order measures of social progress.

5) <u>The passive solar alternative.</u> Passive solar heated buildings use the design, orientation and landscaping of the structure to store and ventilate solar heat. Instead of using energy resources or renewable energy systems they use building materials and engineering and design skill. They can be cheap, they work and they may well prove to be the most economic way to build solar homes.

6) <u>Conservation and renewable energy systems.</u> There is no purpose in solar heating an uninsulated building. We can't feed excessive energy appetites with any energy source. As we note in the Introduction, the use of solar and other renewable energy systems goes hand in hand with strategies for conserving energy and, indeed, is the most important measure for making them practical.

Perhaps more important than any changes in technology is our need to change our social and economic relations. The poor in the third world and the poor in America will be the hardest hit by rising costs for fuel because they have the least money to pay and their demand for it will not fall much as prices rise. Technology alone will not make them better off. Indeed, technology is neither the saviour nor the villian but a tool that can be used to provide more and better from less. But if the control of renewable energy systems is centralized by profit motivated individuals and institutions with no regard for just and equitable distribution of wealth we will have all gained very little. We will have substituted a solar economy for a fossil fuel economy and failed in the meantime to increase the quality of human existence.

(fall, 1977)

Richard Merrill
Thomas Gage

INTRODUCTION

This book was written to fill a void. The groups involved with the *Energy Primer* had been receiving numerous inquiries about methods of supplying energy needs that could be implemented by individuals and small groups. Solar/wind devices and methane digesters seemed to be particularly popular, but there were also the inevitable questions about "organic" gardening and farming, water power, wood and alcohol. There were general questions too: "How can I feed and fuel those around me without relying so much on outside power sources that pollute and that are rising in cost and dwindling in supply?" People wanted to know about food and fuel supplies that came from renewable sources of energy rather than non-renewable oil wells and nuclear reactors.

But it seemed that people were asking much more than just: "*How* can I build a solar collector," or, "*Should* I build a solar collector?" Most questions clearly reflected a growing dissatisfaction with a culture that allowed fewer and fewer options with regard to our control over life's everyday needs. The supermarket and the wall plug are still the major supply sources and there seems no end to it.

So we set out to write a book about renewable energy systems that people could use for themselves; a description of how they work, their limitations and potentials and the hardware and techniques necessary to grow, build and maintain them. But we soon realized that there were limitations. For one thing, the different food and energy systems were in varying states of development and redefinement. New ideas were being offered every day about solar collectors, scaled-down waste systems, wind generators, etc. New companies were starting out all the time in response to the growing demands for devices that harness solar and wind power. Inevitably the book would soon be obsolete. In addition, renewable energy systems were, by their very nature, geared to the local conditions of climate, economy, geography and resources. It made little sense to try and describe endless possibilities and designs that were already described in widely scattered community and regional publications, technical journals, "underground" brochures, survival texts and energy magazines both funky and slick. What we wanted was a sourcebook that brought basic information together, not a cookbook.

And we needed something, too: a perspective to what we were doing.

Energy Scarcity, Conservation and Practical Alternatives:

For generations our culture has been enjoying the benefits of a cheap, easily available energy source . . . fossil fuels. Today, practically all of our luxuries and necessities are totally dependent on coal, natural gas and oil. The fact that reserves of these fossil fuels are limited has become a platform for politicians, a bargaining point for diplomats, a lucrative commodity for the rich and a glaring reality for everyone.

> *Geochemical evidence suggests that every year about 28 million tons of carbon go into the formation of new fossil sediments. Current consumption of fossil fuels is about 6 billion tons of carbon per year. The rate of consumption is therefore over 200 times the rate of deposition. On this basis we can say that fossil fuels are limited and non-renewable.*
>
> —I.T. Rosenqvist
> 1972 Conference on Energy & Humanity

Because we have built our culture and standard of living on easy energy, conventional wisdom pushes harder and harder to find "alternative" energy sources capable of feeding our excessive needs. The most obvious alternative has been the search for more inaccessible deposits of fossil fuels . . . further offshore and deeper in the ground. But very soon the cost of retrieving new reserves will exceed the benefits provided by them. The search for more oil can only be a short term solution with long term implications for the environment.

Further down the line, many see the development of nuclear power as the answer to our energy problems. But we should at least pause for a moment. The development of nuclear power is not only a mistake, but a self-destructive response to a complex problem. In the first place, uranium itself is a limited resource; reserves can only hope to last for a few generations at best. Moreover, there is the strong possibility that the wide-scale use of nuclear power plants will simply inaugurate an era of world wide nuclear armament and global conditions in which no one can afford to make a mistake . . . political or technical. And finally there are the inevitable nuclear wastes generated and *left behind* by nuclear power plants that threaten the very existence of life as we know it. Of most concern is plutonium, a by-product of breeder reactors and the deadliest, and one of the most persistent elements in the known universe. Plutonium has a half-life of 24,000 years, and to our knowledge there is no way of safely containing anything . . . let alone plutonium . . . for 24,000 years (rockets to the sun and salt mines notwithstanding). For a few generations of more easy energy we give to our future generations the consequences and burdens of plutonium and other nuclear wastes. We view this as insanity.

If we dismiss nuclear fusion as being an undeveloped alternative with its own problems of radioactive waste (albeit far less than those of fission), we are left with one alternative energy source that could supply much of our needs . . . solar power and all of its manifestations in the wind, water and plants.

There are a number of ambitious ideas for harnessing solar energy. For example, some people have suggested that we could cover the southwest deserts with solar reflectors or send satellites into orbit to microwave solar energy back to earth. There are also plans for building bigger hydroelectric dams and massive digesters to recycle manure from oversized feedlots. There are even schemes for covering the mid-west plains and offshore areas with wind generators, and for growing high-yield plants in vast areas of marginal land to fuel power stations. Certainly these alternatives are infinitely more desirable than nuclear energy, and there is little doubt that some of them will be developed to help supply future energy needs. However, in perspective, they must be viewed as nothing more than extensions of the growing tendency to centralize the generation and control of energy and to alienate people further from the natural forces that support them. A central power station, regardless of its source of energy, still requires elaborate transmission grids to disperse the energy. To some degree this may be realistic, but as *the only* way of providing energy in an increasingly unstable social order it leaves few options for people to adopt.

More and more the energy resources of the world are coming under the control of international corporations and energy cartels. Growing scarcities of energy and natural resources will continue to determine major political policies in the future. To balance this trend, we need to diversify and disperse the physical energy base and dilute the growing economic power bases . . . a trend that will help us all. *The most obvious way to do this is to develop and adopt scaled-down renewable energy systems that are utilized where they are needed and designed for local environments and requirements.* The new politics of self-sufficiency . . . relying on ourselves and our own decentralized energy resources . . . will conflict with the present politics of centralized institutions and industry. Hopefully, in time, it will come to supplement this tradition rather than conflict with it. The important thing is that we generate as many options as we can for a future whose course grows more uncertain every day.

> *"It's easier and cheaper to save the energy we get from conventional sources than it is to earn (generate) energy from newer, more expensive sources such as the wind and sun. For example, it's cheaper and easier to insulate a home than to produce energy (from renewable sources) to heat an uninsulated dwelling. . . If we consume all of our fossil fuels in our cars and 'space-hippy' vans, we'll never see the solar-based society to which we must move if we want to survive."*
>
> —Lee Johnson and Ken Smith

We should have no illusions about local energy systems. Our exaggerated needs *cannot* be supplied by solar, wind, water and biofuel energy alone. The prerequisite to using any renewable energy system is CONSERVATION. Without conservation, techniques and devices for using renewable energy will always seem impractical and will always make little economic sense.

It's not very glamorous to deal with energy conservation. Most people would rather concentrate on new pieces of hardware than cultivate living habits that make solar collectors feasible. Unfortunately most strategies for conservation are based on simplifying what we have rather than creating radically new living patterns. Car pools, lowered thermostats, simple appliances and efficient insulation are all important methods of reducing energy consumption. But by themselves they mean little unless they become part of a personal plan to change our standard of living, recycle our materials and wastes, and make do with less. We needn't be austere, only sensible about our habits. Then and only then can the energies from the sun, wind and garden begin to support our needs. *First we must minimize our needs,* then we can start changing *our hardware.*

There are other realities too. Fossil fuels are an indispensible ingredient for organic chemicals and a valuable fuel for a few high energy processes like ore smelting. At the present time, coal, oil and natural gas are the only practical energy resources that can provide steel for windmills, digester tanks or circulating pumps, and plastic or glass for solar collectors, greenhouses, etc.

Generally speaking, renewable energy devices are still locked into a costly technology that requires high energy resources like fossil fuels to provide construction materials. This is a reality and shouldn't be hidden in the "something for nothing" attitude that renewable energy systems seem to generate. However this needn't always be so. Future energy policies *can* be redirected so that material and construction designs will be oriented for posterity and not planned obsolescence.

The problem is that we just don't know what the trade-offs really are. Perhaps it will take several years for a wind generator to "pay back" the energy costs of making it, and there is some evidence to indicate that this may be so. At what scale are renewable energy systems really practical. . . the household, village, community or city? These and many more questions remain to be asked. As we observe, think and change, we will often be very much on our own. We will be our own "experts" and we will be our own "scientists." In this process we will develop many values and perfect many processes that can and should be shared. A desire to facilitate this exchange of information has been one of the moving forces behind the creation of this book.

—R.M., —C.M., —T.G., —J.R.B.

ENERGY Conservation

ENERGY AND ITS CONSERVATION

Richard Merrill

An obvious result of the "energy crisis" has been an increased awareness of the need to conserve energy. Yet it seems that the real meaning and implications of energy conservation are still not widely understood. Many people still consider energy conservation in terms of modest inconvenience to their life style . . . like turning down the thermostat, driving in car pools etc. Others feel that incentives and programs for energy conservation will only increase unemployment, halt economic growth and generally put the brakes on our rate of social "progress." This attitude stems from our traditional belief that the more energy we consume, the more affluent we become.

It is true that widespread energy conservation will drastically change the patterns of our lives: the way we eat; the way we travel; the criteria we use to separate needs from desires and austerity from affluence. But what seem like unpopular choices for changing the way we live now will, in fact, become desirable and will improve the quality of our lives in the near future. This is because such changes are inevitable. The era of cheap "limitless" energy is over. Nuclear power and fossil fuels may string us out for a while, but their economic, political and environmental implications are looking less favorable all the time. Energy conservation can provide more people with the amenities of life, help to increase employment, redistribute existing wealth and resources, and make possible the integration of solar energy into our lives.

This is easily said, but as we shall see, there is growing evidence to support such claims. First we need to understand what energy is and that it is not a quantity that can be "saved." Certain limited energy resources like fossil fuels can be conserved; but what energy conservation really means in its broadest sense is that we reduce the rate at which we consume energy and generate waste heat. Towards this end there are three broad strategies we need to implement: 1) a change in our habits of energy consumption to reduce our irrational need for energy, 2) an increase in the efficiency at which we obtain "available work" from our technologies and life support systems, and 3) an increase in the use of renewable energy systems. To make these points more clear, let's take a brief look at some basic concepts of energy.

CONCEPTS OF ENERGY

Energy Defined

We tend to use several terms rather loosely when we talk of energy. "Force," "work" and "power," for example, have very precise technical meanings related to energy in spite of their popular usage. Defining them will help us define energy.

The most fundamental unit of energy is force, of which there are four kinds: gravity, electromagnetic (X-rays, visible light, infra-red heat waves, radio waves etc.), weak interactions (responsible for radioactive decay) and strong interactions (that bind atomic nuclei together). Newton was the first to define force as the mass of a body times its acceleration (rate at which it goes faster). We experience force usually as a push or pull. We can feel the force of a magnet, although we take for granted the forces that bind atoms together and prevent us from falling through the floor. The most familiar kind of force is gravity which gives us weight (the mass of our bodies times the acceleration due to gravity).

Having defined force, the rest of the terms can be defined rather easily (Table I).

TABLE I	DEFINITIONS AND EXAMPLES OF THE TERMS: WORK, POWER AND ENERGY.		
		EXAMPLE	MEASURE
work = force x distance (or energy)		275 lb stone lifted 2 ft off the ground	2 ft x 275 lbs = 550 foot-pounds (ft-lbs)
power = force x velocity (rate of work)		225 lb stone lifted 2 ft off ground in 2 seconds (1 ft/second)	550 ft-lbs x 1 ft/second = 1 horsepower 1 horsepower = 745 watts (mechanical power) = (electrical power)
energy = power x time (potential for work)		225 lb stone lifted off ground at 1 ft per second for 1 hour	1 horsepower x 1 hour = 1 horsepower-hour 1 horsepower-hour = 745 watt-hours (mechanical energy) = (electrical energy)

We can see from Table I that power is the rate at which energy flows or is used. A machine with a low power rating can often do the same job as one with a high power rating, but it will take a longer time. A man, a horse or a tractor can plow a field, but the tractor will do it in the least time and consume the most energy per unit time (power). We can look at this another way. Consider the power output (in kilowatts) of various "prime movers" (any device that converts food, fuel or force to work or power) throughout history (1). First there was human power (0.1 kw), then the horse (0.5 kw), windmill (15 kw), waterwheel (300 kw), steam engine (2,000 kw), internal combustion engine (10,000 kw), gas turbine (80,000 kw), water turbine (100,000 kw), and steam turbine (1,000,000 kw). Clearly the increasing rate at which we have been able to use energy is a trend deeply rooted in our history.

Energy can now be defined as the measure of the ability of a device or system to do work. Hence it is not surprising (but often confusing) that energy and work are measured in similar units and used interchangeably. However, unlike work, energy comes in several forms and both English and metric equivalents. There are, as a result, several units of energy. The common units of work are the foot-pound (English units) and the Joule (metric system). The common units of energy are: British Thermal Unit or BTU (Heat and chemical energy); horsepower-hour, foot-pound and joule (mechanical energy) and the kilowatt-hour (electrical energy). For more on the units of work, power and energy see the Table of Conversions at the end of the ENERGY PRIMER.

The Laws of Thermodynamics

As noted above, energy can be defined as the capacity to do work. This seems a rather narrow definition since, as William Blake has noted: "Energy is eternal delight"; there are, no doubt, many forms of energy, both cosmic and spiritual, that we have yet to fully recognize (see, for example, ref. 2). But whatever form energy takes, it is, as far as we know, governed by the laws of thermodynamics . . . two of the fundamental principles of our physical existence. We cannot escape the implications of these laws, although much of human suffering has been the result of a failure to recognize and appreciate the limitations they set on human activities. They can be stated in several ways:

FIRST LAW: Energy cannot be created or destroyed . . . it can only change from one of its forms such as heat, light, mechanical, chemical or electrical to another (see Tables II and V).
SECOND LAW: All energy in the universe is constantly being degraded into a lower form of heat energy.

Taken together the two laws of thermodynamics state that although the quantity of energy in the universe is constant, the quality of energy is changing into a less useful form, the random disorderly motion of molecules (or simply heat). In other words, a system PLUS ITS SURROUNDINGS tends toward increasing disorder (unless, of course, it is fed with outside energy). The word "entropy" is used to identify this increasing disorder and universal degradation into heat. We can now state the laws in another way:

FIRST LAW: The total energy of the universe is constant.
SECOND LAW: The total entropy of the universe is increasing.

What the Second Law really says is that no conversion of energy from one form to another is 100% efficient, some energy is always wasted, i.e., lost irretrievably to the surroundings. For example, running a car on 100 units of gasoline energy will not give us 100 units of equivalent work. Most of the available energy in the gasoline is lost to the environment as heat from the friction of tires and from the engine (Table IV). We might be able to tap some of the exhaust heat in a useful way, BUT WE COULD NEVER USE ALL OF IT. The important point is that in the process of running our car we have not <u>lost</u> any energy; it has only become degraded into a less useful form . . . heat.

We can now state the Laws of Thermodynamics in a more frivolous way (3):

FIRST LAW: You can't get something for nothing, you can only break even.
SECOND LAW: You can't even break even . . . you can only lose.

There are strong implications to the Laws of Thermodynamics. For one thing the more we try to support increasing numbers of people at higher and higher levels of energy consumption, the more disorder there will be in our life-support systems (e.g. pollution). For another thing it should be clear from the Second Law that, unlike matter, energy cannot be recycled. It passes through and around us only once, perhaps taking several forms, but eventually dissipating to the cosmos as heat. Thus, as noted earlier, we cannot really "conserve" energy in the strict sense; we can only slow down the <u>rate</u> at which we produce waste heat or entropy. This is a subtle but important point which is discussed next when we consider what is meant by "efficiency."

Energy: Forms and Conversions

Energy comes in several forms: gravitational, mechanical, thermal, chemical, electrical, radiant, nuclear etc. The most abundant form of energy in the universe is gravity, which produces such diverse sources of power as supernovae (gravitational collapse of certain stars) and water running downhill. Chemical energy is found in all products of photosynthesis such as wood, food and fibre crops, methane, alcohol, and fossil fuels. Large amounts of energy are available when mass is released as energy in nuclear reactions. In fact, as we proceed from physical to chemical to nuclear forms of energy — that is, as forces act at ever closer distances (physical space, molecular space, nuclear space) — larger amounts of energy <u>and</u> <u>pollution</u> are released. As noted above we can't get something for nothing. As we adopt more intensive forms of energy, we pay a higher price for that convenience and power.

Although energy cannot be created or destroyed it can change form. Table II lists some common examples of direct energy conversions. However, most useful forms of energy are the result of a <u>series</u> of energy conversions (or energy "system"). For example, consider the many ways that solar energy can be converted into thermal energy (say of hot water):

TABLE II. EXAMPLES OF DIFFERENT KINDS OF ENERGY CONVERSIONS.

From	To MECHANICAL	THERMAL	CHEMICAL	ELECTRICAL	LIGHT
MECHANICAL	Bicycle Flywheel Water wheel Wind mill	Friction Heat pump	Detonation of nitroglycerine	Wind generator Water turbine	Friction (sparks)
THERMAL	Engines Wind Falling water	Heat exchanger (eg: radiator)	Pyrolysis Endothermic reaction	Thermocouple Thermionic generator	Luminescence
CHEMICAL	Muscle contraction Firecracker Jet engine	Food Fuel Compost Match	Metabolism Anaerobic digestion	Fuel cell Battery	Candle Bioluminescence
ELECTRICAL	Motor Solenoid	Resistor Spark plug	Electrolysis Battery	Transformer Inverter	Light bulb Lightning
LIGHT	Photoelectric cell (eg: door opener)	Solar collector Greenhouse	Photosynthesis	Solar cell	Fluorescence

Types of Efficiencies

We all use the term "efficiency" in a very loose way — to indicate superior performance, to describe something as being "better" etc. There are, in fact, several kinds of efficiency.

<u>Energetic Efficiencies:</u> The First Law of Thermodynamics tells us that as energy is converted from one form to another (i.e., as work is performed) some of the available energy is lost as heat. The amount of waste heat is one measure of the efficiency of the conversion or system.

(1) First Law Efficiency = $\dfrac{\text{Energy in a desired place or form}}{\text{Total energy actually used}}$

TABLE III — SOME PATHWAYS FOR CONVERTING SOLAR ENERGY INTO THERMAL ENERGY (OF HOT WATER).

FOSSIL FUEL— ELECTRIC—THERMAL

solar (photons) → fossil fuel chemical → boiler thermal → steam turbine mechanical → generator electrical → resistance heating thermal

SOLAR—ELECTRIC THERMAL

solar (radiant) → reflector thermal → absorber thermal → steam turbine mechanical → generator electrical → resistance heating thermal

HYDRO—ELECTRIC—THERMAL

solar (radiant) gravity → falling water mechanical → water turbine mechanical → generator electrical → resistance heating thermal

WIND—ELECTRIC—THERMAL

solar (radiant) → wind mechanical → wind turbine mechanical → generator electrical → resistance heating thermal

WOOD—THERMAL

solar (photons) → wood chemical → combustion thermal

SOLAR—THERMAL

solar (radiant) → solar collector thermal

This is a simple in-and-out energy accounting. It is the usual way we think of physical or technical efficiency. Because it measures the changes in the quantity of useful energy it is called the "first-law" efficiency. For further discussions of first-law efficiencies in the ENERGY PRIMER see pgs. 48-49 (solar collectors); 156-157 (biomass production); 162-164 (agricultural systems); and 220-221 (wood heaters).

Another way of stating first-law efficiencies is in terms of "net energy"; the total energy available, minus the energy used to find, concentrate and deliver the energy to its intended place. Many energy resources have a low net energy because it takes a large amount of energy to make them available. For example, the net energy of fossil fuels mined and shipped from remote places (e.g., north slope of Alaska) or inaccessible resources (e.g., tar sands or oil shale) is much less than those used near the place of extraction or drilled from shallow wells. The net energy of processed foods is less than foods grown in your backyard and the net energy of nuclear power plants is very low when we consider the energy costs of mining and processing uranium fuels, building the reactor, transporting and storing nuclear wastes, and meeting complex safety regulations. (4)

In discussing first-law efficiencies (F.L.E.'s) one should be careful to distinguish between the efficiency of a particular device and the efficiency of the entire system of which it is a part. For example, the F.L.E. of a gas stove and an electric stove are around 37% and 75% respectively (5); but the system efficiencies are reversed: gas stove = 33% and electric stove = 17%. This difference is due to the fact that a great deal of energy is lost at the electric power plant (nearly 70%), whereas natural gas is delivered directly to the house with little loss in available work. Similarly, the F.L.E. of an internal combustion engine is about 30% (the ratio of work on the pistons to potential energy in the gasoline). But this should not be considered as the overall efficiency of the automobile. If we note all of the energy losses from drilling and transporting the crude oil at the well head to the work of moving the automobile, the first-law <u>system</u> efficiency of the automobile is quite low (Table IV), without even including the energy needed to produce the automobile itself.

TABLE IV

FIRST-LAW SYSTEM EFFICIENCY OF THE AUTOMOBILE (FROM REF. 41) NOT INCLUDING "AUTOMOBILE PRODUCTION EFFICIENCY." To find the efficiency of an entire energy system, multiply the efficiencies of each conversion step within a subsystem and then multiply the final products of the subsystems.

PROCESS	LOSSES	PROCESS EFFICIENCY	CUMULATIVE SYSTEM EFFICIENCY
Gasoline efficiency			
production of crude oil	at well head	96%	96%
refining of crude oil	at refinery	87%	84%
transportation of gasoline	vehicles	97%	81%
Automobile Efficiency			
engine thermal efficiency	heat losses from engine	29%	29%
engine mechanical efficiency	accessories (fan, generator, etc.) misc. heat (exhaust, radiator)	71%	21%
rolling efficiency	heat from tires	30%	6%
System Efficiency		81% x 6% =	5%

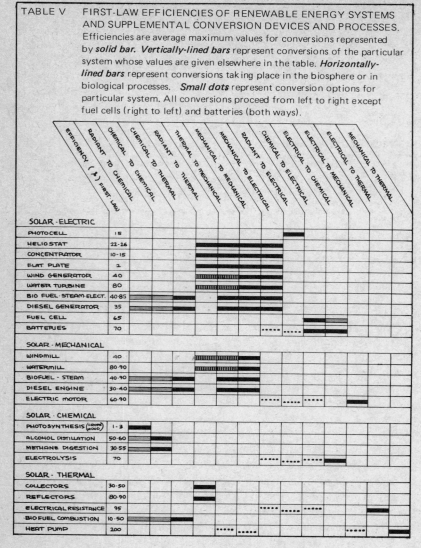

TABLE V FIRST-LAW EFFICIENCIES OF RENEWABLE ENERGY SYSTEMS AND SUPPLEMENTAL CONVERSION DEVICES AND PROCESSES. Efficiencies are average maximum values for conversions represented by *solid bar*. *Vertically-lined bars* represent conversions of the particular system whose values are given elsewhere in the table. *Horizontally-lined bars* represent conversions taking place in the biosphere or in biological processes. *Small dots* represent conversion options for particular system. All conversions proceed from left to right except fuel cells (right to left) and batteries (both ways).

	EFFICIENCY (%) FIRST-LAW
SOLAR - ELECTRIC	
PHOTOCELL	15
HELIO STAT	22-26
CONCENTRATOR	10-15
FLAT PLATE	2
WIND GENERATOR	40
WATER TURBINE	80
BIO FUEL - STEAM ELECT.	40-85
DIESEL GENERATOR	35
FUEL CELL	65
BATTERIES	70
SOLAR - MECHANICAL	
WINDMILL	40
WATERMILL	80-90
BIOFUEL - STEAM	40-90
DIESEL ENGINE	30-40
ELECTRIC MOTOR	60-90
SOLAR - CHEMICAL	
PHOTOSYNTHESIS (biomass/wood)	1-3
ALCOHOL DISTILLATION	50-60
METHANE DIGESTION	30-55
ELECTROLYSIS	70
SOLAR - THERMAL	
COLLECTORS	30-50
REFLECTORS	80-90
ELECTRICAL RESISTANCE	95
BIO FUEL COMBUSTION	10-50
HEAT PUMP	200

Other examples can be drawn from various solar-based energy systems (Table V).

First-law efficiencies are useful in keeping track of energy flows and comparing devices and processes of a particular type under specific conditions. "How much of the solar energy falling on a given collector reaches the water in the storage tank?" or "How much of the torque in a windmill shaft reaches the water pump?" etc. However F.L.E.'s are inadequate as indicators of the potential for fuel savings. This is because losses of energy quality, in addition to losses of energy quantity, are inherent in any process. As a result, F.L.E.'s have several drawbacks: 1) Their maximum value depends on the system and on temperatures and may be greater than, less than, or equal to 1. For example, household furnaces typically have F.L.E.'s of 0.6 meaning that 60% of the heat of combustion of the fuel is delivered as useful heat to the space. This measure implies that a 100% efficient furnace would be the best possible. But this is incorrect since a heat pump, which extracts heat from outdoors and transfers it at a higher temperature to the interior space, can make available as heat more than 100% of the electrical energy it consumes. Similarly, a typical air conditioner (essentially a heat pump operating in reverse) has an efficiency (Coefficient of Operation or COP) of 200%, or COP = 2, a measure which provides no hint of the maximum possible performance.

2) First-law efficiencies do not adequately emphasize the actual role of the second law of thermodynamics in determining the possible efficiency of energy use. The F.L.E. of fossil-fuel power plants, for example, is about 40% (fuel energy to electrical energy). However, the maximum theoretical efficiency is less than 100% because of the constraints of the second law of thermodynamics (irretrievable heat energy losses). In this case 40% is only a partial measure of performance since 100% efficiency cannot be achieved.

A more useful measure of efficiency, therefore, would take into account both quantity and quality losses and would show how well a particular energy conversion system performed relative to an ideal one in which there is loss of neither quantity nor quality (6) . . . in other words a measure of performance relative to the optimal performance permitted by both the first and second laws of thermodynamics. This "second-law" efficiency (S.L.E.) can be defined as follows:

(2) Second-Law Efficiency = $\dfrac{\text{Minimum amount of energy required by second law for specific task}}{\text{Total energy actually used}}$

The minimum (theoretical) amount of energy required by any task depends on the source of energy and the end to which it is used (Table VI).

The second-law efficiency allows one to determine the quality of performance of any task relative to what it could ideally be. It shows how much room there is for improvement in principle. More importantly, it measures the worth of a task and not a device, and it implies that for maximum thermodynamic efficiency we need to match the task we wish to perform with an appropriate energy source. Thus, it makes little sense to heat a house to 70°F. with a 10,000°F. thermonuclear reactor; or, to cite a popular phrase: "Why cut butter with a chain saw?"

For most activities, S.L.E.'s are less than 10% (7), suggesting that energy is being used very inefficiently today. For example, a typical domestic oil furnace has an S.L.E. of only 5% compared with its F.L.E. of 60%. In this case, to maximize the task of heating a house with fuel, a furnace should be replaced by a fuel cell and a heat pump.

Economic efficiency is measured by comparing the dollar cost of energy conserving strategies to the dollar cost of energy consumption without energy conservation. These comparisons are called life cycle cost analyses, and they are done with a consideration given to initial investment, reliability of the energy source and varying forecasts of the price of energy. For example, an uninsulated house that loses 111 million BTU's of heat per year (see following article) would, at present fuel prices of natural gas, represent an annual cost of around $300. If the owner spent $500 on insulation and weatherstripping so that the house only lost 21 million BTU's per year, the annual heating bill would only be $60 and the weatherizing of the house would pay for itself in 2 years (assuming no increase in gas prices). Economic efficiency is a relative measure, but the inevitable rise in fossil fuel prices will continue to make the seemingly high initial costs of energy conservation measures more realistic as time goes by.

Social efficiency attempts to measure the social implications of energy use which cannot be assigned a dollar value. The physical efficiency of an automobile changes very little when five people are riding instead of one, but the social efficiency changes a great deal. Some important considerations here are exertion, time, convenience, risk, pleasure or nuisance (8). The automobile is very inefficient regardless of how you look at it, but its level of convenience is very high . . . hence it is socially efficient . . . at the present. Hopefully this will change as people come to understand the total environmental price we pay for the "convenience" of the automobile (9).

Much of the social inefficiency of energy use in the industrialized countries is the result of careless habit, fostered by cheap energy. Rising energy and resource costs may alter those habits, but it will not be a simple process. The contrast between Sweden and the United States in the economic efficiency of energy use reflects a difference in social efficiency which lies at the heart of the problem. Sweden, with a per capita income similar to that of the U.S., uses only a little more than half as much energy per person . . . despite a harsher climate that requires more energy for heating and transportation (10). The greater efficiency of the Swedes is due to an organization of society and an attitude toward thrift and waste that promote the efficient use of energy and materials. Small cars, public transport, compact urban areas and an awareness of conservation result

TABLE VI

SECOND-LAW EFFICIENCIES are calculated by knowing theoretical minimum amount of energy required for specific task (see Eq. 2), which, in turn, depends on the source of energy and the end use of the energy produced. For actual formulae of each kind of task see Ref. 9.

END USE	SOURCE WORK IN	FUEL (HEAT OF COMBUSTION)	HEAT FROM A RESERVOIR
work out	e.g., electric motor; wind-mill.	e.g., power plant, steam engine	e.g., geothermal plant; solar-boiler electric system
heat added to warm reservoir	e.g., electrically driven heat pump	e.g., wood stove, engine-driven heat pump	e.g., solar hot water heater; solar cooker
heat extracted from cool reservoir	e.g., electric refrigerator	e.g., gas-powered air conditioner	e.g., absorption refrigerator

in a much lower energy consumption than in the United States (11) where large cars, meager mass transit systems, sprawling cities, and a throwaway economy have inflated the consumption of energy. Other industrialized countries, with a high "standard of living," also consume a great deal less energy than the U.S. (12, 13)

Renewable and Non-Renewable Energy Resources: A Deeper Distinction

Although there are many forms of energy, we can distinguish between two basic types of energy resources (Table VII): 1) Income or renewable energy resources which are produced continuously in nature and are essentially inexhaustible, at least in the time framework of human societies, and 2) capital or non-renewable energy resources which have accumulated over the ages and are not quickly replaceable when they are exhausted.

TABLE VII ENERGY RESOURCES CLASSIFIED ACCORDING TO WHETHER THEY ARE RENEWABLE OR NON-RENEWABLE.	
INCOME OR RENEWABLE ENERGY RESOURCES	**CAPITAL OR NON-RENEWABLE ENERGY RESOURCES**
DIRECT SOLAR (heat and light)	FOSSIL FUELS
BIOLOGICAL (photochemical)	gas (natural gas)
crops and livestock	liquid (petroleum, tar sands, shak)
wood	solid (coal)
organic wastes	NUCLEAR
biofuels	fission (U-235, U-239, thorium 232)
animal and human power	fusion (deuterium, lithium 6)
INDIRECT SOLAR	GEOTHERMAL (heat traps)
water or hydro	
wind	
waves	
thermal gradients	
tidal	
GEOTHERMAL (heat flow)	

There is, however, a more profound distinction to be made. Although renewable energy resources represent a limitless supply of energy (the sun), there is a definite limit to the rate at which that energy can be obtained (the solar constant). Just the opposite is true of non-renewable energy resources which can be used at an unlimited rate, but which are limited in supply. To press this important distinction, consider the following: Imagine a small group of people living on an island surrounded by an immense fresh water sea. In the middle of the island there is a fresh water lake. The island inhabitants have the option of either drawing their water from the lake with several powerful pumps, or drawing their water from the surrounding sea with a limited supply of buckets. Using all the pumps would allow the islanders to prosper and multiply at a rapid rate . . . but only for a while . . . until the lake emptied. Using the buckets, on the other hand, would provide the islanders with their water supply at a much slower rate, but for a greatly extended period of time. The point here is that the future of a society is greatly determined by the nature of the energy resources that it uses. This fact is all the more obvious when we consider the implications of a solar-based as compared to a nuclear-based society.

PATTERNS OF ENERGY CONSUMPTION

History is Energy Changing

In many ways human history has been determined by the way various cultures have manipulated and used energy. For centuries preceding the industrial revolution people relied on the chemical energy of plants and animals and the natural forces of wind and water power to provide the necessities of life. As more efficient ways were discovered to exploit these renewable energy resources, gradual changes took places in the way people lived. But beginning around the 18th century the discovery of power devices able to convert steam and, later, fossil fuels into useful work produced a phenomenal growth in energy consumption and rapid social changes unprecedented in human history. Such things as the switch from wood to coal and whale oil to petroleum and the development of the steam engine, which expanded the geographical limitations of water power; the internal combustion engine, which broadened our mobility; electricity, which greatly increased available work and communications; powerful steam, gas and water turbines for generating power, and, finally, nuclear

energy — all these "discoveries" had a profound influence on the course of human history and helped to usher in an era of rapid energy consumption (Fig. 1). Since 1850 the world energy demand per person has increased from about 10 kilowatt-hours per day to nearly 50 kilowatt-hours, and continues to increase at about 3% per year. Most of the increase has been due to the activities of industrialized countries. Thus, during the period 1850-1970 the per capita energy demand in India increased from 6 to 9 kilowatt-hours per day whereas in the United States it went from 100 to over 200 kilowatt-hours per day. (16)

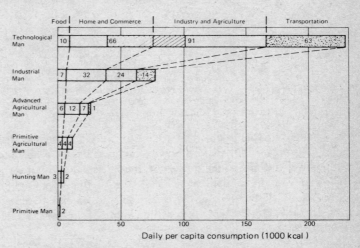

FIG. 1. DAILY CONSUMPTION OF ENERGY PER PERSON FOR SIX TYPES OF HUMAN CULTURE. From Ref. 14 and 15.

Most of the increase in energy consumption has been due to a radical shift in the nature of the energy base of our society — from dispersed natural forces available to large numbers of people to limited reservoirs of intensive chemical energy (fossil fuels) controlled by a few corporations. In 1850, 90% of the U.S. energy was supplied by renewable energy sources (wood, wind and water). Currently more than 75% is supplied by non-renewable hydro-carbon fuels controlled entirely by energy cartels. No doubt the inevitable future changes to new energy resources (nuclear, solar, coal) will cause major changes in the way we live and influence greatly the degree to which people have control over their energy resources.

Global Energy Consumption

There are two major patterns to the way energy is consumed on a global scale: 1) About 80% of the world's energy comes from fossil fuels, about 20% from dung and vegetable wastes, about 1% from water power (mainly hydro-electric) and minor amounts from nuclear, solar, geothermal and wind power (Table VIII). 2) About 75% of the world's energy is consumed by a few rich countries representing less than 30% of the global population. Hence it is not surprising to see a strong correlation between a nation's gross national product (GNP) and its energy consumption (Fig. 2); the more energy a country consumes, the richer it becomes. This has prompted many people to criticize energy conservation strategies because

TABLE VIII WORLD CONSUMPTION OF ENERGY (From Ref. 17)		
SOURCE	ENERGY CONTENT $(10^{15}$ kcal)	% TOTAL
Non-Renewable		
Crude oil	22.7	34.8
Coal and lignite	17.1	26.2
Natural gas	10.9	16.7
Uranium	0.55	0.84
Renewable		
Plant food and feed	6.2	9.5
Dried dung	3.3	5.1
Wood fuel	1.6	2.5
Vegetable refuse	1.6	2.5
Falling water	1.1	1.7
Fish	0.06	0.09
Geothermal power	0.04	0.06
	65.2	100.0%

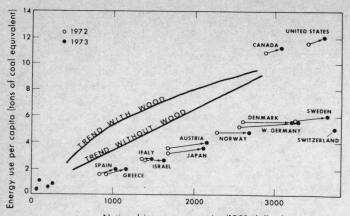

FIG. 2. PER CAPITA ENERGY USE AND NATIONAL INCOME OF VARIOUS NATIONS. Data do not include the large amount of wood used, primarily in Third World Countries. From Ref. 19.

they would lower our standard of living. However, this argument is highly misleading: 1) Among industrialized countries there is only a very loose relationship between GNP and the quality of life. 2) As pointed out earlier, many countries with a standard of living similar to that of the United States consume much less energy. They are able to do this because of cultural habits and value systems that permit a more efficient use of the energy that is available.

U.S. Energy Consumption

Today, the U.S. with 6% of the world's population uses 35% of the total energy. (15) This oft-quoted statistic tends to be taken for granted, but it is always worth repeating. Fig. 3 shows the distribution of energy use in the U.S. (1972) by sector and by end use within each sector.

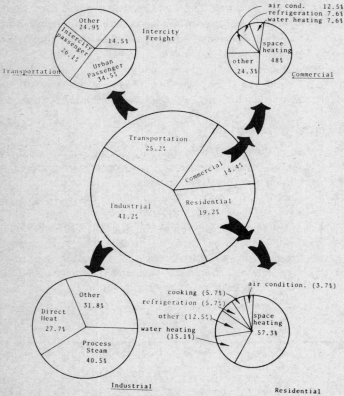

FIG. 3. UNITED STATES ENERGY USE (1972) BY SECTOR AND END USE. From Ref. 5.

Transportation (fuel), space heating (residential and commercial), process steam (industrial) and direct heat (industrial) account for over 70% of the total energy consumed; residential space heating accounts for the largest single end use within any sector.

We can summarize the gross input-output statistics by the energy flow diagram in Fig. 4. Before the early 1970's very few people had ever even considered U.S. energy use in terms of energy efficiency. The 1971 study of Cook (15) was the first detailed analysis of the overall energy efficiency of the U.S. economy. His figure of 50% was later revised to the 36% efficiency shown in Fig. 4. which also shows that we waste largely in the generation of electricity and in transportation.

There has been a great deal of debate as to how fast energy consumption will continue to increase in the U.S. Energy planners in the 1960's projected growth in energy use on the basis of past experiences alone. All concluded that the increase in energy consumption would continue to grow faster than the population—much faster. In the years before the oil embargo of 1973 most energy-use studies (e.g., 21-25) virtually ignored energy conservation and improved technical efficiency in their forecasts of growth (usually 3-5%). However, beginning with the important Ford Foundation Study in 1973 (26), which helped to establish respectability for energy conservation, recent studies have revealed an enormous potential for energy savings in a variety of ways throughout society (4, 6, 8, 27-29). Low-energy growth forecasts are well on their way to attaining the status of conventional wisdom.

ENERGY CONSERVATION

Energy conservation can be defined as the act of adjusting energy use and habits so as to reduce the use of energy per unit of output (work or well being) while maintaining or reducing the total cost of the system (19). In other words, strategies for energy conservation do not just save *quantities* of energy; they must also improve the *quality* of energy use within economic constraints. Because the price of energy is now increasing faster than the price of capital or labor, it is obvious that energy conservation will continue to become more practical as time goes by.

Kinds of Energy Conservation and the Potential for Savings:

There are several ways to conserve energy. Table IX lists six possible strategies (19).

Schipper (8) has reviewed over 45 major studies on energy conservation and has summarized the potential savings in different sectors of the U.S. economy. He suggests that up to 35% of the current U.S. energy use could be saved by energy conservation without major disruptions in the U.S. economy, the life style of people or employment. Ross and Williams concluded that 42% of our energy could thus be saved (6), and a recent forecast from ERDA (30) notes that U.S. energy needs in the 1990's could be 20-40% below what was previously expected in light of higher energy prices and more appropriate technologies. Established policies towards the potentials of energy conservation are changing, bringing to mind a statement by Paul Valery: "The trouble with our time is that the future is not what it used to be."

Energy Conservation and Employment:

We have long been told that increasing energy production/consumption is the key element in producing a sound national economy and more jobs. Corporations and government have always stressed that as energy production increases so does economic growth and employment. In fact, faith in the relationship between energy consumption and social well being is so ingrained in our cultural and political tradition that energy use *per se* has become a gauge of social progress. However, it is becoming more

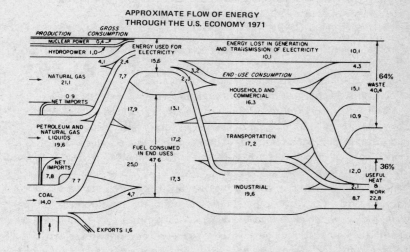

FIG. 4. ENERGY FLOW DIAGRAM OF UNITED STATES ECONOMY (1971) showing an overall first-law efficiency of only 36%. From Ref. 20.

TABLE IX   TYPES OF ENERGY CONSERVATION STRATEGIES.

IN EXISTING SYSTEMS   (From Ref. 19)

Leak Plugging
Reducing heat and cooling losses in life-support systems; adjusting energy systems that are not running at design efficiency. Leak plugging techniques are generally implemented once and then remain passively effective. Examples: building insulation and weatherproofing, heat recovery in industry.

Energy Management (Belt Tightening)
Energy management, unlike leak plugging, must be actively pursued and thus may cause minor changes in life style. Examples: Turning off lights, heat or cooling; changing thermostat settings; driving more slowly; car-pooling.

Mode Mixing
Changing the mix of transportation to utilize modes requiring less energy per passenger mile. Examples: Mass transit, improved freight transportation.

IN NEW SYSTEMS

Thrifty Technology
Introduction of innovative technology not in common usage in any energy system to increase the useful output per unit of energy consumed. Examples: heat pumps for space and industrial heat, electric ignition of gas water heaters, new propulsion systems in transportation.

Input Juggling
Substituting various indirect forms of energy (materials, labor, capital, design and machines) for direct energy use. Example: Recycling programs substitute capital and labor for the extra energy lost through throwaway and obsolete items. Solar energy substitutes capital and labor for heating.

Output Juggling
Changes in life-style, consumer preferences, or shifts from manufacturing to services in the community, which lead to lower energy requirements from the output of a system or service. Driving smaller cars.

American consumption since 1960 should be reflected in a lowering of the rate of unemployment; but the exact opposite has been the case. The tenet, "more energy leads to more jobs," is, in fact, a myth.

Several examples can be cited as evidence:

• In the steel industry from 1959-1969, employment declined from 450,000 to 100,000 as production increased 45%, and energy use increased (32).

• According to the Bonneville Power Administration, the aluminum industry in the Pacific Northwest consumes 25% of the region's electricity but only provides one-half of one percent of the total jobs in that region (33).

• In all, the major energy-producing and energy-using industries consume over 30% of the nation's energy, yet they only provide about 10% of the nation's jobs (26).

• From 1961-1973, electric utilities increased their power output by 130%, their revenues by 260%, their construction costs by 340%, but employment in electric utilities increased only 21% (8).

• In agriculture the use of energy—for fertilizer, chemicals and automated equipment—has increased 4 times since 1920. During that time there has been a steep decline in farm employment by more than 90%.

There are many more examples, but the basis of the argument is simple; sizeable sums of money are being committed to expand energy production. Latest estimates are that it will take at least a trillion dollars to build 500 nuclear power plants currently forecast for the year 2000. The current price tag for a commercially-useable breeder reactor is $10 billion, and the cost of each coal plant is around $1 billion. However as more money is being spent for sophisticated energy production, less money is available for investments in other areas which serve people's needs, provide more jobs per dollar, consume less energy and create fewer environmental and public health hazards (8, 19, 31, 32). In fact high impact energy conservation programs, including retrofitting existing buildings, the manufacturing of high efficiency tools and durable items plus the construction of solar energy systems could increase employment substantially (31, 34, 35).

and more obvious that this long held belief is, in fact, a myth. A great deal of evidence to support this claim is outlined in an excellent paper, JOBS AND ENERGY (31). The authors Grossman and Daneker point out that it is difficult to see a correlation between energy consumption and employment at a time when energy consumption is at an all time high and unemployment is at its highest level since the depression. If, as we are told, unemployment can be ended only by stepping up energy development with the largest systems possible, then the phenomenal increase in

References Cited

(1) Steinhart, C. and J. Steinhart. 1974. ENERGY. Duxbury Press, North Scituate, Massachusetts.

(2) Chesterman, J. et al. 1974. ENERGY AND POWER: AN INDEX OF POSSIBILITIES. Pantheon Books, New York.

(3) Miller, G.T. 1971. ENERGETICS, KINETICS, AND LIFE. Wadsworth Publ. Co., Inc., Belmont, Caliofrnia.

(4) Lovins, A.B. 1975. WORLD ENERGY STRATEGIES. Friends of the Earth International, Ballinger Publ. Co., San Francisco.

(5) PATTERNS OF ENERGY CONSUMPTION. Stanford Research Institute, Menlo Park, CA 1972. U.S. Gov't. Printing Office.

(6) Ross, M.H. and R.H. Williams. 1977. THE POTENTIALS FOR FUEL CONSERVATION. Technology Review: Feb. 1977.

(7) Carnahan, W., et al. 1975. EFFICIENT USE OF ENERGY, A PHYSICS PERSPECTIVE. In: "Efficient Energy Use," American Physical Society, 1975, American Institute of Physics, New York.

(8) Schipper, L. 1976. ENERGY CONSERVATION: ITS NATURE, HIDDEN BENEFITS AND HIDDEN BARRIERS, Energy Communications, 2(4):331-414.

(9) Illich, I. 1974. ENERGY AND EQUITY. Harper and Row, N.Y.

(10) Schipper, L. and A.J. Lichtenberg. 1976. EFFICIENT ENERGY USE AND WELL-BEING: THE SWEDISH EXAMPLE. Science, 194:1001-1013, 3 Dec., 1976.

(11) Doernberg, A. 1975. COMPARATIVE ANALYSIS OF ENERGY USE IN SWEDEN AND THE UNITED STATES. Brookhaven Nat'l. Lab. Rep. BNL-20539.

(12) The Work Group of the International Federation of Institutes for Advanced Study. 1976. ENERGY IN DENMARK—1900/2005 Rept. 7, The Niels Bohr Institute, The University of Copenhagen, Copenhagen, DK-2100, Denmark.

(13) Goen, R. and R. White. 1975. COMPARISON OF ENERGY CONSUMPTION BETWEEN WEST GERMANY AND THE UNITED STATES. Gov't Printing Office, Washington, D.C.

(14) O'Toole, J. 1976. ENERGY AND SOCIAL CHANGE. MIT Press, Cambridge, Massachusetts.

(15) Cook, E. 1971. THE FLOW OF ENERGY IN AN INDUSTRIAL SOCIETY. Scientiflc American, Sept. 1971.

(16) Starr, C. 1971. ENERGY AND POWER. Scientific American Sept. 1971.

(17) Cook, E. 1976. MAN, ENERGY, SOCIETY. W.H. Freeman & Co., San Francisco.

(18) Morrison, W.E. 1968. AN ENERGY MODEL FOR THE UNITED STATES. U.S. Bureau of Mines Circular 8384, Washington, D.C., U.S. Gov't Printing Office.

(19) Schipper, Lee. 1976. RAISING THE PRODUCTIVITY OF ENERGY UTILIZATION. Annual Review of Energy, Vol. 1, pg. 455.

(20) E. Cook. 1973. In: "Resource Conservation, Recovery, and Solid Waste Disposal," Senate Committee on Public Works.

(21) UNDERSTANDING THE NATION'S ENERGY DILEMMA, Joint Committee on Atomic Energy, U.S. Congress, July 1973. U.S. Gov't Printing Office.

(22) Dupree, W. and J. West, 1972. U.S. ENERGY THROUGH THE YEAR 2000. U.S. Dept. Interior, Washington, D.C.

(23) 1972 ANNUAL REPORT, Federal Power Commission, Washington, D.C., 1973.

(24) POTENTIAL NUCLEAR POWER GROWTH, Washington, 1098, U.S. Atomic Energy Commission, Washington, D.C. 1971.

(25) ENERGY OUTLOOK IN THE UNITED STATES to 1985, Chase Manhattan Bank, New York, 1973.

(26) A TIME TO CHOOSE, Energy Policy Project of the Ford Foundation, Ballinger Publishing Co., Cambridge, Massachusetts, 1974.

(27) Erickson, L. 1974. "A Review of Forecasts for U.S. Energy Consumption." IN: ENERGY AND HUMAN WELFARE, B. Commoner et al (ed), MacMillan Co., N.Y.

(28) Ross, M.H. and R.H. Williams. 1975. ASSESSING THE POTENTIAL FOR ENERGY CONSERVATION. Institute for Public Policy Analysis, Dept. 75-02, State Univ. of New York, Albany, N.Y.

(30) U.S. Energy Research and Development Administration. 1975. A NATIONAL PLAN FOR ENERGY RESEARCH, DEVELOPMENT, AND DEMONSTRATION. Publ. No. ERDA 48, Washington, D.C.

(31) Grossman, R. and G. Daneker, 1977. JOBS AND ENERGY. Environmentalists for Full Employment. 1985 Massachusetts Ave., N.W., Washington, D.C. 20036.

(32) Grossman, Richard. 1977. CREATING JOBS THROUGH ENERGY SELF-RELIANCE. Self-Reliance, No. 7, May-June 1977. Institute for Local Self-Reliance, Washington, D.C.

(33) Natural Resources Defense Council. 1977. AN ELECTRICAL ENERGY FUTURE FOR THE PACIFIC NORTHWEST: AN ALTERNATIVE SCENARIO, Palo Alto, California.

(34) Laitner, S. 1976. THE IMPACT OF SOLAR ENERGY AND CONSERVATION TECHNOLOGIES ON EMPLOYMENT. Critical Mass., Box 1538 Washington, D.C. 20013.

(35) THE USE OF SOLAR ENERGY FOR SPACE HEATING AND HOT WATER SYSTEMS. Energy Policy Office. State of Massachusetts, Boston: April 1976.

HEAT PUMPS
Robin Saunders and Harry Whitehouse

Operation

A heat pump is a device which transfers thermal energy or heat from a "cooler" region to a "warmer" one; this is contrary to the normal flow of heat which naturally flows from a warmer region to a cooler region. We can reverse this natural tendency, if we supply some additional energy (usually in the form of work).

Any refrigerator acts as a heat pump. Heat is withdrawn at a low temperature and is rejected at a higher temperature level. In cases where only refrigeration is needed, the rejected heat is usually wasted (place your hand at the back or bottom of your refrigerator sometime to feel this rejected heat). A heat pump is essentially a refrigerator where the heat rejected by the condenser is used for a useful purpose.

The attractiveness of a heat pump arises from the small amount of work required to transfer a given amount of thermal energy (or heat). Under favorable conditions, three or four units of energy can be transferred for every unit of energy supplied as work. This means that we can supply a dwelling with three or four units of heat by investing only one unit of work—say from a hydro unit, windmill, or utility line.

Suppose you lived in a small, one-room dwelling exposed to outside air temperatures of 40°F. In principle, you could heat that room using a modified refrigerator. First, remove the refrigerator door. Then place the refrigerator in an open door or large, open window with the "cold space" exposed to the outside air and the heat coils on the inside. If you sealed up the cracks properly and supplied the appropriate electrical input, the room temperature would rise.

To understand why the room is warmed, we need to understand how the working fluid (often freon) and the "refrigeration cycle" operate. First, freon gas is compressed to a moderately high pressure which produces an accompanying temperature increase. It is during this compression process that the work for the cycle must be supplied. The high temperature freon gas (typically around 140°F) is then circulated through pipes which are exposed to the room air. Since the freon is hotter than the room air, heat energy is transferred into the room until the freon gas cools and condenses to a liquid. As we said before, the amount of energy transferred into the room is several times that put into the compressor. To find out where this "extra" energy comes from, we must follow the cycle further.

Immediatley after condensation, the freon (now a liquid) is still at a high temperature and pressure. The temperature of this fluid can be reduced significantly (below 0°F) by reducing the pressure suddenly. This pressure reduction is accomplished with an inexpensive expansion valve. The low pressure freon, which is now a mixture of liquid and gas, is then circulated through a pipe exposed to the outside air. Here the freon takes energy from the "cold" outside air. It can do this easily since the freon temperature is well below 0°F! This is where the "extra" energy is obtained. The key point is that the outside air contains energy which can be used to heat the room—you merely need a device to collect it.

The cycle is completed as the cold freon collects enough energy from the outside air to vaporize completely. The resulting gas is then fed back into the compressor to start a new cycle. It is often useful to think of this gas as a "carrier" for the energy collected from the outside air, and that the compressor "conditions" the fluid so that the energy can be delivered to the room at a higher temperature. It should be realized that the energy used to operate the compressor eventually enters the room as heat. Thus, if three units of energy are collected from the outside air, and one unit of energy is supplied from a mechanical or electrical source to run the compressor, four units of energy will be transferred into the house.

Coefficient of Performance

The term Coefficient of Performance (COP) is often used in heat pump and refrigeration literature. This is simply the ratio of energy delivered to the room divided by work input to the compressor. The higher the COP, the better. In the example the COP would be 4.0, since four units of energy were supplied for each unit of work inputed. A COP of two to four

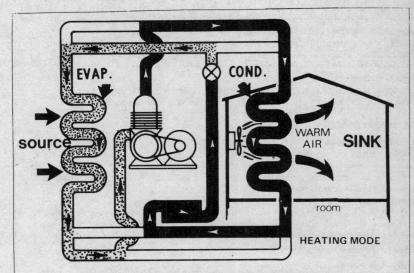

OPERATION OF A HEAT PUMP

During the heating mode, low temperature (liquid) freon is pumped through the evaporator where it absorbs heat from the <u>source</u> (air, water, earth) and gasifies. It is then compressed to a moderately high temperature and pressure, and forced through the condensor where it releases its heat to the room as it returns to a liquid state to begin the cycle again. A fan blowing air over the condensor circulates warm air through the room. During the cooling mode, this process is reversed. The room becomes the source from which heat is extracted and the outside becomes the sink. The evaporator, of course, is cold and so the fan circulates cool air.

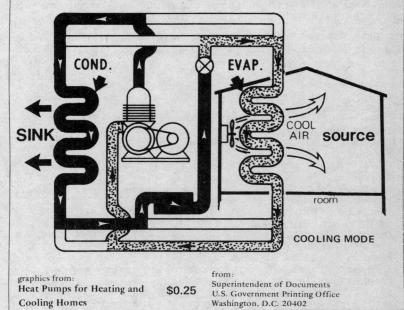

graphics from:
Heat Pumps for Heating and Cooling Homes $0.25

from:
Superintendent of Documents
U.S. Government Printing Office
Washington, D.C. 20402

is common for units operating with moderate outside temperatures (30 to 40°F). However, the COP decreases as the outside temperature falls, and manufacturers' specifications should be consulted to obtain the exact COP for the outside temperatures anticipated in a particular locality.

Considerations

A new heat pump, which can be used both to heat and cool a living space costs about $2500. Typically such units include automatic controls for humidity and temperature. The main disadvantage of a heat pump is, obviously, the high initial cost. In the long run however, heat pumps will save money since fuel bills will be significantly lower than those incurred with gas or electric heat. This is especially true in light of the most recent increases in the cost of natural gas.

Heat pumps suffer from two major problems. First, the COP is reduced significantly in severe climates. Sometimes this can be remedied by switching to another chemical in the freon family which has different thermal properties. However, one should obtain qualified help before embarking on such an expensive proposition. A second problem involves the tendency for ice to form on the "cold" coils and severely degrade the operation. Manufacturers claim to have solved this problem on newer units, but again, you should seek a qualified opinion if you live in a particularly humid location.

To summarize, a heat pump is a fairly sophisticated device which can save considerably on fuel costs for heating. The unit can easily be reversed for cooling in the summer. Commercial units have a high initial cost, but are economical in the long run due to fuel savings. In principle, heat pumps can be constructed with parts from commercial-size refrigeration units, but considerable expertise is required. Home construction is not a job for beginners!

Straight refrigeration systems are much more common and can easily be adapted to particular situations. Cheap refrigeration systems can be obtained from restaurant supply houses or institutions. These usually have belt driven compressors (originally driven by an electric motor, of course). These motors are usually in the range of ½ to 4 horse power. Considering the usual efficiencies of electric generators and motors, it would take 1.5 to 2 times that size in waterwheel generated electric power. Even if an electric generating installation is being built, the saving in power by attaching a direct drive refrigeration system is significant enough that it should be given serious consideration, particularly with the larger units.

Finally, we should note that there are advantages to hooking the compressor in a heat pump or refrigerator directly to a water turbine or windmill. The inefficiency of turning mechanical energy into electric and then the electrical energy back into mechanical seems to be self-evident. However, the advantages of using a heat pump should be considered regardless of whether the compressor receives its energy input from renewable or non-renewable resources.

* * * * * * * * * * * * * * * * * *

Heat pumps are very efficient devices for heating space since they produce more heat energy output than the energy input required for their operation. Heat pumps can be powered mechanically or electrically by conventional fossil fuel devices or by renewable wind or water systems. They can also be made even more efficient by utilizing solar collectors to preheat air or water that the heat pump uses in conventional heat distribution systems (fan-coil or hydronic). Solar-assisted heat pumps (see Figure on pg. 60) consume less energy than conventional heating systems (Table I).

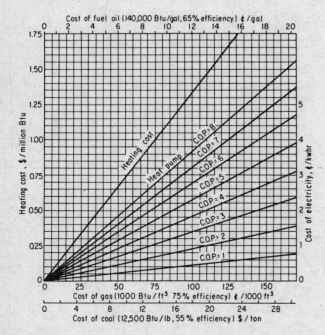

Relative heating costs for different fuels. Thus fuel costs for natural gas at $1 per 1000 ft³ = costs of fuel oil at 12¢ per gallon = costs of coal at $18 per ton = electric resistance heater (C.O.P. = 1) at 45¢ per Kwhr = electric-driven heat pump (C.O.P. = 4) and electricity costs of 1.8¢ per Kwhr. Source: Threlkeld, THERMAL ENVIRONMENTAL ENGINEERING (Prentice-Hall).

TABLE I	ANNUAL FUEL CONSUMPTION OF VARIOUS SPACE HEATING SYSTEMS IN MILLION BTU's. Source: Denton, J.C. 1974. Integrated Solar-Powered Climate Conditioning Systems, Univ. of Pennsylvania, Center of Energy Management.

SPACE HEATING SYSTEM	MILLION BTU's/YR
Solar Heat Pump System[1]	91
Direct Solar Heating[2]	110
Solar Heat Pump System[3]	130
Conventional Air to Air Heat Pump	141
Combustion Furnace (60% efficient)	167
Electric Resistance Heat	325

[1] Solar-assisted heat pump system with "bypass" around heat pump when solar collectors produce enough heat.

[2] Solar space-heating system with no heat pump.

[3] Solar-assisted heat pump system (without bypass)

Heat Pump Manufacturers

ADDISON PRODUCTS CO.
Box 63
Addison, Mich. 49220

AMANA REFRIGERATION, INC.
Amana, Iowa 52203

AMERICAN AIR FILTER CO.
215 Central Ave.
Louisville, KY 40208

BARD MFG. CO.
Box 607
Bryan, Ohio 43506

BRYANT AIR CONDITIONING
7310 W. Morris St.
Indianapolis, Ind. 46231

CARRIER AIR CONDITIONING
Carrier Pky.
Syracuse, NY 13201

COMMAND-AIRE CORP.
3221 Speight Ave., Box 7916
Waco, Tex. 76710

BRYANT DAY & NIGHT
 AND PAYNE CO.
855 Anaheim-Puente Rd.
City of Industry, CA 91744

DUNHAM-BUSH, INC.
177 South St.
West Hartford, Conn. 06110

FEDDERS CORP.
Woodbridge Ave.
Edison, NJ 08817

FHP MFG. CORP.
610 Southwest 12th Ave.
Pompano Beach, Fla. 33060

FRASER & JOHNSTON CO.
2222 Grant Ave.
San Lorenzo, CA 94580

FIEDRICH AIR CONDITIONING
 & REFRIGERATION CO.
4200 North Pan Am (IH 35 Expy.)
San Antonio, Tex. 78295

GENERAL ELECTRIC CO.
Appliance Park
Louisville, KY 40225

GOETTLE BROS. METAL
 PRODUCTS., INC.
2005 East Indian School Rd.
Phoenix, Ariz. 85016

HEAT EXCHANGERS, INC.
8100 N. Monticello
Skokie, IL. 60076

HEIL-QUAKER CORP.
647 Thompson Lane
Nashville, Tenn. 37240

HENRY FURNACE CO.
Medina, Ohio 44256

LENNOX INDUSTRIES, INC.
200 S. 12th Ave.
Marshalltown, Iowa 50158

LUXAIRE, INC.
West of Filbert St.
Elyria, Ohio 44035

MUELLER CLIMATROL CORP.
255 Old New Brunswick Rd.
Piscataway, NJ 08854

RHEEM MFG. CO.
5600 Old Greenwood Rd.
Fort Smith, Ark. 72901

THE SINGLE CLIMATE CONTROL
 DIV.
1300 Federal Blvd.
Carteret, NJ 07008

STEWART-WARNER CORP.
320 N. Patterson St.
Lebanon, Ind. 46052

TAPPAN AIR CONDITIONING
 DIV.
206 Woodford Ave.
Elyria, Ohio 44035

THE TRANE CO.
3600 Pammel Creek Rd.
La Crosse, Wis. 54601

WESTINGHOUSE ELECTRIC CO.
Box 2510
Staunton, VA 24401

WHIRLPOOL HEATING & COOLING
647 Thompson Lane
Nashville, Tenn. 37204

WILLIAMSON CO.
3500 Madison Rd.
Cincinnati, Ohio 45209

YORK DIV. of BORG-WARNER
Box 1592
York, PA 17405

BUILDING HEAT LOSSES
Michael Riordan

**Adapted in part from THE SOLAR HOME BOOK
by Bruce Anderson with Michael Riordan**

There is a tremendous potential for energy conservation in both new and existing buildings. According to the American Institute of Architects, energy savings of 30% in existing buildings and 60% for new structures are regarded as "very conservative" estimates of present possibilities.[1] In more candid moments, they will quote figures of 50% and 80% possible energy savings. Energy is used in buildings for a multitude of purposes—for space heating and cooling, for water heating, lighting, cooking, and refrigeration, and for running a burgeoning list of gadgets and appliances. But the lion's share of energy goes to keep buildings warm in winter and cool in summer. Currently in the United States, space heating and cooling account for about 60% of both residential and commercial energy use.[2] All told, 18% of the nation's energy consumption is devoted to space heating, and another 3% goes for air-conditioning. *So measures to improve the thermal efficiency of buildings are a priority for our energy conservation efforts.*

Heat Flow

Heat energy is simply the motion of the atoms and molecules in a substance—their twirling, vibrating, and banging against each other. Unfortunately, heat energy flows naturally from warm areas to cold, and unless there are provisions for retarding this flow, heat will quickly escape from a building. The actual mechanisms of heat flow are numerous, and so are the methods of retarding them. Therefore, let's review briefly the three basic methods of heat flow—conduction, convection and radiation.

Conduction is the transfer of energy between adjacent molecules in a solid, liquid or gas. Molecules that are in a greater state of agitation transfer some of this energy to nearby molecules. In this way, the handle of an iron skillet left on a hot stove soon becomes too hot to touch. But the rate of heat flow to the handle of a copper skillet is even faster, because copper has a higher *conductivity.* A "conductor" is a material of high conductivity, like copper, and an "insulator" is a material of low conductivity, like wood.

Convection heat flow occurs through the movement of fluids—liquids or gases. A warm fluid can move or be moved to a cooler area where it transfers its heat to warm that area. In a kettle of water on a stove, the heated water at the bottom rises and mixes with the cooler water above, spreading the heat and warming the entire volume much more quickly than could have been done by heat conduction alone. As a fluid is warmed, it expands and becomes less dense, making it buoyant in the surrounding cooler fluid. The warm fluid rises and the cooler fluid that flows in to replace it is warmed in turn. The warm fluid moves to a cooler place, where it surrenders its heat and cools, becoming more dense and settling back to the bottom. This fluid movement is known as *natural,* or *gravity convection.* When we want more control over this heat flow, we use a pump or blower to move the heated fluid in what is called *forced convection.* For example, the heat from a warm air furnace is frequently distributed to the rooms of a building using a fan or blower to circulate the warm air.

Radiation heat flow is the transfer of heat energy through open space by electromagnetic waves. This flow occurs even in a vacuum—just as sunlight can leap across outer space. Objects that stop the flow of light also block radiation heat flow. All objects give off some *thermal radiation* at a rate which depends on their temperature and surface characteristics. Even though we cannot see it, we can feel this thermal radiation when we stand in front of a hot stove. Certain films can "see" this radiation by photographing only (infra-red) heat waves.

In general, the *rate* of heat flow is proportional to the temperature difference between the source of heat and the object or space to which it is flowing. Heat flows out of a building at a faster rate on a cold day than on a mild day. This flow of heat out of a building is called its *heat loss.* During the summer the outside temperature is greater than that inside, and we speak of the heat flow into a building as its *heat gain.* Since we can do little about the temperature difference between inside and outside, most of our effort goes into increasing a building's resistance to heat loss or heat gain. Fortunately, most of the measures that reduce the heat loss during winter also limit heat gain during the summer. Of the three basic

contributions to heat loss, radiation is the most difficult to calculate at the scale of a building. So the total heat loss is taken as the sum of the infiltration heat loss through open doors or cracks around window frames, and the transmission (conduction) heat loss through the exterior skin of the building (the roofs, walls, windows, and floors). The transmission heat loss is usually the largest contribution to the total building heat loss and also the easiest to calculate.

Calculating Transmission Heat Losses

The materials used in the construction of a building all have some ability to retard the transmission of heat. A measure of this ability is called *thermal resistance,* or *R-value,* of the material. The *thermal resistance R* of a slab of building material is equal to the slab thickness

TABLE I	R-VALUES FOR COMMON BUILDING MATERIALS (from the *ASHRAE Handbook of Fundamentals* — 1972)	
	R-values[a]	
Material	per inch	total
WALL & CEILING		
Hardwood	0.91	——
Softwood	1.25	——
Particleboard	1.85	——
Plywood (1/2")	——	0.62
Gypsumboard (1/2")	——	0.45
Insulating board sheathing (1/2")	——	1.32
Wood shingles, lapped	——	0.87
Wood bevel siding, lapped	——	0.81
FLOOR		
Wood subfloor (25/32")	——	0.98
Cork tile (1/8")	——	0.28
Tile (asphalt, linoleum, etc.)	——	0.05
Carpet with rubber pad	——	1.23
Carpet with fibrous pad	——	2.08
Building Paper	——	0.06
Polyethylene sheet	——	0.00
COVERING		
Plaster	0.20	——
Acoustic tile (3/4")	——	1.78
Asphalt roll roofing	——	0.15
Asphalt shingles	——	0.44
Wood shingles	——	0.94
GLASS		
Flat glass (1/8")	——	0.88[c]
Insulating glass, double (1/4" air gap)	——	1.54[c]
Storm windows, double (1—4" air gap)	——	1.79[c]
Plastic bubble (skylight)	——	0.87[c]
BRICKS-CONCRETE		
Brick, common	0.20	——
Concrete or stone	0.08	——
Concrete block (solid, oven-dried)	0.11	——
Concrete block (8" thick, 2 cores)	——	1.04
Concrete block (8" thick, filled cores)[b]	——	1.93
INSULATION		
Mineral wool batt (3—3-1/2")	——	11.0
Mineral wool batt (5-1/4—6-1/2")	——	19.0
Fiberglass board	4.35	——
Corkboard	3.75	——
Expanded polyurethane	6.25	——
Expanded polystyrene, white	4.05	——
Expanded polystyrene, molded beads	3.75	——
Urea Formaldehyde	5.50	——
Loose fill insulation:		
Mineral wool	3.80	——
Cellulose fiber	4.05	——
Perlite	2.85	——
Vermiculite	2.20	——
AIR		
Inside (still)	——	0.68
Outside (7-1/2 mph)	——	0.25
(15 mph)	——	0.17

[a] For average temperatures of $50°$ F.

[b] Filled with perlite, vermiculite, or mineral wool loose-fill insulation.

[c] Includes effects of inside and outside air films.

divided by the conductivity of the material—the thicker the slab, the higher its R-value. Materials with high R-values, like polyurethane foam, we call "insulators"; those with low R-values are called "conductors." Two sets of R-values are provided in Table I—those for 1" thick slabs and those for commonly available thicknesses. R-values also depend on the temperature of the material—particularly for most insulators. Since the change is slight, the R-values given in Table I are good for most calculations.

The rate of heat transmission through an exterior surface of a building is given by the general formula:

$$H_{tr} = \frac{A \times (T_i - T_o)}{R_t}$$

where

(H_{tr}) = heat transmitted in Btu/hr

A = area of building surface in square feet

$T_i - T_o$ = difference between indoor and outdoor air temperatures in °F

(Eq. 1)

R_t = total thermal resistance of the entire building section

R_t, the "total thermal resistance" of an entire building section (a wall, floor, ceiling, or other building surface viewed on edge), is just the sum of the R-values of its component materials, including R-values of inside and outside air films and of air gaps within the section.

For convenience, when dealing with an entire building section, we often use the *coefficient of heat transmission*, or *U-value*, which is just the inverse of R_t, or $U = 1/R_t$. A U-value generally applies only to an entire section of wall, not to the individual components. The U-value of a building section is a measure of how well it transmits heat—the lower the U-value, the less heat conducted per hour. Numerically, U is the rate of heat loss in Btu per hour through a square foot of surface when there is a 1°F temperature difference between inside and outside air (Btu/hr - ft² - °F). Consequently, the rate of heat transmission (H_{tr}) through a building surface of area (A) is:

(Eq. 2) $$H_{tr} = U \times A \times (T_i - T_o)$$

For example, the heat transmitted through a 50-square-foot wall section with a U-value of 0.12 in which the inside temperature was 65°F and the outside temperature was 40°F is:

$$H_{tr} = 0.12 \times 50 \times (65 - 40)$$
$$= 150 \text{ Btu/hr}$$

Computation of the U-value of a building section involves adding up all the R-values of the components and taking the inverse of this sum.

As an example, the U-values of two typical *walls*—one insulated and the other uninsulated—are calculated in Table II. Note that the uninsulated wall conducts heat about three times more rapidly than the insulated wall. For a 50-square-foot wall, this is the difference between 2300 Btu and 690 Btu lost over a period of 8 hours.

A sample calculation of the U-values for two typical *floors* is presented in Table III. Both floors sit above a ventilated crawl space; they differ only in the addition of foil-faced 3½" batts of mineral wool insulation to the second floor. The decrease in U-value is dramatic. Note that a ¾" air gap between the subfloor and insulation has a fairly large R-value. Convection heat flow doesn't work too well in a downward direction (across this gap) because warm air rises and the foil surface of the insulation reflects most of the thermal radiation back to the subfloor above. If the occupants put a carpet with a rubber pad (R = 1.23) on the hardwood floor, the U-value of the uninsulated floor would drop from 0.36 to 0.25. But with the insulated floor, the U-value drops only from 0.058 to 0.054. Adding more insulation to an already-well-insulated building section is usually a waste of money.

In Table IV, we compare the U-values of uninsulated and insulated *ceilings*. Both ceilings are exposed to the outdoor air through a ventilated attic, but the attic vents are much smaller and protected from winter winds in the insulated case. In this case, the attic air space and roofing materials contribute additional resistance to the total. The drop in U-values is even more dramatic here—more than eightfold!

TABLE II   U-VALUES OF TYPICAL FRAME WALLS — INSULATED AND UNINSULATED (See Table I for R-values.)

WALL CONSTRUCTION COMPONENTS	R-VALUES	
	UNINSULATED	INSULATED
Outside air film (15 mph wind)	0.17	0.17
Wood bevel siding, lapped	0.81	0.81
1/2" Insulating board sheathing	1.32	1.32
3-1/2" Air gap	1.01	––
3-1/2" Mineral wool batt	––	11.00
1/2" Gypsum board	0.45	0.45
Inside air film (still)	0.68	0.68
TOTALS (R_t)	4.44	14.43
U-VALUES (U = $1/R_t$)	0.23	0.069

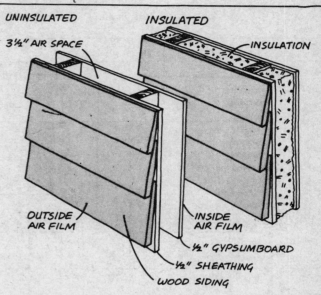

FIG. 1   CROSS SECTIONS OF INSULATED AND UNINSULATED WALLS

TABLE III   U-VALUES OF TYPICAL FLOORS (insulated and uninsulated)

FLOOR CONSTRUCTION COMPONENTS	R-VALUES	
	UNINSULATED	INSULATED
Outside air film (15 mph wind)	0.17	0.17
3-1/2" Mineral wool batt	––	11.00
3/4" Air gap, foil on one side	––	3.55
25/32" Wood subfloor	0.98	0.98
Vapor permeable felt	0.06	0.06
3/4" Hardwood floor	0.68	0.68
Inside air film (still)	0.92	0.92
TOTALS (R_t)	2.81	17.36
U-VALUES (U = $1/R_t$)	0.36	0.058

TABLE IV   U-VALUES OF TYPICAL CEILINGS (insulated and uninsulated)

CEILING CONSTRUCTION COMPONENTS	R-VALUES	
	UNINSULATED	INSULATED
Outside air film (15 mph wind)	0.17	0.17
Wood shingles	––	0.94
Building paper	––	0.06
1/2" Plywood	––	0.62
Attic air gap	––	0.85
6" Mineral wool batt	––	19.00
1/2" Gypsumboard	0.45	0.45
3/4" Acoustic tile	1.78	1.78
Inside air film (still air)	0.61	0.61
TOTALS (R_t)	3.01	24.48
U-VALUES ($1/R_t$)	0.33	0.041

Calculating Infiltration Heat Losses

A second type of heat loss in buildings is from the *air infiltration* through openings in walls and through cracks around doors and windows. Warm air is lost by these routes and all the cold outdoor air that enters to replace it must be heated to room temperature for maximum comfort. The amount of air infiltration is fairly predictable. It depends upon wind speed and upon the linear footage of cracks around each window or door. The rate of infiltration heat loss is given by the general formula:

$$H_{in} = C \times L \times Q \times (T_i - T_o)$$

where C = the heat capacity of air (0.018 Btu/ft³/°F)

L = the total crack length in feet

Q = the rate of air leakage in cubic feet per hour
(Eq. 3)   per foot of crack (see Table V)

TABLE V AIR INFILTRATION THROUGH WINDOW CRACKS		Air Leakage (Q)[1] at Wind Velocity (mph)				
Window Type	**Remarks**	**5**	**10**	**15**	**20**	**25**
Double-hung wood sash	Average fitted,[2] non-weather stripped	7	21	39	59	80
	Average fitted,[2] weatherstripped	4	13	24	36	49
	Poorly fitted,[3] non-weatherstripped	27	69	111	154	199
	Poorly fitted,[3] weatherstripped	6	19	34	51	71
Double-hung metal sash	Non-weatherstripped	20	47	74	104	137
	Weatherstripped	6	19	32	46	60
Rolled-section steel sash	Industrial pivoted[2]	52	108	176	244	304
	Residential casement[4]	14	32	52	76	100

[1] Air leakage, Q, is measured in cubic feet of air per foot of crack per hour.
[2] Crack = 1/16 inch.
[3] Crack = 3/32 inch.
[4] Crack = 1/32 inch.

SOURCES: *The Solar Home Book,* 1976
ASHRAE — Guide and Data Book, 1965

For example, with 15 mph winds beating against an average double-hung, non-weatherstripped, wood-sash window, the rate of air leakage (Q) is 39 cubic feet per hour for each foot of crack. If the total crack length is 16 feet and the temperature is 65°F indoors and 40°F outdoors, the infiltration heat loss is:

$$H_{in} = 0.018 \times 16 \times 39 \times (65 - 40)$$
$$= 281 \text{ Btu/hr}$$

This is about the same as the transmission heat loss through a 50-square-foot uninsulated wall under the same conditions! If the window was weatherstripped (Q = 24 instead of 39), the infiltration heat loss would drop to 173 Btu/hr. To get the total infiltration heat loss, it is important to note that outside air entering in one side of a building will push air through the other side. Because of this, we usually use only ½ of the total crack length for the value of "L" for each kind of crack.

In addition to the heat lost by infiltration through the cracks, there is also a convection heat loss through open doors and windows. Generally these losses are offset by the heat generated by electric lights and appliances and by the occupants themselves.

Heat Load Calculations for a Typical Uninsulated House

We can now calculate the overall building heat loss, or its *heat load.* We take the building heat load to be the sum of the transmission heat losses through exterior surfaces and the infiltration heat losses through cracks around windows and doors.

There are two important quantities we'll need to determine—the *design heat load* and the *seasonal heat load.* The design heat load (Btu/hr), is the hourly heat loss of the house at the lowest expected outdoor temperature *(design temperature).* The design temperature and design heat load determine the size of the heating equipment needed to keep a house warm during the worst cold spells. On the other hand, the *seasonal heat load* is the overall heat loss of a building (in Btu's) during an entire heat season (the period from October to May when a house needs heat). To calculate the seasonal heat load, you need to understand the concept of *degree-days.* If the outdoor temperature is 1°F below the indoor temperature for one day, we say one degree-day has accumulated. Standard practice uses 65°F as the base indoor temperature, so that if the outdoor temperature is 40°F for a full day, we say 65 - 40 = 25 degree-days accumulate. The total number of degree-days in a single heating season is useful because it affects how often and how much the outdoor temperature falls below the comfort level. It determines just how much fuel you are likely to burn during this period to keep your house warm. Local oil dealers, propane distributors, and heating contractors can usually provide the degree-days and design temperatures for your local area.

To calculate the design and seasonal heat loads of a house, you merely add the losses through all exterior surfaces and cracks. The design load can be calculated by substituting the design temperature into equations (2) and (3). To get the seasonal heat load, you merely substitute 24 times the number of degree-days in place of $(T_i - T_o)$ in both equations. For example let's calculate the design and seasonal heat loads of a drafty, uninsulated, wood-frame house in the San Francisco Bay area, where the design temperature is about 35°F and the number of degree-days is around 3000.

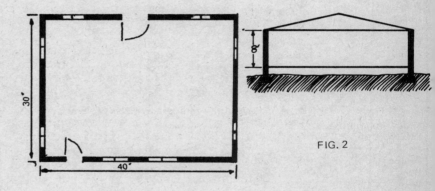

FIG. 2

Assume the house is 40 feet long and 30 feet wide (Fig. 2). It has uninsulated frame walls and a hardwood floor above a ventilated crawl space. The uninsulated ceiling has acoustic tile and sits below a ventilated attic and a low, pitched roof of plywood and wood shingles. The house has 16 single-pane, double-hung, wood-sash windows (each 4 feet high by 2.5 feet wide) and 2 solid oak doors (each 7 feet by 3 feet). The doors and windows are somewhat weathered and fit their frames rather poorly.

First, we need the U-values of each kind of exterior surface. From Table II, we know that an uninsulated frame wall has a U-value of 0.23. Table I gives us R_t = 0.88 for a single pane of glass (U = 1/0.88 = 1.13). Also from Table I, we get R = 0.91 for a 1" thickness of oak. Adding the resistances of inside and outside air films, we get R_t = 1.76 or U = 0.57 for the two doors. A U-value of 0.36 for bare hardwood floors above a crawl space comes directly from Table III, but about half the floor area of this house is covered with carpets (U = 0.24). Finally, the uninsulated ceiling has a U-value of 0.33 (Table IV). The transmission heat losses through each of these surfaces are calculated using equation (2). They are summarized in Table VI.

Infiltration heat losses are also calculated in Table VI by using appropriate Q-values from Table V. Double-hung, poorly-fitted, wood-sash windows have a Q-value of 111 ft³/hr in a 15 mph wind. But around poorly-fitted doors, the infiltration rate is almost twice that: 220 ft³/hr for each crack foot. And there is still some infiltration through cracks around window and door frames as well—with a Q-value of 11. These Q-values are then multiplied by the heat capacity of a cubic foot of air (0.018 Btu/ft³/°F), and ½ the total length of each type of crack to calculate the design and seasonal heat losses due to infiltration.

Under design conditions of a 15 mph wind and 35°F air temperature, this California house loses a total of 46,263 Btu/hr (= 35,336 Btu/hr transmission losses + 10,927 Btu/hr infiltration losses). Its heating system must be large enough to supply about 46,000 Btu/hr, or its temperature indoors will fall below 65°F *during the severest weather.* In colder climates, the same house would have even greater design heat loads—for example, 77,000 Btu/hr in Chattanooga, Tennessee, where the design temperature is 15°F, and 123,000 Btu/hr in Sioux Falls, South Dakota, where the design temperature is -15°F. At $8.00 per million Btu, a typical price for electric heat, that's almost a dollar an hour to heat this house under design conditions in Sioux Falls.

The seasonal heat load of our California house (Table VI) is 111 million Btu, or a total cost of about $300 at present prices for natural gas and heating efficiency (65%). Natural gas prices are sure to double in the near future, and the home owner would be wise to invest now in insulation, weatherstripping, storm windows, or shutters.

One of the primary reasons for making these heat load calculations is to discover for yourself just where your house is losing the most heat. You will note from Table VI that relatively little heat is being lost through and around the doors. But a great deal of heat is escaping through the ceiling, floors and windows. With this kind of knowledge, you can make much wiser and more economical choices for weatherproofing the house.

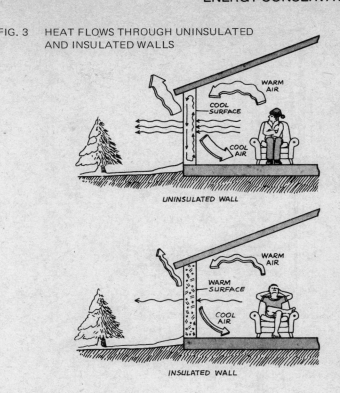

FIG. 3 HEAT FLOWS THROUGH UNINSULATED AND INSULATED WALLS

TABLE VI	DESIGN AND SEASONAL HEAT LOAD CALCULATIONS — UNINSULATED HOUSE			
			Transmission Heat Losses	
Surface	Area (Ft^2)	U-value[a] (Btu/hr/ft^2/°F)	Design (Btu/hr)	Seasonal (million Btu's)
Walls	918	0.23	6,334	15.20
Windows	160	1.13	5,424	13.02
Doors	42	0.57	718	1.72
Bare floor	600	0.36	6,480	15.55
Carpeted Floor	600	0.25	4,500	10.80
Ceiling	1200	0.33	11,800	28.51
TOTAL TRANSMISSION HEAT LOSSES			35,336	84.80
			Infiltration Heat Losses	
Cracks around:	Length (ft)	Q-value[b] (ft^3/hr/ft)	Design (Btu/hr)	Seasonal (million Btu's)
Window sash	128	111	7,672	18.41
Door	20	220	2,376	5.70
Window & Door Frames	148	11	879	2.11
TOTAL INFILTRATION HEAT LOSSES			10,927	26.22
BUILDING HEAT LOADS			46,263	111.02

[a]See Table II, III and IV. [b]See Table V.

Insulation and Weatherproofing

Only a summary of the many methods of retarding heat loss or heat gain is possible here. These methods include insulation, storm windows and doors, insulating shutters, weatherstripping, and wind protection. In most cases, the money you spend on such measures can be recouped in a matter of a few years through lower costs for heating and cooling. In new buildings, the initial cost of heating and cooling equipment are diminished because much smaller units can be installed. And in new houses, it becomes easier to use wood stoves, solar heating, or natural cooling methods to provide a large part of your comfort needs.

Insulation retards the flow of heat across air gaps in walls, floors and ceilings. Most common forms of insulation are fibrous or cellular materials that trap air in tiny pockets and prevent air circulation in the space they occupy. Often one or both sides of the insulation material has a shiny foil surface that reflects thermal radiation—increasing its ability to retard the flow of heat. In addition, the layer of insulation keeps the interior surfaces warm (closer to the indoor air temperature). Then the rooms are more comfortable because radiative and convective heat flows between these surfaces and the inhabitants are drastically reduced (see Fig. 3). Often the thermostat can be set much lower in winter.

There are a number of different forms of insulation on the market (see Table I for R-values).

Blankets or *batts,* usually of fiberglass or mineral wool, fit between the joists in wood-frame construction. Often one side of the blanket or batt is faced with aluminum foil to provide extra thermal resistance.

Boards of cork, fiberglass, foamed glass, polyurethane, or polystyrene come in pre-cut sizes. They are frequently used as perimeter insulation around foundations or added to existing walls during renovation.

Blown Insulation—polyurethane or urea formaldehyde—are spray foams that are applied under pressure by skilled contractors.

Loose-fills such as polystyrene beads, fiberglass, perlite, vermiculite and cellulose fiber are loose granular materials that can be poured into place in stud walls. They are also ideal for insulating existing masonry walls made with hollow concrete blocks or cavity-wall construction.

References 3-5 at the end of this section will aid you in the selection of insulation appropriate to your purposes and pocketbook. You should first insulate ceilings and roofs, because warm air collects at the top of rooms and heat flows upward. Six inches of fiberglass batt insulation (R = 19) has been the standard for ceilings and roofs in cold climates, while 3½ inches (R = 11) has been used in walls. These standards are being accepted in mild climates like California and greatly upgraded in colder areas, where R-values of 20 in the walls and 25 to 30 in the ceilings are not excessive. Often these R-values can only be attained using polystyrene, polyurethane or urea formaldehyde insulation. (See Table I.)

The next place to stem heat flows is at the windows. Under the same indoor and outdoor conditions and for the same surface area, a single pane of glass will conduct 115 Btu/hr, double glass will conduct 60 Btu/hr and a wall insulated to R-11 standards will conduct 7 Btu/hr. You lose the same amount of heat through a single-pane window 2½ feet wide and 4 feet high as you do through an insulated wall 20 feet long and 8 feet high! Addition of a storm window will cut these conduction losses by almost a factor of two. Depending upon their R-values and tightness of construction, an insulating curtain or shutter can reduce conduction losses through a window even more . . . by a factor of two to ten. They will also reduce radiation and convection heat transfers from warm bodies in the room to the cold window surface.

Insulating curtains or shutters need not be elaborate. A tightly-woven material lined with "fiberfill"—a loose stuffing or blanket-type insulating material—can be suspended from a curtain hanger or attached directly to the top window frame (see Fig. 4). Velcro fasteners (like those on down jackets or sleeping bags) allow a tight seal and permit you to open the curtain during times of winter sun. You can fashion cheap, effective window shutters from 1'' or 2'' thick polystyrene board insulation. Measure and cut the boards so that they fit snugly into the window frame and block the flow of cold air from the window surface into the room. A

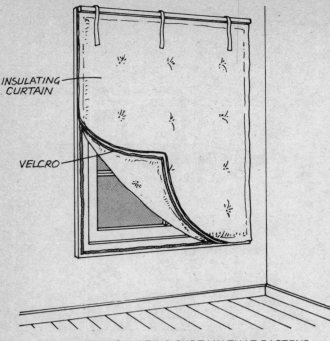

INSULATING CURTAIN

VELCRO

FIG. 4   AN INSULATING CURTAIN THAT FASTENS
TIGHTLY TO THE WINDOW FRAME

recent window shutter invented by Zomeworks Corporation is the "Night-wall"—styrofoam panels attached directly to the glass surface with magnetic clips. The panels are light enough to stay firmly in place against the glass but can be easily removed and stored during the day. Get the styrofoam from your local building supplier and Zomeworks will sell you the magnetic clips.

On north windows, the insulating curtains or shutters can be left in place for most of the heating season. For east, west and especially south windows, they should be closed only at night or during sunless winter periods. These same windows can be shuttered on sunny summer days, with the north windows left uncovered for indirect lighting.

As you reduce heat transmission losses, the penetration of outdoor air becomes a large part of the remaining heat loss. Air infiltration can account for 20 to 55 percent of the heat loss in most homes. Caulking and weatherstripping are the most straightforward, traditional solutions to this problem. The addition of storm windows and doors or insulating curtains and shutters helps restrict infiltration through cracks around the door or window sash. You can also install tighter-fitting windows or plant trees for wind protection, but these measures involve more planning and expense.

Weatherstripping helps to check infiltration where edges of doors and windows meet their frames. From Table V, you can see that weatherstripping improves the performance of any window—particularly in high winds. Fixed windows save the most energy. You need a few operable windows for ventilation, but the rest can usually be caulked *shut* for the winter. All others should be weatherstripped. With double-hung windows, it's best to use spring-bronze or felt-hair weatherstripping. Gasket, compression-type weatherstripping works best on hinged doors and casement or awning-type windows. A good discussion of weatherstripping, caulking and sealants can be found in reference 6.

Strong winds blowing directly upon a building can severely increase infiltration losses. And they also increase the conduction heat flow through the exterior surfaces—particularly the windows and doors. In 20 mph winds, an average house loses *twice* as much heat as it does in a 5 mph breeze. With good wind protection on three sides of a house, however, fuel expenses can be cut by as much as 30 percent. New buildings should be oriented away from prevailing winter winds or screened from them by densely planted evergreens and proper landscape (see references 7-9). Because winter winds come mostly from the north and west, entrances should not be located on these sides and the number of windows should be kept to a minimum. Storm windows and doors are recommended for those that must be located there. Ventilated attics and crawl spaces can be shut off during winter, or at least protected from the wind by baffle plates.

Heat Load Calculations for an Insulated House

Now let's examine the design and seasonal heat loads of a house that has been well-insulated, with doors and windows shuttered and weather-stripped. For ease of comparison, let's use our California frame house (see Fig. 2). In this case, however, the walls are insulated with 3½" batts of fiberglass insulation, so that their U-value is 0.069 instead of 0.23. Six inches of insulation has been added to the ceiling and the attic protected from the winds, so that U = 0.041 instead of 0.33 (see Table IV). Under the floors, foil-faced 3½" fiberglass batts lower the U-value of the bare floors to 0.058 and that of carpeted floors to 0.054. Storm windows placed on all the windows cut their U-value to 0.60, and insulating curtains that are drawn at night lower it further to 0.20, for a daily average of 0.40. The storm doors over both entrances cut their U-value to 0.34. Transmission heat losses through all these surfaces are once again calculated using equation (2), and the results are displayed in Table VII.

TABLE VII   HEAT LOAD CALCULATIONS — INSULATED HOUSE

Surface	Area (ft^2)	U-value ($Btu/hr/ft^2/°F$)	Transmission Heat Losses Design (Btu/hr)	Seasonal (million Btu)
Walls	918	0.069	1900	4.56
Windows	160	0.40	1920	4.61
Doors	42	0.34	428	1.03
Bare Floor	600	0.058	1044	2.51
Carpeted Floor	600	0.054	972	2.33
Ceiling	1200	0.041	1476	3.54
TOTAL TRANSMISSION HEAT LOSSES			7740	18.58

Cracks around	Length (ft)	Q-value ($ft^3/hr/ft$)	Infiltration Heat Losses Design (Btu/hr)	Seasonal (million Btu)
Window sash	64	24	829	1.99
Door	20	34	367	0.88
TOTAL INFILTRATION HEAT LOSSES			1196	2.87
BUILDING HEAT LOADS			8936	21.45

Note that the total transmission heat losses are more than a factor of 4 smaller than those of the uninsulated house (see Table VI). Even larger reductions occur in the infiltration losses. With the windows on the north and west caulked shut, we need only consider the infiltration through the south windows, which have been weatherstripped to a Q-value of 24. The addition of weatherstripped storm doors (Q = 34) and the elimination of leakage by caulking around window and door frames cuts even further into the infiltration losses.

The design heat load of this house is down to 9000 Btu/hr, or less than 20% of that in the drafty, uninsulated house. The seasonal heat load of 21.45 million Btu costs only $60 at present gas prices—a savings of $240 per year. This dividend will only increase as fuel prices rise in the coming years.

References

[1] A.I.A. Energy Steering Committee, 1975. ENERGY AND THE BUILT ENVIRON-MENT. The American Institute of Architects, Washington, D.C.

[2] Stanford Research Institute, 1971. PATTERNS OF ENERGY CONSUMPTION IN THE UNITED STATES. Menlo Park, California.

[3] Eccli, Eugene, 1976. LOW-COST, ENERGY-EFFICIENT SHELTER. Rodale Press, Emmaus, Pennsylvania.

[4] HUD, 1975. IN THE BANK . . . OR UP THE CHIMNEY? Chilton Book Company, Radnor, Pennsylvania.

[5] NAHB Research Foundation, 1971. INSULATION MANUAL: HOMES/APARTMENTS. NAHB Research Foundation, Inc., Rockville, Maryland.

[6] Eugene Eccli, *op. cit.*

[7] Robinette, G.O., 1972. PLANTS, PEOPLE AND ENVIRONMENTAL QUALITY. U.S. Government Printing Office, Washington, D.C.

[8] "The Technology of the Cooling Effects of Trees and Shrubs," Robert Deering, HOUSING AND BUILDING IN HOT CLIMATES, Building Research Advisory Board, Report No. 5, 1952. National Academy of Sciences, Washington, D.C.

[9] White, Robert. 1954. EFFECTS OF LANDSCAPE ON NATURAL VENTILATION. Texas A & M Research Report No. 45. Texas A & M, College Station, Texas.

ENERGY CONSERVATION REVIEWS
ENERGY

The energy crisis engulfs us, and so do books about energy; books that describe what energy is, how it is converted and measured, what it can do, how its use influences our social structure, how its abuses have determined our history, and how energy policies

affect our lives and living habits. We have found the following books to be especially valuable in helping us to learn about energy and to sort out the complexities of our energy problems and the practical options open to us.

MAN, ENERGY, SOCIETY

Earl Cook is a geographer/geologist and has contributed many important papers to the field of energy and mineral uses. His book MAN, ENERGY, SOCIETY might be considered as a sequel to Cottrell's book, ENERGY AND SOCIETY in that Cook tries to give a social, historical and geographic view of energy as a background for understanding contemporary problems while discussing how the use of various energy resources has affected social evolution. Cook brings this concept up

to date by suggesting that the current energy crisis is caused by "an impaired availability of energy to (industrialized) countries . . . and not a worldwide shortage of energy." With excellent graphics, Cook describes the availability and potential of various energy resources . . . but always with an historic, social and geographic perspective. The chapter on renewable energy resources is superficial but contains some very important concepts as to the social implications of renewable energy systems. This book is highly recommended.

— R.M.

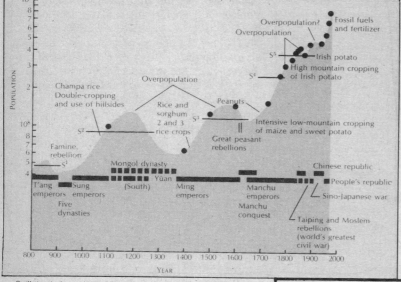

MAN, ENERGY, SOCIETY
Earl Cook
1976; 478 pp.
$7.95

from:
W.H. Freeman
San Francisco, CA

Oscillations in the population of China appear to result from introduction of new food plants, expansion of cropped land, increase in food supply, and subsequent overpopulation, soil erosion, famine, and warfare. Note the logarithmic vertical scale. The horizontal lines marked S^1, S^2, S^3, S^4, and S^5 represent estimated population levels that would have been stable with contemporary food supply. Present vigorous efforts to reduce China's population growth rate may emerge from an estimate of S^6, not shown here. (Basic data from Ping-Ti Ho, 1959.)

ENERGY AND SOCIETY

This work traces man's use of energy, from readily available but low-yield sources . . . to more complicated but high-yield forms . . . The thesis is that the amounts and types of energy employed condition man's way of life materially, and set somewhat predictable limits on what he can do and on how society will be organized.

With that, Cottrell set out on an elegant description of history as moved by the energy sources people have used. His chapter "The Industrialization of Agriculture" was one of the first convincing analyses showing modern agriculture to be an energy sink rather than an energy producer. By using energy as the common denominator for history, Cottrell succeeded in predicting 20 years ago most of today's dilemmas.

—R.M.

Energy and Society
William F. Cottrell
1955; 330 pp
$13.00

from:
Greenwood Press Publishers
51 Riverside Avenue
Westport. Connecticut 06880

or WHOLE EARTH TRUCK STORE

THE POVERTY OF POWER

THE POVERTY OF POWER has received wide acclaim as being one of the most important analyses of our energy problems to date. To me it is a very enlightening book . . . as far as it goes. To be sure, it is a lucid description of the relationship between the production and economic systems of our society, and the ecological systems of nature. Commoner notes that the objectives of these systems are usually at odds with one another, so that "efforts to solve one crisis seem to clash with the solutions of the other . . . resulting in the confusion and gloom that besets the country . . . [However] what confronts us is not a series of separate crises, but a single basic default" . . . According to Commoner, this means a lack of appreciation of the Second Law of Thermodynamics in major policy issues. He then discusses in a most imaginative way the meaning of the Second Law and its relationship to order, time, probability, information and entropy. Commoner gets his inspiration from the 1974 report of the American Physical Society describing the need for a more general measure of "efficiency" . . . one based on the Second Law (energy quality), rather than the First Law (energy quantity) of Thermodynamics. This is followed by a thought

ENERGY FOR SURVIVAL

This is one of the most important energy books of the early 1970's. Wilson Clark has done a superb job of assembling up-to-date energy information and using it to tell a comprehensive story on a largely non-technical level. Many readers will feel the main value of the book lies in its ecological views of energy use and development. It is also an extensive information source, as well as a guide to sources. About one thousand references are given. The book covers the history, sociology, politics, and corporate structure of energy and its use, as well as technical and environmental aspects.

One of the really outstanding sections of this book is *Electricity from Nuclear Fission*. After reading Clark's well documented account, it's hard to see how anyone could feel secure about present government-industry safeguards. In fact the real value of Clark's evaluations lies more in the questions he raises than in his conclusions. This book is a veritable who's-who and what's-what of "alternative" energy technology. Fossil-fuel and energy conversion processes are extremely well covered. It also contains one of the best discussions of solar, wind and biofuel energy systems now in print. Highly recommended.

— Roger Douglass
— Richard Merrill

Energy for Survival
Wilson Clark
1974; 652 pp.
$4.95

from:
Doubleday & Co., Inc.
501 Franklin Avenue
Garden City, NY 11530

or WHOLE EARTH TRUCK STORE

provoking analysis of how our fossil fuels are being misused, why they seem to be in short supply, and why nuclear power is such a disaster.

Commoner implies throughout that if we are to base economic and ecological decisions on the common good, we must use the Second Law of Thermodynamics as a guideline. This means, among other things, that a technology must be matched thermodynamically to the task being performed (heating a house to 70°F with a 10,000°F thermonuclear reactor is absurd, etc). The book is incomplete in that it fails to deal precisely with the possibility that thermodynamic efficiency may have more to do with the way society is organized to distribute energy resources, than it does with the way a machine or technological system performs.

— R.M.

The Poverty of Power
Barry Commoner
1976; 297 pp
$2.50

from:
Bantam Books, Inc.
414 E. Golf Rd.
Des Plaines, IL 60016

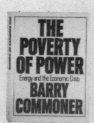

The Laws of Thermodynamics mold our lives . . . often without us realizing it and always in spite of our activities. The Laws tell us that the quantity of energy in the universe can't change (First Law), whereas its quality must change . . . (Second Law). Because it is qualitative, the Second Law of Thermodynamics tends to be interpreted according to one's discipline and inclinations. Biologists talk of losses in information, ecological successions and the Second Law as a (only apparent) barrier to biological organization. Physicists and chemists describe kinetics, the one-way flow of physical reactions and the Second Law as "times arrow." Economists and social thinkers point to the "entropy trap," "Second Law efficiencies" and the inevitability of no-growth economics. The common denominator here, of course, is the unifying principle of "entropy" . . . a concept of such fundamental importance that it resists analysis and, hence, traditional understanding. Three books to ponder:

— R.M.

ENERGY AND ECOLOGY

Four lectures delivered in 1966 at the University of Chicago. Probably the most lucid account of recent ideas in ecological theory. Margalef explores and attempts to integrate the dynamics of ecological feedback systems, biological diversity and stability, energy flows, information theory, evolution, the organization of ecosystems, and the effect humans have on these systems. He concludes by describing the changes of natural communities as they move through time in the short run (succession) and the long run (evolution). The result is that this book is an elegant description of the laws of thermodynamics from a biological point of view. It just doesn't provide another intellectual branch to cling to . . . it projects one into a totally different frame of ecological reference and energy consciousness.

— R.M.

Evolution cannot be understood except in the frame of ecosystems. By the natural process of succession, which is inherent in every ecosystem, the evolution of species is pushed — or sucked — in the direction taken by succession, in what has been called increasing maturity. The implication is that in general the process of evolution should conform to the same trend manifest in succession. Succession is in progress everywhere and evolution follows, encased in succession's frame. As a consequence, we expect to find a parallel trend in several phylogenetic lines which can also be recognized as a trend realized in succession. For instance, a decrease in the ratio of production to biomass and an increase of the efficiency and specialization should be common trends in phylogenetic lines.

Perspectives in Ecological Theory
Ramon Margalef
1968; 111 pp
$2.45

from:
University of Chicago Press
11030 S. Langley Avenue
Chicago, Illinois 60628

or WHOLE EARTH TRUCK STORE

ENERGY, KINETICS AND LIFE

This book explains in a semi-technical manner the concept of entropy and its relation to the basic processes of chemistry and biology. Specifically, Miller discusses entropy as it relates to the human condition and our every-day experiences, and presents a critical analysis of the many "technology-will-solve-our-problems" scenarios (he calls them myths) now abounding, explaining their limitations in terms of thermodynamics. His last chapter, *Concluding Unscientific Postscript*, is a series of unique essays relating entropy to concepts of evil, ethical values, consciousness and priorities of social action.

— R.M.

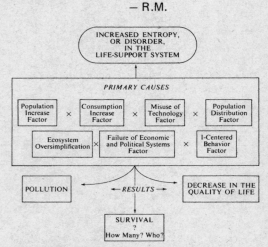

A Very Crude Model of Major Causes and Results of the Ecological Crisis on Spaceship Earth.

Energetics, Kinetics, and Life
G. Tyler Miller, Jr.
1971; 360 pp
$6.95

from:
Wadsworth
Publishing Co., Inc.
10 Davis Drive
Belmont, CA 94002

or WHOLE EARTH TRUCK STORE

ENERGY AND ECONOMICS

It has been said that ecology is nothing more than long-term economics. If this is so, then the essence of today's human problems has a great deal to do with our transition from short term expediency to long term consequences . . . from the painful conflicts of growth, to the cultural symbiosis of global equilibrium. Such a transition will require not only new policies and technologies, but a fundamental reorganization of economic premises, social priorities, cultural traditions and political strategies. This book attempts to describe thermodynamics in economic terms. It is a challenging book that relates entropy to economic value, industrial development, the _real_ limits of growth and to social conflict.

— R.M.

The Entropy Law and the Economic Process
Nicholas Georgescu-Roegen
1971; 457 pp.
Harvard University Press   **$5.95**
Cambridge, Massachusetts

LOVINS ON ENERGY

Since the early 1970's, Amory Lovins has been writing about the "soft" energy path we need to follow in dealing with our energy problems. His articles and books have given intellectual support to what seems to be the central issue of the real energy crisis . . . the long-term political and social implications of our energy options. On the one hand we have a widening of our traditional "hard energy path" . . . centralized energy production from non-renewable resources (nuclear and fossil fuel) powering esoteric technologies controlled by a few corporations, and creating convenience at the expense of freedom and environmental quality. On the other hand we can choose the "soft energy path" . . . decentralized energy production from solar and other renewable resources, produced where they are needed, motivated by grass roots politics and controlled by local associations and regional needs. These are the extreme choices and the future will (hopefully, at least) be somewhat of a middle ground.

Lovins' writings present a straightforward global energy analysis, clear and precise (see especially WORLD ENERGY STRATEGIES). His critique of nuclear energy is particularly devastating (NON-NUCLEAR FUTURES). But the real value of his works lies in his elegant rationale and demonstrations that: (1) industrialized countries can develop a future that is relatively free of fossil fuels and nuclear power by using solar, hydro, wind and liquid organic fuels and by mixing these resources to the specific situation needed (Second Law efficiencies); (2) the soft and hard energy technology options are mutually exclusive . . . once we commit ourselves to one path we cannot turn around; (3) the widespread utilization of renewable energy systems can help create a more humane and equitable society; and (4) underlying the difference between the soft and hard energy paths is a fundamental different perception about how society should be organized.

— R.M.

It is encouraging, then, that the concept of a soft energy path brings a broad convergence which . . . cuts across traditional lines of political conflict. It offers a potential argument for every [point of view] : civil rights for liberals, States' rights for conservatives, availability of capital for business-people, environmental protection for conservationists, old values for the old, new values for the young, exciting technologies for the secular, and spiritual rebirth for the religious.

THE LOVINS TRILOGY

World Energy Strategies: Facts, Issues and Options
1973; 131 pp
$4.95

Non-Nuclear Futures
1975; 151 pp.
$5.95

Soft Energy Paths
1977; 231 pp.
$6.95

all from:
Friends of the Earth
529 Commercial Street
San Francisco, CA 94111

ODUM ON ENERGY

Environment, Power and Society bent my mind the first time I read it. It is difficult to sum up swiftly, but you might think of it as an energy I Ching. The book is packed with unique energy flow diagrams, tables of energy data and flashes of inspirational prose. With further readings, however, you get the idea that Odum tends to stretch the analogy between physical and biological systems a bit too much, and we are left with life forces being managed like machines and reduced to little more than kilocalories.

Read Cottrel first and then read this . . . carefully:

In recent years studies of the energetics of ecological systems have suggested general means for applying basic laws of energy and matter to the complex systems of nature and man. In this book, energy language is used to consider the pressing problem of survival in our time . . . the partnership of man in nature. An effort is made to show that energy analysis can help answer many of the questions of economics, law and religion, already stated in other languages.

Energy Basis for Man and Nature is a more lucid book than *Environment, Power and Society.* It outlines the concepts of net energy and Odum's unique energy language/symbolism much more clearly than in his earlier book. It also elaborates on Odum's thesis that:

Everything is based on energy. Energy is the source and control of all things, all values, and all actions of humans and nature . . . the options available for the future are set out for us by the laws of energy. As patterns of energy change, so do human roles. When national governments are confused about the future, the reason is that (they) fail to understand the primary role of energy . . . how it controls our economy, our international relationships, our standard of living and our culture.

P.S. Odum has also written two papers that are perhaps the most concise examination of the global energy situation to date. 1) *Energy, Ecology and Economics* (written for the Royal Swedish Academy of Sciences), and 2) *More Perspectives on World Energy Relationships.* Both are reprinted in *The Mother Earth News* #27 and the *Whole Earth Epilog.*

— R.M.

Environment, Power, and Society
Howard T. Odum
1971; 331 pp.   from:
    John Wiley and Sons, Inc.
$6.95   One Wiley Drive
    Somerset, New Jersey 08873

Energy Basis for Man and Nature
Howard T. Odum and Elisabeth C. Odum
1976; 297 pp   from:
    McGraw-Hill Book Co.
    1221 Avenue of the Americas
$7.95   New York, NY 10020

EARTH, ENERGY AND EVERYONE

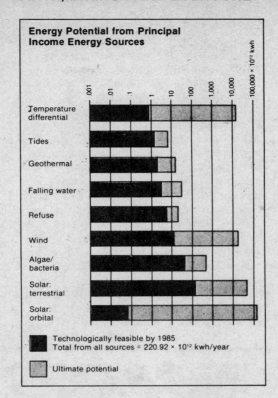

Energy Potential from Principal Income Energy Sources

Sources (top to bottom): Temperature differential, Tides, Geothermal, Falling water, Refuse, Wind, Algae/bacteria, Solar: terrestrial, Solar: orbital

Scale: .001, .01, .1, 1, 10, 100, 1,000, 10,000, 100,000 × 10¹² kwh

■ Technologically feasible by 1985
Total from all sources = 220.92 × 10¹² kwh/year

▨ Ultimate potential

EARTH, ENERGY AND EVERYONE documents a World Game Workshop sponsored by the research and education organization, Earth Metabolic Design of New Haven, Connecticut. It is a unique energy book in that it documents in outline form the history, advantages and disadvantages of 21 major renewable and non-renewable energy resources. More importantly, it graphs the potential development for most of the energy resources according to their scale of application: Single Dwelling Unit, Community Unit, Industrial Unit, Regional Sub-System, Global System. Finally, as noted by Buckminster Fuller in the Foreword:

This book makes it incontestably clear that it is feasible to harvest enough of our daily income of extraterrestrial energy as well as of [geothermal power] all generated at an inexorable, nature-sustained rate to provide all humanity and all their generations to come with a higher standard of living and greater freedoms than ever have been experienced by any humans and to do so by 1985, while completely phasing out all future use or development of fossil fuels, atomic and fusion energies.

The basic strategy for accomplishing this end is a hydrogen-based energy system fueled by the energy income of solar, wind, hydro- and biofuel energy resources. With so much potential for a sane energy future, this book also makes it clear that the "energy crisis" is really an "energy-policy crisis."

— R.M.

Energy, Earth and Everyone
Medard Gabel, et al (ed)
1975; 192 pp.
$4.95

from:
Simon and Schuster
One W. 39th St.
New York, NY 10018

ENERGY: SOURCES, USE AND ROLE IN HUMAN AFFAIRS

This is one of the best popular overviews of energy now available. The Steinharts discuss in plain language the potentials and drawbacks to fossil fuels and nuclear energy. They also give a relatively good outline of solar power and other renewable energy systems . . . an area glossed over in most energy overviews. One chapter, *The Energy We Eat*, is perhaps the definitive textbook treatment of the energy budget of our food system. The description of energy conversions and thermodynamics is superficial . . . but this detracts little from an otherwise fine book.

— R.M.

Energy: Sources, Use and Role in Human Affairs
Carol and John Steinhart
1974; 362 pp.

$5.95

from:
Wadsworth Publishing Co.
Duxbury Press
North Scituate, Massachusetts

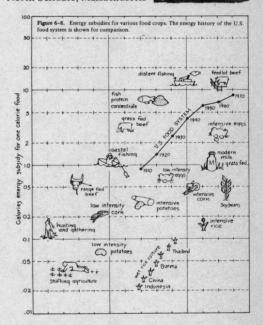

Figure 6–8. Energy subsidies for various food crops. The energy history of the U.S. food system is shown for comparison.

ENERGY

With all the talk about "energy crisis" these days, few of us have a good theoretical understanding of what "energy" is all about. Most of us equate an "energy crisis" with long lines at the gas station, but do we know the finer points of the subject? What's the difference between potential and kinetic energy? What relation does heat, radiation and electricity have to the concept of energy? What are the basic units of measurement we use when discussing energy? The author takes us through the complexities of this broad subject as only a master of the subject can—simply, clearly and directly.

— C.M.

Energy
Bruce Chalmers
1963; 289 pp

$9.25

from:
Academic Press
111 Fifth Avenue
New York, NY 10003

or WHOLE EARTH TRUCK STORE

DIRECT ENERGY CONVERSION

Understanding the theoretical basis of energy conservation means understanding thermodynamics and the physics of energy conversions and storage. This is a highly technical book for those with a strong background in the physical sciences. Most of the text is a discussion of thermodynamic theory applied to solar-thermal electricity; photovoltaics; fuelcells; plus thermoelectric, thermionic and magneto-hydrodynamic generators. For understanding these processes at this level it is an excellent book.

—Larry Lewis

Direct Energy Conversion
Stanley W. Angrist
1971; 488 pp

$15.95

Allyn and Bacon, Inc.
Boston, MA.

ENERGY FUTURES IN INDUSTRY

ENERGY FUTURES is the result of three years of investigation of 142 corporations involved in energy research and development (renewable and non-renewable sources). As a simple survey, it stands on its own. The research programs of all the large coprorations are outlined and the technologies of the energy systems are discussed. Smaller companies are not reviewed except in the wind section. What is most significant about the book is the over-all picture it gives of the relationship between industry and government. Nearly every corporation states that they will not or cannot engage in solar development and related technologies without ERDA and other government backing. The reasons for this state of affairs are many, but basically it reflects, in part, the capital shortage in the U.S.; the fact that corporations are mostly interested in short term profits and not the long term picture; and that for many large corporations, alternative energy sources compete with their existing technologies.
— Donald Marier

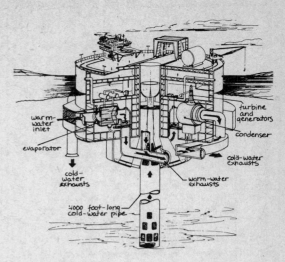

TRW's proposed ocean thermal electric powerplant

Energy Futures: Industry and the New Technologies
S.W. Herman and
J. S. Cannon
1976; 752 pp.

$10.00
(summary copy)
$256.00
(complete report)

from:
The National Center
for Community Action
1711 Connecticut Ave.
Washington, D.C. 20009

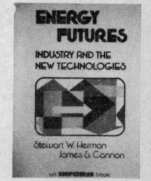

"SCIENCE" MAGAZINE, ENERGY ARTICLES

SCIENCE magazine, the official weekly publication of the American Association for the Advancement of Science, often contains excellent articles on subjects of energy conservation and technology. Scientifically objective and full of data, the articles generally represent the state-of-the-arts of the scientific establishment. ENERGY: USE, CONSERVATION AND SUPPLY is a reprinted collection of energy articles that appeared over a period from March 1973 to July 1974. They are placed under the headings of: *People and Institutions; Energy and Food; Oil, Coal, Gas and Uranium;* and *Developing Technologies* (Solar, Wind, Methanol, Geothermal, Hydrogen, etc.). In addition, the entire April 19, 1974 issue of SCIENCE (available in any library) is devoted to energy, and includes many articles not contained in the book above. Since 1974, energy articles in SCIENCE have been common but scattered.

— R.M.

Energy: Use, Conservation and Supply
Philip H. Abelson (ed)
1974; 154 pp

$4.95

from:
The American Association for the Advancement of Science
1515 Massachusetts Avenue
Washington, D.C. 20005

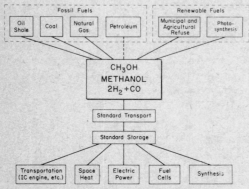

Sources, transport, and possible applications of methanol.

ENERGY AND POWER

This is a classic popular primer for the "energy crisis." It consists of eleven essays that first appeared in the September 1971 SCIENTIFIC AMERICAN. The essays have been editorially composed into a gallery of energy portraits which illustrate key energy concepts and processes. The book is very comprehensive. Freeman Dyson's *Energy in the Universe* could give a person religion; whereas Earl Cook's *The Flow of Energy in an Industrial Society*, Roy Rappaport's *The Flow of Energy in an Agricultural Society*, and Claude Summers' *The Conversion of Energy* have become modern day classics. ENERGY AND POWER takes a holistic view of our energy resources and our manner of using them. It will be useful long after other energy books have become obsolete.
— R.M.

Energy and Power
A Scientific American Book
1971; 144 pp

$3.50

from:
W. H. Freeman and Co.
660 Market St.
San Francisco, CA 94104

KILOWATT COUNTER

If you need to learn about energy from scratch and in practical terms, this is the best place to start. KILOWATT COUNTER discusses the fundamental concepts of energy (efficiency, units of measure, conversions, net energy, patterns of use, etc.), and their application to our everyday lives. Numerous examples are given that help you to survey energy use in your home, and allow you to make value judgements about your habits of energy consumption. There is also a brief overview of renewable energy resources. This would be an excellent primer for school use.
— R.M.

Kilowatt Counter
Gil Friend and David Morris
1975; 36 pp.

$2.00

from:
Alternative Sources of Energy
Rt. 2, Box 90A
Milaca, MN 56353

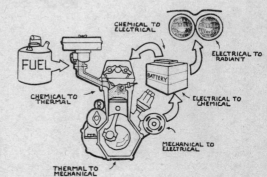

THE AUTOMOBILE: A SYSTEM OF ENERGY CONVERSIONS

ENERGY CONSERVATION

ENERGY AND EQUITY

Illich is a social critic who has taken stabs at the medical establishment, education and at energy consumption. His basic hypothesis in this brief book is that we need to establish limits in our industrialized societies which will help create a more humanistic culture that is not exploitative, over-consumptive or wasteful. The focus for the analysis is transportation or traffic; the vast amounts of energy, time and money we use getting around. Equity is at the heart of the issue because, for example, the "jet-setter" and the bicyclist are unequal and for various reasons have valued their time in different proportions—the "jet-setter" is a consumer of energy rather than of time.

If there is fault to be found with this book it is in its simplicity. More empirical work with better thermodynamic analysis has been done by Odum, Lovins and others.

— T.G.

Energy and Equity
Ivan Illich
1974; 83 pp.
$0.95

Perennial Library
Harper & Row, Publishers, Inc.
10 East 53rd St.
New York, N.Y. 10022

MISCELLANEOUS ENERGY BOOKS

Energy
Gerard M. Crawley
1975; 337 pp
Macmillan Publishing Co., Inc.
New York
$10.95

A fine elementary textbook for discussing energy concepts, resources, conversions and potentials. Solar energy is touched on lightly.

Energy and the Environment
John M. Fowler
1975; 496 pp
McGraw-Hill Book Co.
New York
$8.50

A good general overview of energy concepts, traditional "alternative" energy options and efficiencies of current practices and technologies. Lots of data, but deals primarily with megascale systems. Poor coverage of renewable energy.

Explaining Energy
Lee Schipper
(Energy Resources Group, U.C. Berkeley)
1976; 71 pp
National Technical Information Service
LBL-4458
Springfield, VA 22161
$6.00

Over priced, but an excellent non-technical description of energy and various energy technologies. Booklet includes reproductions of most of the important energy diagrams, tables and figures scattered throughout the literature. Over 800 energy references are also listed.

ENERGY: THE CASE FOR CONSERVATION

This is a brief overview of energy conservation and a fine critique of the many myths preventing its wide adoption. Hayes concludes from his analysis:

> More than one-half the current U.S. energy budget is waste. For the next quarter century the United States could meet all its new energy needs simply by improving the efficiency of existing uses. The energy saved could be used for other purposes and relieve us of the immediate pressure to commit enormous resources to dangerous energy sources before we have fully explored all alternatives. Energy derived from conservation would be safer, more reliable, and less polluting than energy from any other source. Energy conservation could reduce our vulnerability in foreign affairs and improve our balance of payments position. Moreover, a strong energy conservation program would save consumers billions of dollars each year.

Hayes makes it clear that, contrary to popular opinion, a dollar invested in energy conservation often makes more net energy available than a dollar invested in developing new energy resources. Furthermore, a reduction of energy use does not demand a reduction in economic production. But Hayes implies much more and joins a growing number of analysts who are suggesting that energy conservation will not only save energy . . . but will also improve the quality of our lives.

— R.M.

Energy: The Case For Conservation
Denis Hayes
1976; 77 pp.
$2.00

from:
Worldwatch Institute
1776 Massachusetts Ave., N.W.
Washington, D.C. 20036

KEEPING WARM FOR HALF THE COST

This is by far the best homeowner's guide for reducing energy use in the home since Eugene and Sandy Eccli's *Save Energy, Save Money*. Belts are tighter in England, and they have a stock of houses several hundred years old that require thoughtful techniques for insulation, so the British have a lot of experience we can learn from. The book's best new information seems to be a section on clear how-to information for making insulating window shutters and storm windows, and ideas for commercial products available in England that could be usefully produced here: sheetrock laminated to styrene insulation for direct application to existing walls; rigid foam insulating tiles to apply to ceilings where access to rafter space is difficult; foam panels covered with masonite for insulating existing concrete floors; kits for glass storm windows. Good detailed instructions, clear illustrations and lots of practical know-how for dealing with difficult or unusual situations. Insulation levels suggested should be at least doubled for the U.S. (and probably for England), but otherwise pretty directly applicable here.

—Tom Bender

HOMEOWNERS' GUIDE TO SAVING ENERGY

This is a very practical and well-illustrated book. Photos and detailed graphics explain the fundamentals of home weatherizing ("R" and "U" values, etc.) how to insulate and weatherstrip your home, and how to save energy by the proper maintenance of heating systems, air conditioners, refrigerators, freezers, water heaters and other large appliances. There is also a section on home wiring and lighting techniques designed to save energy.

— R.M.

Homeowner's Guide to Saving Energy
Billy L. Price and James T. Price
1976; 288 pp.
$8.95

from:
Tab Books
Blue Ridge Summit, PA 17214

350 WAYS TO SAVE ENERGY IN YOUR HOME AND CAR

There are scores of books describing the many ways one can save energy around the home. This book describes a variety of rather conventional and obvious techniques for home insulation, for the maintenance of heating units and large appliances, and for breaking bad habits that consume energy. There is also a nice section on driving for energy economy.

— R.M.

350 Ways to Save Energy (and Money) in Your Home and Car
Henry R. Spies et al
1974; 198 pp.
$3.95

from:
Crown Publishers, Inc.
419 Park Avenue South
New York, NY 10016

This top-hung single shutter is easily constructed and installed. A batten is screwed to the ceiling joists to hold a metal plate for the magnetic mortice catches which are let into either end of the shutter. Inset: The shutter is lifted and lowered using a hook screwed into a pole and which engages screw eyes in the shutter's lower edge.

Keeping Warm for Half the Cost
Phil Townsend and John Colesby
1975; 91 pp
$3.00 (US)

from:
Conservation Tools & Technology
143 Maple Road, Surbiton
Surrey KT6 4BH ENGLAND

AMERICAN INSTITUTE OF ARCHITECTS: ENERGY CONSERVATION AND BUILDING.

all books available from: American Institute of Architects 1735 New York Avenue, N.W. Washington, DC 20006

ENERGY AND THE BUILT ENVIRONMENT: A GAP IN CURRENT STRATEGIES

Energy conservation and on-site generation in buildings is an enormous potential "energy-source" that policy-makers have (until recently) completely ignored in their rush to increase energy supplies. This report, which has catalyzed much of the AIA's current thinking on energy conservation in building design, concludes that we could be saving the equivalent of 12.5 million barrels of oil per day by 1990. This amount is about equal to the projected energy supply from any one of the prime fossil or nuclear sources: domestic oil, coal, domestic and imported natural gas, or nuclear energy. The document is extremely valuable because it ties the problems and potentials of energy conservation in buildings to the larger picture of national energy supply and demand. A must reading for the serious design professional and interested layperson.

Energy and the Built Environment: A Gap in Current Strategies
Leo A. Daly, FAIA
1974; 18 pp   **free**

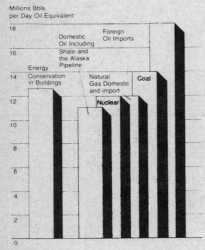

Figure 13: Energy Conservation in Buildings as a Substitute for Supply—1990

ENERGY AND THE BUILT ENVIRONMENT: A NATION OF ENERGY-EFFICIENT BUILDINGS BY 1990

This follow-up report outlines a national energy program to achieve the potentials of energy-efficient buildings described in its predecessor. It shows how the program can be made feasible, and presents a series of recommendations for immediate action.

**Energy and the Built Environment:
A Nation of Energy-Efficient Buildings by 1990**
AIA Energy Steering Committee
1975; 20 pp
free

ENERGY CONSERVATION IN BUILDING DESIGN

This comprehensive survey of energy-conserving options in building design is addressed primarily to practicing architects. The heart of the publication outlines those areas of the building process where energy savings are possible and attempts to describe the trade-offs that must be considered when choosing a particular measure. The discussion is geared to large commercial buildings and the book is not much help to a designer, builder or owner of single-family dwellings. Basically, it's a good survey for the architect just beginning to worry about energy-consumption in the buildings he or she designs.

Energy Conservation in Building Design
AIA Research Corporation
1974; 156 pp
$5.00

NEW DESIGN CONCEPTS FOR ENERGY CONSERVING BUILDINGS

In the near future, energy-conscious design must become a standard practice of the architectural profession. This book is a compendious account of one attempt to stimulate such practice in professional schools of architecture. Sponsored by the AIA Research Corporation in the Spring of 1976, the National Student Competition in Energy-Conscious Design attracted hundreds of entries from across the United States. Of these, 103 design projects are represented here, and 12 Entries of Distinction are discussed in detail using the students' own comments and working drawings. Although some of these projects strike me as awkward and unrealistic, it's intriguing to glimpse the incredible diversity of building forms created in response to a few common, energy-wise aims.

—Michael Riordan

New Design Concepts for Energy Conserving Buildings
AIA Research Corporation
1976; 120 pp.
$9.95

YOUR ENERGY-EFFICIENT HOUSE

Although this book has a very comprehensive table of contents, its treatment of those contents is rather simplistic. It takes a crack at owner-design of homestead and house, both new and old, but just doesn't seem to follow through. There's some good general material on natural ventilation and the use of plants for wind control, but the sections on energy conservation and natural fuel sources are no match for other publications that present clearer and more directly usable information for about the same price. However, this is a reasonable introductory book for people interested in energy-conservation in their homes.

—Lynn Nelson

Sometimes we need simple books to get us going. Useful for grade school/high school classes in energy conservation and building.

—R.M.

Your Energy-Efficient House
Anthony Adams
1975; 118 pp
$4.95

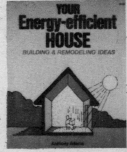

from:
Garden Way Publishing
Charlotte, VT 05445

or WHOLE EARTH TRUCK STORE

SAVE ENERGY, SAVE MONEY

This booklet should be subtitled "Energy Conservation for Those With a Low Income." It's full of nifty little tips and ideas that cost only a few pennies but save you dollars in the long run. The Ecclis are continually stuffing old newspapers or rags into crevices and bags of leaves under crawl spaces or slapping old blankets over doors and windows in a feverish effort to keep the heat in. Their solutions may not strike the middle-class homeowner as being very pretty, but if you're cold *and* poor, you've little time for aesthetics. There isn't much detail given in the 40 typewritten pages, but there are a goodly number of illustrations that make their point well. You should read this booklet carefully, though, because there are a few errors. Otherwise, it's a highly recommended publication.

—Michael Riordan

Save Energy, Save Money
Eugene Eccli
and Sandy Fulton Eccli
1974; 40 pp
free

from:
Community Services Agency
Washington D.C.
(OEO pamphlet 6143-5)

IN THE BANK . . .
OR UP THE CHIMNEY?

If you're ready to *do* something about the amount (and cost) of the energy it takes to heat and cool your home, but are a little hazy about the choices and their costs, this is the manual for you. A how-to-go-about-it gem, this well-illustrated guide leads you step by step through the process of determining just what energy-saving measures your house needs, their approximate cost and your potential savings. Then comes the actual how-to information. For each measure selected, the guide tells you (1) how hard it is so you can decide whether to do it yourself or should hire a contractor; and (2) how to get the job done. If you prefer the do-it-yourself approach, you're told what tools, types and amounts of needed materials, and you're given safety tips and step-by-step installation instructions. Each step is illustrated so you're not left guessing how to get from step 3 to step 4. If you opt for contractor installation, there are pointers to help you make sure the job gets done right.

The manual also includes other practical information, left out of many "save energy" publications, such as buying the insulation, choosing a contractor, getting financing, and maintaining and repairing your house's heating and air-conditioning equipment. All in all, it'll give you confidence to tackle energy-saving home improvements and empower you to do a lot of the work yourself. It's one of the best books around on its subject, and at $1.95 it also seems to be the cheapest.

—Lynn Nelson

In The Bank . . . Or Up The Chimney?
U.S. Department of Housing and Urban Development
1976; 73 pp

$1.95

from:
Chilton Book Company
Radnor, PA 19089

or WHOLE EARTH
TRUCK STORE

KEEPING THE HEAT IN

Conserving energy in buildings is not a simple task. Proper applications of insulation, use of storm windows, blocking of air drafts and using energy conserving building materials takes planning, money and work. This Canadian Government report is one of the best "how to do it" books available. Insulation of basements, attics, walls, floors, crawl spaces and more is covered in simple, direct language and excellent diagrams. A special section has estimates of fuel savings when insulation is applied to certain parts of a house in different Canadian climates.

—T.G.

Keeping the Heat In
Office of Energy Conservation
Energy, Mines and Resources, Canada
1976; 106 pp

INSULATION GUIDES

MAKING THE MOST OF YOUR ENERGY DOLLARS

The homeowner's guide to selecting energy conservation measures for the home based on the technical report, *Retrofitting Existing Housing for Energy Conservation.* Both the technical report and the homeowner's guide are still the "best of the bunch" of all the insulation guides we've seen. Insulation values are keyed to climate and energy costs—you pick what you think energy will cost through the life of your home. They also give economics for storm windows, weatherstripping, floor insulation and other conservation measures.

—Lee Johnson
RAIN

Making the Most of Your Energy Dollars
National Bureau of Standards Consumer Information
Series 8
1975; **$0.70** from:
U.S. Government Printing Office
Washington, DC 20402

or WHOLE EARTH TRUCK STORE

MINIMUM DESIGN STANDARDS FOR HEAT LOSS CALCULATIONS

A concise and practical guide. Resistance values for common building materials are given, as are instructions for calculating heat transmission through a building section. Appendix 4, "Overall Heat Transmission Coefficients of Typical Wall, Ceiling and Floor Sections," eliminates the need to calculate the coefficient if the building section is identical to one of the many sketched. It also gives instructions for calculating the heat loss from an entire building.

—Gail Morrison

Minimum Design Standards for Heat Loss Calculations
1973; 66 pp from:

Free U.S. Dept. of Housing and Urban Development
451 Seventh Street SW
Washington, D.C. 20410

or your regional HUD office

HOW TO INSULATE HOMES FOR ELECTRIC HEATING AND AIR CONDITIONING

A small booklet that shows how to install insulation for the greatest effectiveness. It is similar to the illustrated how-to- pages in the NAHB Research Foundation's *Insulation Manual.*

—C.M.

How to Insulate Homes for Electric Heating and Air Conditioning
1973; 32 pp **Free**
from:
National Mineral Wool Insulation Association, Inc.
211 East 51st Street
New York, NY 10022

RETROFITTING EXISTING HOUSING FOR ENERGY CONSERVATION: AN ECONOMIC ANALYSIS

An excellent technical analysis comparing the economic desirability of adding insulation, storm windows and weatherstripping to existing houses. The first study available which analyzes for a wide range of energy costs as well as climatic conditions. It also contains a model that can be used to calculate what combinations will give homeowners the greatest savings in investing different amounts of money in energy conservation measures for their homes.

—Lee Johnson
RAIN

Retrofitting Existing Housing for Energy Conservation: An Economic Analysis
National Bureau of Standards Building Science
Series 64
1975; from:
U.S. Government Printing Office
$1.35 Washington, DC 20402

or WHOLE EARTH TRUCK STORE

INSULATION MANUAL: HOMES/APARTMENTS

A manual directed toward builders and contractors interested in keeping construction costs down (who isn't) and increasing the sales appeal of their product. It's slanted toward mineral wool and fiberglass insulation, but one glance at the list of sponsors tells why. All in all, though, this is a very useful and enlightening report. Direct, concise and informative, it has that no-fooling-around, let's-get-down-to-the-facts approach I like. There's an excellent section on installing blanket insulation and detailed reference sections for calculating heat losses and gains. Unfortunately, the cost comparison data is already out-of-date because it's based on 1970 fuel prices—a revised edition is sorely needed. —Michael Riordan

Insulation Manual: Homes/Apartments
NAHB Research Foundation, Inc.
1971; 44 pp from:
$4.00 NAHB Research Foundation, Inc.
P.O. Box 1627
Rockville, MD 20850

or WHOLE EARTH TRUCK STORE

1946-60 · TORONTO/MONTREAL · 1500 SQ. FT.

Original Condition	Condition After Upgrade	Fuel Saved
No Insulation Storms	Basement R 8 Ceiling R 20–30	45%
No Insulation No Storms	Ceiling R 20–30 Basement R 8 Storms	48%

free

from:
Book People
2940 Seventh St.
Berkeley, CA. 94710

EARTH-COVERED BUILDINGS:

We've come a long way from the days when Malcolm Wells was a lone voice crying in the desert, warning us to bury our buildings in the earth. The benefits of using the earth to cover roofs and walls are becoming more obvious every day. Energy is saved in the operation of an earth-covered building because of the insulating value of the soil and the fact that the average soil temperature is lower than outdoor air temperatures in summer and higher in winter (see graph below). Some estimates put the possible energy savings at 30%. Energy is also saved in the construction phase because much less is needed to fabricate the building materials and transport them to the site. And as an extra bonus, the noise levels inside an earth-covered building are severely reduced, while exterior surfaces can be used for trees, shrubs, and other vegetation.

For a thoughtful, entertaining critique of earth-covered architecture, read "Underground Architecture," by Malcolm Wells in the Fall 1976 issue of CoEvolution Quarterly. Popular Science ran a good survey of current happenings: "Underground Houses," in their April 1977 issue. Watch for Malcolm Wells' Underground Designs, soon to be published.

— Michael Riordan

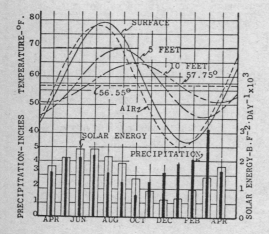

annual variation of soil temperature, mean air temperature, solar energy and precipitation at a selected site.

from: *The Use of Earth Covered Buildings*

EARTH INTEGRATED ARCHITECTURE

This publication is a comprehensive workbook about earth-covered building in the vicinity of Phoenix, Arizona. It's addressed primarily to the problems of underground building in arid soils and climates, but much of the material can be generalized to other areas. Though it suffers from shoddy editing and is repetitious in places, the book is a pretty thorough catalog of the considerations that enter into the design of a building integrated into the earth. Particularly cogent are two sections detailing soil properties and landscaping requirements. Other sections allude to the compatibility of earth-integrated architecture with solar heating and natural cooling, but they shy away from in-depth discussions.

— Michael Riordan

Earth Integrated Architecture
James W. Scalise, editor
1975: 314 pp.   **$10.00**

from:
Architectural Foundation Publications
College of Architecture
Arizona State-University
Tempe, Arizona 85281

THE USE OF EARTH COVERED BUILDINGS

This publication contains the Proceedings of the July 1975 Fort Worth, Texas, conference on underground buildings. This is the most comprehensive and up-to-date source available on underground building. "Down-to-earth" information on legal, economic, insurance, structural, psychological, historical, and energy considerations. Life-cycle costs show underground building is becoming more viable as energy costs to operate buildings increase. Very comprehensive bibliography, list of people actively working on underground building, etc.

—Tom Bender

The Use of Earth Covered Buildings
National Science Foundation
1975; 353 pp.
$3.25

Superintendent of Documents
U.S. Government Printing Office
Washington, DC 20402

U.S. DEPARTMENT OF HOUSING AND URBAN DEVELOPMENT: EARTH SHELTERS

The Office of International Affairs of the U.S. Department of Housing and Urban Development makes available a wide variety of free publications on low-cost, self-help housing. Send for their "Selected Publications Checklist" to get a listing of what's available. As you'd expect, many of these books and pamphlets have a strong international messianic flavor (What can I do to help my downtrodden brethren?), but there's a lot of solid, well-reasearched information available, and the price is right.

Earth for Homes
Ideas and Methods Exchange #22
This manual attempts to identify those methods of earth-building that are most durable and amenable to owner-building. Topics covered include soil composition, earth-wall construction and stabilization, earth floors and roofs, finishing materials, and the design of earth homes. Adobe and rammed-earth construction methods are particularly well covered.

Handbook for Building Homes of Earth
This booklet covers much of the same ground as EARTH FOR HOMES, but it is much more systematic in its treatment. It has less to say about the pros and cons of the various types of earth construction and provides more detailed descriptions of the techniques involved. A very valuable book for the owner-builder committed to building his home with earth.

— Michael Riordan

Order from:
from: U.S. Dept. of Housing & Urban Dev.
Washington, DC 20410

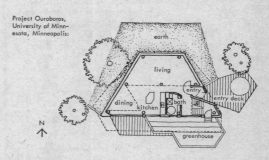

Project Ouroboros,
University of Minnesota, Minneapolis:

ENERGY CONSERVATION IN THE COMMUNITY

The structure of a community—its commercial buildings and their relation to residences, its arteries and its land use patterns—can have a critical impact upon energy consumption. The urban sprawl of many cities in the American Southwest, for example, places undue emphasis on automobile transportation and the consequent waste of ever more expensive gasoline. More compact urban arrangements, with mixed *residential and commercial areas, are much more conducive to mass transit systems and pedestrian and bicycle traffic. People are just beginning to come to grips with the question of how urban and regional planning policies can affect the energy consumption of a community.*

— Michael Riordan

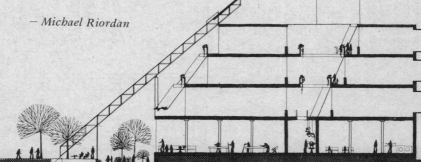

from "Winona"

A LANDSCAPE FOR HUMANS

To me this seems to be one of the most ignored books of the 70's, by one of the true pioneers of decentralist planning and so-called "appropriate technology." Van Dresser proposes a plan of regional development quite unlike traditional concepts. He suggests, rather, (1) that there should be a redistribution of population, of production and of patterns of trade, for greater regional self-sufficiency, (2) that this smaller range of urban places should be functional economic and cultural elements, (3) that more of our social effort should be directed away from massive transport and towards the enrichment and diversification of localized production within efficient smaller communities, (4) that there should be a production technology that utilizes renewable resources on a small-scale, skill-intensive basis, and (5) that there should be an ecological science of community design that lives in symbiosis with the soil, climate and biotic regimes of an area. Heavy idealistic stuff, but this book attempts to deal with the practicality of these alternatives in terms of the culture and environment of a specific regional community in the US (southern Rocky Mountains of northern New Mexico), which, Van Dresser believes, typifies both classical patterns of rural decay and an unusual potential for regeneration. A LANDSCAPE FOR HUMANS should be required reading for every county and regional planning department in the country.

—Richard Merrill

A Landscape for Humans
Peter van Dresser
1972; 128 pp.
$3.00

from:
Biotechnic Press
P.O. Box 26091
Albuquerque, NM 87125

GRASSY BROOK VILLAGE

When complete, Grassy Brook Village will consist of 20 housing units in two clusters, each with its own solar heating and waste handling systems. With the help of HUD financing for the solar heating system, construction of the first 10-unit cluster began in the Fall of 1976. Completion is scheduled for September 1977. Another cluster is planned for the future, as well as a wind energy system to provide electricity for all 20 homes.

—Michael Riordan

For more information contact:
Richard D. Blazej
R.F.D. No. 1
Newfane, Vermont 05345

WINONA

Another of Dennis Holloway's multi-disciplinary student teams at the University of Minnesota has examined the application of energy conservation and alternative energy sources to a total existing community — the river town of Winona in southeastern Minnesota. Their project, which illustrated how the community as a whole might become energy self-sufficient by the year 2000, has evolved into a traveling exhibition, and a companion volume crammed full of ideas and designs. There are suggestions for restructuring neighborhoods by clustering the housing while providing more in-city space for gardening. Solar collectors are on rooftops, and windmills dot the land.

There's a proposal to renovate old downtown buildings for combined commercial and residential use — heated of course by solar energy. These and many other designs are illustrated with very detailed drawings and photographs of the models used in the traveling exhibition. One wishes, however, that the verbal descriptions had gone into similar detail instead of skimming the surface and describing only possibilities. Nevertheless, this is an important and unique book, for it illustrates how community cooperation and planning can be crucial elements in attaining true energy self-sufficiency.

—Michael Riordan

Winona
Huldah Curl, editor
1975, 122 pp.
$5.00 Postpaid

from:
Publications: Winona
University of Minnesota
2818 Como Avenue, S.E.
Minneapolis, MN 55414

BUILDING VALUE

A comprehensive checklist of new guidelines for determining the value of a building design — in light of increasing fossil fuel costs and scarce natural resources. Prepared for the California Office of the State Architect, this manual examines building values from several perspectives, including its value as an energy structure and its impact upon public services and the community as a whole. Most of the document is devoted to an extensive resource section which contains some of the best annotated bibliographies I've seen in this area.

—Michael Riordan

Unlike the case of the private builder, both the cost of state buildings and the associated costs of energy supply, transportation, water and waste treatment, health, and other support services come out of the same pocket. It is therefore appropriate to examine and evaluate these systems together in order to minimize total costs. The concept of valuating building projects as public investments addresses reduction of the impacts on utility, transportation, safety service and other public services. In light of this kind of accounting, the future value of increased self-sufficiency in buildings is clear, particularly if it is acquired by the use or reuse of energy or other resources acquired or retained on site (e.g. solar energy and heat recovery).

Building Value: Energy Design Guidelines for State Buildings
Tom Bender and Lane deMoll
1976, 99 pp.
$3.00

from:
Office of the State Architect
P.O. Box 1079
Sacramento, CA 95805

A STRATEGY FOR ENERGY CONSERVATION

A study of energy conservation in Davis, California, this landmark report reached some conclusions that were quite startling at the time of its publication but have become common knowledge nowadays. For example, with some very simple measures, the energy consumption in Davis homes and apartments could be cut by 50%. The report goes on to propose a number of detailed building standards and neighborhood planning requirements to bring this reduction about. Most noteworthy are suggested building code modifications to govern the window area and orientation, exterior color, and the heat storage capacity of a building. The city of Davis has taken many of these proposals to heart and become a model city for energy conservation. Many other communities would do well to follow suit and this report can help show the way.

—Michael Riordan

A Strategy for Energy Conservation
Jonathan Hammond et al.
1974; 51 pp.
$5.00

from:
Living Systems
Route 1, Box 170
Winters, CA 95694

PLANNING FOR ENERGY CONSERVATION

Whereas a few communities are just now beginning to talk about energy conservation, the California city of Davis has been doing things for a few years. With a new solar building code, Davis is now reexamining its standards and practices in neighborhood planning and urban design that could lead to needed energy conservation. One of the groups behind this effort, Living Systems, has produced a manual full of resolutions that the city might adopt to curb growing energy use and encourage the substitution of renewable energy sources in place of fossil fuels. This remarkable document contains many of the nitty gritty details of how you can get city hall to back such an effort. It's thorough, well-researched, well-documented and concise.

I especially liked the section on *Street Design for Energy Conservation.* By building narrower streets and shading them well, local air temperatures could be reduced by 10°F on summer days — with drastic reductions in electrical consumption for air conditioners. What's more, narrow streets are cheaper to build, safer for residents and children, and more conducive to bicycle traffic. Urban planners across the country should read the manual just for this section alone.

—Michael Riordan

Planning for Energy Conservation: Draft Report
1976, 83 pp.
$6.50

from:
Living Systems
Route 1, Box 170
Winters, CA 95694

Also ask about: **The Davis Energy Conservation Program,** 1977

SOLAR

SOLAR RADIATION AND ITS USES ON EARTH

John I. Yellott

Professional Mechanical Engineer
Visiting Professor in Architecture, Arizona State University

Astronomers tell us that our sun is only one of untold billions of similar stars, in varying stages of their progression from incomprehensible origins to ultimate extinction. Our particular star is thought to have originated between eight and ten billion years ago and its present rate of energy output, approximately 3.8×10^{23} kilowatts, is caused by the conversion of mass into energy at the rate of some 4.7 million tons per second. It is expected to continue to emit radiant energy at this rate for another four billion years, so, for all practical purposes, it is the only perpetually renewable source of energy which the planet Earth possesses.

It is high time that we get on with the task of learning how to use the massive amounts of energy which the sun gives us each day, and that is what this chapter is all about. Appendix I is a glossary of the words and symbols used in heliotechnology, while Appendix II contains a more in depth mathematical and conceptual approach for the serious student of solar technology. It gives the information needed to make quantitative estimates of how big a solar device must be in order to achieve a desired result. Methods of estimating solar heater performance are in Appendix III, and resources for further work are referred to in Appendix IV.

We will now turn our attention to the uses to which we can put this highly variable but vitally important energy resource, the sun's radiation.

FUNDAMENTAL PRINCIPLES OF SOLAR ENERGY UTILIZATION

There are three primary processes by which the sun's radiation can be put to technical use: (1) Heliochemical, (2) Helioelectrical and (3) Heliothermal. These are derived from the Greek word Helios, meaning "the sun." Other terms often used are: heliostatic, referring to devices which are stationary with respect to the sun or which make the sun appear to stand still; heliotropic, referring to devices which follow or track the sun, and heliochronometry, which means the telling of time by means of the sun. The latter, discussed in a later section, is probably the most ancient of the subjects to be covered in this chapter, since people have been using the shadows cast by various devices to tell time since the days of ancient Eygpt.

Heliochemical

From our point of view, the first of these is by far the most important, since it is the heliochemical process called photosynthesis (App. IV, No. 1) which enables certain wavelengths from the solar spectrum to cause carbon dioxide and water to unite with nutrients from the soil and thus create the plants and oxygen by which we live. All of the coal, oil and natural gas that our planet has ever possessed has come from photosynthesis in ages past, and the food on which we exist today comes from that same process. We are just beginning to understand photosynthesis, but even the greatest laboratory in the world cannot equal a single blade of grass or the leaf of a tree in bringing about this remarkable process.

Man has not yet made much use of heliochemical processes, except perhaps in the field of photography, and so we will turn our attention to the remaining two processes, helioelectrical and heliothermal.

Helioelectrical

Helioelectrical devices are entirely man-made, since nature has not produced anything which converts solar radiation directly into useable electricity. These are also the most recent of the sun-powered devices to make their appearance. Really significant helioelectrical apparatus dates back only two decades to the invention of the silicon solar battery. Silicon solar cells utilize the ability of solar radiation to dislodge electrons from properly treated silicon and thus to cause an electric current to flow.

These were the first sun-powered devices to attain real and unqualified success; the exploration of space and the landing of men on the moon would not have been possible without the arrays of silicon cells which have powered virtually all of our space vehicles. It is to be hoped that the heliothermal apparatus which is about to be discussed will make an equally rapid and significant impact as that made by silicon solar cells.

Heliothermal

Heliothermal devices absorb solar radiation on blackened surfaces and convert it into heat. The black surface will attain a temperature at which equilibrium is established between the rate at which energy is being absorbed and the rate at which the absorbed energy is being lost to the surrounding atmosphere and put to useful service.

The blackened sheet of metal, heavily insulated on the side away from the sun and covered with a sheet of glass to trap the absorbed solar heat, may approach 300°F. The heat absorbed by the metal may be carried away by water circulating through tubes attached to the metal plate, or by air blowing over the sun-heated metal. The glass cover plays a vital part by transmitting as much as 90 per cent of the solar radiation to the plate and by refusing to transmit back to the atmosphere the longwave radiation (heat) which is emitted by the plate. The glass also makes a major reduction in the amount of convective heat loss, i.e., the carrying away of heat by movement of the air at the surface of the plate.

Heliothermal devices have been used with varying degrees of success for more than a century to produce temperatures ranging from below the freezing point of water to nearly 6000°F. We will discuss them in order of the temperature at which they operate, excepting the production of ice. This will be covered in a later section, because even though ice can be produced from solar radiation, it can only be done chemically by using the process of absorption refrigeration which requires a relatively high temperature. Our discussion of heliothermal devices will begin with the solar still, which does its work at temperatures very close to the ambient air temperature.

SOLAR STILLS AND SOLAR DRYERS

Probably the oldest intentional use of the sun's ability to produce heat came when primitive man allowed pools of salt water to evaporate to produce salt essential to the human diet. This process is still in use, as is demonstrated by the great salt works near San Francisco.

Solar Distillation

Controlled evaporation to produce fresh potable water from salt or brackish sources is a much more recent development, but, like most of the heliothermal devices which will be described in this chapter, the basic ideas can be traced back for at least a century. The first large solar still was built at Las Salinas in the Andes Mountains of Chile by J. Harding in 1872, to provide drinking water for the men and animals working in a copper mine. It was nearly an acre in extent, made of wood and glass, and it could deliver some 6000 gallons per day of pure water from a very brackish source.

Its principle of operation is exactly the same as that of the many stills which have been built in the past few years in Australia and on the arid Greek islands in the eastern Mediterranean Ocean. Fig. 1 shows the general idea of these simple devices. A water-tight compartment, made of wood or concrete, is painted black to absorb the solar radiation which enters through the glass roof of the still. Salt or brackish water, or even sewage effluent, is allowed to flow into the channel or box to a depth of four to six inches. The incoming solar radiation heats the water, causing some of it to evaporate and this condenses on the inner surface of the glass roof.

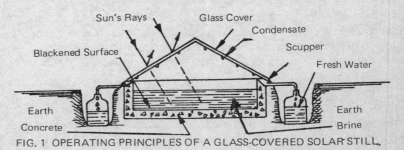

FIG. 1 OPERATING PRINCIPLES OF A GLASS-COVERED SOLAR STILL.

Since water will "wet" (App. I) clean glass, the condensation takes place in the form of a very thin film which flows down by gravity into the scupper which leads the condensate off to a suitable container. The water channel is insulated by the earth. After some of the water has been distilled off, the brine which is left is flushed out at intervals. This prevents the concentration of salt from building up to the point when it will cover the the bottom of the still with reflective white crystals.

The product of such a still is distilled water which can be used for drinking, for filling storage batteries or for any other purpose for which pure water is needed. Attempts have been made to use plastic films as still covers, to eliminate the weight and cost of the glass roofs, but the plastic films have not proved to be as reliable as glass. Most films are not "wet" by water and so the condensation takes place in droplets which tend to reflect the sunlight instead of transmitting it into the basin. Also, few films are completely resistant to deterioration from the ultraviolet component of the solar radiation. They are really only useful in stills which are expected to be used for a short period of time.

Fig. 2 shows the desert survival still devised by R. D. Jackson and C. H. Von Bavel of the Water Resources Laboratory in Phoenix. The kit includes only a sheet of transparent plastic and a tin can, since the other needed materials can be found in the desert. Even in the driest earth there is always some moisture and it can be distilled out by creating a heat trap, as shown in Fig. 2, by using a transparent plastic cover, a rock to weight the cover down in its center and a can to catch the droplets of moisture as they trickle down the inner surface of the plastic. This is the desert version of the ocean rescue still, developed in 1940 by Dr. Maria Telkes, using inflatable plastic bags which, in an emergency, can be used to make drinking water from ocean water.

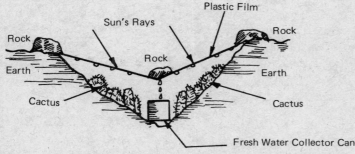

FIG. 2 DESERT SURVIVAL SOLAR STILL

Dr. M. Kobayashi, formerly Managing Director of Nippon Electric Co. of Tokyo, proved that water can be extracted from virtually any kind of soil by the earth-still shown in Fig. 3. He used a typical still construction, complete with cover glass and reflector to increase the amount of solar radiation reaching the earth. He has tested his still at the top of Mt. Fujiyama where the soil is volcanic ash, and in the arid deserts of Pakistan, and he has never failed to produce water that is pure and potable.

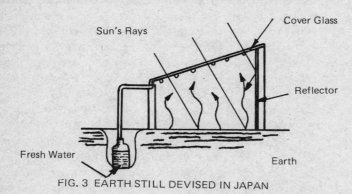

FIG. 3 EARTH STILL DEVISED IN JAPAN

The Brace Research Institute of McGill University has published plans for a simple solar still (App. IV, No. 2) which uses a tray made by folding up the edges of a flat sheet of galvanized steel. The corners are soldered to make the tray water-tight and a drain pipe is soldered to the bottom of the tray. The tray is quite shallow since it is intended to hold less than an inch of water. It is painted black with a flat black waterproof plastic paint after a suitable priming coat has been applied.

The bottom of the still is made of hardboard sheet (Masonite) 1/8 inch thick and the space between the pan and the bottom is filled with wood shavings to act as insulation. In the U.S. glass wool would probably be used as the insulating material, since it is readily available and relatively inex-

pensive. Fig. 4 shows a cross section of the Brace still with enough of the details to convey the essential ideas. The ends of the still are elongated triangles, made of plywood, and the sides are made of single pieces of good quality soft pine, 2 inch by 4 inch, and long enough to suit the dimensions of the tray.

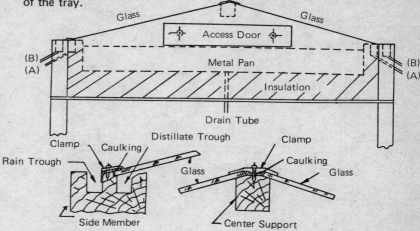

FIG. 4 CONSTRUCTION DETAILS OF A GLASS-COVERED SOLAR STILL, DESIGNED BY BRACE RESEARCH INSTITUTE.

The two troughs serve to carry off the distillate "A" and any rain "B" which may fall on the still (Fig. 4). The cover glasses are supported at an angle of about 15 degrees to the horizontal, with a 1 inch square supporting member which is appropriately beveled. The glass is given a water tight caulking at the upper and lower edges with Dow Corning No. 780 Building Sealant or its equivalent, and a galvanized sheet steel clamp is attached to the supporting member by flat-head screws.

The still is mounted so that the tray is level, with the long dimension running east and west, in a location which is unshaded all day long. The still is filled to a depth of about 1 inch with fresh water each morning. The glass must be thoroughly cleaned before it is installed and the outer surface must be kept clean so that the collected rainwater can also be used.

Where there is adequate sunshine and plenty of non-potable water available, a solar still such as that shown in Fig. 4 is a practical means of turning that water into drinking water at the rate of about one gallon per day for each 10 to 12 square feet of area. Larger stills will produce proportionally greater quantities of distillate, comparable to the 6000 gallons per day produced by the great stills at Las Salinas, Chile, and Coober Pedy, Australia. In summer, temperatures within the tray can be expected to reach $140°F$, with considerably lower temperatures attained during the winter months.

There is a very extensive bibliography devoted to solar distillation. App. IV, No. 3 will lead the reader to much of the literature devoted to this subject, including references to the work of the Commonwealth Scientific and Industrial Research Organization (CSIRO). Most of the successful solar stills in use today use long concrete troughs, painted black, covered with tilted glass roofs similar to that shown in Fig. 4. Careful attention to details, particularly to the prevention of leaks, is essential to good performance.

SOLAR CROP DRYERS

Sunshine has been used since time immemorial for the purpose of drying crops. This has traditionally been accomplished by simply exposing the crop to the sun's rays and hoping that no rain would fall until the drying process had been completed. A more sophisticated technology has been evolved during the past two decades to make more effective use of the sun's thermal power and to minimize the usual contamination associated with natural dehydration. The most common sources of contaminants are:

(a) airborne dust and wind-blown debris, such as leaves;
(b) insect infestation and presence of larvae, etc.;
(c) animal and human interference.

To minimize contamination and to maximize the effectiveness of the sun's rays, it is essential that the drying area be covered by a transparent material. Glass is preferred because of its resistance to deterioration, while plastic films such as Mylar have merit because of their low cost and freedom from breakage due to stones, hail, etc. The Brace Research Institute has published (App. IV, No. 4) a very good description of a solar drying unit which can be made by anyone who has access to simple carpentry tools. This dryer is essentially a solar hot box, which can be used to dehydrate fruit, vegetables, fish, or any other product which needs to be dehydrated for preservation. It consists of a solar "hot box," with two layers of transparent covering, which can be either double strength glass or plastic film (Fig. 5).

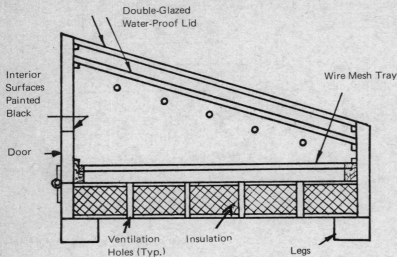

FIG. 5 SECTION THROUGH TRAY-TYPE SOLAR DRYER
(adapted from Brace Research Institute design)

The box itself is generally made of plywood, although larger, permanent dryers may be made of adobe, brick, stone or concrete. The insulation should consist of any materials, such as wood shavings or sawdust, which are locally available and inexpensive, although these are subject to infestation by ants and termites and would have to be treated to prevent this problem. Glass wool is the preferred insulating material since it can survive any temperature which the hot box will attain, and will not support insect life.

The operation of the dryer is quite simple. The ventilation holes at the bottom and the top of the dryer allow air to enter and to carry away the moisture which is removed from the vegetables or other materials by the heat of the sun. The access door in the rear panel enables the materials to be placed on the drying tray, and to be removed when they are dry. The interior of the cabinet should be painted black, while the exteriors of the side and rear panels should be painted with aluminum paint. The number of ventilation holes is determined by the amount of moisture which must be removed from the materials which are being dried.

The angle of slope of the dryer cover should vary with the latitude, but in general an angle of the local latitude minus 10 degrees will be found to be satisfactory. The use of aluminum foil within the cabinet to reflect some of the transmitted rays onto the material may be helpful. The drying trays are made of galvanized wire mesh (i.e., "chicken wire"). Where electricity is available, the use of a small fan to draw air through the dryer would be helpful, but it is not necessary.

The dryer should be glazed, preferably with two layers of glass, fitted in with adequate room for thermal expansion. Ventilation is essential so that the moisture which is "distilled" out of the agricultural material can be removed from the cabinet. This means that screened air holes in the bottom, sides, and the back of the cabinet are essential. One of the most valuable assets of such a dryer is its ability to keep produce dry during rain storms, so the glazed top should be watertight.

Experience with a cabinet dryer of this type indicates that there are upper temperature limits which should not be exceeded for most products. A hot box of this type can readily attain temperatures above 200°F if it is not ventilated, so some degree of care should be taken to prevent overheating of the materials which are to be dried.

SOLAR WATER AND AIR HEATERS

The heating of water and air by solar radiation is one of the oldest and most valuable applications of heliothermal technology. The variety of water heaters and air heaters is almost endless and they all operate on much the same principle.

Solar Water Heaters

Fig. 6 shows a typical flat-plate grid collector for heating water, and Fig. 7 shows the components in an exploded cross-section. The components, progressing inward from the top, are:

(a) The *glazing*, which is usually double strength (1/8 inch) window glass, although 3/16 or even 1/4 inch glass may be used if the size of the pane is so large that the wind loading requires that thickness. Tempered glass is being suggested because of its resistance to breakage, but its extra cost may not be justified in most applications. The glass must be installed with gaskets or caulking which provide enough flexibility to allow the glass to expand when it gets hot and to contract again when it cools off at night. More details on the behavior of glass and other possible glazing materials will be given later in this section.

(b) The *water tubes*, which, in the first generation of heaters built in the 1930's, were generally arranged in a zig-zag pattern. Today's collectors generally use a grid pattern. The metal was formerly copper, but now aluminum and steel are being more widely used for economic reasons. The tubes are usually 1/2 to 3/4 inch inside diameter, with 1 to 1-1/4 inch pipe used for the inlet and outlet headers.

(c) The *flat plate*, which may be any metal—copper, aluminum or steel—that has good thermal conductivity and is reasonable in cost. The term "flat plate" is used to distinguish this type of collector from the concentrator-type which will be discussed later when we deal with high-temperature collectors. The surface of the plate may be corrugated or vee-ed, or fins may be used which are perpendicular to the tubes. The major problem is finding an inexpensive way of fastening the tubes to the plate with a good thermal bond. Thermal cements (App. I) are an efficient way of doing this.

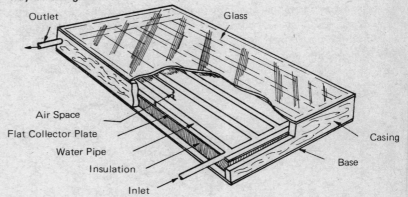

FIG. 6 TYPICAL FLAT-PLATE SOLAR COLLECTOR
(for heating water)

The metal plate must be coated with a radiation-absorbing paint or plating; if painted it should be properly primed before the black coating is applied. There are special coatings available which can absorb most of the sun's rays and re-radiate very little longwave radiation, but these finishes, called "selective surfaces" (App. IV, No. 5) generally require very special equipment and they are really needed only when the collector is intended to operate at quite high temperatures. Flat black paint, properly applied to prevent peeling and cracking, will do a good job for ordinary domestic solar water heaters.

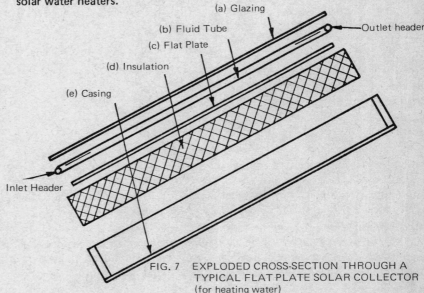

FIG. 7 EXPLODED CROSS-SECTION THROUGH A TYPICAL FLAT PLATE SOLAR COLLECTOR
(for heating water)

(d) The *insulation*, which may be any low-conductivity material that is available and can withstand temperatures up to 200°F. In India, dried palm leaves have been used, but in the U.S., glass wool is the most widely used insulator because it has a low thermal conductivity (see App. I for definitions and values of this important quantity) and it is available at moderate cost in a wide range of widths and thicknesses. Remember that foil-clad insulation only makes use of the radiation-reducing qualities of the foil when the shiny surface has an air space between it and the adjacent surface. Foamed insulation is being used to some extent in collectors and it can add structural strength. Care must be taken to avoid the high temperatures which will cause the foam to melt when a double-glazed collector is exposed to the sun without any water flowing through it.

(e) The *casing*, which holds the collector together and, in combination with the glazing, makes it water- and dust-proof. A simple wooden box, adequately painted and fitted with a hardboard (Masonite) base, will do. Factory built collectors usually have casings made of sheet metal, rust-proofed and painted to resist deterioration.

The flow of water through flat-plate collectors can occur in many ways, as indicated by Fig. 8.

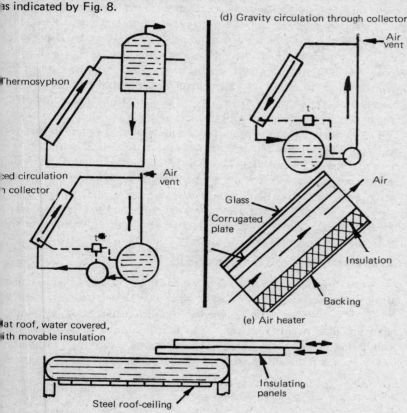

FIG. 8 SOLAR WATER AND AIR HEATERS.

Fig. 8(a) shows the typical thermosyphon system, which will be described in considerable detail in the next section because it is by far the most widely-used heliothermal device. Fig. 8(b) shows a forced-circulation system, with a pump to take water from the lower part of the storage tank and force it upward through the heater to the top of the tank. An extension to the vertical down-comer is shown to represent an air-vent, which will cause the collector to be drained of its water whenever the pump stops. This is necessary to prevent freezing in regions where the outdoor air temperature falls below 32°F for any protracted period of time. An alternative is to use a mixture of water and antifreeze (ethylene glycol is most widely used and readily available), but this requires the use of a heat-exchanger (App. I) between the water circuit and the storage tank. In addition, any antifreeze liquid is certain to be relatively expensive and it may be highly poisonous.

This system would use a thermostat, designated "t," to sense the temperatures of the water in the tank and the surface of the collector. When the sun is shining brightly enough to heat the collector surface to a temperature above that of the water in the tank, the pump is started, and it runs until the sun has moved to a position in the sky where its rays will no longer produce enough heat to enable the collector to operate.

Fig. 8(c) shows the "Skytherm" system invented by Harold Hay of Los Angeles, in which a steel ceiling-roof is covered with water which is confined in shallow plastic bags (e.g., water beds). The insulating panels shown above the water bags can be moved away horizontally by pulling on a cable (not shown) so that the sun's rays can warm the water during sunny winter days. The water will in turn warm the metal ceiling-roof which then becomes a radiant heater for the room below. At night the panels are moved back to cover the water bags and eliminate loss of heat to the cold night air.

During the summer the operation is reversed; the panels remain over the bags during the day and they are rolled back at night so that the water can lose heat to the night sky. During the hottest part of the summer, water can be sprayed on the bags. Some of this exposed water will evaporate, cooling the enclosed water and thus causing the ceiling to function as a radiant cooler.

The "Skytherm" system was tested extensively for eighteen months in Phoenix during 1967-68. It kept a small, one-room structure above 68°F in winter and below 80°F in summer for all except 1% of the 13,140 hours of test conditions. A full-scale "Skytherm" residence is currently in operation at Atascadero, California.

Fig. 8(d) shows a gravity system in which the pump raises the water to a distribution pipe along the top of the collector and gravity makes the water flow downward over the black flat plate in the collector and back to the tank. This is the system which Dr. Harry Thomason uses in his "Solaris" homes. The advantage to this system is that there are no pipes to form a grid in the collector, thus reducing the cost and the weight of the unit. A thermostat performs the same function here that it did for type (b) and the vent at the top of the piping causes the water to drain back to the tank whenever the pump stops.

The Thermosyphon Water Heater

The simplest and most reliable heliothermal device is the thermosyphon solar water heater which is shown in Fig. 8(a), and in more detail in Fig. 9. This combines a flat plate collector with an insulated water storage tank mounted high enough above the collector so that the cold water in the downcomer pipe will displace by convection the sun-heated water in the collector tubes. This causes a slow but continuous circulation of water downward from the bottom of the tank, upward from the collector, and back to the upper part of the tank. The action will continue as long as the sun shines on the collector. In a good sunny location, free from shadows throughout the entire day, a 4 foot x 8 foot collector will give 40 to 50 gallons of hot water per day. The temperature of the water in the upper part of the tank will vary from 165°F on a hot summer day to 115°F on a cold winter day.

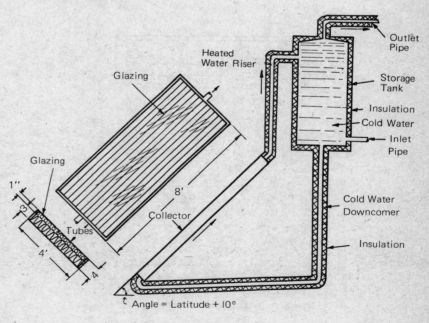

FIG. 9 TYPICAL THERMOSYPHON SOLAR WATER HEATER

A single layer of glass is generally adequate to reduce the heat loss from the surface of most collectors, but in very cold areas double-glazing is desirable. Care must be taken in applying glass, to make sure that it is free to expand or contract, because the inner glass will get very hot by radiation and convection from the collector plate, while the outer glass will remain relatively cool due to the loss of heat to the outdoor air.

The bottom of the storage tank should be at least 1 foot above the top of the collector and the connecting pipe from the downcomer to the bottom header should slope downwards towards the collector.

All piping should be insulated and the tank should be completely surrounded with as much insulation as possible, up to 6 inches in thickness. This should be covered with a thin sheet-metal jacket painted in some dark color to take advantage of the sun's ability to help keep the tank warm during the day.

There are many other ways (Fig. 10) to produce water heater collector plates. The Australians have traditionally used copper tubing soldered to copper sheet as was used in thousands of solar water heaters built a generation ago in the U.S. Now copper is too expensive for large scale use, although its resistance to corrosion and its ease of forming and soldering give it a major advantage over steel and aluminum.

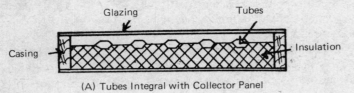

(A) Tubes Integral with Collector Panel

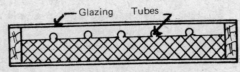

(B) Tubes Bonded to Collector Panel

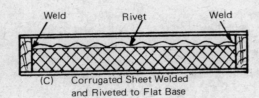

(C) Corrugated Sheet Welded and Riveted to Flat Base

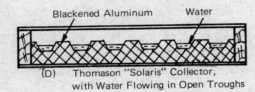

(D) Thomason "Solaris" Collector, with Water Flowing in Open Troughs

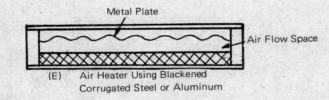

(E) Air Heater Using Blackened Corrugated Steel or Aluminum

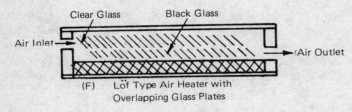

(F) Lof Type Air Heater with Overlapping Glass Plates

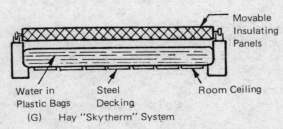

(G) Hay "Skytherm" System

FIG. 10 FLAT PLATE COLLECTOR TYPES

A factory-made collector plate is shown in Fig. 10(a), representing an integral tube-in-sheet which is made in the U.S. under the Roll-Bond® patents held by the Olin Brass Co. Their process involves the printing (using a very special "ink") of the desired tube pattern on one flat sheet of aluminum. A sandwich is then made by putting a second sheet of aluminum over the first and bonding the two together with heat and pressure over their entire surfaces, except for the printed areas. The tubes are then created by inserting a special needle into an unbonded portion of the edge and inflating the tube pattern with a pressurized fluid. This is the

process by which virtually all refrigerator freezing compartments are made, and patterns of extreme complexity can be produced on panels of relatively large size (up to 3 or 4 feet by 12 feet).

The one major problem encountered with aluminum tube-in-sheet is the corrosion which generally occurs when untreated water comes into contact with bare aluminum. The Showa Aluminum Co. of Japan has a U.S. patent on the use of zinc powder in the special "ink" used in the Roll-Bond® process and this produces the equivalent of a galvanizing action on the water passages which, according to Showa, makes them entirely resistant to corrosion. When these panels become available in the U.S., they will be very valuable components of the large number of collectors which are going to be needed here.

The actual temperature of the water used for domestic purposes in the U.S. ranges from $100°$-$110°F$ for bathroom purposes, which is about as high as the human skin can tolerate. Dish washing, to remove grease, may require higher temperatures, although proper detergents can take care of that problem in the $100°F$ range quite nicely. Institutions are required by law to maintain $180°F$ in their dishwashers for sterilization, and this is above the limit attainable by a single-glazed collector during most of the year. For such purposes, the solar water heater would be used as a pre-heater, thus conserving a substantial part of the energy that would otherwise be required.

The Japanese have developed a wide variety of solar water heaters; new types are also being announced in Australia and the U.S. with each new issue of *Popular Science*. Some of these will undoubtedly prove to be successful, but others will lack either the long life or the high performance which a successful installation must have. Beware of the low-cost plastic variety, which are likely to have a very short life and to be susceptible to the high temperatures which are reached very quickly when un-cooled black surfaces are exposed to the sun.

Solar Water Heating on a Large Scale

The heating of large quantities of water can be done by using a number of individual collectors connected in parallel. The Australians (App. IV, No. 7) who have had much experience with such installations, believe that not more than twenty-four riser tubes should be used in parallel coming from a single header. They have demonstrated that very large batteries of collectors can be made to work satisfactorily by using the arrangement shown in Fig. 11.

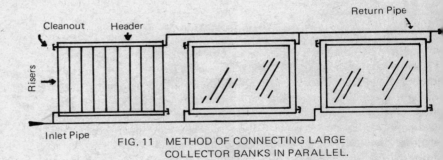

FIG. 11 METHOD OF CONNECTING LARGE COLLECTOR BANKS IN PARALLEL.

They have also shown that relatively high temperatures can be attained for large quantities of water for industrial and commercial purposes by using three types of collector batteries in series. The first, receiving the coldest incoming water, are not covered with glass and hence are relatively inexpensive; the second battery is comprised of single-glazed collectors which are more expensive and also more efficient with warmer water. The third battery, which elevates the water to its final temperature, is double glazed; these are the most expensive units, but they are also the most efficient at high operating temperatures.

Solar Air Heaters

Fig. 8(e) shows a simple and inexpensive air heater, consisting of a cover glass (plastic film may be used, too), a corrugated plate of sheet steel or aluminum painted black, a space through which the air can flow, a layer of insulation and a Masonite or plywood backing to keep the assembly water-proof. The air can be made to flow by means of a fan or blower, or if the system is properly designed, it will rise due to convection (the "chimney effect") because the heated air is lighter than the cold air outside.

Air heaters are much less expensive than water heaters because there is no need to worry about freezing, and any leakage which occurs will not cause the kind of damage which water can create. The drawback to air systems is the fact that fans are larger, more expensive and more power-consuming than the small pumps used with some solar water heating systems [Fig. 8(b) and (d)]. Also, the ducts which are used to carry the air are much larger and more costly than the pipes used with water systems. Each type has its advantages and disadvantages, and each prospective installation must be given careful thought to make sure that the best possible choice is made.

The simplest of all solar air heaters, and one which is in increasingly wide use now, employs a heavy southfacing vertical concrete wall (Fig. 12) which is painted a dark color and covered with a sheet of glass. An air space runs between the concrete and the glass, and the chimney effect causes the heated air to rise. Openings at the bottom and top of the wall allow the cold air along the floor to enter the air space and similar openings at the top provide an opportunity for the warm air to re-enter the room behind the wall. The air then circulates through the room, warming the walls and the occupants. Small electric baseboard heaters are used to provide heat during long periods of bad weather.

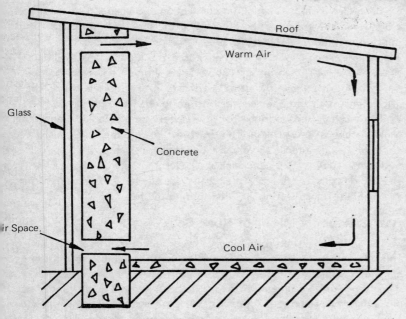

FIG. 12 TROMBE-TYPE SOLAR SOUTH WALL
AIR HEATER USED AT ODEILLO, FRANCE

The overlapping glass plate air heater shown in Fig. 10(f) was invented during World War II by Dr. George Löf, who uses this system in his own home in Denver. The plates are actually divided into two segments, with the upper piece being clear and the lower piece being black. The division was found to be necessary when the first prototype was tested using single sheets of glass with the lower half painted black. These promptly broke from thermal shock as soon as they were exposed to the sun, because the lower part immediately became hot, while the upper part remained cool. Dividing the glass sheets into two parts cured this problem completely, and the Löf house in Denver has been operating satisfactorily since the late 1950's.

Flat metal plates, painted black and covered with glass or plastic glazing, will do a good job of heating air, as Dr. Telkes proved in the solar house which she built at Dover, Massachusetts, in 1939. However, the heat transfer from the metal to the air is improved if the plates are corrugated to increase the amount of surface available for heat transfer. The Australians are making good use of thin copper air heaters, with the copper corrugated into small vees and treated on its outer surface to become selective in its ability to absorb solar radiation and yet not re-radiate much longwave radiation.

Suggestions for Building Solar Water and Air Heaters

There are no hard and fast rules for building satisfactory solar heaters, nor are there any precise dimensions which must be followed. Generally speaking, there should be a single glazing at least, spaced about 1 inch above the collector plate. Glass will generally give much longer life than plastic glazing, but if a shortened life is acceptable, either Mylar or Tedlar (both are duPont films) will do almost as good a job as glass except at very high temperatures.

There are no precise dimensions which must be specified for either air or water heaters. Metal sheets, plywood, Masonite and similar materials are generally available in 4 foot by 8 foot dimensions. Aluminum, galvanized steel, and copper are usually available in 2 to 3 foot widths and almost any specified length. Glass wool insulation is available in thicknesses ranging from 1 to 4 inches. Glass can be purchased in a wide range of dimensions, but individual sheets larger than 4 feet square are difficult for one person to handle.

A 4 foot by 4 foot module, as used in Australia, is a very good choice for a water heater, since it will weigh close to 100 lbs. and that is as much as two people can handle readily, particularly when they are working on a rooftop with a steep slope. For air heaters, the weight is much less and 4 foot by 8 foot units can be conveniently carried and installed by two people. Corrugated aluminum is much lighter than corrugated sheet steel of the same dimensions and it generally costs little if any more per square foot since the thickness of the aluminum can be less than the steel.

Dr. Thomason gives quite detailed instructions for building the aluminum water heater which is the main feature of his "Solaris" house (App. IV, No. 8). The construction details of most of the other solar houses which have been built during the past three decades are to be found in the articles published by their builders and many of them are well described in *Introduction to the Utilization of Solar Energy* by Zarem and Erway (App. IV, No. 3).

Factory-produced water heaters are now available from a number of manufacturers, and more will soon be on the market. The only type which cannot be built by the builder of average skills is the Roll-Bond® variety which is made of aluminum and requires special welding equipment.

A quite satisfactory water heater can be made using 1 inch galvanized steel pipe as the headers and 1/2 inch galvanized tube as the risers. Thin galvanized sheet steel can be used as the collector plate; this is easier to handle if it is used in the form of strips about 6 inches wide. Drill holes large enough to accept the 1/2 inch tubes at 4 inch centers in the headers and solder the tubes into the headers after the headers have been threaded on both ends.

This will produce a grid of pipe and tube which will weigh less than 30 pounds; the addition of the galvanized steel strips will add another 10 to 12 pounds of weight. Let the steel strips overlap lengthwise and solder or cement the tubes to the strips.

Mount the collector in a box with wooden sides and a Masonite bottom. A soft Neoprene gasket and an aluminum cover strip will give the glass the necessary room to expand and contract, while you retain the water-tight feature which is essential. Three inches of glass-wool insulation should be adequate and a few small holes may be drilled along the bottom of the collector box to allow it to "breathe" as its temperature changes. A flat black paint should be applied to the collector plate and the tubes, after they have been primed with the material which is recommended by the paint manufacturer.

The water inlet connection may be made to either end of the bottom header and the other end should be closed with an ordinary 1 inch standard pipe cap. This cap can be removed at appropriate time intervals to remove dirt and scale from the header. The outlet connection should be made at the opposite corner of the upper header, so there will be approximately equal resistance to water flow across the collector grid.

The use of a water softener ahead of the solar collector will greatly extend the useful life of the collector by preventing the formation of scale on the inner surface of the piping in the hot portion of the collector.

Air heaters are much easier to make, since they require only an air-tight box with a glazing material on the top; this may be either glass or a Tedlar or Mylar film. The heat-absorbing panel may be corrugated or rolled sheet metal, preferably aluminum, with a "dead air space" about 1 inch wide between the glazing and the collector. The air-flow space will be along the back of the aluminum, and both sides of the metal should be painted black. The top or upper surface must be black to absorb the solar radiation which will be transmitted through the glazing. The lower surface should also be black so that it can radiate heat to the covering over the glass wool insulation and allow that surface to help in the air heating process.

Unless you are expert at cutting glass, have this job done for you at the hardware store where you buy the glass. A badly-cut piece of glass is likely to have defects along the edges and when temperature differences begin to occur across the glass, as they will when the plate begins to be heated by the sun, thermal stresses may be set up which can break the glass. Four foot by four foot lights (panes) of glass are also much easier to replace if breakage does occur due to hail stones or other objects which may strike the glazing. Breakage of the glass covers due to hail or stones can be lessened by the use of 1/2 inch wire mesh, supported on an angle-iron framework several inches above the glass. The use of this screen reduces effective absorber area by about 15%, so the total absorber area should be increased to compensate for this reduction.

SOLAR SPACE HEATING AND HEAT STORAGE

Solar space heating may be accomplished in many ways and again there are no hard and fast rules. Before undertaking to heat a building by solar energy, a competent estimate must be made of the amount of heat which the structure will need under adverse winter conditions and at night. The solar heater must be able to provide not only the heat which will be needed hour by hour during a sunny day but it must have additional capacity so that heat can be stored during the day for use at night or during cloudy days.

There are three methods of heat storage which are in use today; the first two representing relatively low technology but complete feasibility. The third, although it has been under study for many years, is still in the developmental stage.

Heat Storage in Water Tanks

Large tanks, filled or almost filled with water, represent the best method that is presently available to store large amounts of heat or cold.

Every substance possesses its own "specific heat," which is the amount of heat in Btu's needed to raise the temperature of that substance 1 degree F. On a pound-for-pound basis, one can store more heat in a substance of higher specific heat than in a substance of lower specific heat. Water has a specific heat of 1.0 Btu/lb./°F, an antifreeze such as ethylene glycol has a specific heat of 0.6, but almost all solid materials from rocks to iron and aluminum have a specific heat of 0.2 Btu/lb./°F.

Water weighs 62.4 pounds per cubic foot or 8.34 pounds per gallon. If we heat water 10 degrees F above the temperature at which we need the heat, we can store 624 Btu in a cubic foot or 83.4 Btu in a gallon of water. Assume that it is house heating which we wish to accomplish and that by using a very efficient system of heat distribution we can use water as cool as 90°F to heat the house.

If we store hot water at 150°F in a cylindrical steel tank, as in Fig. 13(a), we can store 62.4 lb./cu.ft. x (150°F - 90°F) = 3744 Btu in a cubic foot of water or 8.34 lb./gal. x (150°F - 90°F) = 500.4 Btu in a gallon. A 2000 gallon tank of water will thus be able to store about 1,000,000 Btu's at a temperature difference of 60°F.

If we insulate the tank with 6 inches of glass wool or, even better, 6 inches of urethane foam, we will lose only about 1000 Btu per hour through the glass wool and half that amount through the urethane (assuming that the tank is 6 feet in diameter and 10 feet long).

The storage of cold is not quite so simple, since we cannot use such a large temperature range without running into the complication of freezing the water. We may well want to do just that, as we will see in the heat-of-fusion section upcoming, but assuming that we want the water to remain in the liquid state and that we want to use it for cooling at 55°F, we have only 23 degrees of temperature at our disposal (water freezes at 32°F). The 2000 gallon tank contains 8.34 lb./gal. x 2000 gal. = 16,680 lbs. of water and can thus store about 16,680 lbs. x 23°F = 383,640 Btu's of "coolth," as compared with nearly three times that much when we are storing "warmth."

Now we see that, if we can allow a 60°F temperature rise and fall, we can store about 1,000,000 Btu in a 2,000 gallon tank but if we restrict ourselves to the 23°F temperature rise and fall for cooling, we will need a tank nearly three times as large, or almost a 6,000 gallon tank. For a 6 foot diameter tank, the length will stretch out to about 30 feet. Obviously, if we want to store very large amounts of "coolth" in a tank of water, the tank is going to have to be very large.

The tank should have an outlet connection at one end, near the bottom, from which the water can be pumped to the solar collector, and a return inlet, above the water level, thus leaving some air space in the tank. An access hole, large enough for someone to get inside the tank, will be a wise precaution (not shown in Fig. 13(a)). There should be two connections on the far end, one above the center of the tank to supply water to the distribution system and one near the bottom to receive the return water.

The mechanics of storing heat in water are simple; in addition, water is available almost everywhere. The disadvantages are the space, weight (8.3 tons for 2,000 gallons, 25 tons for 6,000 gallons), and the cost of a steel tank with a 1/4 to 3/8 inch thick shell. Despite these disadvantages, water storage remains as the only practical system when we are using water as the medium for collecting and distributing the heat.

Heat Storage by Means of Rock Beds

The ability of a solid to *transfer* heat is a function of its surface area per unit volume; the more surface area, the greater the ability to transfer heat per unit volume. The ability of a material to *store* heat is a function of its density per unit volume and its specific heat. A solid cubic foot of basalt has a density of about 184 pounds, but a surface area of only six square feet, making it a poor candidate for a heat transfer system, even though it is a good means of heat storage. If this big cube is broken into many 1/2 inch cubes, we will have 13,824 small cubes and their total surface area will be 144 square feet, giving us a much better heat transfer situation. However, in trying to pack the cubes helter-skelter into the original one cubic foot container, we will find, because of air spaces, the cubic foot will only hold about 89 pounds. This will still give us about 70 square feet of heat transfer surface, compared to the 6 square feet of the original block.

Going back to 23°F as the temperature range over which we were operating with the water tank, 1 cubic foot of basalt pebbles can store 89 lbs. x 0.2 Btu/lb./°F x 23°F = 409 Btu. If we are concentrating on heating and can allow a 60°F temperature difference, we can store almost three times as much heat; 1,068 Btu per cubic foot or 24,000 Btu per ton. To store 1,000,000 Btu with the 60°F temperature range, we would need 41-2/3 tons, which is a formidable pile of rocks.

The practical way to handle a rock pile storage situation is to use the arrangement shown in Fig. 13(b), with the rocks contained in a cylinder, and their weight supported by a steel grill at the bottom of the container. A central pipe with a damper will allow the rocks to be by-passed when the heat is needed directly in the structure.

The air-heated house erected by Dr. George Löf in Denver in 1959 uses two large cylinders, 3 feet in diameter and 18 feet high in the center of the building, to store 12 tons of 1 to 1½ inch rock with a bulk density of 96 pounds per cubic foot. The Thomason houses in Washington, D.C., use 25 to 50 tons of "fist-sized" rocks to provide supplemental heat storage as well as heat transfer surface. Two demonstration houses erected by the writer for the U.S. government in 1958, one in Casablanca and one in Tunis, used the system shown in Fig. 13(b), with a vertical rock pile contained in a Sonotube cylinder.

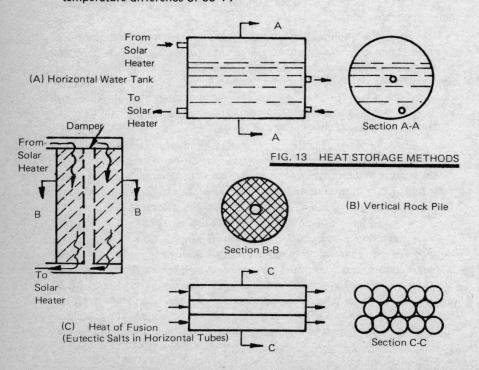

(A) Horizontal Water Tank

From Solar Heater

To Solar Heater

Section A-A

From Solar Heater

Damper

To Solar Heater

(C) Heat of Fusion (Eutectic Salts in Horizontal Tubes)

Section B-B

(B) Vertical Rock Pile

Section C-C

FIG. 13   HEAT STORAGE METHODS

Rock piles have the advantage that they can neither freeze nor leak, but their capacity is limited. However, they can be safely used under a building since, once they are put in place, very little harm can come to them. The Australians have made particularly effective use of rock beds in the device called the "Rock Bed Regenerator," which will be discussed later.

Heat of Fusion Storage

When a tank of water is cooled by refrigeration, it will give up 1 Btu per pound for each degree that it is cooled until it gets to $32°F$. At that temperature, ice will begin to form and there will be no further change in temperature until another 144 Btu (called the "heat of fusion") have been removed. One hundred forty-four Btu per pound must be added to the ice-water mixture to melt the ice and return it to liquid form again. Its limitation is that $32°F$ is the only temperature at which water will freeze into ice and then melt back again to water.

There are other substances which freeze and melt at more convenient temperatures; the most widely known being Glauber's salt, sodium sulfate decahydrate ($Na_2SO_4 \cdot 10H_2O$). This melts and freezes again at $88°$ to $90°F$, with a heat of fusion of 108 Btu per pound. There are many problems connected with this and similar materials. Their potential value in the storage of warmth and coolth lies in the fact that 1,000,000 Btu's can be stored in only 5 tons of salt as compared with 25 tons of water and 125 tons of rocks (with a $20°F$ temperature swing).

Dr. Telkes of the Institute of Energy Conversion at the University of Delaware has devised ways of adding small amounts of other substances to heat of fusion materials to make them behave properly. The major difficulty appears to be that the heat of fusion materials, unlike water, sink when they freeze and it is not a simple matter to find ways to make them become homogeneous liquids again when they melt. The first attempts to use Glauber's salt in relatively large drums were unsuccessful and now Dr. Telkes has turned to long slim plastic tubes (Fig. 13(c)), similar in shape and size to fluorescent lamps, to provide both storage for the materials and heat transfer surface for the University of Delaware Solar House, "Solar One." Her latest combinations will provide storage of warmth at $120°$ to $126°F$ and coolth at $55°F$.

SOLAR AIR HEATING SYSTEMS

Fig. 14 shows a simple solar air heating system which can be used to provide heat and store excess energy during the day by heating the gravel in the storage tank. One fan and five dampers are needed to enable air to be drawn through the air heater and then to be sent either through the storage cylinder and back to the collector, or diverted directly to the house on cold days. At night, the collector may be by-passed and all of the house air may be circulated through the storage cylinder. For use in most parts of the U.S., an auxiliary electric heater, probably in the form of a grid of resistance wire, would have to be used to guarantee comfort during prolonged periods of bad weather. A fireplace with a Heatalator [see BioFuels—Wood] would do equally well in rural areas where firewood is available.

For heating small, single-story houses, air heaters have a number of advantages, including simplicity and low initial cost. However, the duct-work for the movement of air requires careful planning. Air ducts are many times larger than water pipes in cross-section, and problems can easily develop when one tries to run the ducts in walls and around structural members. Duct-work generally must be purchased from sheet-metal shops, where some pieces probably will have to be custom-made. Fortunately, the temperatures involved in most solar air heaters are such that the air may safely be conveyed in insulated passages between floor or roof joists and through the air spaces which exist in stud walls. Care must be taken to ensure equalized flow through all parts of the air heater. This will generally involve some balancing of the air flow by means of dampers after the system is built.

In summary, an air heating system needs the four elements shown in Fig. 15: (a) is the collector, facing south, steeply tilted and insulated so that the collected heat will go into the air and not be wasted. A fan (b) is needed to circulate the air. A filter on the fan will minimize the dust and pollen problem within the house. The rock bed (c) provides storage of heat in winter and storage of coolth in other seasons when the air at night is cold enough to cool the rocks so that they in turn cool the house during

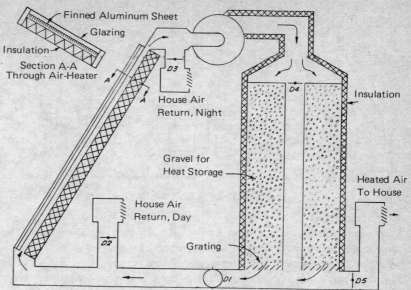

FIG. 14 SOLAR AIR HEATER SYSTEM WITH VERTICAL ROCK BED HEAT STORAGE

the day. An auxiliary heating system in the form of the electric grid (e) can be used to heat the house directly on cloudy, cold winter days or to store heat in the rock bed by properly setting the dampers. Fig. 15 has a maximum of flexibility because its eight dampers permit outdoor air to be brought in, blown through the rockbed and then out again. The rockbed can thus be cooled down at night and the house air can be cooled during the day by blowing it through the cooled rocks.

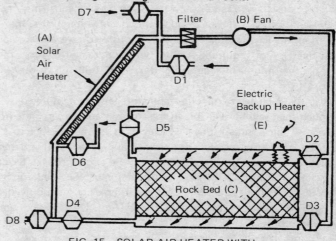

FIG. 15 SOLAR AIR HEATER WITH ROCK BED HEAT STORAGE (8 dampers)

South Wall Air Heaters

A number of houses have been built in southern France using the concept shown in Fig. 12. This is an outgrowth of work done during the past decade by Dr. Felix Trombe of the Solar Energy Laboratory at Odeillo. The major activity at Odeillo is centered around the great solar furnace, but the building which houses the nine-story solar concentrator, and many of the staff homes in the vicinity, use south wall heaters during the winter months, with electric auxiliary heaters for long periods of low insolation.

This idea is just beginning to be used again in the U.S., although the Telkes-Raymond house in Dover, Mass. used vertical south-wall heaters, and "Solar One" at the University of Delaware is employing this system, too.

Solar Water Heating Systems for Space Heating

The basic principles of a solar space heating system using water instead of air as the heat transfer fluid are given in Fig. 16. A collector similar to Fig. 6, or Fig. 10 (a)—(d), is mounted on a south-facing exposure or a flat roof of a structure. Heat storage is provided by an insulated tank which can hold from 2,000 to 20,000 gallons of water depending upon the size of the building and the length of time that heat must be provided from storage.

The house heating system works independently of the solar collecting and storage system (Fig. 16), since the solar-system pump, P1, goes to work whenever the solar thermostat, T1, senses that the temperature of the collector panel is warmer than the water in the storage tank. The pump runs at constant speed and so the temperature of the water leaving the top of the collector will vary from a few to many degrees above the

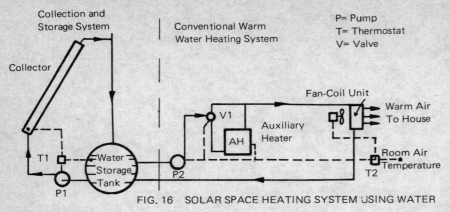

FIG. 16   SOLAR SPACE HEATING SYSTEM USING WATER

tank temperature.

When the thermostat senses that the collector panel is falling below the tank temperature, the pump will stop and all of the water in the collector panels will drain back into the tank. The system is thus "fail-safe," because it will be protected against freezing since the system will drain itself whenever the pump stops because of power failure or from a signal from the thermostat. There will be enough inertia in the system to prevent it from stopping if a passing cloud happens to cross the sun, but it will stop and remain shut down if prolonged cloudiness sets in.

The actual heating of the building is done by any conventional warm water system (e.g., water circulation pipes and radiators) into which the circulating pump P2 will supply warm water from the storage tank. When the house thermostat T2 calls for heat, it starts P2, and valve V1 directs the warm water into the house heating system, indicated in Fig. 16 as a fan-coil unit. This consists of a fan or blower, driven by an electric motor, and a heater containing finned tubes through which the water flows. In many homes, the fan in such a system runs continuously to give air circulation even when heat is not required. A filter will add to the comfort of the occupants of the house by removing dust and pollen at very little added cost.

Generally, the solar heat collection and storage system will be designed so that enough heat will be available in the storage tank to carry the house through several cold, sunless days. However, it is not feasible to store enough heat for every possible contingency and so an auxiliary heater must be provided. Shown in Fig. 16, this may be a fuel-fired heater or an electric resistance heater. If an auxiliary heater must be used for many hours per year, or if the region is one where summer cooling is needed as urgently as winter heating, then a heat pump (App. I) is the economical answer to this problem.

The heat pump can use cool or even cold water from the storage tank as its heat input and use a small amount of electric power to raise the temperature of the water up to the point where it can do the heating job. This system is discussed in detail in the articles cited in App. IV, No. 9.

The system shown in Fig. 16 has been used in many buildings, including the MIT solar houses which were built in the 1930's. These used large south-facing collectors which constituted the south-facing portion of the roofs, and large water tanks in the basement. Several experimenters have used bare metal roofs with tubes integral with the metal roofing (similar to Fig. 10(a)) as their heat collectors, with water storage tanks and heat pumps to raise the temperature of the circulating water to the point where it can warm the building. Prominent among these experimenters were Mr. Yanagimachi of Tokyo and Mr. and Mrs. Raymond Bliss of Tucson, who designed the University of Arizona Solar Laboratory which was in use for several years until it was torn down to make way for the University's Medical School.

There are two other water heating systems which should be mentioned in this section, the first being the "Solaris" houses built by Dr. Thomason in a suburb of Washington, D.C., in 1959, 1960 and 1963. The Thomason houses all use the open-flow system shown in Fig. 10(d). The collectors are mounted on steeply-pitched south-facing roofs and they receive their water from a perforated pipe which runs along the ridge of the roof. A large (1,600 gallon) water tank in the basement provides the major heat storage and this is surrounded by 50 to 60 tons of river rock. These rocks provide additional heat storage and the heat transfer surface which is needed to warm the air which is circulated through the house by a small blower.

The collectors are covered by a single layer of glass and they drain automatically whenever the pump is shut off by the thermostat which senses

the temperature of the collector surface. Corrugated aluminum roofing, primed and painted black, is used as the collector surface. This is probably the lowest cost collector panel in use today. App. IV, No. 8, gives additional details on the "Solaris" system.

The "Skytherm" solar heated and naturally cooled system (Figs. 8(c) and 10(g)), invented by Mr. Hay, now has its first full-scale application. A building of this design is now in regular use at Atascadero, California. Professors Kenneth Haggard and Philip Niles of California State Polytechnical University did the architectural design and engineering. Technical assistance came from Professor Philip Niles on the instrumentation phase. Funds for building the house were put up by the inventor, Mr. Hay. HUD (the Department of Housing and Urban Development) contracted with Cal Poly to instrument and monitor the operation of the house.

The "Skytherm" system is particularly well adapted to the southwest desert, where the skies are generally clear both winter and summer, and where the summer humidity is low. Daytime temperatures may be very high (110°F and above) but the "Skytherm" system has demonstrated its ability to cope with them.

Another solar-water heated house which should be mentioned is the residence of Mr. and Mrs. Steve Baer, near Albuquerque, New Mexico which uses steel barrels filled with water, behind single glazing, to absorb the winter sun on the southern exposure of the house. An insulating panel, hinged at the bottom, is lowered when the sun begins to strike the south wall, admitting the sun's rays to warm the blackened ends of the barrels. The barrels serve as radiators to warm the building, and an ingenious device allows shutters to open automatically whenever the house needs to be cooled. The exterior insulating panel is raised at night to retain the heat which the barrels have collected during the daylight hours.

SOLAR COOLING

Solar cooling is something of a misnomer, since there is actually no way by which we can use the heat of the sun directly to produce cooling. However, we can use the heat to produce hot water or steam, and with that we can refrigerate, using the process known as absorption refrigeration. There are natural cooling methods which are treated in the next section which might be considered to be sun-related but they do not actually use solar radiation as such.

Heliothermal Cooling Systems

The process by which the sun's radiant heat is used indirectly to produce cooling is known as "absorption refrigeration." This process was discovered almost accidentally by the British scientist, Michael Faraday, in 1824. Faraday was trying to make liquid ammonia, using the apparatus shown in Fig. 17. It consisted of a U-shaped test tube, with silver chloride saturated with ammonia in the left leg, and nothing, at first, in the right leg. A glass of water, or some similar simple cooling means, was provided for the right leg, and a source of heat, shown as a Bunsen burner on the left. As Faraday applied heat gently to the saturated silver chloride, the ammonia gas was driven off, and it condensed as a liquid on the right side.

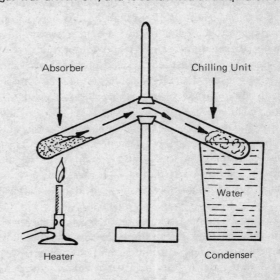

FIG. 17   FARADAY'S 1824 DEMONSTRATION OF ABSORPTION REFRIGERATION

Faraday then proceeded to cut off the heat, and go on about his business. When he returned, he found, much to his surprise, that the liquid ammonia had disappeared from the right side of the apparatus, but the outer surface of that side was covered with ice. He reasoned, quite correctly, that the ammonia had evaporated, taking heat away, freezing the water as the ammonia went back into absorption in the silver chloride. This was the first absorption refrigeration process, and, about one hundred years later, this principle was used for household refrigerators.

The ammonia-water absorption system was discovered somewhat later. Here the ammonia is the refrigerant and the water, instead of silver chloride, is the absorber. Water can absorb very large quantities of ammonia when it is at a low temperature, but, when the water is heated under high pressure, it will give off the ammonia as a gas. This ammonia can then be condensed back to liquid form and allowed to expand through a valve, where some of the ammonia will evaporate and cool down the rest of it.

This form of absorption refrigeration has been very widely used, with low pressure steam as the heat source. The only mechanical work involved is the power needed to run a small pump to raise the pressure of the ammonia-water. In a high pressure "generator," heat is then applied to drive the ammonia away from the water. Cooling must also be applied at this point to cool the water down again and to condense the ammonia. So, both heat and cooling water, such as that obtained from a cooling tower or an evaporative cooler, must be provided.

The ammonia-water system is also under pressure, and building codes require that such apparatus be used in outdoor locations. The chilled fluid which results from this process, generally cold water, can be taken indoors and used to cool air. This system is widely used where there is a large quantity of low pressure steam available which would otherwise be wasted or where very hot water is available.

The absorption process which is being studied in a number of experimental buildings uses lithium bromide as the absorbant and water as the refrigerant. Water has been used as a refrigerant for many years, using centrifugal compressors or steam jets to reduce the pressure on a tank full of water. This causes enough of it to evaporate to chill down the remainder. The compressed water vapor is then condensed by the use of water from a cooling tower or a similar source. It then goes back to the tank to complete the cycle and to be re-used. Attempts have been made to use steam generated in solar boilers to power such a system, but they have not been successful.

The lithium bromide system uses the ability of that substance to absorb large quantities of water when it is relatively cool, and to give up the water vapor when it is heated. The temperature to which the heating must proceed is about 200° to 210°F in conventional commercial units, such as those manufactured by the Arkla-Servel Company. Newer versions of the lithium bromide-water systems use solar-generated hot water at 180° to 200°F to regenerate the strong lithium bromide-water solution, and thus to drive off the water, which must be cooled before it can go back into the chiller tank to complete the cycle and continue the process.

The major difficulty with solar-powered absorption refrigeration at the present time is the temperature at which it has to operate and the relatively low Coefficient of Performance. The latter is simply the ratio of the cooling effect to the amount of heat that must be generated to produce this effect, frequently measured in "tons of refrigeration" (App. I). The Coefficient of Performance of a typical lithium bromide absorption system is no higher than 0.6, as opposed to the 3.0 or better attained by even a small compression refrigeration system driven by an electric motor.

Temperatures in the 200°F range are not easy to obtain with solar collectors, even if two cover plates and a selective surface are used. The efficiency of collection is likely to be no higher than 50% at noon and an hour or two before. After that time, the efficiency falls off rapidly, as the sun moves away from its noon position, and the day-long efficiency of a collector operating at 200°F is likely to be no more than 30%.

There are many other combinations, aside from ammonia-water and lithium bromide-water, which can be used in absorption systems, and these are being carefully reviewed now to see whether a system can be developed which will operate at lower temperatures, with a higher Coefficient of Performance than those which characterize today's absorption systems. Dr. Erich Farber and the Solar Energy Laboratory, University of Florida (App. IV, No. 10) are doing significant work in solar cooling.

Natural Cooling Methods

There are two natural cooling methods which have been used with considerable success in appropriate climatic conditions. These are (1) night sky radiation and (2) evaporation.

Almost everyone has observed the effect of night sky radiation when they have noticed frost on the ground early in the morning, even when the air temperature has not dropped to 32°F. The reason for the cooling of the ground to the freezing point is that depending largely upon the amount of water vapor in the atmosphere, the sky is a good absorber of radiation. When the sky is very dry, the apparent temperature of the sky, as far as ground heat losses are concerned, may be very low.

The term "nocturnal radiation" is again a misnomer, because radiation from surfaces on the earth to the sky proceeds all day long, but it is generally masked by the incoming solar radiation during the daytime. At night when this source of heat is not present, the cooling effect of radiation to the sky can be very marked indeed. It is most noticeable when the sky is clear and cloudless, and the relative humidity is low. It is least noticeable on an overcast night, because then the clouds tend to radiate back to the earth almost as much heat as they receive from the ground and the grass and buildings which cover it.

Raymond Bliss, when he was planning the solar energy laboratory for the University of Arizona, made a careful study of the water vapor in the atmosphere. He found that the rate of heat loss from a black surface on the ground, at the temperature of the air near the ground, ranged from 20 to 40 Btu (App. I) depending upon the amount of moisture in the air. The dew point of the air at the surface has been found to be a good measure of the total amount of moisture in the atmosphere in any given location, and the Bliss curves (App. IV, No. 11), show that there is very little change in the "nocturnal radiation" rate as the air temperature changes, but there is change from 20 to 40 Btu as the relative humidity of the air changes from 80% to 10%.

Evaporative cooling is the cooling of air by the evaporation of moisture into that air. The process of passing air through pads which are saturated with water, and evaporating some of the moisture picked up by the air, has been used for many years as a means of reducing the temperature of desert air to a tolerable level. The difficulty with this very simple process is that the moisture which reduces the temperature also raises the relative humidity, until the result is cold, clammy air which is almost as uncomfortable as the hot, dry air which entered the evaporative cooler. These devices are frequently called "swamp coolers" in the southwest where they have been widely used for several generations. Today, they are employed most widely to cool cattle sheds, manufacturing operations and large areas where so much heat must be removed that conventional refrigeration would be far too costly.

A modification of the simple evaporative cooler has been developed in Australia, and given the name "Rock Bed Regenerative Air Cooler (RBR)." One such device is shown in Fig. 18, and hundreds of units using this simple principle are in operation in Australia. The system uses two beds of rocks, set side-by-side and separated by an air space in which a damper is located. Water sprays are mounted close to the inner surface of each bed of rocks, and two fans are used. In Fig. 18, the outdoor air is being drawn into the rock bed on the right, which has just been evaporatively cooled, while the rock bed on the left is undergoing evaporative cooling by having the indoor air flow through it on its way outward to the atmosphere.

At the beginning of each cycle, water is sprayed into the rock bed for 10 to 15 seconds, thoroughly saturating the rocks. The air from the house,

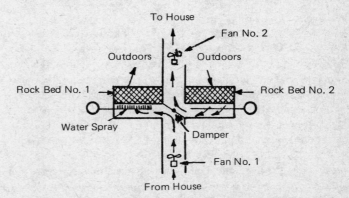

FIG. 18 ROCK BED REGENERATOR

blowing through the rock bed, evaporates the moisture away from the rocks and thereby cools them and the air, increasing the humidity of the air as well. This does no harm, since the air simply goes back to the atmosphere. On the other side of the system, air is being drawn into the house through rocks which have just been cooled. Only a very small amount of moisture is added to the air since it is the rocks which have been cooled, and the only moisture remaining is the small amount which adheres to the surface of the rocks.

The first prototypes in the U.S. are now under construction in Arizona. They will be very useful adjuncts to solar heating systems, since the amount of power that they require is only about 10% of that used by a mechanical refrigeration system, and the amount of water needed is very small. The evaporation of 1.5 gallons of water per hour is equivalent to 1 ton of refrigeration, and the cost of the water is negligible.

The Australians have found that the rock beds can also be used to conserve heat from outgoing ventilation air in the winter. They operate during the winter without the water sprays. The rock bed will be heated almost to the exhaust air temperature in the exhaust portion of the cycle, and the incoming cold outdoor air will then be preheated almost to the indoor temperature by passing through the heated rocks. The RBR system has tremendous promise for both heating and cooling, and it is likely to find wide use in many parts of the United States.

SOLAR COOKING

Cooking by the use of solar heat is a very old art. The first solar cooker was probably that built in Bombay in 1880, and several other ingenious ovens, including that shown in Fig. 19, have originated in India.

Solar Ovens

One of the best solar ovens, little known in this country, was developed by Dr. M. K. Ghosh of Jamshedpur. Shown in Fig. 19, the Ghosh heater consists of a simple wooden box, about 24 inches square and 12 inches deep. It is lined with 2 inches of insulation (glass wool would be very satisfactory here) and it contains a blackened metal liner in which the cooking is done. The interior insulation and the hinged top are lined with shiny aluminum foil, the former to reduce radiation from the oven to the sides and bottom of the box, and the latter to reflect the sun's rays into the oven through the double cover glasses. Since much of India lies near the equator and the sun is nearly overhead at midday, the Ghosh oven does an excellent job of cooking the noon meal.

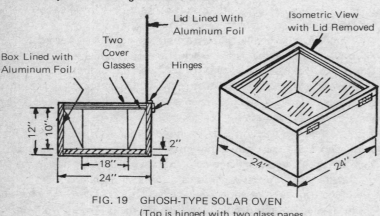

FIG. 19   GHOSH-TYPE SOLAR OVEN
(Top is hinged with two glass panes
spaced 3/4" apart, sides and bottom
of 3/8" plywood).

Another type of oven, originated by Dr. Telkes, uses a metal box with double glazing over its open end and a hinged, insulated door at the other end. It is contained in a casing which may be made of metal or plywood, and the whole affair is arranged so that it can turn and tilt to follow the sun. Reflecting wings made of shiny aluminum (Alzak® is the best brand) are attached at an angle of 60 degrees to the plane of the cover glasses. These nearly double the amount of solar radiation which can enter the oven. On a bright sunny day, the Telkes ovens can get to 400°F from about 9:00 a.m. to 3:00 p.m.

A simplified version of the Telkes oven is described by Dan Halacy on pp. 51-62 of his excellent do-it-yourself handbook, *Solar Science Projects* (App. IV, No. 6). His oven uses the Telkes principle, but it sets the cover

glasses at an angle of 45 degrees to the oven so that it does not need to be tilted.

The basic principle of operation of a reflector-type solar cooker is shown in Fig. 20, where a simple parabolic reflector, arranged on a sun-following mount, concentrates the sun's rays on a grill which can support a pot or a frying pan. This type of cooker will become hot just as soon as it is adjusted to face the sun and, according to Dr. Farrington Daniels, a 4 foot diameter reflector-type cooker will deliver the equivalent of a 400 watt electric hot plate under bright sunshine. He gives detailed instructions for the construction of a plastic shell of the proper shape which can be lined with aluminized Mylar. App. IV, No. 5, Chapter 5, pp. 89-103 covers the topic of solar cooking; pp. 102-103 gives an excellent bibliography on the subject.

Blackened Pot

Sun's Rays

(A) Plano-Convex Lens or   (B) Fresnel Lens
    "Magnifying Glass"

FIG. 20   REFLECTOR COOKER   FIG. 21   SOLAR CONCENTRATORS
(Must be adjustable to
follow the sun).

Dan Halacy shows how to make an even simpler reflective cooker out of corrugated cardboard and kitchen-type aluminum foil on pp. 14-27 of *Solar Science Projects* (App. IV, No. 6). He advises the use of sun glasses when you are cooking with his aluminized broiler because the reflectance can be dazzling. The reflecting surface does not get hot, but the grill area does, so be careful when you put on the pots and pans. Also, Brace Research Institute, in *How to Make a Solar Steam Cooker*, Do-It-Yourself Leaflet L-2 ($1.00), describes a steam cooker which uses the simplest of materials for making a broiler. See App. IV, No. 2 for ordering information.

SOLAR FURNACES

While we are dealing with some of the higher temperature solar devices, mention should be made of solar furnaces which can produce small but extremely hot images of the sun. Fig. 21(a) shows the familiar magnifying glass which can be used to start a fire whenever the sun is shining. The diameter of available lenses is not great enough to do much real work with a burning glass. The Fresnel (pronounced Frā-nĕl') lens shown in Fig. 21(b) is available from Edmund Scientific Co. (App. IV, No. 12) in the form of a 12 inch square plastic sheet and you also can buy a booklet entitled *Fun with Fresnel Lenses*. Dan Halacy's book, *Solar Science Projects*, gives explicit details for making a mounting which will convert the Fresnel lens into a miniature solar furnace.

A concentrating curved-mirror solar furnace can reach extremely high temperatures (to 6,000°F, depending on size). Control of the temperature of the target at the focal spot (App. I) may be accomplished by using a diaphragm to "stop down" incoming rays, or by using a cylinder which can be moved to or fro along the axis of the concentrator to regulate the amount of radiation which reaches the target. Military surplus searchlights are the best source of high-precision reflectors at reasonable prices, and Edmund Scientific Co. frequently has some for sale.

The largest solar furnace now in existence is located at Odeillo in the French Pyrenees. There, starting in 1957, Dr. Felix Trombe and the National Center for Scientific Research have constructed a gigantic paraboloid which is nine stories high, uses 63 heliostats, each of which is 24 feet square, to produce an image which is a perfect circle, 12 inches in diameter. The furnace can concentrate 1,000 kilowatts of thermal energy into this small area and temperatures pushing up towards 6,000°F have been reached on sunny days.

ELECTRICITY FROM THE SUN

Photovoltaic Devices

About a century ago, when the scientific foundations were being laid for today's heliotechnology, a Frenchman, Becquerel, found that sunlight could produce minute amounts of electricity when it entered a very special kind of "wet cell" battery. Later on, other workers found that sunlight could change the resistance of certain metals and that very small amounts of electricity would be generated when sunlight illuminated discs of selenium or certain types of copper oxide. These devices were very useful as light meters but they could not produce enough power to do anything more strenuous than moving the pointer on a meter or activating a very sensitive relay.

In May, 1954 a one-page report, authored by D. M. Chapin, C. S. Fuller, and G. L. Pearson, appeared in the *Journal of Applied Physics* (Vol. 25, No. 5, p. 676). In brief terms they announced the discovery of a new treatment for ultra-pure silicon, which gave it the property of generating electricity from sunlight with a conversion efficiency of 6%. This was ten times better than any previous efficiency for the direct conversion of sunlight into electricity (the previous record was held by Dr. Telkes with a battery of thermocouples). Drs. Fuller, Chapin and Pearson immediately applied their invention to a small transistorized radio transmitter and receiver.

It was not until the U.S. embarked upon its space program late in 1957 that a unique application was found for the silicon solar battery. There was no other source of power for the satellites which we were planning to launch. NASA was persuaded to put silicon cells on its first permanent satellite, Vanguard One, which was orbited in 1958. They worked so well that all but one of the satellites which have been orbited since that time have been powered by increasingly complex arrays of silicon solar cells. Communication satellites use tens of thousands of silicon cells and Skylab I produces some 20 kilowatts from its array of solar panels.

The details of the theory and construction of silicon cells are beyond the scope of this chapter, so it must suffice to say that, under bright sunlight, each 2 centimeter square cell will produce a short circuit current of 0.10 amperes (100 milliamperes) and an open circuit voltage of 0.60 (600 millivolts). Under optimum load, they produce about 90 milliamps at 400 millivolts, or approximately 36 milliwatts. This is certainly a tiny amount of power when it is compared with today's gigantic steam turbo-generators; but silicon cell technology has advanced so rapidly that tens of thousands of individual cells can be connected together. This can be done rapidly and reliably so that today the communications satellite has become the standard means of intercontinental communication for voice, television and even computer language.

Although the cost is still too high for the installation of great panels of solar cells on every rooftop, great strides have been made and the cost has been reduced from $1000 per watt to $20 per watt, with more reductions on the way. Ways of producing far less expensive silicon cells are now under intensive study. Another approach is to concentrate ten times as much sunshine onto the cells as they would normally receive. The main problem here is to remove the heat which will build up if the cells are not cooled effectively. The more optimistic silicon cell specialists think that, within another decade, the costs will come down to $2 per watt, which will be competitive with conventional methods of power generation.

Although making silicon cells is not a do-it-yourself project, from time to time bargains are available in the way of space-reject cells (Herbach and Rademan of Philadelphia (App. IV, No. 13) is an excellent source), and these are well worth considering for powering radios and other apparatus which require very small amounts of power. Fig. 22 shows a method of mounting solar cells on a sun-tracking device to obtain more power than can be gained from a fixed array. The circuit diagram indicates a series-connected battery of cells which can produce any desired voltage (just use twice as many cells as you need volts, since under load each cell will give just about 1/2 volt). A number of such series-connected groups connected in parallel can produce a respectable amount of power and keep a 12-volt storage battery charged for running a radio or even a small TV set. We are not yet at the point where solar cells will run air conditioners or other large motor loads, but Dr. Böer believes that he can produce another type of cell, using cadmium sulfide rather than silicon, which can be produced at far lower cost than the silicon cells and still retain adequate efficiency to do the job.

Heliothermal Generators

There have been many attempts to produce steam from heliothermal

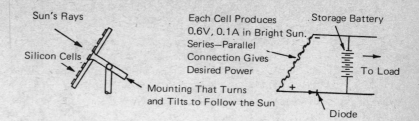

FIG. 22   SOLAR ELECTRIC GENERATOR USING SILICON CELLS

devices, beginning with John Ericsson's work in New York just after the Civil War and reaching its high point early in the 1900's with the building of great solar engines in Arizona and California by A. G. Eneas and in Egypt by F. Shuman and C. V. Boys. None of these attempts succeeded for reasons which are too numerous to be mentioned here.

Flat plate collectors cannot attain temperatures much above 200°F and engines operating at temperatures in that range and discharging their waste heat to air at 80° to 90°F offer little prospect of generating useful amounts of power. Concentrating collectors which do not yet exist, except for laboratory prototypes, are needed to reach temperatures high enough to make engines or turbines operate with competitive efficiencies. When these become available, they will probably use vapors of the Freon family rather than water and they will use the cycle pioneered a century ago by an Englishman, Rankine. For a more detailed account of the various heat-engine cycles, read Chapter 7, *Introduction to the Utilization of Solar Energy* (App. IV, No. 3).

Today's research efforts in the field of heliothermal power generation fall into two categories: (A) small engines using Freon vapor to produce 5 to 10 horsepower to drive pumps, generators or refrigeration compressors and (B) very large installations using vast areas of concentrating collectors to produce enough steam to run a boiler mounted on a tower of moderate height. This is surrounded by a great field of slightly curved mirrors which turn and tilt to reflect the sun's rays into the boiler and thus to produce steam.

The storage of heat to run the solar-power engine at night or on partially cloudy days is a problem which is not even close to solution, so present thinking is that solar steam stations will be tied in with the existing electric power grids. Solar plants will contribute power to the grid when the sun is shining, thus conserving fuel which would otherwise have to be burned or water which would have to run through turbines. The solar plants would be idle at night, when the demand for power in today's air conditioned world is low.

SUNDIALS AND THE TELLING OF TIME

The use of shadows cast by the sun on the earth to designate the passage of time is so ancient that no one really knows who built the first sun dial. They were certainly in use by 1500 B.C. (see App. IV, No. 14 for a comprehensive study of the science of "dialing") and a bewildering variety of sundials is to be found in Europe and Asia. The simplest and most widely-used type is the horizontal dial.

The dial plate is the base upon which the hour lines are inscribed and the "gnomon," attached to the plate and pointing true north (not compass north!), is the element which casts the shadow. The angle between the gnomon and the base must be exactly equal to the local latitude and then, for reasons which an astronomer can readily explain, the gnomon will be aligned exactly with the earth's axis. The hour lines for a properly constructed horizontal dial will always denote the number of hours away from solar noon, and so corrections must be made to account for the Equation of Time and departure of the local longitude from the Local Standard Time Meridian. These corrections are explained in App. II, and the details for calculating hour line angles are given in App. IV, No. 14.

CONCLUSIONS

The purpose of this chapter has been to give the reader the basic principles upon which the rapidly growing field of heliotechnology is based. For those who want to learn more about this fascinating field, the list of resources given in App. IV will prove to be exceedingly helpful. Regular reading of *Solar Energy*, the Journal of the International Solar Energy Society, will help, too.

APPENDIX I

Glossary

A — Symbol used here to denote the solar azimuth, which is the angle between the south-north line at a given location and the projection of the earth-sun line in the horizontal plane. It is ∠HOS in Fig. II.4.

absorptance — The ratio of absorbed to incident solar radiation.

ASHRAE — Acronym denoting the American Society of Heating Refrigerating and Air Conditioning Engineers, 345 E. 47th St., New York, NY 10017. ASHRAE Handbooks are the source of most of the basic data on heating and air conditioning.

B — Symbol used here to denote the solar altitude above the horizontal plane. It is ∠HOQ in Fig. II.4.

Btu — British thermal unit, which is the amount of heat required to raise the temperature of one pound of water one degree Fahrenheit. (No longer used in England, since they have gone metric!)

B̲t̲u̲ — Symbol used here to denote "Btu per hour per square foot."

C — Designates the Celsius temperature scale where water freezes at 0°C and boils at 100°C. To convert from degrees C to degrees F multiply the temperature in deg. C by 9/5 (or 1.8) and add 32. Example: convert 18°C to degrees F.; $(18^\circ$C x 1.8) + 32° = 64.4°F. To convert from degrees F to degrees C reverse the process and subtract 32 from the Fahrenheit temperature and then multiply by 5/9. Example: convert 112°F to degrees C.; $(112^\circ - 32^\circ)$ x 5/9 = 44.4°C.

coolth — A moderate degree of coolness (antonym of warmth).

D — Symbol used here to denote the solar declination, which is the angle between the earth-sun line and the earth's equatorial plane, as shown in Fig. II.2. It varies day by day throughout the year from $+23.47^\circ$ on June 21 to -23.47° on Dec. 21.

declination — See the definition of symbol D.

deg. — Abbreviation used for degrees of arc and also degrees of temperature.

emissivity — The property of emitting heat radiation; possessed by all materials to a varying extent. "Emittance" is the numerical value of this property, expressed as a decimal fraction, for a particular material. Normal emittance is the value measured at 90 deg. to the plane of the sample and hemispherical emittance is the total amount emitted in all directions. We are generally interested in hemispherical, rather than normal, emittance. Emittance values range from 0.05 for brightly polished metals to 0.96 for flat black paint. Most non-metals have high values of emittance.

F — Designates the Fahrenheit temperature scale, where water freezes at 32°F and boils at 212°F. See the definition of symbol C for the equation to convert degrees F to degrees C.

focal spot — The rays of the sun are not precisely parallel, despite its 93,000,000 mile distance, and so they are concentrated not at a point but on a spot, the diameter of which is given by the following formula:

$$\text{Diameter of the focal spot} = \frac{\text{focal length}}{107.3}$$

The number 107.3 is one-half of the reciprocal of the tangent of 16 minutes of arc, which is in turn one-half of the angle which the sun's disk subtends when it is observed from the earth.

G — Symbol used here to denote the wall-solar azimuth, the angle in the horizontal plane between the solar azimuth line and the line normal to a specified wall. It is ∠HOP in Fig. II.4.

H — Symbol used here to denote the sun's hour angle, equal to the number of minutes from local solar noon, divided by 4 to convert to degrees of arc.

heat exchanger — A device used to transfer heat from a fluid flowing on one side of a barrier to a fluid flowing on the other side of the barrier. Quite often this is done by running a coil of pipe through a tank.

heat pump — A device which transfers thermal energy or heat from a relatively low-temperature reservoir to one at a higher temperature. Heat normally flows from a warmer region to a cooler region; this process is reversed when we supply additional energy to a heat pump.

heliochemical — Process which uses the sun's radiation to cause chemical reactions.

heliochronometry — Telling of time by means of the sun.

helioelectrical — Process which uses the sun's radiation to produce electricity.

heliostatic — An adjective referring to devices which make the sun appear to stand still.

heliothermal — Process which uses the sun's radiation to produce heat.

heliotropic — An adjective referring to devices which track the sun, following its apparent motion across the sky.

hour angle — See the definition of symbol H.

I — Symbol used to denote solar radiation intensity in B̲t̲u̲.

I_d — Symbol used to denote diffuse or sky radiation reaching a specified surface, in B̲t̲u̲.

I_{DN} — Symbol used here to denote direct beam solar irradiation on a surface normal to the sun's rays, in B̲t̲u̲.

$I_{D\theta}$ — Symbol used here to denote direct beam solar irradiation on a surface with an incident angle Θ between the sun's rays and the normal to the surface, in B̲t̲u̲.

I_L — Symbol used here to denote total insolation on a south-facing surface tilted upward from the horizontal at the angle of the local latitude, in B̲t̲u̲.

I_{sc} — Symbol used here to denote the solar constant, the intensity of solar radiation beyond the earth's atmosphere, at the average earth-sun distance, on a surface perpendicular to the sun's rays. The value for the solar constant is 1,353 W/m^2, 1.940 $cal/cm^2/min$, 429.2 Btu/sq. ft./hr., 125.7 W/sq. ft., or 1.81 horsepower/sq. meter.

$I_{t\theta}$ — Symbol used here to denote total insolation, direct plus diffuse, on a surface with an incident angle Θ between the sun's rays and the normal to the surface, in B̲t̲u̲.

incident angle — The angle between the sun's rays and a line perpendicular (normal) to the irradiated surface.

insolation — Solar irradiation.

k — See the definition of thermal conductivity.

kilowatt — One thousand watts. A watt is a unit of power equal to one joule per second. Power is the rate at which work is done. A kilowatt hour is the total energy developed when the power of one kilowatt acts for one hour. This is the common unit of electrical power consumption.

L — Symbol used here to denote local latitude in degrees.

latitude — The angular distance north (+) or south (-) of the equator, measured in degrees of arc.

Langley — The meteorologist's unit of solar radiation intensity, equivalent to 1.0 gram calorie per square centimeter, usually used in terms of Langleys per minute. 1 Langley per minute = 221.2 Btu per hour per sq. ft.

m — Symbol used here to denote air mass, the scientist's way of expressing the ratio of the mass of atmosphere in the actual earth-sun path to the mass which would exist if the sun were directly overhead. In space, beyond the earth's atmosphere, m = 0.

micron — Unit of solar radiation wavelength, equal to one-millionth of a meter.

normal — In geometry, a word which means perpendicular.

photosynthesis — The process by which the sun's radiation, in certain specific wavelengths, causes water, carbon dioxide and nutrients to react, thus producing oxygen in plants.

pyranometer — A solar radiometer which measures total insolation, including both the direct and the diffuse radiation.

radiometer — An instrument which measures the intensity of any kind of radiation. The adjective "solar" is needed to denote instruments, like the pyrheliometer and the pyranometer, which measure insolation.

selective surface — A surface which can absorb most of the sun's shortwave radiant energy, 41% of which is visible, but which re-radiates very little longwave (infrared) radiant energy.

solar constant — See the definition of symbol I_{sc}.

T — Tilt angle, upward from the horizontal.

Tabor surface — A black nickel selective surface coating, invented by Dr. Harry Tabor, which absorbs 90% of the incoming solar energy but re-radiates only about 10% as much longwave radiation as would be emitted by a coat of flat black paint.

thermal cement — A material which can be used to make metal-to-metal seals which have both high mechanical strength and good thermal conductivity. Generally, the basic composition of such cements is a liquid, such as sodium silicate (water glass), in which a large amount of iron filings is suspended. When exposed to the atmosphere, this mixture hardens and the iron filings give it a thermal conductivity comparable to steel itself. There are many commercial sources for thermal cements, including epoxies, etc.

thermal conductance — The amount of heat in Btu which can be conducted through a particular solid material, one foot square, which is 1-inch thick and has a temperature difference of 1°F maintained between its two surfaces. "k" is the symbol used here to designate thermal conductance. Most metals are good conductors and their conductivity varies with their temperature. At 212°F (100°C), silver has the highest conductivity, 2,856 B̲t̲u̲/($^\circ$F/in.), while mild steel has a con

thermal resistance	The reciprocal of thermal conductance. The thermal resistance of a material is its thickness, ℓ, in inches divided by its thermal conductivity, k. The resistance of a series of different materials, all in thermal contact with each other, is the sum of the individual resistances.
thermocouple	A thermoelectric device which has a combination of two dissimilar wires with their ends connected together. A millivolt meter is connected in the circuit to measure the voltage which is generated when the two junctions are at different temperatures. If one junction is kept in a bath of ice and water, at 32°F, the voltage generated (measured in millivolts) is a measure of the temperature of the other junction above the 32°F reference point.
thermopile	A large number of thermocouples connected in series with all of the hot junctions located in one region and all of the cold junctions located in another. It is used to measure very small temperature differences with a high degree of accuracy.
thermosyphon	The convective circulation of fluid which occurs in a closed system when less-dense warm fluid rises, displaced by denser, cooler fluid in the same system.
theta	See the definition of symbol Θ.
Θ	Symbol for the Greek letter "theta," used here to designate the angle of incidence between a solar ray and the line "normal" to the surface which is being irradiated.
Θ_L	Symbol used here to denote the incident angle in degrees for solar rays falling on a south-facing surface tilted up from the horizontal at an angle equal to the local latitude, L.
Θ_v Θ_h	Symbols used here to denote the incident angles on vertical and horizontal surfaces, respectively, in degrees.
ton of refrigeration	One ton of refrigeration means the removal of heat at the rate of 12,000 Btu per hour. This unit comes from the fact that melting a ton (2,000 lbs.) of ice requires 2,000 x 144 Btu, which if done over a 24 hour period, requires a heat removal rate of 12,000 Btu per hour.
transmittance	The ratio of radiant energy transmitted to energy incident on a surface. In solar technology it is often affected by the thickness and composition of the glass cover plates on a collector, and to a major extent by the angle of incidence Θ between the sun's rays and a line normal to the surface.
U	Symbol for the heat loss coefficient, in Btu/°F.
wall azimuth	The line in the horizontal plane that is perpendicular to the receiving surface. It is ∠SOP in Fig. II.4.
wall-solar azimuth	See the definition of symbol G.
warmth	A moderate degree of heat.
"wet surface"	A surface upon which water will condense in a thin, even sheet, as opposed to forming large, distinct drops. Plastic surfaces can be treated to become "wettable."

APPENDIX II

THE SUN AND ITS RELATION TO THE WHOLE EARTH

The Sun's Radiant Energy

The energy radiated from the sun's outer surface (called the photosphere) travels in spheres of ever-increasing diameter. The intensity of radiation on a unit of area, such as a square foot, or a sq. meter, or a sq. centimeter, varies inversely as the square of the distance from the sun. The intensity of solar radiation at the edge of the Earth's atmosphere, at the average Earth-sun distance, measured on a surface perpendicular to the solar rays, is called the solar constant, I_{sc} (App. IV, No. 15). The average Earth-sun distance, approximately 92,956,000 miles, is called an Astronomical Unit (AU), and at this distance the Solar Constant, I_{sc}, is 429.2 Btu per hour per square foot. ("Btu per hour per square foot" will be used so frequently that we will use the symbol Btu to denote this unit.) Space scientists generally use the International system of units and they say that I_{sc} = 1,353 watts per sq. meter while meterologists use still another system of terminology, in which the solar constant is 1.940 Langleys per minute.

The Earth's orbit is almost exactly circular but the sun is somewhat off-center and so we are further from the sun on July 1 (94,482,000 miles) than we are on January 1 (91,325,000 miles). This variation is great enough to make a detectable change in the intensity of solar radiation at the Earth, but it is more important to space technicians than to others of us on Earth. Here, other and far greater variations exist because of the manner in which the Earth rotates about its own axis.

For nearly two thousand years, between the time of the Greek philosopher Aristarchus of Samos (300 BC) and the Polish astronomer Copernicus (1473-1543), virtually everyone believed that the Earth was the center of the universe and that the sun and all of the other heavenly bodies revolved about our planet. Thanks to Copernicus, Kepler, Galileo and other courageous men of that era, we now know that the apparent motion of the sun across the sky is actually the result of the Earth's own rotation. We spin at the rate of 360.99 degrees in 24 hours and so the sun appears to move across the sky at the rate of 15.04 degrees per hour.

Time and the Turning of the Earth

Our changing seasons are caused by the fact that, in our annual journey around the sun (it actually takes 365.25 days, which is why we need a "Leap Year" every four years), our rotational axis is tilted at 23.47 degrees with respect to the "plane of the ecliptic" which contains our orbit. Fig. II.1 is the classic method of representing the Earth-sun relationship as it would be viewed by an observer in far outer space. It shows that on December 21 the sun is shining primarily on the southern hemisphere, with the northern half of the globe tilted away. On June 21, the situation is reversed and we receive most of the sunshine, while winter prevails south of the equator.

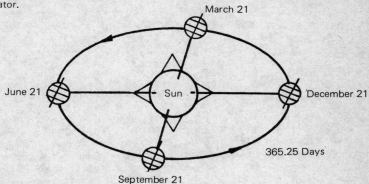

FIG. II.1 ANNUAL MOTION OF THE EARTH AROUND THE SUN.

Fig. II.2 shows the apparent annual motion of the sun with respect to the Earth. The angle made by the Earth-sun line and the equatorial plane is called the sun's "declination" and it varies day by day throughout the year from +23.47 degrees on June 21 to -23.47 degrees on December 21. On March 21 and September 23, the Earth-sun line coincides with the equator and, since the day and the night are each 12 hours long on these two days, they are called the spring and fall equinoxes. We generally think of the Earth as it appears in Fig. II.2, with its axis pointing towards the North Star, Polaris, and a series of circular lines of latitude which are parallel to the equator. Two of these lines, shown on Fig. II.2, are designated as Tropics, with the Tropic of Cancer, at +23.5 degrees North, denoting the most northerly position of the Earth-sun line and the Tropic of Capricorn at -23.5 degrees South, performing the same function in the southern hemisphere.

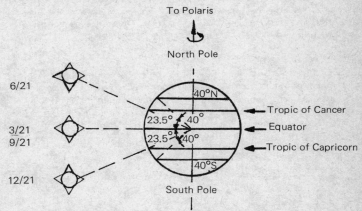

FIG. II.2 APPARENT ANNUAL MOTION OF THE
SUN WITH RESPECT TO THE EARTH
(caused by the 23.5° tilt of the Earth's axis).

The lines of latitude customarily shown on maps are actually designations of the angle between the equator and a line from the center of the Earth to its surface, as shown by the solid lines on Fig. II.2 which denote 40 degrees North and South latitudes, and the dashed lines which show how the 40 degree angular distance is measured. Latitudes vary from 0 degrees at the equator to 90 degrees at the poles. It is interesting to note that some 75% of the world's population inhabit the land areas between 40 degrees North and 40 degrees South, with 60% living north of the equator.

The declination of the sun for any given date is determined by the year day, starting from January 1, and Table II.1 gives the declinations for the 21st day of each month for the year 1974. For any other year, there will be a very minor difference between the values given in Table II.1 and those given in The Old Farmer's Almanac or some similar reliable sources which obtain their data from the U.S. Naval Observatory in Washington, D.C.

We are primarily interested in the position of the sun with respect to any given location on the Earth at any particular time, but that time must be reckoned by the sun's own peculiar system of chronometry, which differs in some important details from our standard or daylight saving time. First, the velocity of the Earth varies as it pursues its orbital path, because, as Kepler discovered four centuries ago, we speed up as we swing further away from the sun and slow down as we draw nearer. This gives rise to a series of variations, known as "the equation of time," between noon as given by a man-made clock which keeps time uniformly, and solar noon, which is the time when the sun is directly overhead at any given location. Table II.1 gives values of "the equation of time," and its sign is positive (+) when the sun is "ahead" of the clock and negative (-) when the sun is "behind" the clock.

Date	1/21	2/21	3/21	4/21	5/21	6/21
Declination, degrees	-19.9°	-10.6°	0.0°	+11.9°	+20.3°	+23.45°
Eq. of Time, minutes	-11.2	-13.9	-7.5	+1.1	+3.3	-1.4
Date	7/21	8/21	9/21	10/21	11/21	12/21
Declination, degrees	+20.5°	+12.1°	+0.5°	-10.7°	-19.9°	-23.45°
Eq. of Time, minutes	-6.2	-2.4	+7.5	+15.4	+13.8	-1.6

TABLE II.1 DECLINATIONS AND EQUATION OF TIME FOR THE
21st DAY OF EACH MONTH FOR THE YEAR 1974

For purposes of determining time, the imaginary circle which girdles the Earth at the equator has been divided into 24 segments of 15 degrees each and circular lines of longitude extend from the North Pole through each of these division points to the South Pole. Time is reckoned as beginning at the longitude of Greenwich, England and the sun appears to move around the Earth at the rate of 15 degrees per hour or 4.0 minutes per degree.

The world has been divided into 24 Time Zones and those which affect the United States are Eastern, starting at 75 degrees West longitude, Central at 90° West, Mountain at 105° West, Pacific at 120° West, Yukon at 135° West and Alaska-Hawaii at 150° West. At 180° West the International Date Line is reached and the longitude diminishes to reach 0 degrees again at Greenwich.

At any given location, solar time is found by starting with the local standard time (one hour slower than the Daylight Saving Time which, as of 1974, is in force throughout the year in all states except Arizona and Hawaii). Corrections must then be made for the local longitude, since the sun needs 4.0 minutes to traverse each degree between the Standard Time longitude and the local longitude. This correction must be made first and the equation of time must then be added, taking proper account of its algebraic sign.

As an example, we will find the solar time at noon, Pacific Daylight Saving Time, for Palo Alto, California on February 21, when the equation of time is -13.9 min. The longitude of Palo Alto is approximately 122 degrees West, so the time zone correction is 4.0 min./deg. x (122° - 120°) = 8.0 min. The solar time is thus:

Solar time = (12.00 - 1:00) - 8.0 - 13.9
= 11:00 - 21.9
= 10:38.1 a.m.

Solar time may also be expressed in terms of the Hour Angle, H, which is equal to the number of minutes from local solar noon, divided by 4 to convert to degrees. The concept used here involves the imaginary circles shown in Fig. II.3 in which the sun appears to move as our earth turns beneath it. The sun completes its daily journey in 24 hours, so each hour involves a 15 degree progression upward from sunrise in the morning and downward along the circle in the afternoon. The hour angle, H, for the Palo Alto solar noon situation is:

$$H = \frac{11 \text{ hr. } 60 \text{ min. } - 10 \text{ hr. } 38.1 \text{ min.}}{4}$$

$$= \frac{1:21.9}{4} = \frac{81.9}{4} = 20.475 \text{ deg.}$$

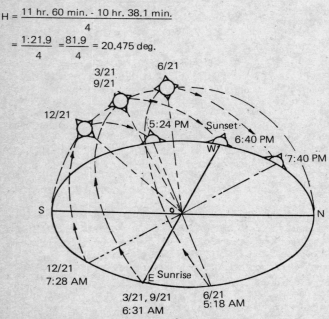

FIG. II.3 APPARENT MOTION OF THE SUN AS SEEN   (latitude = 33°30').
BY AN OBSERVER AT PHOENIX, ARIZONA

Time and the Position of the Sun

The intensity of the sun's radiation at a given time and place and the angle at which its direct rays strike a surface are dependent upon the sun's position in the sky, the season of the year, the degree of cloudiness and the tilt of the surface with respect to the horizontal. The position to the sun is defined by its altitude above the horizon, which in scientific terms is denoted by the Greek letter Beta (β), and by its azimuth.

We will use the letter B to symbolize the solar altitude and the letter A to denote the azimuth, which is the angle between the projection on the horizontal plane of the Earth-sun line and the north-south line in that same plane. A is measured from the south, for our purposes, and so the azimuth line is eastward in the mornings and westward in the afternoons. Fig. II.4 shows the sun in a typical afternoon situation; the surface on which it is shining is a vertical wall, facing south-east.

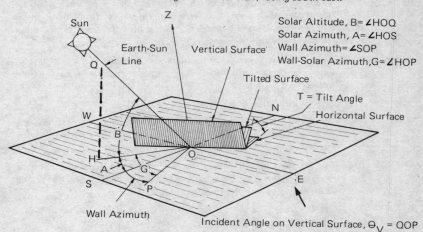

Solar Altitude, B= ∠HOQ
Solar Azimuth, A= ∠HOS
Wall Azimuth= ∠SOP
Wall-Solar Azimuth, G= ∠HOP
T = Tilt Angle
Incident Angle on Vertical Surface, Θ$_V$ = QOP

FIG. II.4 SOLAR POSITION ANGLES FOR AN AFTERNOON CONDITION,
WITH THE SUN SHINING ON A VERTICAL SURFACE
WHICH FACES SOUTH-EAST.

Using the symbol ∠ to designate an angle, and defining the angles by the letters used in Fig. II.4, the solar altitude B in this case is ∠ QOH and the solar azimuth A is ∠HOS. When we know the latitude L and the longitude for the location in which we are interested, and the date and time are specified so that we can determine the declination D and the hour angle H, we can refer to one of the standard texts on this subject (App. IV, No. 16) and find the trigonometrical equations by which the altitude and azimuth may be found.

These are: Eq. II.1

| Sine of Solar Altitude | = | [Cosine of Local Latitude x | Cosine of Day's Declination x | Cosine of Hour Angle] | + | [Sine of Local Latitude x | Sine of Day's Declination] |

Put into mathematical terms, this is simply:

$$\text{Sin B} = \text{Cos L Cos D Cos H} + \text{Sin L Sin D} \qquad \text{Eq. II.1a}$$

After the altitude has been found, the azimuth comes from:

| Sine of Solar Azimuth | = | [Cosine of Day's Declination x | Sine of Hour Angle] | ÷ | Cosine of Solar Altitude | Eq. II.2

which is:

$$\text{Sin A} = \frac{\text{Cos D Sin H}}{\text{Cos B}} \qquad \text{Eq. II.2a}$$

To show that these calculations are not really too complicated in this day of electronic computers, let's find the solar altitude and azimuth for Palo Alto at noon, PDST, on February 21. We have already found that the hour angle H is 20.475 degrees and Table II.1 gives the declination D as -10.6 degrees. The latitude L is found from an atlas to be 37.5 degrees. We can now tabulate the values which we will need, using any convenient set of trigonometrical tables, a slide rule, or a calculator such as the Hewlett-Packard 45 to do the heavy work. For example, we know:

EXAMPLE:

Latitude (L)   = 37.5 deg; Sin L = 0.609; Cos L = 0.793
Declination (D) = -10.6 deg; Sin D = -0.184; Cos D = 0.983
Hour Angle (H) = 20.475 deg; Sin H = 0.350; Cos H = 0.937

Using Eq. II.1a, we find:

Sin B = Cos L Cos D Cos H + Sin L Sin D
= 0.793 x 0.983 x 0.937 + 0.609 x (-0.184)
= 0.730 - 0.112
= 0.618

Then B, equal to the "angle whose sine is 0.618," equals 38.17 degrees.

Cos B = 0.786, so we can then find the azimuth from Eq. II.2a:

$$\text{Sin A} = \frac{\text{Cos D Sin H}}{\text{Cos B}} = \frac{0.983 \times 0.350}{0.786} = 0.438$$

The azimuth, equal to the "angle whose sine is 0.438," equals 25.96 degrees east.

Normally, this kind of mathematical exercise is left for students to do as homework problems. The experienced solar technologist knows that he can find the altitude and azimuth already tabulated for him for the 21st day of each month, at latitudes from 24 to 64 degrees North, in App. IV, No. 16, pp. 388-392. In App. IV, No. 17, pp. 59.5 and 59.6, B and A are given for 40 degrees North. Table II.2 gives in simplified form data similar to those given in App. IV, No. 17.

TABLE II.2 SOLAR POSITION ANGLES, ALTITUDE (B) AND AZIMUTH (A) FOR 32, 40 AND 48 DEG. NORTH LATITUDE; DIRECT NORMAL IRRADIATION IN Btu (I_{DN}); TOTAL INSOLATION (I_L) IN Btu AND ANGLES OF INCIDENCE (θ) FOR SOUTH-FACING SURFACES TILTED UPWARD AT ANGLE L, EQUAL TO THE LOCAL LATITUDE.

Jan 21, Nov 21 Dec.=-19.9	θ_L = Incident Angle at Lat.														
	32° North Lat.					40° North Lat.					48° North Lat.				
Solar Time	Alt.	Azim	I_{DN}	I_L	θ_L	Alt.	Azim	I_{DN}	I_L	θ_L	Alt.	Azim	I_{DN}	I_L	θ_L
6 6	--	--													
7 5	1.5	65.4	2	0	75.9										
8 4	12.7	56.6	196	104	62.0	8.2	55.4	136	72	62.0	3.6	54.7	36	19	62.0
9 3	22.6	46.1	263	190	48.4	17.0	44.1	232	167	48.4	11.2	42.7	179	129	48.4
10 2	30.8	33.2	289	252	35.5	24.0	31.0	268	233	35.5	17.1	29.5	233	202	35.5
11 1	36.2	17.6	301	291	24.8	28.6	16.1	283	273	24.8	20.9	15.1	255	245	24.8
am 12 pm	38.2	0.0	304	304	20.0	30.2	0.0	288	287	20.0	22.2	0.0	261	259	20.0
Day-Long Total, Btu/Ft²			2406	1980				2128	1778				1668	1448	

Feb 21, Oct 21 Dec.=-10.7	θ_L = Incident Angle at Lat.														
	32° North Lat.					40° North Lat.					48° North Lat.				
Solar Time	Alt.	Azim	I_{DN}	I_L	θ_L	Alt.	Azim	I_{DN}	I_L	θ_L	Alt.	Azim	I_{DN}	I_L	θ_L
6 6															
7 5	6.8	73.1	99	32	75.2	4.5	72.3	48	15	75.2	2.0	71.9	4	1	75.2
8 4	18.7	64.0	229	128	60.5	15.0	61.9	204	113	60.5	11.2	60.2	165	91	60.5
9 3	29.5	53.0	273	208	45.9	24.5	49.8	257	195	45.9	19.3	47.4	233	176	45.9
10 2	38.7	39.1	293	269	31.5	32.4	35.6	280	257	31.5	25.7	33.1	262	239	31.5
11 1	45.1	21.1	302	307	18.0	37.6	18.7	291	295	18.0	30.0	17.1	274	277	18.0
am 12 pm	47.5	0.0	304	320	10.0	39.5	0.0	294	308	10.0	31.5	0.0	278	291	10.0
Day-Long Total, But/Ft²			2696	2208				2454	2060				2154	1860	

March 21 Sept 21 Dec. = 0.0	θ_L = Incident Angle at Lat.														
	32° North Lat.					40° North Lat.					48° North Lat.				
Solar Time	Alt.	Azim	I_{DN}	I_L	θ_L	Alt.	Azim	I_{DN}	I_L	θ_L	Alt.	Azim	I_{DN}	I_L	θ_L
6 6															
7 5	12.7	8 19	163	56	75.0	11.4	80.2	149	51	75.0	10.0	78.7	131	44	75.0
8 4	25.1	73.0	240	141	60.0	22.5	69.6	230	134	60.0	19.5	66.8	215	124	60.o
9 3	36.8	62.1	272	215	45.0	32.8	57.3	263	208	45.0	28.2	53.4	251	197	45.0
10 2	47.3	47.5	287	273	30.0	41.6	41.9	280	265	30.0	35.4	37.8	269	254	30.0
11 1	55.0	26.8	294	309	15.0	47.7	22.6	287	301	15.0	40.3	19.8	278	289	15.0
am 12 pm	58.0	0.0	296	321	0.0	50.0	0.0	290	313	0.0	4210	0.0	280	302	0.0
Day-Long Total, Btu/Ft²			2808	2308				2708	2228				2568	2118	

April 21 Aug 21 Dec.=12.1	θ_L = Incident Angle at Lat.														
	32° North Lat.					40° North Lat.					48° North Lat.				
Solar Time	Alt.	Azim	I_{DN}	I_L	θ_L	Alt.	Azim	I_{DN}	I_L	θ_L	Alt.	Azim	I_{DN}	I_L	θ_L
6 6	6.5	100.5	59	7	90.0	7.9	99.5	81	9	90.0	9.1	98.3	99	10	90.0
7 5	19.1	92.8	190	69	75.3	19.3	90.0	191	69	75.3	19.1	87.2	190	67	75.3
8 4	31.8	84.7	240	144	60.7	30.7	79.9	237	141	60.7	29.0	75.4	232	137	60.7
9 3	44.3	75.0	263	212	46.2	41.8	67.9	260	207	46.2	38.4	61.8	254	201	46.2
10 2	56.1	61.3	276	264	32.0	51.7	52.1	272	259	32.0	46.4	45.1	266	252	32.0
11 1	66.0	38.4	282	298	18.9	59.3	29.7	278	292	18.9	52.2	24.3	272	285	18.9
am 12 pm	70.3	0.0	284	309	11.6	62.3	0.0	280	303	11.6	54.3	0.0	274	296	11.6
Day-Long Total, Btu/Ft²			2902	2296				2916	2258				2898	2200	

May 21 July 21 Dec.=20.5	θ_L = Incident Angle at Lat.														
	32° North Lat.					40° North Lat.					48° North Lat.				
Solar Time	Alt.	Azim	I_{DN}	I_L	θ_L	Alt.	Azim	I_{DN}	I_L	θ_L	Alt.	Azim	I_{DN}	I_L	θ_L
6 6	10.7	107.7	113	14	90.0	13.1	106.1	138	17	90.0	15.2	104.1	156	18	90.0
7 5	23.1	100.6	203	75	75.9	24.3	97.2	208	75	75.9	25.1	93.5	211	75	75.9
8 4	35.7	93.6	241	143	62.0	35.8	87.8	241	142	62.0	35.1	82.1	240	140	62.0
9 3	48.4	85.5	261	205	48.4	47.2	76.7	259	203	48.4	44.8	68.8	256	199	48.4
10 2	60.9	74.3	271	254	35.5	57.9	61.7	269	251	35.5	53.5	51.9	266	246	35.5
11 1	72.4	53.3	277	285	24.8	66.7	37.9	275	281	24.8	60.1	29.0	271	276	24.8
am 12 pm	78.6	0.0	279	296	20.0	70.6	0.0	276	292	20.0	62.6	0.0	272	286	20.0
Day-Long Total, Btu/Ft²			3012	2250				3062	2230				3158	2200	

June 21 Dec.=23.45	θ_L = Incident Angle at Lat.														
	32° North Lat.					40° North Lat.					48° North Lat.				
Solar Time	Alt.	Azim	I_{DN}	I_L	θ_L	Alt.	Azim	I_{DN}	I_L	θ_L	Alt.	Azim	I_{DN}	I_L	θ_L
6 6	12.2	110.2	131	16	90.0	14.8	108.4	155	18	90.0	17.2	106.2	172	19	90.0
7 5	24.3	103.4	210	76	76.3	26.0	99.7	216	77	76.3	27.0	95.8	220	77	76.3
8 4	36.9	96.8	245	143	62.7	37.4	90.7	246	142	62.7	37.1	84.6	246	140	62.7
9 3	49.6	89.4	264	204	49.6	48.8	80.2	263	202	49.6	46.9	71.6	261	198	49.6
10 2	62.2	79.7	274	251	37.4	59.8	65.8	272	248	37.4	55.8	54.8	269	244	37.4
11 1	74.2	60.9	279	282	27.6	69.2	41.9	277	278	27.6	62.7	31.2	274	273	27.6
am 12 pm	81.5	0.0	280	292	23.4	73.5	0.0	279	289	23.4	65.5	0.0	275	283	23.4
Day-Long Total, Btu/Ft²			3084	2234				3180	2224				3312	2204	

Dec. 21 Dec.=-23.45	θ_L = Incident Angle at Lat.														
	32° North Lat.					40° North Lat.					48° North Lat.				
Solar Time	Alt.	Azim	I_{DN}	I_L	θ_L	Alt.	Azim	I_{DN}	I_L	θ_L	Alt.	Azim	I_{DN}	I_L	θ_L
6 6	--	--													
7 5	--	--													
8 4	10.3	53.8	176	90	62.7	5.5	53.0	89	45	62.7					
9 3	19.8	43.6	257	180	49.6	14.0	41.9	217	152	49.6	8.0	40.9	140	98	49.6
10 2	27.6	31.2	288	244	37.4	20.7	29.4	261	221	37.4	13.6	28.2	214	180	37.4
11 1	32.7	16.4	301	282	27.6	25.0	15.2	280	262	27.6	17.3	14.4	242	226	27.6
am 12 pm	34.6	0.0	304	295	23.4	26.6	0.0	285	275	23.4	18.6	0.0	250	241	23.4
Day-Long Total, Btu/Ft²			2348	1888				1978	1634				1444	1250	

Interpolating among the tabulated values is generally quicker and quite sufficiently accurate for most solar purposes, where the other values which are going to be employed later are not precise enough to justify more than slide rule accuracy at most.

There is one more angle which is very important in solar technology and this is the angle between the direct rays of the sun and a line perpendicular (normal) to the surface which is trying to absorb or reflect the sun's rays. This is called the <u>angle of incidence</u>, and it is shown in Fig. II.4 as ∠QOP. It is generally symbolized by the Greek letter theta, θ, and we will use that symbol here because it is both distinctive and easily typed by combining a zero, 0, and a dash, –.

The incident angle can be calculated by using another trigonometrical equation and another angle, G, the wall-solar azimuth, which is the angle in the horizontal plane between the sun's azimuth line, OH in Fig. II.4, and the wall's azimuth, OP, which is the line (in the horizontal plane) that is perpendicular to the wall. The symbol generally used for the wall-solar azimuth is the Greek letter Gamma (γ), but we will use the letter G instead. In Fig. II.4, the wall azimuth is ∠SOP and the wall-solar azimuth G = ∠HOP.

The incident angle can be found from the equation:

$$\text{Cosine of Incident Angle} = \left[\text{Cosine of Solar Altitude} \times \text{Cosine of Wall-Solar Azimuth} \times \text{Sine of Tilt Angle}\right] + \left[\text{Sine of Solar Altitude} \times \text{Cosine of Tilt Angle}\right] \quad \text{Eq. II.3}$$

In mathematical terms.

$$\text{Cos } \theta = \text{Cos B Cos G Sin T} + \text{Sin B Cos T} \quad \text{Eq. II.3a}$$

The <u>tilt angle</u>, T, is measured from the horizontal and it can vary from 0 degrees, when the surface is actually horizontal, to 90 degrees, when the surface is vertical. In that case, a look at the basic trigonometrical triangle shows us that Sin T = 1.0 and Cos T = 0.0, so the incident angle for any vertical surface becomes simply:

$$\text{Cosine of Incident Angle for Any Vertical Surface} = \text{Cosine of Solar Altitude} \times \text{Cosine of Wall-Solar Azimuth} \quad \text{Eq. 11.4}$$

or

$$\text{Cos } \theta_v = \text{Cos B Cos G} \quad \text{Eq. II.4a}$$

For a horizontal surface, the tilt angle T = 0 degrees, and so Sin T = 0.0 and Cos T = 1.0. Thus for any horizontal surface such as a flat roof, Eq. II.3 tells us that:

$$\text{Cosine of Incident Angle for Any Horizontal Surface} = \text{Sine of Solar Altitude} \quad \text{Eq. II.5}$$

or

$$\text{Cos } \theta_h = \text{Sin B} \quad \text{Eq. II.5a}$$

By inspection of Fig. II.4, it can be seen that:

$$\text{Incident Angle for Any Horizontal Surface} = 90 \text{ degrees} - \text{B} \quad \text{Eq. II.6}$$

If we assign some values to the angles shown in Fig. II.4, we can see how these equations actually work. Given that:

Example:

Local Latitude (L) = 40 degrees
Solar Time = 1:00 p.m.
Date = September 21

From Table II.2, we find that the solar altitude B = 47.7 degrees and the solar azimuth = 22.6 degrees west. If the wall faces south-east, its azimuth is 45 degrees east, so the wall-solar azimuth G = 22.6° + 45° = 67.6 degrees. Let the surface be a vertical wall, so the tilt angle T = 90 degrees and we can use Eq. II.4a, which becomes:

$$Cos\ \theta_v = Cos\ B\ Cos\ G$$
$$= Cos\ 47.7^\circ\ Cos\ 67.6^\circ$$
$$= 0.673 \times 0.381$$
$$= 0.256$$

and in this particular example, then θ_v is the "angle whose cosine is 0.256," which turns out to be 75.14 degrees.

There are some useful short-cuts which can simplify the finding of some solar angles. At solar noon, the solar azimuth is always 0.0 degrees and the solar altitude is:

$$B_{noon} = 90\ \text{degrees} - L + D \qquad \text{Eq. 11.7}$$

To return to the Palo Alto example used earlier, at solar noon on February 21, the solar altitude is:

$$B_{noon} = 90\ \text{degrees} - 37.5^\circ + (-10.6^\circ) = 41.9\ \text{degrees}$$

On June 21 in Palo Alto, when the declination is $+23.5^\circ$

$$B_{noon} = 90\ \text{degrees} - 37.5^\circ + 23.5^\circ = 76\ \text{degrees}$$

Time and the Intensity of Solar Radiation

We have already learned that the intensity of the sun's rays on a surface perpendicular to those rays, located at the average earth-sun distance and outside of the Earth's atmosphere, is 429.2 Btu. April 4 and October 4 are approximately the dates when the average Earth-sun distance is attained (the exact date changes slightly from year to year). The maximum Earth-sun distance occurs on July 4, and then the surface beyond the Earth's atmosphere receives only 415.2 Btu; on January 4, we reach our closest proximity to the sun and the extra-terrestial intensity becomes 443.6 Btu. For those who want to learn more about the sun's radiation in space, App. IV, No. 18, is an excellent and authoritative source of information.

In passing through the atmosphere to reach the surface of the Earth, solar radiation undergoes a number of changes. Ultraviolet radiation which is less than 0.3 microns in wavelength is all absorbed in the upper atmosphere, primarily by the ozone which abounds there. The longwave infrared beyond about 2.6 microns is all absorbed by the water vapor in the lower atmosphere. In between, there are numerous absorption bands caused by other components of the atmosphere, including carbon dioxide.

Some of the incoming radiation is scattered in all directions by the air molecules themselves and this scattered radiation is primarily in the blue portion of the visible spectrum, thus causing the blue color which characterizes a clear sky.

The direct beam radiation is considerably reduced in intensity as it passes through the atmosphere and the amount of the reduction depends upon the length of the atmospheric path which it must traverse. Meteorologists describe this in terms of the air mass, m, which is the ratio of the actual path length OQ to the path length OZ which would exist if the sun were directly overhead. This ratio evidently depends upon the sun's altitude above the horizon, B, and a glance at Fig. 11.5, a much simplified version of the actual situation, will show that:

$$m = \frac{OQ}{OZ} = \frac{1}{\text{Sine of the Solar Altitude}} = \frac{1}{Sin\ B} \qquad \text{Eq. 11.8}$$

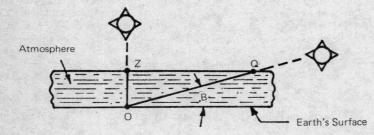

FIG. 11.5 SIMPLIFIED DRAWING OF THE PATH OF THE SUN'S RAYS THROUGH THE EARTH'S ATMOSPHERE (air mass, m, OQ/OZ = 1/Sin B).

To return to the Palo Alto example, on February 21, the solar altitude at noon is 41.9 degrees and so the air mass is $1.0/Sin\ 41.9^\circ = 1.0/0.67 = 1.5$. On June 21, the noon solar altitude is 76 degrees and the air mass at that time and date is $1.0/Sin\ 76^\circ$ $1.0/0.97 = 1.03$.

The intensity of the direct beam depends upon the solar altitude, since that determines how much atmosphere the rays have to traverse, and it also depends upon the amount of water vapor, dust particles and man-made pollutants which the atmosphere contains. Water vapor is the primary factor which determines how much direct radiation will reach the earth's surface on a clear day and there is a very marked variation in water vapor content throughout the year.

The probable values of the direct normal radiation intensity on clear days throughout the year and their variation with solar altitude are shown in Fig. 11.6. These values (see App. IV, No. 16, for a more detailed explanation) take into account the annual variation of humidity in the United States. The term "direct normal irradiation" means the intensity of the direct radiation falling on a surface perpendicular or "normal" to the rays, and it is symbolized by I_{DN}.

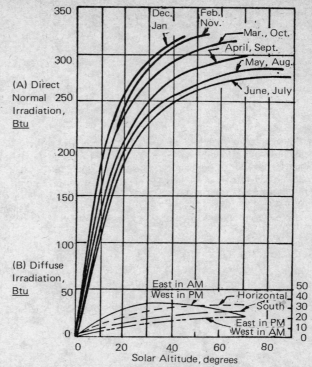

FIG. 11.6 (A) VARIATION OF DIRECT NORMAL IRRADIATION WITH SOLAR ALTITUDE THROUGHOUT THE YEAR; (B) DIFFUSE RADIATION VS. SOLAR ALTITUDE FOR VARIOUS SURFACE ORIENTATIONS.

Most of the surfaces in which we are interested are not normal to the sun's direct rays and so we must return to the angle of incidence, θ between the incoming rays and a line that is normal to the surface, by using the following equation to find the intensity on any surface for which θ is known:

| Direct Solar Irradiation on a Surface with an Incident Angle θ | = | Direct Normal Irradiation | x | Cosine of Incident Angle, θ | Eq. 11.9 |

or

$$I_{D\theta} = I_{DN}\ Cos\ \theta \qquad \text{Eq. 11.9a}$$

The physical explanation of this very valuable equation (which is the reason for all of the trigonometry in which we have indulged up to this point) is shown in Fig. 11.7. Here we have a beam of direct sunshine which is exactly 1 foot square falling on a horizontal surface with an incident angle θ which is shown by $\angle$ MOZ. Because the horizontal surface is not normal to the sunbeam, the energy contained in the 1 square foot, represented by the letters MN, is spread over a larger area represented by the letters ON and so the intensity, or the amount of energy per square foot of area is:

$$I_{D\theta} = I_{DN} \times \frac{MN}{ON} = I_{DN}\ Cos\ \theta \qquad \text{Eq. 11.9b}$$

The laws of geometry tell us that $\angle$ MNO is equal to the incident angle $\angle$ MOZ, and the ratio of the two sides of the triangle, MN/ON, is the cosine of $\angle$ MNO, which is also the cosine of the incident angle θ.

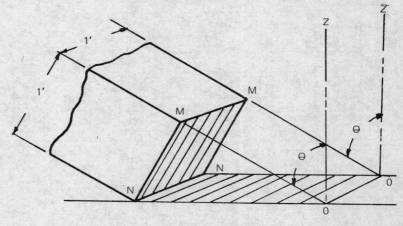

FIG. 11.7 EXPLANATION OF THE COSINE LAW ($I_{D\theta} = I_{DN} \times Cos\ \theta$).

Most of the surfaces in which we are interested in our study of solar technology will be tilted with respect to the horizontal, but we have learned, thanks to Eq. 11.3, how to find θ for any surface. Because of the sun's apparent daily motion across the sky, shown in Fig. II.3, the incident angle is constantly changing. If we restrict our attention to surfaces which face the south, the incident angle for any particular surface with a fixed tilt will depend only on the time of day and the date, since they determine the sun's altitude and azimuth. Using a surface at 40 degrees north latitude, tilted at 40 degrees upward from the horizontal with July 21 as the date, we find that incident angle varies from approximately 90 degrees at 6:00 a.m. (we will use the 24 hour clock and call that 0600 hours) to a minimum of 20 degrees at noon (1200 hours) and back to 90 degrees again at 1800 hours.

Fig. II.8 shows the hourly variation of θ and it is immediately seen that, for a south-facing surface, the incident angle is symmetrical about solar noon. This means that θ will be the same for a given number of hours away from solar noon in the afternoon as it is for the same hour angle in the morning. The only difference will be that the sun's rays will come from the east in the morning and from the west in the afternoon.

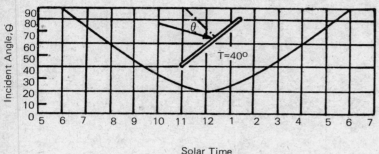

FIG. II.8 HOURLY VARIATION OF INCIDENT ANGLE ON SOUTH-FACING SURFACE
(July 21, 40° N latitude, receiving surface tilted upward from horizontal 40°).

Thus far, we have considered only the sun's direct radiation and we appear to have forgotten about the radiation which, in an earlier paragraph, was scattered by the air molecules and the atmospheric dust. Part of this radiation, which is called "diffuse" since it comes from all parts of the sky, reaches the earth's surface and so the total irradiation or "insolation," $I_{t\theta}$, is the sum of this diffuse radiation, which we symbolized by I_d, and the direct radiation, $I_{D\theta}$. In equation form we would say that:

| Total Solar Irradiation | = | [Direct Normal Solar Irradiation | x | Cosine of Incident Angle, θ |] | + | Diffuse Solar Irradiation | Eq. II.10 |

or

$$I_{t\theta} = I_{DN} \cos \theta + I_d \qquad \text{Eq. II.10a}$$

There is no easy way to estimate the intensity of diffuse radiation, since it too depends upon the amount of dust, moisture and the degree of cloudiness of the sky. On a completely cloudy day, the only radiation that we receive is the diffuse component, while on a clear day it depends upon just how clear the sky really is and how much invisible water vapor it contains. The lower curves in Fig. II.6 give typical values of diffuse radiation for various surfaces in terms of the solar altitude.

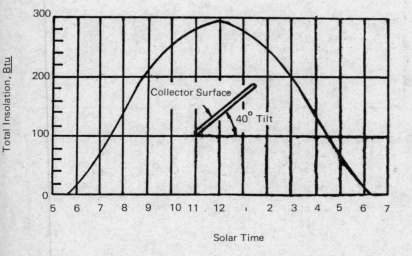

FIG. II.9 HOURLY VARIATION OF TOTAL INSOLATION ON SOUTH-FACING SURFACE
(July 21, 40° N latitude; receiving surface tilted upward from horizontal 40°).

The University of Florida has developed a computer program (App. IV, No. 19) by which the total insolation can be computed from ASHRAE data for surfaces with any orientation and any degree of tilt from the horizontal. Table II.2 gives selected values of the solar data which are to be found in App. IV, No. 16 and No. 17. Space does not permit the use of all of the available data and so only a single surface is considered in Table II.2, facing south and tilted at the angle of the local latitude.

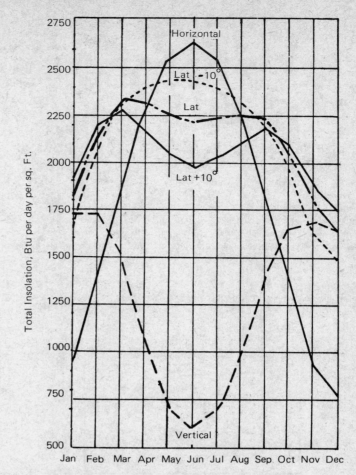

FIG. II.10 VARIATION OF DAY-LONG TOTAL INSOLATION THROUGHOUT THE YEAR FOR SOUTH-FACING SURFACES WITH VARYING TILT ANGLES (on the 21st day of each month).

An extension of the University of Florida program, produced by Miss Ceresse Nada of Alhambra High School, Phoenix, gave insolation data for surfaces facing in directions other than due south. For 40 degrees North latitude, the loss in day-long insolation on January 21 is 1% for a 10 degree deviation, 4% for 20 degrees and 8% for 30 degrees. Thus exact adherence to the general rule about facing due south is not essential for good performance of a solar device. Fig. II.9 shows the hourly variation in total insolation on July 21, for a south-facing surface tilted upward at 40 degrees from the horizontal, at 40 degrees North latitude.

The angle of tilt is very important in the performance of solar collectors, as Fig. II.10 shows. A horizontal surface receives more irradiation in mid-summer and less in mid-winter than any of the other surfaces shown in Fig. II.10. A vertical surface receives a significant amount of irradiation during the winter months when the solar altitude is low, but it receives very little during the summer months when the sun is high in the sky. A surface tilted at the angle of the local latitude (L) will do a good job all year long, but its performance will be exceeded during the summer by a surface tilted at L - 10°, and in winter by one which is tilted at L + 10°. In general, the complexity involved in making a collector adjustable in its tilt angle is not compensated by improved performance. It is preferable to select a suitable angle, based on the function which the collector is intended to serve, and let it remain fixed in place.

Thus far, no reference has been made to radiation which may be reflected from surfaces which are on the south of the collector surface, but this can be significant in the winter when the ground is covered with snow which is a very good reflector. For details on this subject, see App. IV, No. 17, p. 59.9.

Conclusion

The rather tedious mathematical approach contained in the foregoing pages will give to the serious student of solar technology the information which one needs to make quantitative estimates of how big solar devices must be in order to achieve a desired result. The highest intensity attained by the sun's direct rays at sea level does not exceed 304 Btu at 32 degrees North latitude, which means that the atmosphere has filtered out some 30% of the solar radiation even on a clear day. In mid-summer, the highest observed intensity at sea level will be about 280 Btu, indicating that the summer atmosphere absorbs even more of the incoming radiation because of its higher humidity.

We have spoken of clear days and sea-level conditions, and we will find higher intensities at higher elevations and in regions where the humidity is exceptionally low, as it is in the arid southwest. To correct for these situations, App. IV, No. 16, gives a map with "Clearness Numbers," which show that the values given in Table II.2 and Fig. II.10 will be exceeded by as much as 10% in the Rocky Mountain states, while the Gulf Coast and most of Florida will have lower values than the so-called standard, by as much as 10 to 15%.

Clouds present a major problem to the solar technologist, because it is difficult to predict when they will appear and how long they will persist. The local Weather

Bureau (now operated by the National Oceanic and Atmospheric Administration) has long records of the hours of sunshine, or at least the percentage of possible sunshine, on a month by month basis for all major cities. Maps are available in App. IV, No. 20, which show monthly variation of insolation on horizontal surfaces, percent of possible sunshine, number of hours of sunshine per year, and other information which can be used to estimate how much solar radiation can be expected to be available at any given place in an average year.

APPENDIX III

METHODS OF ESTIMATING SOLAR HEATER PERFORMANCE

The technical aspects of collector design and performance estimates are discussed in detail in the publication listed in App. IV, No. 21. The fundamental principles will be treated briefly and simply in the following section.

The first decision which must be made is the angle of tilt of the collector panels. In App. II we have learned that in the northern hemisphere the collectors should face the south while "down under" they should face towards the north. A few degrees deviation from a true south exposure will do very little harm to the collector's performance.

The collector plate can only make use of the heat which actually reaches it, so the ability of the glazing material to transmit the sun's radiation is very important. This property is called transmittance and it depends upon the thickness of the glazing, its composition, and to a major extent upon the angle of incidence θ between the sun's rays and a line normal to the surface (Fig. III.1).

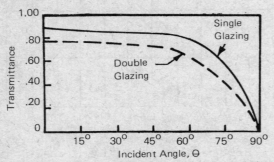

FIG. III.1 TRANSMITTANCE VS. θ FOR CLEAR GLASS.

The equations for finding θ are given in App. II and this angle is very important in determining the intensity of the direct solar radiation falling on a surface tilted away from normal incidence. Table II.2 (App. II) gives data for both the total insolation $I_{t\theta}$, in Btu, and the incident angle θ, for south-facing surfaces tilted upward at the angle of the local latitude, for the 21st day of each month at 40 degrees North latitude. App. IV, No. 17 and No. 19, give similar information for other latitudes. Both $I_{t\theta}$ and θ change minute by minute throughout the day, and so does the collector performance!

The objective of the collector is to heat a fluid and so the following equation is used to see how well the collector is doing its job:

$$\text{Efficiency of Collector, Percent (\%)} = 100 \times \left[\frac{\text{Rate of Fluid Flow} \times \text{Specific Heat of Fluid} \times \text{Fluid Temp. Rise}}{\text{Total Insolation with Incident Angle } \theta, I_{t\theta}}\right] \quad \text{Eq. III.1}$$

The units of measurement are:

$$\text{\% Efficiency} = 100 \times \left[\frac{\dfrac{\text{lb.}}{\text{hr.ft.}^2} \times \dfrac{\text{Btu}}{\text{lb.}^\circ\text{F}} \times {}^\circ\text{F}}{\dfrac{\text{Btu}}{\text{hr.ft.}^2}}\right] \quad \text{Eq. III.1a}$$

The rate of fluid flow is measured in pounds per hour, the specific heat is a quantity of heat in Btu needed to warm a pound of that fluid 1 degree F, and the temperature rise is given in degrees F.

The collector efficiency can also be expressed in terms of what happens to the incoming solar radiation, since we can also say that:

$$\text{Total Solar Radiation in BTU's} = \text{Heat Carried Away by Working Fluid in BTU's} + \text{Heat Loss From All Collector Surfaces in BTU's} \quad \text{Eq. 111.2}$$

The units of measurement are:

$$\text{Btu} = \text{Btu} + \text{Btu} \quad \text{Eq. III.2a}$$

The collector has six surfaces, but its insulation can easily be made so effective that very little heat is wasted through the back or sides. It is the sun-facing glazed or bare surface which is the hottest and largest in area and so it is responsible for most of the lost heat.

Heat always flows from high temperature to low temperature and the rate of heat flow is equal to the product of the area involved, the difference of temperature between the collector surface and the surrounding (ambient) air, and a factor which the engineer calls the "heat loss coefficient," symbolized by the letter U. The heat loss

coefficient U depends upon (a) the number of cover glasses employed, which may range from zero for a bare collector to three when very high temperatures are needed, (b) the nature of the collector surface and its ability to emit longwave radiation, and (c) the amount of the temperature difference which is causing the heat to flow. App. IV, No. 21, deals with these matters in great detail and Fig. III.2 is adapted from Dr. Austin Whillier's chapter in that reference (p. 32) which gives a great mass of engineering data, most of which is beyond the scope of this chapter. The units for U are Btu (Btu per square foot per hour) per degree F, or $\text{Btu}/{}^\circ\text{F}$.

The general equation for the collector's performance is:

$$\text{Heat Carried Away by the Fluid} = \left[\text{Radiation Transmitted by Glazing} \times \text{Absorptance of Collector Surface}\right] - \left[\text{Heat Loss Coeff., U} \times (t_s - t_o)\right] \quad \text{Eq. III.3}$$

The units of measurement are:

$$\text{Btu} = \left[\text{Btu} \times \frac{\text{Btu (absorbed)}}{\text{Btu (incident)}}\right] - \left[\frac{\text{Btu}}{{}^\circ\text{F}} \times {}^\circ\text{F}\right] \quad \text{Eq. III.3a}$$

t_s and t_o are the symbols used here to designate the average temperature of the collector (t_s) and the ambient air temperature (t_o).

Fig. III.2 shows how the heat loss coefficient varies with the number of glass cover plates and with the difference in temperature between the collector plate and the ambient air, starting with an average plate temperature of 100°F and going up to 240°F, which is just about as hot as a flat plate collector is likely to get. The chart is based on the use of a non-selective black coating on the collector plate, with an emittance (App. I) of 0.95. Page 34 of App. IV, No. 21, gives the remaining assumptions, all of them quite reasonable, upon which Fig. III.2 is based.

The ability of glass or other transparent materials to transmit solar radiation depends upon the thickness and quality of the glass and upon the angle of incidence between the glass and the sun's rays. Table 1 gives values of the transmittance for one, two and three cover glasses for incident angles from zero to 90 degrees. The absorptance for flat black paint is also given for the same incident angles and finally the first two terms of Eq. III.3 are multiplied together to give the decimal fraction of the incoming radiation which reaches and is absorbed by the collector plate.

Incident Angle, θ	0°	10°	20°	30°	40°	50°	60°	70°	80°	90°
Transmittance										
n = 1	0.87	0.87	0.87	0.87	0.86	0.84	0.79	0.68	0.42	0.0
n = 2	0.77	0.77	0.77	0.76	0.75	0.73	0.67	0.53	0.25	0.0
Absorptance										
	0.96	0.96	0.96	0.95	0.94	0.92	0.88	0.82	0.67	0.0
Transmittance X Absorptance										
n = 1	0.84	0.84	0.84	0.83	0.81	0.77	0.70	0.56	0.28	0.0
n = 2	0.74	0.74	0.74	0.72	0.71	0.67	0.59	0.43	0.17	0.0

TABLE III.1   VARIATION WITH INCIDENT ANGLE θ OF: TRANSMITTANCE FOR SOLAR RADIATION OF n = 1 OR 2 DOUBLE STRENGTH GLASS COVER PLATES; ABSOPRTANCE FOR FLAT BLACK PAINT; PRODUCTS OF TRANSMITTANCE X ABSORPTANCE FOR n = 1 OR 2 GLASS COVERS.

We are often interested in knowing what the efficiency of collection is likely to be under a given set of circumstances and Eq. III.1 gave one way to find this. However, we are more likely to know the quantities given in Eq. III.3 and this can be revised to give another way of estimating the efficiency of collection:

$$\text{Efficiency of Collector, Percent (\%)} = 100 \left[\text{Transmittance x Absorptance} - \frac{\text{Heat Loss Coeff., U} \times (t_s - t_o)}{\text{Total Insolation, } I_{t\theta}}\right] \quad \text{Eq. III.4}$$

The units of measurement are:

$$\text{\% Efficiency} = 100 \left[\left(\frac{\text{Btu (transmitted)}}{\text{Btu (incident)}} \times \frac{\text{Btu (absorbed)}}{\text{Btu (incident)}}\right) - \frac{\dfrac{\text{Btu}}{{}^\circ\text{F}} \times {}^\circ\text{F}}{\text{Btu}}\right] \quad \text{Eq. III.4a}$$

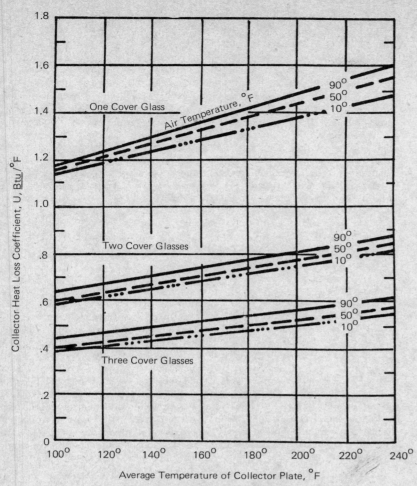

FIG. III:2   VARIATION OF HEAT LOSS COEFFICIENT, U.
WITH COLLECTOR PLATE TEMPERATURE
(non-selective surface).

EXAMPLE:

To see how this equation works, let us assume that we have a south-facing collector, tilted upward at 90 degrees from the horizontal, at 40 degrees north latitude at 2:00 p.m. solar time, on January 21, at a place where the outdoor temperature is 50°F. We would like to have the collector run at 110°F to provide domestic hot water; Fig. III.2 tells us that the heat loss coefficient, U, will be 1.13 Btu/°F for one cover plate, 0.62 for two and 0.41 for three cover plates.

We must next find the total insolation and the incident angle for this condition and Table II.2 tells us that $I_{t\theta}$ is 237 Btu/ft² while θ = 35.9 degrees (by calculation, using Eq. II.3a). From Table III.1, we find the following values for the product of the transmittance and the absorptance at 40 degrees incident angle: 0.81 for n = 1; 0.71 for n = 2; 0.60 for n = 3.

By substituting these values into Eq. III.4 we can easily find the probable value of the efficiency of this collector on a clear winter day, with a 60 degree F temperature difference between the collector plate and the ambient air. For n = 1 (one cover glass) the efficiency is 51%; if we add a second cover glass, the efficiency rises to 55.3% but if we add a third glass, the efficiency drops to 49.6% because the reduction in transmittance more than offsets the reduction in the heat loss coefficient.

For south-facing collectors, tilted at the angle of the local latitude, the collection efficiency is generally highest in the early afternoon, when the incident angle is still below 40 degrees and the outdoor air is at its maximum temperature, thus reducing the heat loss as far as possible with the specified number of cover glasses.

We can easily find the maximum temperature which a particular collector will attain when no heat is being withdrawn from it in the form of hot water or air. In this case the efficiency will be zero and the collector will attain what is called its "equilibrium temperature." For the case just considered, with two cover glasses, the equilibrium temperature is found to be approximately 250°F.

If we are trying to heat the fluid to an average temperature of 150°F, the efficiency with two cover glasses will drop to 42%, but if we are content to heat the fluid (let's assume that it is water) to 75°F, the efficiency would be 69% with one cover glass and 66% with two glasses.

There is one other case which should be considered and that is the minimum rate of insolation which must be attained before the collector can begin to operate. This can also be found from Eq. III.4 by letting the efficiency again become zero and finding the insolation rate. Assuming again that we are dealing with a double-glazed collector, with a relatively low incident angle so that the product of the transmittance times the absorptance remains near 0.71 and the plate-to-air temperature difference is still 50°F, the minimum insolation which will cause the surface temperature to rise to 110°F will be approximately 43 Btu/ft² which, on a clear January morning, would occur before 8:00 a.m. solar time.

At 8:00 a.m., the air temperature is more likely to be 35°F than 50°F so the surface-to-air temperature difference would be 75°F, the heat loss coefficient would be 0.60 and the necessary insolation would be 64 Btu/ft² which would be reached just before 8:00 a.m.

Now, refer to Fig. III.3. At very low temperature differences, the flat black collector will do a slightly better job because its absorptance (0.96) will generally be higher than that of the selective surface. As the temperature rises, the heat loss for the collector plate also rises (Eq. III.3) and the lower longwave emittance of the selective surface soon begins to show its superiority.

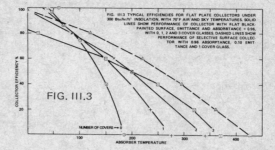

Increasing the number of cover plates from zero to 1 makes a dramatic improvement in collector performance by cutting the heat loss coefficient by a factor of two; and the addition of a second cover glass gives an additional improvement at the higher levels of surface-to-air temperature difference. At the lower temperature differentials, the single glass is better because, as Table III.1c shows, the product of the transmittance times the absorptance falls off sharply as the number of cover glasses increases.

As the time from solar noon increases, there is always a decrease in the insolation of a south-facing collector and an increase in the incident angle which cause the transmittance times absorptance product to fall off.

Generally, the sun must rise to an elevation of 10 to 15 degrees in the sky before its rays are strong enough and the incident angle is favorable enough to permit effective heating. In mid-summer, the sun rises north of east and, as Table II.2 shows, the direct rays of the sun will not reach a south-facing collector until nearly 9:00 a.m. and they will move away again after 3:00 p.m. Thus, depending upon the slope of the collector, there will generally be between 6 and 8 hours of collectable sunshine per day.

Note: There is a difference between INSTANTANEOUS EFFICIENCY of a collector, and the TOTAL DAY-LONG EFFICIENCY. That is the total amount of heat collected and stored for one day with clear sky, compared to the efficiency at solar noon. There are many variables to be considered, and we can be mislead by instantaneous collector efficiencies.

APPENDIX IV

RESOURCES

No. 1   Virtually every aspect of the subject "Energy" is discussed at great length in the April 19, 1974 issue of Science, published by the American Association for the Advancement of Science, Vol. 184, No. 4134. Photosynthesis is discussed in detail by Dr. Melvin Calvin, University of California, Berkeley, in his article "Solar Energy by Photosynthesis," pp. 375-381 of that publication. Scientific American has had excellent articles on energy in a number of recent issues. In September, 1970, Vol. 223, No. 3, G. M. Woodwell discusses photosynthesis very effectively in "The Energy Cycle of the Biosphere," pp. 64-97. In the September, 1969 issue, Vol. 221, No. 3, John D. Issacs discusses "The Nature of Oceanic Life," pp. 146-165, and he shows that much of the photosynthetic activity on our globe takes place in the oceans. In September, 1971, Vol. 224, No. 3, the entire issue was devoted to a superb survey of the energy situation.

No. 2   Solar still construction can be done in many ways. A good set of directions for making a simple solar still is to be found in Simple Solar Still for the Production of Distilled Water, Technical Report No. T 17, Revised, September, 1967 by T. A. Lawand. This booklet is available for $1.00 plus $0.25 check or mail order handling fee from Brace Research Institute, MacDonald College of McGill University, Ste. Anne de Bellevue, Quebec, Canada H9X 3M1.

No. 3   Distillation of salt or brackish water by stills is discussed in detail by E. D. Howe in Chapter 12, pp. 301-307, of Introduction to the Utilization of Solar Energy, edited by A. M. Zarem and D. D. Erway (McGraw-Hill: 1963) and now available from Xerox University Microfilms. [See Index for page number of this book's review. Ed.] Page 312, references 12-26, covers the subject of solar distillation from the time of Harding and Las Salinas (1872) to the reports issued by the Office of Saline Waters, U.S. Department of the Interior, and the CSIRO, Australia, in the mid-1950's. For later work, the serious student should consult sources such as the Proceedings of the Conference on New Sources of Energy (United Nations: 1964, and recently reissued), held at Rome in 1961, and the publication Solar Energy, issued quarterly by the International Solar Energy Society.

No. 4   Brace Research Institute, Canada, has published How to Make a Solar Cabinet Dryer for Agricultural Produce, Do-It-Yourself Leaflet L-6, Revised, March, 1973 ($1.00), which gives construction details for a simple cabinet-type solar dryer. See Resource No. 2 for ordering information. Much more extensive work on drying, using separate air heaters, has been carried out by the CSIRO in Australia.

No. 5   Selective surfaces possess the property of absorbing a large portion of the shortwave solar radiation which strikes them, while re-radiating only a small fraction of the longwave radiation which a black surface would emit at the temperature attained by the surface. This concept was pioneered by Israeli scientist

Harry Tabor, and his first paper on this subject appeared in the Transactions of the Conference on Scientific Uses of Solar Energy (University of Arizona Press: 1958), held at Tucson in 1955, Vol. 2, Part 1-A, pp. 23-40. In the same publication, J. T. Gier and R. V. Dunkle published their findings on the same subject, pp. 41-56.

Many other papers on the subject have appeared since that time, including specific instructions for preparing selective surfaces on copper sheet, given by Dr. Farrington Daniels in his book Direct Use of the Sun's Energy (Yale University Press: 1964), which was recently reissued by Ballantine Books (No. 23794, $1.95). Since that time, many papers have appeared in Solar Energy Journal, dealing with various aspects of selective surfaces in temperature control for space craft.

No. 6   Thermosyphon water heaters are the most widely used solar devices in the world. They vary in complexity; the very simple zig-zag type is described in detail by D. S. Halacy, Jr., in Chapter 6 of Solar Science Projects (formerly Fun with the Sun) available for $0.75 from Scholastic Book Services, 900 Sylvan Avenue, Englewood Cliffs, NJ 07632.

Brace Research Institute, Canada, has published How to Build a Solar Water Heater, Do-It-Yourself Leaflet No. L-4, Revised, February, 1973 ($1.00), which gives detailed instructions for the building of a corrugated plate absorber and making it into a thermosyphon system. Alternative methods of construction are given, including directions for making a grid-type collector in which a flat metal sheet is wired to the pipes to establish the thermal contact. The use of thermal cement would improve the performance of such a unit. See Resource No. 2 for ordering information.

A detailed study of the performance of thermosyphon heaters with selective surfaces is given in "An Investigation of Solar Water Heater Performance," by J. I. Yellott and Rainer Sobotka. The paper was presented at the Summer Annual Meeting of ASHRAE in 1964, and it is to be found in ASHRAE Transactions, Vol. 70, 1964, p. 425.

No. 7   Flow distribution in solar absorber banks has been studied carefully in Australia, and Paper No. 4/35, presented at the 1970 International Solar Energy Conference by R. V. Dunkle and E. T. Davey of the CSIRO, covers it in considerable detail. Large batteries of solar collectors have not yet been installed in the United States, but experience in this field will soon be gained, and anyone who is planning to make use of more than a few collectors should certainly obtain and study this paper.

No. 8   Dr. Harry Thomason's solar-heated buildings near Washington, D.C. have been described in many papers, including "Three Solar Homes," ASME paper 65 WA/SOL-3, and in Solar Energy Journal, Vol. IV, No. 4, October, 1960. A booklet describing his system, entitled Solar House Plans, is available from Edmund Scientific Co., 701 Edscorp Building, Barrington, NJ 08007 (Catalog No. 9440, $10.00), and a license to build under the Thomason patents (of which there are many) may also be obtained through Edmund (Catalog No. 9441, $20.00).

No. 9   The first office building in the U.S. to be equipped with a solar-assisted heat pump was built in 1956 by Messrs. Bridgers and Paxton, a consulting engineering firm in Albuquerque, NM. Their 4300 square foot building used a large expanse of south-facing wall to support 790 square feet of single-glazed copper-and-aluminum collectors, with a 6,000 gallon storage tank and an electric-powered heat pump. This building is about to be re-activated with new and improved collectors as part of the National Science Foundation's solar energy application program. It is described in detail in Heating, Piping and Air Conditioning for November, 1957, Vol. 27, p. 165 and in the ASHRAE Transactions for 1957, also.

No. 10   The University of Florida, under the leadership of Dr. Erich Farber, has been a leader in solar energy research for the past twenty years. A good reference to their work is "Solar Energy Conversion and Utilization" by E. A. Farber, Building Systems Design, June, 1972, or "Solar Energy, Its Conversion and Utilization," by E. A. Farber, Solar Energy Journal, Vol. 14, No. 3, February, 1973, pp. 243-252. The Solar Energy Laboratory, University of Florida, Gainesville, Florida 32601, can supply reprints of many of their papers, including their outstanding work on solar refrigeration. Write for a list of their publications and current prices.

No. 11   The Bliss curves are described in Solar Energy Journal, Vol. 5, No. 3, 1961, p. 103.

No. 12   Edmund Scientific Co., 701 Edscorp Building, Barrington, NJ 08007, is an excellent source of solar energy devices, and they will gladly send their latest catalog if you simply ask to be put on their mailing list.

No. 13   Silicon solar cells have been used in very large numbers in the space program. Many surplus cells have found their way to the "space surplus" market, and they are available in many forms, from single cells to complete modules. Convenient sources are Herbach and Rademan, Inc. 401 East Erie Avenue, Philadelphia, PA 19134, Edmund Scientific Co. (Resource No. 12), and electronic and radio shops. Silicon cell radiometers called "Sol-A-Meters," calibrated and ready for use, may be obtained from Matrix, Inc., 537 South 31st Street, Mesa, AZ 85204 ($195.00 and up).

No. 14   (a) The best reference on the subject of dials and dialing is Sundials by R. N. and M. L. Mayall (Charles T. Branford, Boston: 1958), now out of print. This 200 page volume tells how to construct dozens of different kinds of sundials, gives values of the solar declination and the Equation of Time for every day in the year, and it is an invaluable aid to anyone who wants to build a sundial. The Encyclopaedia Brittanica also has a good article on this subject.

(b) To calculate the angle between the 12 o'clock line, which always runs exactly true north, and the hour lines, we must make our last excursion into trigonometry and use the following equation:

$$\text{Tangent of Angle } X = \text{Sine of Local Latitude} \times \text{Tangent of Hour Angle} \qquad \text{Eq. IV.1}$$

X is the angle between the north-south line and the hour lines; the hour angle is simply the number of hours from solar noon which applied to each of the time lines, divided by 15 degrees per hour (just as we endeavored to explain in App. II). The local latitude (L) must be known exactly and a local surveyor or a good map will give you that information; the sine needs to be found only once.

For the great sundial at Carefree, Arizona, which has a gnomon which is 42 feet long and 4 feet wide (it also carries a solar water heater on its surface), the local latitude is 33 degrees 50 minutes, so the Sine is 0.5567. The angles for the hour lines, measuring southward towards the west for the morning hours and towards the east for the afternoon hours, were calculated as follows:

Time	H, Hour Angle	Tangent of Hour Angle	Tan H x Sin L	X = Angle whose Tangent = Tan H x Sin L
12	0	0.0	0.0	0
11, 1	15	0.2679	0.1492	8.489 = 8 deg. 29 min.
10, 2	30	0.5774	0.3214	17.818 = 17 deg. 49 min.
9, 3	45	1.0000	0.5567	29.105 = 29 deg. 06 min.
8, 4	60	1.7321	0.9642	43.957 = 43 deg. 57 min.
7, 5	75	3.7321	2.0776	64.298 = 64 deg. 18 min.
6,6	90	Infinite	Infinite	90.000 = 90 deg. 00 min.

No. 15   The most definitive work on this subject is NASA Report SP-8005, Solar Electromagnetic Radiation, May, 1971 Edition, by M. P. Thekaekara. This report is available for $3.00 from National Technical Information Service, U.S. Dept. of Commerce, P.O. Box 2553, Springfield, Virginia 22151. The earlier works of Dr. Charles G. Abbot, published by the Smithsonian Institution, are valuable in relating historical steps towards today's accurate evaluation of the solar constant. Also see Resource No. 18.

No. 16   Solar altitude and azimuth are given in the ASHRAE Handbook of Fundamentals, 1972 Edition, Chapter 22, and in

No. 17   ASHRAE Handbook of Utilization, 1974 Edition, Chapter 59, pp. 59.2 - 59.7. These publications give altitude and azimuth for latitudes from 24 degrees to 64 degrees north, by 8 degree intervals. Values by increments of 1 degree are given in Bulletin No. 214, Volumes 2 and 3, Tables of Computed Altitude and Azimuth, U.S. Government Printing Office, Washington, D.C. 20402. Note that these are intended for navigation purposes, and the azimuth is measured from the north, with noon values being 180°, instead of from the south, with noon values being 0 degrees.

No. 18   Solar radiation in space is well covered by a number of publications written by M. P. Thekaekara. Among these are Solar Electromagnetic Radiation (see Resource No. 15), and Solar Energy Monitor in Space (SEMIS), available free from Dr. M. P. Thekaekara, Code 322, NASA Goddard Space Flight Center, Greenbelt, MD 20771. Similar material by Dr. Thekaekara is to be found in "Solar Energy Outside the Earth's Atmosphere," Solar Energy Journal, Vol. 14, No. 2, January 1973, pp. 109-127. Also see "Evaluating the Light from the Sun," Optical Spectra, March, 1972, pp. 32-35, and "Proposed Standard Values of the Solar Constant and the Solar Spectrum," Journal of Engineering Science, Vol. 4, September-October, 1970, p. 609. For possible variations in the solar constant, see the work of Dr. Abbot and also "Extraterrestrial Solar Energy and Its Variations," by Dr. Thekaekara, in Solar Energy Monitor in Space (SEMIS).

No. 19   Solar irradiation of south-facing surfaces on earth is covered by the work of various ASHRAE committees, for which Resources No. 16 and No. 17 are helpful. The most comprehensive publication is ASHRAE Paper No. 825, "Development and Use of Insolation Data for South-Facing Surfaces in Northern Latitudes," by C.A. Morrison and E.A. Farber of the University of Florida, Gainesville.

No. 20   For actual solar intensities, sunshine hours and percentage of possible sunshine, see Climatic Atlas of the United States, 1968 Edition, U.S. Government Printing Office, Washington, D.C. 20402 ($4.25).

No. 21   Low Temperature Engineering Application of Solar Energy is a 78 page monograph with a somewhat misleading title, since the words "low temperature" were intended to distinguish between the moderate temperatures used in space heating and water heating, and the very high temperatures attained in solar furnaces. Produced by the Technical Committee on Solar Energy Utilization under the editorship of Richard C. Jordan, this volume was published in 1966 by ASHRAE. It was re-issued in 1974, and a new version is now in preparation. This is a highly technical but extremely valuable summary of heliothermal technology as it was in 1966.

Dr. Tabor's chapter on selective surfaces gives specific directions for producing such surfaces on aluminum, galvanized steel, and copper. Dr. Austin Whillier has an extremely valuable chapter on the factors which influence collector performance. Other chapters cover such topics as "Availability of Solar Energy for Flat Plate Collectors," "Measurement of Solar Radiation," "Use of Flat Plate Collectors in Tropical Regions," and "Solar Water Heaters." This booklet is highly recommended to anyone who wants to make a serious study of heliotechnology. It is available from ASHRAE Sales Department, 345 47th Street, New York, NY 10017 for $9.00.

SOLAR AVAILABILITY: QUANTITATIVE DATA ON SOLAR ENERGY

by Matthew P. Thekaekara,
Goddard Space Flight Center,
Greenbelt, MD

Extraterrestrial Solar Energy

Quantitative data on available solar energy is an essential parameter for the design of solar systems and for computing their efficiencies. Two quantities of prime importance in this regard are the *Solar Constant* and the *Extraterrestrial Solar Spectrum*. The solar constant is the amount of energy incident on a unit area exposed normally to the sun's rays at the average Sun-Earth distance, in the *absence* of the Earth's atmosphere. The currently accepted values of the solar constant and solar spectrum are those published by the American Society of Testing and Materials as the ASTM Standard E 490-73a (Ref. 1). They also form the Design Criteria for NASA space vehicles (Ref. 2).

The value of the solar constant is 1353 Watts/m^2, which is equivalent to 125.7 Watts/ft^2, 1.94 cal/cm^2/min, 1.94 Langleys/min, 429.2 Btu/ft^2/hr, or 1.81 horsepower/m^2. The estimated error of this value is ±1.5 percent. These values are based on measurements made by different observers, mainly from high altitude jet craft. References 3 and 4 should be consulted for detailed information on the derivation of the ASTM standard.

The solar constant is defined for the average Sun-Earth distance. As the Earth moves in its annual elliptical orbit around the sun the total solar energy received varies by ±3.5 percent. There are also small and undetermined variations due to cyclic or sporadic changes in the sun itself. These variations are more significant in certain portions of the spectrum than in others.

Solar Irradiance at Ground Level

The solar energy received on a surface at ground level has two components, that received directly from the sun and that diffused by the sky. A spectral curve of the direct solar radiation is shown in Fig. 1. The total direct solar energy transmitted by the atmosphere in this case is 956.2 Watts/m^2 or 70.7 percent of that received above the atmosphere. As solar zenith angle (the angle between the sun's rays and the local vertical) increases, the transmitted energy decreases. Table I gives data on the total irradiance at ground level on a surface exposed normally to the sun's rays, for four values of solar zenith angle, and for four levels of *Atmospheric Pollution* or *Turbidity*. It is significant that as air mass increases or turbidity increases, the relative amount of energy in the IR increases and that in the visible and UV decreases.

AIR MASS	SOLAR ZENITH ANGLE (DEGREES)	TURBIDITY FACTORS α	TURBIDITY FACTORS β	TOTAL IRRADIANCE W m-2	RATIO OF TOTAL IRRADIANCE TO SOLAR CONSTANT %	UV, λ<0.4μm %	VISIBLE 0.4μm<λ<0.72μm %	INFRARED λ>0.72μm %
0	0			1353.0	100.0	8.7	40.1	51.1
1	0	1.30	0.02	958.2	70.7	4.8	46.9	48.3
4	75.5	1.30	0.02	595.2	44.0	1.23	44.2	54.5
7	81.8	1.30	0.02	413.6	30.6	0.35	39.4	60.3
10	84.3	1.30	0.02	302.5	22.4	0.102	34.7	65.2
1	0	1.30	0.04	924.9	68.4	4.6	46.4	49.0
4	75.5	1.30	0.04	528.9	39.1	1.04	42.1	56.9
7	81.8	1.30	0.04	342.0	25.3	0.26	35.9	63.8
10	84.3	1.30	0.04	234.5	17.3	0.065	30.3	69.6
1	0	0.66	0.085	889.2	65.7	4.7	46.4	48.9
4	75.5	0.66	0.085	448.7	33.2	1.14	42.4	56.5
7	81.8	0.66	0.085	255.2	18.9	0.30	36.3	63.4
10	84.3	0.66	0.085	153.8	11.4	0.08	30.7	69.2
1	0	0.66	0.17	800.2	59.1	4.5	45.4	50.1
4	75.5	0.66	0.17	303.1	22.4	0.88	38.3	60.8
7	81.8	0.66	0.17	133.3	9.85	0.14	30.0	69.8
10	84.3	0.66	0.17	63.4	4.69	0.039	22.9	77.1

TABLE I   SOLAR IRRADIANCE ON UNIT AREA EXPOSED NORMALLY TO THE SUN'S RAYS, COMPUTED FOR DIFFERENT AIR MASS VALUES, U. S. STANDARD ATMOSPHERE, 2cm OF H$_2$O AND 0.34cm OF OZONE.

Direct solar irradiance (I_{DN}) discussed above is for a surface normal to the sun's rays. For a horizontal surface or a surface with any other slope, the energy (E) received per unit area (E_s) is: $E_s = E_{normal} \cos \theta$ where θ is the angle between the solar rays and the surface normal (see Appendix III, pg. 48 EP). For optimum efficiency of the collector plate, it is advantageous to have low values of θ, especially around solar noon when the air mass is least. For year-round operation a surface facing south with a slope equal to the latitude of the place is desirable. For winter heating, but no air conditioning, a greater slope is desirable and for summer heating (of swimming pools, for example) a smaller slope is indicated.

Direct solar irradiance is the major component in the total energy on a surface, followed by diffuse radiation due to the sky. It is always available during day time, even when direct solar radiation is blocked by clouds. A representative figure for diffuse sky irradiance on a horizontal surface is between one-fifth and one-sixth of the direct solar irradiance, but it can be as much as 1/3 at higher latitude.

Another major area where precise quantitative prediction is desirable but almost impossible is the solar energy available at any location over an extended period of time and the relative number of days of sunshine, cloud-cover or rain. "As unpredictable as the weather" is a common expression; statistical data based on previous measurements will have to serve as a guide for the future. An estimate of the energy available at 61 different locations may be made from Fig. 2. Station 4, Inyokern, CA, has the maximum value of 569 cal/cm^2 per day (2381 joules/cm^2 per day) and Seattle University, Station 24A, has the minimum value of 269 cal/cm^2. These values of annual means of daily insolation should not be confused with what will be available on any given day and location. [In fact, daily total insolation has a wider range of variation from day to day at any one location than the yearly averages of such totals from one location to another, ed.]

An area of one cm^2 exposed normally to the sun's rays outside the

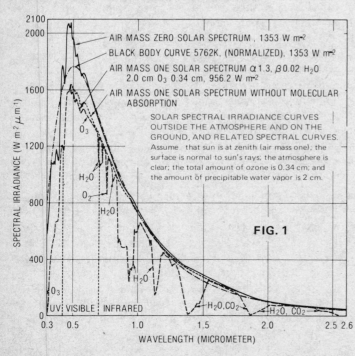

FIG. 1

SOLAR SPECTRAL IRRADIANCE CURVES OUTSIDE THE ATMOSPHERE AND ON THE GROUND, AND RELATED SPECTRAL CURVES. Assume that sun is at zenith (air mass one); the surface is normal to sun's rays; the atmosphere is clear; the total amount of ozone is 0.34 cm; and the amount of precipitable water vapor is 2 cm.

AIR MASS ZERO SOLAR SPECTRUM, 1353 W m⁻²
BLACK BODY CURVE 5762K, (NORMALIZED), 1353 W m⁻²
AIR MASS ONE SOLAR SPECTRUM α 1.3, β 0.02 H₂O 2.0 cm O₃ 0.34 cm, 956.2 W m⁻²
AIR MASS ONE SOLAR SPECTRUM WITHOUT MOLECULAR ABSORPTION

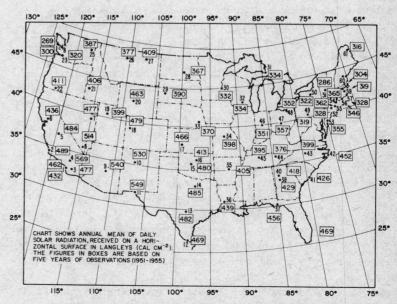

FIG. 2 ANNUAL MEAN OF DAILY INSOLATION RECEIVED ON A HORIZONTAL SURFACE AT 61 LOCATIONS IN CONTINENTAL UNITED STATES

CHART SHOWS ANNUAL MEAN OF DAILY SOLAR RADIATION, RECEIVED ON A HORIZONTAL SURFACE IN LANGLEYS (CAL CM⁻²). THE FIGURES IN BOXES ARE BASED ON FIVE YEARS OF OBSERVATIONS (1951-1955)

atmosphere at the mean Sun-Earth distance would receive in 2 hours and 19 minutes the same amount, 269 cal/cm^2, as the average for Seattle University for a whole day. In most areas of the south-west U.S. the daily insolation is considerably greater than in the north-west. Not all locations in the U.S. are equally cost-effective in solar energy conversion.

Detailed information on solar energy available on the ground can be obtained from the National Oceanic and Atmospheric Administration (NOAA) which maintains an extensive network. There is also a "sunshine switch" network with over 160 stations. Efforts are now being made by NOAA to upgrade and expand the national network for monitoring solar irradiance at ground level. The repository and distribution center for information about solar irradiance data is the National Climatic Center, Federal Building, Asheville, NC 28801.

Conclusion

The Earth-atmosphere system receives energy from the sun at the rate of 5.445 $\times$ 10^{24} joules or 1.513 $\times$ 10^{18} KW-hrs. per year. The total output of all man-made energy producing systems in the world in 1970 was less than 0.004 percent of this, 2 $\times$ 10^{20} joules. Of this energy 35 percent was produced in the United States, and all but 4 percent from a rapidly dwindling supply of fossil fuels (which after all are stored solar energy). Using today's solar energy for today's needs is a challenge greater than the conquest of space. Knowing how much solar energy is available is part of meeting this challenge.

References

1. Anon: Standard Specification for Solar Constant and Air Mass Zero Solar Spectral Irradiance, ASTM Standard, E-490-73a, 1974 Annual Book of ASTM Standards, Part 41, Philadelphia, PA 1974.
2. Anon: Solar Electromagnetic Radiation, NASA SP8005, Washington, D.C., May 1971.
3. Thekaekara, M.P., ed., The Solar Constant and Solar Spectrum Measured from a Research Aircraft, NASA TR R-351, Washington, D.C., October 1970.
4. Thekaekara, M.P., Solar Energy Outside the Earth's Atmosphere, Solar Energy, Vol. 14, No. 2, 1973, pp. 109-127.
5. Campen, C.F., et al., ed., Handbook of Geophysics (AFCRL), MacMillan Co., New York, pp. 15-16, 1960.

SOLAR COMMENTS
Chuck Missar, Solar Co-Editor, 1974

The information presented here on solar energy is just a hint of what is available. For a technology that has had so little practical application, there is ample information available to educate the experimenter and builder. Solar devices for distilling water, heating domestic and swimming pool water, cooking food, and melting materials have been in use for years, the technology is well known and the results are reasonably predictable.

Let's look at solar energy in historical perspective. Really active research and experimentation on the subject started in the late 19th century. Periodically since then (maybe on 10-15 year cycles), interest in it has risen and then fallen. With oil prices so low for so long and with our "oil will last forever" mentality, we could not justify the expense of solar hardware. Under these conditions of oscillating interest in solar technology, research has been carried out by a small group of dedicated people who believed in the environmental and long term social benefits that could come from the use of solar energy. Maybe that is why this group of pioneering "eccentrics" has so much in common with the burgeoning group of solar advocates today. Together they know that solar energy needs to be developed for the safety and comfort of future generations.

Many large, well established firms are becoming quite involved with solar energy research and hardware. These firms will be the leaders in many solar energy activities, since they have the capital, momentum, and reputation to carry out successful research and development. Most of us interested in doing-it-ourselves will live off the technological fall-out from these firms. By paying close attention to new patents, government information from the National Technical Information Service and the National Aeronautics and Space Administration, trade journals, and our own research, we can learn how to incorporate the latest engineering achievements in our small scale systems without having to support directly large, centralized industries. An appropriate phrase often heard in the hall at Portola Institute is "keep on livin' in the cracks." This applies to solar energy as much as it does to economics or shelter.

Solar energy is well beyond the vague generality stage. The solar chapter attempts to illustrate this by directing you to sources of ideas, plans and hardware from which you can build or buy your own equipment. The beauty of solar energy is that anybody can build a simple system that will decrease demands on our dwindling fossil fuel reserves. A simple cooker or a scrap-fabricated water heater are things you can build tomorrow, if you have sufficient interest. And, as we try to make clear elsewhere in the ENERGY PRIMER, more self-reliance and the understanding of our immediate world that this entails are an absolute necessity in our chaotic world.

Self-reliance can be expensive. Solar collectors use a lot of metal and glass, they require electronic controls to work efficiently, and maintenance is an ever present reality. We cannot expect large segments of the population of the United States, much less the other nations of the world, to use solar energy at the same rate they have been using fossil fuel energy. Besides, the "costs" of solar energy resources and hardware might use up more non-renewable energy than renewable energy gleaned from the sun. Who knows? If we really look at this, I think we will confirm what we've suspected all along—the answer to most of our problems is to use less, and in the process, get more.

Back to what we can do. First, get out there and try to build some solar devices and begin integrating them into your living habits. Start with a small solar still or dryer, for instance, and work up into something you really feel proud of.

Second, take safety precautions. A collector, left to heat in the sun with water in it but not circulating, can generate enough steam to blow the collector up if there is no means of pressure relief. Many commercial antifreezes are poisonous and should not be used where children or animals might drink the water in which they are mixed. A concentrating collector can blind you almost instantly if you deliberately or accidently look into it, or it can start a fire if it is carelessly aimed. The addition of collector assemblies can alter a structure's deadweight and wind loading characteristics. In short, consider the safety aspects of what you are doing. Dress appropriately for the work to be done and think through each project you undertake.

Third, keep records of mistakes and successes, and share with others the results of your research and building. There's no better turn-on to solar energy than to see a cooker cook or a heater heat. We must learn to discuss the attributes and shortcomings of solar assemblies just as well as we've been doing with automobiles and the other "necessities" of our age. Be prepared for lots of visitors and small talk if you build something, since this is one way people become comfortable with the ideas and hardware involved with solar energy.

We haven't done much to stress foreign vendors of solar hardware. Excellent equipment is available from Israel, Australia, Japan and elsewhere. However, with present transportation charges and exchange rates, it hardly makes sense to feature these items. In keeping to what is available in the United States, we have tried to review some of the bad with the good to give you an idea of the range of publications and hardware available.

Prices are changing rapidly. Don't be surprised if book and hardware prices are higher on items you order. Also, because solar energy technology is changing so fast, some of the information presented here will soon be obsolete. Read *Alternative Sources of Energy* Newsletter, *Solar Energy Digest*, *Solar Age* or *Solar Engineering* for the newest information.

When you send off for "free" information, *please* send a stamped, self-addressed envelope. Many of the vendors and authors listed here are small operations. Ask for information judiciously and help pay for it if you need it so badly. If phrases are in this typeface, it's a quote from the book or the descriptive literature of the project or hardware being reviewed.

SOLAR COMMENTS
Berkeley Solar Group, Solar Co-Editors, 1977

The solar energy field has changed a great deal during the two and one-half years since the first ENERGY PRIMER was put together. Berkeley Solar Group was organized late in 1974, so we've been in operation during the entire life of the original ENERGY PRIMER.

Over this period we've grown from two people waiting for the phone to ring to a group of seven struggling to keep up with the workload. We have worked on numerous solar houses and buildings, done a variety of feasibility studies, written a book on solar retrofitting, taught classes, and developed specialized computer programs for solar and energy conservation design.

From the beginning, we've been guided by two principles:

1. Conserve energy first, then use the sun.
2. Use passive solar heating wherever possible, supplemented as necessary by active systems.

The first principle can be justified purely on economic grounds. Energy conservation practices often cost nothing (for instance, it doesn't cost anything to lower your thermostat) and virtually never cost as much as elaborate solar heating systems. Even expensive energy conserving equipment, such as shuttered windows, pays for itself more quickly than solar equipment. And energy conservation makes sense from many non-economic points of view.

The second principle is less widely accepted than the first. Active solar heating, using flat-plate collectors, pumps or fans, and other mechanical equipment is standard practice for most applications. But in many cases, especially in new construction, there is the passive alternative — using simple, direct solar heating of buildings and taking advantage of natural radiant and convection heat transfer that occurs in buildings. There are at least two reasons for this "active bias." First, some of the initial experiments with passive systems in the early 1950s produced overall efficiencies of under 30%. Discouraged, the investigators turned to active systems and stuck with them, even though their overall efficiencies were still about 30%. As a result, there is a great deal more information available about active system design and performance than there is about passive systems.

The other reason for the active bias is that the prevailing approach to building (as well as to almost everything else) is to subdivide, specialize, and mass produce rather than to integrate. Active systems fit well with this method because they are composed of separate pieces of hardware which can usually be developed, manufactured, and marketed separately. The pieces can be assembled to meet the requirements of any project. The hardware section illustrates how well this approach works for manufacturers (compare it to the solar hardware section of the first ENERGY PRIMER to see how far we've come).

Passive and semi-passive systems require a much more integrated approach. Passive design is essentially architecture. A passive building must not only maintain thermal comfort by responding to and cooperating with the local climate, it must respond to human needs as well. It must be oriented properly for adequate heat gains, it must have good lighting and acoustics, have a sense of being a good place, and, at the same time, have the bathroom in the right location. Those of us who have dabbled in architectural integration know that all this is much more complicated than sizing flat-plate collectors.

A consequence of the active bias is that it is much easier to design and build an active system. Local plumbing stores are beginning to carry flat-plate collectors. Readily available books give reliable engineering data for active systems but ignore or waffle on passive systems. Many designers have more experience and confidence with active systems, so they steer their clients toward them. It is also true that many people building houses are a little nervous about using solar at all and therefore want to stick to what's "proven."

However, the payoff from using passive solar heating is potentially large. Passive systems are in general cheaper, quieter, and more reliable than active systems. In many cases, they produce a more comfortable thermal environment than do active systems. Some passive systems provide cooling in addition to heating, while active solar cooling, using absorption chillers, is still being developed and may never become economically practical.

So we urge you, if you are considering a solar project, to take the extra time to learn about passive solar systems. Read what you can, visit passive houses in your area (your local solar energy association can help you find them), and talk to their designers. Place active and passive on an equal footing and make an honest choice for your situation.

Finally, something about attitude. Solar energy has become something of a fad — all the rock stars and movie producers are going solar these days. However, if solar energy is going to live up to its promise, it must become more than a fad. We must make it real for everybody. This requires a kind of solar pragmatism which asks, "What is needed that solar energy can provide better than any other alternative? What is the best system to do the job? How can we get the system implemented?" An attitude like this places workable, sensible systems above those that enjoy short term popularity and it is essential to the development of solar energy as a real resource.

THE RYAN HOUSE

The Ryan House, located north of Jenner, California, is an example of a combined passive and active system. The Berkeley Solar Group advised the architect, Paul Tyrrell of Vallejo, California, on the design of the passive aspects of the house, and did all the mechanical design for the active system. Construction was completed in late 1976.

The passive system uses a simple, direct gain approach. Four hundred square feet of double glazed windows face south. There are few windows on the other sides of the house. Overhangs are carefully sized to shade the glass during the summer while admitting full sun in the winter. A masonry floor and a central masonry wall absorb the solar heat and keep the house warm in the evening.

The active system is based on 400 square feet of south-facing flat-plate collector mounted 30 feet in front of the house. These panels heat water. An 1100 gallon concrete tank in the house serves as storage. The heat delivery system is a radiant floor slab. Domestic hot water is heated with a coil in the storage tank. Auxiliary heat is provided by fireplaces, a wood stove, and some electric heating units.

Computer modeling that we did while designing the system indicates that 40% of the heating energy for the house comes from the windows, 40% from the active solar, and 20% from the auxiliary. This performance hasn't been verified, but when the system was first completed, the house was operated for several months without auxiliary heat. The house cooled only after long cloudy periods, so a high solar percentage was being achieved.

Photovoltaics or solar cells convert sunlight <u>directly</u> into electricity. Presently, the most practical solar cells are made from silicon . . . the second most abundant element on earth. It is quite appealing, therefore, to consider the possibilities of producing electricity from two of our most reliable resources . . . sunlight and silicon. However, there is a growing debate concerning the future potential of solar cells. On the one hand there are those who believe that the high cost of production will always prevent the widespread use of solar cells, and that even improvements in production will never make them economically competitive with fossil fuels, nuclear power or even other solar-electric systems (wind, solar-boilers, etc). On the other

hand, many people, especially those in the solar cell industry, believe that costs will be significantly lowered over the next few years to the point where they will be economically feasible and politically desirable.

The following articles reflect these two opinions. The first article is by Marshal Merriam, Professor of Materials Science and Engineering, University of California, Berkeley. The second article is excerpted from: SOLAR CELLS: POWER TO THE PEOPLE by David Morris of the Institute of Local Self Reliance (see page 247) which appeared in the ILSR Newsletter (November 1976).

—R.M.

PHOTOVOLTAICS: PROCESS AND POTENTIAL

Marshal Merriam

Physical Principle of the Photovoltaic Effect

The conversion of sunlight to electricity with photovoltaic devices, i.e. solar cells, is a proven technology. The best available devices have satisfactory efficiency, proven reliability, low weight, long life and high cost. The form of the delivered energy (low voltage direct current) is well suited for many applications. If the high cost could be reduced sufficiently there is little doubt that solar cells would be widely used. However, the cost reduction required is so great there is doubt it will ever be achieved.

Sunshine comes in little bundles called photons. When a photon enters a semiconducting material and is absorbed it often happens that one of the primary electronic bonds is broken. The electron evicted from its customary bond can travel through the entire crystalline solid and respond to electric fields and other influences. The bond from which the electron was ejected is short one electron (there are two electrons per bond), and this electron shortage is also mobile. It is customary to refer to the electron shortage as a 'hole,' and to say that the photon has created an 'electron-hole pair.' . . . the electron being negatively charged and the hole being positively charged.

The electron and hole created by the photon will eventually disappear by recombining with each other unless they are physically separated into different regions of the material. This separation is accomplished in photovoltaic devices by the use of a p-n junction. Positive-negative or p-n junctions can be produced in semiconductor crystals through extremely careful control of chemical purity during crystal growth, followed by deliberate introduction of a very small amount of a selected impurity. Introduction is done by diffusion from one edge during later processing. The boundary of the region where the deliberately added impurity (called a 'dopant') exists constitutes the p-n junction. The key characteristic of a p-n junction from the standpoint of understanding the photovoltaic effect is the electric field which exists within it. This electric field originates in the (very slight) chemical difference between the material on the two sides of the junction. The dopant and the residual impurities cannot move around in the material at the temperatures of solar cell operations; thus the electric field, which exists only in the immediate vicinity of the junction, is built-in and permanent.

Electrons and holes, being of opposite charge, will be pushed in different directions by the field if they come into the region near the p-n junction. One side of the junction is called the p-region; the other side is called the n-region. The electric field pushes the holes into the p-region and the electrons into the n-region. Thus the p-region becomes positively charged and the n-region becomes negatively charged. If an external load is connected, this charge difference will drive a current through it. The current will flow so long as the sunlight keeps generating the electron-hole pairs. Fig. 1 is a schematic picture of a photovoltaic cell. Note that the top surface contact, being opaque to sunlight, must cover as little of the surface area of the cell as it can without introducing excessive electrical resistance.

Use of Photovoltaics

The great majority of photovoltaic panels in use today are used for charging batteries. It is commonly said that various navigation beacons, radio repeater stations, etc. are powered by solar cells, but a more exact statement would be that they are operated from batteries, the batteries being charged from solar cells. Battery charging has the good feature that the cells in the panel operate at nearly constant voltage automatically, the charging current they deliver varying according to the light level.

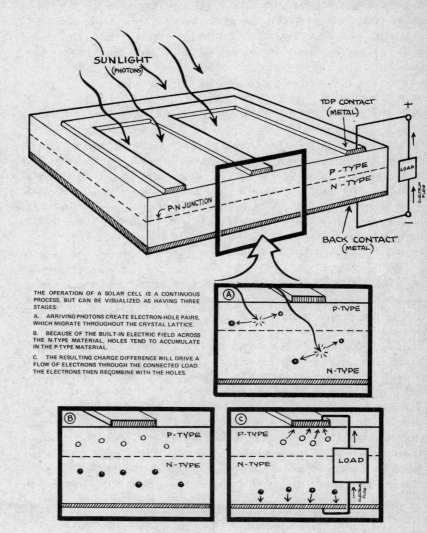

THE OPERATION OF A SOLAR CELL IS A CONTINUOUS PROCESS, BUT CAN BE VISUALIZED AS HAVING THREE STAGES:

A. ARRIVING PHOTONS CREATE ELECTRON-HOLE PAIRS, WHICH MIGRATE THROUGHOUT THE CRYSTAL LATTICE.

B. BECAUSE OF THE BUILT-IN ELECTRIC FIELD ACROSS THE N-TYPE MATERIAL, HOLES TEND TO ACCUMULATE IN THE P-TYPE MATERIAL.

C. THE RESULTING CHARGE DIFFERENCE WILL DRIVE A FLOW OF ELECTRONS THROUGH THE CONNECTED LOAD. THE ELECTRONS THEN RECOMBINE WITH THE HOLES.

FIG. 1 THE WORKINGS OF A PHOTOVOLTAIC CELL

Any applications in which the amount of power needed is very small and connection to the regular electric power net is not convenient is a possible use for solar cells. Situations where storage is not required, either because the power use correlates with insolation or because the load is interruptible, are especially favored. Use of a photovoltaic panel to operate a very small pump for a pumped solar hot water system has been proposed by various investigators. In this case, the photovoltaic cells provide not only the power but also the control. Another use is as a proportional control element in solar heating systems, where the capability of providing a current that is linearly proportional to light intensity is exploited.

A number of government demonstration programs in the USSR and elsewhere have made use of photovoltaic cells to operate loads which were not small, such as irrigation pumps or lights. These cannot be regarded as viable uses because the economics are so unfavorable.

Economics of Terrestrial Use

The present cost of solar cells is about $30 per peak watt when purchased from a retail supply house by an individual and about half that when purchased directly from the factory by competitive bidding in million dollar lots by the government. "Per peak watt" means per watt of output into an optimally matched load at 20°C in strong sunshine. To convert "peak watt" into average watt, where "average" means averaged over the 8760 hours in a year, a factor of 5 is conventional. Thus, enough cells to deliver 24 KWH per day would cost about $150,000 at current retail prices. The value of 24 KWH purchased as commercial electric power is less than $1.50. Obviously, the economics of solar cells for large scale terrestrial use, in competition with commercial electricity, can only be described as hopeless. Proponents who see future large scale photovoltaics are forced to postulate drastic cost reductions. The reductions are postulated to occur because of

- large scale production,
- new photovoltaic materials, and
- use of concentrators.

Arguments of major cost reductions for photovoltaics are, unfortunately, uncertain. Most persons agree that some cost reduction can be expected if large scale production can fill a vastly expanded market; if a continuing R&D program develops improved materials; and if the problems associated with use of concentrators are worked out. The disagreement is about whether the amount of the cost reduction can even approach those now existing for what is required to bring electricity produced by other means, even other solar means (e.g. wind and solar-boiler). Unfortunately, the strongest proponents of the view that costs can be reduced by the required factor of 50 or so are those people in the industry who have a strong vested interest. The solar cell industry is entirely dependent on government support for both its present existence and its future expansion.

Even at today's prices, photovoltaics are economically feasible for use in situations where commercial power is not available, where the power requirement is measured in watts or milliwatts, not kilowatts, and where the power has a high value because of its use — e.g. communications (re: earlier editorial note).

One thing which seems probable is that the economics of terrestrial use will continue to favor decentralized application over centralized ones.

Prophets with visions of square miles of Arizona covered by silicon blankets seem less credible than those who picture a photovoltaic panel on every rooftop — though the situation is anything but clear, and neither prospect is realistic unless drastic price reductions occur. With a decentralized energy source and decentralized consumption, and a cost of electricity which is very nearly directly proportional to the area of sunshine intercepted, it is hard to see any substantial economy of scale in a large photovoltaic power plant.

Summary and Conclusions.

At present prices, photovoltaics are useful only for applications requiring small amounts of power in situations where it would be too costly to bring in conventional electricity. Navigational aids, radio repeater links, remote monitoring instruments are examples of such applications. When properly engineered and installed, and protected from hazards, photovoltaic installations of this type have performed reliably for many years with minimal maintenance.

A widespread use of solar cells on a scale sufficient to make any measurable contribution to meeting human energy needs will depend upon drastically lower prices for the cells. Price reductions to one or two percent of present levels are required, and it is not at all clear that these will ever occur. Cost reduction is the central issue; improvements in cell performance are unimportant by comparison. Photovoltaic technology differs from most other solar energy conversion technologies, even the other solar electricity producing technologies, in that the cost reduction required for large scale practicality is extreme, and depends on breakthroughs in materials and technology which cannot be foreseen with confidence at this time.

PHOTOVOLTAICS: ECONOMY AND POLITICS
David Morris

Economics and Solar Cells

Throughout the 1960's, the price of solar cells remained very high. They were used exclusively in space satellite systems, and the few manufacturers maintained a stable price of about $200 per peak watt.

In the early 1970's, however, a number of terrestrial applications were found for solar cells. Prices dropped from $100 per peak watt in 1970 to $30 in 1973 to $17 in 1975 and most recently to $10. This means that the current cost is about 20 times that of nuclear power.

Manufacturers, researchers, and government officials agree that the high cost of solar cells will drop dramatically if production is scaled up to permit automation. Currently the silicon wafers are cut by hand, the cells are etched and tested by hand, the electrodes are soldered by hand, etc. Automobiles produced this way would cost twenty times more than cars produced on the assembly line. Solar cells, though, lend themselves easily to mass production techniques since they are electronic devices. The solar cell is actually simpler to produce than an integrated circuit. Like all solar energy devices, solar cells require storage systems. At present, storage costs are only about 2.5 to 3 cents per kwh; but as the price for the cell itself drops, the storage system will become a significant cost item. Current technologies utilize lead acid batteries. New types of batteries are being developed and it is expected that advanced battery systems with significantly longer lives and lower costs will come on line within five years.

The industry stands on the verge of new technological breakthroughs, both in materials and in production processes. Price reduction, however, is not dependent upon such breakthroughs. As the Solar Energy Task Force Report of Project Independence noted, "By just extending conventional silicon crystal growing and slicing techniques, and not counting on any major new technology advancements, we are able to project solar cell array costs to about 75 cents per peak watt."

Politics and Solar Cells

Presently the federal government is doing little to support the development of solar cells. The solar cell industry is relatively new, and therefore has little clout with Congress. ERDA was only a short time ago the Atomic Energy Commission; it has a built-in bias in favor of nuclear energy.

Also, since the federal government believes that only big business can significantly affect future energy supplies, it does not actively support the small manufacturers who dominate solar cell production.

The role of the Defense Department is also important. In the early 1950's there was a need in the military for a light-weight electronic replacement for vacuum tubes. As a result, the Department of Defense underwrote its development and, within a few years, the price of $25 per transistor had dropped to 25¢ per transistor. Now, seeing little military use for solar cell electricity, the military has remained aloof from its development.

The result is a disproportionate federal program. In fiscal 1975, nuclear power was provided $1.5 billion and solar cells only $8 million. In the proposed 1977 budget, almost 40 times more money is being spent on nuclear energy than on the total solar electric program (wind and thermal electric). Indeed, the federal government is planning on spending twice as much in protecting nuclear plants against sabotage and Americans against exposure to nuclear wastes than it is to develop the entire solar energy program.

Yet, despite this neglect, the prices continue to fall. With the cost of nuclear plants rising by 15-20% per year and the cost of solar cells dropping by an equal amount each year, it is a matter of only a few years before the one is competitive with the other. Even the solar cell division of ERDA predicts the cost lines will cross in the mid-1980's.

Which brings up an interesting point. Currently, it takes ten years from the time a nuclear plant is first proposed before it begins to produce electricity. Solar cells can begin producing electricity within a few months. This means that by the time a nuclear reactor comes on-line, it will produce more expensive electricity than will solar cells. In addition, solar cells can be put on-line in modular form. Thus the generating capacity can be expanded as the need arises. With nuclear reactors, future consumption must be predicted accurately; and, as the collapse of the nuclear industry in the last two years has shown, such predictions are very shaky. Finally, solar cells require relatively short term capital investment, whereas nuclear requires long term capital financing. An investment in nuclear will not return any profit for a decade.

As a result, we can expect that private capital will shortly forsake the nuclear area for solar cells. And when this occurs we will have to develop entirely new concepts of utility structures and energy generation. For then our homes and neighborhoods will become electricity producers rather than consumers.

SOLAR BOOK REVIEWS ———————— BASIC SOLAR ENERGY

THE COMING AGE OF SOLAR ENERGY

This book will broaden your knowledge and perspective on what has already been done with solar energy (few current developments are really new) and what solar energy might do in the future. The book is short and easy reading, but the author knows his stuff and there's enough substance to the book that you'll be referring back often. Halacy is a good science writer, not of the gee-whiz variety, but one who explains in understandable terms what's really important.

The first half of the book is related to the history and uses of solar energy. The second half of the book details several major current proposals for large-scale solar powerplants, and the last chapter briefly describes others. There's very little space given to solar houses, but the choice of Thomason's and Hay's houses as examples was excellent—they are among the very few solar houses that have been cheap and fully successful.

This book isn't as exhaustive in its coverage as Daniels' DIRECT USE OF THE SUN'S ENERGY, and it doesn't give references, but it's certainly worth adding to your bookshelf.

— Roger Douglass

The Coming Age of Solar Energy
D.S. Halacy, Jr.
1973; 231 pp

$7.95

from:
Harper & Row
Keystone Industrial Park
Scranton, Pennsylvania 18512

or WHOLE EARTH TRUCK STORE

DIRECT USE OF THE SUN'S ENERGY

This is a book to own; there's so much in it, you'll want it handy. Farrington Daniels wrote much of the book from his own first-hand experience as Director of the Solar Energy Laboratory of the University of Wisconsin. He covers all aspects of solar energy research and application. Without stressing mathematical or engineering details (though including complete references to this material) he describes the full range of work on solar collectors, cooking, heating, agricultural and industrial drying, distillation, storage of heat, solar furnaces and engines, cooling and refrigeration, photochemical conversion, photo- and thermoelectric conversion, and many other uses of solar energy.

Dr. Daniels was concerned with developing low-cost solar applications, principally for the benefit of non-industrialized peoples. He wrote this book to interest others in such possibilities. It's by far the best available introduction to the subject.

— Roger Douglass

Direct Use of the Sun's Energy
Farrington Daniels
1964; reprinted 1975

$1.95 paperback

from:
Ballantine Books, Inc.
457 Hahn Road
Westminster, MD 21157

$12.50 hardcover

from:
Yale University Press
92A Yale Station
New Haven, CT 06520

SOLAR ENERGY — THE AWAKENING SCIENCE

This book is an interesting, inspirational and educational book. The style is anecdotal, based on visits to a large number of active solar R&D centers all over the world and interviews with a great number of the leading people. If solar energy ever lacked for "human interest," Behrman's book should set that straight. In addition to the anecdotes there is plenty of pro-solar, small-is-beautiful, ideological conviction expressed by the author, along with corresponding negative views about nuclear power and big energy corporations.

There are no equations, no graphs, and rather few numbers. The book is not for those who wish to learn how to do things in solar technology. However it's highly readable, and full of entertaining stories about the characters and plot of the solar energy drama. Politically oriented solar enthusiasts will enjoy it. There are many fine photographs and a good index. There is no other book like it, and much of the material is written down nowhere else.

—Marshal Merriam

Solar Energy — The Awakening Science
Daniel Behrman
1976; 408 pp

$12.50

from:
Little Brown & Co.
200 West St.
Waltham, MA 02154

SOLAR ENERGY TECHNOLOGY AND APPLICATIONS

According to the preface, this "introduces the various techniques for utilizing solar energy and brings you up to date on work to the present time on the broad spectrum of solar energy systems." Succinctly written for the technical person, it is readily understood by anyone with no background in the field. It is also recommended as a supplementary text for energy related courses.

The book briefly covers the use of solar energy for heating and cooling of buildings, water heating, central station power generation, solar furnaces, crop dryers and stills, fuels from grown and waste organic materials, the ocean thermal gradient power concept, the satellite solar-cell power plant, and wind power. Its extreme brevity is perhaps the book's main drawback. Students of design, architecture or environmental studies should consider Bruce Anderson's book instead. They will get much more for their money.

— Roger Douglass

Solar Energy Technology and Applications
J. Richard Williams
1974; 126 pp

$6.95

from:
Ann Arbor Science
P.O. Box 1425
Ann Arbor, MI 48106

or WHOLE EARTH TRUCK STORE

SOLAR ENERGY FOR MAN

This would be a good book for the kind of ecology-environment-energy students who are common in the undergraduate bodies of American universities today and are often very interested in solar energy. The mathematics is easy, the orientation is toward understanding and appreciating the situation, and the facts are correct. The style is reminiscent of Daniels' "Direct Use of the Sun's Energy," but Brinkworth is no substitute for that book. Daniels is more comprehensive and is fully documented for those who may wish to go to the roots of a question.

There are equations, diagrams, graphs, and plenty of them, and the text is laced with facts. However the scientific and engineering arguments are developed from a basic knowledge of physics and chemistry, such as might be acquired in a top-level secondary school course. It is not necessary to be a graduate engineer to handle this book. No attempt is made to teach technological design; rather the direction is toward understanding and appreciation of how things work.

The book is based solidly on facts, and has no economics or policy content, which means it's like to remain useful for many years. Some of the topics discussed at length are not in the mainstream of solar energy work today, because they do not appear economically promising. There are, for instance, long sections on thermionic emission, thermoelectric conversion, photochemistry, distillation and solar cookers. On the other hand, ocean thermal gradient conversion, power towers, shallow solar ponds and absorption chillers are not mentioned at all.

—Marshal Merriam

Solar Energy for Man
B. J. Brinkworth
1973; 264 pp

$9.95

from:
John Wiley & Sons, Inc.
1 Wiley Drive
Somerset, NJ 08873

or
1530 South Redwood Rd
Salt Lake City, UT 84104

or WHOLE EARTH TRUCK STORE

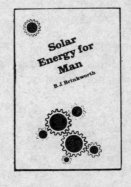

SUNSPOTS

Steve Baer subtitles this book "Collected Facts and Solar Fiction." It consists of somewhat reworked writings that first appeared in the TRIBAL MESSENGER and other magazines and journals. The underlying point of most of the material seems to be that if we thought a little harder about things, we could get by with less energy, less equipment, and fewer government boondoggles. Other material does not seem to have an underlying point, but it is very entertaining anyway.

A few quotes:

You can see that the sun lingers at its highest position, hardly changing in the sky for two months; then rushes through the fall towards winter where it will again linger . . . The sun follows a giant spiral in our sky. Each day it cuts a new thread winding its way up or down. The threads are closest together at the sun's upper and lower limits and farthest apart midway between . . . Almost anything that fluctuates between two extremes lingers at the extremes and rushes between them.

* * *

(6) & (7) - Sometimes (6) is better than (7). Sometimes (7) is better than (6). Generally, if the collector temperature is more than 100° hotter than outside (6), double glazing is best.

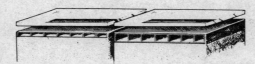

(7) & (8) - Generally (8) is better than (7) since there is liquid circulating directly behind the collecting plate everywhere.

(9) & (10) - Generally (9) will be a better collector than (10). This is because collectors grow cold or fall asleep when the sun goes behind a cloud or during the night. A heavy collector with a large mass of water has a great deal of heat to lose as it assumes the temperature of the air around it. Then when the sun comes up it takes a long time to wake up, for its mass must be warmed up above the useful collecting temperature before you circulate the liquid through it.

Surfaces with different orientations have different climates. Heat collectors copy their orientations from warm parts of the planet. A south wall collector in Albuquerque is parallel to the earth's horizon at a point about 700 miles north of the Antarctic . . . A collector tilted 45 degrees in Albuquerque is parallel to the horizon at a point 1200 miles north of Easter Island in the Pacific.

* * *

Like a parent checking the health of a child by feeling its forehead and looking in its eyes, one can judge a heat collector by looking for glare and feeling for hot glass. An efficient collector is dark and its outer glazing is cool.

— Chip Barnaby

Sunspots
1975; 115 pp.

$3.00

from:
Zomeworks Corporation
P.O. Box 712
Albuquerque,
NM 87103

or WHOLE EARTH TRUCK STORE

A FLORIDIAN'S GUIDE TO SOLAR ENERGY

This book was put together by the Florida State Energy Office. It is an effort to fill the "information gap" so that members of the public, particularly Floridians, can make intelligent decisions about using solar energy. The book contains a lot of information, but it is presented in a poorly organized yet highly structured outline form that is very hard to read.

Sections include energy conservation, solar system operation, collectors, heat storage, and "characteristics of reputable manufacturers of solar energy systems." An appendix gives solar data for Florida.

There are some interesting notes on the solar water heater industry that flourished in Florida during the 30's and 40's. Cheap electricity and corrosion problems halted installations in the 50's. As of 1966, though, there were still 16,000 solar water heaters in the Miami area.

—Chip Barnaby

A Floridian's Guide to Solar Energy
Florida State Energy Office
1976; 120 pp.

$1.50

from:
Florida Solar Energy Center
300 State Road 401
Cape Canaveral, Florida 32920

or WHOLE EARTH TRUCK STORE

PUMPED SOLAR WATER HEATING SYSTEM

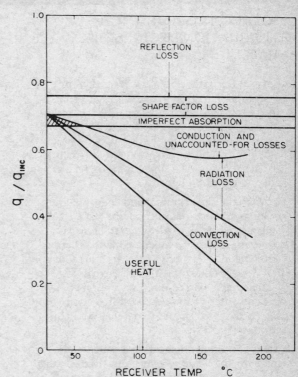

- solar heat collector
- automatic air discharge valve
- cold water to collector
- hot water from collector
- hot water to house
- air inlet drain cock
- gate valve
- cold water from street
- pump controls
- solar storage tank
- water drain cock
- check valve
- gate valve
- circulating pump

SOLAR SYSTEM DESIGN AND ENGINEERING

SOLAR ENERGY THERMAL PROCESSES

SOLAR ENERGY THERMAL PROCESSES was the first solar engineering book to appear, and is still one of the best. As the title implies, the subject matter is limited to thermal applications, and, particularly, relatively low temperature applications—in heating water and space. There is a chapter on absorption cooling, but none on swimming pools. The great bulk of the material however is not directly a discussion of applications but a treatment of fundamentals such as solar radiation, the solar constant, direct and diffuse radiation, and the calculation of insolation on tilted surfaces in the absence of the atmosphere. (Though it is an exactly solvable problem, the formula giving the insolation on a surface tilted at arbitrary angle, at any location, any time of day and year is written down in surprisingly few places. It is given here, with the five relevant angles defined clearly and precisely—an example of the kind of factual orientation that pervades the book.)

There is a chapter on heat transfer, with application to solar collectors, and one on emissive, absorptive and reflective properties of surfaces. The theory of flat plate collectors

is developed in terms of efficiencies, transmittance-absorptance product, etc., so that the performance of a particular collector design can be described and calculated. There are also chapters on focusing collectors, thermal storage, and computer modeling. In each case the relevant configurations are discussed in terms of engineering parameters, with numerical examples where appropriate, rather than in general descriptive terms.

This is a book by engineers for engineers. Others with engineering or science training will find it manageable and very useful in solar work. Contractors, builders, and architects, may be able to find what they want, with a struggle, but the book is not intended primarily for them.

—Marshal Merriam

Solar Energy Thermal Processes
John A. Duffie and W. A. Beckman
1974; 386 pp

$18.00

from:
Wiley InterScience
One Wiley Drive
Somerset, N.J. 08873

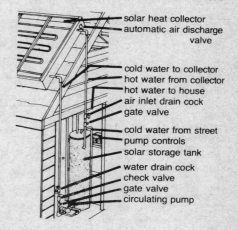

Figure 8.8.3 Distribution of incident energy for the 1.89 m reflector, 6 cm diameter receiver as a function of receiver surface temperature.

SOLAR HEATING AND COOLING

This book is intended for engineers and technical types designing solar heating and cooling systems. The authors have surveyed and condensed much of the literature produced in the solar community prior to 1975 and put it together to make a good introduction to the basics of active solar systems. Chapter headings include *Fundamentals of Heat Transfer, Methods of Solar Energy Collection, Solar Heating of Buildings,* and *Solar Cooling of Buildings.* Extensive appendices include data on properties of materials, climate, energy conservation, and sun rights legislation. The book is full of diagrams, tables, and formulas useful to designers Discussions of concentrating collectors and active solar cooling seem more complete than most.

SOLAR HEATING AND COOLING has a number of weaknesses, however. There is a very strong mechnical bias in the entire book, so much so, that although passive cooling systems are mentioned, passive heating is relegated to a two page appendix which cautions against the use of large glass areas. An uninformed reader would conclude from this book that the solar system designer determines heating and cooling loads by asking the architect and then automatically proceeds to design an active heating and cooling system using either flatplate or concentrating collectors. Also, the collector sizing procedure, although workable, is extremely oversimplified and unsophisticated compared to other more recent statistically based methods.

This is a useful reference book on active systems, but is not a good source for anyone trying to get a general solar education.

—Bruce A. Wilcox

Solar Heating and Cooling
J. F. Kreider and F. Kreith
1975; 342 pp.
$22.50

from:
McGraw-Hill
New York, N.Y.

or WHOLE EARTH TRUCK STORE

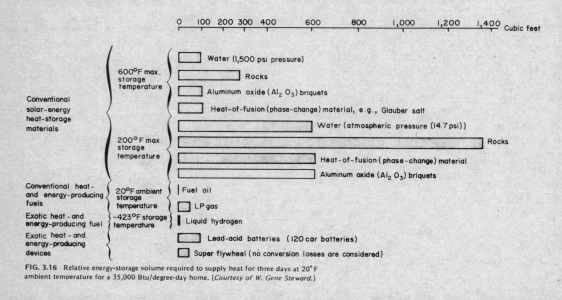

FIG. 3.16   Relative energy-storage volume required to supply heat for three days at 20°F ambient temperature for a 35,000 Btu/degree-day home. *(Courtesy of W. Gene Steward.)*

INTRODUCTION TO THE UTILIZATION OF SOLAR ENERGY

Starting in 1954-55 the UCLA Engineering Extension offered a course by this title, to which many solar energy authorities were invited to lecture. This book contains rewritten and expanded versions of some of these lectures, and reflects the technical state-of-the-arts of the early 1960's.

Following an introduction by Farrington Daniels and a section entitled *Energy Sources of the Future,* there are chapters on *Availability of Solar Energy, Properties of Surfaces and Diathermanous Materials, Flat Plate Collectors* (by H.C. Hottel and D.C. Erway)—*Concentration of Solar Energy, Conversion of Solar to Mechanical Energy, Direct Conversion of Solar to Electrical Energy,* and *Photochemical Processes for Utilization of Solar Energy.*

These excellent chapters on the technical aspects of solar energy are followed by a good discussion of the complex *Economics of Solar Energy.*

The next chapter, by George Löf, is on *Heating and Cooling of Buildings with Solar Energy.* The book concludes with a brief chapter on *Distillation of Sea Water and other Low-Temperature Applications of Solar Energy* (water heating, stoves, greenhouses), a chapter on solar furnaces for materials research and processing, and a chapter on *Space Applications of Solar Energy.*

This book costs $.05 per page, so if you can find the library copy it might be cheaper to do your own duplication of sections of interest to you.

—Roger Douglas

Introduction to the Utilization of Solar Energy
A.M. Zarem and Duane D. Erway
1963; 398 pp.
$20.50 hardcopy
$6.85 microfilm
plus tax and shipping

from:
Xerox University Microfilms
300 North Zeeb Road
Ann Arbor, Michigan 48104

SOLARON CORPORATION APPLICATION ENGINEERING MANUAL

Solaron Corporation, founded in 1974 by solar pioneer Dr. George Löf, makes probably the best packaged solar heating system on the market. It is an air system that is well engineered and well built. Solaron has set up a distributor network around the country and is installing an increasing number of systems on houses and commercial buildings.

To help their distributors and clients plan systems, Solaron has put together this engineering manual. Although the book is obviously primarily concerned with Solaron equipment, it contains a wealth of information useful to anyone building air systems. For instance, the manual discusses in some detail the construction of rockbed storage units. Rock size, bed depth, air flow, and enclosure construction are covered.

There are also worksheets and calculation methods for sizing systems, tabulated weather data, and detailed information about the Solaron collector. Many of the calculation methods are tailored to Solaron equipment, so one must be careful when using them in other situations.

— Chip Barnaby

Solaron Corporation Application Engineering Manual
Solaron Corporation
1976; 85 pp.
$10.00

from:
Solaron Corporation
300 Galleria Tower
720 So. Colorado Blvd.
Denver, CO 80222

or:
Local Solaron dealer
(check your Yellow Pages)

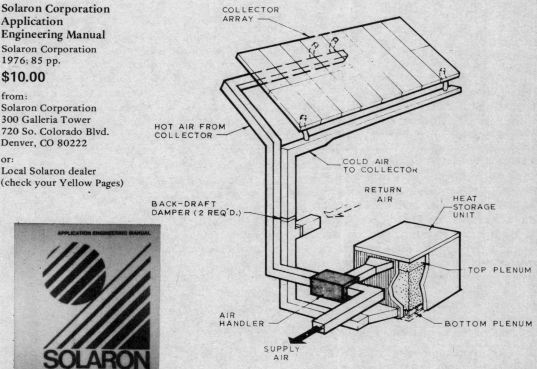

SOLAR ENERGY: A PRACTICAL GUIDE

Lufkin's book really lives up to its name. It's a very readable discussion of the basics of active, flat-plate solar systems. It's written for the technically competent layman who has a calculator and a working knowledge of trigonometry and algebra. Chapters include: heating load, solar radiation availability, practical collectors, storage and controls, heat pumps, and equipment manufacturers.

Lufkin explains complex problems in a clear, logical style that gets right to the point, particularly in discussions of the relative merits of various collector designs. He also throws in some goodies I've never seen elsewhere, such as a simple regression equation for calculating average energy needs from fuel bills and degree days. Unfortunately, he never presents a comprehensive design procedure for sizing collector systems. This is still one of the best books I've seen for the self education of designers and do-it-yourselfers.

— Bruce A. Wilcox

Solar Energy: A Practical Guide
D. H. Lufkin
110 pp.   from:
$10.50   D. H. Lufkin
               303 West College Terrace
               Frederick, Maryland 21701

SOLAR HEATING SYSTEMS DESIGN MANUAL

In 1975, Bell and Gossett converted their training center to solar heating and cooling. As a result they started getting hundreds of requests for system design information. They have responded by assembling this manual.

This manual is the best single starting point for anyone interested in the detailed mechanical design of complicated active systems. The material on system sizing is not as good as some of the more recent simplified methods. However, the Systems chapter gives schematics and design considerations for several drain back and antifreeze system arrangements—information that is not available anywhere else.

Since Bell and Gossett is in the hydronic heating equipment business, it is not surprising that the systems they recommend involve rather elaborate plumbing. Unfortunately, however, water based active systems are necessarily complicated, especially when freeze protection is involved. If you are designing hydronic solar systems, I recommend that you read this book.

—Chip Barnaby

Solar Heating Systems Design Manual
ITT Training and Education Department
1976; 130 pp.

$2.50

from:
Fluid Handling Division
8200 N. Austin Avenue
Morton Grove, IL 60053

ERDA'S PACIFIC REGIONAL SOLAR HEATING HANDBOOK

One of the most useful solar energy engineering books available today, particularly for west coast designers. For the book, Douglas Balcomb and James Hedstrom at LASL ran a computerized solar system simulator for active domestic water heating and space heating systems using weather data for six western U.S. cities (Phoenix, Santa Maria, Fresno, Medford, Seattle, and Bismark, ND). The result of this work is a set of graphs which give the solar designer an indication of the effects of changes in major design parameters on the performance of a solar system. It is so comprehensive and well produced that it is probably the best introduction to active system design theory available anywhere.

The book presents graphs showing the effects of changing collector area, tilt, orientation, number of glazings, glass transmission, surface absorbtivity and emissivity, coolant flow rate, collector heat transfer, collector insulation, collector heat capacity, pipe insulation, heat exchanger effectiveness, storage size and heat loss, and design water temperature on the performance of liquid space heating systems. The data are presented for two system sizes, 40% and 75% solar heat at all six locations.

If there's any complaint to be made, it's that this book concentrates on active systems because there is not comparable information for passive systems. Doug Balcomb and many others are working to solve this problem.

— R.M.

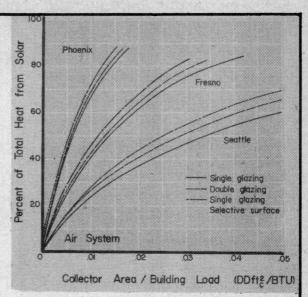

ERDA's Pacific Regional Solar Heating Handbook
San Francisco Operations Office (ERDA)
and Los Alamos Scientific Laboratory
1976 2nd edition; 108 pp

$3.50

from:
U.S. Government Printing Office

DESIGN MANUAL FOR SOLAR HEATING OF BUILDINGS AND DOMESTIC HOT WATER

This book is adapted from a design manual the author wrote for the Navy. It explains, in a very straight-forward way, how to size active space and water heating systems. The method used is based on the f-chart technique developed at the University of Wisconsin. The manual provides step-by-step instructions and clear worksheets on how to use the f-chart method for both air and water systems. Worksheets are also provided for doing life-cycle cost calculations. Filled out worksheets are included as examples.

This manual is just the thing for anyone with some engineering skill who wants to size active systems. The presentation style and nomenclature are fairly technical, however, so people who have not had much experience doing engineering calculations may have trouble using the book. Also, there is necessarily a fair amount of calculation drudgery involved in this method. So far though, this is the easiest method that can be done without a computer.

— Chip Barnaby

Design Manual for Solar Heating of Buildings and Domestic Hot Water
Richard L. Field
1976; 81 pp.   **$5.95**
from:
SOLPUB Co.
1831 Weston
Camarillo, CA 93010

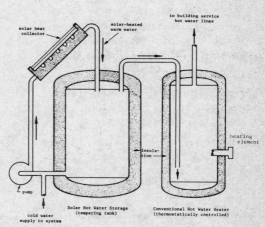

Schematic of potable hot water heating system using solar storage (tempering) tank ahead of conventional fueled or electric service water heater.

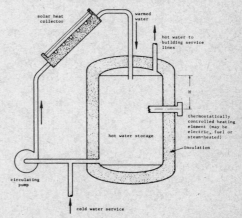

Single tank system for hot water storage and heating system.

LOW TEMPERATURE ENGINEERING APPLICATION OF SOLAR ENERGY

This short book contains six articles by leading solar engineers. Each article is an authoritative technical treatment of a particular subject relevant to the use of solar energy at low temperatures—i.e., up to boiling water temperatures. The six chapters are: Availability of Solar Energy for Flat Plate Collectors, The Measurement of Solar Radiation, Design Factors Influencing Collector Performance, Selective Surface for Solar Collectors, Potential Utilization of Flat Plate Collectors in Tropical Regions, Solar Water Heaters.

The material is not heavily dated, but much of it can be found in more recently published sources, such as Duffie and Beckman's *Solar Energy Thermal Processes*. Some of the articles, especially the first and third, are quite condensed and must be read with persistence.

— Marshal Merriam

Low Temperature Engineering   **$8.50**
Application of Solar Energy
Richard C. Jordan (ed.)
1967; 78 pp

from:
ASHRAE
345 E. 47th St.
New York, NY 10017

OTHER ASHRAE PUBLICATIONS

In their publications list, ASHRAE offers additional solar publications, including:

- SOLAR ENERGY AND THE FLAT-PLATE COLLECTOR, by Francis de Winter. An annotated bibliography.
- APPLICATIONS OF SOLAR ENERGY FOR HEATING AND COOLING OF BUILDINGS, edited by R.C. Jordan and B.Y.H. Liu. This is a revised version of LOW TEMPERATURE ENGINEERING . . . reviewed above. I have not seen it yet but it has been well reviewed.
- SOLAR ENERGY REPRINTS. Fourteen articles reprinted from the November, 1975 ASHRAE Journal.

Given the interest in solar energy, ASHRAE will probably publish more solar material in the future. Write to them for their publication list and prices.

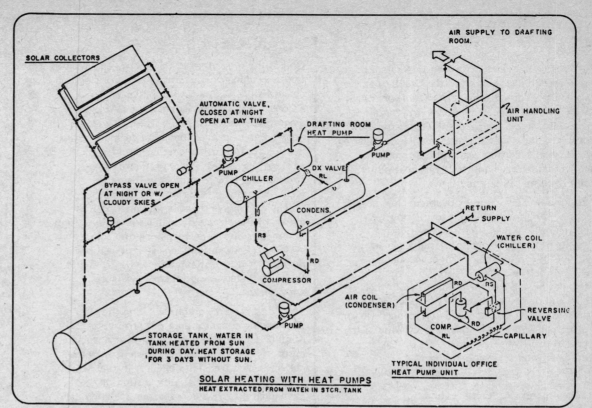

SOLAR HEATING WITH HEAT PUMPS
HEAT EXTRACTED FROM WATER IN STOR. TANK

ASHRAE HANDBOOK AND PRODUCT DIRECTORY (4 VOLUMES)

Large, definitive books published by the American Society of Heating, Refrigerating, and Air-conditioning Engineers. These are the standard reference volumes for engineers working in "energy related" areas. The books are technical and contain numerous references into the engineering literature for those who want more information. They are the definitive books for most solar designers.

One of the four volumes is revised each year or so in rotation so no volume is more than four or five years old. Each volume includes an up-to-date product directory that lists and categorizes manufacturers of heating, cooling, refrigeration, and related equipment. One can look up "Valves, Thermostatically Controlled," for instance, and find the names of 22 manufacturers.

from:
ASHRAE (Sales Dept.)
345 East 47th Street
New York, NY 10017

The four volumes are:

- HANDBOOK OF FUNDAMENTALS (1972; $32.00)
  Theory, general engineering data, basic materials, load calculations, duct and pipe sizing, and weather data. Best single source available on heat loss and gain calculations and insulation data. A new FUNDAMENTALS will be out this year.
- APPLICATIONS (1974; $40.00)
  Air-conditioning and heating, food refrigeration, and industrial applications. Chapter 59, "*Solar Energy Utilization for Heating and Cooling,*" was written by John Yellott and is a more comprehensive version of his article in this book.
- EQUIPMENT (1975; $40.00)
  Air handling, refrigeration, heating, general components (such as pumps and motors), and unitary equipment.
- SYSTEMS (1976; $40.00)
  Air-conditioning and heating systems, industrial ventilation, and refrigeration.

UNIFORM SOLAR ENERGY CODE

THE UNIFORM SOLAR ENERGY CODE is one of the first published attempts at providing "a safe and functional solar energy system with minimum regulations." The book is organized into two major parts: the first is devoted to an ordinance-type format suitable for adoption by local agencies, and the second is almost entirely an appendix on "Collector Sizing and Calculations."

The book begins with a model ordinance that has little to do with solar energy and the code. It is mainly a restatement of conventional plumbing practices. Not only does it fail to provide guidelines, it adds confusion to an issue where very precise performance guidelines are required.

The second half of the book (Appendix A) finally deals directly with solar systems, and is based entirely on material submitted by the Copper Development Association. It covers simplified methods for sizing a solar heated domestic hot water or space heating system. The sizing calculations and examples are done in sufficient detail to allow the do-it-yourselfer to use this particular sizing approach.

The overall value of this book is limited at best. It does give a fairly concise view of conventional plumbing practices and some sizing methods for two common solar applications but never really addresses the issue of solar energy components and the code. It is almost exclusively oriented toward water systems, with little or no mention of air or passive systems. It also tends to discourage owner-built components and systems.

Although the task of a uniform solar energy code is a large one, this is one attempt that never really hit the target. Instead it has the appearance of having been put together by a committee at the last minute.

—Dean Anson

Uniform Solar Energy Code
International Association of Plumbing and Mechanical Officials (IAPMO)
102 pp; 1976 edition.

$6.95   paperback

from:
IAPMO Headquarters
5032 Alhambra Avenue
Los Angeles, CA 90032

SOLAR DESIGN AIDS AND METHODS

Designing any solar system requires a fair amount of calculation. Some of the calculations, such as those required to size a collector, are almost impossible to do without a computer. Others, such as for duct sizing, are tedious at best. Various design aids and methods are available that speed up this type of work. Here are some we have found useful.

DUCTULATOR AND SYSTEM SYZER

The Ductulator is a handy device for sizing ducts put out by the Trane Company. If you know how much air you want to move from here to there, the Ductulator will tell you what size round or rectangular duct you need. This can save you a lot of drudgery when designing air solar systems or forced air heating systems.

The System Syzer does for pipes what the Ductulator does for ducts, plus a lot more. The scales on the calculator give you friction loss in various types of pipe, flow velocities, and flow rate as a function of pressure.

—Chip Barnaby

System Syzer
$2.50

from:
Bell & Gossett
8200 N. Austin Ave.
Morton Grove, IL 60053

Ductulator
$1.00

from:
The Trane Company
La Crosse, WI 54601

or: **Add 50¢ to either price for postage.**

Check local distributors of Trane and Bell and Gossett.

BENNETT SUN ANGLE CHARTS

Large format, easy to use charts showing the sun's path across the sky during each month of the year as a function of both solar azimuth and altitude (expanding parabolas). Charts available for latitude 24°N through 52°N in 2°increments.

— Richard Merrill

Bennett Sun Angle Charts
$2.00 per chart

from:
R.T. Bennett
6 Snoden Rd.
Bala Cynwyd, PA 19004

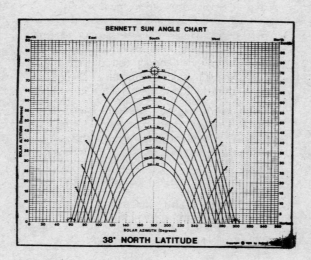

BENNETT SUN ANGLE CHART

38° NORTH LATITUDE

SUN ANGLE CALCULATOR

The Sun Angle Calculator is a nicely packaged set of curves and plastic overlays that will tell you where the sun is in the sky and other useful solar angles. The Calculator can save literally hours of trigonometric agony when you're trying to figure out sunshine availability or shading.

Besides the position of the sun, the Calculator can tell you the profile angle (useful for sizing sunshades on buildings), the angle of incidence of the solar beam on a wall, and daylight availability. Because the Calculator is intended for window design, it does not give angles of incidence on sloping surfaces (such as solar collectors). Also, it has charts for 4° increments of latitude, so if you want high accuracy, you may have to interpolate between charts.

All in all, though, the Sun Angle Calculator is very useful. It comes with a very clear instruction book.

—Chip Barnaby

Sun Angle Calculator
Libbey — Owens — Ford
$4.00

from:
Libbey — Owens — Ford
Toledo, Ohio

or WHOLE EARTH TRUCK STORE

DESIGN METHODS AND COMPUTER PROGRAMS

SIMPLE METHODS

Several simplified methods for estimating the performance of active solar heating systems have been developed in the last two years. The methods are simple enough to do by hand with the aid of a scientific calculator. However, so far they all require month-by-month calculation. The procedure must therefore be run through 12 times and the results summed up. I can assure you that the novelty of pushing buttons on a calculator wears off by the time you get to about April. A programmable calculator or a microcomputer could handle these jobs, though. The two most widely known methods are:

f-Chart. This method was developed at the University of Wisconsin with the aid of 300 simulation runs on the program TRNSYS (see below). The result is a chart that allows the user to determine the fraction f of the heating energy required during each month of the year. Space and water heating loads, monthly total solar radiation, and monthly average temperatures must be known.

The most convenient way to use f-Chart is to get the DESIGN MANUAL FOR SOLAR HEATING OF BUILDINGS AND DOMESTIC HOT WATER by Richard L. Field (reviewed elsewhere in this section). Worksheets and explanations are provided in the DESIGN MANUAL that are extremely helpful. Also, life-cycle cost calculation methods are provided for economic evaluation of proposed solar heating systems.

Simplified Method. The simplified method developed by the Los Alamos Scientific Laboratory is somewhat similar to f-Chart in that it derives an "X-factor" for each month which approximates the solar fraction of the heating load. The "X-factors" have many standard collector and system parameters built in (such as the number of glazings and the heat removal efficiency). This means that the method cannot be applied to "non-standard" collectors such as trickle collectors, while f-Chart can.

The most straightforward description of the simplified method is found in ERDA's PACIFIC REGIONAL SOLAR HEATING HANDBOOK, reviewed elsewhere in this section.

SIMULATION

Detailed design and research studies of active solar systems are done with simulation computer programs. The programs model the behavior of the system at short time intervals, such as half an hour, throughout an entire year. The best known solar simulator is TRNSYS, developed at the University of Wisconsin. To use TRNSYS, the user specifies the components in the solar system and how they are connected.

Detailed weather data must also be supplied. The computer program models the operation of the system and prints out temperatures, energy transfer, and other information at various levels of detail. TRNSYS has proved to be accurate and very useful for studying proposed solar systems on a research basis. It is, however, too expensive to use routinely for system design.

BUILDING SIMULATORS

The simulation method can be applied to buildings as well as active solar systems. Building simulation programs tell you the amount of energy required to heat and cool a building. Since the programs take account of the solar energy coming into windows, some of them can be used for passive solar design. Effort is underway now to increase the passive modeling capabilities of some of the programs. In a few years, this work will yield simple passive design methods analogous to the ones described above for active systems.

Two building simulation programs are worth mentioning by name:

CAL/ERDA is a program being developed for the State of California and the Federal Energy Research and Development Administration. CAL/ERDA has numerous features that make it easy to use. It calculates building energy requirements and fuel costs. CAL/ERDA is in the public domain and will become widely available over the next four years.

NBSLD is a program written at the National Bureau of Standards several years ago that uses an extremely accurate calculation method. Although it is expensive to run, NBSLD is very useful because it produces detailed information about the thermal behavior of a building. Berkeley Solar Group has developed a revised and extended version of NBSLD (which we call NBSGLD) which we use in our work on energy conserving buildings.

—Chip Barnaby

SOLAR RADIATION AND CLIMATE

WORLD DISTRIBUTION OF SOLAR RADIATION

This is a dated but somewhat useful solar data source that was done in 1965 as background information for "appraising the economics of a proposed solar energy application in a particular area . . ." The goal was to provide maps of radiation isolines on a global scale and present the information in a way that would allow the user to interpolate from the map for any given location.

The data are in the form of monthly averages of daily totals, and are presented both as a table with values for individual stations, and also as 12 monthly maps of global solar radiation intensity. The data for particular stations is useful beyond the actual values listed—it also indicates the data source and type of instrument used so that the user can make evaluations as to reliability. The monthly maps are preceded by a location map of the original data points. These are keyed to indicate where the data have been modified. When extracting values for a particular location from the isoline map, one can easily refer back to see if there is an actual data point nearby.

The maps were drawn with an emphasis on three ordered criteria: long-term averages, local influences, and relationship to climatic/vegetative associations. Although the original data are presented in an uncritical manner and come from a wide variety of sources, information such as type of instrument used and data modifications are presented in such a way as to be useful for solar applications.

—Dean Anson

World Distribution of Solar Radiation
George O. G. Löf, John A. Duffie, Clayton O. Smith

$6.00

from:
Report No. 21, Solar Energy Laboratory
The University of Wisconsin
1966; 85 pp

or WHOLE EARTH TRUCK STORE

SOLAR AND TERRESTRIAL RADIATION

The author has spent many years studying the characteristics of solar radiation, especially the measurement of intensity and polarization, and is presently Professor of Meteorology at the University of California, Davis. SOLAR AND TERRESTRIAL RADIATION is an excellent book which will be of value to a wide range of workers including those concerned with the science of solar radiation plus those trying to design innovative solar devices or systems.

The emphasis throughout is on measurement—what has been measured, how it was measured, what the results were. The characteristics and working principles of nearly all types of pyranometers are covered in some detail.

Chapter headings include: *Principles of Radiation Instruments, Radiation Sensors and Sources, Solar Radiation: Direct Component,*

CALIFORNIA SOLAR DATA MANUAL

The DATA MANUAL pulls together and tabulates all reliable solar data for the State of California. The book was assembled by Lawrence Berkeley Laboratory and represents a new standard of how to present climate information. Highly readable discussions cover the data sources used, the errors involved, error correction methods, and the consequences of using erroneous data. There are also sections on simplified solar performance calculation methods, general solar information, and estimating unmeasured radiation quantities.

The actual climate and radiation data are presented for 15 climate zones. Calculated radiation levels for tilted surfaces are also given, which is very helpful for using the simplified performance methods that are becoming available. My only criticism here is that the zones are not small enough to represent the rapid climate changes that occur as one moves inland from the California Coast.

I recommend this book to people doing solar work in California and to people working elsewhere who want to learn more about getting the most from the scanty sunshine data available.

— Chip Barnaby

California Solar Data Manual
Energy and Environment Division
Lawrence Berkeley Laboratory
1977; 301 pp.
from:
California Energy and Resources Conservation
Alternative Implementation Division
1111 Howe Ave.
Sacramento, CA 95285

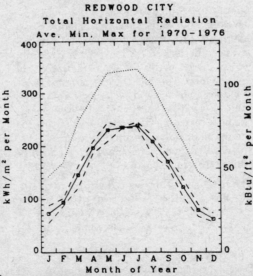

REDWOOD CITY
Total Horizontal Radiation
Ave. Min. Max for 1970-1976

Solar Radiation: Diffuse Component, Ultraviolet Radiation from the Sun and Sky, Illumination, Polarization of Light in the Atmosphere, Duration of Sunshine, The Solar Constant, Terrestrial Radiation: Field Characteristics, Terrestrial Radiation: Methods of Measurement.

For any serious student of solar energy Coulson's book will be extremely useful. It complements, extends and updates Robinson's earlier book (Robinson, N. 1966. SOLAR RADIATION, Elsevier, N.Y.), and has a greater emphasis on measurement. Highly recommended.

—Marshal Merriam

CLIMATE

We all talk about the weather a lot, but how many of us know where to get good information concerning it? First we can turn to the National Climatic Center in Asheville, North Carolina. NCC is the largest climatological data facility of the Environmental Science Services Administration (ESSA), which in turn is part of the National Oceanic and Atmospheric Administration (NOAA), which in turn is part of the U.S. Department of Commerce.

NOAA publications are detailed in *Selective Guide to Climatic Data Sources.* I would recommend this book to people trying to determine what weather information is available for their area. It is confusing in its details at first, but persevere and it will eventually make sense. If you order this book, ask for an updated price list. And if you order *from* this book, send your order for everything but the *Climatic Atlas of the United States* (see review in the wind chapter) to NOAA, even though it tells you to mail orders to the Superintendent of Documents, US Government Printing Office. NOAA has been fast and efficient in responding to my requests while the GPO hasn't always been so good.

For a good, cheap overview of solar radiation in Langleys by months for the entire U.S., order *Mean Daily Solar Radiation, Monthly and Annual.* It's a double-sided map with one large summary chart of numerical solar radiation data. A sampling of NOAA publications by title includes: *Climatological Data; Climatological Data, National Summary; Local Climatological Data; Hourly Precipitation Data; Monthly Climatic Data for the World; Climatic Guide; Average Precipitation in the United States; Daily Normals of Temperature and Heating Degree Days; Heating Degree Day Normals; Sunshine and Cloudiness at Selected Stations in the United States.*

— C.M.

Selective Guide to Climatic Data Sources
1969; 94 pp

$1.00

Mean Daily Solar Radiation, Monthly and Annual
1964; 16" x 22"

$0.10

from:
Environmental Data Service
National Climatic Center
Federal Building
Asheville, North Carolina 28801

Solar and Terrestrial Radiation
Kinsell L. Coulson
1975; 322 pp.

$27.00

from
Academic Press
111 5th Avenue
New York, NY 10003

SOLAR BUILDING APPLICATIONS

SOLAR FOR YOUR PRESENT HOME

This book emphasizes in a logical and easy-to-follow format, realistic and practical solar information for applications in the San Francisco Bay. By substituting local climate data, much of this material can be readily used for other locations. Considering the potentials of the solar retrofit market, it is surprising there is so little specific information on the subject.

The book begins with several retrofit examples, some fundamentals, and "things to watch out for," e.g., common materials like copper and glass are compared with aluminum and plastic in terms of dependability, longevity, and cost.

Methods for estimating energy use are presented in a checklist and worksheet approach in the next section. This includes details for analyzing individual retrofit possibilities in terms of where to locate collectors and storage units, and a step-by-step discussion of how to do it. Twelve types of solar systems are used as

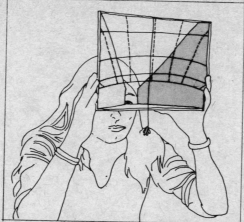

Figure 3-10: Using the Solar Viewer. Note that eye level is even with the bottom of the Viewer. Face south and hold the Viewer level, as indicated by the string-and-paperclip level indicator. Move your eye around to check shading at different times of the day and year.

examples. This gives a picture of what to expect from a particular system, type of building, and energy use. The exercise of filling in the worksheet is educational, and lessens the possibility of forgetting a particular item which could easily lead to an unrealistic and uneconomic decision. The book suggests to retrofit now, since prices of solar systems aren't likely to come down.

— Dean Anson

Solar for Your Present Home
Charles Barnaby, Phillip Caesar, and Lynn Nelson
1977

from:
California Energy Resources Conservation
and Development Commission
1111 Howe Avenue
Sacramento, CA 95285

For more on building applications see the Solar Architecture section of the ENERGY PRIMER (p. 84).

SOLAR WATER HEATING, DISTILLATION AND DRYING

SOLAR WATER HEATING, FOR THE HANDYMAN

This is a matter-of-fact book for the person with some tools and modest building skills. Paige addresses the book to those who want to build a solar water heater with existing plumbing and not over-elaborate "systems." He stresses designs of the thermosyphon and batch collectors using simple, easily available materials. However, with several free hand drawings this book is more of an inspirational guide than a how-to manual . . . but that's what "handy" people need anyway.

—Richard Merrill

Solar Water Heating for the Handyman
Steven Paige
1974; 31 pp
$4.00

from:
Edmund Scientific Co.
Barrington, NJ 08007

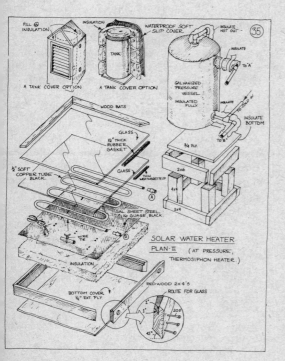

SOLAR WATER HEATING IN SOUTH AFRICA

This is an important report for anyone in the business of developing, manufacturing, or testing solar water heaters. For those building only one or two units the value will not be great enough to justify the trouble involved in obtaining a copy.

Thermosyphon circulation and reverse circulation are considered in mathematical terms. The relative performances of nine different absorber designs are evaluated on the basis of a three week performance testing program. Other tests are described which show the influence on efficiency of climate variation, cover transparency, insulation, storage tank height, flow inlet connection height and forced circulation. Pipe lengths and pipe diameters are discussed. There are 48 figures and 31 tables, which gives some idea of the thoroughness.

This is a top quality engineering report, with considerable relevance to the water heater scene in the United States today.

—Marshal Merriam

Comparison of average efficiencies of different absorbers

Type of absorber	Average efficiency (%)
B: commercial radiator	56,4
C: two corrugated galvanized steel sheets	50,8
D: corrugated galvanized steel sheet on flat galvanized steel sheet	50,4
E: aluminium tube-in-strip	57,1
F: galvanized steel pipe framework on copper strips	54,6
G: low-cost unit, galvanized steel	55,9
G: low-cost unit, fibre-glass	55,6
G: asbestos-cement, uninsulated	34,7
G: asbestos-cement, insulated	44,0
H: copper tube-in-strip	49,2
I: two flat steel plates	48,9
J: black polyethylene piping	53,5

Solar Water Heating in South Africa
D.N.W. Chinnery
1971; 79 pp
$6.00

from:
South African CSIRO
P.O. Box 395
Pretoria, South Africa

DESIGN AND BUILD A SOLAR POOL HEATER

This book is about the detailed design and economics of solar pool heating systems using copper flat-plate collectors. Now for most people swimming pools are a luxury . . . unnecessary and otherwise. But DeWinter has assembled information useful elsewhere. Especially valuable are: (1) the many matrix graphs showing simultaneous relationships between several design variables; (2) description of copper-soldering techniques; and (3) a useful discussion of life cycle costing and solar economics. Also, the price is right.

— Richard Merrill

How to Design and Build a Solar Swimming Pool Heater
Francis de Winter
1974; 54 pp
free

from:
Copper Development Association
405 Lexington Avenue
New York, NY 10017

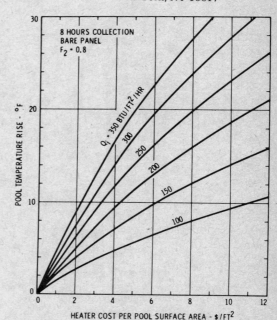

How the rise in pool temperature produced by a bare solar collector varies depending on solar input and the investment in the collector per sq. ft. of pool area.

BREADBOX SOLAR WATER HEATER

Zomeworks does it again. A solar water heater that needs no "collector." Steve Baer writes, "We have done a great deal of work on this kind of water heater and feel it is a good system for someone to build himself. They are bulky and heavy and are not very suitable for shop production and sale, but the simple construction methods make them right for backyard construction." Basically, the heater is a black water tank in an insulated box equipped with two glass walls to admit sunlight, and two insulated lids that reflect sun into the box during the day and close to keep in the heat at night. The plans are big, clear, and hand drawn in Steve's own inimitable style.

— J. Baldwin

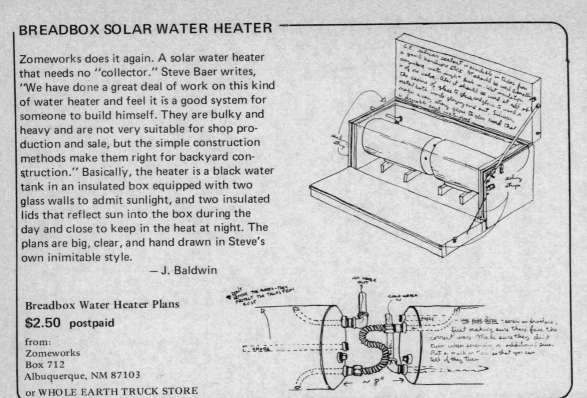

Breadbox Water Heater Plans

$2.50 postpaid

from:
Zomeworks
Box 712
Albuquerque, NM 87103

or WHOLE EARTH TRUCK STORE

DO-IT-YOURSELF BOOKLETS

Hot Water
Scott Morgan et al.
1975; 31 pp.

from:
HOT WATER
350 E. Mountain Drive
Santa Barbara, CA 93108

$2.00

Excellent easy-to-follow booklet describing the construction of open and closed solar hot water systems from readily available materials. Book includes instructions for retrofitting with existing hot water tank; conversion of gas water heater into closed system using gas flame core; and construction of "stack coil water heating system (heat exchangers in wood stove pipes).

Build Your Own Solar Water Heater
1974; 25 pp.

from:
Florida Conservation Foundation
935 Orange Ave.
Winter Park, FL 32789

$2.50

Instructions for building a solar water heater with commonly available materials, 43 illustrations and step-by-step directions. Very practical and handy from the region where solar water heaters have been in operation the longest.

SMALL-SCALE SOLAR DISTILLATION

A well-designed, durable, basin-type solar still will cost in the neighborhood of $1 per square foot, produce 0.1 gallon of fresh water per square foot on a good day, and average 25 gallons per square foot per year production in a favorable climate. This book describes designs of large stills that have been built, discusses materials and economics, and gives detailed mathematical theory of operation. There are 30 figures, 6 tables, 29 references and a bibliography. Personal survival stills are not discussed. Combinations of solar with other heat input and fresh water production with other functions receive very limited treatment. *Solar Distillation . . .* was written by solar still experts from around the world.

— Roger Douglass

Solar Distillation As A Means Of Meeting Small-Scale Water Demands
1970; 86 pp
Sales No. E.70.11.B.1

$2.50

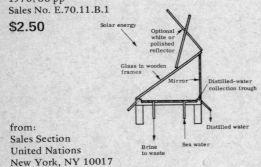

from:
Sales Section
United Nations
New York, NY 10017

BRACE RESEARCH   SOLAR STILLS

How to Make a Solar Still (Plastic Covered)
1973; 13 pp; Do-It-Yourself-Leaflet L-1 **$1.00**

Hard facts, bill-of-materials, detailed drawings for a four foot wide by any length solar still. Basic, good.

Simple Solar Still for the Production of Distilled Water
1967; 6 pp; Technical Report No. T 17 **$1.00**

This is a description of a simple solar distillation unit designed primarily for use in service stations with the object of providing distilled water for automobile batteries. Distilled water is very necessary for battery maintenance, especially in warm and arid regions. The still should produce under normal operating conditions an <u>average</u> of three litres of distilled water per day.

Plans for a Glass and Concrete Solar Still
1972; 8 pp + 2 drawings:
Technical Report No. T 58 **$4.00**

In 1969, Brace built a 200 gallon per day still in Haiti. This report is the plans and specifications for that still. If you're in the market for a 50 foot by 75 foot still made from concrete, butyl, rubber, glass and silicone glass sealant, here's the answer. Very clear and understandable drawings.

from:
Brace Research Institute
Macdonald College of McGill University
Ste. Anne de Bellevue
Quebec, Canada H9x 3M1

MANUAL ON SOLAR DISTILLATION OF SALINE WATER

This compendium on solar stills appears to be exhaustive and accurate, and is likely to be extremely helpful to anyone planning to build a solar distillation plant, either large or small. The information was assembled by the authors while working for Batelle Memorial Institute under contract to the Office of Saline Water. It is an absolutely first class job.

The history of solar distillation is a history of problems and failures—not necessarily performance failures, but failures in that continuous, substantial and expensive maintenance was required to keep them performing. Consequently solar distillation today is an abandoned technology. It was not so at the time this manual was compiled. Of the 27 large stills discussed in detail, 20 were operating in 1970. Only a very few, perhaps none, are operating in 1977. No doubt there are still situations where solar distillation is still the best of bad alternatives, and builders of stills in such places can usefully benefit from the experiences of past builders, like those discussed in this book. The performance specifications and problems encountered, on essentially every large still ever built, are collected here, together with the relevant details on a great many smaller ones. There are 23 tables, 152 figures and 529 references, a 26 page chapter on economics of solar distillation, and exhaustive discussion of design and technology.

—Marshal Merriam

Manual on Solar Distillation of Saline Water
S.G. Talbert, J.A. Eibling and G.O.G. Löf
1970; 263 pp

$2.00

from:
Superintendent of Documents,
Washington, D.C.

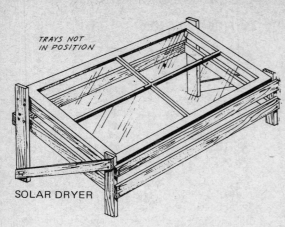

TRAYS NOT
IN POSITION

SOLAR DRYER

HOW TO BUILD A SOLAR CROP DRYER

This is a pamphlet put together by the New Mexico Solar Energy Association. It gives step-by-step instructions for construction of a simple, $60 air-convection crop dryer based on a design by Peter van Dresser. Good diagrams and a short discussion of food drying are included. A bargain.

— Chip Barnaby

How to Build a Solar Crop Dryer
NMSEA
1976; 10 pp.   from:
NMSEA
$.50   P.O. Box 2004
Santa Fe, New Mexico 87501

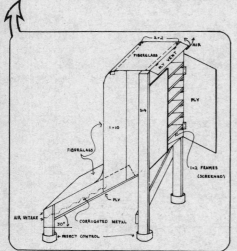

SOLAR ENERGIZED FOOD DEHYDRATOR

Most of the really useful solar hardware is coming from small outfits. There seems to be a distinct aversion by big companies to get into a business where they can't sell or control the monopolistically critical components (hee hee). Anyway, here's a nifty set of plans that seem to be well thought out. At the very least, they'll get you started on experiments of your own. These are *very* clear and nicely drawn. Most other plan-sellers, please take note.

— J. Baldwin

Solar Energized Food Dehydrator
$5.00 plans, postpaid
$200.00 complete unit, F.O.B.

from:
Solar Survival
Cherry Hill Rd.
Harrisville, NH 03450

A SURVEY OF SOLAR AGRICULTURAL DRYERS

A collection of 25 case studies of solar drying methods from around the world. Brace Research Institute collected this information by sending a questionnaire to people working on solar drying. The responses are written up in consistent form and illustrated with diagrams and photographs.

The survey focuses on small-to-medium scale dryers that are adaptable to rural areas of developing countries. A representative, not exhaustive, sample of such dryers is included. The sample is broken down into 5 categories: *Natural and Sun-dryers, Direct Solar Dryers, Mixed Mode Solar Dryers, Indirect Solar Dryers* (with forced ventilation), and *Solar Lumber Dryers.*

Each case study includes climatological, performance, and cost information. Names and addresses of contact people are also given.

— Chip Barnaby

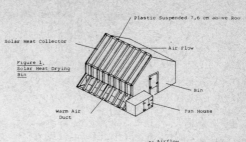

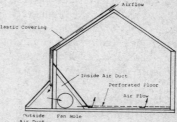

A Survey of Solar Agricultural Dryers
Brace Research Institute
1975
$5.00   from:
Brace Research Institute
MacDonald College of McGill University
Ste. Anne de Bellevue
Quebec, Canada

BOOKS ON FOOD DRYING AND PRESERVATION

Home Storage of Vegetables and Fruits
Evelyn V. Loveday
1973; 152 pp

Another book on canning, freezing, curing and drying. While the drying chapter is not as complete as that of the other two drying books mentioned here, it does give basic plans for various types of dryers—solar, oven, stove, and electric.

$3.00

from:
Garden Way Publishing Company
Charlotte, Vermont 05445

or   WHOLE EARTH TRUCK STORE

Stocking Up
1973; 351 pp

A thorough book on all aspects of food preservation. Written with the usual Rodale care for the life-giving values of food. The drying section is basic, simplified, with some good charts. With some adapting, it could be a guide for solar drying.

$8.95

from:
Rodale Press, Inc., Book Division
33 East Minor
Emmaus, Pennsylvania 10849

or WHOLE EARTH TRUCK STORE

How to Make a Solar Cabinet Dryer for Agricultural Produce
1973; 11 pp
Do-It-Yourself Leaflet L-6

Hands-on plans for a small dryer. See John Yellott's article for further details.

$1.00

from:
Brace Research Institute
Macdonald College of McGill University
Ste. Anne de Bellevue 800
Quebec, Canada

Dry It—You'll Like It
G. MacManiman
1973; 64 pp

A recipe book of drying fruits & vegetables.

The authors contend that 110°F is the maximum drying temperature for any plant drying task. Written for use with a dryer utilizing a thermostatically-controlled, wall-type heater with a circulating fan, the information presented could be adapted to your solar dryer needs.

$3.95

from:
Living Foods Dehydrators
P.O. Box 546
Fall City, Washington 98024

or WHOLE EARTH TRUCK STORE

Putting Food By
R. Hertzberg, B. Vaughan and J. Greene
1973; 372 pp

In addition to thorough discussions of canning, freezing and curing, *Putting Food By* has an excellent chapter on drying. Procedures, equipment, a few paragraphs on solar dryers, and lots of practical tips on drying fruits, vegetables, herbs and meats. This is the only book that talks about bacteria growth and control in the food drying process. It thoroughly evaluates dryer tray materials, including the problems presented by food acidity.

$2.50

from:
Bantam Books

or WHOLE EARTH TRUCK STORE

For more information on hardware, see the 5 articles on solar drying in Vol V, UN *New Sources of Energy* (1961), the article by G.O.G. Lof in Vol. 6, No. 4 (1962) of *Solar Energy* Journal, and occasional other articles in the same publication.

PHOTOVOLTAICS

SOLAR CELLS

This volume in the IEEE Selected Reprint Series consists of reprinted classic papers dealing with photovoltaic theory and applications from 1954 through late 1973. The papers range from discovery to development, production and application of the cells. Also included is a very extensive bibliography. The entries are arranged by subject, and an author index is provided. Believed to be the most complete bibliography existing, it consists of over 1200 entries and occupies 78 pages.

An effort has been made to address the problem of cost reduction for terrestrial applications by including the discussions of working groups from a 1973 Workshop on Photovoltaic Conversion. These include opinion and speculation, and critical questions from peers. They give an insight into the thinking of the experts in the field, at least as of 1973.

This book is recommended for those who prefer to get their information from the source. In the photovoltaic field such an approach has much to recommend it; predigested summaries are sometimes unreliable, misleading and wrong. (Press releases are even worse!) For the researcher, serious student, or development engineer this is the right book. The hobbyist, tinkerer, or journalist will find it heavy going. Unique in its breadth and depth, *Solar Cells* is an extremely valuable addition to the literature on photovoltaics.

—Marshal Merriam

Solar Cells
Charles E. Backus (ed.)
1976; 504 pp

$24.95 (cloth)

$12.45 (paper)

from:
The Institute of Electrical and
    Electronics Engineers
345 East 47th Street
New York, NY 10017

SOLAR CELLS

This is the best and most up-to-date book on photovoltaics. The author is a research scientist with IBM who has been a well-known figure in solar cell research for some years. Thus the book is scholarly, very accurate, quantitative and clear. It will be used for many years as a basic source.

SOLAR CELLS, however, is not a design manual for the hobbyist, or a useful source for the person who wants to speculate about cost of photovoltaic in future decades without coming to a complete understanding of how solar cells work and what limits them today. The author states that the book is intended as a background and general reference source primarily for the non-expert and for those interested in entering the field, and this it certainly is. Anyone with a knowledge of the physics of semiconductors who has a desire to learn and a will to persist can master the book, and many less well prepared can gain a great deal from it.

All types of cells are covered, and the factors affecting performance in both the space and terrestrial environments are discussed in some detail. Important results reported through mid-1975 are included. The index is complete and several hundred references to the original literature are given.

— Marshal Merriam

Solar Cells (Vol. II, Semiconductors and Semimetals)
Harold J. Hovel
1975; 250 pp

$14.50

OTHER BOOKS (TECHNICAL)

Sunlight to Electricity
Joseph A. Merrigan   **$10.00**   The MIT Press
1975; 163 pp   Cambridge, MA

Solar Photovoltaic Energy Hearings
J.J. Loferski, 1974
Subcommittee on Energy of the Committee on Science and Astronautics, U.S. House of Representatives, June 6 and 11, 1974.

OTHER BOOKS (POPULAR)

The Dawning of Solar Cells
David Morris   **$1.50**
1975; 16 pp
An excellent review of the rather biased information from manufacturers and government reports indicating that solar cells have the potential of providing us with all of our electrical needs by 2000. Paper details the economic and technical steps necessary to make photovoltaics a reality . . . and also their political implications.

Solar Cells and Photocells
Rufus P. Turner   **$3.95**
1976; 96 pp

Intended for experimenters, technicians, and science fair participants. Practical applications of photocells and solar cells include light meter, light relays, and control circuits, etc. With complete schematics.

BOTH OF ABOVE   EARS Reprint Service
BOOKS AVAILABLE   2239 East Colfax
FROM: Denver, CO 80206

Photovoltaics
Pietro Widmer
1977; 12 pp
  .50

Solar Energy Society of Canada
BC Chapter
1271 Howe Street
Vancouver, BC V6Z 1R8

A well illustrated, and simply written companion to David Morris' book.

Energy from the Sun
D.M. Chapin
1962

Edmund Scientific
701 Edscorp Bldg.
Bamhston NJ 08007

Oriented to the high school student, this is an excellent layperson's guide to the science of solar cells. The booklet accompanies a kit with which the student can make solar cells.

The Photovoltaic Generation of Electricity
Bruce Chalmers
in Scientific American
October 1976

This is a semi-technical article written to be understood by the layperson, describing the operation and efficiency of photovoltaic cells.

SOLAR PERIODICALS

The best way to maintain contact with the solar community and keep in touch with state-of-the-art technical developments is to subscribe to one of the many solar-oriented periodicals. Most of these are directed toward a fairly general reader while at the same time providing quite detailed technical information.

Magazines like POPULAR SCIENCE and SUNSET regularly feature articles on solar energy, as well as having reprints available for a nominal charge. Note that many groups listed throughout the ENERGY PRIMER have their own periodicals relating to solar energy.

1) **Solar Age.** Published monthly by SolarVision, Inc., 200 East Main Street, Port Jervis, N.Y. 12771. $12/6 mos., $20/yr. Regularly features articles on active and passive systems, with frequent coverage of ''indirect'' solar energies such as wind power and wood burning.

2) **Solar Energy Digest.** Published monthly by William B. Edmondson, P.O. Box 17776, San Diego, CA 92117. $28.50/yr. A newsletter with up-to-date information on new products and developments. Contains frequent editorial comments. About 12 pages.

3) **Solar Engineering.** Published monthly by Solar Engineering Publishers, Inc., 8435 N. Stemmons Freeway, Suite 880, Dallas, Texas 75247. $10/yr. Strong emphasis toward commercial and manufacturing market, with much product information. Official publication of the Solar Energy Industries Association.

4) **Solar Heating and Cooling.** Published bi-monthly by Gordon Publications, 20 Community Place, Morristown, N.J. 07960. $6/yr. Includes much product information and orientation toward commercial reader.

5) **Sunworld.** Published quarterly by the International Solar Energy Society, 320 Vassar, Berkeley, CA 94708. $12/yr. (free to ISES members). A nice blend of the literary and the technical—with solar articles from all over the world.

6) **The Solar Energy Intelligence Report.** Published bi-weekly by Business Publishers, Box 1067, Silver Springs, MD 20910. $75/yr; $40/6 mos. Features government activity; funding, granting, and research.

7) **Solar Outlook.** Published weekly by Observer Publishing Co., Canal Square, Washington D.C. 20007. ''Only weekly periodical of the solar industry,'' focuses on solar developments at the Federal level.

—Dean Anson

SOLAR EDUCATION

LEARNING ABOUT SOLAR ENERGY

A few years ago, it was difficult to find courses and workshops on solar energy. Now the problem is selecting the right one out of the many that are offered. To find out about courses in your area, we suggest the following:

- *Call or write the extension or continuing education offices of colleges and universities. Solar courses are most often categorized as engineering.*
- *Call people working in solar energy. They often are the ones who teach the courses given at universities, colleges, etc., or they know who is teaching them. Some areas have a solar energy listing in the Yellow Pages, so check there to get some names.*
- *Check the listings of "open education" organizations. In the San Francisco area, for instance, several solar courses are offered by that sort of group.*

Get enough information on a course's contents to assure yourself that it will match your needs. Some of the courses offered are extremely technical, others are introductory. There is also a wide range of prices for the courses.

SOLAR ENERGY

This is a fine book on all phases of solar energy, written on a junior high school level. The book starts with a discussion of what energy is and the implications of the earth's limited reserves. This is followed by a simple, but accurate, discussion of how the sun works, what sunlight is, and the radiation balance of the earth.

The rest of the book discusses solar collectors, solar heating of buildings, controlled photosynthesis as a source of food and fuel, solar furnaces, coolers, and stills, photovoltaic and thermoelectric generators, and power from the wind and ocean thermal gradients. Although this book is 17 years old, little could be added to it today. This book should be in every school library.
— Roger Douglass

Solar Energy
Franklyn M. Branley
1957; 117 pp
$3.95

from:
Thomas Y. Crowell Co.
666 Fifth Avenue
New York, NY 10019

or WHOLE EARTH TRUCK STORE

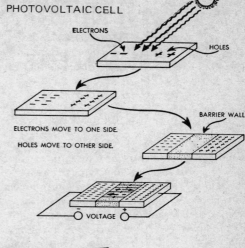

OPERATION OF PHOTOVOLTAIC CELL

ELECTRONS / HOLES
ELECTRONS MOVE TO ONE SIDE.
HOLES MOVE TO OTHER SIDE.
BARRIER WALL
VOLTAGE

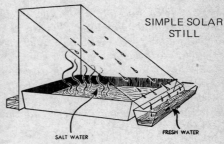

SIMPLE SOLAR STILL

SALT WATER / FRESH WATER

SOLAR ENERGY EXPERIMENTS

There are many books these days that *explain* the principles and applications of solar energy. However, there are very few books that outline techniques for *teaching* solar energy . . . especially at the high school and college level. As with any natural science course, the goal should be to take students patiently through natural processes familiar to them in their daily activities, rather than try to overload them with facts, data and models that may or may not have meaning in their lives. This book comes very close to that goal. As the authors note:

This is a . . . manual of 18 experiments and 8 class-room activities. The 18 experiments are of varying difficulty and cover the important aspects of solar energy utilization. Each experiment is self-contained with its own introduction and background information, making this book a suitable solar energy curriculum outline. . . Among the topics discussed are altitude and azimuth of the sun, radiation characteristics, energy collection with converging lenses, air and water solar collectors, and energy storage in gravel beds and in salt hydrates . . . Many experiments are directly applicable to existing physics, general science and environmental science curricula . . .

My only criticism of the book is that it needs more refined classroom activities concerning energy conservation (as prelude to solar energy utilization), and still others concerned with methods of calculating seasonal heating loads of (the students') houses. Otherwise very good.
— Richard Merrill

Solar Energy Experiments for High School and College Students
T.W. Norton, D.C. Hunter and R. J. Cheng
1977; 129 pp
$4.95

from:
Rodale Press
Emmaus, PA 18049

or WHOLE EARTH TRUCK STORE

SOLAR ENERGY EXPERIMENTS FOR SECONDARY SCHOOL STUDENTS

When did you last see a procedure for measuring the solar collection efficiency of a closed automobile? This book contains one, along with 17 other experiments and 8 classroom activities that cover just about every aspect of solar energy.

Typical experiments are "Energy Consumed by a Hot Bath" and "Measuring the Solar Radiation Spectrum." Each experiment write-up includes clear instructions, plus background and discussion information. Emphasis is placed on using simple equipment and a little physics to learn as much as possible about solar energy and household energy use.

The classroom activities involve more calculation study than the experiments, but they have the same practical flavor. For example, "Energy Consumption" has the students plot out their families' fuel use from daily gas and electric readings. The consciousness raising value of this exercise is immense.

This is an excellent book for anyone involved in solar and energy conservation education. Some of the experiments seem rather advanced for high school students, but they are all clearly presented and well thought out. The appendices contain handy reference data.
—Chip Barnaby

Solar Energy Experiments for Secondary School Students
T. W. Norton, D. C. Hunter, and R. J. Cheng
1975; 158 pp

from:
Atmospheric Sciences Research Center
State University of New York at Albany
ASRC - SUNY Pub. No. 353

SOLAR SCIENCE PROJECTS FOR A CLEANER ENVIRONMENT

The world could use more books like this one! The back cover promises "7 Solar Energy Projects You Can Build and Use" and that's exactly what the book consists of. The seven projects are: reflector cooker (parabolic mirror of cardboard and aluminum foil), still, furnace, oven, water heater, solar motor and solar radio (the last two operated by photovoltaics). Each chapter, except the first and last, consists of detailed how-to-do-it plans and directions for a solar energy device. Interspersed with the instructions are notes of history, large scale application, and general explanatory material, which have the effect of keeping the experimenter's interest up, and also providing something for the armchair enthusiast.

The level and style is about right for the Boy Scout/Girl Scout market, and in fact this book could be the basis of an excellent Merit Badge Pamphlet. However, it's a fun book for children of all ages. The book was published first in 1959, during a previous solar energy age under the title *Fun With the Sun*. However, it's not dated in the least. You can't lose with this one.
—Marshal Merriam

Solar Science Project for a Cleaner Environment
D. S. Halacy, Jr.
1974; 96 pp
$0.85

from:
Scholastic Book Services
900 Sylvan Ave.
Englewoods Cliffs, N.J. 07632

or WHOLE EARTH TRUCK STORE

FILMS & VIDEO

If one good picture is worth a thousand words, then a good movie, video tape or slide show is worth a zillion words. We've found some of these "visual aids" to be very valuable in introducing and reinforcing the concepts of renewable energy systems. Obviously there are many sources we haven't listed, and many new films are being made all the time. For example, almost any good television station has run a series on solar energy, and many local TV stations have no doubt done programs on local solar energy activities. Furthermore, any solar energy teacher or school program worth its salt will have assembled its own slide shows about solar energy. Remember, if you do rent a film, do so early, since demand is high, and waits can be long.

The following films, and video-tapes are but a sampling of the available audio-visual information on solar energy. To keep up-to-date, consult the following sources: (1) Index to 16 mm Educational Films, Nat'l. Information Center for Educational Media, University of Southern California, University Park, Los Angeles, CA 90007; (2) Educational Guide to Free Films, Educators Progress Service, Inc., Randolph, WI 53956; (3) Solar Age magazine, Solar Vision, Inc., 200 East Main St., Port Jervis, NY 12771; (4) Solar Engineering magazine, SEIA, 1001 Connecticut Ave., N.M., Washington D.C. 20036.

—C.M., R.M.

FILMS: 16mm

ACI Productions, 36 West 45th St., 11th Floor, New York, NY 10036.

SOLAR ENERGY—TO CAPTURE THE POWER OF SUN AND TIDE. 21 min., color, 1975. Predicts that fossil fuels will be consumed by the end of this century and that the creation of more sources of energy is in order.

ENERGY—TOWARD THE AGE OF ABUNDANCE. 20 min., color, 1975. Explains how solar energy can serve the world through an orbiting satellite of solar cells.

Almanac Films, Inc., 915 Broadway, New York, NY 10010.

SOLAR ENERGY. 23 min., B&W, 1956. Demonstrates industrial potential of solar energy, solar batteries and their uses.

Association-Sterling Films, 866 Third Ave., New York, NY 10022

RIVER TO THE SUN. 19 min., color, 1974. L.C. No. 74-702509. Shows development of power sources in area served by Minnesota Power and Light. Examines use of coal, nuclear and solar energy for electric power.

Association-Sterling Films, 6644 Sierra Lane, Dublin, CA 94566.

JAPAN AND THE SUN. 14 min., color. Focuses on life of Japanese in rural agricultural communities, fishing villages, and urban areas, finding a way to utilize solar energy effectively.

Benchmark Films, Inc., 145 Scarborough Rd., Briarcliff Manor, NY 10510.

THE NEW ALCHEMISTS. 29 min., color, 1975. LC no. 75-704279. Explains use of organic fertilizers, solar heat, and windmill for energy at experimental plant and fish farm near Falmouth, Massachusetts.

CCM Films, Inc., 866 Third Ave., New York, NY 10022.

THE SUN SERVES NIGER. 18 min., color, 1969. Use of solar batteries in this sub-Saharan country, an example of the potential in developing countries.

Counselor Films, Inc., 2100 Locust St., Philadelphia, PA 19103.

THE SEARCH FOR NEW ENERGY. 15 min., color. Covers breeder reactors, fusion, solar power systems, windmills and tidal power.

Doubleday Multimedia, 1371 Reynolds Ave., Santa Ana, CA 92705.

PUTTING THE SUN TO WORK. 4½ min., 1974, color. On producing electricity directly from solar energy.

Energy Research and Development Administration, Film Library, PO Box 62, Oak Ridge, TN 37830.

PUTTING THE SUN TO WORK. 5 min., color, 1974. Discusses developments in solar energy, and use of solar collectors to produce electricity and heat homes.

Environmental Action Reprint Service (EARS), 2239 East Colfax, Denver CO 80206.

THE AGE OF THE SUN. 21 min., color. Specific examples of solar technology—included are anaerobic fermentation, wind energy conservation, ocean thermal conversion, thermal electric generation, solar space satellites, solar heating and cooling and photovoltaics.

ENERGY: NEW SOURCES. 20 min., color. Surveys briefly: wind, tides, methane from trash, thermal gradients in the oceans, geothermal fusion and solar energy.

Great Plains Instructional TV Library, University of Nebraska, PO Box 80669, Lincoln, NB 68501.

THE DAY AFTER TOMORROW. 30 min., color, 1974. LC No. 75-704149. Discusses use of solar power, geothermal resources and thermonuclear fusion as energy sources.

Handel Film Corp., 8730 Sunset Blvd., West Hollywood, CA 90069.

HOW TO MAKE A SOLAR HEATER. 20 min., color. Shows step-by-step construction of solar heater from recycled materials as a classroom project.

Indiana University Audio-Visual Center, Bloomington, ID 47401.

ALTERNATIVE ENERGY SOURCES, 30 min., color. Examines coal conversion, geothermal, nuclear power, and experimental solar energy farms.

Montage Films, P.O. Box 38128, Hollywood, CA 90038.

SOLAR ENERGY: UNLIMITED POWER. 13½ min., color, 1977. Discusses how sun creates energy and weather, early solar experiments, solar water and space heating and air conditioning, solar architecture, high temperature solar conversion, ocean thermal, bio-conversion and photovoltaic.

Nasa Public Affairs Office, (place nearest you) Washington, D.C. 20409; Greenbelt, MD 20771; Hampton, VA 23365; Kennedy Space Center, FL 32899; Marshall Space Flight Center, AL 35812; 21000 Brookpark Road, Cleveland, OH 44135; Houston, TX 77058; Moffett Field, CA 94035.

GREAT IS THE HOUSE OF THE SUN. 21 min., color, order no. HQ 144. University of Hawaii researchers study solar radiation in space, while others prepare experiments for space satellites to study radiation.

ECLIPSE OF THE QUIET SUN. 27 min., color, order no. NAV 010. Solar radiation effects on Earth. Rare study of a total eclipse in Canada.

Office of Counsellor (Scientific), Embassy of Australia, 1601 Massachusetts Ave., Washington, D.C. 20036.

WATER FROM THE SUN. 15 min., 1967, color. Shows construction of solar still in South Australia built in 1966.

SOLAR WATER HEATING. 13 min., color, 1964. Excellent film for classroom or solar energy workshop.

DESIGN FOR CLIMATE. 21 min., color, 1967. Designing for control of natural "elements" including solar heat, sun and sky glare, wind, rain and noise.

R.H.R. Filmedia, Inc., 48 West 48th St., New York, NY 10036.

HERE COMES THE SUN. 15 min, color, 1974. Students, teachers, and communities find solar heating an important asset to help relieve energy shortage.

Stuart Finley, Inc., 3428 Manfield Rd., Falls Church, VA 22041

THE SOLAR GENERATION. 21 min., color, 1976. Ranges from Archimedes to today. Looks at solar cell research program, solar furnace, and prototype solar-thermal electric plant in Genoa, Italy.

University of Colorado, Educational Media Center, Boulder CO 80309.

SOLAR POWER—THE GIVER OF LIFE. 29 min., color. Examines solar stills, cookers and collectors and the impact of commercial solar installations on the environment.

Xerox Films, 245 Long Hill Rd., Middletown, CT 06457.

POWER WITHOUT END. 16 min., color, LC No. 76-700254. Examines possible uses and limitations of power from the sun, wind, geothermal energy, tides and flowing water.

VIDEO-TAPES

Instructional Services Department, Wayne State University, 2978 West Grand Blvd., Detroit, MI 48202.

ENERGY, TECHNOLOGY AND SOCIETY. Series of 55 half-hour programs available for credit—write for information. Five courses specifically dealing with solar energy.

Solar Engineering Magazine, 8435 N. Stemmons Freeway, Suite 880, Dallas, TX 75247

Cassette No. 1. Pres. of Solar Energy Industries Association discusses potential market for solar heating and air conditioning (15 min.)

Cassette No. 2. HUD executive is interviewed on Solar Demonstration Program of the Dept. of Housing and Urban Development (15 min.)

Cassette No. 3. Visual demonstration of the proper operation of controls of a solar system. (9 min.)

Solarvision, Inc., Box A, Hurley, NY 12443.

SOLAR REALITY — A DIALOGUE WITH JOHN YELLOTT. The solar pioneer talks about the past, present and future of solar energy.

ARCHITECTURE, BUILDING, AND THE SUN— THE KAREN TERRY HOUSE. An account of the conception, construction, and performance of one of the most sophisticated "passive" projects thus far.

Time-Life Films, Time and Life Bldg., Rockefeller Center, New York, NY 10020.

SUNBEAM SOLUTIONS. 38 min., color. BBC-TV documentary on solar and geothermal power, electric and hydrogen-powered autos, burning garbage for electricity, extracting methane from sewage.

CONFERENCE PROCEEDINGS

VARIOUS SOLAR ENERGY CONFERENCE PROCEEDINGS

There are many technical reports on solar energy that have been compiled from various meetings and conferences. To some degree these represent the state-of-the-art progress in solar work, while on an individual basis, they may exhibit a good deal of variation in quality. The availability and cost of these publications is frequently a deterrent to many users, so the best recommendation is to check them out from an engineering library.

1. **Proceedings of the United Nations Conference on New Sources of Energy**
United Nations, 1964 $16.
Volumes 4, 5, and 6 are on solar energy.

2. **Technical Program and Abstracts. International Solar Energy Society**
Proceedings of conference held in Fort Collins, Colorado, 1974.

3. **Solar Use Now — A Resource for People**
International Solar Energy Society. Extended abstracts from conference held in Los Angeles, California, 1975.
CONTENTS:
series 10—economic and social aspects; developing countries; general papers.
series 20—solar radiation; photovoltaic, photo-chemical, photobiological processes; solar furnaces.
series 30—materials, flat plate collectors, energy storage.
series 40—solar heating and cooling of buildings; drying and distillation.
series 50—focusing collectors; solar thermal power.

4. **Sharing the Sun: Solar Technology in the Seventies**
American Section of the International Solar Energy Society and Solar Energy Society of Canada. Proceedings of joint conference held in Winnipeg, Canada, 1976. 10 volumes.
CONTENTS:
volume 1—international agencies; intergovernmental agencies; ERDA program; solar flux.
volume 2—flat plate collectors; focusing collectors.
volume 3—cooling methods; heating and cooling; heat pumps.
volume 4—passive systems; retrofit systems; simulations studies; design methods; general systems.
volume 5—general aspects; low temperature thermal energy systems and applications; intermediate temperature thermal energy systems; high temperature thermal energy systems; ocean thermal.
volume 6—photovoltaics; materials.
volume 7—agricultural and industrial process applications; bioconversion; wind; new developments.
volume 8—solar storage; chemical storage; solar heat for buildings; education and training; data dissemination and communications; solar water heaters.
volume 9—socio-economic aspects; socio-cultural implications.
volume 10—business and commercial implications; posters; miscellaneous.

$10/vol; $80 for all 10
from:
The American Section of the International Solar Energy Society
300 State Road 401
Cape Canaveral, Florida 32920

5. **Solar Cooling and Heating: A National Forum**
Energy Research and Development Administration and the School of Continuing Studies, University of Miami. Proceedings of condensed papers presented in Miami Beach, Florida, 1976.
CONTENTS:
session 1A—architectural considerations
session 1B—energy storage
session 2A—flat plate collectors
session 2B—solar buildings
session 3A—heating systems
session 3B—concentrating collectors
session 4A—cooling systems
session 4B—system simulation and control
session 5A—heating and cooling systems
session 5B—economics, incentives and legal aspects
$45.00 paperback
from:
Tony Pajares
School of Continuing Studies
Conference Services
University of Miami
P.O. Box 248005
Coral Gables, Florida 33124

6. **Workshop on Passive Systems. Energy Research and Development Administration. Proceedings of Conference held in Albuquerque, New Mexico, 1976.** (see review on pg. 69.)

7. **Proceedings of the Solar Energy — Fuel and Food Workshop. The Utilization of Solar Energy in Greenhouses and Integrated Greenhouse—Residential Systems.** April 5-6, 1976. Tucson, AZ.
Most of the papers given are concerned with solar heating commercial greenhouses. Some, however, deal with residence-greenhouse arrangements.

— Dean Anson

OLDIES BUT GOODIES

Solar Heating and Cooling for Buildings Workshop
Redfield Allen, Editor, 1973; 226 pp; Order No. 223-536 $3.00
from: Nat. Technical Information Service, U.S. Dept. of Commerce, P.O. B. 1553, Springfield, VA 22151

Wind and Solar Energy: Proceedings of the New Dehli Symposium
1956; out of print

Solar and Aeolian Energy: Proceedings of the International Seminar at Sounion, Greece, 1961
ed by A. G. Spanides and A.D. Hatzikakidis
1964; 491 pp; out of print

Transactions of the Conference on the Use of Solar Energy: The Scientific Basis
1955; out of print

Solar Energy Research
ed. by F. Daniels and J. A. Duffie
1955; 290 pp; out of print

Proceedings of the World Symposium on Applied Solar Energy
1956; 304 pp; $15
from: Johnson Reprint Corporation, 111 Fifth Ave., New York, NY 10003

SOLAR DIRECTORIES

U.S. Federal Energy Administration, 1976. BUYING SOLAR, by Joe Dawson. Washington, D.C.: FEA, 71 pp. Free. Operation of solar systems, comparison of air and water systems, economics, and discussion of performance guarantees. No discussion of passive systems.

Southern California Solar Energy Association. 1976. WESTERN REGIONAL SOLAR ENERGY DIRECTORY. San Diego: San Diego Urban Observatory, 54 pp. $2.00. (Order from: SCSEA Directory, 202 "C" Street, 11B, San Diego, CA 92101). A listing of 500 companies and individuals involved in the solar energy industry in California and other western states. Includes architects, engineers, contractors, builders, solar consultants, manufacturers and distributors.

Solar Energy Industries Association, Inc., 1975. SOLAR ENERGY INDUSTRY DIRECTORY AND BUYER'S GUIDE. Washington, D.C. SEIA, Inc., $2.00. (Order from SEIA, Inc., 1001 Connecticut Avenue, N.W., Washington, D.C.)

Pesko, Carolyn. 1976. SOLAR DIRECTORY. Ann Arbor: Ann Arbor Science Publishers, Inc., $20.00.

Shurcliff, William A. 1975. INFORMATION DIRECTORY OF ORGANIZATIONS AND PEOPLE INVOLVED IN THE SOLAR HEATING OF BUILDINGS. Cambridge, MA: Shurcliff, $5.00 - $7.00.

DIRECTORY OF THE SOLAR INDUSTRY. Hampton, NH: 1976. $7.50 (Available from Solar Data, 13 Evergreen Road, Hampton, NH 03842.)

U.S. Congress. Committee on Science and Technology. 1975. SURVEY OF SOLAR ENERGY PRODUCTS AND SERVICES. Washington, D.C.: Library of Congress, Science Policy Research Division, $4.60.

SOLAR INDUSTRY INDEX

The Solar Energy Industries Assocation (SEIA) is a group that represents commercial solar energy organizations. The SOLAR INDUSTRY INDEX is a listing of all SEIA members with a geographical breakdown by state, plus a product and services index of member and non-member firms who paid to be listed. This index is broken down into a number of categories, including solar heating systems, passive systems, contracting, consultants, photovoltaics, wind power, bioconversion and power generation. An extensive cross reference system is provided so that readers can find companies which offer any of the above products or services in any state. There is also an outline of basic solar energy systems and solar system economics.

The index is a basic tool for locating products and services, particularly for people in the solar business. Where else can you find the aluminum extrusion manufacturers on the West Coast who make collector panel components? The 1977 first edition seems to list only a fraction of the firms active in the field, and they are primarily large corporations. As additional companies find this a good means of advertising, the index will grow and become even more useful.

—Bruce A. Wilcox

Solar Industry Index
1977 (Annual); 400 pp.

from:
Solar Energy Industry Association
1001 Connecticut Avenue N.W. Suite 632
Washington, D.C. 20036

SOLAR HARDWARE

It is important to point out a few things about the hardware listings on the following pages.

• Almost all of the equipment listed is oriented toward active systems. This is not because we think active systems are better than passive. On the contrary, we encourage people to use passive systems whenever feasible. It's just that passive systems don't require as much specialized hardware.

• It is virtually impossible to present accurate cost information about solar systems and components because the prices are so variable. There are at least three basic reasons for this: (1) The effects of the current economic inflation, (2) the fact that many solar companies, still in the experimental stages of development, are trying to recoup their initial research costs, and (3) the fact that the average consumer is basically unaware of the tradeoffs in design and materials of solar hardware. This allows solar companies to sell similar products at different prices with few of the usual competitive checks and balances present in a more stable marketplace. As consumers learn more about solar energy systems and about the types and designs of solar devices available, the solar industry will have to respond with fair prices and even better quality products.

The tables in this section are intended to give you an idea of what sort of solar hardware is currently available (as of July 1977) and where you might go to get it. Note that many solar companies are beginning to work through established or growing distribution networks. In your inquiries to solar companies ask for the names of possible local dealers (e.g., plumbing supply houses, hardware stores, etc.). The primary purpose of the solar hardware listings, however, is to give you a rough idea of the types of components and materials currently being used in the manufacturing of solar products together with their relative prices. We have tried to include a representative sample of manufacturers; this is definitely not an exhaustive list. For more complete and up-to-date lists of companies dealing in solar products see reviews of SOLAR DIRECTORIES (pg. 69).

In the tables, quotation marks indicate that we've quoted right out of the manufacturer's literature. Blank spaces are left where information was unavailable.

GLAZING

Glass is probably the most reliable glazing material available. It maintains its properties essentially forever. Ordinary window glass transmits about 90% of the visible solar radiation incident upon it, and low-iron, clear float or water-white glass transmits even more. Glass also reduces radiative heat loss from the absorber panel by blocking infrared radiation.

Other materials are often used in place of glass because of lower weight, greater resistance to impact, lower price, and greater ease of installation. Fibreglass reinforced polyester (fibreglass) degrades after prolonged exposure to sunlight, but can be coated with DuPont Tedlar which protects it from such degradation. Tedlar-coated fibreglass has been used successfully in greenhouse applications for years. Fibreglass degrades at high temperatures (over 200°F or so, depending on the specific material) and therefore should not be used as the inner glazing on a double-glazed collector.

Tedlar can be used alone as glazing but is more subject to wind damage than the rigid glazing materials. As an inner glazing on double-glazed collectors it can be very effective and its light weight is certainly an advantage.

Polycarbonate plastics should not be used on collectors, but they are suitable for use on greenhouses. It is good to bear in mind that all plastics are derived from organic materials and in time will decay, especially in direct sunlight.

Lifetime estimates on glazing materials for solar collectors are not particularly reliable, due to the fact that these materials have only been used in such applications for a few years. So far they seem to be doing an acceptable job.

Cheap sources of ordinary glass are used windows from torn-down buildings, and window and sliding glass door seconds. If you scrounge around a bit you can save money.

Company	Product Name	Materials	Transmissivity	Max. Temp.	Weight per Ft^2	Estimated Lifetime	Cost per Ft^2	Comments
Window Glass Suppliers	glass	glass; low-room tempered water-white, float	86-91% 78% for ¼"	400°F and up	about 1 lb.	forever		the basic glazing material.
ASG Industries Kingsport, TN	Sunadex™ Lustraglass® Starlux®	tempered water white glass low iron sheet glass clear float glass	86-91%	400°F and up	about 1 lb.	forever		
3M Company 3M Center, Bldg. 236-1 St. Paul, MN 55101	"Flexigard"	Polyester Films	85-90%	300°F	.5 oz.	over 10 yrs.	furnished on request to Mfgr.	3M makes a variety of films and tapes for Solar Glazing.
E.I. DuPont Co. Film Dept. Wilmington, DE 19898	Tedlar® Teflon FEP	polyvinylfluoride film flourocarbon film	93-98%	225°F, up to 400°F	0.11 oz.	more than 15 yrs.	$0.02- $0.16 $0.17	high tensile strength, but fatigues in wind; excellent as inner glazing
Filon Div. Vistron Corp. 12333 S. Van Ness Ave. Hawthorne, CA 90250	Filon®	Tedlar®-coated fibreglass reinforced polyester	75-90%	250°F	4-8 oz.	5-20 yrs.	$0.60- $1.70, depending on quantity	
Glasteel 1516 Santa Anita Ave. So. El Monte, CA 91733	z-lok Kristal Clear	Tedlar®-coated fibreglass; reinforced polyester	85-90%		4-5 oz.	10-20 yrs.	$0.43	
Kalwall Corp. 88 Pine St. Manchester, NH 03103	Sun-lite®	fibreglass reinforced polyester	75% double 85-90% single	225°F	2.8-4.7 oz.	7 yrs. (Premium II) 20 yrs.	$0.50	
Lasco Industries 1561 Chapin Rd. Montebello, CA	Lascolite-T®	Tedlar®-coated fibreglass; reinforced polyester	85-95%		4-5 oz.			
Rohm and Haas Port Plastics Inc. 1047 North Fair Oaks Ave. Sunnyvale, CA 94086	Tuffak-Twinwall	twin-walled, hollow channeled polycarbonate sheet.	80% in visible	140°F	4 oz.			plastic doubleglazing for low-temperatures (e.g. greenhouses)

ABSORBER PLATES

Listed below are several companies which sell bare absorber plates from which you can fabricate your own complete collectors. Tranter, Olin, Ilse, and Kennecott sell only the absorber plate. The others market pre-fabricated collectors as well, and their addresses and absorber plate characteristics can be found in the Flat-Plate Collectors chart. It is likely that most flat-plate collector companies would be willing to sell their absorber component by itself. Write and ask . . . otherwise build it yourself.

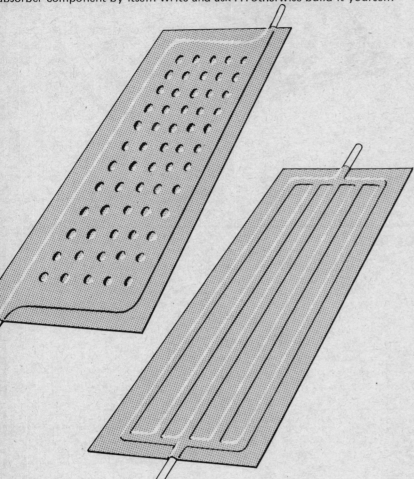

SOME PROPERTIES OF ABSORBER METAL MATERIALS

	Copper	Steel	Aluminum
Thermal Conductivity (Btu-in/hr-sq.ft.-°F)	220	26	120
Specific Gravity (lb/in³)	0.32	0.28	0.10
Tap Water Compatible	excellent	fair	poor
Corrosion Resistance[a]	excellent	good	fair
Energy Cost (Btu/lb)	47,400	11,600	116,690
Plate Thickness for Equal Performance (in)	0.016	0.064	0.032
Cost for Equal Performance ($/ft²) (6" tube spacing for CU and AL: 4" for steel)	2.88	0.90	0.58
Energy Cost for Equal Performance (Btu/ft²)	47,000	45,200	70,000
Weight for Equal Performance (lb/ft²)	1	3.9	0.6
Relative Costs for Equal Strength	35	1	5.6
Coatings (absorptivity/emissivity)	Flat Black .95/.85 Naturally oxidized .82/.58 Solarsorb® .94/.56 'Copper Black' .89/.17 Ebanol .90/.16 Black Chrome .95/.10 Polished Copper .35/.039	Solarsorb® .94/.56 many others	Black Paint .96/.88 Solarsorb® .94/.56 PbS crystals .89/.20 Polished AL .10/.12 CuO coating .85/.11

[a] If properly protected by the use of inhibitors (in closed systems), general corrosion is usually insignificant and the metals should last 100 years or more. Corrosion increases with heavy metals, acids and high water velocities.

COMPANY	ABSORBER	SIZE (INCHES)	APPROX. COST	COMMENTS
Berry Solar Products P.O. Box 327 Edison, NJ 08817	selective black chrome on thin copper sheet		$0.90 - $1.80 per sq. ft.	"Solar Foil" (.0014" & .0028" thick) "Solar Strip" (.005" & .013" thick)
Energy Systems Inc. 634 Crest Dr. El Cajon, CA 92021	aluminum fin (with non-selective coating) press fitted around parallel copper tubes	36 x 78 (¾" header) . . . 48 x 72 (1½" header) . . . 48 x 120 (1½" header) . . .	$5.90/ft² $7.00/ft² $5.50/ft²	
Ilse Engineering 7177 Arrowhead Rd. Duluth, MN 55811	steel sandwich panels	48 x 144 max.	$3.25 per sq. ft.	maximum operating pressure, 25 psig.
Kennecott Copper Corp. 128 Spring St. Lexington, MA 02173	all copper (square) tube in (thin) plate; flat black coating	22 x 96; 34 x 76 and 34 x 94	$40; $60	maximum operating pressure, 25 psig.
Olin Brass East Alton, IL 62024	aluminum or copper "Roll-Bond"	copper 34 x 96 max; aluminum 36 x 110 max		
Tranter 735 E. Hazel St. Lansing, MI 48909	all carbon-steel, spot-welded, quilted internal tubes; stainless steel available	34 x 92 standard, others by special order	$2.60 - $3.10 per sq. ft.	used with closed systems; 90% internal wetting
Alten Corp. **Raypak Inc.** **Reynolds Aluminum** **Revere** **Solar Development Inc. (SDI)** **Sunburst Solar Energy, Inc.** **Sunworks** many others	See Flat-Plate Collectors Low and Mid-Temperature			

━━━━ LOW TEMPERATURE FLAT-PLATE COLLECTORS ━━━━

The collectors in this chart are all suitable for low temperature applications: swimming pools, aquaculture systems and other situations requiring large volumes of water at a relatively low temperature (70°F - 100°F). With improvements in efficiency, most of the collectors can also be used for water heating and space heating (fan-coil). Likewise, many of the mid-temperature collectors listed in the next chart can also be used for low temperature applications. In fact, many companies sell their mid-temperature collectors with large headers or without glazing and insulation for this purpose.

Plastic collectors should not be glazed because of the possibility of high stagnation temperatures melting them. If heating is only required during the warmer months, the low cost of plastic or rubber collectors often makes them desirable, although there is some debate as to the durability of plastic collectors over long periods of time. If large volumes of warm water are needed all year, glazed metal collectors are necessary to reduce winter heat losses in most locations.

The heating requirements of swimming pools and aquaculture ponds can be greatly reduced if pool covers are used. There are several companies selling such covers, and effective ones can be made from readily available materials. If the heating load is reduced, the collector area (and thus the cost of the system) will be reduced accordingly. An extra advantage of pool covers is a reduction in the loss of water (from evaporation).

━━━━ PLASTIC OR RUBBER ━━━━

COMPANY	DESCRIPTION	SIZE (IN.)	COST	COMMENTS
ALCOA H.C. Products Co. Box 68 Princeton, IL 61559	ABS plastic tube-on-fin with protective coating	47 x 47		modular size adapts to easy installation
AquaSolar Inc. 1234 Zucchini Ave. Sarasota, FL 33577	ABS plastic pipe in serpentine grid. internal fins in pipe.	variable	$2.60 per sq. ft.	with remote control sensors and other accessories
Burke Rubber Co. 2250 S. Tenth St. San Jose, CA 95112	"bag" trickle collector of DuPont Hypalon® synthetic rubber or EPDM	96 x 96 96 x 144 96 x 192	$2.50 - 3.50 per sq. ft.	Hypalon®, certified potable for higher temps (e.g., batch collectors); May need separate return line to pool
Fafco 1050 B. East Duane Sunnyvale, CA 94086	¼" parallel square tubes in polyolefin plastic panel	48 x 120	$3.60 per sq. ft.	lightweight, unglazed, uninsulated
Sunburst Box 2799 Menlo Park, CA 94025	400 ft. of coiled UV resistant plastic pipe	.84" I.D. 32 ft² coil	$2.50 - $3.00 per sq. ft.	unglazed uninsulated
Sundu Co. 3319 Keys Lane Anaheim, CA 92804	5/16" ABS rectangular tubes, parallel in plastic panel	18 x 34 48 x 96 48 x 120	$1.85 per sq. ft.	"units in operation for 10 years show minimum deterioration."
	━━━━ METAL ━━━━			
Alten Associates 2594 Leghorn St. Mountain View, CA 94043	#120 U — extruded AL plate around serpentine CU tube #110F — with fibreglass glazing and insulation	48 x 10 ft. 48 x 10 ft. 48 x 20 ft.	$5.00/ft² $10.00/ft² $9.50/ft²	unglazed uninsulated
Falbel Energy Systems 472 Westover Rd. Stamford, CT	See Flat Plate Collectors Mid. Temperature			
Raypak 31111 Agoura Rd. West Lake Village, CA 91359	See Mid-Temperature Collectors	44 x 96 44 x 120	$5.75/ft² $5.00/ft²	with controls unglazed, uninsulated
Solar Applications 7926 Convoy Ct. San Diego, CA 92111	See Mid-Temperature Collectors	34 x 76		1¼" manifold collector suitable for low temp. applications
Solar Energy Research 1228 5th St. Denver, CO 80202	"Thermo-Spray™" trickle collector on CU absorber plate; sunlite glazing	24 x 72 to 24 x 192	$10 per sq. ft.	very efficient trickle collector with other uses
R-M Products **Solar Development, Inc.**	see Flat Plate Collectors Mid. Temperature			advertised for use as pool heaters.

MID-TEMPERATURE FLAT-PLATE COLLECTORS

MID TEMPERATURE COLLECTORS operate best in the 110-180°F range and are suitable for heating water, heating space with air or water, for drying foods, for operating solar-assisted heat pumps, etc. More efficient mid-temperature collectors (those with tube-in-plate and/or selective absorbers, proportional controls, multiple glazing, etc.) generate higher temperatures suitable for absorption cooling, steam, etc.

The following is an explanation of those columns which are not self-explanatory:

• Type. Collectors are listed in sub-tables according to whether they use liquid or air as a heat transfer. A few companies sell both types.

• Glazing. "No." is the number of glazings; "Type" is what they are made of. Refer to the GLAZINGS TABLE for details.

• Absorber. "Material:" Aluminum or steel absorbers have potential corrosion problems. They should only be used with closed systems. Care should be taken to avoid electrolytic corrosion between dissimilar metals in all systems. Many of the absorbers are both copper and aluminum. This seems to be perfectly acceptable so long as water only comes in contact with the copper. For a detailed and up-to-date discussion of the tradeoffs or different absorber plate materials (copper, aluminum, steel) see SOLAR AGE, Vol. 2, May 1977. Also see the ABSORBER METALS TABLE on page 71. "Type:" Tube-in-plate absorbers are able to withstand much higher pressures than Roll-Bond® or spot-welded absorbers. "Coating:" Selective surfaces (e.g., copper oxide, black chrome) absorb most of the sunlight which hits them and reradiate only a small portion of the infrared radiation which a non-selective absorber would emit at the same temperature. They are more efficient than non-selective surfaces (see Fig. 111. 3 pg. 49), though the difference in efficiency may not always warrant the added cost.

• Insulation. See pg. 16 for heat retaining values of various insulation materials.

A WORD ON COLLECTOR PERFORMANCE:
Energy collected by solar collectors is a function of many parameters, including environmental conditions, and hardware   characteristics. Important environmental conditions are:

• Amount of solar energy incident on collectors

• Outside air temperature

• Wind velocity

Important hardware characteristics are:

• Number of glazings

• Absorber surface properties (absorbtivity, selective vs. non-selective surface)

• Plate construction

• Amount of edge and back insulation

The most important system characteristic is the temperature of the fluid entering the collector.

The amount of solar energy incident on the collector strongly affects the collector efficiency (collector efficiency is defined as the ratio of energy collected to energy incident on collector surface, times 100%). The larger the amount of solar energy, the higher the efficiency, other things being equal.

The collector looses heat to the outside air. The larger the temperature difference between collector absorber and outside air, the more the heat loss and therefore the lower the efficiency. To a lesser extent, wind velocity over the glazing affects efficiency. The higher the wind velocity, the greater the heat loss.

Appendix III of John Yellott's article explains the effect that the number of glazings (selective and non-selective) has on collector efficiency.

Plate construction is another important factor. If there is not a good bond between the tubes and the plate, the heat transfer from the plate to the fluid will be impaired. Similarly if the tubes are placed too far apart heat transfer becomes less efficient. However, if they are placed too closely, the slight increase in efficiency will usually not be worth the additional cost. Also important is the fluid flow within the collector. Hot spots are areas where flow is limited and where heat is lost. Ideally the whole plate should be at the same temperature for maximum efficiency. Roll-Bond® collectors are good in terms of uniform plate temperatures, though they have other disadvantages such as low maximum pressures. Parallel tube-in-plate collectors tend to be better than serpentine, but you pay for the extra connections.

Edge and back insulation serve the obvious purpose of keeping the heat from escaping the collector.

Finally, collector efficiency depends on the temperature of the fluid (or gas) entering the collector. The hotter the transfer medium, the more heat it loses to the outside. Given the same amount of incoming energy, a collector running at a high temperature will realize a lower net energy gain than it would at a lower temperature.

This table only lists a small proportion of the over 80 "Collector Sub-System Manufacturers" listed in the SOLAR INDUSTRY INDEX. (see pg. 69) There are probably more than 120 companies presently marketing mid-temperature solar collectors in the U.S. (July 1977).

As usual we encourage people to learn and build their own collectors whenever possible. The books listed in the SOLAR BOOK SECTION will serve as guideposts.

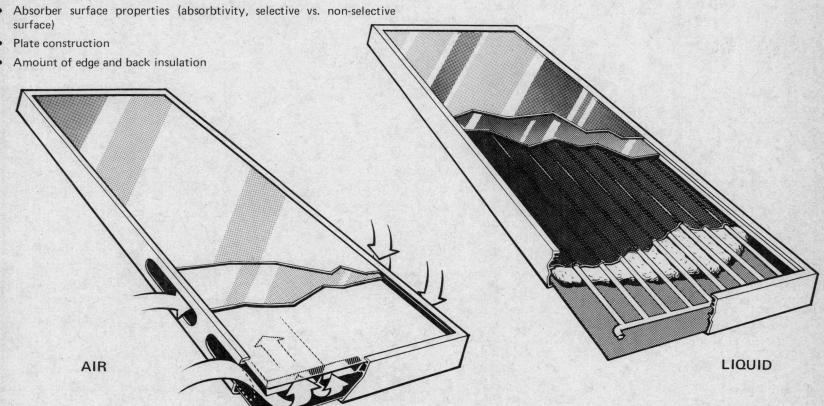

AIR

LIQUID

MID-TEMPERATURE COLLECTORS

COMPANY	GLAZING		ABSORBER			INSULATION	BOX	DIMEN. (INCH)	APPROX. COST
	No.	Type	Material	Type	Coating				

LIQUID

COMPANY	No.	Type	Material	Type	Coating	INSULATION	BOX	DIMEN. (INCH)	APPROX. COST
Alten Corporation 2594 Leghorn St. Mountain View, CA 94043	1 or 2	glass	CU tube; AL plate	serpentine tube in finned plate; silicone bond	non-selective	fibreglass	aluminum	47 x 93 49 x 243	$13 - $15 per sq. ft.
American Heliothermal Corp. 3515 S. Tamarac Denver, CO 80237	1	water white glass	all steel	parallel steel, pipes compression bonded between plates	selective	"fibrous mineral wool"	steel	38 x 73 42 x 80	
AMETEK Power Systems Group One Spring Ave. Hatfield, PA	2	low-iron glass or plastic	all CU	Roll-Bond®	selective	fibreglass	Cor-Ten steel	26 x 100	
Bay Area Solar Collector 3068 Scott Blvd. Santa Clara, CA 95050	1 or 2	Tedlar-coated fibreglass	all CU or all AL	Roll-Bond®	non-selective	Technifoam (iso-cyanurate)	aluminum	22 x 96 34 x 96	
Berry Solar Products P.O. Box 327 Edison, NJ 08817	1	low-iron glass; Lexan or fibreglass	all CU	parallel tube-in-plate; solder and epoxy bond	selective	fibreglass and polyurethane	aluminum	28 x 98	
Chamberlain Mfg. Co. 845 Larch Ave. Elmhurst, IL 60126	1 or 2	low-iron glass	steel CU	seam-weld, pressure expanded tube-on-plate	selective	polyurethane	galvanized steel	36 x 72 36 x 84 36 x 96	
Columbia Solar Energy Div. 55 High Street Holbrook, MA 02343	1 or 2	fibreglass	all CU (Kennecott absorber plate)		selective, non-selective	fibreglass	fibreglass	37 x 101	$13 - $15 per sq. ft.
Daystar Corp. 90 Cambridge St. Burlington, MA 01803	1	low-iron glass	all CU	serpentine tube-in-plate	selective: "folded polymer structure"	urethane	aluminum	44 x 72	$11 - $13 per sq. ft.
Energy Systems, Inc. 634 Crest Dr. El Cajon, CA 92021	1 or 2	glass	CU tube AL plate	parallel tube-in-plate; mechanical bond	non-selective	polyurethane		48 x 72 48 x 120	$11 - $13 per sq. ft.
Grumman Energy Systems Sunstream 4175 Veteran's Memorial Highway Ronkonkoma, NY 11779	1 2	acrylic glass	all AL, or CU tube, AL plate	Roll-Bond®, or parallel tube-in-plate, mechanical bond			enameled aluminum	36 x 108	
Halstead and Mitchell Div. of Halstead Industries Scottsboro, AL 35768	2	low-iron glass	CU tube, AL fin	finned coil	non-selective	Foamglas	aluminum	35 x 77	
Lennox Industries, Inc. P.O. Box 250 Marshalltown, IA 50158	1 or 2	low-iron glass	CU tube, steel plate	parallel tube-in-plate, solder bond	selective	fibreglass board	"galv-alume"M steel	36 x 72	$15/ft²
Lof Solar Energy Systems Libbey-Owens Ford Co. 1701 E. Broadway Toledo, OH 43605	2	glass	all CU	parallel tube-in-plate, solder bond	non-selective	fibreglass	aluminum	36 x 84	
PPG Industries, Inc. Solar Systems Sales One Gateway Center Pittsburg, PA 15222	1 or 2	glass	all CU or all AL	tube-in-plate, solder bond (CU) or Roll-Bond®(AL)	non-selective	fibreglass	galvanized steel	34 x 76	$11/ft²
Raypak, Inc. 31111 Agoura Rd. Westlake Village, CA 91361	1 or 2	low-iron glass	CU tube AL plate	parallel tube-in plate, mechanical bond	non-selective	fibreglass	galvanized sheet metal	37 x 82	$16 - $19 per sq. ft.
Revere Copper & Brass, Inc. Solar Energy Dept. P.O. Box 151 Rome, NY 13440	1 or 2	glass	all CU	integral "tube in strip"	non-selective		aluminum	36 x 78	$11 - $13 per sq. ft.
Reynolds Metals Co. Box 27003 Richmond, VA 23261	2	Tedlar®	all AL	integral tube in fin (extruded)	non-selective	"closed cell foam material"	aluminum	48 x 96	
Solar Applications, Inc. 7926 Convoy Ct. San Diego, CA 92111	1 or 2	glass	all CU	parallel, tube-on-plate, solder bond	non-selective	fibreglass	plastic	34 x 76	$14 - $15 per sq. ft.

COMPANY	GLAZING		ABSORBER			INSULATION	BOX	DIMEN. (INCH)	APPROX. COST
	No.	Type	Material	Type	Coating				
Solar Energy Products, Inc. 1208 N.W. 8th Ave. Gainesville, FL 32601	1	glass	CU tube, AL plate	parallel tube-in-plate, mechanical bond	non-selective	iso-cyanurate	aluminum	38 x 98 48 x 98	$9 - $10 per sq. ft.
Solar Engineering, Inc. Box 1358 Boca Raton, FL 33432	1	plexiglass	all CU	serpentine tube-on-plate, mech. bonded	selective	"rigid foam"	fibreglass	50 x 99	
Solar Innovations 412 Longfellow Blvd. Lakeland, FL 33801	1	Kalwall Sun-Lite®	all CU	Roll-Bond®	non-selective	polyurethane	aluminum	23 x 96	$13/ft²
Solar King 281 Gould St. Reno, NV 89502	2	water-white glass	all CU all CU			Technifoam (iso-cyanurate)	redwood and plywood	48 x 96	$10/ft²
Sunburst Solar Energy, Inc. P.O. Box 2799 Menlo Park, CA 94025	1	Tedlar-coated fibreglass	CU tube, AL plate	parallel tube-in-plate, mechanical bond	non-selective	"rigid foam"	aluminum	48 x 84 48 x 124	
Universal Solar Energy Co. (USECO) 1802 Madrid Ave. Lake Worth, FL 33461	1 or 2	glass	all CU	serpentine tube-in-plate		polyurethane	aluminum with enamel	48 x 96 or 48 x 120	$12 - $15 per sq. ft.
Western Energy, Inc. 454 Forest Ave. Palo Alto, CA 94302	1	glass	all CU	serpentine tube-in-plate; solder bond	non-selective; selective optional	polyurethane	aluminum	35 x 93	$8.40 /ft²

AIR

COMPANY	GLAZING		ABSORBER			INSULATION	BOX	DIMEN. (INCH)	APPROX. COST
Div-Sol, Inc. Box 614 Marlboro, MA 01752	1 or 2	Tedlar®	all AL	corrugated sheet	selective	buyer supplied	buyer supplied		$5 - $6 per sq. ft.
Hoffman Products, Inc. Box 975 Willmar, MN 56201	2	low-iron glass	steel	1 sheet	non-selective	fibreglass	steel	36 x 78	$15.25 /ft²
Kalwall Corp. Solar Components Div. 88 Pine St. Manchester, NH 03103	1	Sunlite®	all AL	flat or corrugated	non-selective	buyer supplied	aluminum		$5/ft²
Solaron Corp. 720 S. Colorado Blvd. Denver, CO 80222	2	low-iron glass	all steel	steel sheet with air channels	non-selective ceramic	fibreglass	painted steel	36 x 72	$17/ft²

LIQUID AND AIR

COMPANY	TYPE	GLAZING		ABSORBER			INSULATION	BOX	DIMEN. (INCH)	APPROX. COST
		No.	Type	Material	Type	Coating				
Northrup Inc. 302 Nichols Dr. Hutchins, TX 75141		1 1	Tedlar® glass	CU tube; AL fin	parallel tube-in-plate mechanical bond	non-selective non-selective	fibreglass fibreglass	steel steel	35 x 101 35 x 101	$8/ft² $9.75/ft²
	LIQ.	1 2	glass glass	all CU all CU	Roll Bond® Roll Bond®	selective selective	fibreglass fibreglass	steel steel	35 x 101 35 x 101	$15.50/ft² $16.40/ft²
	AIR	2	Tedlar®	steel	sheet steel	non-selective	fibreglass	steel	33 x 96	$11/ft²
R-M Products 5010 Cook St. Denver, CO 80216	LIQ.	1	low-iron glass	all CU	parallel tube-in-plate, solder bond	selective	"foam and fibreglass"	AL and steel	24" x 48" to 25'	$15/ft²
	AIR	1	low-iron glass	all CU	8" plenum chambers with duct header	selective	"foam and fibreglass"	AL and steel	24" x 48" to 25'	$15/ft²
Solar Development 4180 Westroads Dr. West Palm Beach, FL 33407	LIQ.	1 or 2	Kalwall Sunlite®	all CU	parallel or serpentine tube-in-plate, solder bond	non-selective	Technifoam (Iso-cyanurate)	aluminum	48 x 120	$9 - $10 per sq. ft. $5.60/ft²
	AIR	1 or 2	Kalwall Sunlite®		"vitreous screening with baffle"	non-selective	Technifoam (Iso-cyanurate)	aluminum	48 x 96	$5.60/ft²
Sunworks Box 1004 New Haven, CT 06508	LIQ.	1	low-iron glass	all CU	parallel tube-in-plate, solder bond	selective	Technifoam (Iso-cyanurate)	aluminum	36 x 84	$10/ft²
	AIR	1	low-iron glass	all CU	copper sheet with air chamber, mech. bond	selective	Technifoam (Iso-cyanurate)	aluminum	36 x 84	$11/ft²

HIGH TEMPERATURE COLLECTORS

The efficiency of low and mid-temperature flat-plate collectors decreases as the absorber temperature rises (see pg. 49). There are three *general* types of solar collectors that can obtain high temperatures efficiently (greater than about 180°F); (1) STANDARD FLAT-PLATE COLLECTORS WITH STRUCTURAL, CHEMICAL OR ELECTRO-PLATED SELECTIVE SURFACES. Some of these are listed with the "Mid-Temperature Collectors" in the previous listing. (2) EVACUATED TUBES. These are collectors in which the absorber is contained in an evacuated tube. The vacuum serves to reduce convective heat losses, thus allowing higher temperatures. The absorber may be metal, glass tubes or fins that transfer heat to a liquid or gas, and the tubes may contain or be backed by a reflecting surface. Evacuated tubes are effective in the 180-260°F range that serves absorption cooling, finned-tube (hydronic) baseboard heating systems, process water, sterilization, etc. (3) CONCEN-TRATING COLLECTORS. Concentrating collectors, or simply con-centrators, use curved reflecting surfaces or lenses to focus sunlight collect-ed from a large area onto a smaller area. They may be shaped like troughs or deep dishes, parabolic, or circular (see below).

Concentrators are capable of reaching much higher temperatures (200-300°F) than flat-plate collectors. In order to attain these higher temperatures the area of sunshine collection (the reflecting surface or lens) must be greater than the area of the heat absorber (the tube or plate onto which the reflected light is concentrated). With flat-plate collectors these areas are the same. For more information on the theory and application of concentrators see: CAPTURE THE SUN: THE PARABOLIC CURVE AND ITS APPLICATION. Enterprises Unlimited, Star Route, Fern-dale, CA 95536.

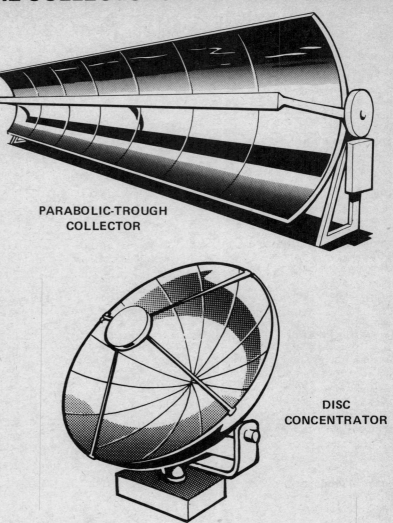

PARABOLIC-TROUGH COLLECTOR

DISC CONCENTRATOR

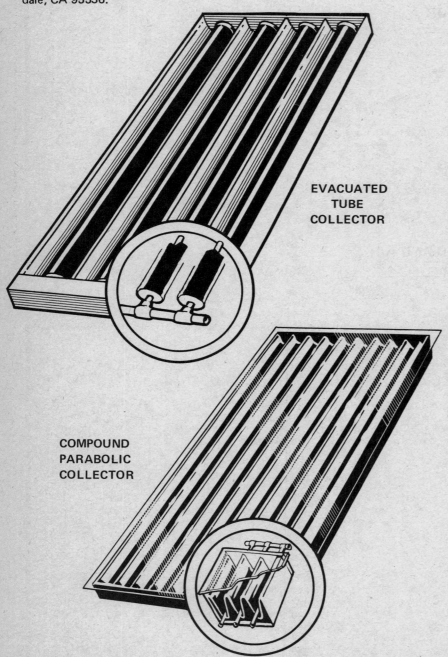

EVACUATED TUBE COLLECTOR

COMPOUND PARABOLIC COLLECTOR

Today commercial concentrators are being used primarily for industrial applications. For example, steam is being produced to sterilize cans in a Campbell soup cannery in Sacramento, California. Some industries are using concentrators to prewarm viscous fuels for easier transport. Other industries are using concentrators for absorption cooling (air conditioning), and a few companies are selling concentrators for producing domestic hot water and operating agricultural irrigation pumps.

In general concentrators cost only slightly more than flat plates ($10-$20/ft^2). Others are quite expensive. Higher costs are due to high tempera-ture piping and tracking devices. Concentrators can only make use of the sun's *direct* rays and most of them must follow ("track") the sun to operate. Flat-plate collectors can collect both direct and diffuse sunlight and thus can afford to be stationary. There are a few concentrators that do not need to track.

Below is an explanation of the columns in the chart.

- Type. Single-axis tracking collectors follow the sun on one axis. No dual-axis trackers are listed. Stationary concentrators do not track the sun, but are designed to be able to concentrate its rays even at a reasonably large angle to the normal.
- Concentration ratio. This is the ratio of effective collection area to the area of the absorber that is losing heat.
- Concentrator. This is the part of the collector that focuses the collected sunlight onto the absorber surface. Many types of materials can be used for this.
- Absorber. The absorber is located at the focus of the concentrator. It absorbs the concentrated sunlight and transfers its heat to the fluid.
- Tracking controls. On the trackers, there are various ways of driving the collectors. Those listed here use a photovoltaic device to activate the steering mechanism. Obviously the stationary concentrators do not need tracking controls.
- Effective collection area. This is the area of sunshine intercepted by the concentrator (length × top width) and not the surface area of the concentrator.

Check with the manufacturers for further details (weight, absorber description, controls, prices, etc.).

COMPANY	TYPE	CONC. RATIO	CONCENTRATOR	ABSORBER	TRACKING CONTROLS	EFFECTIVE COLLECTION AREA
Acurex Corp. 485 Clyde Ave. Mountain View, CA 94042	single-axis tracking; liquid or air	37:1	thin, reflective aluminum parabola, held by edge clamp plates	flattened copper tube, insulated above; glazing beneath	1/12 hp gear motor, shadow band sun sensor	40 ft²
General Electric Co. Solar Heating Mktg. P.O. Box 13601 Philadelphia, PA 19101	stationary evacuated tube concentrator		reflective tray under each tube which carries liquid	evacuated glass tube contains metal tube which carries liquid	not applicable	
Hexcel 1711 Dublin Blvd. Dublin, CA 94566	single-axis tracking. air, oil, water transfers		parabolic trough; acrylic-aluminum reflector	copper-steel tube with pyrex glazing	"shaded photo-transistor sensor"	in 20 ft lengths
KTA Corp. 12300 Washington Ave. Rockville, MD 20852	stationary "glass tube concentrator"		glass tubes silvered on bottom	copper tube, black copper oxide coating; inside second glass tube within concentrator.	not applicable	
Northrup Inc. 302 Nichols Dr. Hutchins, TX 75141	single-axis tracking		acrylic Fresnel lens; fiberglass insulated galvanized steel trough beneath	copper tube, selective black coating	silicon sun-sensor, small electric motor	39 ft²
Owens-Illinois Richard E. Ford P.O. Box 1035 Toledo, OH 43666	stationary evacuated tube concentrator		aluminum with weather resistant surface; cylindrical concentrator under each tube	evacuated glass tube contains feeder-tube	not applicable	27.4 ft²
Polisolar Ltd. P.O. Box 228 3000 Berne 32 Switzerland	single-axis tracking. various heat transfers for specific temperature	6-10:1	parabolic trough glass silvered mirrors	glass tube contains black star-shaped absorber, or black fluid	sun-sensor, 60 watt motor, .03 Kwh/day.	20 ft² per element
Solerjy 150 Green St. San Francisco, CA 94111	stationary concentrator	3.5-5.5:1	inward spiralling cylindrical reflecting concentrator	copper tubing at focal point of spiral	not applicable	16 ft²
Special Optics Box 163 Little Falls, NJ 07424	precision paraboloid and ellipsoid reflectors	many	nickel-plated brass, many coatings available	non supplied	non supplied	specify
Sunpower Systems Corp. 2123 S. Priest Rd. Tempe, AZ 85282	single axis tracking domestic hot water					

— WATER TANKS —

Included in this table are both storage tanks and electric hot water heaters. Systems which run potable water through the collectors need a tank with "solar connections"—that is, an outlet near the bottom of the tank and an inlet near the top to connect to the collector array—in addition to the regular connections to the city water system. Antifreeze (closed) solar systems, in which the collector fluid is entirely contained, use a tank with a heat exchanger. The heated fluid from the collector runs through the heat exchanger and heats the water in the tank. If an electric element is present, it should be at the top of the tank so that it doesn't preheat the fluid going to the collectors.

Many solar hot water heating systems utilize only one storage tank, often the existing conventional tank, complete with gas or electricity for a backup. Other systems use an additional tank, solar heated, which serves as a "preheater" for the conventional tank. While the initial cost of a one-tank system may be less, it has been shown by computer simulation (University of California, Santa Cruz) that the two-tank system will use far less back-up fuel. For one-tank electrical systems, the back-up fuel requirements can be reduced by only using the upper heating element and disconnecting the lower element.

Large storage tanks for space heating systems are available from local tank suppliers, or you can build your own out of concrete or wood. If you do build your own, try to consult a structural engineer.

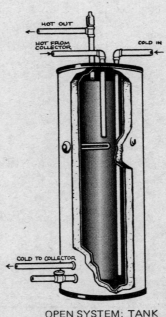

OPEN SYSTEM: TANK WITH SOLAR CONNECTION

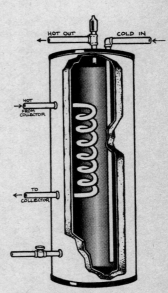

CLOSED SYSTEM: TANK WITH HEAT EXCHANGER

WATER TANKS FOR SOLAR USE

COMPANY	MODEL	GALLONS CAPACITY	DIMENSIONS inches height	diam.	MATERIALS lining	tank	insulation	APPROX. COST	COMMENTS
A. O. Smith Corp. Consumer Products Div. P.O. Box 228 Kankakee, IL 60901	storage tanks STJ-30 to 120 elec. heaters ESH-30 to 120	30 - 120	49 to 62	20 29	glass	steel	"R-10"	$140 - $375	solar connections for thermo-siphon or pumped systems; both storage tanks and electric heaters available.
Ford Products Corp. Ford Products Road Valley Cottage, NY 10989	Aqua-Coil, Aqua-Booster	40, 65 (80 & 120)	53 55	20 24	stone	steel	fibreglass	$140 - $270	finned copper coil heat exchanger for use with anti-freeze; electric element and circulator available.
Heliotrope General 1869 Hidden Mesa Rd. El Cajon, CA 92020	Storex, Storex Electric	Storex: 40, 65, 120 Electric 65, 120	54 56 68	20 24 28	stone	steel	fibreglass	$140 - $300	finned copper coil heat exchanger for use with anti-freeze, both storage tanks & electric heaters available.
Rheem Water Heater Div. City Investing Co. 7600 S. Kedzie Ave. Chicago, IL 60652	Solaraide	66, 82, 120	58 59 62	22 24 28	glass	steel	fibreglass		solar connections for thermo-siphon or pumped systems; available with or without electric heating element.
Solar Systems Sales 180 Country Club Dr. Novato, CA 94947		82			glass	steel	fibreglass		copper heat exchanger, electric element; also tanks of 500, 1200, 2000 & 3000 gallons.
State Water Heaters Ashland City, TN 37015	SCV-82 SCV-120	82 120	56 75	28 28	glass	steel	fibreglass		four internal heat exchanger channels; electric element, built in thermostat
W. L. Jackson Mfg. Co. P.O. Box 11168 Chattanooga, TN 37401	S 080 1 S 100 1 S 120 1	80 100 120	64 57 67	24 28 28	glass	steel	fibreglass		solar connections for thermo-siphon or pumped systems.

FANS AND PUMPS

Active liquid solar systems require a pump to circulate the heat collection fluid. Until recently it was difficult to find a small (1-5 gallon per minute) durable pump, but they are becoming more and more available. Systems exposed to tap water require a bronze or stainless steel pump. Cast iron can be used in recirculating (closed) systems where corrosion can be more easily controlled.

Just as active water systems require a pump, active air systems require fans. Selection of fans is very important because their motors can use quite a bit of energy. Several manufacturers make energy saving models which are good for solar systems. A good fan is probably worth the extra money it will cost.

There are a lot of different kinds of fans, so unless you're pretty familiar with them you should consult an engineer before you buy. The Trane Company (see Fan Listing) publishes some good educational material on the operation and selection of fans. For more on forced air systems see the publications of the Solaron Corp. (pg. 58). Listed below are a few companies which manufacture suitable fans for air systems, along with two who specifically provide solar air handlers.

PUMPS

COMPANY	TYPES OF PUMPS	MATERIALS
Bell and Gossett 8200 N. Austin Ave. Morton Grove, IL 60053	complete line of pumps that can be used for domestic hot water, space heating, and hydronic systems; suitable for large systems.	cast iron or bronze body
Grundfos Pumps Corp. 2555 Clovis Ave. Clovis, CA 93612	pumps for use in large or small domestic hot water or space heating systems; some models have adjustable flow rates, usable with proportional controllers.	cast iron body with stainless steel impeller or aluminum body with stainless steel pump chamber and impeller
March Mfg. Co. 1819 Pickwick Ave. Glenview, IL 60025	good for domestic hot water systems	bronze body
Taco, Inc. 1160 Cranston St. Cranston, RI 02920	good for domestic hot water systems.	cast iron or bronze body
Thrush Products, Inc. P.O. Box 228 Peru, Indiana 46970	complete line of pumps that can be used for domestic hot water, space heating and hydronic systems.	cast iron or bronze body
W.W. Grainger, Inc. 5959 W. Howard St. Chicago, IL 60648	pumps for use in large or small domestic hot water or space heating systems. (Teel brand)	cast iron, bronze, or plastic body

FANS

COMPANY	COMMENTS
The Trane Company La Crosse, WI 54601	expensive, good quality fans
Carrier Corporation Carrier Parkway Syracuse, NY 13201	expensive, good quality fans
W.W. Grainger, Inc. 5959 Howard St. Chicago, IL 60648	cheaper, a wide range of quality.
Solar Control Corp. 5595 Arapahoe Rd. Boulder, CO 80302 80302	Solar Air Mover; provides system air flow and control in one package.
Solaron Corp. 300 Galleria Tower 720 S. Colorado Blvd. Denver, CO 80222	air handling unit similar to the above.

SOLAR CONTROLS

Differential thermostat controls are a necessary part of active solar space and water heating systems. Their basic function is to turn the system on when there is useful heat to be collected, and off when there is not. Sensors are mounted in the hottest part of the collector and the coldest part of the water tank or rock bed. When the collector is several degrees hotter than the storage, the pump or fan is turned on. When the collector temperature approaches the storage temperature, the pump or fan is turned off. Some controls have preset temperature differentials, others can be set to the buyer's specifications, and still others are adjustable.

Proportional controls vary the speed of the pump or fan in proportion to the temperature difference between the collector and the storage. These are slightly more efficient because they reduce power consumption and pump or fan wear.

Some controls for liquid systems can protect the system from freezing by turning the pump on when the collector temperature approaches freezing, or from overheating by turning the pump off when the tank gets too hot. Freeze protection doesn't work if the electricity fails.

Solid state controls may be more reliable than those that use relays because they have no moving parts. For more on differential thermostats see: SOLAR ENGINEERING, November 1976, pg. 27 and February 1977 pg. 20.

COMPANY	MODEL	APPROX. COST	AVAILABLE FEATURES AND COMMENTS
Del Sol Control Corp. 11914 U.S. 1 June, FL 33408	02A Control with or without 809 circulator	$60; $85 with circulator	freeze protection available; all solid state
Hawthorne Industries, Inc. Solar Energy Div. 1501 South Dixie West Palm Beach, FL 33401	H-1500 through H-1512	$40 to $60, depending on features	freeze protection; fixed or proportional control; dual outlet to control two pumps; upper temperature limit of 150° or 165°F, or 87° for pools; all solid state.
Heliotrope General 3733 Kenora Dr. Spring Valley, CA 92077	Delta-T many models	$35 to $50, depending on features	automatic off when collector temperature is below 80°F; freeze protection; upper temperature limit of 160°F; bypass switch; all solid state except one 120 V, 10 amp, 1/3 HP relay.
Natural Power, Inc. New Boston, NH 03070	SC100	$400	Solid state sensor and controller. Differential, low limit, and high limit control settings adjustable over wide range. Temperature read-out meter.
Penn Division Johnson Controls, Inc. 2221 Carden Ct. Oak Brook, IL 60521	series R34, A41W		all solid state controller.
Rho Sigma, Inc. 15150 Raymer St. Van Nuys, CA 91405	500, 102, 104, 106, 240	$40 to $115 depending on features	Upper temperature limit of 140°F; freeze protection; pool controller (240) has adjustable thermostat; model 500 is proportional and solid state, others use relays.
Robertshaw Controls Co. 100 W. Victoria St. Long Beach, CA 90805	SD-10 Solar Commander	dealer $80 - $90	freeze protection, upper temperature limit, high temperature sensors, all solid state.
Solar Control Corp. 5595 Arapahoe Rd. Boulder, CO 80302	74-140, 75-160, 75-176, and 76-100		controls auxiliary system as well as solar; for liquid or air systems; models 75-176 and 76-100 programmable; all solid state.
Sunwater Company 1112 Pioneer Way El Cajon, CA 92020	Suntroller	$50	all solid state

SOLAR WATER HEATING SYSTEMS

Complete solar water heating systems generally consist of collectors, controls, expansion tanks, filters, valves and one or more pumps and water storage tanks. Some of the available systems listed here do not offer complete packages, and many of the companies manufacture one or more of the components (usually collector or tank) making up the rest of the package from other manufacturers' products.

We have tried to make brief informative comments in those cases where the systems were in some way non-standard. We have used the following terms for brevity.

• Closed systems recirculate distilled water, antifreeze and/or corrosion inhibitors through the collectors and storage tanks in a closed loop. Heat picked up from the collectors by the transfer fluids is transferred to the water in the storage tanks through the closed loop or other heat exchangers. In locations where water is particularly corrosive or where freezing temperatures occur often, closed solar water heating systems must be used.

• Open systems circulate usable water directly through the collectors and store it in one or more tanks. Open systems must be drained, heated or circulated by "freeze-protection" controls and pumps during freezing weather.

• Thermosiphon systems utilize the natural upward movement of heated fluids to circulate water from the collectors to the storage tank without the use of pumps and other components of "active" systems. To do this the tank(s) must be located above the collectors.

More detailed information on the collectors in many of the packages is in the Flat-Plate Collector tables.

Solar water heating packages range in price from $500 for retrofit packages, adaptable to existing plumbing (usually with no tank), to $2000 and more for systems complete with 2 tanks, high-efficient collectors, controls, heat exchangers, etc. The SOLAR INDUSTRY INDEX (1977 edition) lists 100 companies currently marketing solar hot water systems. There are undoubtedly many more than this.

Alten Corporation
2594 Leghorn St.
Mountain View, CA 94043

standard complete two-panel system (except piping to and from panels); automatic and manual freeze protection; adjustable panel mounting supports.

American Heliothermal Corp.
3515 S. Tamarac Dr.
Denver, CO 80237

closed single tank system, steel collectors with selective surface, pump, expansion tank, & controls; electric back-up.

American Sun Industries
P.O. Box 263
Newbury Park, CA 91320

open and closed systems; with and without electric back-up; collectors, pump, tank, controllers, etc.

Atlantic Solar Products, Inc.
11800 Sunrise Valley Drive
Reston, VA 22091

closed system; copper Roll-Bond® collectors.

SOLAR WATER HEATING SYSTEMS

Cole Solar Systems, Inc.
440 A East St. Elmo Rd.
Austin, TX 78745

Double Kalwall® collector, pump, controls, one 82 gal tank with elect. back-up; freeze protection.

Columbia Solar Energy Div.
55 High Street
Holbrook, MA 02343

closed, single tank (80 gal.) system; copper absorbers, controls, pump

Conserdyne Corporation
4437 San Fernando Rd.
Glendale, CA 91204

closed, 2-tank system; steel/glass collector with pump and controls

Daystar Corporation
90 Cambridge St.
Burlington, MA 01803

copper/glass collector

Grumman Energy Systems
4175 Veterans Memorial Hwy.
Ronkonkoma, NY 11779

closed or open system available; 2 tanks with controls, expansion tank, etc.

Jackson Manufacturing Co., Inc.
200-26 E. 40th St.
Chattanooga, TN 37401

closed single tank system with electric back-up; AL/glass/selective collector, freeze protection.

McCombs Solar Company
1629 K Street NW
Suite 520
Washington D.C. 20006

closed or open system with freeze protection; electric back-up; 2 tank systems with 65 or 120 gal. storage.

Northrup, Inc.
302 Nichols Dr.
Hutchins, TX 75141

thermosiphon or pumped system available with no freeze protection.

R.M. Products
5010 Cook St.
Denver, CO 80216

closed or open 2 tank system with freeze protection; copper collectors, pump, controls

Raypak, Inc.
31111 Agoura Rd., P.O. Box 5790
Westlake Village, CA 91359

2 tank systems, pumped or thermosiphon, freeze protection; complete with installation and trouble shooting instructions.

Revere
P.O. Box 151
Rome, NY 13440

closed system, copper/glass collectors, tank, pump, controls

SAV
Fred Rice Productions, Inc.
P.O. Box 643/48
780 Eisenhower Dr.
La Quinta, CA 922

"High Speed Cylindrical Solar Water Heater System" collector and storage in one. 12 gal/cylinder

Solar II Enterprises
21416 Bear Creek Rd.
Los Gatos, CA 95030

pump or thermosyphon copper/glass collector

Solar-Aire
82 S. Third St.
San Jose, CA 95113

air collector with blower, air/water heat exchanger, tank

Solar Development, Inc.
4180 Westroads Dr.
West Palm Beach, FL 33407

you supply the (only) tank

Solar Energy Products, Inc.
1208 N.W. 8th Ave.
Gainesville, FL 32601

package (panel, controls, pump, hardware) for converting existing hot water heater into solar storage tank.

Solar Innovations
412 Longfellow Blvd.
Lakeland, FL 33801

adjustable support structures available.

Solar Products, Inc.
614 NW 62nd St.
Miami, FL 33150

open single tank system; thermosiphon or pump, collectors with selective coating and Kalwall® glazing.

Solar Research Div.
525 N. Fifth St.
Brighton, MI 48116

closed system, with or without collectors or storage tank. Used as booster in cold climates.

Solar Structures, Inc.
7 Sundance Rd.
La Grangeville, NY 12540

closed single tank (82 gal.) system; pump, controls, expansion tank; Revere collectors.

Solar Systems Sales
180 Country Club Drive
Novato, CA 94947

closed single tank system with electric back-up; pumps, controls, expansion tank, no collectors.

Solaron Corporation
300 Galleria Towers
720 S. Colorado Blvd.
Denver, CO 80222

air collector with air to water heat exchanger, freeze safe, 2 tank system.

Solarcoa, Inc.
2115 E. Spring St.
Long Beach, CA 90806

complete system for easy retrofit; collectors, 52 or 82 gal tank, pump, controls.

Solargizer Corporation
220 Mulberry St.
Stillwater, MN 55082

closed system, 2 tanks, CU/AL fin absorber. fiberglass glazing, pumps and controls

State Industries, Inc.
Ashland, TN 37015
Henderson, NY 89015

closed single tank (82 or 120 gal.) systems; pump, controls, double glazed steel collectors

Sunearth
R.D. 1 Box 337
Green Lane, PA 18054

closed or open system with freeze protection; AL/CU collector with double plastic glazing.

Sunspot
5511 Ekwill St.
Santa Barbara, CA 93111

closed system; pump, controls, CU/AL fin absorbers.

Sunworks
PO Box 1004
New Haven, CT 06508

one tank system; collectors, pump, controls.

Taco, Inc.
1160 Cranston St.
Cranston, RI 02920

controls, two pumps, and heat exchanger in one unit; you supply collectors and tank.

Universal Solar Energy Co.
1802 Madrid Ave.
Lake Worth, FL 33461

you supply the tank.

Wallace Company
831 Dorsey St.
Gainsville, GA 30501

all copper absorber with glass, tank, controls.

SIMPLIFIED SOLAR WATER HEATING SYSTEM

SOLAR SPACE HEATING SYSTEMS

Space heating packages are less standard than water heating packages, because buildings are far less standard in their space heating than in their water heating needs. Here are a few companies that offer interesting packages. The SOLAR INDUSTRY INDEX (1977 edition) lists 96 companies offering solar space heating packages (virtually all of them active systems).

Alten Corporation
2594 Leghorn St.
Mountain View, CA 94043

water collectors
water storage
hydronic distribution

Champion Homebuilders Co.
5573 E. North Street
Dryden, MI 48428

self contained outside unit for retrofit to existing homes, mobile homes and forced-air systems.

Columbia Solar
High Street
Holbrook, MA 02343

water collectors . . . selective copper panels
water storage . . . tie in for wood stove back-up

Div-Sol, Inc.
P.O. Box 614
Marlboro, MA 01752

air collector
rockbed or water storage with heat exchanger
do-it-yourself system

Northrup, Inc.
302 Nichols Dr.
Hutchins, TX 75141

Sun Duct® air system with no storage for daytime heating

Solaron Corporation
300 Galleria Tower
720 S. Colorado Blvd.
Denver, CO 80222

air collectors
rock storage
choice of fan coil or hydronic distribution

PASSIVE HEATING

Passive systems are looking more and more like the best way to go for space heating. They have certain limitations in that they don't lend themselves as easily to retrofitting, and they require more participation on the part of the owner than do active systems; however they tend to be less expensive, less hardware oriented, and often more energy conserving than active systems.

Attached solar greenhouses can also offer inexpensive solar heat, as well as providing a great place to grow winter tomatoes. (See pgs. 94 and 180 for references to solar-heated greenhouses.)

We haven't listed much in the way of passive space heating hardware. This is because materials for passive systems are largely ordinary building materials, along with some common items like 55-gallon drums. Some non-standard things like moveable insulation are needed, and these are beginning to become available. Skylids and Beadwall, and the insulating window shade plans listed here, are moveable insulation.

For more on passive systems see the Solar Architecture section, pages 84-87, 94-95 and 99.

Zomeworks Corporation
P.O. Box 712
Albuquerque, NM 87103

Skylids: Freon-actuated automatic insulating louvers; open when sunny, close when cloudy or at night. Beadwall: fire-retardent, polystyrene beads to be blown between two layers of glazing for insulation, or sucked out to let sun in. Bead storage containers, blower motors, sock valves, anti-static agent for Beadwall systems; plans and information for various systems also available.

Rainbow Energy Works
3765 17th St. No. B
San Francisco, Ca 94114

plans for an insulating window shade of fabric and polyester bat; rolls up inside wooden box. $2.50

The Solar Room Company
Box 1377
Taos, NM 87571

attached solar greenhouse kit; redwood and steel framing, double plastic glazing treated to resist UV, small fan; available in many sizes, about $2.50 per sq. ft.

MISCELLANEOUS HARDWARE

REFLECTING MATERIALS

All sorts of materials can be used to build reflectors and concentrating collectors. The simplest is household aluminum foil. Some others:

RAM PRODUCTS

Ram sells Plexi-View® acrylic mirrors in two thicknesses, 1/8″ and 1/4″. Clear mirrors start at $2.39 per sq. ft. (1/8″ thick) and $2.94 per sq. ft. (1/4″ thick) in 4′ X 8′ sheets. Many people are testing Plexi-View® mirrors for solar installations, but results are unknown.

Ram Products
1111 North Centerville Road
Sturgis, Michigan 49091

D.J. AND A. REFLECTORS

The aluminum parabola they sell is 48″ in diameter and has a focal-length of approximately 18″. The steel parabola is 49″ in diameter. Aluminum units are $48; steel, $38. F.O.B. Minneapolis, Minnesota. Prices subject to change (aren't they all?).

D.J. & A. also sells a complete cooker, consisting of a Scotchcal covered aluminum reflector, grill, and stand for $95.

Donald Johnson & Associates
2523 16th Ave. S.
Minneapolis, MN 55403

TALBERT REFLECTORS

Large optical mirrors. Reflecting surface supported by aluminum honeycomb-epoxy-fibreglass substrate.

Talbert Reflectors
42 Alcatraz Ave.
Oakland, CA 94609

ALZAK ALUMINUM

Alzak is a brand name for an aluminum reflector material manufactured by Alcoa. It comes in a highly polished or a diffuse finish. The highly polished variety costs at least $1.50 per sq. ft. in 2′ X 6′ X .020″ sheets, and that's for large quantities (180 sheets).

Aluminum Co. of America
Lighting Industry Sales
1501 Alcoa Building
Pittsburgh, Pennsylvania 15219

"SCOTCHCAL" REFLECTING ACRYLIC

"Scotchcal" Brand Series 5400 film is a metallized acrylic film with a pressure-sensitive adhesive and a nominal (thickness) of 0.005″. Average reflectivity of 80-90% solar radiation. From:

Decorative Products Division
3M Center, Bldg. 209-1
St. Paul, MN 55101

ALUMINIZED MYLAR

Aluminized Mylar is commercial polyester or aluminized polyethylene terephthalate film. Basically, aluminized film is used in cryogenic (super-insulation systems) or other applications where high reflectivity is desired. The reflectance of the decorative or general purpose type aluminized Mylar is approximately 85%. The aluminum coating is vacuum deposited onto the film is a vacuum chamber. The actual metal deposit is 99.9% pure aluminum.

Prices start at about $0.20/sq. ft. for .00025 inch thick aluminized mylar in small quantities. On the West Coast, it's distributed by:

Transparent Products Corp.
1727 W. Pico Blvd.
Los Angeles, CA 90015

or 1328 Mission St.
San Francisco, CA 94103

HEAT TRANSFER FLUIDS

Several manufacturers market heat transfer fluids that are currently being sold for use in solar systems. The following is a partial list of these manufacturers:

E.I. duPont, de Nemours & Co.   "Freon" fluorocarbons
"Freon" Products Division
Wilmington, Del. 19893

Monsanto Industrial Chemicals   Therminol® Heat Transfer
800 N. Lindberg Blvd.   fluids 44, 55, 60, 80.
St. Louis, MO 63166

Resource Technology Corp.   Sun-Temp
151 John Downey Dr.   non-aqueous, non-glycol, non-corrosive,
New Britain, CT 06051   non-toxic, inert, stable -40°/500°F

Solar Energy   Dow Corning Silicone Products
Dow Corning Corp.   no tar or acid buildup, non-corrosive
Midland, MI 48640   -50°/450°F

Sunworks   Sunsol 60*
P.O. Box 1004   propylene glycol with inhibitors
New Haven, CT 06508   -55°F/240°F

Union Carbide Corp.   Preston II*
Old Saw Mill River Road   ethylene glycol with
Tarrytown, NY 10591   silicone-silicate inhibitor
or local dealer

Glycol Solutions are usually used for freeze protection. They <u>must be</u> inhibited with a pH buffer to prevent the formation of highly corrosive organic acids. <u>Degrades</u>.

THERMAL CEMENTS

Thermal cements allow you to make a reasonably good heat connection without soldering or welding. They have been used (with varying success) to bond tubes onto absorber plates in flat-plate collectors and to bond pipes to the outside of tanks. Hardening and non-hardening formulas are available. The former work better but don't cope well with thermal expansion. Two manufacturers we know of (there are probably others):

Chemax Corporation   **Thermon Manufacturing Co.**
211 River Road   100 Thermon Drive
New Castle, Delaware 19720   San Marcos, Texas 78666

INSTRUMENTATION

Most measuring that is required in water solar systems can be done with a few thermometers, a calibrated bucket, and a watch. Measuring things in air systems is harder, but sometimes you can borrow an air flow meter from a heating contractor. Measuring sunshine intensity is impossible without a specialized instrument.

For instruments used to measure solar radiation:

Dodge Products   **Spectran Instruments**
Box 19781   P.O. Box 891
Houston, Texas 77024   La Habra, California 90631
A $50 hand-held meter for   High quality and expensive
quick field measurements.

The Eppley Laboratory, Inc.   **Spectrolab, Inc.**
12 Sheffield Avenue   12500 Gladstone Avenue
Newport, Rhode Island 02840   Sylmar, California 91342
High quality and expensive   High quality and expensive.

For general weather instruments, write for the catalogs of:

Weather Measure Corporation   **Texas Electronics, Inc.**
P.O. Box 41257   P.O. Box 7225 K
Sacramento, California 95841   Dallas, Texas 75209

Science Associates, Inc.
Box 230
230 Nassau Street
Princeton, NJ 08540

For temperature or pressure test plug and gauges

Peterson Equipment Company, Inc.
P.O. Box 217
Richardson, Texas 75080

or your local representative.

SOLAR COOLING

If you <u>really</u> want to do absorption cooling, ARKLA has a line of chillers designed to run on "low temperature" hot water (195° F).

Write for info:

Arkla Industries
P.O. Box 534
Evansville, Indiana 47704

PHOTOVOLTAICS

Listed below are several manufacturers of silicon photovoltaic cells. The cells can be arranged to obtain various currents and voltages. Usually the cells are available in either N on P or P on N configuration (see section on photovoltaics), in large arrays or as single cells. They have a practically unlimited lifetime if properly protected from impact and corrosive environments. Contact manufacturers for details and prices.

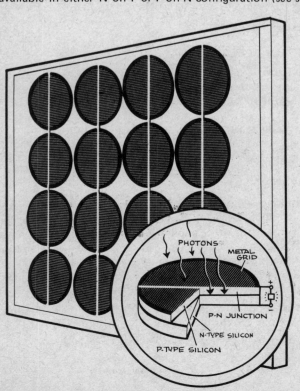

Company	Comments
Optical Coatings Laboratory Photoelectronics Group 15251 E. Don Julian Rd. City of Industry, CA 91746	individual cells, several shapes and sizes; generally for use in electronics applications.
Sensor Technology, Inc. 21012 Lassen St. Chatsworth, CA 91311	individual cells; top contact along edge of cell, gridded also available; several shapes and sizes, P-on-N or N-on-P.
Solarex Corp. 1335 Piccard Dr. Rockville, MD 20850	makes a variety of products; available thru Edmunds Scientific.
Solar Energy Systems 1 Tralee Industrial Park Newark, DE 19711	manufactures cadmium sulfide cells; financed by Shell.
Solar Power Corp. 5 Executive Park Drive Billerica Industrial Park North Billerica, MA 01862	various mounting structures and cell configurations. Subsidiary of Exxon
Spectrolab, Inc. 12500 Gladstone Ave. Sylmar, CA 91342	modular arrays, various mounting structures; complete systems with voltage regulator and storage batteries. Part of Hughes Aircraft

A SOLAR WHITE HOUSE

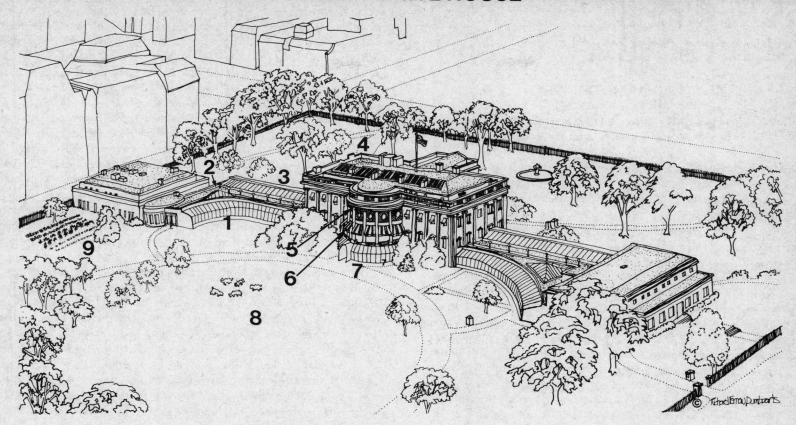

Richard Fernau, a Berkeley, California, designer interested in energy conserving buildings, has made a modest proposal for the Executive Mansion at 1600 Pennsylvania Avenue. The idea of a solar heated White House was initiated by the Friends of the Earth and this drawing was first published in their newspaper "Not Man Apart." The scheme is in the original Palladian spirit of the White House and incorporates greenhouses reminiscent of the conservatories installed by President Harrison in the 1890's. Fernau developed the design with the help of Berkeley Solar Group and Bruce Corson.

The drawing shows the following solar heating and energy saving features:

1. Solar heated greenhouses enclosing symmetrical colonnades of water-filled steel columns.

2. Skylights in greenhouses to capture additional solar heat.
3. Skylights in corridors and offices for solar heating and lighting.
4. Flat-plate collectors to heat 550 gallons of water per day for household use.
5. Sleeping porch to provide a naturally air-conditioned place to spend warm summer evenings.
6. Red, white, and blue awnings to cut summer heat gain through south facing windows.
7. Entry enclosure allows the President to greet visiting diplomats in the comfort of a solar heated enclosure.
8. Sheep mow the White House lawn, saving 350 gallons of gasoline annually.
9. A vegetable garden allows the Presidential family to produce a portion of their food locally.

SOLAR ARCHITECTURE

SOLAR ARCHITECTURE
Michael Riordan

**Adapted in part from THE SOLAR HOME BOOK
by Bruce Anderson with Michael Riordan**

The decision to use solar energy sources for heating and cooling has many implications for architecture. The orientation and composition of roofs and exterior walls, their relation to site and surrounding vegetation, the local landscape and even the interior layout of a building are all influenced by such a choice. *In a rational solar architecture, building form reflects not only the needs of inhabitants and site constraints but also the cyclic energy flows of the natural environment.*[1] Such variable climatic factors as the amount of solar radiation, position of the sun, direction of prevailing winds, air temperature, and humidity all find expression in the geometry and structure of buildings. On one level, solar architecture can focus on the need to incorporate large solar collectors and heat-storage devices into building design. But in a broader sense, the building *itself* can be designed to collect and store solar energy in its walls, roof, and floors. Natural cooling by means of shading, ventilation, or thermal radiation to the night sky can also be included. In such "passive" approaches, the building *becomes* the heating and cooling system—rather than relying on extra mechanical equipment to accomplish these ends.

There are a growing variety of methods for trapping solar energy to heat and cool. These methods can be grouped into three broad categories[2]—direct, integrated, and indirect (see Fig. 1). In *direct* methods, the sun's rays penetrate into the interior of a building. Massive internal structures, such as concrete slab floors or adobe walls, absorb the solar heat generated and release it when the sun isn't shining. Since the sun's energy is absorbed *inside* the house right where the heat is needed, there is little need for a separate system to distribute this heat to the rooms. In an *integrated* system, the sun's rays are intercepted and absorbed by a massive exterior wall or roof—which usually doubles as a heat storage container. Heat flows to the rooms by conduction, natural convection, or thermal radiation. Both of these approaches to solar heating are termed *passive* because the heat flows by natural means (present terminology seems to favor the term "direct" to describe all "passive" systems). Of course, fans or blowers can be used to aid natural convection and provide better control of the heat, but the priority of passive systems is to allow natural patterns of heat flow to move uninhibitedly.

Indirect systems gather solar energy in flat-plate or concentrating collectors mounted on the roof or walls, or even standing apart from the building. Pumps or fans are needed to circulate liquids or air through the collectors and back to a heat storage container—often a tank of water or a bed of rocks. When heat is needed, air or water from the building's heating system is warmed by this stored heat and circulated to the rooms. Indirect systems are also called *active* systems because of their reliance on mechanical power to move the collected solar heat.

The three approaches to solar heating put different constraints on building design. For example, with direct methods and integrated systems, a building should not be too deep along the north-south axis, and its interior layout must be open to encourage ready circulation of the collected heat. By contrast, the major design problem with an indirect system may be the need to incorporate a large collector surface into the south side of a building in a low-cost but aesthetically pleasing fashion. Of course, there are many constraints that are common to all approaches. Among them are the need to provide a large southern exposure and to insulate exerior surfaces well.

The right approach to solar heating and natural cooling will depend upon climate, site, and the needs of the inhabitants. Direct methods and integrated systems work well in the American Southwest, with its abundance of direct sunshine, but they may fare poorly in cloudy regions. Indirect systems are preferred in densely-packed neighborhoods where rooftops have the only unobstructed southern exposures. And the *size* of a building is important, too. Passive systems work best in small buildings like single-family homes, where the solar heat doesn't have to travel very far from the point of collection. With larger buildings, however, additional mechanical power must be summoned to help move the heat to the point where it's needed. Active systems begin to make much more sense in factories, schools, and office buildings. But regardless of the approach used, there are sizeable cost reductions to be realized by integrating solar heating and cooling with the other purposes and functions of a building. *The intelligent designer owes it to himself and to his clients to remain open to the many possibilities of solar architecture.*

Direct Solar Heating

One of the best ways to use solar energy for heating is to design a building to collect and store the energy itself. To achieve this end, the building should satisfy three criteria:[3]

- It must let sunlight in when it needs heat and keep it out when it doesn't; it must also let coolness in as needed.
- It must store the heat (or coolness) for times when the sun isn't shining (or it's too hot out).
- It must trap this heat (or coolness), letting it escape only very slowly, and distribute it throughout the building.

The first aim is accomplished by orienting and designing the building so as to let sunshine penetrate through walls and windows during winter and by using shading to keep it out during summer. Buildings made of heavy materials such as stone or concrete will satisfy the second aim by storing the accumulated heat or coolness over longer periods of time. Finally, well-insulated, weatherstripped, shuttered buildings with open interior designs can keep the heat or coolness inside and let it flow to all occupied areas.

The direct rays of the sun strike differently oriented surfaces with varying intensity and a building should have an *orientation* that can receive this heat in winter and shed it in summer. In the temperate zones of the Northern Hemisphere, the principal facade of a building should face within 30° of due south. In this configuration, the south-facing walls absorb the most radiation from the low winter sun while the roof—which can reject excess heat more easily—catches the brunt of the intense summer sun. Olgyay[4] suggests that a building be oriented somewhat east of south to catch the early morning sun when heat is most needed. It should face away from the west and southwest—the direction of principal summer heat gain.[5] In general, the optimum *shape* for direct solar heating is a form elongated along the east-west direction. And for equivalent floor areas, two or three-story buildings fare better than those with only a single floor. Of course, other factors such as site constraints and personal needs will influence the shape of a building, but these can often be combined with the requirements of direct solar heating.

Although a building's orientation and shape are important, the *size and placement of its windows* are the most significant factors in capturing the

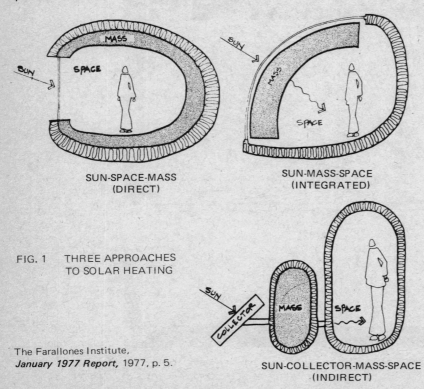

SUN-SPACE-MASS
(DIRECT)

SUN-MASS-SPACE
(INTEGRATED)

FIG. 1   THREE APPROACHES
TO SOLAR HEATING

SUN-COLLECTOR-MASS-SPACE
(INDIRECT)

The Farallones Institute,
January 1977 Report, 1977, p. 5.

| TABLE I | SOLAR HEAT GAIN FACTORS FOR 40°N LATITUDE. Values given represent the clear day solar heat gain, in Btu/ft²/day, through a single layer of clear, double-strength glass at the orientation listed. |

		WINDOW ORIENTATION				
		Vertical				
21st Day Of:	Horiz.	South	SE/SW	E/W	NE/NW	North
January	706	1630*	1174	508	127	118
February	1092	1626*	1285	715	225	162
March	1528	1384*	1318	961	422	224
April	1924	978	1199*	1115	654	306
May	2166	712	1068	1173*	813	406
June	2242	622	1007	1200*	894	484
July	2148	694	1047	1163*	821	422
August	1890	942	1163*	1090	656	322
September	1476	1344*	1266	920	416	232
October	1070	1566*	1234	694	226	166
November	706	1596*	1151	504	132	122
December	564	1482*	1104	430	103	98

*Vertical orientation which receives the most solar heat gain in any particular month.

SOURCE: ASHRAE, *Handbook of Fundamentals*, 1972.

sun's energy. Solar heat gains thru south-facing glass can be substantial in winter (see Table I). Studies by F. W. Hutchinson[6] showed that more than twice as much solar energy is transmitted during that season than in the summer. In all the cities studied, the solar heat gain through double-pane, south-facing windows was more than enough to offset the heat conduction losses back through the glass. But there are a few drawbacks. Large expanses of south glass can overheat the rooms on clear mid-winter days when the sun arcs in a low path across the sky. The solar heat must be absorbed and stored in massive interior structures so that it can be used at night rather than vented by day to keep the rooms comfy. Insulating curtains or shutters help stretch the lifetime of this stored heat. In fact, *movable insulation* is one of the rapidly advancing techniques that make passive solar heating possible. You can also minimize summer heat gains with intelligent placement and shading of windows. South glass can be readily shaded with a fixed overhang (see Fig. 2) or movable awning. East and west windows are very difficult to shade with fixed overhangs because the sun is low in the sky both early morning and late afternoon. Deciduous trees and hedges are a good bet on these sides—especially on the west where summer heat gains can be most oppressive. Reflecting and heat-absorbing glass can also be used.[7] At any rate, the amount of glass on the north, east, and west sides should be kept to the minimum necessary for natural lighting. Windows on these sides don't provide much solar heat gain in winter, but they can be a source of very large heat losses then.

A building with lots of south glass needs some means of storing the solar heat so that rooms don't overheat on sunny winter days but stay warm at night. The building materials themselves—the roof, walls, and floors—can provide effective heat storage. In particular, heavy building materials can store large quantities of heat without getting too warm. As the inside air cools, the warmed materials replace the lost heat, keeping the rooms warm. Well-insulated buildings with really massive interiors can stay warm for a few days without needing supplementary heat from fires or furnaces. During the summer, a massive building can also store coolness at night for use during the hot day. Ventilation of cool night air will cool all

the materials inside. The next day they will absorb excess heat from the room air—keeping it cool.

To store large amounts of heat (or coolness) in a fixed volume, use materials which have high *heat capacity*. In general, heavy materials like concrete or stone are your best bet. Over a temperature rise of 10°F, a cubic foot of concrete absorbs and stores 320 Btu while a cubic foot of pine timber stores only 180 Btu. Concrete, stone, and adobe walls insulated on the *outside* are far superior "heat sinks" than wood-framed walls. Another excellent heat sink is a concrete floor slab with insulation on the underside. Containers of water placed within the building confines can store more heat per unit volume than any other material (see Table II).

Interior heat sinks can do the most good with maximum exposure to direct sunlight. This means buildings that are shallow in the north-south dimension, or tall relative to their depth. Either way, midwinter sunlight penetrates through south windows all the way to the back of the building and the absorbed solar heat can more readily reach all parts of the house by natural convection and thermal radiation.

An open interior plan—free of rooms closed off from the south side—further encourages this natural circulation. Still, those areas closest to the glass surfaces will feel the greatest daily temperature fluctuations. These areas are best used for active functions like working and playing. More passive functions like reading or sleeping can be relegated to the north side of a house. In a two-story house, you'd probably put the bedrooms on the second floor or in a loft to take advantage of the warm air that collects there. Early risers who cherish the first rays of dawn would probably want their bedrooms facing east. There are many other examples, too many to enumerate here, of the ways interior living spaces can be organized around the cycles of natural energy patterns.

TABLE II	SPECIFIC HEATS AND HEAT CAPACITIES OF COMMON MATERIALS		
Material	Specific Heat (Btu/lb/°F)	Density (lb/ft³)	Heat Capacity (Btu/ft³/°F)
Water (40°F)	1.00	62.5	62.5
Steel	0.12	489	58.7
Cast Iron	0.12	450	54.0
Copper	0.092	556	51.2
Aluminum	0.214	171	36.6
Basalt	0.20	180	36.0
Marble	0.21	162	34.0
Concrete	0.22	144	31.7
Asphalt	0.22	132	29.0
Ice (32°F)	0.487	57.5	28.0
Glass	0.18	154	27.7
White Oak	0.57	47	26.8
Brick	0.20	123	24.6
Limestone	0.217	103	22.4
Gypsum	0.26	78	20.3
Sand	0.191	94.6	18.1
White Pine	0.67	27	18.1
White Fir	0.65	27	17.6
Clay	0.22	63	13.9
Asbestos Wool	0.20	36	7.2
Glass Wool	0.157	3.25	0.51
Air (75°F)	0.24	0.075	0.018

FIG. 2   SHADING A SOUTH WINDOW WITH A FIXED OVERHANG. FOR METHODS USED TO CALCULATE SHADING ANGLES AND SIZES OF SHADING DEVICES SEE *SOLAR HOME BOOK*, PG. 90.

DECEMBER 21

MARCH 21 OR SEPTEMBER 21

JUNE 21

SOURCE: Anderson and Riordan *The Solar Home Book*, pg. 90.

Diurnal Heat Flows in a Typical Solar House

A specific example of direct solar heating will help illustrate the principles outlined here. Consider the 1200 ft^2 California house described in the Energy Conservation section (pg. 18). This house has been insulated and weatherstripped to the same standards as before, but the double-glazed windows have *all* been moved to the south wall to maximize the solar heat gain. Relocating the windows doesn't change the heat load calculations much; under design conditions the house still loses about 9000 Btu/hr or 216,000 Btu per day. However, on a clear winter day, the solar heat *gain* through the 160 square feet of south glass can replace this loss (Table I, Fig. 3).

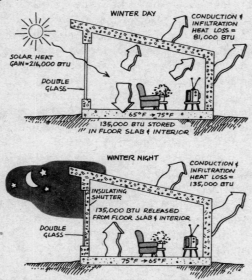

FIG. 3   INVENTORY OF SOLAR HEAT GAIN, STORAGE, AND HEAT LOSS IN A TYPICAL HOUSE HEATED BY DIRECT METHODS.

But that's only half of the story. Unless the house can store all this solar heat for later use, most of it will have to be rejected. The house needs an hourly solar heat gain of only 9000 Btu under the worst conditions. But an average of 24,000 Btu (= 216,000 Btu ÷ 9 hrs.) penetrates the south glass each hour of a clear day—an excess of 15,000 Btu/hr. To absorb all this excess heat, we must make one more design change. Let's put the house on a concrete slab floor 4 inches thick insulated from direct contact with the outside. Some of the 135,000 Btu (= 15,000 Btu/hr × 9 hrs.) excess is absorbed in the interior walls and furniture. But most of it goes into the concrete slab, raising the slab temperature about 10°F—say, from 65°F at sunrise to 75°F by sunset. By sunrise the next morning, all 135,000 Btu (= 9000 Btu/hr × 15 hours) will have escaped. Room and slab temperatures will be back to 65°F.

So if the local climate were a sequence of clear days accompanied by design conditions, this California dwelling could remain fairly comfortable using direct solar heating as its only source of heat. Room air temperatures would swing about 10°F daily, but the inhabitants would probably become accustomed to this temperature swing and organize their activities and dressing habits to accomodate it. Unfortunately, nature and the weather are rarely ever so predictable. Cloudy periods of intense cold are inevitable, and some form of backup heating—a small gas furnace or wood stove—is needed to make life bearable during these times. Still, 50 to 80 percent solar heating is possible using only these direct methods.

This apparent miracle was accomplished with three fairly straightforward design changes: insulating very well, shifting all the windows to the south side, and providing a concrete floor slab. Each measure is essential to the overall design. The uninsulated California house discussed earlier needs more than a million Btu to keep it warm on a design day. Solar heating can't match such extravagant demands. But with all the windows moved to the south side of the *insulated* house, the solar heat gain just about matches the design heat load. Of course, the house could face within 25° of due south and still get more than 90% of the maximum solar heat gain (Table I). Finally, the concrete slab is needed to control the absorbed heat so that the house remains comfortable both day and night. Interior temperatures have to swing about 10°F so that heat can be stored inside the building and released as needed. Such a temperature

swing cannot be eliminated in direct heating methods, but it can be smaller in more massive houses.

The ardent reader should be cautioned against taking the numbers calculated in this example too seriously. All sorts of simplifications and assumptions were necessary to keep the treatment straightforward. Room and slab temperatures are never uniform, for example, and outdoor conditions don't remain fixed very long. It takes a sophisticated computer program to accurately model the thermal response of a building to the vagaries of climate. Even then, we shouldn't try to specify the building performance too exactly, because the weather itself is too unpredictable and will always find ways to confound our equations. A good backup heater is the best insurance against capricious weather patterns and less-than-perfect designs. Nevertheless, the above example illustrates that direct solar heating *can* provide a large portion—certainly more than half—of the winter heat demand of a well designed dwelling.

Integrated Systems

In an integrated solar heating system, the sun's rays do not penetrate to the interior of the building. Instead, they are absorbed by a massive exterior structure that replaces the roof or south wall and doubles as the heat storage container. Some form of transparent material covers this mass to retard heat losses to the outdoor air. The absorbed heat flows to the rooms by natural means—usually thermal radiation or natural convection. Sometimes a fan or blower is used to supplement this flow. In such approaches, the solar "apparatus" is an intrinsic part of the building surface—hence the name, "integrated" system. Materials are usually kept simple; little or no electrical power is used to distribute the solar heat; and integrated systems are frequently cheap and reliable.

The three most familiar examples of integrated solar heating systems are:[8]

- the concrete south-wall collector, as pioneered by Felix Trombe and Jacques Michel in Odeillo, France;
- the "water-wall" collector whose best example is the Drumwall used by the Baer house in Corrales, New Mexico; and
- the "Sky Therm" roof-pond collector developed by Harold Hay in Phoenix, Arizona, and used by a house in Atascadero, California.

All three approaches permit a welter of variations that suit them to a range of climates and purposes.[9] Movable insulation over the collector surface helps retard nightly heat losses in winter. Similarly with the collection surface covered by movable insulation during summer days, the storage mass can absorb the excess room heat and cool the interior by radiating outside when the insulation is removed at night.

The first actual dwelling to make significant use of concrete wall collectors was built in Odeillo in 1967.[10] The operation of this collector is shown schematically in Figure 4. When sunlight strikes the rough blackened surface, the concrete becomes warm and heats the air in the space between wall and glass. Some of the solar heat is carried off by the air, which rises and enters the room, but a large portion of this heat migrates slowly through the concrete. The wall continues to radiate heat into the house well into the night. In the first two Odeillo houses, the

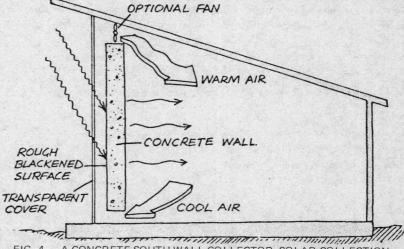

FIG. 4   A CONCRETE SOUTH WALL COLLECTOR. SOLAR COLLECTION, HEAT STORAGE, AND HEAT DISTRIBUTION ARE COMBINED IN ONE UNIT.
SOURCE: Anderson and Riordan, *The Solar Home Book*, 1976.

concrete wall was 2 feet thick and covered by double panes of glass 4 inches away. Solar heat migrates through this wall with a time delay of 10 to 15 hours. On a clear day, as much as 70 percent of the solar heat eventually reaching the rooms follows this path, while the remainder is brought in by natural convection. Other houses using concrete wall collectors in conjunction with sophisticated systems of venting and movable insulation are described elsewhere.[11]

Water is an even better heat storage medium than concrete. A cubic foot of water can store 624 Btu for a $10°F$ temperature rise, compared to only 320 Btu for concrete. Convection currents in water rapidly transfer the solar heat from the collection surface to the rest of the water volume and thence to the rooms. The collection surface remains cool, and a water wall collector loses less heat to the outdoors than a concrete wall collector. The major problem with water is that it is a liquid. The structures needed to confine water are often the major expenditure in the construction of a water wall collector.

Perhaps the best example of a water-wall collector is the Drumwall installed by Steve Baer and Zomeworks Corporation in his Corrales home.[12] The four south-facing walls of this "zome" structure are arrays of 55-gallon oil drums filled with water and stacked horizontally as shown in Figure 5.

FIG. 5   SOUTH FACE OF STEVE BAER'S ZOME. INSULATING SHUTTERS ARE LOWERED ON WINTER DAYS TO EXPOSE THE DRUMWALLS.
Courtesy of Zomeworks Corporation.

At night or on cloudy days, these collectors are covered by movable insulating shutters hinged at the base of each wall. Each shutter lies flat on the ground during sunny winter days with its aluminum inner surface reflecting additional sunlight onto the drums. The shutters are used accordingly to heat or cool the house as described earlier.

Ease of construction is one important reason for the emphasis on vertical wall collectors. Glass is much easier to install, weatherproof, and maintain in a vertical orientation. There are fewer structural complications with walls than with roofs, and you needn't worry about hail, snow, or freezing rain. The total amount of clear day solar heat on a south wall (see Fig. 6) follows the seasonal need very closely. That is, during the winter, when the sun is low in the sky, heat gains on a vertical wall are actually greater than in the summer when the sun is higher in the sky (see Fig. 2). With an additional 10 to 30 percent more sunlight reflected onto a south wall from fallen snow or other surfaces, it can actually receive *more* solar heat gain than a tilted surface.

In spite of the advantages of vertical wall collectors, there has been work done with flat roof collectors. Harold Hay and John Yellott pioneered this approach, placing containers of water on flat roofs to collect and store the sun's energy.[13] The solar heat collected in such roof ponds is radiated directly to the rooms below. Because the insolation on a horizontal plane is small in winter, these ponds must cover most of a roof. But very even heating is possible, and water temperatures as low as $70°F$ can still be used to heat the house.

Roof pond collectors are better suited to lower latitudes—those between $35°S$ and $35°N$, where the sun climbs high in the sky on a winter day. They are extremely well suited to summer cooling, which is more important than winter heating at these latitudes. With the collector on a flat roof, there's little need to orient the building towards the south. Freed from this constraint, the building can assume an external appearance similar to other flat-roofed structures in its immediate area.

Direct solar heating methods are still important for buildings which use integrated systems. Often the best approach is to combine a few south windows with wall collectors or roof ponds. Open interior layouts aid natural distribution of the heat, but a manually operated fan will help get more heat to the north side of a building. With a few more dampers, the fan can also cool the collector surface and send excess heat to a gravel or rock bed beneath the floor. As always, ample insulation and weatherproofing are mandatory.

As the functions of solar heat collection, storage, and distribution become even more separated, you must summon more and more external power to move the heat around. Expensive ductwork or piping is needed to control and divert this heat. The larger a building, the more likely it will need such an active or indirect system to supply its needs. Truly passive solar heating and cooling systems make sense only in small dwellings or in larger structures whose geometry insures that collection sites are close to all areas needing heat. The real beauty of passive solar architecture lies in its ability to liberate the inhabitants from dependency on external power sources. People can return to a closer relationship with their environment, which they discover to be much more beneficial than they had ever imagined.

References

[1] Knowles, Ralph. 1974. ENERGY AND FORM. Cambridge, Massachusetts: The MIT Press.

[2] Anderson, Bruce, with Michael Riordan. 1976. THE SOLAR HOME BOOK: HEATING, COOLING AND DESIGNING WITH THE SUN. Harrisville, New Hampshire: Cheshire Books.

[3] Ibid, pp. 78-79.

[4] Olgyay, Victor. 1963. DESIGN WITH CLIMATE. Princeton, New Jersey: Princeton University Press, p. 60.

[5] More comprehensive tables of solar heat gain, including hourly data and other latitudes, can be found in ASHRAE, HANDBOOK OF FUNDAMENTALS, 1972.

[6] Hutchinson, F. W. 1947. "The Solar House: Analysis and Research," in PROGRESSIVE ARCHITECTURE 28, May 1947.

[7] Anderson, op. cit., p. 87.

[8] Moorcraft, Colin. 1973. "Solar Energy in Housing," in ARCHITECTURAL DESIGN 42, October 1973, pp. 634-658.

[9] Anderson, op. cit., pp. 121-141.

[10] F. Trombe, et al. "Some Performance Characteristics of the CNRS Solar House Collectors," in PASSIVE SOLAR HEATING AND COOLING (Proceedings of and ERDA Conference held May 18-19, 1976, in Albuquerque, New Mexico), pp. 201-222.

[11] Anderson, op. cit., pp. 124-125.

[12] Baer, Steve. 1975. SUNSPOTS. Albuquerque, New Mexico: Zomeworks Corporation. See also 1973. "Solar House," in ALTERNATIVE SOURCES OF ENERGY 10, p. 8.

[13] Hay, H. R. and J. I. Yellott. 1970. "A Naturally Air-conditioned Building," in MECHANICAL ENGINEERING 92, no. 1, pp. 19-25.

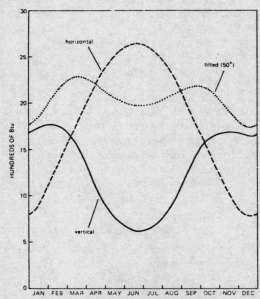

FIG. 6   CLEAR DAY INSOLATION ON HORIZONTAL SURFACES, AND ON SOUTH-FACING VERTICAL AND TILTED SURFACES. REFLECTED RADIATION NOT INCLUDED.

SOLAR ARCHITECTURE REVIEWS

DESIGN WITH CLIMATE

All the recent books on climate and architecture refer back to this one—the pivotal work in the field. Olgyay's explanations of the human comfort zone, climatic need timetables, sun path diagrams, shading effectiveness, and wind-breaks are the best work on these subjects I've seen. Since it first appeared, this book has been the basis for a lot of thinking and planning in architecture. It would be well worth your time to read it.

—C.M.

The desirable procedure would be to work with, not against, the forces of nature and to make use of their potentialities to create better living conditions. The structure which in a given environmental setting reduces undesirable stresses, and at the same time utilizes all natural resources favorable to human comfort, may be called "climate balanced." Perfect balance can scarcely be achieved except under exceptional environmental circumstances. But it is possible to achieve a house of great comfort at lowered cost through reduction of mechanical conditioning. We will do well to study the broad climate layout, then apply the findings, through a specific region, to a specific structure. And one must be ever alert to regional variations.

Design With Climate
Victor Olgyay
1963; 190 pp
$28.50

from:
Princeton University Press
Princeton, NJ 08540

or WHOLE EARTH TRUCK STORE

THE OWNER BUILT HOME

This book is the bible of low cost building and design principles for the owner-builder. It is divided into four sections: "Site and Climate," "Materials and Skills," "Form and Function," and "Design and Structure." The ideas that Kern presents deal with how to make your house as efficient as possible, in terms of construction methods, use, and energy consumption. The section on "Site and Climate," especially, discusses how to heat, cool, and ventilate your house naturally, without the use of gadgets. This idea of using the house itself to produce the desired climate is as important and revolutionary as developing new, improved windmills and solar collectors. This is not a construction manual; the detail needed for such a book is lacking. It is a book on design principles, and much of the information is not available anywhere else.

—Larry Strain

The Owner Built Home
Ken Kern
1975; 374 pp
$6.95

from:
Charles Scribner's Sons
Vreeland Avenue
Totowa, N.J. 07512

THE ARCHITECTURE OF THE WELL-TEMPERED ENVIRONMENT

A brilliant and lucid historical treatise on the past hundred years of architecture. Human beings altering and creating their own environment. Banham's hypothesis is that mechanical innovations such as air-conditioning and lighting have had a profound effect upon architecture. Many amazing architectural schemes are reviewed and illustrated including solar walls and natural ventilation. This book should be required reading for both the owner-builder and the architect.

—T.G.

The Architecture of the Well-Tempered Environment
Reyner Banham
1969; 295 pp
$5.95

from:
The University of Chicago Press
11030 South Langley Avenue
Chicago, Illinois 60628

or WHOLE EARTH TRUCK STORE

ENERGY AND FORM

With sources of cheap fossil energy dwindling rapidly, the architectural dictum "form follows function" begins to lose ground to a renewed awareness that building form must also reflect environmental stresses. This highly erudite work is a searching analysis of the implications natural energy flows have for the geometry of human settlements. Author Knowles takes his cue from the adaptive responses of plants, animals, and aboriginal peoples to the recurring flows of sunlight, wind and water. He argues that man-made structures and urban arrangements must respect similar principles in an energy-poor future. But his central premise is surprisingly hopeful—that an energy-conserving architecture responsive to natural variation will have the diversity of form essential for a rich and humane community life.

This is an important book. It is lavishly illustrated and extremely well-written, the culmination of ten years work by one of the most original thinkers in the field of modern architecture. Highly recommended.

—Michael Riordan

(The integration of climate and architecture) . . . requires recognition of the basic fact that, in general, the desirable internal state of a building is steady while the outside environment goes through cyclic changes. For this reason, while a static description is required and is certainly employed by designers today, the cyclic or dynamic character of the environment must also be recognized if the building form itself is going to help support a steady state.

Energy and Form
by Ralph Knowles
1974; 198 pp
$27.50

from:
The MIT Press
28 Carleton Street
Cambridge, MA 02142

or WHOLE EARTH TRUCK STORE

ARCHITECTURE AND CLIMATE

CLIMATE AND ARCHITECTURE BIBLIOGRAPHY

Several books are available which give a general introduction to the design of buildings that are responsive to climatic conditions. *Design With Climate* (reviewed above) is by far the best, with excellent overall graphic approach to form and micro-climate design of housing. *Man, Climate and Architecture* contains the best summary of climate design information for engineers and builders but is not as graphic as Groff Conklin's book, *The Weather-Conditioned House*. Jeffrey Aronin's *Climate and Architecture* is especially strong on biological climate modification and is less centered on sun control than other books. All cover generally similar ground, with some variation, and are worth reading for their different insights.

The best climate information source for designers is the excellent *Regional Climate Analyses* published in the *AIA Bulletin* in 1949-52 (reviewed below). *The Climatic Atlas of the U.S.* by Stephen Vishner is an excellent general source of information from 150 reporting stations but is graphically difficult to decipher and contains few design considerations for builders. Rudolf Geiger's *The Climate Near the Ground* has the most comprehensive coverage on the effects of sun, wind, vegetation and topography on micro-climatic conditions. It also covers some effects on dwellings and behavior.

—Tom Bender

Aronin, Jeffrey, 1953. CLIMATE AND ARCHITECTURE, Reinhold, out of print.
Conklin, Groff, 1958. THE WEATHER-CONDITIONED HOUSE, Van Nostrand Reinhold, out of print.
Geiger, Rudolf, 1950. THE CLIMATE NEAR THE GROUND, $18 from: Harvard University Press, 79 Garden Street, Cambridge, MA 02138.
Givoni, Baruch, 1950. MAN, CLIMATE AND ARCHITECTURE, $48.55 from: Applied Science Publishers, Ripple Road, Barking, Essex, England.
Vishner, Stephen, 1954. CLIMATIC ATLAS OF THE U.S. $13.50 from: Harvard University Press.

REGIONAL CLIMATE ANALYSES AND DESIGN DATA

An extremely useful and usable guide to macro-climates in 15 representative U.S. cities and their surrounding climatic regions. Average temperature profiles, solar radiation, wind speed and direction, precipitation and humidity are tabulated for each month in very readable graphs. What's more, the implications for building design are summarized in accompanying charts. The price is a rip-off (not the AIA's fault), so you might try to locate a friend's copy and Xerox just the pages you need for your area. Or look for back issues of the *Bulletin of the A.I.A.*, 1949-52, where all the analyses were originally published. A must for the serious designer.

—Michael Riordan

Regional Climate Analyses and Design Data
American Institute of Architects
1950; 200 pp
from: **$15.00** (pbk.)
Xerox University Microfilms   **$5.00** Microfilm
300 North Zeeb Road
Ann Arbor, MI 48106

SOLAR CONTROL AND SHADING DEVICES

There's a lot of new information these days about using building elements to collect winter sunlight, but what about keeping unwanted sunlight out during the summer? Fortunately, the definitive work on this subject has just been reprinted in paperback by its publisher. By far the most detailed examination of sun control I've seen, this book goes beyond the merely descriptive to provide the reader with excellent design tools that he or she will find invaluable in practical work. It explains the use of sun-path diagrams and shading masks to model shading conditions at different times of the day for every month of the year. The last half of the book is devoted to more than 50 real-life examples of shading devices designed with this method. All are accompanied by excellent drawings and photographs.

—Michael Riordan

Solar Control and Shading Devices
Aladar Olgyay and Victor Olgyay
1957; 202 pp.

$7.50

from:
Princeton University Press
Princeton, NJ 08540

THE AUTONOMOUS HOUSE

The authors candidly admit that an autonomous house is not the answer for everyone; nor is it *the* answer for our energy needs on a large scale. It is however, a novel idea that can provide people who are interested in self-sufficiency with a logical and workable model that can be applied to almost any climate. The autonomous house is not the drop-out, back to the land house but is a sophisticated, ingenious dwelling space that integrates the best renewable energy systems to fulfill total energy needs.

Sun, wind, heat pumps, waste recycling, water purification, batteries, fuel cells, and heat storage are the topics of the first seven chapters. Each area is covered in an overview approach on research in that field throughout the world. Many of the examples are from the U.S. and present a survey of the experimental solar and wind energy systems research in this country. The last chapter briefly integrates the systems discussed and applies them to a model house in England. Much of the research for this book was conducted at the Cambridge University School of Architecture.

— T.G.

The Autonomous House
Brenda and Robert Vale
1975; 224 pp.

$10.00

from:
Universe Books
381 Park Avenue South
New York, NY 10016

CLIMATE AND HOUSE DESIGN

Here's a condensed guidebook to intelligent shelter design, particularly in arid or humid tropical regions. Climatic factors are involved in all three steps of the design process, i.e., the sketch design stage, the plan development stage, and the element design stage. It's a very direct monograph on how to use solar, wind, and humidity information wisely. It has some short cut graphs to determine the sun path diagram, and a shadow angle protractor has been included. Olgyay's *Design With Climate* was the source of much of the material presented.

—C.M.

This neglect of the traditional designs and building methods, according to this group of experts, is partially due to the fact that virtually all professional training is oriented along the lines of the western countries, even in non-western societies. Not many architects from the developing countries would advocate their own countries' traditional designs and construction methods, national pride notwithstanding. On the contrary, it was mentioned that design solutions and building codes emanating from a by-gone colonial era and representing an entirely different cultural and climatic background are prevalent, and even promoted, in developing countries today. The minimum standards specified in many developing countries are hardly relevant to the situation of the masses of their populations who cannot afford even these minimum, which are not only of foreign inspiration but also urban in conception, whereas the populations concerned are mostly rural in character. It is far more practical, it was stressed, to start from what is found and what is feasible and improve upon the shanties and hovels that exist.

Climate and House Design
1971; 93 pp
Sales No. E.69.IV.11

$3.00   from:
United Nations
Sales Section
New York, NY 10017

PLANTS, PEOPLE AND ENVIRONMENTAL QUALITY

The best overview around on the incredibly varied uses of plants in human settlements. Although written for practicing landscape architects, Robinette's work is extremely useful for anyone planning a home, designing a small homestead, or using plants to control the local environment of a building to promote natural heating, cooling, and ventilation. For design work, it presents excellent guidelines for using plants as screens, walls, for noise reduction and air purification, to give green rather than glare to the eye, and for controlling such microclimatic variables as solar radiation, wind, and air temperature. With its wide scope, depth of treatment, and excellent use of graphics, the book is unique in its field. You can practically skip from one chart to the next collecting good ideas on how to use plants for landscaping needs. Much of the material in the book comes from hard-to-get scientific papers, foreign sources and university research reports. Robinette has done a great service by presenting all this information in such a readable, usable, and visually pleasing manner.

—Jerry Yudelson

Plants, People, and Environmental Quality
G. O. Robinette
1972; 139 pp

$4.35
   from:
U.S. Government Printing Office
Washington, DC 20402

or WHOLE EARTH TRUCK STORE

STREETS WITH TREES

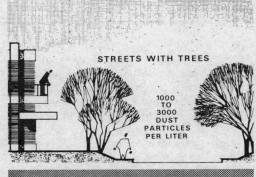

1000 TO 3000 DUST PARTICLES PER LITER

10,000 TO 12,000 DUST PARTICLES PER LITER

STREETS WITHOUT TREES

ENVIRONMENTAL TECHNOLOGIES IN ARCHITECTURE

Planning before building is a must. This is certainly true for structural considerations—for load, materials, stress etc., but it is equally true for environmental considerations—for heating and cooling, lighting, sanitation and noise control. This book discusses the environmental considerations in a rigorous technical manner with numerous tables and graphs mostly taken from the ASHRAE Guide and Data Book 1961. It is aimed at the architect and engineer rather than the owner/builder though the latter may, with some study, get something out of this text. Considerable time is spent with examples and the tables and graphs enable one to calculate heating requirements and heat loss, acoustic requirements, sanitation needs, lighting needs and designs, and electric power use, distribution and wiring. Only passing mention is made of "future" energy sources.

—T.G.

Environmental Technologies in Architecture
B. Y. Kinsey Jr., H. M. Sharp
1951, 1963; 788 pp

$17.95

from:
Prentice-Hall, Inc.
Box 500
Englewood Cliffs, N.J. 07632

SOLAR ARCHITECTURE

THE SOLAR HOME BOOK

I like to think of this book as the next progression in sophistication from the general overview offered in the ENERGY PRIMER. It presents an in depth look at solar energy applications to housing. Strong emphasis is placed on energy flow basics and passive solar energy systems and designs. Most homeowners interested in a real understanding of "how to do it" should start here. Architects, engineers and builders will find many of their most fundamental questions answered after reading this book. Extremely well illustrated, with a number of useful tables and diagrams, this is by far the best of a great number of books published on the subject.

—T.G.

This book stands a chance of becoming the definitive work in the field. Beautifully illustrated and perfectly organized, it introduces the reader to solar energy in homes with a descriptive cataloguing of those classic examples that have been around for many years—demonstrating diverse approaches and the fact that solar energy has been a practical resource for a long time.

—Donald Aitken

Of the numerous solar book titles appearing on the market these days, it is difficult for those of us who are in daily contact with solar material to be satisfied with the new additions. It is therefore refreshing to see a work like THE SOLAR HOME BOOK which not only simply states its purpose in the title, but delivers so completely throughout the text.

In summary, a very fine solar book, and more importantly a very useful solar book, and finally a solar book I can recommend to those requests for information that start out with— "Tell me everything you know about solar energy!".

—Dean Anson

Solar heating system in the Davis House. There are two distinct flow loops: collector-to-storage and storage-to-house.

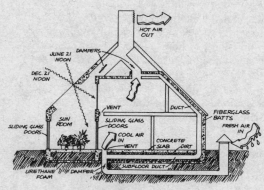

Cross section of the Jackson House—summer operation.

The Solar Home Book
Bruce Anderson, Michael Riordan
1976; 297 pp
$7.50   from:
Cheshire Books
Church Hill
Harrisville, NH 03450

or WHOLE EARTH TRUCK STORE

DESIGN FOR A LIMITED PLANET

DESIGN FOR A LIMITED PLANET is a smorgasbord of nearly 40 "alternative energy houses," most of them solar. Unlike the standard "solar book" approach that emphasizes system types—this book focuses on the designers and/or inhabitants of the homes. Each section is titled with the name(s) or the designer/inhabitant(s), and the text is laced with their own descriptions of what the house design expresses and their experience of living in it. The book's special delight—numerous photographs of house interiors and exteriors, and the people involved with them—gives the reader a good feeling for solar homes as living environments.

The houses themselves range from the very down-home to the very sophisticated, from the owner-builder and group projects to the offerings of architects and designers. The authors seem to go on a materials binge occasionally (three beer can houses?), and some of the homes seem to have made it into the book on the strength of far-out appearance rather than performance. There's also no hesitation to include those with a saga of malfunctions and impressive cost overruns.

Overall, however, the book provides an important visual sense of solar homes lacking in other books. And the words of the designers and inhabitants about living with natural processes, and the sense of sun, place and self they've gained from their involvement, is worth the price of the book alone.

—Lynn Nelson

Design for a Limited Planet
Norma Skurka and Jon Naar
1976; 215 pp.
$5.95

from:
Ballantine Books
457 Hahn Rd.
Westminster, MD 21157

or WHOLE EARTH TRUCK STORE

SOLAR DWELLING DESIGN CONCEPTS

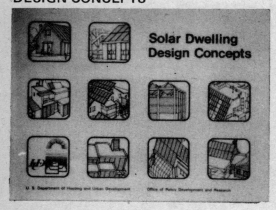

Solar Dwelling Design Concepts
AIA Research Corporation
1976; 146 pp.
$2.30   from:
U.S. Government Printing Office
Washington, DC 20402

One of the best AIA publications I've seen, this volume is a collection of recent work by the AIA Research Corporation and eleven subcontractors that explores the building design considerations involved in solar heating and cooling. The front end of the book is devoted to a quick survey of historical examples, a summary of available solar components and systems, and a discussion of the climatic factors involved in solar house design. Then an excellent chapter on site planning for solar energy utilization forms the core of the work. To close, the book examines the impact of solar energy use on traditional dwelling designs and explores a wide variety of innovative designs submitted by the contractors. There's a bit too much about active systems for my tastes. But it's illustrated well and the price is right—one of the few real bargains available in the field.

—Michael Riordan

THERMAL ENVIRONMENTAL ENGINEERING

This comprehensive technical text covers the basic principles of psychrometrics, moist air thermodynamics, solar radiation calculations for buildings plus heating and cooling via mechanical systems. It also includes information on how to figure thermal lag in heat flows through walls and roofs. The best general treatment available until someone writes a book focusing on the thermodynamics of direct solar heating.

—Lee Johnson
RAIN

Thermal Environmental Engineering
James L. Threlkeld
1970; 495 pp.
$16.15   from:
Prentice-Hall
Englewood Cliffs, NJ 07632

SOLAR ENERGY HOME DESIGN IN FOUR CLIMATES

This book presents solar house design methods and four examples of applications of the methods. Bruce Anderson and TEA approach solar design with an intelligent mix of energy conservation and architecture. As a result, the solar houses they design are both nice places to live and use less than half energy per square foot than required by conventional construction.

The strength of SOLAR ENERGY HOME DESIGN is that it presents an entire design process for solar houses. Building sites in Phoenix, Minneapolis, Boston, and Charleston are used as examples for the methods discussed. Site and climate analysis are emphasized. A final design for each city is presented, complete with notes on why certain design decisions were made.

I can quibble with many of the calculation methods that TEA recommends. However, they consider everything that should be considered and the calculations are valid first approximations. You're clearly better off with the TEA methods then you would be with no methods at all.

So, SOLAR ENERGY HOME DESIGN is well worth reading to see how one group of designers approach their work.

—Chip Barnaby

Solar Energy Home Design in Four Climates
Bruce Anderson and Total Environmental Action
1975; 198 pp

$12.75 paperback

from:
Total Environmental Action
Church Hill
Harrisville, NH 03450

ENERGY, ENVIRONMENT AND BUILDING

A somewhat outdated overview of energy conservation, solar space and water heating applications, wind and water power, water conservation, waste management and integrated natural energy systems for autonomous houses. It's an ambitious but somewhat scattered book, but its historical sense is good, with many examples of natural energy projects dating back to the 1930's. Other good features include the coverage of English, European, Commonwealth, and Japanese experiments with solar heated buildings and autonomous houses.

The book is somewhat difficult to follow, due to the profusion of topics, lack of a coherent organizing principle, and a rather forbidding layout. Nevertheless, designers of solar homes and natural energy environmetns will probably want to scan the book for such details as building layouts and floor plans for a number of solar dwellings and offices, heliothermic site planning, heat recovery systems, and autonomous house flow charts. There are some very extensive bibliographies which include many items written before 1970. If you're really into building integrated energy systems, it's probably worth having in your library.

—Jerry Yudelson

Energy, Environment and Building
Philip Steadman
1975; 287 pp.

$5.95

from:
Cambridge University Press
32 East 57th Street
New York, NY 10022

or WHOLE EARTH TRUCK STORE

SOLAR HEATED BUILDINGS — A BRIEF SURVEY

After years of heroic effort trying to keep us supplied with accurate descriptions of solar heated buildings that keep popping up everywhere, Bill Shurcliff is throwing in the towel. But he goes out in style—his 13th and final edition of SOLAR HEATED BUILDINGS is a classic. No brief survey, it's the most thorough compendium of information available on existing solar buildings and their heating and cooling systems. There are 306 pages crammed full of the usual detailed, down-to-the-last-Btu descriptions of the solar equipment in 319 (count 'em) houses, schools, and commercial buildings located in 38 states and 11 foreign countries. And there are even 24 pages of photographs! All in all, it's a *tour-de-force* of self-publishing and destined to become a collector's item. No serious designer, architect, or engineer should be without a copy.

—Michael Riordan

Fig. 2. Cumulative number of solar buildings in USA at end of indicated year.
A: Buildings completed.
B: Buildings completed, or under construction, or to be under construction within a few months.

**Solar Heating Buildings —
A Brief Survey (Final Edition)**
William A. Shurcliff
1977; 330 pp.

$12.00 prepaid

from:
Wm. A. Shurcliff
19 Appleton Street
Cambridge, MA 02138

SUN EARTH

Here is the coffee-table book on solar energy. This outsized and overpriced volume contains a lot of Richard Crowther's excellent ideas and approaches to the passive use of natural energies. But it lacks the systematic development of basic themes and the hard numerical data one needs in order to *do* anything with these ideas. Although the book is very well illustrated, the writing leaves much to be desired—often jumping from one topic to another without a hint to the reader. One wishes that the publishers had spent less on the reflective cover and all the pretty yellow suns inside and had used the money for a good editor.

—Michael Riordan

Sun Earth
Richard L. Crowther, AIA
1976; 232 pp.

$12.95

from:
Crowther/Solar Group
310 Steele Street
Denver, CO 80206

or WHOLE EARTH TRUCK STORE

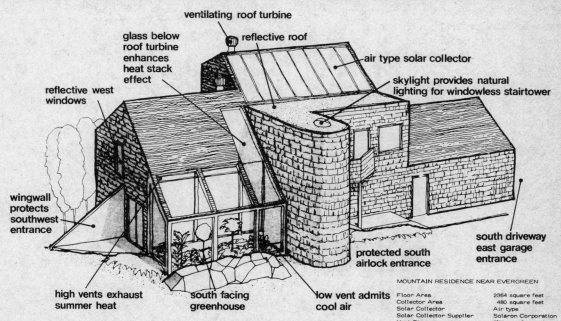

MOUNTAIN RESIDENCE NEAR EVERGREEN

Floor Area	2364 square feet
Collector Area	480 square feet
Solar Collector	Air type
Solar Collector Supplier	Solaron Corporation
Heat Storage	9½ tons of gravel

HOMEOWNER'S GUIDE TO SOLAR HEATING AND COOLING

Written in straight-forward, non-technical style, this book is intended to help the homeowner make decisions about using active solar heating and cooling. It goes into some detail on system components and their operation, and this is very useful for the person without a great deal of experience with mechanical or solar equipment. Included are large drawings of individual components, such as a tempering valve, and brief explanations of their function. Also included are descriptions of system operation, with some good material on solar swimming pool heating.

The book goes on to give performance and economic calculation methods. The approaches used are not very accurate and there are some errors and omissions in the information presented.

All in all, HOMEOWNER'S GUIDE is useful as a general guide, but should be read carefully to catch errors. I would suggest comparing information with THE SOLAR HOME BOOK or DESIGNING AND BUILDING A SOLAR HOUSE before making any final decisions about a system.

— Chip Barnaby

Homeowner's Guide to Solar Heating and Cooling
William M. Foster
1976; 196 pp.

$4.95

from:
Tab Books
Blue Ridge Summit, PA 17214

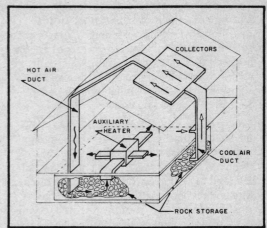
Fig. 2-7. A room heating system using solar-heated air, rock storage, a circulating blower, and an auxiliary heater.

SOLAR ENERGY FOR PACIFIC NORTHWEST BUILDINGS

This "regional" book contains a collection of information about using solar energy in the Pacific Northwest. This part of the country is often written off as hopeless for solar applications because of the area's persistent cloud cover and incredibly cheap hydro-power. One of the points made here is that the coldest days of the year are the clear ones, so solar may prove useful as a peak-saving energy source.

SOLAR ENERGY FOR PACIFIC NORTHWEST BUILDINGS has three sections: *Solar Data*, *Northwest Solar Houses*, and *Other Influences on Regional Architecture*. The *Solar Data* section includes handy radiation maps of Oregon; calculated insolation on tilted surfaces; graphs relating insolation to electrical demand in Eugene, Oregon; and sun angle charts for 44°N. A good discussion deals with analyzing a site for solar suitability. A calculation method is included for preliminary collector sizing.

The section on Northwest Solar Houses concentrates on the Mathew house in Coos Bay, Oregon and the Hoffman house in Surrey, B.C. Diagrams and cost analyses are included.

The final section of the book, *Other Influences on Regional Architecture*, makes some nice points about shading and sun control. It also includes climatological "timetables" that show expected weather conditions by time of day and month.

— Chip Barnaby

Solar Energy for Pacific Northwest Buildings
John S. Reynolds
1974 (revised 1976); 70 pp.

$4.00

from:
The Center for Environmental Research
School of Architecture and Allied Arts
University of Oregon
Eugene, OR 97403

SOLAR COOLING

These two conference proceedings contain a vast amount of information on the principles, modeling, functioning, and general status of solar driven cooling systems. Much of this information is available nowhere else. No professional directly concerned with solar cooling of buildings can afford to be without them. The curious engineer or architect will not find a ready answer here, though he may find it informative and interesting to read the various reports of what is going on in the field. He will probably come to a fuller appreciation of the complexities and problems; solar cooling systems have plenty of both.

These workshops were held in conjunction with ASHRAE and ISES and sponsored by NSF-RANN and ERDA. Consequently every ERDA or NSF-RANN supported R&D project is reported on; it was a command performance.

The principal systems of course are absorption chillers, mostly LiBr-H$_2$O. However Rankine cycle systems, in which a heat engine drives a vapor compression refrigerator, and also desiccant systems and miscellaneous other topics are discussed.

The second book (the 1975 workshop) is not only longer, but more tightly edited and substantive than the first. (By then there was more to report.) However both are well worth having.

—Marshal Merriam

Solar Cooling for Buildings (Proceedings of Los Angeles Workshop, February 1974)
Francis de Winter (ed.)
1975; 231 pp

$3.00

from:
Superintendent of Documents
U.S. Government Printing Office
Washington, D.C.
(stock 3800-00189)

The Use of Solar Energy for the Cooling of Buildings (Proceedings of Los Angeles Workshop, July 1975)
Francis and J.W. de Winter (eds.)
1976; 382 pp

$10.75

from:
National Technical Information Service
Springfield, VA
(Doc. No. SAN/1122-76/2)

SOLAR HEATED HOUSES FOR NEW ENGLAND AND OTHER NORTH TEMPERATE CLIMATES

This book is a study of the performance and cost effectiveness of various solar heating systems built into standard house types. Water, air, and passive systems are analyzed. Cape, saltbox, and townhouse designs are considered. The climate at Blue Hill, Massachusetts (near Boston) was used for the study. The conclusions are therefore applicable to many temperate locations acorss the country.

In preparing this book, MASSDESIGN used a simple computer program that simulates the thermal performance of a house and solar heating system through a weather year. The characteristics, predicted performance, and cost analysis for each system/house combination are presented.

Because the study was limited to more-or-less "traditional" house designs, heated to traditional temperatures (68°F day and 60°F night), the passive systems investigated did not turn out to be cost effective. More recent work by Doug Balcomb and others shows that passive systems in fact are probably more cost effective than active ones in many climates.

So, if you are interested in how active solar systems can be installed on conventional houses, you should take a look at SOLAR HEATED HOUSES. Other books that deal with both building form and solar systems are usually more useful, however.

—Chip Barnaby

Solar Heated Houses for New England and Other North Temperate Climates
MASSDESIGN
1975; 68 pp.

$7.50 paperbound

from:
MASSDESIGN
18 Brattle Street
Cambridge, MA 02138

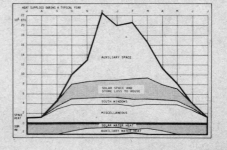

DESIGNING AND BUILDING A SOLAR HOUSE

This is an excellent book. The author is an architect who has been a designer or consultant for over 80 solar houses. The strength of the book is that nearly all the material presented is based on real experience—rather than saying "this seems like a good idea," Watson says "so and so did it this way, and this is what happened." He illustrates many of his statements with photographs.

Sections of the book cover solar energy history and principles of operation; passive systems; active systems; energy conservation; solar house design; and building a solar house. Appendices list references; solar houses in the U.S.; solar designers, architects, and engineers; and solar design calculations. The calculation method is presented complete with clear worksheets and thorough discussion of the approaches used. While the method recommended is not particularly accurate, it is easy to do by hand and provides good guidance when designing a house.

In the sections covering solar house design, Watson stresses energy conservation and balanced use of active and passive solar heating. He presents six solar heating alternatives, ranging from domestic hot water heating through large-capacity solar space heating. The six alternatives are subjected to economic analysis under different assumptions. The upshot is that large-capacity solar systems are economically recommended only when fuel prices are high, but that some degree of solar heating is always justified in northern climates. Watson therefore suggests an incremental approach—design for energy conservation and limited solar use for the present, but allow for installing a larger system in the future.

—Chip Barnaby

Figure 6.4. House designs based on combined climatic conditions.

Cool Climate
Maximum thermal retention
Maximum solar heat gain
Minimum wind resistance

Temperate Climate
Moderate thermal retention
Moderate solar heat gain
Slight wind exposure (humidity control)
Moderate internal air flow

Hot Dry Climate
Minimum solar heat gain
Moderate wind resistance (dust)
Moderate internal air flow

Hot Wet Climate
Maximum wind exposure
Maximum internal air flow

Cool climates. Partly underground massing, designs that enable snow accumulation to stand against walls or on roofs, and internal heat zoning especially with "air-locks" at entry ways, all help to reduce heat losses in cool-climate houses. These solutions have been used for centuries in traditional dwellings throughout Northern Europe, Scandinavia and North America.

Temperate climates. In temperate climates, the humidity which builds up inside the building and in the building materials needs to be controlled by exposure to the drying effect of sun and wind. This usually argues against the underground massing that could be used in colder climates. Greenhouses and other passive solar-heat devices can be used to advantage in temperate zones, while cooling needed for only a small portion of the summer can be provided by roof monitors or other ventilators which minimize additional energy requirements.

Hot-dry climates. Natural cooling and heating designs, interior courtyards (for sun and wind control), and partly underground placement of the building (for reduced heat gain in summer) all are appropriate for hot-dry climates where night temperatures fall well below day averages, even in summer. The glare and intensity of direct sunlight in hot-dry climates require sun controls at the windows, and in dry, desert areas, protection is needed against wind-blown dust and sand.

Hot-wet climates. The major objective in hot-wet climates is to maximize the effect of natural cooling by exposure to breezes. Walls with windows, screens and ventilating louvers will provide storm protection, insect control, privacy and variable light conditions. Interior courtyards and roof monitors can be used to produce natural chimney effects for ventilation. In some instances, a building may be raised to increase its exposure to the wind, as well as for insect and humidity control.

The generalized design recommendations illustrated here must be adjusted carefully to the specific conditions of building sites. This is the great opportunity in designing a home—to take advantage of the particular climate conditions at the building site to increase comfort by natural means.

Designing and Building a Solar House
Donald Watson
1977; 281 pp.
$8.95

from:
Garden Way Publishing
Charlotte, VT 05445
or WHOLE EARTH TRUCK STORE

THE FUEL SAVERS; A KIT OF SOLAR IDEAS FOR EXISTING HOMES

Prepared for a New Jersey community action program, THE FUEL SAVERS gives a brief but effective introduction to "winterization," and then launches into 20 projects for retrofitting homes to use solar energy, and to cut heat loss. Projects are generally chosen for their low cost and do-it-yourself potential, and range from insulating curtains and windowbox collectors to free-standing solar structures and freeze-proof solar water heaters. Each project has its own section that includes a basic description and possible variations.

Unfortunately, it's a bit of a struggle to get at the information due to poor visual presentation. However, as the title states, the projects are presented as "ideas" rather than as detailed how-to instructions, and THE FUEL SAVERS certainly accomplishes its purpose by providing a good introduction to the numerous approaches to low-cost energy retrofitting.

— Lynn Nelson

The Fuel Savers; A Kit of Solar Ideas for Existing Homes
Dan Scully, Don Prowler, Bruce Anderson, with Doug Mahone
1976; 60 pp
$3.00

from:
Total Environmental Action, Inc.
Church Hill
Harrisville, NH 03450

BASICS OF SOLAR HEATING AND HOT WATER SYSTEMS

This is one of the best visual introductions to solar energy systems available. Material is presented in a series of illustrated matrices showing various solar water and space heating systems in different modes of operation. This

form is also used to describe the BASICS OF SOLAR UTILIZATION (Climate, Radiation, Collector Orientation, etc.) and SOLAR COMPONENTS DESCRIPTION. For a broad yet succinct introduction to solar water and space heating systems this book is as good as they come.

— Richard Merrill

Basics of Solar Heating and Hot Water Systems
AIA Research Corporation
1977; 46 pp.
$4.00

from:
AIA Research Corp.
1735 New York Ave., N.W.
Washington, D.C. 20006

** STORAGE - DISTRIBUTION DIAGRAMS **

4. RADIANT HEAT DISTRIBUTION

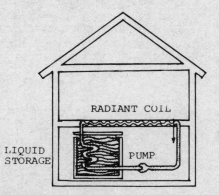

5. HEAT PUMP ASSISTED DISTRIBUTION

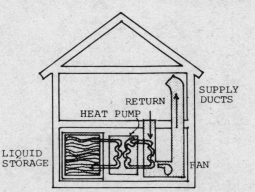

6. WARM AIR DISTRIBUTION FROM ROCK STORAGE

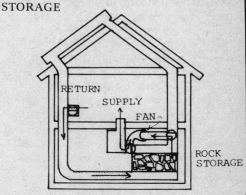

SIMULATION ANALYSIS OF PASSIVE SOLAR HEATED BUILDINGS

A fairly technical paper that tries to determine the optimum size and configuration for a passive solar heating system. With an extremely simple computer model, the authors study passive system responses to the climate of Los Alamos, New Mexico. They find that 70% solar heating is possible with a 6" thick water wall sitting behind double glass. Even better performance is possible with a thicker wall, the use of insulation over the glass at night, or tolerance of daily temperature swings of more than $10°F$. An important and seminal work.

—Michael Riordan

Simulation Analysis of Passive Solar Heated Buildings
LA-UR-76-1719, by J. Douglas Balcomb, James C. Hedstrom, and Robert D. McFarland
1976, 16 pp.

free with large self addressed stamped envelope

from:
Los Alamos Scientific Lab
Solar Energy Lab (Mail Stop 571)
Los Alamos, NM 87544

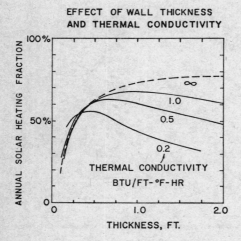

EFFECT OF WALL THICKNESS AND THERMAL CONDUCTIVITY

Yearly performance of a passive masonry wall as a function of thickness for various thermal conductivities.

PASSIVE SOLAR HEATING AND AND COOLING CONFERENCE

On May 18 and 19, 1976, most of the U.S. experts on passive solar heating converged on Albuquerque to compare notes on their work. The variety and success of their projects served to underscore the participants fervent belief that passive approaches to solar heating and natural cooling will become a dominant force in the near future. The conference PROCEEDINGS contain a wealth of information you couldn't hope to find elsewhere. In addition to descriptions of quite a few functioning passive solar houses, there are performance calculations for 6 buildings. Economics, computer simulation, and terminology were discussed in parallel sessions. Truly a landmark event and publication.

—Michael Riordan

Passive Solar Heating and Cooling
Conference and Workshop Proceedings
May 18-19, 1976, Albuquerque, New Mexico
1977, 345 pp.

price unknown

from:
Solar Energy Lab (Mail Stop 571)
Los Alamos Scientific Laboratory
Los Alamos, NM 87544

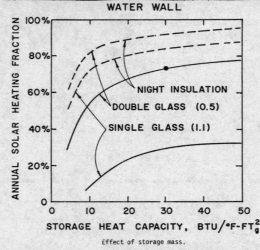

WATER WALL

Effect of storage mass.

SOLAR ORIENTED ARCHITECTURE

A fairly extensive survey of existing solar dwellings done by John Yellott and co-workers for the AIA Research Corporation. A total of 70 solar homes are described in tabular form with an illustration of each provided. Climatic and building data are given, as well as the method of solar collection and storage, heat distribution and auxiliary heating. At the close of this report, detailed descriptions and drawings of ten residences are used to illustrate four principal categories of solar dwelling design. The design considerations and trade-offs involved in each category are treated in some detail, too. A well-organized and very instructive work.

—Michael Riordan

Solar Oriented Architecture
Solar Energy Applications Team,
Arizona State University
1975; 142 pp.

$12.50

from:
Architectural Foundation Publications
College of Architecture
Arizona State University
Tempe, AZ 85281

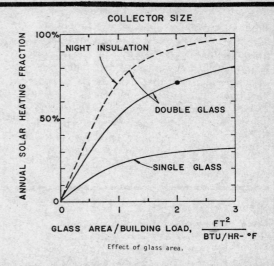

COLLECTOR SIZE

Effect of glass area.

SOLAR GREENHOUSES

SOLAR ENERGY-FOOD AND FUEL WORKSHOP

On April 5 and 6, 1976, the Environmental Research Laboratory (University Arizona), in cooperation with ERDA and the USDA, sponsored a workshop/symposium on "The Utilization of Solar Energy in Greenhouses and Integrated Greenhouse-Residential Systems." To date it is the definitive summary work on the solar heating and cooling of commercial greenhouses. Articles include heating methods involving: solar ponds; non-integral and integral flat-plate collectors (both fan-coil and hydronic systems); deep mine air; various hybrid systems plus techniques for energy conservation (thermal blankets, self-fogging roofs, liquid foam insulation, etc.) There are also a few good articles on residential heating with solar greenhouses. Throughout the Proceedings report special attention is given to performance data and economic payback. This is an excellent sourcebook.

— Richard Merrill

The solar pond is shown to the right of the two-module Agricultural Engineering greenhouse. Heat from the pond will be discharged in the covered and nearest greenhouse module. The adjacent module will be covered and heated conventionally.

Solar Energy Food and Fuel Workshop
Merle Jensen (ed)
1976; 261 pp.

$5.00

from:
Environmental Research Labs
University of Arizona
Tucson International Airport
Tucson, AZ 85706

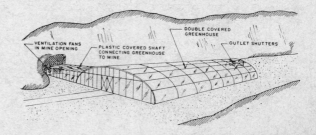

THE FOOD AND HEAT PRODUCING SOLAR GREENHOUSE

If you already own a home and are looking for an affordable way to solar heat is, THE FOOD AND HEAT PRODUCING SOLAR GREENHOUSE is a must. This highly readable gem of a publication will tell you all you need to know to design, build and operate a low-cost solar greenhouse attached to your home that also doubles as a year-round food-growing room.

The book begins with a clear discussion of basics, demonstrating how an "attached solar greenhouse" can be designed to produce enough heat for itself and part of the adjoining house as well. The actual greenhouse design presented—with the combined attributes of low cost, sturdy construction and good solar performance in a variety of regions—makes the book a special bargain. What follows is equally valuable—an extended section that takes you step by step through actual construction of your attached solar greenhouse. The chapter on greenhouse food production details what to plant when and where, plant care, and includes a good layperson's guide to natural control of greenhouse garden pests.

What's particularly reassuring about the book is that it's based on the authors' own extensive experience with solar greenhouses. Both authors have built numerous low-cost solar greenhouses from the Southwest to the Pacific Northwest as part of the Solar Sustenance Project.

Given the numerous examples of solar greenhouses (both attached and free-standing) in the book's "State of the Art" chapter, it appears that utilization of solar greenhouses is booming. If so, it's fortunate—and a tribute to the authors—that the first book out on the subject is such an excellent one.

— Lynn Nelson

The Food and Heat Producing Solar Greenhouse: Design, Construction, Operation
Rick Fisher and Bill Yanda
1976; 161 pp.

$6.00

from:
John Muir Publications
P.O. Box 613
Santa Fe, NM 87501

or WHOLE EARTH TRUCK STORE

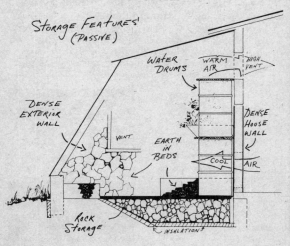

NOTI SOLAR GREENHOUSE

This report outlines briefly the performance and design of a free-standing, low-cost solar greenhouse designed for Northwest conditions. Located at Noti, Oregon the greenhouse was designed by students at the School of Architecture, University of Oregon. It is constructed of 2 X 4 framing and fir pole post and beam. Thermal mass is basalt stone serving as a north retaining wall and placed in a cavity between outside earth berm and inside chicken wire. Windows and doors are recycled, and the passive system is backed up with a wood stove. Glazing is double Filon. Performance, climatic and horticultural data are given. Now for the other regions and climates. . .

—Richard Merrill

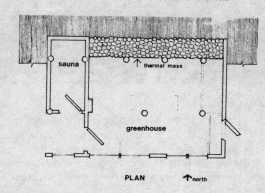

Noti Solar Greenhouse: Performance and Analysis
Eric Hoff, David Jenkins and Jim Van Duyn
1977; 32 pp

$2.00

from:
Center for Environmental Research
School of Architecture and Allied Arts
University of Oregon
Eugene, OR 97403

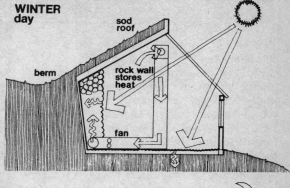

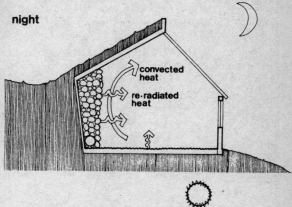

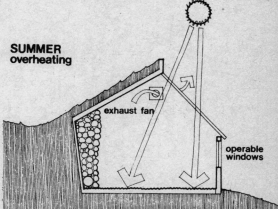

SOME OTHER SOLAR GREENHOUSE BOOKS

For a more detailed list of solar greenhouse references see pg. 181 in the Energy Primer.

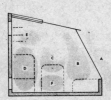

Profile: SPRING—SUMMER—FALL
A. Shading plants.
B. Fruiters. Tomatoes, cucumbers trained up twine. Trim foliage, squash, melons.
C. Seedlings, herbs, fruiters. Hydroponic table.
D. Low light, coolest greens. In late summer new fruiters can go here. Climbers, flowers.
E. Flowers, shade lovers.
F. Berries, shade lovers.

An Attached Solar Greenhouse
W.F. and Susan Yanda
1976; 18 pp

The Lightning Tree
Box 1837
Santa Fe, NM 87501
$1.50

A bilingual (Spanish), step-by-step manual for the construction of low cost attached solar-heated greenhouses. A really fine how-to manual.

The Passive Solar Greenhouse and Organic Hydroponics: A Primer
Rick Kasprzak
1977; 80 pp.

R.L.D. Publications
Box 1443
Flagstaff, AZ 86002
$5.00

Some good information on solar greenhouse design and organic hydroponics.

Solar-Heated Greenhouse
Bill and Marsha Mackie
1977; 16 pp

Sun Experimental Farms
835 Fleishauer Lane
McMinnville, OR 97128
$2.00

Describes construction of a 11 x 16', free-standing solar greenhouse with oil drums and graywater-fishtanks as thermal mass, fibreglass insulation and glazing, etc. Photos and diagrams.

Two Solar Aquaculture-Greenhouse Systems for Western Washington
Woody and Becky Deryckx
1976; 43 pp
$3.00

Hunter Action Center
Evergreen State College
Olympia, WA 98505

Describes the integration of small fish cultures into two free-standing solar greenhouses.

The Survival Greenhouse
James B. DeKorne
1975; 165 pp.

The Walden Foundation
Box 5
El Rito, NM 87530
$7.50

Describes the philosophy, design and management of a free-standing (pit) solar greenhouse that integrates hydroponic crops, small livestock, fish culture and wind power as a greenhouse ecosystem.

GROUPS AND DESIGNERS

OUROBOROS

One of the most active university goups in the area of energy conservation and architecture is at the University of Minnesota. Called Ouroboros, the project includes the design, construction and continuing study of two residences in the Minneapolis area. Both structures were built with student labor and local funding.

The first residence, Ouroboros South, is a 2000 square foot, two story house with energy conserving design features: a sod roof, well insulated walls, natural ventilation, south facing glass, a greenhouse, and a roof-mounted collector. Other features include a 5 kw Aero wind generator, a Clivus Mulstrum composting toilet, and an enclosed sewage system for recycling water and wastes.

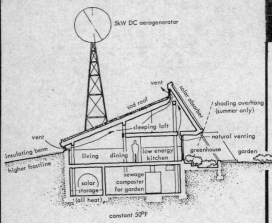

The second project, Ouroboros East, involves the retrofitting of a typical older urban home in St. Paul. When complete, the project will demonstrate methods of reducing energy consumption of an existing house. Water conservation measures are also included, and wastes are composted for re-use in an attached greenhouse and backyard garden.

In both projects, there has been a strong emphasis on local funding and community involvement. Ouroboros East now doubles as a living quarters for students and a community education center where people can learn about energy-conserving measures in classes or hands-on workshops

For an extension of these projects to a community scale, see the review of WINONA (pg.29).

— Michael Riordan

A book, OUROBOROS/EAST, describing plans for that project was published in early 1974. The proceeds from the book's sale will help finance the remodeling. It has loads of ideas, drawings, sources, and references and is highly recommended for natural and low-energy architecture freaks. In addition, it includes information on how to grow plants and keep bees—there's something for everyone.

— C.M.

Ouroboros/East
1974; 209 pp
$5.50

from:
University of Minnesota
School of Architecture
110 Architecture Building
Minneapolis, Minnesota 55455

ACCESS

So far, work at the Milwaukee test house, 2915 N. First Street, has concentrated on insulation, weatherproofing, shutters and storm windows. Together with thermostat set-backs and furnace modifications, they have achieved a 50% reduction in energy consumption. Work is presently underway for the installation of rooftop solar collectors, and experiments in water conservation, food production, composting toilets; wind energy systems are planned for the future. ACCESS is also involved, together with the School of Architecture at the University of Illinois at Chicago, in the planning of another demonstration project at 128 South Waller in Chicago. ACCESS Manuals No. 3 and No. 4 describe these projects in more detail.

— Michael Riordan

For information, write:
ACCESS Program
School of Architecture and Urban Planning
University of Wisconsin—Milwaukee
Milwaukee, WI 53201

CAMBRIDGE UNIVERSITY

The Department of Architecture at Cambridge publishes an amazing set of pamphlets dealing with autonomous housing, renewable energy sources and agriculture. Occasionally the information presented lacks depth and sometimes the geographical area discussed only applies to Britain. These papers, however, present one of the best places to start reading and understanding about houses which are energy self-sufficient (autonomous).

— T.G.

Although there are voluntary movements away from consumption growth there has so far been little analysis of the consequences; saving in one resource may lead to increased consumption of others, as seems the case with substituting autonomous plant and ambient resources for the central and non-renewable varieties. Using the household as a model of the world it seems clear that the most energetic and fortunate members have a duty to help the remainder, not only through technological and financial aid but by lowering the goals of attainment and redistributing the means to attain them.

Publications List Available Free from:

University of Cambridge
Department of Architecture
Technical Research Division

1 Scroope Terrace
Cambridge CB2 1PX England

McGILL UNIVERSITY:
Minimum Cost Housing Project

The objective of the Minimum Cost Housing program, initiated in 1971, is to produce original and contributive work in the context of world housing problems. Since this requires relatively long periods of time, we have introduced the concept of on-going research projects, within which students can make original contributions. Case studies complement workshop activities in these projects. Research is in progress in the areas of: BUILDING MATERIALS (Sulphur building, Re-use of waste materials, Garbage housing, Earth), SANITATION IN HOT AND COLD CLIMATES (Low-cost sanitation, Composting, Reduced-water washing, Wastewater treatment and use), ECO-DEVELOPMENT (Appropriate technologies, Rural housing, Cooperatives, Small-scale production).

Available books include 3 valuable additions to AT information:

- **ROOFTOP WASTELANDS**
  A description of a two-year project in community rooftop gardening, includes sections on organics, solar cold frames and container gardening. 1976, 32 p. **$2.00**

- **STOP THE FIVE GALLON FLUSH!** — **A Survey of Alternative Waste Disposal Systems**
  A classification and description of sanitation methods and a catalog of 66 systems. New chapter on composting. A very important book. 1976, 82 p. **$4.00**

- **THE ECOL OPERATION** — **Ecology + Building + Common Sense**
  Describes the experience of developing and building a habitable low-cost house which incorporates asbestos super-tiles of sewer pipes, sulphur lock-blocks, modular logs, roof-top solar stills and a re-cycling toilet. 1975, 128 p. **$5.00**

from:
The Secretary
School of Architecture
McGill University
3480 University St.
Montreal, Quebec
Canada
H3A 2A7

DESIGN GROUPS

The following are among the most well-established and reliable firms with broad experience in energy-efficient architecture and engineering. Contact them for technical help on a specific project, but try not to bother them with inquiries for free information—they're probably quite overworked and underpaid right now. This list is by no means comprehensive and does not constitute a warranty of their services.

Sky Therm Processes and Engineering Co.
2464 Wilshire Boulevard
Los Angeles, CA 90057

Harold Hay's firm is the place to go for the best advice on roof pond solar heating and natural cooling systems. Be prepared to pay for the information you seek.

Dubin-Bloome Associates
42 West 39th Street
New York, NY 10018

Fred Dubin and cohorts have engineered a number of medium-to-large scale energy-efficient and solar heated buildings—principally in the Northeast. Included in their projects are Grassy Brook Village, Cary Arboretum, and the GSA Building in Manchester, New Hampshire.

Zomeworks Corporation
P.O. Box 712
Albuquerque, NM 87103

Perhaps the most truly innovative solar design firm in the country. Besides turning out solar components and devices that astound you with their simplicity, Steve Baer and others in this company have designed some of the most effective passive solar homes in the Southwest.

The Solar Group, Inc.
2830 East Third Avenue
Denver, CO 80206

Richard Crowther, author of SUN/EARTH, is the principal of this design group. Very skilled and competent in all phases of active and passive solar system design.

Total Environmental Action, Inc.
Church Hill
Harrisville, NH 03450

A multi-disciplinary team of architects, designers and engineers "dedicated to providing ecologically sound environments that fulfill real human needs." Good in many areas of active and passive solar system design and very strong in building design. President Bruce Anderson wrote the manuscript that became THE SOLAR HOME BOOK.

Solar Service Corporation
306 Cranford Road
Cherry Hill, NJ 08003

When they can find time away from their many publications, Malcolm Wells and Irwin Spetgang are doing some of the most innovative work in the country in the realm of underground buildings and environmentally-sensitive architecture.

Living Systems
Route 1, Box 170
Winters, CA 95694

Jonathan Hammond, David Bainbridge, and others specialize in passive solar heating and natural cooling designs—especially for small-scale, residential buildings in warm-temperate climates. Also, they do a lot of work in neighborhood and urban planning for energy conservation.

For a more complete listing of architects and engineers with experience or interest in solar design services, see: LIST OF PROFESSIONALS ACTIVE IN THE DESIGN OF SOLAR-HEATED FACILITIES, from AIA/Research Corporation, 1735 New York Ave., N.W., Washington D.C. 20006.

SITE PLANNING

Often the character of a site — its topography, vegetation and soil composition — has a strong influence on the structures built thereon and the activities occurring therein. Perhaps the dominant theme of this excellent book is this responsiveness to the essential character of a site and its use, in designs for human habitation. While not strictly an "energy" book per se, SITE PLANNING contains much worthwhile advice for the energy-conscious designer. Building arrangements that complement an existing landscape will use much less fossil energy than those competing with it.

Lynch has a very democratic approach to environmental design. He sees it as a continuously evolving process involving many participants, including those who live and work in a building. The function of a site planner is to build freedom and flexibility into the initial design while ensuring that the essential character of the site is preserved.

—Michael Riordan

Site Planning
Kevin Lynch, 2nd Edition
1971, 384 pp.
$13.30

from:
The MIT Press
28 Carleton St.
Cambridge, MA 02142

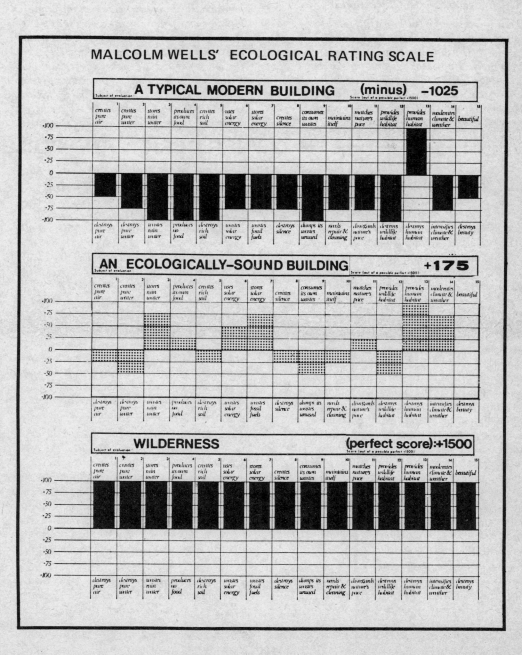

MALCOLM WELLS' ECOLOGICAL RATING SCALE

A TYPICAL MODERN BUILDING (minus) −1025

AN ECOLOGICALLY-SOUND BUILDING +175

WILDERNESS (perfect score):+1500

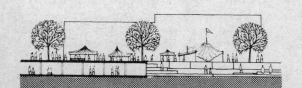

BUILDING LAWS AND SOLAR HEATING

There are several types of building laws: different codes which regulate the procedures and materials used in construction; housing codes that attempt to establish minimum standards for occupancy, sanitation, ventilation, etc.; zoning laws that govern the size and use of buildings; and, most recently, energy use standards for various kinds of buildings. In general, building laws are written and enacted from altruistic motives; the first building code provided minimum standards for New York tenements.

Building codes are the most visible kind of building laws. There are three major codes enforced in the United States. In all but a few states, it is left to local governments to decide which code, if any, will be used. Recently, building codes have been criticized for unnecessarily raising construction costs and hindering various energy saving practices. Although the codes are frequently updated, it is usually for the purposes of making them more restrictive. They often fail to note "new" construction techniques until they are quite outdated. The codes usually require maximum construction rather than necessary construction. They make it extremely difficult to use recycled materials, and are highly restrictive to experimental techniques.

Building laws make the construction of energy-saving homes more difficult. Alternatives to the flush toilet are almost universally prohibited. Graywater systems are similarly discouraged. Even when the owner is prepared to deal with miscalculations, many state and local ordinances require expensive conventional back-ups to solar or wood

heating systems. Ironically, it is the energy-conservation laws which can be the most restrictive. In California, for example, a new state law sets maximum limits on the amount of glass area in new homes, thus making it more difficult to design a passive home. Sometimes, the laws are ludicrous. A proposed law in California would require double glazing in all homes except those that are solar heated, presumably on the assumption that solar heat need not be conserved.

Building laws have a particularly harsh effect upon the person who wants to build his own home. The building codes only deal specifically with certain time-honored techniques particularly suited to the mass builder. If one chooses to use some other technique, particularly if it is an unusual but sensible approach like passive solar heating, one is certain to encounter unpleasant sessions with the building inspector. Construction procedures that use native or recycled materials are similarly discouraged. The codes (particularly electrical and plumbing) are written in complex, legal terminology. Probably the biggest gripe that the owner-builder will have with the code is that he must spend a lot of time, money, and energy complying with sections of the code that will never affect anyone but himself.

Building laws undoubtedly have their place in protecting homeowners from unscrupulous builders and from uncaring neighbors, but in their present form the laws are unnecessarily restrictive to homeowners willing to accept the consequences of their own mistakes. More emphasis should be placed on performance standards rather than specification standards. People ought to be allowed to build (or hire someone to build) their own

houses in any manner they choose, as long as it doesn't threaten the health or safety of the community.

—Alan Edgar

ENERGY CONSERVATION BUILDING CODE WORKBOOK. The City of Davis, California. 1976.

FREEDOM TO BUILD. J.F.C. Turner & R. Fichter, Eds., 1972. 301 pp.

HOUSING, Lane S., "Housing the Underhoused" 1970-1971. Sternlied and Sagalya, Eds., AMS Press, NY.

HOUSING CODE STANDARDS—THREE CRITICAL STUDIES. U.S. Gov't. Printing Office, Washington, D.C. Mood, E.W., B. Lieberman, O. Sutermeister.

THE OWNER-BUILDER AND THE CODE. Kern, Ken, T. Kogan, R. Thallon, 1976. Owner-Builder Publications, Oakhurst, California.

Information or copies of the major building codes can be obtained from:

National Building Code, American Insurance Association, 85 John Street, New York, NY 10038.

Uniform Building Code (ICBO), International Conference of Building Officials, 5360 South Workman Mill Road, Whittier, CA 90601.

Southern Standard Building Code, Southern Building Code Conference, 3617 8th Avenue South, Birmingham, Alabama 35222.

Basic Building Code, Building Officials Conference of America (BOCA), 1313 E. 60th Street, Chicago, IL 60637.

OWNER DESIGN AND BUILDING

Today, one of the biggest drawbacks to solar heating is its cost. Although the pro-rated cost of a solar system over its lifetime may be less than conventional heating, the initial cost of an active system is undeniably high. Similarly, a passive solar home designed by an architect will be expensive. The alternative is to design, and possibly build, your own solar-heated house.

From an architect's viewpoint, the chief difference between active and passive solar heating is that with active systems, the design can be modified and adapted to suit the layout of the house (heat can be forced to go anywhere); with a passive system, it is the house itself which must be adapted to the heating design (heat will only flow naturally to certain places). The passive solar house is essentially a single, large solar collector in which people live. These people demand that the house-collector be neither too hot nor too cold, that it be attractive and easy to clean, etc. The function of the architect, in this case, is to reconcile the needs and desires of the house's occupants with its function as a solar collector.

This design procedure requires a great deal of interaction between the architect and the future inhabitants of a house. He must become sufficiently familiar with their life-

style to know where to place the various living areas of the house in relation to the location of the heat. He must educate the family in the basics of passive design and in the everyday habits necessary to the maintenance of passive heating devices (window shades, insulating curtains, vents, etc.). Since there isn't a large body of passive design experience to draw upon (compared to conventional houses), the architect must design a totally custom and somewhat experimental house. This procedure is very expensive. The architect makes his living by charging a percentage of the estimated cost of the finished house. Since his expense in time will be large, his fee must be too; there is no incentive to design low-cost passive homes.

Faced with this reality, the prospective owner of a solar-heated home must either be prepared to spend a lot of money, or do the design work (and possibly the building) himself. Although a great deal of time is involved, the design of a passive solar home, or any house, does not require a great deal of technical skill or knowledge. Whatever information is needed is available to one who has the time and patience to look for it.

Having designed a passive home, the future homeowner can reap even greater savings and satisfaction by actually building it.

Although building a house doesn't require any special skills or experience that can't be learned on the job, it does require a commitment to spend at least several months (bare minimum) steadily building. A listing of some books (there are many more) of value to the owner-designer/builder follows.

One cautionary note: at this time there doesn't seem to be any book that really deals with the needs and specific problems of the urban (or suburban) owner-builder. Most of the books geared to the amateur house-builder focus on techniques suited to rural areas and recommend practices that may not comply with local building laws. Books that do cover standard construction practices are mostly designed as texts or trade manuals and are more technical and large-scale than the owner-builder requires. What is needed are books for urban owner-builders faced with hostile officials, indifferent lumber yards, and amused neighbors. Everyone can't move to the country. But designing and building a solar tempered house can be a viable route to low-cost, energy-conscious housing for anyone able to acquire a building site. Hopefully, more and more people will find this out.

— Alan Edgar

DESIGN AND PLANNING

*Most amateur housebuilding books advise you
not to attempt your own planning. If you're
primarily interested in a 1950's style tract
house or a "vacation" home, then certainly
investigate the many companies selling sets
of ready-drawn plans; their catalogs can be
found on any newsstand. If you have access
to or can afford an architect, you'll be months
ahead using his services. If neither of these
fit your circumstances, the following books
will help you design your house and draw your
own plans. They did for me.*
— *Alan Edgar*

**Do-It-Yourselfers Guide to Home Planning
and Construction**
William Clarneau
Tab Books
Blue Ridge Summit, PA 17214
$4.95

Although this book is by no means what its title
claims, it is still well worth reading. It's a highly
readable account of the author's experiences building his own home (in a city, no less).

Dwelling
River
Freestone Publishing Co.
Albion, CA 95410
$5.00

An interesting review of rural housing alternatives;
interview format explores the thought processes
behind houses; good discussions of basic shelter and
building codes; good to get you thinking about your
individual shelter needs.

Shelter
Lloyd Kahn, ed.
Random House
Westminster Maryland 21157
$6.00

Another book to start you thinking; a photo book
of how people everywhere shelter themselves; if
you're tempted to buy a kit or set of plans, look at
this book and read the following book first.

Your Engineered House
Rex Roberts
M. Evans and Company, Inc.
New York, N.Y.
$4.95

One of the first owner-builder books, it reflects its
age in a general lack of concern for energy/materials
conservation (some sections also have a markedly
sexist tone); still, this book is the best to check your
preconceived notions against: it punches holes in a
lot of long-accepted practices.

Owner Built House
Ken Kern
(see review, pg. 88)

Valuable chiefly in areas without building codes.

Sunset Books
Lane Publishing Co.
Menlo Park, CA
$2.45 (generally)

These books are filled with good ideas; much specific
information useful in laying out rooms; nice graphics.

From the Ground Up
John Cole and Charles Wing
Little, Brown & Co.
$7.95

This book is a "must-read" before you start. It's
full of confidence-building information of value
throughout the design and construction stages of
your house. The authors assume that you'll do all
your own work. While the book is primarily aimed
at rural builders, it will be of value to any owner-
builder. The section on stresses and load calculations
is particularly useful and not to be found in any
other amateur book. This and the following book
deal specifically with solar tempered houses.

Low Cost, Energy Efficient Shelter
Eugene Eccli, ed.
Rodale Press, Inc.
Emmaus, PA
$5.95

This is another book that's valuable throughout the
whole housebuilding process. Emphasizes energy-
efficient design . . . often a bit austere. Very good
section on sealing up your house. Goes into more
detail than most books on financing and codes. Very
much into small houses.

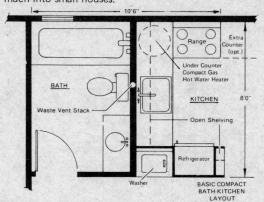

BASIC COMPACT
BATH-KITCHEN
LAYOUT

Drawing Plans for Your Own Home
Jane Curran
McGraw-Hill
New York
$19.95

Package containing a text, 6 drawing templates, and
some paper; very basic approach presupposes a total
lack of drafting experience; enough information to
get you started but probably not enough advanced
instruction to get you through a full set of plans.

Architectural Drawing and Planning (2nd ed.)
William Goodban and Jack Hayslett
McGraw-Hill
New York

The best text on architectural drafting. It assumes
some knowledge of basic mechanical drawing. Lots
of handy information in appendices; a must if
you're drawing your own plans.

National Construction Estimator
Gary Moselle
Craftsman Book Co.
Solana Beach, CA
$7.50

This book will save you many phone calls and trips
to the store as you work out your house plans; gives
prices of most materials and labor involved in
residential construction. Great for quickly comparing
the costs of different solutions to a problem.

PASSIVE SOLAR HEATING DESIGN

*There are only a few books presently available
that present passive solar design concepts for
the layman.*

The Solar Home Book, pg. 90
Passive Solar Heating and Cooling Conference,
pg. 94
Solar Energy—Food and Fuel Workshop, pg. 94
The Food and Heat Producing Solar Greenhouse,
pg. 95

From the Ground Up (see above)
Low Cost, Energy Efficient Shelter (see above)

BOOKS ON BUILDING

Illustrated Housebuilding
Graham Blackburn
Overlook Press, Woodstock, N.Y.
$4.95

Very clear, readable instructions on building simple
houses. Good line drawings make construction procedures quite clear.

The House Building Book
Dan Browne
McGraw-Hill, New York
$5.95

This book is a guided tour with a carpenter through
every step involved in conventional frame construction. A very useful book as well as being enjoyable
to read. A good companion to the next book.

How to Build a Wood-Frame House
L.O. Anderson
Dover Books, New York
$3.50

Methodical description of all procedures involved
in wood-frame construction; clear line drawings
make up for lack of detail in text and are well
worth the low price of the book; reprint of USDA
Forest Products Lab handbook; one of the best.

Step-by-Step Housebuilding
Charles Neal
Stein and Day, New York
$6.95

This book uses photographs to illustrate its text;
unfortunately, they're not as clear as good line
drawings; however, they are useful to show how the
finished product actually looks.

Build Your Own Home
Robert C. Reschke
Structures Publishing Co.
Farmington, Michigan
$6.95

Very much into pre-drawn plans, pre-fabricated
roof trusses, how to build a tract house. If that's
what you want, this is the book.

Modern Carpentry
Willis H. Wagner
Goodheart Willcox
South Holland, IL
$11.92

A textbook intended to supplement a live teacher;
very clear, non-technical but not a how-to source.
This book's chief value is in acquainting the owner-
builder with professional procedures and terminology.

Practical House Carpentry
J. Douglas Wilson
McGraw-Hill
$3.95

This book is probably more detailed than most
owner-builders need. Good reference if your design
has any complex shapes.

Dwelling House Construction
Albert G.H. Dietz
MIT Press
Cambridge, MA
$3.95

An old classic that's probably of more value to
someone restoring an old house than building a new
one.

WATER

HARNESSING THE POWER OF WATER
Robin Saunders
Mechanical Engineer

Most everyone has witnessed the destruction caused by torrential floods, the subtleties of weathering or erosion, the power of wave motion, the strength and mystique of grand rivers, or the gentleness and swiftness of small streams. The power of water has the capacity for destruction and useful work.

Essentially water power is a form of solar energy. The sun begins the hydrologic cycle by evaporating water from lakes and oceans and then heating the air. The hot air then rises over the water carrying moisture with it to the land. The cycle continues when the water falls as precipitation onto the land, and the potential energy of the water is dissipated as the water rushes and meanders its way back to the lakes and oceans.

The potential of water at an elevation above sea level is one of the "purest" forms of energy available. It is almost pollution free (when not contaminated) and can provide power without producing waste residuals. It is relatively easy to control and produces a high efficiency. From 80% to 90% of controlled water energy can usually be converted to useful work. This is dramatic when compared with the 25% to 45% efficiencies of solar, chemical and thermal energy systems. As a result, large and small rivers around the world have been dammed and waterwheels and water turbines installed to capture the energy of water.

Here, we will be concerned with small hydro-plants that can service the needs of individuals and small communities. In many cases even very small streams can be harnessed to produce power. The power that can be developed at a site is calculated as the rate of flow of water (measured in cubic feet per minute) multiplied by the "head" or vertical distance (measured in feet) the water drops in a given distance. It is these two quantities which must first be measured to see if they are adequate to develop a hydro-plant.

Most hydro-power installations will require the construction of a dam. A dam can increase the reliability and power available from a stream. It can also provide a means by which to regulate the flow of water and can add to the elevation of the water (by making it deeper) thus providing greater head to operate the wheel or turbine.

Both waterwheels and turbines deliver their power as torque on a shaft. Pulleys, belts, chains or gear boxes are connected to the shaft to deliver power to such things as grinding wheels, compressors, pumps or electric generators.

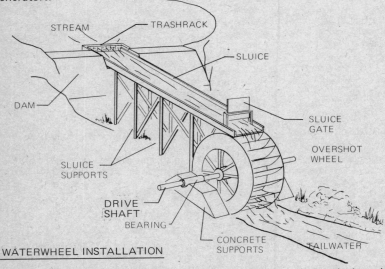

WATERWHEEL INSTALLATION

Waterwheels are the old-style, large diameter, slow turning devices that are driven by the velocity of the weight of the water. Because they are slow turning they are more useful for producing mechanical power for grinding and pumping than electrical power. The mechanical power can be connected by pulleys and belts to drive saws, lathes, drill presses and other tools. Waterwheels can provide a small amount of electric power (up to 10 HP) but it involves complex and expensive "gearing up" to produce the necessary speeds to actuate an electric generator.

Water turbines are preferable for producing electricity from the stream. Turbines generally are small diameter, high rpm devices that are driven by water under pressure (through a pipe or nozzle). When coupled with a generator, even relatively small turbines can provide electricity for most homestead needs.

Economics

Perhaps the greatest stumbling block to utilizing water power will be cost. Purchasing a manufactured waterwheel or turbine will obviously be more expensive than building your own. (In the Reviews/Water section a few companies and their relative costs are reviewed.) However, only a few of the many different types of waterwheels and turbines should be attempted by the home-builder. This includes most of the lower technology waterwheels but only one of the turbines. The Banki Turbine can with some time and perhaps some technical assistance be home-built. The Pelton Wheel and the Reaction Turbines are probably best left to the manufacturer. A home-built unit could be inefficient as well as downright dangerous. The cost of the waterwheel or turbine is only a portion of the over-all cost of installing a hydro-power system. Building dams also costs money. Here, the expense can be small if you use indigenous materials but if concrete and steel reinforcing are used the cost can be much greater. Then comes the expense for pipe or sluice to carry the water to the wheel or turbine. Depending upon the amount of flow and the distance involved this could be a very minor expense or a considerable one.

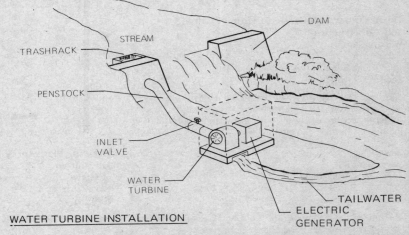

WATER TURBINE INSTALLATION

Probably the largest expense besides the wheel or turbine itself will be the gears or pulleys, the shafting and the electrical generator. A 1 KW generator should be obtainable for under $200 new and considerably less if surplus or used. A new 10 KW unit should start in the $800 range. The pulleys, drive belts and chains or gears needed to "gear up" the power of the waterwheel or turbine should be rather inexpensive.

The cheapest electrical generating hydro-power system that could be installed, dam, generator, turbine and all would cost a minimum of $500, and more likely $1,000. On the other end of the scale, a manufactured turbine, with dam and piping being constructed by the owner could cost from $3,000 to $10,000.

It is difficult to determine the long range cost of building or buying a hydro-plant, as opposed to hooking up with the local utility company. At present it is probably cheaper to go with the utility. But as the age of cheap utilities and the promise of something-for-nothing from nuclear power is disappearing, building an independent power source will become a more economical investment.

Environmental Considerations

Waterwheels and turbines in and of themselves have a negligible effect on the environment. However, the damming of a river or stream, a necessity with most installations, has an important and sometimes irrevocable effect upon the long-term ecological balance of that particular environment. Certainly dams can create a better environment for some animals and plants, and they can and do prevent natural disasters such as floods and severe erosion. But it is important to know that by building a dam, you are also creating a pond or lake where a stream or river used to exist; that you are flooding an already existing river ecosystem, encouraging the accumulation of silt, and perhaps providing a breeding ground for mosquitoes. The resulting pond or lake behind a dam also usually raises the water table behind the dam (as a result of seepage) and lowers it below the dam. Innumerable other changes are effected by the construction of a dam, and it is generally fair to say that the larger the dam the greater the changes. It is therefore of primary importance to foresee the ecological impact of installing a hydro-plant, and if necessary, to forego that particular site plan or the entire project.

WATER RIGHTS

Water is subject to a complex array of laws and regulations concerning its use. Wherever you may settle, some rules and regulations about water will surely follow. This is particularly true in the west and southwest states where arid conditions mean that almost all water that flows needs to be used by someone. Generally in the U.S. there are two types of water rights associated with property: Riparian rights and Appropriative rights.

RIPARIAN RIGHTS were originally brought to the United States through English common law. These laws were easily adapted as water laws for the eastern and midwestern U.S., where the supply of water was similar to that in England. Riparian rights are usually defined in the following way: The owner of any land that contains or is adjacent to a water course has a common law right to "reasonable use" of the natural flow. This right is shared equally by all other properties along the water course. And each is subject to the equal rights of all other riparian properties, or in some states, the higher priority of upstream users. This "reasonable use" means sharing the available water for domestic use, and then using excess water for irrigation of commercial crops or watering livestock.

The usual distinction between riparian and appropriative users is the concept of equality among riparian users as compared to the time priority "first come, first served" among appropriative users. APPROPRIATIVE RIGHTS evolved in the arid portions of the country where little rainfall and periods of drought forced a different tactic for controlling water use. In these places, water that is used for other than domestic purposes has to be claimed through a legal process to establish appropriation use. This means that the first person to make a claim on the water can have the exclusive use of as much as he/she needs; later settlers up and down stream have the right to the use or claim of any water left over. The theory is that water is so limited that there is not enough to share.

Appropriative rights are the exclusive law in Montana, Idaho, Wyoming, Nevada, Utah, Colorado, Arizona, New Mexico and Alaska. The states neighboring this drier region have a mixture of both riparian and appropriative rights; Washington, Oregon, California, North and South Dakota, Nebraska, Kansas and Texas.

The usual procedure for appropriative rights (assuming there is enough extra water available) requires public notice and perhaps a public hearing. If your claim is challenged, and it is important enough to warrant going to battle over the issue, a lawyer may be necessary (be sure to find one who knows about these matters; not many do).

Some older properties have mining or manufacturing claims that are listed by use of the water, rather than by a specific quantity of water appropriated. These may be listed as simply the amount of water required to operate a certain number of machines, or for certain processes that require water for washing or sluicing. In some areas these rights may still be on the books and tied to the property rights. To properly research these rights is a job for an historian (this should be part of the title search when purchasing a piece of property).

The local water districts and state water resource departments each has its own set of rules and rituals to obey in obtaining a dependable supply of water. Study them carefully; in many cases they can be circumvented though it is usually best to follow the proper procedures and establish a legal claim for your supply of water.

Once the claim has been filed, you must show ability to exercise the claim for appropriation (similar to mining and homesteading claims). This is usually done by filing a statement with the state's Department of Water Resources, of quantities and schedules of water removed. It may even be wise to file a report of how much is removed under the riparian rights; this can insure the supply in case someone upstream should attempt to cut it off.

Any local soil conservationist or Resources Conservation District (an agency of the U.S. Department of Agriculture) can usually be helpful in supplying information and assistance, not only concerning water rights problems, but also about dams, reservoirs and soils in the area. In addition, you should check with your local authorities and possibly file plans for your dam with them.

Water rights laws are some of the *most* involved around. There have been volumes of court discussions concerning the relative priorities of public, private, industrial, irrigation, hydro-electric, mining and navigational needs. Not only are they complicated, but water rights are often highly localized . . . differing considerably from state to state, county to county and city to city. In California, for example, there are such additional rights, besides the riparian and appropriative as: ground water, prescriptive, developed, surface, consumer, Pueblo, spring water, combinations, governmental, and used water. The rule then should be to investigate the laws and the local regulations before planning use of any water, either for irrigation, general water supply, or power.

MEASURING AVAILABLE WATER POWER

Assuming that you have overcome any legal, economic or environmental problems on your property you can now begin planning your hydro-power plant by calculating how much power is available from your stream. The amount of horse-power possible is determined by; (1) what quantities of water are available (the flow), and (2) what the drop or change in elevation (the head) along the water course is.

Measuring Flow

The volume of water flowing is found by measuring the capacity of the stream bed and the flow rate of the stream. Accurate measurements of the water flow are important for decisions about the size and type of water power installation. You will want to know; (a) normal flow and (b) minimum flow. If a unit is built just on the basis of normal flow measurements, it may be inefficient or even useless during times of low flow. If it is built just on the basis of minimum flow measurements, which usually occur during the late summer, the unit may produce considerably less power than possible. All water courses have a variation of flow. There are often daily as well as seasonal differences. The more measurements you make throughout the year, the better estimate you will have of what the water flow really is. Once you know the stream's varying flows, a system can be built to operate with these flows.

The Small Stream

$$Q = \frac{V}{S}$$ where
Q = flow rate in cubic feet per second (cfs)
V = volume of bucket in cubic feet
S = filling time in seconds   (Eq. 1)

The easiest way to measure the flow rate of a small creek or stream with a capacity of less than one cubic foot per second, is to build a temporary dam in the stream. Channel all of the water flow into a pipe or trough and catch it in a bucket of known volume. Measure the time it takes to fill the bucket and use Eq. 1 to find the flow rate.

The Medium Stream: The Weir Method

$Q = T \times W$ where
Q = flow rate in cubic feet per second
T = flow value in cubic feet per second *per foot of width of the weir*, read from a standard table (Table I)
W = width of the weir in feet   (Eq. 2)

The weir method is used for measuring flows of medium streams with a capacity of more than one cubic foot per second. Basically a kind of water meter, a weir is usually a rectangular notch of definite dimensions, located in the center of a small dam. By measuring the (1) depth of water going over the weir and referring to standard tables, and measuring (2) the weir width, the volume of flow can be accurately calculated.

In order for standard tables to apply, the weir must be constructed of standard proportions. Before you build the dam, measure the depth of the stream at the site. The depth of the weir notch "H" (see Fig. 1A) must equal this. The weir notch must be located in the center of the weir dam (see Fig. 1B) with its lower edge at least a foot above the downstream water level. The opening must have a width at least three times its height (3H), and larger if possible. The notch will then be large enough for the water to pass through easily to measure the large flows, and yet not too large so that it cannot also accurately measure the small flows.

The three edges of the opening should be cut or filed on a 45° slant downstream, to produce a sharp edge on the upstream side of the weir. The sharp edges keep the water from becoming turbulent as it spills over the weir, so that measurements will be accurate.

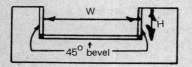

FIG. 1A WEIR

W SHOULD BE ABOUT 3 TIMES H. THE THREE SIDES ARE CUT TO A 45° BEVEL; SHARP EDGE PLACED UPSTREAM

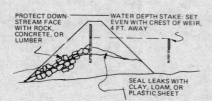

FIG. 1C WEIR DAM CROSS-SECTION

The weir dam need not be permanent, although if left in, it is convenient for continuously monitoring stream flow. The temporary dam can be made simply from logs, tongue and groove lumber, scrap iron, or the like. The dam must be perpendicular to the flow of the stream. And, when the weir is installed, be certain that the sides are cut perpendicular to the bottom, and that the bottom is perfectly level.

All water must flow through the weir opening, so any leakage through the sides and bottom of the dam must be sealed off. Side and downstream leakage can be stopped with planks extended into the banks and below the bed of the stream. Upstream leakage can be sealed off with clay or sheet plastic.

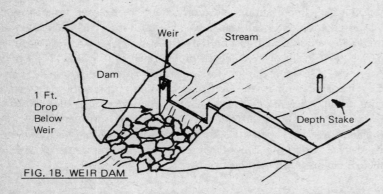

FIG. 1B. WEIR DAM

The accepted method of measuring the water depth over the weir notch (in lieu of putting a ruler in the weir notch) is to drive a stake in a spot accessible from the bank and at least four feet upstream from the weir. The reason for placing the stake at this distance is that the level of the water begins to fall as it nears the weir, where the water forms a crest. Four feet is a safe distance away from the weir to avoid measuring this lower water.

Pound the stake down until the top of the stake is exactly level with the bottom of the weir opening. A level can be established by placing a plank between the bottom of the weir opening and the stake, and using a carpenter's level. To measure the "head" (depth) of water flowing over the weir, allow the stream to reach its maximum flow through the weir, and place a ruler on the stake. Then directly measure the depth in feet of water over the stake.

Having measured the depth of the water, refer to the Flow Rate Weir Table (Table I) for the flow rate for that depth labelled "H" for Head. The flow rate in Table I is given in cubic feet per second for each foot of width of the weir. It is necessary to multiply the flow rate by the width of the weir in feet, to find the actual flow rate. For example: If the depth (H) of the water is 1 foot then the given flow rate from Table I is 3.26. To find the flow over a weir that is 4 feet wide, multiply 4 x 3.26 which equals 13.04 cubic feet per second.

The Large Stream: Float Method

The following method is not as accurate as the previous two. It is impractical to dam a larger stream and measure it for preliminary study, but with large amounts of water, a precise measurement is probably not so important.

TABLE I — FLOW RATE WEIR TABLE

Table for rating the flow over a rectangular weir.
Flow (Q) is in cubic feet per second per foot of width of the weir.
Multiply the flow value given times the width of the weir in feet to find the actual flow rate.

H, head (feet)	Q, flow (cfs)	H, head (feet)	Q, flow (cfs)	H, head (feet)	Q, flow (cfs)	H, head (feet)	Q, flow (cfs)
.05	.037	1.05	3.51	2.05	9.37	3.05	16.66
.10	.105	1.10	3.76	2.10	9.71	3.10	17.05
.15	.193	1.15	4.01	2.15	10.05	3.15	17.45
.20	.297	1.20	4.27	2.20	10.39	3.20	17.84
.25	.414	1.25	4.54	2.25	10.73	3.25	18.24
.30	.544	1.30	4.81	2.30	11.08	3.30	18.65
.35	.685	1.35	5.08	2.35	11.43	3.35	19.05
.40	.836	1.40	5.36	2.40	11.79	3.40	19.46
.45	.996	1.45	5.65	2.45	12.14	3.45	19.87
.50	1.17	1.50	5.93	2.50	12.51	3.50	20.28
.55	1.34	1.55	6.23	2.55	12.87	3.55	20.69
.60	1.53	1.60	6.52	2.60	13.23	3.60	21.10
.65	1.72	1.65	6.83	2.65	13.60	3.65	21.53
.70	1.92	1.70	7.13	2.70	13.97	3.70	21.95
.75	2.13	1.75	7.44	2.75	14.35	3.75	22.37
.80	2.34	1.80	7.75	2.80	14.73	3.80	22.79
.85	2.57	1.85	8.07	2.85	15.11	3.85	23.22
.90	2.79	1.90	8.39	2.90	15.49	3.90	23.65
.95	3.02	1.95	8.71	2.95	15.88	3.95	24.08
1.00	3.26	2.00	9.04	3.00	16.26	4.00	24.52

$Q = A \times V$ where

Q = flow rate in cubic feet per second (cfs)

A = average cross-sectional flow area in square feet
D(depth) $\times$ W (width)

D = the average depth of the stream

$$\frac{(d_0 + d_1 + d_2 + d_3 + \ldots d_n)}{n}$$

which is the sum of the depths at n stations *of equal* width, divided by n

W = the width of the stream

V = the velocity in feet per second of a float

(Eq. 3)

Choose a length of stream that is fairly straight, with sides approximately parallel, at least 30 feet long (the longer the better), that has a relatively smooth and unobstructed bottom. Stake out a point at each end of the length, and erect posts on each side of the bank at these points. Connect the two upstream posts by a level wire or rope (use a carpenter's line level). Proceed the same way with the downstream posts (see Fig. 2).

Divide the stream into at least five equal sections along the wires (the more sections, the better), and measure the water depth for each section. Then average the depth figures by adding each value and dividing by the number of values. For example, if you have 7 readings of equal width, add $depth_0 + depth_1 + depth_2 + depth_3 + depth_4 + depth_5 + depth_6 + depth_7$ and divide by 7. Since Fig. 2 shows d_0 and d_7 at the edge of the stream, their depths are zero, they are not included in the calculation, and the sum of the values is divided by 5. For other situations, Eq. 4 should be used.

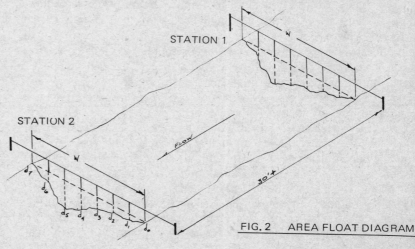

FIG. 2 AREA FLOAT DIAGRAM

$$A = \frac{(d_0 + d_1 + d_2 + d_3 + \dots d_n)}{n} \times (W)$$

where   A = average cross-sectional flow area in square feet
            d = the depth at each reading
            n = the number of readings taken along the
(Eq. 4)   stream's width
            W = the width of the stream

Now to find the stream's cross-sectional area, multiply the average value of the depth times the stream's width (the length of the wire or rope as in Eq. 4.)

Remember that you are trying to find the average area of a section of the stream, so you must take the value of (A) for each station, add them together and divide by two.

Your next step is to measure the stream's velocity in order to determine Q (the stream's flow rate). Make a float of light wood, or use a bottle that will ride awash. A pennant can be put on the float so that its progress can be followed easily. Now set the float adrift, in the middle of the stream, upstream from the first wire. Time its progress down the stream with a stop watch, beginning just when the float passes the first wire, and stopping just as it passes the second wire. Since the water does not flow as fast on the bottom as it does on the surface, you must multiply your calculations by a coefficient to give you a more accurate estimate of the stream's velocity.

$$V = \frac{D}{T} \times .8$$

where   V = velocity in feet per second (fps)
            D = distance in feet
            T = time in seconds
(Eq. 5)   .8 = coefficient

Having determined the area and velocity of the stream, Eq. 6 gives the flow rate of the stream.

$$Q = A \times V$$

where   Q = flow rate (cfs)
            V = velocity (feet per second)
(Eq. 6)   A = cross-section area in square feet

Measuring Head

The "head" or height of fall of the water determines what kind of waterwheel or turbine you will choose. Commonly, this distance is measured in feet. Head produces a pressure . . . water pressure. Basically the weight of the water at a given head exerts a pressure that is proportional to that head . . . the greater the head the greater the pressure. Pressure is usually measured in pounds of force per square inch (psi). Pipes, fittings, valves and turbines may be rated either in head or psi. The relationship between the two, if you need to know head and the equipment is rated in psi and vice versa, is:

1 ft. head = .433 psi   .water weighs 62.4 lbs. per cubic foot
                    .there are 144 square inches (in^2) per square foot
                    .thus $\frac{62.4 lbs/ft^3}{144 in^2/ft^2}$ = .433 psi/ft of head

An elevation of head will produce the same pressure, no matter what the volume or quantity of water is in that distance. If the head or depth of water is 20 feet the pressure of that water will be (20 x .433) = 8.7 psi whether there is a whole reservoir of water 20 feet deep or a 2 inch pipe filled with water 20 feet high.

You can get a rough estimate of head on your land from a detailed topographic map. The U.S. Geological Survey prints topographic maps of the entire U.S., with elevation contour intervals of 40'. They are available directly from the USGS (see page 114) or from some local sporting goods stores. These maps are useful for making note of particularly choice sites. However, since they only give an approximation of slopes and elevations, for a more accurate measure of head it will be necessary to take a level survey.

Level surveying is a relatively easy process. A good description of a poorman's survey using a carpenter's level can be found in *Cloudburst* (see page 114). Also included in that publication is a critique that recommends the all-purpose hand level. The hand level is a metal sight tube with a plain glass cover at each end and a prism, cross hair, and spirit level inside the tube. In most cases the hand level will be the simplest and least expensive way to go. More expensive and elaborate instruments are available (a surveying transit or a surveyor's level) but they will probably not be needed for these basic measurements.

Measuring the vertical difference in elevation between point A and point B.

    1. Basic Eyeball Method with Handlevel: How High is Your Eye?

Measure the distance standing upright from the bottom of your feet to the middle of your eye. Stand at the lower point (point B) and sight through the level at the top of an object or at the ground next to an object so you know where to go next. Go stand at that spot and sight again, always working towards point A. With a pole in the ground at point A the elevation of the final sighting can be marked and subtracted from the total. By simply multiplying the elevation of the eye by the number of sightings (minus the extra elevation in the last sighting) the vertical elevation change from point A to B is determined.

    2. With surveying Rod and Handlevel

Attach a tape measure to a pole with the zero end at the bottom. Have one person hold the rod while another sights through the handlevel. This does require some conversation between the "rod-man" and the "instrument man" about where the "level point" is on the rod and what the value of that point is.

The instrument man must tell the rod man where the level point is on the rod (indicated with a pointing finger) and the rod man must read off and write down the value on the tape of that point. The instrument man stays put while the rod man moves on towards point B to some notable point where a new reading is taken. Then the instrument man "leap-frogs" downhill past the rod for a new reading and so forth. Add up the differences of readings for each time the instrument man moves past the rod man; the total of these differences will be the elevation change from point A to point B.

The head is then the change in elevation between point A and point B. It is desirable to obtain the greatest amount of head possible. This can be accomplished in several ways. First, you can choose a site for the turbine where the greatest drop in the stream occurs in the shortest distance. Secondly, the amount of head can be increased with the construction of a dam (see Dams). Most hydro-power systems will require a dam anyway, and the higher the dam can be built, the greater the head will be. Finally, the head may be increased by the use of a channel or sluice. The channel or sluice will be downstream from the dam to carry the water to a place where a steeper drop occurs. Other determining factors must be considered in trying to realize the greatest amount of head. The most basic are cost, property lines, construction laws, soil condition and pond area in back of the dam.

Calculating Power

We can define power as the ability to do work. In order to determine if the hydro-power installation will meet your needs, the amount of available power must be calculated. The amount of this power is proportional to the head available and the flow rate of the water. Waterwheels and water turbines generate mechanical power (however, turbines are usually hooked up to an electrical generator, and the mechanical power generates electrical power). The mechanical power is usually measured in horsepower.

$$THP = \frac{Q \times H}{8.8}$$

where   THP = theoretical horsepower
            Q = flow rate in cubic feet per second (cfs)
            H = head in feet
(Eq. 7)   8.8 = correction factor for the units

Eq. 7 illustrates the theoretical horsepower available from the head and flow of a stream. Eq. 8 below refines the calculation of this available power. There are losses in the amount of head due to friction in the channel or pipe which carries water from the dam to the wheel or turbine. Thus, actual head is the amount of head loss (due to friction) subtracted from the total head available. A discussion of the head loss in pipes and canals can be found in the Channels, Sluices and Pipes section.

Another loss factor that Eq. 8 accounts for is the efficiency of the devices (turbine, generator, and any mechanical connection between the two: belts, gears, chains, pulleys) used to harness the power. In the case of water power, efficiency is an indicator of the conversion performance of the machinery used to harness the water power. It is usually measured as a percentage. Each machine or device will have its own efficiency percentage, and they must be multiplied together in order to obtain the over-all efficiency:

$$NHP = \frac{Q \times H \times E}{8.8}$$ where

(Eq. 8)

NHP = net horsepower
Q = flow rate in cubic feet per second
H = actual head in feet
E = efficiency of all the devices multiplied
    times one another
    (i.e. 75% X 85% X 80%)
8.8 = correction factor for the units

DAMS

In developing water power, dams are needed for three functions: (1) to divert the stream flow to the waterwheel or turbine, (2) to store the energy of the flowing stream, and (3) to raise the water level (head) to increase the available power. In addition to being used to help develop power, a dam may be useful in providing a pond for watering livestock, for fire protection or for irrigation needs. However, keep in mind the possibility of ecological harm involved in damming a stream, and forego the construction if necessary.

There are four basic criteria for deciding possible sites for the dam and powerhouse: (1) the ease of building the dam, considering the width of the stream and the stability of the soil; (2) maximizing the amount of possible storage volume behind the dam without damaging the ecological balance; (3) minimizing the distance to a good powerhouse site in order to lower the difficulty and the expense of moving the water; and at the same time (4) finding a place where the greatest amount of head is available.

Once a site has been chosen the size and type of dam needed will largely depend upon the stream course and surroundings as well as your needs for power. An assessment must be made of basic requirements for power (both mechanical and electrical) and some estimate of future needs. This could be as simple as adding up the power needs for lighting and refrigeration, but could also include assessing needs for machinery, power tools, or appliances. Essentially then, the size and type of dam will depend upon the size and type of waterwheel or turbine that will be needed to fill power needs, and also upon the flow rate of the stream, the head available, the local restrictions on size and permanency, and the money available.

Diversion Dams

When there is a creek with a continuous flow and a natural drop that together will provide sufficient power for your needs, a small diversion dam will suffice. This can simply be a log placed up against some projecting rocks, with rocks, gravel and earth placed upstream to stop the under-flow. Even a temporary dam of a few rows of sandbags will serve well (at least until the first flood comes) and can be cheaply and easily replaced. All that is necessary is a sufficient dam to divert the water into a sluice or pipe intake which carries the water to the turbine (see Channels, Sluices and Pipes).

Small diversion dams have the advantage of easily "washing out" during large flows thus preventing possible damage downstream which the washing out of a larger more permanent dam might cause. Also, diversion dams may be useful for running some of the stream's water into an off-stream storage location.

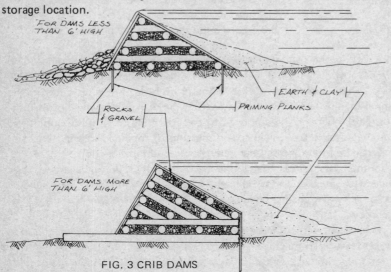

FIG. 3 CRIB DAMS

The need for storage is one of the dam's primary functions, but during times of large flow or flood can cause potentially dangerous situations. Normally the water would be stored directly behind the dam in the stream's course. However, if a diversion were used to divert the water to a side canyon or hollow, the need for storage would be met, while avoiding the danger of a flood washout. When large flows occurred the diversion dam could either be easily taken down or allowed to wash out by the stream's force; both situations leaving the off-stream storage facility full and intact. When the large flow subsided, the dam could simply be rebuilt.

Low Dams of Simple Construction

These can be built by adding to the diversion dam's structure. Instead of one log, use several stacked together log cabin style, or like a corn-crib —hence the term "crib dam." The crib dam (see Fig. 3) consists of green logs or heavier timbers stacked perpendicular to each other, spaced about 2 or 3 feet apart. These should be spiked together where they cross, and the spaces in between filled with rocks and gravel. The upstream side, especially the base, should be covered with planks or sheets of plastic to prevent leakage, and then further covered with earth or clay to seal the edges. Priming planks should be driven into the soil approximately two to three feet deep at the upstream face to limit the seepage under the dam (particularly on porous soils). Priming planks are wooden boards, preferably tongue and groove, with one end cut to a point on one edge. They are driven into the soil so that the long pointed side is placed next to the board that was previously driven. Then as each successive board is driven into the soil it is forced up snug against the preceding board as a result of the angle of the bottom cut.

PRIMING PLANKS

The downstream face of the dam must be protected from erosion or undercutting wherever water will spill over. This is most important at a time of large flows! The spillways can be made of concrete, lumber, or simply a pile of rocks large enough to withstand the continual flow. Crib dams can be built with the lower cross-timbers extended out to form a series of small water cascades downstream. Each cross-timber step should be at least as wide as it is tall.

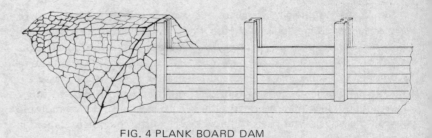

FIG. 4 PLANK BOARD DAM

Earth-Fill Dam

This is the cheapest kind of dam if earth moving equipment is available. Sometimes these can be small gravel dams (under 5 feet) that can wash out with each season's flood, and can be rebuilt when necessary. For larger earth dams (in California anything more than 6 feet high or 5 acre feet of storage) a registered civil engineer will have to be consulted, and some soil studies should be made to determine the method of construction. For more information on the structure, placement and suitability of earth fill dams, see the USDA Handbook No. 387. "Ponds for Water Supply and Recreation," page 67.

Plank Board Dam

Much like a plank-board overflow spillway, a small dam can be constructed from wooden planks supported by posts set in a concrete foundation (see Fig. 4). The posts can be wood 4 x 4's (or larger) with steel channel or angle-iron attached to the sides, or the posts could be steel I-beams set directly in the foundation. The wooden planks can be dropped into the steel slots to form as much of a dam as is needed (up to the height of the posts). The upstream face of a plank board dam will often need to be sealed with plastic sheeting to prevent leakage. The planks (2 x 6 or larger) can be either added or removed to vary the height of the dam and can be completely removed during the flood season.

Rock Masonry Dam

With a plentiful supply of rocks and stone nearby, a good rock masonry dam can be built of uncut rock laid in cement mortar. This style should be built with a base at least 8/10 the height. With a masonry dam more than 8 feet high some engineering consultation would be important (and may even be required). Many soil factors can threaten the stability of the dam.

Concrete Gravity Dam

One of the more common building materials for containing water is good old concrete. A 1925 USDA Bulletin: "Power for the Farm From Small Streams," recommends a mixture of one part Portland cement, 2 parts sand, and 4 parts gravel or broken stone. Large boulders can be thrown in, but they should be well set in the concrete and should not exceed 30% of the total volume. The dam with dimensions shown in Fig. 5 should not be more than 50 feet long.

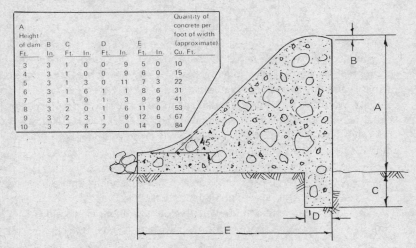

A Height of dam		B		C		D		E		Quantity of concrete per foot of width (approximate)
Ft.	In.	Ft.	In.	Ft.	In.	Ft.	In.	Ft.	In.	Cu. Ft.
3		3	1	0	0	9	5	0		10
4		3	1	0	0	9	6	0		15
5		3	1	3	0	11	7	3		22
6		3	1	6	1	1	8	6		31
7		3	1	9	1	3	9	9		41
8		3	2	0	1	6	11	0		53
9		3	2	3	1	9	12	6		67
10		3	2	6	2	0	14	0		84

FIG. 5 CONCRETE GRAVITY DAM

Concrete Block Dam

For simpler construction, and considerably less volume of concrete mix, a structure can be built of manufactured standard concrete blocks. The blocks will need to be reinforced with re-bar, placed on a firm concrete footing, and protected on the downstream face with a pile of sizable rocks. The construction of this type dam is very similar to retaining walls used to hold back earth for buildings. The re-bar must be placed first, then the footing concrete poured, and the first row of blocks placed on the wet concrete. After that has set, the other rows of blocks can be placed (with mortar joints) around the vertical bars; the horizontal bars can be placed where needed, and then the holes in the block should be filled with poured concrete.

LARGE AND SMALL FLOWS

The process of developing a water power installation is complicated by variations in the quantity of stream flow. Large flows can wash out dams, destroy waterwheels and turbines and create dangerous situations downstream. It is important to know what the possible large flows for the particular stream will be and how best to prepare for them.

The maximum expected flow in a stream cannot always be measured directly; it could happen at 3 am on a cold rainy night. Nor can flows be determined by the regular high-water marks along the banks, or by old timers who remember the "big rains of ought-8." The "Big Flood" for a particular drainage will come as the result of the right combination of strength and intensity of an exceptional storm as well as the soil conditions, moisture content, vegetation, and the physical structure of the drainage area around that stream. Hydrologists, local flood control people, soil conservation agencies, or water districts can often offer assistance in predicting the flood flows for small drainages. Another source, where these agencies are not available, is a US Department of Agriculture handbook (No. 387) "Ponds for Water Supply and Recreation" (see page 67).

For all types of dams there must be some provision to handle the maximum flows that can come with a severe rain storm; otherwise the

sides, foundation, and the structure of the dam can be weakened. This is particularly true for earth dams, since any flow that exceeds the capacity of the spillway will erode the dam with frightening ease. As this would occur at a time when the stream is already at flood stage, the added volume of water from a washed out dam could be exceptionally dangerous downstream and the owner of the dam could be legally liable for the damage.

With small dams, especially earth dams, the best way for large flows to safely pass downstream may be to allow them to wash out the dam. With larger dams this method may be impractical, costly and dangerous. In these situations it will be necessary to use a culvert or weir for the large flows to pass through.

An opening can be made with a large culvert placed through the dam. The *Water Measurement Manual*, page 115, will aid in determining the capacity for a given length and size of pipe. The face of the dam should be protected from the downstream outflow of the culvert, or, if you have one long enough, the culvert can be extended beyond the footing of the dam to the natural stream course. The problem with pipes and culverts is that they are easily plugged by brush and trash. They must be watched almost continuously during storms or the water back-up may overflow the dam. Because of this problem, culverts do not make good spillways; so, there should be some additional spillway provided. However, pipes and culverts can be useful to control the flow rate downstream, a simple "gate valve" (of a sort) can be made by placing a piece of plywood over the upstream end of the pipe. The flow rate can then be controlled by adjusting the plywood to allow the proper opening.

A sufficient flood flow capacity can be provided by one or more rectangular openings in the dam. These could be built like large rectangular weirs to provide a continuous measure of the flow rate, as well. (Use Table I to find the size weir required to handle the flood flows). Such an opening would not be easily plugged by trash, and the flow could be regulated by placing planks flat across the upstream side of the opening to raise the spill level.

Plank board dams are well suited to handle floods. During times of large flow the planks may simply be removed to allow more water to pass downstream (see Dams).

Small flows do not present the potential danger that large flows do. They may, however, be of insufficient quantity to operate the waterwheel or turbine. Although it may be desirable to store water behind the dam in order to gain a larger flow for the wheel or turbine, it may not be possible or necessarily advisable. Water right laws in many cases will prohibit halting the flow downstream even if you plan to let it flow later in the same day. Also, during times of low flow, the intermittent storage and release of water will cause some erosion in the stream course above and below the dam. In short, it is probably best to design your hydro-power installation to operate at the minimum seasonal flow rate, rather than attempting storage during that time. This suggestion should not be taken as advice against storage of water during times of normal or average flow, but here too, care should be taken and the laws should be investigated.

CHANNELS, SLUICES AND PIPES

Every hydro-power installation (with the possible exception of an undershot wheel) will require a means to carry the water from the stream to the waterwheel or turbine. In most cases a dam will have been constructed and the task will be to take all or part of the water and run it to the wheel or turbine. In many cases there will be a considerable distance from the dam to the turbine and it will be important to consider the various possibilities and compare the costs of both materials and labor. Most of the lower technology waterwheels will be best serviced by an open channel or sluice. The Pelton wheel, Banki, Propeller, Kaplan and Francis turbines will almost always require piping. Discussion of how each of these units operate will follow in the Waterwheels and Water Turbines section.

Basically, there are two ways to move water from one location to another—either with unpressured flow along an open channel (e.g. a canal, sluice or a natural stream) or with flow under pressure contained in a pipe. Each has advantages in the appropriate situation.

A canal is simply a ditch dug out along the ground, that will maintain a nearly constant elevation (with a very gradual down grade slope) and carry water at a low velocity with a minimal amount of head loss. The head loss will be the same as the loss of elevation along the canal. The size (i.e. the cross-section area) of the canal that will be required to carry a given amount of flow will depend on the roughness of the canal, the amount of vegetation growing in the canal and the slope of the canal. Canals are more often used for irrigation than for a hydro-power plant, although the two needs, power and agriculture, will often be compatible—a dam and canal can divert water for both purposes. In short, a canal will move water with a minimum of both cost and head loss.

The sluice is really another type of open channel. The difference being that a canal is usually dug out of the ground, whereas a sluice is an elevated structure or open box channel, somewhat like the old Roman aqueduct or gold-miners flume. They are useful for carrying water over or around obstacles where a ditch would be impractical (e.g. rocky ground, side canyons, or very steep hillsides). The overshot water wheel requires a sluice to carry the water out over the structure of the wheel.

The sluice itself can be made from wood or metal. The sluice must be supported by a strong structure. Remember that water weighs 62.4 lbs. per cubic foot, so a 2′ x 4′ box channel would have to carry *500 lbs. per foot of length*, or each four feet of sluice would hold a ton of water!

Channel Flow

There are several objectives in channel design: to (1) move the water with as little head loss as possible; this means a gradual slope, slower velocity, and an adequately large channel; (2) do the job as cheaply as possible, avoiding any undue construction complications, and keeping the channel as small and as short as possible; and (3) move as much water as is available within the above constraints. Some compromises must be made along the way, since it is unlikely that all these criteria can be met in any one design.

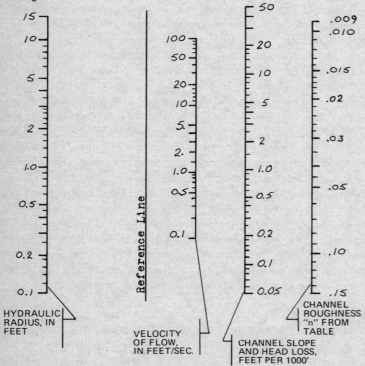

FIG. 6A   CHANNEL FLOW NOMOGRAPH

USE OF THE NOMOGRAPH: Draw a straight line from the appropriate value for "n" through the velocity of flow desired, to Reference line; this sets the pivot point. Then a second line is drawn from the hydraulic radius (determined by the size and shape of the channel) through the pivot point to the head loss scale; this gives the slope required to overcome the head loss in a given channel for water moving at a given rate.

To find the channel size and slope needed to meet the above requirements involves the use of a "Channel Flow Nomograph" (Fig. 6A). The two following methods will describe how to use it. In order to avoid complications in the examples, we are assuming a uniform channel cross-section, a steady flow rate of water, and a constant slope along the channel.

Method Number One involves the use of a chart called a "nomograph" (Fig. 6B) to determine the required slope of the channel. First, determine the amount of water that must be diverted to the channel; this depends on the amount of flow available from the stream and the wheel or turbine requirements (see Waterwheels and Water Turbines). Now consult Table II

to find the maximum allowable velocity for the water to move through the channel. For example, if the flow is 1 cfs, and the channel were dug in soil with about 40% clay content, the maximum water velocity without causing erosion in the channel would be 1.8 feet per second.

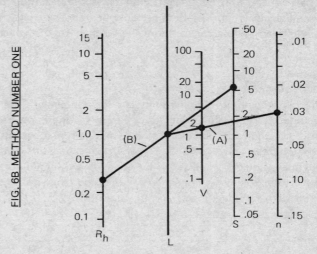

The next value that needs to be known in order to work with the nomograph is the channel roughness (n). This is also read directly from Table II and in this example would be .03.

Last, we must find the hydraulic radius in feet. The hydraulic radius is equal to a proportion of the channel through which the water flows. The first step is to find the minimum channel cross-sectional area that is required to carry the stream's flow (see Eq. 9).

$$A = \frac{Q}{V}$$

where   A = cross-section area of the channel in square feet

Q = flow in cubic feet per second

V = maximum water velocity in feet per second

(Eq. 9)   before erosion begins (from Table II)

Continuing with our "example" stream, using Eq. 9, "A" would equal 1 cfs divided by 1.8 fps or 0.56 ft². Once the channel area is figured, the channel shape is chosen. The shape is largely a function of the materials found or used. Channels made of timber, masonry, concrete or rock should have walls constructed perpendicular to the bottom (see Fig. 7). The water level height should be one half of the width. Earth channels should have walls built at a 45° angle (see Fig. 8). Design them so that the water-level height is one half that of the channel width at the bottom.

CHANNEL ROUGHNESS

TABLE II

Composition of channel wall	Maximum water velocity (ft/sec) before erosion begins	Channel roughness (n)
fine grained sand	0.6	0.030
coarse sand	1.2	0.030
small stones	2.4	0.030
coarse stone	4.0	0.030
rock	25.0	0.033
earth:		
sandy loam, 40% clay	1.8	0.030
loamy soil, 65% clay	3.0	0.030
clay loam, 85% clay	4.8	0.030
soil loam, 95% clay	6.2	0.030
100% clay	7.3	0.030
earth bottom with rubble sides	(use one of above factors for earth)	0.033
concrete with sandy water	10.0	0.016
concrete with clean water	20.0	0.016
wood	25.0	0.015
metal	no limit	0.015

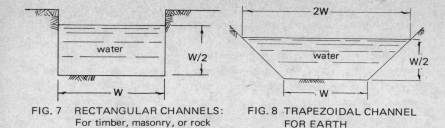

FIG. 7   RECTANGULAR CHANNELS:
For timber, masonry, or rock

FIG. 8 TRAPEZOIDAL CHANNEL
FOR EARTH

Now we must solve for (W) the width across the bottom of the channel which will enable us to find the hydraulic radius, the last value we must know to use the nomograph.

For Rectangular Channels: Timber, Masonry, Concrete or Rock Lined

$$Area = \frac{W^2}{2} \quad so \ W = \sqrt{2A}$$

Hydraulic Radius $(R_h) = (0.25)(W)$ (Eq. 10)

For Trapezoidal Channels: Earth

$$Area = (.75)(W^2) \quad so \ W = \sqrt{\frac{A}{.75}}$$

Hydraulic Radius $(R_h) = (0.31)(W)$ (Eq. 11)

To find the (W) of our example stream, use $W = \sqrt{\frac{A}{.75}}$. We know A = .56 ft², then $W = \sqrt{\frac{A}{.75}} = \sqrt{\frac{.56}{.75}} = 0.86$ ft. Now the hydraulic radius can be calculated using the formula $R_h = (.31)(W)$. $R_h = (.31)(.86) = 0.27$. The hydraulic radius is then .27 ft.

To continue with the example, we have found the flow (1 cfs), the velocity (1.8 fps), the channel roughness (.03), and the hydraulic radius (.27 ft) and now need to find, using the nomograph, the channel slope required to move 1 cfs at a maximum velocity of 1.8 fps.

Using the nomograph (Fig. 6B), locate the point .03 on the *channel roughness* scale (n) and locate 1.8 fps on the *velocity* scale (V). Draw the line (A) through these two points to the *reference line* (L). Then locate the .27 on the *hydraulic radius* scale (R_h). A second line (B) drawn from .27 through the point at which line (A) intersects the reference line will intersect the *channel loss/head loss* scale (S) at approximately 7.6. The channel slope/head loss is then 7.6 feet per thousand feet. Evaluating this outcome to see if this is the channel to construct will depend upon the length of the channel and the amount of head needed to operate the wheel or turbine.

In most cases the head loss and slope found in this manner will be too great for the requirements of the site, and so some redesign will be necessary. To redesign for a lesser, more gradual slope, a larger channel (with a larger hydraulic radius) will be necessary to move the same amount of water since the velocity would then be less.

Method Number Two begins with an assumed velocity much less than the maximum allowable. If we use the same flow rate, 1 cfs, it might be logical to choose a velocity of 0.5 fps. The channel area can then be found using Eq. 9. The example would require an area 2 ft². The corresponding hydraulic radius for a trapezoidal earth channel would be 0.51 calculated using $(0.31) \times \sqrt{2 \ ft^2 / .75}$.

Using the nomograph (Fig. 6C) and plugging in the appropriate values gives a channel loss/head loss of only .25; a drop of about 3 feet per thousand feet of channel length.

If the slope is still too great, the channel design can be changed by: (a) assuming a slower velocity, with either a larger channel or a smaller flow, or (2) reducing the channel roughness by removing vegetation or otherwise "smoothing out" the channel.

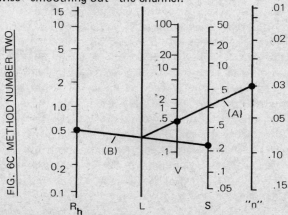

FIG. 6C METHOD NUMBER TWO

Pipe Flow

The other alternative for water flow, besides the open channel flow, is the pressure flow of water contained in a pipe. This type of flow could exist over a very wide range of conditions: from the large volume of low pressure water flowing in a culvert (as under a road) to the small volume, high pressure flow of water in a pipe. Both cases can be analyzed with the pipe flow nomograph (Fig.9). It is important to find the proper diameter and strength of pipe to handle the required flow without undue expense or excessive head loss. Although water pipe is expensive (particularly the high pressure steel pipe) it is the only way to deliver water under pressure.

Pipe flow nomographs are used to find head flow loss in pipes. The use of the nomograph in Fig. 9 is relatively easy. One line is drawn through the proper values on the flow rate (Q) scale and pipe size (D) scale, and will indicate the water velocity in the pipe and the corresponding head loss per 100 feet of pipe.

For example, to use a 4" steel pipe to move a flow of 1 cfs, draw a line from the 4" mark on line "D" on the nomograph through 1.0 cfs on the "Q" line. This will give a value on the head loss line "H_L" of 20 feet of head loss per 100 feet of pipe. In most instances this would be too prohibitive a loss, and so some larger pipe would be required. To use a 6" pipe for the same flow rate (1 cfs) would mean a head loss of only 3 feet per 100 feet. This is a much more reasonable size of pipe for the 1 cfs flow.

This nomograph (Fig. 9) is only for the use of steel pipe. Steel is relatively rough compared with other kinds of pipes, with a roughness coefficient (C) of 100. It is possible to find the head loss and velocity in pipes other than steel (as long as the flow rate "Q" and pipe diameter are known). By finding the C value of the pipe (Table III) and using Eq. 12, the velocity, flow rate and head loss can be found.

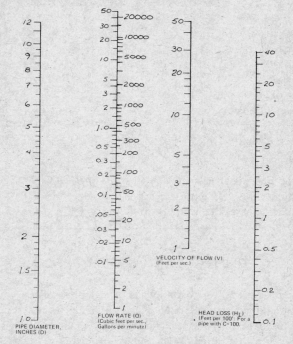

FIG. 9   NOMOGRAPH FOR HEAD LOSS IN STEEL PIPE (C=100)

$V = (V_N)\dfrac{C}{100}$ where

(Eq. 12A)

V = velocity of water through a pipe other than steel

V_N = velocity of water read from nomograph, when Q and D are known

C = pipe roughness coefficient (Table III)

$Q = (Q_N)\dfrac{C}{100}$ where

(Eq. 12B)

Q = flow rate of water (cfs) through pipe other than steel

Q_N = flow rate value read from nomograph

C = pipe roughness coefficient (Table III)

$H_L = (H_{LN})(\dfrac{100}{C})^{1.85}$

where

(Eq. 12C)

H_L = head loss per 100 feet of pipe

H_{LN} = head loss read from nomograph

C = pipe roughness coefficient (Table III)

To take $\dfrac{100}{C}$ to the 1.85 power, it is necessary to use logarithms. Either consult a friend, textbook or slide rule for assistance.

For example, using a new pipe lined with bitumastic enamel, and the same flow rate (1 cfs) and diameter pipe (6 inches) then:

$$V = 5 \text{ (fps)} \left(\frac{145}{100}\right) = 7.25 \frac{\text{feet}}{\text{second}}, \text{ a faster V, as a result of smoother pipe}$$

$$Q = (1 \text{ cfs}) \left(\frac{145}{100}\right) = 1.45 \text{ cfs}$$

$$H_L = (3\text{ft}) \left(\frac{100}{145}\right)^{1.85} = (3)(.512) = 1.54 \text{ feet per 100 feet}$$

TABLE III PIPE ROUGHNESS TABLE
VALUE OF C* FOR USE WITH PIPE-FLOW NOMOGRAPH

Pipe	C*
New tar-coated	130
New cement-lined	150
New cast-iron, pit cast	120-130
New cast-iron, centrifugally cast	125-135
Cement lining, applied by hand	125-135
Bitumastic enamel, hand brushed	135-145
Bitumastic enamel, centrifugally applied	145-155
Ordinary tar-dipped cast-iron, 20 years service in inactive water.	110-125
Ordinary tar-dipped cast-iron after long service with severe tuberculation	30-40
Ordinary tar-dipped cast-iron; average tuberculation;	
new	135
5 years old	120
10 years old	110
15 years old	105
20 years old	95
30 years old	85
40 years old	80
New bituminous enamel-lined	150
Transite	140+
Poly-vinyl chloride (PVC)	130

*C is a smoothness coefficient which gives a numerical value for how smooth the pipe and fittings are. A rougher pipe with a lower value increases the head loss in the pipe.

WATERWHEELS AND WATER TURBINES

So far we have been discussing ways to measure available water power, build dams and construct channels and sluices. Now we shall turn our attention to the actual devices, the waterwheels and turbines that turn water power into useful work. The choice of turbine or wheel is probably the biggest decision in building a hydro-power plant. The main factors that will determine your choice are:

1. *Flow rates:* minimum to be available, maximum to be utilized and maximum to be routed through the dam.

2. *Available head:* elevation difference in feet or meters between head waters and tail waters.

3. *Site sketch:* an elevation or topographic map with dam and power locations indicated.

4. *Soil conditions:* determines the possibility of erosion and the size and slope of a canal; also a consideration in dam construction.

5. *Pipe length:* required from dam or end of canal to powersite.

6. *Water conditions:* clear, muddy, sandy, acid, etc.

7. *Tailwater elevation:* the maximum and minimum water level immediately below the turbine.

8. *Air temperature :* annual maximum and minimum, particularly amount of exposure to freezing temperatures.

9. *Power generation:* a waterwheel if you want mechanical energy, or a turbine for electrical energy, and how much of each type of energy is needed.

10. *Cost/Labor:* pre-packaged vs "home-built."

11. *Materials:* use of native or purchased materials to suit the design and use.

12. *Maintenance:* some assessment of reliability and ease of repair.

We will discuss those waterwheels and water turbines that are appropriate for the small scale projects that individuals and communities might reasonably consider. There are really too many designs, devices and inventions associated with hydro-power to describe in this limited space. Furthur research has led us to conclude that construction instructions for wheels that may be "home-built" are adequately treated in other publications. Consequently we will not take the time here to rehash what is already available, and the reader should refer to p.114 *Reviews/Water* section. For the more technically minded, the discussion of each wheel and turbine has *some* detailed and geometric coverage of construction.

WATERWHEELS

Waterwheels are particularly useful for generating mechanical power. This power is taken off the center shaft of the wheel and usually connected, via belts and pulleys, to machinery. Waterwheels generally operate at between 2 and 12 revolutions per minute (rpm) and are appropriate for slow speed applications such as turning grinding wheels, pumping water and sometimes running lightweight machinery and tools (i.e. lathes, drill presses and saws). Waterwheels may be used to generate electric power but because of the slow rotational speeds, difficulties are often encountered. Waterwheels will operate in situations where there are large fluctuations in the flow rate. The changing flow will bring about a change in the rpm of the wheel.

Undershot Wheel

The most basic design (and simplest concept) in waterwheels is the old-style undershot wheel. The earliest design for the undershot wheel was a simple paddle wheel, immersed in the stream flow, that splashed along with the current. This sort of wheel powered the fountains of Louis XIV's Palace of Versailles. A number of undershot wheels, each 14 meters in diameter, powered the fountain's pumps with an overall efficiency of only 10%.

The refinements that were developed to improve the efficiency of the undershot wheel included better controls on the water in order to increase the velocity of the water as it hit the paddles, and to limit the amount of water so the paddles would not get bogged down in the backwater.

Since the really significant loss of energy in the paddle-type undershot wheels came from the shock and turbulence as the water hit the flat paddles, around 1800 the shape of the paddles was changed to reduce this loss. The end result of this change, the curved blades of the Poncelet undershot wheel, is still the last word in the design of the undershot waterwheel.

Poncelet Wheel

These "low-technology" wheels are best suited for heads ranging from three feet up to ten feet and flows ranging from three cubic feet up to whatever is available. They generally operate at a low rpm (for example 7.4 rpm for a 14 foot diameter wheel with a six foot head) as determined using the appropriate formula from Table IV. Poncelet wheels usually develop a high torque; this coupled with their low rpm make them best suited for mechanical work rather than electrical generation. Poncelet wheels are usually made of wood with reasonably heavy timbers being used for the spokes. The buckets or vanes are usually made of sheet-steel.

A well designed Poncelet wheel utilizes the impulse of the water jet as it strikes the vanes at the bottom of the wheel. The vanes are curved so that they allow the water to enter the buckets with a minimum of shock (which would cause some energy loss). The water then "runs up" the curve of the bucket vane, exerting a force on the wheel. As the energy of the water is transferred to the wheel and the wheel rotates, the water falls back and drops from the wheel to the tailwater with nearly zero velocity. The Poncelet is 70% to 85% efficient.

The diameter of the wheel is largely a matter of preferred use and limitations in materials, usually with a minimum diameter of 14 feet up to a maximum of four times the head. For the same output the smaller wheels will turn much faster and with less torque than the larger wheels.

TABLE IV WATERWHEEL AND WATER TURBINE SPECIFICATIONS

	Wheel or Turbine Type	Range of head (feet)	Wheel or runner diameter (feet)	Optimum rpm	Efficiency %	Ability to handle changing: Q flow	H head	Technology of construction	Materials
WATERWHEELS	Undershot Wheel	6'-15'	3H	$\dfrac{42.1\sqrt{H}}{D}$	35-45%	good	fair	low	metal/wood
	Poncelet Wheel	3'-10'	2H-4H (>14')	$\dfrac{42.1\sqrt{H}}{D}$	60-80%	good	fair	medium	metal/wood
	Breast Wheel	6'-15'	H-3H	dependent on design less than overshot	40-70%	good	fair +20%H	low	metal/wood
	Overshot Wheel	10'-30'	.75H	$\dfrac{41.8}{\sqrt{D}}$	60-85%	good	none	low	metal/wood
WATER TURBINES	Michell (Banki)	15'-150'	1'-3' +	$\dfrac{862\sqrt{H}}{D, \text{ in inches}}$	60-85%	good	good	medium	welded steel
	Pelton Wheel	50'-4000'	1'-20'	$\dfrac{76.6\sqrt{H}}{D}$	80-94%	good	fair	medium/high	steel, cast iron or bronze
	Francis	100'-1500'	1'-20'	dependent on design	80-93%	poor	poor	high	cast or
	Kaplan	14'-120'	2'-30'	50-220	80-92%	poor	good	high	machined
	Propeller	8'-200'	2'-30'	50-220	80-92%	poor	poor	high	steel

H = head in feet
D = diameter of wheel in feet

The bottom of the sluice under a Poncelet wheel must be made in a close-fitting breast for an arc of 30° (15° on either side of bottom-dead center) see Fig. 10. *Breast* or *breast works* may be defined as a structure that is formed to fit close to the rim of the waterwheel, usually intended to help the wheel retain water in its buckets. A breast should be made of concrete or other easily-formed durable materials. It should fit as close as possible to the wheel but not so close that the wheel could rub or bind as it rotates. The tailrace beyond the breast should be deepened and widened to insure that the backwater does not hinder the rotation of the wheel.

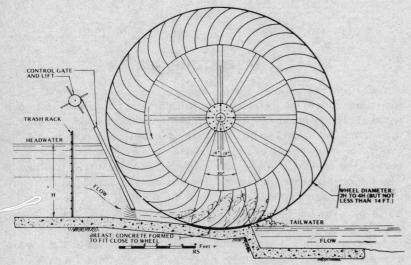

FIG. 10   PONCELET WHEEL

Breast Wheel

Breast wheels can be used for heads between 5 and 15 feet; however, since they are not quite as efficient as overshot wheels for similar heads, they are generally only considered for heads under 10 feet. They are more difficult to construct than the other types of basic water wheels, in that they require a close-fitting breast works (similar to the one pictured in Fig. 10) to keep the water in the buckets along the lower half of the wheel. These breast works further complicate matters since the wheel must be protected from rocks, logs, and debris, to prevent the wheel being jammed and damaged.

When the water enters the wheel at about the elevation of the center shaft it is considered a "breast wheel"; when the entrance is below the center shaft, the wheel is a "low-breast"; when above, it is a "high-breast".

From late 19th and early 20th century records of American and European practice (still pretty much the state-of-the-art for these wheels) it is evident that the efficiency will vary with the type of breast wheel: about 35% to 40% for low-breast wheels, little better than a simple undershot; about 45% to 55% for midrange; and for "high-breast," about 60% to 65%, approaching that of the overshot wheels.

Not only do breast wheels have complicated, curved breastworks and entrance gate, but the buckets must be "ventilated" to allow air to escape to the next higher bucket as each fills. It seems that considering the operating inefficiencies, maintenance problems, and difficulty of construction, it would make more sense to build an overshot wheel. And for the same effort in design and construction, a Poncelet wheel can give a higher efficiency than the medium or low-breast wheels.

Overshot Wheel

Overshot wheels have traditionally been used for falls (heads) of 10 to 30 feet. They have sometimes been used with higher falls, but because of the size of the wheel needed, there are usually more practical ways to harness the energy. In an overshot wheel, a small portion of the power is a result of the impact of the water as it enters the bucket. Most of the power is from the weight of the water as it descends in the buckets. The wheel should be designed so that the water enters the buckets smoothly and efficiently so that the greatest amount of power is produced with the least amount of waste. The efficiency of this wheel, if well constructed, is anywhere from 70% to 80%.

When the supply of water is small, during dry times, the wheel still operates, even though the buckets are only partially filled. This, of course, reduces the power output, and yet the efficiency is increased since less water spills from a less-full bucket. The corollary to this would seem to be that a wheel with an excess carrying capacity would operate at a higher efficiency. However, this increases the construction cost and usually the efficiency gained is not worth the extra expense.

Fig. 11A shows an overshot wheel with curved steel buckets. The water, controlled by the sluice gate at G, flows along a trough or sluice, A, to a drop at the crown of the wheel at C. The end of the sluice is slightly curved toward the wheel and placed such that the water enters at (or slightly after) the vertical center-line of the wheel. The sides of the sluice are extended just enough to fill several buckets at a time, without losing water over the sides of the wheel.

The supply of water as it is regulated by the sluice gate, is usually limited to under an 8 inch depth (see Fig. 11A). In most cases this flow would be controlled by hand, although with a little mechanical ingenuity, the gate could be regulated by an automatic governor.

Overshot Construction

Suppose that the total fall or head available is 20 feet. To insure that the centrifugal force from the water in this head filling the buckets will not cause too much spilling, the rim speed, U, should not be too great. The rim speed is the velocity of the outer edge of the wheel, usually expressed in feet per second. One value for U has been given in an engineering handbook (c. 1930) as U = 2D. With the curved buckets in common usage (as shown in Fig. 11A), a better value of U = 1.55√2D, where D is the diameter of the wheel.

As a first approximation, assume the wheel diameter is 16 feet. Then U = 1.55 (√2 × 16) = 8.8 feet per second. The rim velocity should be between 50% to 70% of the velocity of the water, the best being about 65%. If the rim speed is much faster than this, the back of the buckets will tend to "throw" the water out and away from the wheel.

For our example in which the rim speed is 8.8 feet per second, the velocity of the water entering the buckets would be 8.8 ÷ 65% = 13.5 feet per second.

The headwater behind the sluice gate required to produce this velocity is:

$$h = \frac{v^2}{2g}$$

where h = head in feet (at the sluice gate)
　　v = velocity of water entering buckets in feet per second

(Eq. 13)　g = acceleration of gravity, 32.2 feet per second^2

Then, $h = \frac{13.5^2}{2 \times 32.2} = 2.82$ feet. Because of loss of velocity in the sluice and gate due to friction, another 10% should be added to the 2.82 feet to give a required gate head of about 3.1 feet. At best, only half of this velocity is useful work, i.e. contributing to the torque of the wheel. The rest, say 1.6 feet, is lost in the turbulence as the buckets fill.

At this point the initial estimate of wheel diameter can be evaluated. The sum of gate head (3.1 feet), wheel diameter and tailwater clearance (about .5 feet, sufficient so that the wheel does not "hit" the tailwater and get slowed down while it is rotating), should be the total head difference between headwater and tailwater. In this case the total head of 20 feet, minus the gate head (3.1 feet), wheel diameter (16 feet), and tail water drop (about .5 feet), leaves only .4 foot error in the initial estimate of wheel diameter. This is well within the accuracy of the design.

The 16 foot example wheel would operate best at about 10 rpm (from Table IV). Some care should be taken to keep the wheel running at the design rpm by manually regulating the flow with the sluice gate.

Once the wheel diameter and the rim speed are known then the rpm can be calculated with Eq. 14.

$$rpm = \frac{(U)(60)}{C}$$　•where　rpm = revolutions per minute
　　　　　　　　　U = rim speed in feet per second
　　　　　　　　　C = wheel circumference in feet (π diameter)

(Eq. 14)

It is necessary to determine the volume of the buckets that is available to be filled with each revolution of the wheel. An approximate formula for this is a product of the area of the buckets (the width times the depth) times the circumference of the wheel at mid-depth of the buckets (that is π times the diameter of the wheel minus half the depth of the buckets)

Volume per revolution = $\pi[(D - b)] \times [(\text{width of buckets}) \times (\text{depth of buckets})]$

(Eq. 15)

The buckets are only assumed to be partially full, because a full fill would cause losses due to spilling with no net power gain. Assume then that wooden buckets will be filled approximately 50% and metal buckets approximately 67%. The flow (Q) needed to fill these buckets at the optimum rpm can now be determined using:

Q = (volume per revolution) x (rpm) x (% fill)　where q = flow in cfm

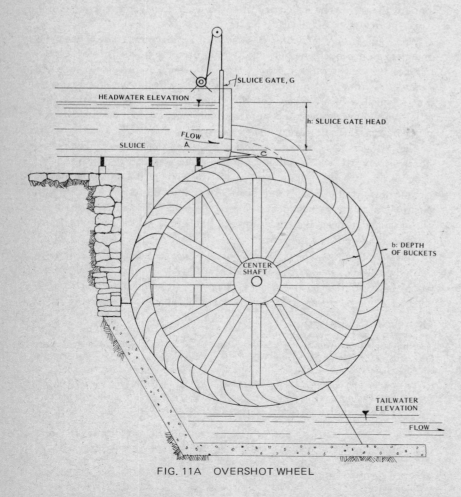

FIG. 11A　OVERSHOT WHEEL

The particular depth chosen for the buckets is rather subjective. One source says the depth should be ".3$\sqrt{D}$ to .5$\sqrt{D}$ where narrower wheels are desired." Another says simply that "spacing and depth should be the same" and the spacing should be ".10 inch to .18 inch."

From the usual practice in the 19th century, it seems that a reasonable number of buckets for a wheel should be about 2.5D (where D is in feet). This has to be a whole number so that the buckets are equally spaced around the wheel. The spacing, s, would be $s = \frac{\pi D}{n}$ where n = number of buckets. Or, to find the angle in degrees between each bucket, divide the full circle, 360° by n.

To find curvature of the buckets for an overshot wheel (see Fig. 11B), set the distance J to K equal to 1/3 b and the distances L to M equal to (1.2) times the circumvential pitch or bucket spacing, S. A bucket chord line is drawn from M to K with a point, R, found near to K (the length of K to R is about 1/4 the length of J to K). The center of the arc from R to M is set at 0 on a line that is offset 15° from the radius. The arc from M to R should be "rounded" into the radial line J.K.

During construction of this wheel, the job of placing the buckets would be much easier with a template or jig already cut to the exact bucket curvature. If the jig were set up to include the center of the wheel, it could be moved rather quickly around the wheel to indicate the placement of each bucket.

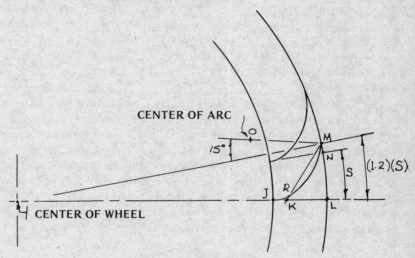

FIG. 11B　OVERSHOT WHEEL: BUCKET CONSTRUCTION

WATER TURBINES

Water turbines are used for producing either direct current (D.C.) or alternating current (A.C.) electricity. Turbines can be classified into two types; (1) impulse, and (2) reaction. Impulse turbines include the Michell (Banki) and Pelton types. They are vertical wheels that utilize the kinetic energy (or momentum) of a jet of water which strikes the buckets or blades. The buckets or blades are so shaped that they turn the flow of water through as near 180° as possible and move at a speed which results in the spent water falling straight to the bottom of the wheel housing, which is mostly full of air. The Michell (or Banki) turbine is simple to construct, while the Pelton wheel is more complicated.

Reaction turbines are horizontal wheels that are moved at high speeds. Unlike the impulse wheels, they are encased in a housing which is completely filled with water, and the continuation of the outlet below the machine results in the formation of a slight vacuum, which both increases the total head and reduces turbulence. These are really too "high-tech" to consider "home-building" and should be purchased from a manufacturer.

Michell Turbine (Banki Turbine)

There is some confusion in this country about the origin of this style water turbine. In the early part of this century an English engineer, A.G.M. Michell, wrote a discussion of the "cross-flow" turbine. Later European references usually mention Michell as the source. It was, however, a paper by Donat Banki, "Neue Wasserturbine," that introduced the concept to America. A study published in 1949 by the Oregon State College (now the University) included a "free translation" of Banki's paper, along with their test results (see page 119). Subsequent references to this type turbine usually have called it a Banki Turbine, although the European manufacturers list their products as the "Michell cross-flow turbine."

The Michell turbine is probably the best choice for most small hydro-power installations. It is fairly simple to construct, requiring some welding, simple machining and a few amenities in the workshop. The steel parts can be cut from stock sheet and standard steel pipe. The design is essentially the same for a very wide range of flows and head. For installations with variations in flow rates, the rotational speed for top efficiency will remain the same (for a constant head) over a range of flows from 1/4 of the flow to full design flow. A 12" diameter wheel with a head of 15 feet will operate at about 280 rpm. One speed change of approximately 6.5:1 will give the rpm necessary to generate A.C. power. A flow of 0.9 cfs would be required for one horsepower output at this head.

The *Michell Construction* section has a brief description of the theory of this unique turbine; further, more technical discussion can be found in the Oregon State study. Basically there are two parts, a nozzle (Fig. 12A) and a turbine runner (Fig. 12B). The drum-shaped runner is built of 2 discs connected at the rim by a series of curved blades. The rectangular nozzle squirts a jet of water at an angle of approximately 16° across the width of the wheel; the water flows across the blades to their inner edge, flies across the empty space within the drum, strikes the blades on the inner side of the rim and exits from the wheel about 180° from the first point of contact. About 3/4 of the power is developed in the first pass through the blades, the remaining 1/4 in the second pass. With a well designed set-up with smooth nozzle surfaces, as well as thin, smooth blades, the maximum possible efficiency would be 88%; the Oregon State test in 1948 with a "home-built" turbine managed a respectable 68%; the German machines are rated at 84%.

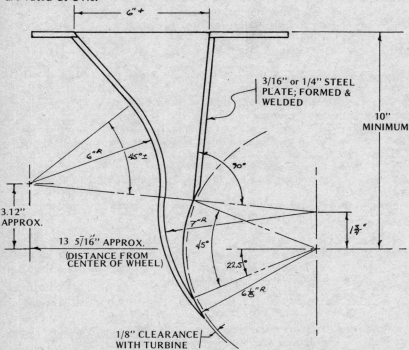

FIG. 12A   NOZZLE FOR MICHELL TURBINE (BANKI)

The cost of materials to construct a Michell turbine is really rather insignificant, especially when compared with the other types of manufactured water turbines. The big expenses will be for the electrical generating equipment, the piping, and, should you need it, the cost of having some machine shop put the wheel together. The Michell turbine is undoubtedly the best power source (where the head is appropriate, in the range of 15 feet to about 100 feet) for the limited budget and those with a desire to be able to do-it-themselves. No other wheel or turbine is quite as versatile, easy to build, and still useful for power generation.

Michell Construction

The following data is for construction of a 12" diameter wheel. The dimensions for wheel diameter and blade curvature can be changed proportionally for larger wheels though the angles will remain the same.

The end plates are 1 foot diameter discs, cut from 1/4 inch sheet steel, with keyed hubs welded in to fit a suitable sized shaft (this is dependent on power requirements). The blades are cut from standard 4 inch steel water pipe (wall thickness is .237 inch). Each blade is cut for a 72° arc; this can be measured at 1/5 the circumference of the pipe (for a 4" pipe the distance along the arc would be 2.83 inches, or .236 feet). Each length of pipe suitable for blade pieces will make only 4 blades (each piece 1/5 the pipe's circumference); since there is some loss of material with each cut, there would not be enough left over after 4 blades and 5 cuts to allow a 5th blade.

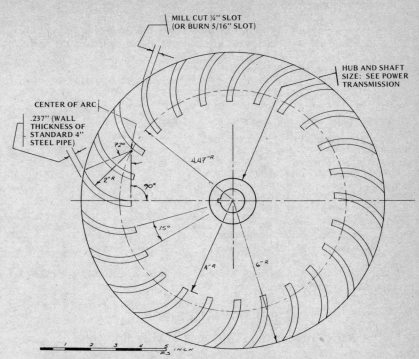

FIG. 12B   12" WHEEL WITH 24 BLADES FOR MICHELL TURBINE

The length of each blade must be 3/4 inch longer than the inside width of the wheel (W_2) to allow enough to stick out beyond each side plate for a 1/8 inch welding tab. The slots on the side plates can be cut with a welding torch. This should require about a 5/16 inch wide cut. For a more accurate job, the slots could be milled out with a .25 inch mill-bit, assuming a milling machine is available. Every center-of-radius for the arc of each cut will fall on a circle of 4.47 inch radius as measured from the center of the wheel (again, this is for a 12 inch wheel). If the wheel is to have 24 blades, each blade will be placed every 15 degrees around the wheel (i.e. every 360/24 degrees); and so, each center-of-radius will be 15 degrees apart around the 4.47 inch radius circle. Once the centers are located for the arc of each blade slot, the arcs are drawn at 2 inch radii, and the slots can be cut.

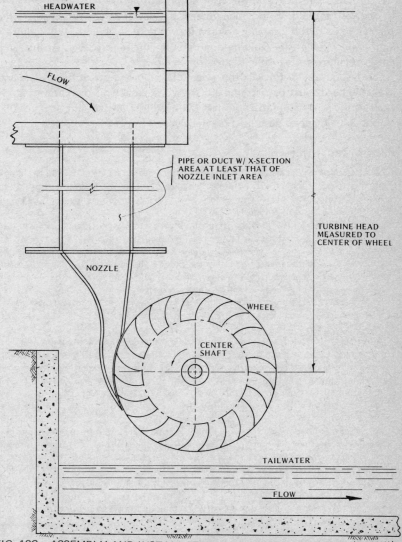

FIG. 12C   ASSEMBLY AND INSTALLATION FOR MICHELL TURBINE (BANKI)

For constant speed regulation (something essential for running an A.C. generator), a slide gate valve would have to be added to the nozzle, plus a centrifugal governor to actuate the valve. (This little mechanism could cost as much as all the other materials combined.) For high heads, the entrance works would be connected directly to a pipe.

Design equations for Michell Turbine:

(Eq. 16)   Width of nozzle = $W_1 = \dfrac{(210)(Q)}{(D)(\sqrt{H})}$

(Eq. 17)   Inside width of turbine = $W_2 = W_1 + 1$ inch

(Eq. 18)   Length of blades = $W_2 + 3/4$ inch

(Eq. 19)   Optimum rpm = $\dfrac{(862)(\sqrt{H})}{(D)}$

where Q = flow rate in cubic feet per second
  D = wheel diameter in inches
  H = head at the wheel in feet
  W_1 = outside width of turbine runner
  W_2 = inside width of turbine runner

The specific details of assembly of the wheel and nozzle can be best understood from the drawings (Figs. 12A, 12B and 12C).

PELTON WHEELS
Written with the Assistance of Guy Immega

The Pelton wheel is an important type of impulse water turbine because with an adequate head it can develop a high rpm useful for A.C. power generation. A Pelton wheel is usually small in diameter with specially shaped bucket cups mounted on the perimeter. Water is directed into the cups by a nozzle. Characteristically, a Pelton wheel requires a very high head of water (over 50 feet), delivered at a small volume of flow through a pipe. This makes the Pelton wheel particularly useful for a small mountain stream. A Pelton wheel turns at high speed (up to 1000 rpm) which makes it attractive for use as a power source for electrical generators (which usually require 1800 or 3600 rpm). The efficiency of a Pelton wheel with polished cups can be up to 93%.

The selection of Pelton wheels is dependent upon the site available, and the power required. With a higher head, the power can be obtained from a smaller wheel, with a smaller volume flow rate. A lower head will require a larger wheel at a greater volume flow rate, for the same power. At the same head, smaller wheels turn at higher rpm's.

To lay out a site for a Pelton wheel, the volume flow of the stream and available head must be determined. A "worst case" approximation for the head at the wheel should allow for 1/3 head loss; the stream flow calculations can be made as described previously.

Pelton Construction

Aside from the wheel itself, the most important part in a Pelton wheel installation is the feed pipe. Generally, to obtain a high head of water, many hundreds of feet of pipe are required. With this length of pipe, a primary concern is the friction of the water flowing in the pipe. Pipe friction is generally expressed in loss of head per 100 feet of pipe. If the pipe feeding a Pelton wheel is too small, then the pressure of water at the nozzle is reduced (effective head reduced).

Friction will be great in a narrow pipe. As the pipe diameter gets larger, the friction in the pipe decreases. But at some point, the high cost of a large diameter pipe outweighs the advantage of less friction. The general rule-of-thumb is that "maximum power-per-dollar invested is extracted from a pipe at 1/3 head loss. . ." This means that a 1/3 head loss gives the highest efficiency at the lowest price. Pipes with diameters allowing less than 1/3 head loss are more "efficient," but are prohibitively expensive.

Head loss may be determined by the use of the pipe flow nomograph (Fig. 9). For instance, suppose the pipe feeding a Pelton wheel must be 100 feet long and will have a static head of 200 feet. (Static head is the pressure inside a pipe filled with water, when the water is not flowing; in this case the pressure is equal to 200 feet of head.) Then the maximum tolerable head loss is 1/3 of 200 feet, or 66 feet. That leaves 134 feet of effective head at full flow. Expressed differently, the head loss must be less than 6.6 feet per 100 feet of pipe.

Suppose also that the stream is less than 6 inches wide and 6 inches deep, and the volume flow is about 32 cfm (.534 cfs), and that the entire stream is fed into the pipe. From the pipe flow nomograph, a 4 inch pipe is found to be more than sufficient to carry the flow, within the allowed 1/3 head loss.

Equation 20 is for finding the speed of the jet of water at the nozzle.

$V = \sqrt{2gH}$ where V = nozzle velocity, when maximum power is taken from
  the feeder pipe
  g = acceleration of gravity, 32 feet per second²

(Eq. 20)   H = effective head at full flow

The nozzle size is set by either the design flow rate of the system or the capacity of the wheel, whichever is less. At any point in the pipe or nozzle, the flow is a product of the velocity of the water through a particular cross-section.

Q = (V)(A)   where Q = flow
  V = velocity
  A = cross-sectional area   (Eq. 21)

Since the nozzle velocity is known from Eq. 20, and the design flow (Q) is known, the theoretical nozzle opening can be found:

T.A. = $\dfrac{Q}{V}$ where   T.A. = theoretical nozzle opening (area) in inches squared
  Q = flow in cfs
  V = velocity in feet per second   (Eq. 22)

With the above example, the area = $\dfrac{.534}{92.6}$ (144 inch²/feet²) = .83 inch²

However, to find the actual flow through a given nozzle the theoretical area must be divided by a "nozzle coefficient" (C_n). C_n is .97 for a plain nozzle without controls. To find the necessary area for delivering the design flow:

Area required = $\dfrac{\text{theoretical area}}{C_n}$

$A = \dfrac{.83}{.97} = .855$ inch²   (Eq. 23)

Then, using simple geometry, the radius and diameter of the proper nozzle can be found.

$r = \sqrt{\dfrac{A}{\pi}}$ where r = radius of the nozzle in inches
  A = area required for delivering the design flow

then $r = \sqrt{\dfrac{.855}{3.14}} = .522$ inch

And since diameter = 2 x radius, D = (2) x (.522) = 1.044 inch   (Eq. 24)

The 4 inch feed pipe we selected on the basis of pipe flow will be more than adequate for this 1 inch diameter nozzle.

When the nozzle is delivering water so that the wheel can achieve maximum power, the rim speed of the wheel will be 1/2 the velocity of the nozzle jet.

U = .5 V   where U = rim speed of the wheel (Pelton wheel)
  V = velocity of the jet of water from the nozzle   (Eq. 25)

In the case of the 12 inch wheel with a 1 inch nozzle, the circumference of the wheel is: C = πD, or C = $\pi\dfrac{12 \text{ inches}}{12 \text{ in/ft}}$ = 3.14 feet.
From Eq. 20 V = $\sqrt{2gH}$, we know that with the available head of 134 feet, the nozzle jet velocity will be 92.6 feet/second. Therefore, from Eq. 24, the rim speed will be half the velocity, or 46.3 feet/second.

To find rpm use Eq. 14 (from Overshot wheels):

rpm = $\dfrac{\text{rim speed}}{\text{circumference}}$ = $\dfrac{46.3 \text{ ft/sec}}{3.14 \text{ ft/revolution}}$ = 14.74 $\dfrac{\text{revolutions}}{\text{second}}$ x $\dfrac{60 \text{ sec}}{\text{minute}}$ = 885 rpm

If the wheel is overloaded, the speed will be slower and the power will drop. If the wheel is underloaded, the speed will be greater, but water will be wasted.

SCALE MODEL OF A PELTON WHEEL: Showing nozzle on right and pulley on left. Normally the cups on the wheel would be notched in the center so that water shooting out the nozzle would hit more than one cup at a time.

Reaction Turbines: Francis Wheel, Kaplan Turbine, Propeller (and others)

Some people may come into possession of one of this class of reaction turbines. These consist of a number of curved and convoluted vanes or runners arranged around a central shaft. The water flowing through these vanes causes the wheel to rotate by the "reaction" or pressure of the water. These turbines require careful engineering to operate properly and must be purchased from a manufacturer rather than be home-built.

Reaction turbines are usually designed for a limited range of flow and head conditions, and are not suitable for other than their design specifications. They are very efficient when properly regulated for load and flows (up to 93%) and turn at a high rpm. The high rpm makes them ideal for driving an electrical generator. Because of their shape and structure they are expensive, but it is possible to find used or surplus turbines at bargain prices (be sure to check the specifications of head and flow as the turbine may be useless if it doesn't fit the situation).

The only knowledge that is needed for installing a turbine of this class is its rating for head and flow rate, operating speed, and of course all the basic information about the site at which the turbine will be placed.

The James Leffel Company manufactures a complete package unit for small hydro-power sites (see page 119). Leffel's design, a Hoppe turbine, a variation of the Francis, is available for heads of 3 to 25 feet with an electrical output of 1 to 10 kw.

POWER TRANSMISSION

In many cases the power available from a hydro-power installation will not supply the amount of power desired. In those cases it will probably be necessary to re-define your needs or fill them on a priority basis. Some needs can also be filled by another source of power. It is thus useful to think in terms of an integrated power systems approach from the outset.

It is probably pretty obvious that the operating speeds of the water wheels and turbines will not exactly match the speeds required to drive the generator or compressor. The common design of A.C. generators requires a particular rpm in order to produce the proper voltage and frequency of electricity. For the 60 cycle frequency common in the U.S. and Canada, the minimum input of a two-pole generator is 3600 rpm. Four pole generators are slightly more expensive, but operate at 1800 rpm. Six and eight-pole generators are also available (with proportionately slower operating speeds of 1200 rpm and 900 rpm) but these are specialty items that can be hard to find and are quite expensive besides. Even the speediest turbine of them all, the Pelton wheel, would require a head of more than 2000 feet to drive a 12 inch wheel at the 1800 rpm that is needed for a four-pole A.C. generator.

At the other extreme in power sources, the overshot wheel would turn at about 10 rpm (for a 16 foot wheel); the gearing up to get the speed necessary for electric generation would require a gear ratio of over 1-to-100. This sort of speed change would, in itself, cause some considerable loss in the overall efficiency of the system.

D.C. generators and alternators are available throughout the world for use in autos. They can operate at most any rpm above their minimum (that is, something faster than about 700 rpm); also, they are usually equipped (in an auto) with some sort of voltage regulator to avoid overcharging the batteries. This sort of flexibility in operation, along with their ready availability and their bargain prices, make them attractive for use in small scale generating plants (for windmills, as well as waterwheels). The D.C. technology developed for automotive systems can be applied for some household uses: D.C. storage batteries, light bulbs, radios, tape players and small motors are available. The biggest problem in converting a household to D.C. use are those motors in appliances and power tools. Most are designed for the common 60 cycle A.C. power source at 120 volts and would be completely useless for a 12 volt D.C. power supply.

Most installations use belts and pulleys to transfer the power from turbine to generator and, at the same time, obtain the needed rpm along the way. Larger installations, or those with high torques (like the water-wheels) need to have some sort of metal gearing or drive chains to handle the loads. Ready-built gear boxes and speed changers are usually available at surplus or used machinery dealers. And, of course, the neighborhood auto junk yard has a plentiful supply of rear axles, already complete with

roller bearings and wheel mounting bolts. These can be arranged to provide a gear ratio of from 3:1 to almost 9:1. (Of the three rotating parts, that is, the two wheels and one drive shaft connection, one must be fixed in order for power to be transferred through the other two.) To check the gear ratio of any gear box in question merely turn one shaft and count the resulting turns of the other shaft.

Glossary

axial flow	A term for hydraulic machinery, pumps, turbines, in which the water flows parallel to the power shaft (axis of rotation) as in a propeller pump.
center of curvature	That spot where the point of the compass is stuck when drawing an arc or circle.
cfm.	Water flow rate, cubic feet per minute.
cfs.	Cubic feet per second.
control gate	See gate.
design flow	That flow rate for which the turbine is designed.
fps = feet/second	Velocity in feet per second. Also: (fps)(60) = feet/minute - feet per minute.
flume	An old term for a wooden or metal box channel: an aqueduct.
gate	In a sluice or canal, a structure of vertical sliding boards or metal that controls the flow of water (as in a "watergate").
gpm.	Gallons per minute (there are 7.48 gallons in each cubic foot).
head	The elevation of water that is available and so, a measure of the energy of the water. In some cases the pressure in a pipe may be indicated by the head in feet of water. The law of conservation of energy in water flow is given by the equation:

$$\frac{v^2}{2g} + \frac{P}{\alpha} + \frac{(\text{elevation})}{\text{in feet}} = \text{a constant along a continuous flow.}$$

v = velocity in feet/second
2g = 64.34 feet per second^2
α = 64.4 lbs. per cubic foot
P = pressure in lbs. per square foot

headwater	The static head (without velocity) usually behind a dam, sluice, or weir.
HP	Horsepower. A measure of power, equivalent to 745 watts.
hydraulic radius	A concept used in analysis of water flow in channels. Equal to the cross-section area of the flow divided by the wetted perimeter.
penstock	A pipe to carry water to the turbine, usually under a high pressure.
percent grade	% Grade. The slope of ground, creek, or canal—in feet per 100 feet or meters/100 meters. (Not the same as the channel flow equations "s.")
psi	Pounds per square inch. A pressure measurement, equivalent to .433 foot of head (of water).
Q	Symbol for flow rate, usually in cubic feet per second or liters per second.
radial flow	For hydro machinery, pumps and turbines; where the water flows radially out from the power shaft; as in a centrifugal pump, or radially as in a Francis wheel.
radius of curvature	The distance from the center of curvature to the arc.
rim speed	The velocity of a point on the rim of a rotating wheel or turbine = (rpm)(2π)(radius).
rpm	Rotation, revolutions per minute.
slope	In channel flow calculations: slope, s = feet of drop per 1000 feet of horizontal distance (or meters drop per kilometers).
tailrace	The channel that carries the tailwater flow.
tailwater	The water surface elevation immediately downstream of a water-wheel or turbine (see various wheel illustrations).
torque	Something that produces or tends to produce rotation or torsion and whose effectiveness is measured by the product of the force and the perpendicular distance from the line of action of the force to the axis of rotation.
tuberculation	The pits and lumps of rust and corrosion in steel and cast-iron pipe.
weir	An exact opening (rectangular, triangular, or trapezoidal) used to accurately measure water flow rates.
wetted perimeter	A concept used in flow analysis. It is that portion or length of the channel cross-section that is in contact with the flow (measured perpendicular to the flow). For a circular pipe flowing full, the wetted perimeter would equal the circumference of the pipe.

WATER REVIEWS

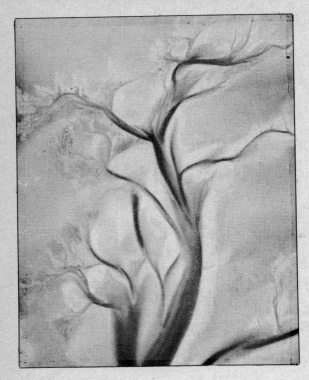

CLOUDBURST

This is a practical handbook for rural survival; a teaching book, with experience and ideas on building tools and using them: saunas, curing fish, domes, compost shredders and more. Forty pages are devoted to waterwheels including one of the more "recent" available articles on the subject—a 1947 *Popular Science* reprint entitled, "Harnessing the Small Stream."

The long-awaited edition of CLOUDBURST II is now available. Similar in format to the first book, it is short of energy information. The exceptions are two articles on hydraulic rams and one on flat-plate solar collectors.

—T. G.

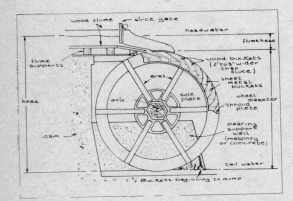

Cloudburst
Vic Marks, editor
1973; 128 pp

Cloudburst II
Vic Marks, editor
1976; 128 pp

$4.95 Each

from:
Cloudburst Press
Box 79
Brackendale, B.C.
Canada

or WHOLE EARTH
TRUCK STORE

SENSITIVE CHAOS

Fluids in motion. The beauty and strength, motion, flow, grace and harmony of water and wind. ALIVE. A theoretical look at the inter-relationship of nature and human beings using movement as the common denominator. Opinionated and presumptuous in spots, Schwenk hammers away at the aliveness and biosimilarity of nature and human beings.

—T.G.

Sensitive Chaos
Theodor Schwenk
1965; 231 pp

$10.00

from:
Rudolph Steiner Press
35 Park Road
London, England N.W.1

or WHOLE EARTH
TRUCK STORE

The ray has assimilated the movement of the waves into its fin movements (after Hesse-Doflein)

VOLUNTEERS IN TECHNICAL ASSISTANCE

VITA is a non-profit corporation that provides information and technical assistance to groups involved with projects around the world. Much of their assistance is directed to unindustrialized nations. Their expertise is "low-tech" engineering technology predominantly in the areas of water and water resources, agriculture and food, home improvement, sanitation, construction and communication. *The Low Cost Development of Small Water Power Sites* pamphlet has basic information on measuring head and flow as well as information on various types of waterwheels and turbines.

—T.G.

VILLAGE TECHNOLOGY HANDBOOK
1970; 350 pp; $9.00

Making Building BLocks with the CINVA-RAM BLOCK
1966; 30 pp; $1.50

Low Cost Development of Small Water Power Sites
1967; 50 pp; $2.50

Solar Cooker Construction Manual
1967; 25 pp; $2.00

Hydraulic Ram
1970; 15 pp; $1.50

Low Cost Windmill for Developing Nations
1970; 45 pp; $2.50

Design Manual for Water Wheels
1975; 80 pp; $4.50

Handpumps for Village Wells
1975; 15 pp; $1.75

Freshwater Fish Pond Culture and Management
1976; 200 pp; $1.50

Using Water Resources
1977; 160 pp; $5.00

Publications List and
Information Available from:

VITA
3706 Rhode Island Ave.
Mt. Rainier, Maryland 20822

WATER INFORMATION CENTER, INC.

These people specialize in publishing technical books about water resources and conducting studies about water around the world. There is no water power information to be found among their long list of publications. However, for the person interested in water shortages and groundwater problems, this is a place to get a good bibliography of reading material that should keep you busy for years. Many of the books they publish can be found in libraries where reading them will put less of a dent in your pocket.

—T.G.

A particularly interesting title they publish is:

WATER ATLAS OF THE UNITED STATES
J. J. Geraghty, D. W. Miller, F. van der Leeden, F. L. Troise
1973; 200 pp; $40.00

Four years in preparation . . . this updated and expanded new atlas now gives you . . . 33 new maps devoted to the nation's water quality and pollution problems. Sample map-titles from this section include: Thermal pollution, Pipeline Spills, Mercury, DDT, Sewage, Hardness, Sodium, Arsenic and Lead, Cadmium and Chromium, Offshore Disposal and Fish Kills to name just a few.

Information from:
WATER INFORMATION CENTER, Inc.
7 High Street
Huntington, New York 11743

U.S. GEOLOGICAL SURVEY

Geological surveys of the U.S. Mostly large scale. Often individual studies of rivers, National Parks, earthquake areas, soil studies, etc. Their standard topographic maps are 7½ minute and 15 minute maps which sell for $0.75 postpaid. They also have maps covering larger areas. Free for the asking are state maps showing what smaller area topo maps are available and a useful pamphlet "Topographic Maps" which explains how to read all the numbers and curved lines.

All maps available from: Distribution Sect., U.S. Geological Survey, Federal Center, Denver, Colo. 80225. Also from: Local centers. Check the phone book in larger cities.

FINDING AND BUYING YOUR PLACE IN THE COUNTRY

After a year of selling books on buying land to people who come to us at the Whole Earth Truck Store, this is the first one I can recommend with a clean conscience. With chapters on looking for land, climate, water rights, legal information, escrow, the contract, financing your purchase and much more. It's a damn good book.

—T.G.

Finding and Buying Your Place in the Country,
Les Scher, 1974; 393 pp

$6.95

from: Collier-Macmillan
Publishing Co., Inc.
866 Third Ave., NY NY
10022
or WHOLE EARTH
TRUCK STORE

SURVEYING

"Surveying is the art of measuring and locating lines, angles and elevations on the earth's surface." It is also the method used to stake out those territorial claims that are so important to the animal in man. **Surveying** is a standard handbook for surveyors; it is also used as a text for college courses and would be sufficient for a self taught course in land surveying.

Surveying
C. B. Breed and A. J. Bone
1942, 1971; 495 pp

$14.25

from:
John Wiley & Sons
605 Third Avenue
New York, NY 10016

or WHOLE EARTH
TRUCK STORE

High quality, reliable equipment together with a sound knowledge of surveying technique, make a survey successful. Most of the companies which make surveying tools make good ones. Two that are well known among land surveyors are Lietz and Berger. Lietz has a complete line of instruments and accessories, including machetes, technical books and drafting equipment. Berger is smaller but has all the basics necessary for survey work.
—Robin Saunders

Berger Instruments
37 Williams Street
Boston, MA 02119

Has free booklet HOW TO USE TRANSITS AND LEVELS

Lietz
829 Cowan Road
Burlingame, CA 94010

WATER RIGHTS

Possibly the most important consideration to buying land is the availability of water. Many first-time land owners don't realize that the presence of water doesn't necessarily guarantee its availability. Laws regulating water rights are complicated, often having more to do with custom and tradition than legal fact. It is possible to have water flowing next to your land and not be able to use it. A basic understanding of water rights is essential when purchasing raw land. Possible sources of information might include a local lawyer, county agricultural office, or neighbors. A good general discussion of water rights is contained in FINDING AND BUYING YOUR PLACE IN THE COUNTRY (see opposite). The following books deal with the subject in greater detail.
— Alan Edgar

Davis, Clifford, 1971. RIPARIAN WATER LAWS: A FUNCTIONAL ANALYSIS. 81 pp. NTIS, Springfield, VA.

Habercorn, Guy E. Jr., 1974. WATER RIGHTS AND WATER LAW: A BIBLIOGRAPHY. 160 pp. NTIS, Springfield, VA.

Kendall, James H., 1967. WATER LAW: STREAM-FLOW RIGHTS IN NEW ENGLAND AND NEW YORK STATE. 47 pp. New England Interstate Water Pollution Control Commission, Boston, MA 02108.

Trelease, Frank J., 1967. WATER LAW: CASES AND MATERIALS. 364 pp. West Publishing Co. St. Paul, Minn.

WATER MEASUREMENT

Since water is such a precious commodity, some sort of accounting is usually necessary for measuring and metering the distribution of the wealth. This is particularly true for those situations where a large volume of water must be purchased from some local water utility, (e.g. irrigation) or for a community of water users to meter each individual outlet. These are needs that require some accuracy in measurement; but there is no need to resort to expensive mechanical devices. These two books discuss those means or "artificial controls" that can be used to measure the water flow rate with a great deal of accuracy and a minimum of expense and maintenance. The World Meteorological Organization book is primarily concerned with stream flow measurement made with open flow weirs and flumes. The Water Measurement Manual is a standard in the water resources business; it catalogues all sorts of possibilities for flow measurement with pipes, gates, and orifices as well as the standard weirs and flumes. For the money and the information the Water Measurement Manual is really the best bargain.
—Robin Saunders

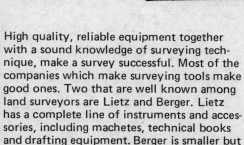

Cipolletti weir in a permanent bulkhead discharging under free-flow conditions.

Use of Weirs and Flumes in Stream Gauging:
WMO 280
1971; 55 pp
$6.23

from: World
Meteorological Org.
Publications Ctr.
P.O. Box 433,
NY, NY 10016

Water Measurement Manual: Bureau of Reclamation, 1967;
323 pp **$6.40**

from: Sup't of Documents
U.S. Gov. Printing Office
Washington, D.C. 20402
or WHOLE EARTH
TRUCK STORE

WATERSHEDS

Water, Earth and Man
Richard J. Chorley
1969; 580 pp; $25
Associated Book Publishers
North Way Andover
Hampshire, England
FP10 5RE
or WHOLE EARTH TRUCK STORE

Vegetation and Watershed Management
E. A. Coleman
1953; 415 pp.
Ronald Press out of print

Seeding and Planting in the Practice of Forestry
James W. Toumey
1916; 455 pp.
John Wiley & Sons out of print

Water
Luna B. Leopold and Kenneth S. Davis
1970; 196 pp; $9.32
Time-Life Books

Watershed Planning: A Selected Bibliography
1976; 57 pp; $5.50
Council of Planning Librarians
P.O. Box 229
Monticello, IL 61856

WINDMILLS AND WATERMILLS

This highly readable book traces the history of watermills from their earliest uses in the Middle East as devices to move water, through the development of the first grain mills in Roman times, to their ultimate refinement in the textile mills of England and North America during the 18th and 19th centuries. The text is illustrated with many excellent photographs and drawings.

Windmills and Watermills
John Reynolds
1971, 196 pp
$8.95

from:
Praeger
111 Fourth Ave.
New York, NY 10003

DOWSING

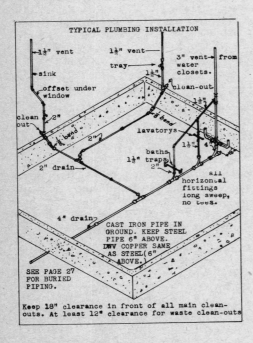

Moses is said to have been the first dowser—with a little help from his friend.

Both *Dowsing* and *The Beginner's Handbook of Dowsing* are "how to" books. *Dowsing* in pamphlet form gives more information especially on different types of dowsing rods. It is also cheaper, and its author has more of a professional dowsing background. The handbook is considerably better illustrated with most of the basic stuff in it. Both sources refer to the American Society of Dowsers Inc., Danville, Vermont 05828, a non-profit educational and scientific organization. They publish a quarterly journal—"The American Dowser."

—T.G.

Dowsing
Gordon Maclean, Sr.
1971; 46 pp

$1.00

from:
Gordon Maclean, Sr.
30 Day Street
South Portland, Maine 04106

The Beginner's Handbook of Dowsing
Joseph Baum
1974; 31 pp

$3.95

from:
Crown Publishers Inc.
419 Park Avenue South
New York, N.Y. 10016

or WHOLE EARTH TRUCK STORE

PLANNING FOR AN INDIVIDUAL WATER SYSTEM

For those who like their education with a bit of visual aids, this book has plenty of multicolored pictures, charts, and even a few cartoons to lighten the subject. In spite of the slick approach, the information is all very relevant for those installing their own water supply system. Much emphasis is placed on water quality, both how to prevent contamination and how to use the methods of water treatment.

—Robin Saunders

Planning for an Individual Water System, 1955, 1973; 155 pp

$6.95

from:
American Assoc. for Vocational Instructional Materials, Engineering Ctr.
Athens, Georgia 30602
or WHOLE EARTH TRUCK STORE

PROFESSIONAL PLUMBING ILLUSTRATED

We all know that plumbers cost a lot by the hour; at those rates it wouldn't take long to pay for the price of this book. It is very complete, not only what to do, but even considerable amounts of practical advice on procedures and potential problems. How to install everything from Roman baths, furnaces and gas pipes to the plumbing system for an entire house. This book written by a former plumbing inspector even includes useful math basics for plumbers.

—Robin Saunders

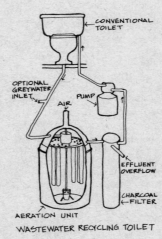

Professional Plumbing Illustrated, and Plumbing Mathematics Simplified
Arthur J. Smith
1959, 1972; 243 pp

$6.00

from:
Mrs. Arthur J. Smith
4037 Golf Drive
San Jose, California 95127

RESIDENTIAL WATER CONSERVATION

Talk of desalinization plants has filled the newspapers and technical journals in recent years. The world is becoming acutely aware that shortages of water are imminent and could be disastrous. Expensive desalinization plants are surely the wrong way to solve the problem at the present time in the U.S.A. We can save money and save water by judicious use of this valuable natural resource.

Both common sense approaches and the latest technologies are explored here. Historical background of the development of urban water and waste systems is reviewed as well as water utility rate structures and almost every aspect of residential water use from drinking to water recycling. This book is a must for every owner-builder and contractor in the country.

—T.G.

Residential Water Conservation
Murray Milne
1976; 468 pp

$7.50

from:
California Water Resources Center
475 Kerr Hall
University of California
Davis, California 95616

WATER SUPPLY AND WATER WELLS

In many areas of the world ground water is in plentiful supply. Searching for it is usually a matter of consulting geologic maps, learning about existing wells in the area and looking for surface evidence such as streams and ponds. Designing, drilling and maintaining the well and water system must be done carefully and with an understanding of the processes. It is not an easy task to develop a useable and ecological water supply.

The *Water Well Manual*, a reprint of a U.S. Department of State publication *Small Wells Manual*, covers the area of finding water, well drilling and maintenance more than adequately.

For more information especially on "power" drilling methods, a useful but out of print publication from the USGPO is *Well Drilling Operations* (check your local library).

A final suggestion is *Water Supply for Rural Areas*, a World Health Organization book. This is probably the most complete book of the three in covering water supply. It discusses many health related aspects of water, hand dug wells, pumping systems and also construction of small dams.

I suggest using the *Water Well Manual* for its information on wells and *Water Supply for Rural Areas* for its information on developing a good overall water supply system. Both do the job.

—T.G.

Water Supply for Rural Areas
E.G. Wagner and J. Lanoix
1959; 339 pp

$17.60

from:
Q Corp; World Health Org. WHO
49 Sheridan Ave.
Albany, N.Y. 12210

Water Well Manual
U.P. Gibson & R.D. Singer
1971; 155 pp.

$6.00

from:
Premier Press
P.O. Box 4428
Berkeley, CA 94704

both available from WHOLE EARTH TRUCK STORE

COMMUNITY WATER SYSTEMS

Some of my friends in Big Sur, California could have used this one when they were expanding an old 1930 water supply system to service a dozen new homes. They miscalculated the pipe size, got a trickle out of the kitchen sink and had to dig up a thousand feet of the wrong size pipe.

Included are basic explanations and practical data on population densities, water use, wells and sources of supply, distribution systems, water quality, and cost accounting for residential as well as industrial applications. Although the approach is mainly to simplify the planning procedure of large and small scale public water systems for those unfamiliar with the workings, the book may even prove useful to those civil engineers working in this field who find themselves too close to the subject to see the easy approximations.

—Robin Saunders

Pipe Discharges in Gallons Per Minute for Different Diameter Pipe for Known Lengths and Under a Pressure of 40 Pounds Per Square Inch

Pipe Size Inches	Length of Pipe in Feet						
	100	200	300	400	500	600	700
1½	100	64	52	44	40	37	36
2	200	130	100	86	76	70	66
2½	300	210	160	140	120	110	105
3	600	410	325	290	250	235	220
4	1,200	730	580	490	410	380	350
6	3,200	2,400	1,700	1,450	1,300	1,180	1,090
8	7,100	4,700	3,600	2,900	2,600	2,300	2,100
10	13,000	8,200	6,200	5,200	4,400	4,000	3,600
12	18,500	11,200	8,800	7,300	6,300	5,600	5,100

Cut-Away of AWWA Valve Showing Wedging Mechanism   Hub Ends   Flanged Ends

Community Water Systems, Joseph S. Ameen, 1960, 214 pp

$7.50

from:
Technical Proceedings
P.O. Box 5041
High Point, N.C., 27262
or WHOLE EARTH
TRUCK STORE

IRRIGATION PUMPING PLANTS

Basically about sizing, designing, and installing pumping plants for irrigation systems. Also, gets into some handy stuff on trouble-shooting and maintenance problems that are part of the regular operation of pumps and their motors.

—Robin Saunders

Irrigation Pumping Plants
U.S.D.A.
1959; 70 pp

$0.45

from:
Superintendent of Documents
U.S. Government Printing Office
Washington, D.C. 20402

PONDS FOR WATER SUPPLY AND RECREATION

The demand for water in the US has increased rapidly over the years. The USDI has put out this handbook to satisfy those farmers, ranchers, and other country folks who want to provide for themselves rather than wait for "that big irrigation project proposed for '84." The stated concerns are for irrigation, fishing, recreation, fire protection, and wildlife habitat; of course, the same sort of dam can be used for a water-power reservoir (the USDI doesn't mention that possibility).

They do, however, include most all the information that you need to know to build your own earth-fill dam (for whatever purpose), including: assessing the needs for water; preliminary site studies; drainage areas required for various size ponds; regional climates and storm run-off; requirements for inlet works and over-flow spillways; various construction techniques; and sealing the ponds to limit seepage. Some of the most useful information is the basic course in do-it-yourself hydrology to estimate the storm run-off and average annual stream flows that the reservoir will have to handle. The discussion is really for large farm ponds and reservoirs and the construction methods are really for those with some heavy equipment that can move a lot of dirt. However, the basic hydraulics discussed are just as important for any size or shape of dam.

—Robin Saunders

From an economic viewpoint, locate the pond where the largest storage volume can be obtained with the least amount of earthfill. A good site usually is one where a dam can be built across a narrow section of a valley, the side slopes are steep, and the slope of the valley floor permits a large area to be flooded. Such sites also minimize the area of shallow water. Avoid large areas of shallow water because of excessive evaporation and the growth of noxious aquatic plants.

**Ponds for Water Supply and Recreation;
Agriculture Handbook No. 387**, Soil Conservation Service, U.S. Dept. of Agriculture, 55 pp

$1.25

from:
Sup't of Documents
U.S. Gov. Printing Office
Washington, D.C. 20402
or WHOLE EARTH
TRUCK STORE

DESIGN OF SMALL DAMS

This is the definitive text on the design and construction of earth fill dams. The dams discussed and illustrated are medium sized or large by most standards. With considerable information on ecological impacts, soil geology, soil placement, construction techniques and the like this book has become part of the reference library of most civil engineers working in the field.

—Robin Saunders

Design of Small Dams
Dept. of the Int.
1973; 816 pp

$12.65

from:
Sup't of Documents
U.S. Gov. Printing Off.
Washington, D.C. 20402

(A) WITH ROCK FILL TOE

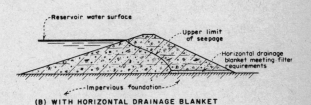

(B) WITH HORIZONTAL DRAINAGE BLANKET

Figure 114. Modified homogeneous dam.

SMALL EARTH DAMS

Dam building is a very involved undertaking. This circular endeavors to supply you with practical suggestions which may be of assistance to you. It is up to you to fit them together as they apply to your situation.

About 20,000 small earth dams have been built on California farms in the last 25 years and more are being built all the time. While many of the early dams were built primarily as stock-watering ponds, farmers are also deriving benefits from their reservoirs in the way of irrigation water and even recreational purposes.

But dams have disadvantages too. They are expensive to build; they require labor and more expense to maintain properly; they may increase the nearby mosquito population.

**Small Earth Dams
Circular 467**
Lloyd N. Brown
1965; 23 pp.

FREE
from:

California Agricultural Extension
90 University Hall
University of California
Berkeley, CA 94720

HYDRAULIC RAMS—

HYDRAULIC RAM: AN ENGINEERING BELIEVE-IT-OR-NOT

The principles behind the successful operation of a hydraulic ram are engineering facts of life that I always have found difficult to believe. By means of air pressure, built up by the weight and velocity of the water, a stream of falling water manages automatically to pump a good part of itself to a height of 25 feet for each foot that it falls. Given a flow of not less than 1½ gallons per minute from a spring, creek, or artesian well, and—for example— a fall or "head" of four feet, you could elevate part of the flowing water to a tank 100 feet above the ram. These facts suggest that hydraulic rams ought to be an important factor in water-supply systems in many rural areas—and they are.

The Rife Company manufactures three different types of rams. Costs run between $300 and $3000 (FOB factory) with drive or intake pipe sizes from 1¼" to 8". Rife claims to have received more than 10,000 inquiries from having been reviewed in the Last Whole Earth Catalog. We at the Truck Store haven't heard any complaints.

Green and Carter Ltd. manufactures the Vulcan ram. With drive pipe size from 1¼" to 7" their rams cost a bit less than Rife's but the shipping expense from merry ole' England will certainly make up the difference.
—T.G.

Information free

Rife Hydraulic Engine Mfg. Co.
Box 367
Millburn, N.J. 07041

Green & Carter Ltd.
Vulcan Iron Works
Kingsworthy
Winchester
Hampshire
England

HANDBOOK OF APPLIED HYDRAULICS

If you were going to build Hoover Dam or supply Chicago with water this is the book you would use to find out how.
—T.G.

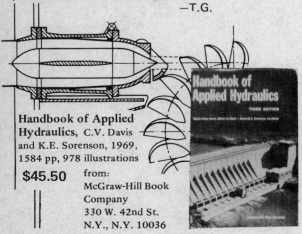

Handbook of Applied Hydraulics, C.V. Davis and K.E. Sorenson, 1969, 1584 pp, 978 illustrations

$45.50 from:
McGraw-Hill Book Company
330 W. 42nd St.
N.Y., N.Y. 10036

TIDAL POWER

The book is compiled from the contributions to the 1971 International Conference on the Utilization of Tidal Power. It is perhaps the most comprehensive statement of state-of-the-art harnessing of tidal power on a large scale; and, as such it is a highly technical tome. The subjects covered include construction techniques, corrosion problems, environmental problems, pumped storage operations, math-models of operations, sedimentation problems and the economics of it all. As one contributor puts it in his introduction, "Tidal power should be recognized as a special source of energy, requiring special technology for its development. Approaching tidal power as saltwater river hydro-power is bound to cripple the development of the appropriate technologies." A point that is repeatedly made in the papers presented is that often tidal projects are rejected as being too expensive, and too "unsure," when the real problem is the lack of information and research. There are, of course, certain requirements of geology and geography that have to be met for any tidal power location to be feasible. But even with obvious restraints it is estimated that Canada could supply as much as 15% of its total national power needs with renewable tidal power.
— Robin Saunders

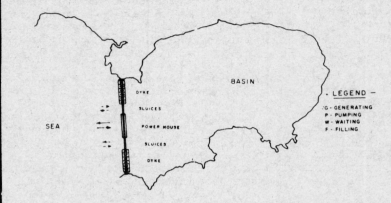

Tidal Power
T. J. Gray & O.K. Gashus
1972; 630 pp

$28.00

from:
Plenum Press
227 West 17th Street
New York, NY 10011

PRIMARY SOURCES (PUBLICATIONS) FOR ADDITIONAL INFORMATION ON WATER POWER

	Head and Flow	Dam	Channels/Sluices	Poncelet	Breast	Overshot	Michell (Banki)	Pelton	Reaction Turbines	Power Transmission
LOW-COST DEVELOPMENT OF SMALL WATER-POWER SITES. From: VITA (see page 114)	Yes	Yes	Yes	Minimal	No	Minimal	Yes	Minimal	Minimal	Minimal
CLOUDBURST (see page 114)	Yes*	Yes*	No	Yes	Yes	Yes	Yes†	No	No	Minimal
YOUR OWN WATER-POWER PLANT. from: Popular Science 1947. Reprinted in the Mother Earth News issues No. 13 & No. 14 (see page 249)	Yes	Yes	No	No	No	Yes	No	Yes	No	Minimal
WATER POWER FOR YOUR HOME. from: Popular Science May 1977	Yes	No	No	No	No	No.	No	Yes	Yes	Yes
HYDROPOWER from: Wadebridge Ecological Centre 73 Molesworth St. Wadebridge, Cornwall, England ± 1.50 postpaid	Yes	Yes	Yes	Yes	No	Minimal	No	Yes	No	Minimal

*reprinted from: YOUR OWN WATER-POWER PLANT (Pop. Sci.)
†reprinted from: LOW-COST DEVELOPMENT OF SMALL WATER-POWER SITES (VITA)

Water in the Service of Man
H.R. Valentine
224pp; 1967
Penguin Books (out of print)

Windmills and Watermills
John Reynolds
196pp; 254 illus; 1971; reprint 1974
Praeger Publishers
111 Fourth Ave.
New York, N.Y. 10003
$8.95

Power From Water
T.A.L. Paton & J.G. Brown
210pp; 1961
Leonard Hill (books) London
(out of print)

Hydroelectric systems come in three sizes; micro, small and large. The micro systems are the size that is applicable to a small community, homestead or house. Most units produce DC power which is then transformed into AC current. As of Spring 1977 we knew of only the following companies which manufactured micro hydropower systems.

— T.G.

INDEPENDENT POWER DEVELOPERS

In the past three years IPD has installed approximately 75 hydropower units. They manufacture a small Pelton wheel and package a complete system including generator, batteries and piping connections. IPD will install a unit anyplace and does consulting work also. A brochure with information is available for $2.

Independent Power Developers, Inc.
Box 1467
Noxon, Montana 59853

THE BANKI WATER TURBINE

This paper covers a study made in 1948 for a "typical" installation of the Banki or Michell cross-flow turbine. The test on this simply constructed water turbine produced 2.75 horsepower at 280 rpm using 2.22 cfs at 16ft. of head: a resulting efficiency of 65%. As the authors state this is only a first try, and by no means the best efficiency for this type of turbine (though quite respectable). The first half of the article is a loose translation of the original paper by Donati Banki "Neue Wasser turbine" — an involved discussion of the theory of hydraulics of the turbine, including the vector analysis of the water flow, blade angles, curvature, and speed (strictly technical stuff).

The second half is a brief discussion of the test set-up and the results, including variations of horsepower and efficiency at different speeds and flow rates.

This is apparently the only complete discussion of the Banki or Michell turbine available in the states. It is interesting to note the prediction that, "...there is a distinct place for the Banki turbine in the small turbine field" has not been realized in this country in the 25 years since the paper was published.

— Robin Saunders

The Banki Water Turbine: Bulletin 25
C.A. Mockmore, Fred Merryfield
1949; 30 pp.

Free single copy
$0.40 additional copies

from:
Oregon State University
Engineering Experimental Station
Corvallis, OR 97331

CANYON INDUSTRIES

Canyon manufactures a small propellor type water wheel. It produces twelve volt D.C. power.

Canyon Industries
P.O. Box 2543
Bellingham, WA 98225

SMALL HYDROELECTRIC SYSTEMS AND EQUIPMENT

William Kitching and Associates manufacture and install three different sizes of Pelton wheels. 4½", 9", and 18" wheels in either cast aluminum alloy or in manganese bronze (the bronze units are more expensive and more durable). Complete units are available as well as just the wheels themselves or any component parts. A brochure is available for $5.

Small Hydroelectric Systems and Equipment
Box 124
Custer, WA 98240

OSSBERGER-TURBINENFABRIK

Seems to be the only company in the world which manufacturers the Michell (Banki) cross-flow turbine. They have units producing from 8.5 KW up to 220 KW and build each unit in accordance with the local conditions. These units are probably better suited for the small community rather than the individual. Share the cost and the power.

— T.G.

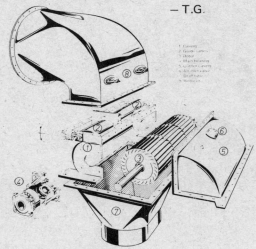

Information Free
from:
Ossberger-Turbinenfabrik
D-8832 Weissenberg i. Bay, P.O. Box 425
Bayern, Germany

or from:
F.W.E. Stapenhorst
285 Labrosse Ave.
Pointe Claire, Quebec
H9R 1A3 Canada

OSSBERGER 542

ELEKTRO G.m.B.H.

Elektro, which is famous for their wind generators, also makes a very high quality small scale water turbine. They do not have any US distributors and their price is very high.

Elektro G.m.b.H.
St. Gallerstrasse 27
Winterthur, Switzerland

PUMPS, PIPES AND POWER

This is a very small company that is primarily in the irrigation business. But they manufacture and install a small hydro system.

Pumps, Pipes and Power
Kingston Village
Austin, Nevada 89310

THE JAMES LEFFEL COMPANY

As far as we can tell Leffel is the only company in the USA that has been manufacturing small hydroelectric units for any length of time (over 100 years). They also manufacture big units.

Their small units, called Hoppes, run on a maximum of 25 ft. of head and a maximum output of 10 KW. With those maximum specifications the unit costs approximately $7200 including turbine and generator but no piping, F.O.B. factory. Cost increases as the amount of head decreases; so to get 10 KW out of 15 ft. of head would cost considerably more.

When we made our inquiry, they had a small Pelton wheel "in stock." Using a head between between 200 and 300 ft. the unit cost approximately $3000.

In requesting information, ask for a Pamphlet A which explains how to measure head and flow, etc., and Bulletin H-49 which describes the workings and different models of the Hoppes unit.

—T.G.

Information Free:

The James Leffel Company
Springfield, Ohio 45501

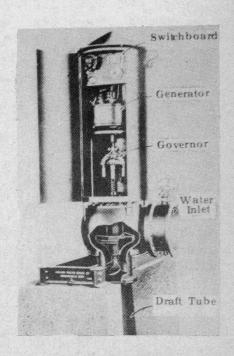

WIND

Since the first printing of the Energy Primer in 1974, we have seen enormous changes in this country, and it seems for the better. Our environmental and energy concerns are beginning to be clearly understood as essential components to our way of life. The catastrophic effect we have had on the environment has made it clear that the total cost of the energy and materials invested in a new technology must be weighed against that system's ability to produce or transform useful energy.

Another thing has become apparent; one of the most important factors pertaining to the energetics of any system or device is its scale in relation to its task. Hence we begin to see that with wind energy conversion systems (WECS) the energy is still very expensive at the individual-user scale, and that because of the liabilities still inherent in the current technology, larger WECS are more appealing on a first cost and maintenance basis. Two things come to mind: (1) the key to the success of WECS is that they be matched very closely to their load; and (2) wind systems are very site-sensitive and each case must be handled with care given to all local parameters as discussed in the following section. However, it is also my feeling that, with the development of a new American wind industry, the problems with scale, storage, and load matching may soon be resolved . . . perhaps within two to three years.

The small WECS that are widely available are for the most part designed from technologies from the 1950's and 1960's, with very little refinement. Much research and development on WECS still remains to be done, but they have a high potential for making a significant contribution to the goal of meeting our future energy needs through the use of a clean and inexhaustible source of energy, the wind.

Wind is another form of energy, created by the sun; the heating of our atmosphere during the day and its absence cooling the night sky. It's like breathing—that's it, the earth breathes. Wind is the reaction of our atmosphere to the incoming energy from the sun. Heat causes low pressure areas and the lack of heat results in high pressure areas. This process causes the wind.

It seems ironic that probably the oldest and most constant character of the universe, e.g., massive movements of energy, heating to cooling (entropy), the motion of our atmosphere, is suddenly rediscovered as a "new source" of energy. History tells us that next to agriculture, it is very possible that wind may have been one of the first sources harnessed by man.

Our main concern regarding wind energy is that it is not as constant and/or predictable as, say, the sun. There are many solutions to this problem, but usually the situation is managed by a *storage system* designed to have the energy available at the time it is needed or desired. Yet, on the other hand, one might look at this concern in a different perspective and not see it as a problem at all, but simply as a challenge to our ability to adapt. If we are truly aware of our capabilities to adapt or adjust, then we also realize our limitations. It should be noted that we have adapted to our lifestyles most effectively considering we inhabit the planet in so many numbers. So that the problem with wind is not predictability, but it is our ability to respond.

Our dependent, real and intimate relationship with the biosphere can no longer afford to be overlooked. Wind systems are visual indicators of amounts of energy used, and therefore assist us in understanding this *environmental relationship*. The right perspective of this situation is important, for then and only then are we able to design our part of the environment, without selling short our individuality or our abilities.

All things in life change, as does the Sun, the Wind, the Water, and all living things in accordance with them.

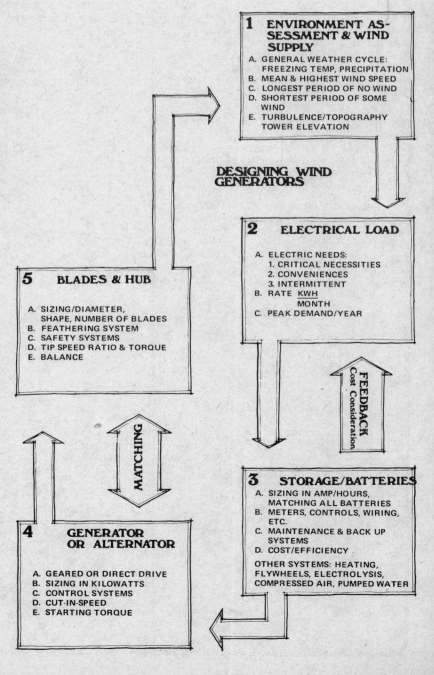

1 ENVIRONMENT ASSESSMENT & WIND SUPPLY

A. GENERAL WEATHER CYCLE: FREEZING TEMP, PRECIPITATION
B. MEAN & HIGHEST WIND SPEED
C. LONGEST PERIOD OF NO WIND
D. SHORTEST PERIOD OF SOME WIND
E. TURBULENCE/TOPOGRAPHY TOWER ELEVATION

DESIGNING WIND GENERATORS

2 ELECTRICAL LOAD

A. ELECTRIC NEEDS:
1. CRITICAL NECESSITIES
2. CONVENIENCES
3. INTERMITTENT
B. RATE $\frac{KWH}{MONTH}$
C. PEAK DEMAND/YEAR

5 BLADES & HUB

A. SIZING/DIAMETER, SHAPE, NUMBER OF BLADES
B. FEATHERING SYSTEM
C. SAFETY SYSTEMS
D. TIP SPEED RATIO & TORQUE
E. BALANCE

FEEDBACK Cost Consideration

MATCHING

3 STORAGE/BATTERIES

A. SIZING IN AMP/HOURS, MATCHING ALL BATTERIES
B. METERS, CONTROLS, WIRING, ETC.
C. MAINTENANCE & BACK UP SYSTEMS
D. COST/EFFICIENCY

OTHER SYSTEMS: HEATING, FLYWHEELS, ELECTROLYSIS, COMPRESSED AIR, PUMPED WATER

4 GENERATOR OR ALTERNATOR

A. GEARED OR DIRECT DRIVE
B. SIZING IN KILOWATTS
C. CONTROL SYSTEMS
D. CUT-IN-SPEED
E. STARTING TORQUE

WIND ENERGY RULES OF THUMB

1. Air is approximately 800 times less dense that water. This means that you've got to pass a lot more air through the blades of a wind energy device as compared to the amount of water you would need to pull through a water turbine to obtain the same amount of energy.
2. A blade is always spilling power.
3. It is important to remember that the wind blows parallel to the ground; — *not* perpendicular to gravity.
4. One should always remember that siting the wind device is the most important consideration and that trees can only grow taller and therefore a wind system should not be placed near them.
5. In certain climates, icing of the blades is a major problem and sometimes motoring of the generator helps take care of this.
6. The bottom tip of the blades of your wind mill should be fifteen feet higher than the tallest object within 50 feet.
7. The base of a 40 to 80 foot tower should be secured in the ground and concreted to a minimum of 5 feet.
8. Plumb your tower perfectly level and square. Leaning towers are dangerous.
9. An overdesigned tower is better than an underdesigned tower.
10. When using bearings use tapered roller bearings or thrust bearings for the "turn table" of the generator.
11. Remember, vibrations kill bearings.
12. It is a good idea never to use bearings or gears in a feathering mechanism.
13. Periodic maintenance of a wind system is an investment in its longevity. Grease is always cheaper than steel.
14. In a conventional generator the size of the armature is related to amperage output and the size of field poles is related to voltage output.
15. The back-up system should have the same power output as the wind generator.
16. You should be able to get by with an inverter rated between one-quarter and one-half the wattage rating of the wind generator.
17. Never let the rate of discharge of a battery in amperes exceed 15% of the amp-hour rating.
18. A general rule is that batteries should never be discharged below 20% of rated charge.
19. The quality of a battery is the function of the softness (paste to lead ratio) of the plates.

A partial list of wind generator companies in the U.S.A. 1910-1970.

Air Electric	Wind Power	Air Charger	Allied
Wind Charger	Rural Lite	Delco	Miller
Parris-Dunn	Air King	Western Electric	Aerodyne
Jacobs	Wind King	Nelson	Zenith

WIND ENERGY CONVERSION

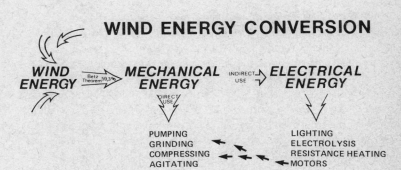

WIND ENERGY — Betz Theorem 59.3% → MECHANICAL ENERGY — INDIRECT USE → ELECTRICAL ENERGY

DIRECT USE ↓

PUMPING
GRINDING
COMPRESSING
AGITATING

LIGHTING
ELECTROLYSIS
RESISTANCE HEATING
MOTORS

WIND PLANT DESIGN OPTIONS

WIND PLANT

- **MATERIALS**
  FABRIC
  WOOD
  METAL
  FIBREGLASS
  PLASTIC
  RUBBER
- **CONSIDERATIONS**
  STRESS
  STRENGTH
  TEMPERATURE
  EFFICIENCY
  COST
  MAINTENANCE
  VIBRATION
  FRICTION
  DEGRADATION

The items discussed in this list may be used either in the building, designing or purchasing of wind driven generators; it is suggested that any additions one finds could be written in, thus making the list more useful.

BLADES

- **DESIGN**
  LIFT VECTOR
  PITCH
  CHORD
  ANGLE OF ATTACK
  LENGTH
  CENTER OF GRAVITY
  TWIST
- **TYPE**
  PROPELLER
  DARRIEUS
  SAVONIUS
  MAGNUS
- **WIND SPEED**
- **TIP SPEED**
- **DIAMETER**
- **BALANCING**
- **TORSION**
- **BENDING MOMENT**
- **TRACKING**

ELECTRICITY GENERATION

- **D.C. (GENERATOR)**
  BRUSHES
  NUMBER OF
  FIELD POLES
- **A.C. (ALTERNATOR)**
  SLIP RINGS
  PHASES (1, 2, 3)
  A.C. TO D.C. WAVE
  RECTIFICATION
- **"CUT IN SPEED"**
- **SIZE AND SHAPE**
- **PERMANENT MAGNET**
- **INDUCTION**
- **VOLTAGE RATING**
- **WATTAGE**
- **AMPERAGE**

CONTROLS

- **HOUSING**
- **DIRECTING INTO WIND**
  VANE (UP-WIND)
  COWLING (DOWN-WIND)
  WIND ROSE
- **SPEED GOVERNING**
- **BRAKES**
- **AIR BRAKES**
- **FLYBALLS**
- **BLADE**
- **FEATHERING**
  TEETERING
  GIMBALED
  HINGELESS
  FULLY ARTICULATED
  HELICAL
  RADIAL
  CONING

STORAGE BATTERIES

- **RATING (AMP HOUR)**
- **CHEMICAL AND CHARACTERISTICS**
- **LEAD ACID**
- **NI-CAD**
- **NUMBER OF CYCLES**
- **RATE OF CHARGE & DISCHARGE, DEPTH OF CHARGE**
- **MATCHING SETS**
- **OTHER STORAGE SYSTEMS**

HARDWARE

- **GEARS**
  SIZE OF FACES
  RATIOS
- **BEARINGS**
- **BUSHINGS**
- **DIRECT DRIVE**
- **LUBRICANTS**
- **BELTS, CHAINS**
- **SPRINGS**
- **CABLE**
- **CAMS**
- **ROLLERS**

CONTROLS

- **REGULATOR**
- **BREAKERS & FUSES**
- **GROUNDING**
- **LIGHTNING ARRESTERS**
- **WIRING**
  SIZING
  A.C. & D.C.
  SYSTEM
  DIFFERENCES
- **SLIP RINGS**

TOWERS

- **DESIGN**
  3 LEGS
  4 LEGS
  OCTAHEDRON
  HELICAL
- **MATERIALS**
  ANGLE STEEL
  PIPE
  WOODEN
  GUIDE WIRES
  CEMENT BASE
  CONCRETE BLOCK
  CONCRETE
- **ALIGNMENT**
- **JOGGING**

INVERTERS D.C. TO A.C.

- **ROTARY OR STATIC**
- **WAVE FORM**
- **CONTINUOUS DUTY**
- **IDLING AMPS**
- **EFFICIENCY**

WIND DRIVEN GENERATORS
BY
James Sencenbaugh

INTRODUCTION

Perhaps the primitive horizontal windmills of 10th century Persia were the first attempt at harnessing wind. This mill, with its sails revolving on a vertical axle mounted in a square tower, was used to grind corn. Diagonally opposed slots in the walls ducted air to the enclosed sail assembly. Tradition tells of the prisoners of Genghis Khan introducing the mills into the East. Horizontal mills became commonplace throughout China, where they were used primarily for irrigation. At the end of the 12th century, mills could be found throughout Northern Europe. By the late 13th century they were in use in Italy, but almost 200 years later the windmill was still unknown in Spain. German crusaders driving through Asia Minor probably instituted the technique in this region. Design from this period on varied greatly and improvements developed independently in many countries.

Unlike windmills which use the wind directly for mechanical energy, a modern wind driven generator extracts energy from the wind and converts it into electricity. A complete wind driven system consists of a: (1) tower to support the wind generator, (2) devices regulating generator voltage, (3) the propeller and hub system, (4) the tail vane, (5) a storage system to store power for use during windless days, and (6) an inverter which converts the stored direct current (D.C.) into regulated alternating current (A.C.) if it is required. An optional backup system, such as a gas or diesel generator, is used to provide power through extremely long calm periods.

With the invention of automobiles and the development of their electrical systems, small D.C. generators of low output and moderate speed input became available on a large scale. Changes in wiring enabled these early generators to be used in the first wind driven designs. At the same time, the rapid increase in aeronautical research led to intensive investigation on airfoil and (propeller) blade design. With this background, the wind driven generator came into its own as the first form of free private electrical power generation. It was first sold as an accessory to battery powered radios. The Zenith radio corporation offered a small 200 watt windplant at a reduced price when bought with one of their battery powered radios. It was during this period that the Wincharger Co. of Sioux City, Iowa was reported to have been turning out 1000 units per day. The wind driven generator field bloomed in the late 1920's and 30's and at its peak over 300 companies were formed throughout the world. A large number of varied designs were available, from a down-wind 1800 watt Win Power to the 3000 watt Jacobs unit. The introduction of the Rural Electrification Agency (REA) in the United States brought a cheaper, more convenient method to have larger amounts of electricity, and wind plants all but disappeared. Jacobs finally closed its doors around 1956 and Wincharger now only makes the original 200 watt 12 volt D.C. model designed in the 1930's.

At present there are only six *major* manufacturers of wind driven generators in the world. Two are in the United States: Dynatechnology (Wincharger) and Bucknell Engineering (bought by Precise Power Corporation). Both build very small 200 watt units on a limited production basis. Dunlite of Australia (which builds plants marketed by Quirk's) manufactures two basic models: a D.C. generator type and a brushless alternator type. Elektro G.m.B.H. of Winterthur, Switzerland offers a complete line of windplants from 50 to 6000 watts. Aerowatt of Paris, France, builds a number of excellent wind plants designed for commercial and marine applications, but their prices are extremely high. Last is Lübing Maschine Fabrik of Barnstorf, West Germany. They primarily build water pumping windplants, but do offer a small 400 watt, 24 volt unit.

The most economical windmill is one which furnishes the kilowatt-hr at the lowest cost. The production of energy by windmills at a favorable cost is made difficult by the fact that the wind is an intermittent source of energy. During a large part of the time it blows too little to produce any useful output and other times it is of such velocities as to cause potential damage to the windplant.

The actual power available from the wind is proportional to the cube of the windspeed. In other words, if the windspeed is doubled, you will

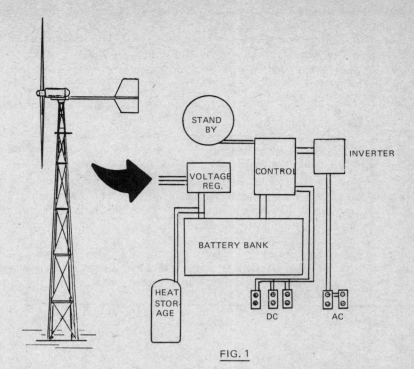

FIG. 1

IN THIS SYSTEM, THE WIND DRIVEN GENERATOR CHARGES A BATTERY BANK, from which DC power is taken directly for use or inverted, making AC for appliances like T.V., and radios. Excess power runs a heating storage system.

get eight times as much power (Cube Law). Another fundamental principle governing windmill design is that it is theoretically impossible in an open-air windplant to recover more than 59.26% (A. Betz) of the kinetic energy contained in the wind. If the prop itself is 75% efficient, and the generator 75% efficient, then 33.34% of the kinetic energy of the wind may be converted into electricity. The other important factor to point out is that the amount of energy captured from the wind by a windplant depends on the amount of wind intercepted; that is, the disk area swept by the blades. A well designed windplant irrespective of the number of blades decelerates the whole horizontal column of air to one-third its free velocity. These facts and the nature of wind currents generally restrict the designer to the most common wind velocities of 3 to 10 meters per second (6.7 to 22.3 mph).

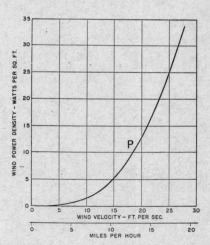

FIG. 2. THEORETICAL POWER DENSITY OF WIND
$$P = (K \cdot A \cdot V^3) \cdot (.5926)$$

Unit of Power P	Unit of Area A	Unit of Velocity V	Value of K
Kilowatts	Square feet	Miles per hour	0·0000053
Kilowatts	Square feet	Knots	0·0000081
Horse power	Square feet	Miles per hour	0·0000071
Watts	Square feet	Feet per second	0·00168
Kilowatts	Square metres	Metres per second	0·00064
Kilowatts	Square metres	Kilometres per hour	0·0000137

WIND FORMULA TABLE 1: Where P= Power in Kilowatts, K= Constant (Air density and other conversion factors), A= Area swept, and V= Wind Velocity.

Table 2 will give an appreciation for the principles involved in wind-plant design. A 6 foot diameter prop, operating at 70% efficiency in a 20 mph wind, can produce 340 watts. This shows the relationship between windspeed and output. If the wind speed is doubled, you will get eight times as much power. Also note the relationship between propeller diameter and output. Keeping the results for the 6 footer in mind, let us double the prop diameter to 12 feet and note the output at 10 and 20 mph. At 10 mph the 12 footer can produce up to 170 watts and at 20 mph, 1360 watts. Hence the power output from the 12 feet diameter prop is 4 times that of the 6 footer, or *power is proportional to the square of the diameter of the prop.* Double the size of the propeller and the power output will increase by a factor of four.

Propeller Diameter in Feet	Wind Velocity in mph					
	5	10	15	20	25	30
2	0.6	5	16	38	73	130
4	2	19	64	150	300	520
6	5	42	140	340	660	1150
8	10	75	260	610	1180	2020
10	15	120	400	950	1840	3180
12	21	170	540	1360	2660	4600
14	29	230	735	1850	3620	6250
16	40	300	1040	2440	4740	8150
18	51	375	1320	3060	6000	10350
20	60	475	1600	3600	7360	12760
22	73	580	1940	4350	8900	15420
24	86	685	2300	5180	10650	18380

TABLE 2. WINDMILL POWER OUTPUT IN WATTS,
assuming $P = (K \cdot A \cdot V^3) \cdot (.5926) \cdot (.70) \cdot (.70)$

ENVIRONMENTAL ASSESSMENT

It is not enough just to know these fundamental principles before deciding to build or buy a windplant. Your choice of site must be assessed to see if a windplant will give you equitable returns, in addition to the positive environmental effects.

Consider the following conditions at the site on a *frequency* and *intensity* basis: rain, freezing temperatures, icing, sleet, hail, sandstorms, and lightning. The life and longevity of a wind plant, as well as its structural design and cost, depend on the completeness of your weather assessment. (Note: Most manufactured wind systems are "tropic-proofed"; be sure to check this if buying a system.)

In order to determine the usable output (Kw-hrs. per month) produced by a particular size wind driven system, the output characteristics of the windplant and the average yearly wind speed at the site must be known. The average yearly winds at a location can be obtained from the National Weather Bureau records center, U.S. Weather Bureau, Federal Building, Asheville, N.C. 28801. They carry statistical data for the U.S. for the last 50 years. Although they might not have information for your exact location, they should have records of a city or area very near. This information is a good start for estimating if the winds in your area are suitable for wind power. Another source of wind information will be your local airport. The next step should be the purchase of an anemometer (wind gauge) to estimate local wind conditions. A small hand held unit is available from Dwyer for $6.95 (see page 146). There is also a more expensive remote reading anemometer by Stewart Northwind for about $90.00 (see page 147). The transmitter assembly can be mounted on a T.V. type mast at the height the wind plant is to be installed.

Another means of assessing local wind velocities is by using the Beaufort scale (Table 3). If readings are taken with regularity and then compared to the local Weather Bureau data, the scale is a very accurate and inexpensive way of measuring wind speed. Readings should be taken every day at the same times for accurate results (typically four times a day). Data should be taken for at least one month, and preferably longer, to determine mean average wind speed. Determining the longest period of no wind and the shortest period of some wind annually are two calculations that will be very helpful in figuring storage systems and back-up system requirements. If the test results show that there is over a 10 mph wind average on an norm of 2 to 3 days per week, you have an adequate site for wind power. The site findings should be compared to official Weather Bureau data for that area.

sites nearby to see if they correlate with the 10 year monthly averages for that area.

Effect Caused by the Wind On Land	At Sea	Beaufort Number	Description	Speed (m/sec)	Speed (miles/hr)
Still; smoke rises vertically	Surface mirror-like	0	Calm	0—0.2	0—1
Smoke drifts but vanes remain still	Only ripples form	1	Light air	0.3—1.5	1—3
Wind felt on face, leaves rustle, vane moves	Small, short wavelets, distinct but not breaking	2	Light breeze	1.6—3.3	4.7
Leaves and small twigs move constantly, streamer or pennant extended	Larger wavelets beginning to break, glassy foam, perhaps scattered white horses	3	Gentle breeze	3.4—5.4	8—12
Raises dust and loose paper, moves twigs and thin branches	Small waves still but longer, fairly frequent white horses	4	Moderate breeze	5.5—7.9	13—18
Small trees in leaf begin to sway	Moderate waves, distinctly elongated, many white horses, perhaps isolated spray	5	Fresh breeze	8.0—10.7	19—24
Large branches move, telegraph wires whistle, umbrellas hard to control	Large waves begin with extensive white foam crests breaking, spray probable	6	Strong wind	10.8—13.8	25—31
Whole trees move; offers some resistance to walkers	Sea heaps up, lines of white foam begin to be blown downwind	7	Stiff wind or moderate gale	13.9—17.1	32—38
Breaks twigs off trees; impedes progress	Moderately high waves with crests of considerable length; foam blown in well-marked streaks; spray blown from crests	8	Stormy wind or fresh gale	17.2—20.7	39—46
Blows off roof tiles and chimney pots	High waves, rolling sea, dense streaks of foam; spray may already reduce visibility	9	Storm or strong gale	20.8—24.4	47—54
Trees uprooted, much structural damage	Heavy rolling sea, white with great foam patches and dense streaks, very high waves with overhanging crests; much spray reduces visibility	10	Heavy storm or whole gale	24.5—28.4	55—63
Widespread damage (very rare inland)	Extraordinarily high waves, spray impedes visibility	11	Hurricane-like storm	28.5—32.6	64—72
	Air full of foam and spray, sea entirely white	12	Hurricane	32.7—36.9	73—82

TABLE 3. THE BEAUFORT SCALE

The relative wind velocity prevailing in any location determines what size of wind generator is best suited for that region. Table 4 shows a sample region having annual average velocities of 10 mph or greater. Note that the 10 mph average wind is made up of many low winds and a few high winds. A 6 mph wind is generally considered to be the lowest wind for any practical use. In terms of energy available, 12 to 25 mph is the range of higher winds providing good power conversion. The relationship between the wind and power available to a wind generator can be exemplified by the detailed study that was made of wind records at Dayton, Ohio. Data was taken from 1936 to 1943, covering a 7 year period. In each month two groups of winds exist. First, there are the frequent or *prevalent winds* ranging from 5 to 13 mph. Second there are the *energy winds* which blow less frequently, ranging from 13 to 23 mph. A quick glance at Table 4 will give a clue to the energy winds. The prevalent winds blow 2—1/2 times more frequently than the more vigorous energy winds; for example, 5 days prevalent as opposed to 2 days of the energy winds. But because the energy varies with the cube of the velocity, the energy winds produce 3/4 of the total power. A windmill utilizing the prevalent winds must be twice the diameter of a windmill running *only* on the energy winds, if each is to produce the same amount of electricity per month.

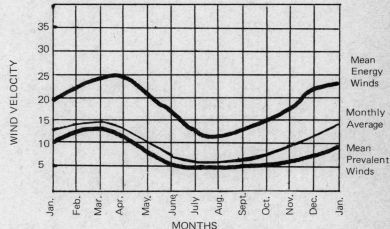

TABLE 4. The curves represent averages of 13 "frequency" and "energy" charts compiled from Weather Bureau records.

Tower design and installation are as important as site selection. A support must be built that is strong enough to handle loads from the dead weight of the generator assembly itself, as well as the thrust loads developed from the propeller at the highest anticipated windspeed. Most commercial towers are designed for a wind loading of at least 140 mph, and then a generous safety factor is added. Manufacturers discourage the installation of these larger units on home roofs because the loads on a 12 or 14 foot diameter prop are so powerful in winds over 45 mph that they could cause serious structural damage to the rafters, or even send the plant crashing down through the roof. A six or eight foot diameter prop on a wood frame structure can cause noise to be transmitted throughout the structure. even though the plant is balanced and running smoothly. This noise, which resembles a low howl or groan, can bother even the most sound sleeper.

The best location for a wind plant is as high as economically possible to reach undisturbed air. Placing a windplant a minimum of 30 to 40 feet above the ground (not on a roof!) will greatly increase the amount of power available to the swept area. Ideally the plant should be placed 15 to 20 feet above all obstacles within a 500 foot radius because surrounding objects have a very disturbing effect on the air and cause whirling eddy currents that greatly effect plant performance.

TURBULENCE/TOPOGRAPHY → TOWER ELEVATION

A hill or ridge of high ground lying in the path of the wind will have a considerable influence on the wind. Remember, also, that the winds blow *parallel* to the ground, not perpendicular to gravity. Obtaining a topographical map from the U.S.G.S. will enable you to estimate any turbulence due to topography. Tall trees behind a wind plant, as well as trees in front, interfere with a wind plant's operation. Most commercial wind plants are designed so that they can be installed 500 to 600 feet away from the point where power is required, so there is leeway for avoiding obstacles.

LOAD

The next assessment which needs to be made is how much power your appliances will actually require. The more accurately you figure these needs, the lower your storage costs become. The storage system expense is a direct result of the load. Figure *all* electrical needs:

1. Critical needs (e.g. refrigerator)
2. Convenience needs (e.g. electric blanket)
3. Intermittent needs (e.g. power saw)

Construct a chart, as shown below, listing the devices in use, hours per day each is in use, and the number of watt-hours each device requires.

Appliances	Watt Rating	Hours/Day in Use	Watt-Hr./Day
4 light bulbs	100 each	6	2400
1 stereo	80	4	320
1 percolator	480	1	480
1 sewing machine motor	30	1	30
		total	3230

Nominal Output Rating of Generator in Watts	Average Monthly Wind Speed in mph					
	6	8	10	12	14	16
50	1.4	3	5	7	9	10
100	3	5	8	11	13	15
250	6	12	18	24	29	32
500	12	24	35	46	55	62
1,000	22	45	65	86	104	120
2,000	40	80	120	160	200	235
4,000	75	150	230	310	390	460
6,000	115	230	350	470	590	710
8,000	150	300	450	600	750	900
10,000	185	370	550	730	910	1090
12,000	215	430	650	870	1090	1310

TABLE 5. AVERAGE MONTHLY OUTPUT IN KILOWATT-HOURS

1. From the table find the average wind speed vs. generator rating to determine Kw-hr./month output. Let's assume the wind-generator is operating in an area with 10 mph average winds and that the generator rating is 4 Kw: from the table we find the expected monthly output to be about 200 Kw-hr./month (the table gives the figure 230, but we'll work with a more conservative figure). To find out how many Kw-hr. per day of electricity we could use (i.e. the *use rate*) with a 200 Kw-hr./month supply from the wind, divide the power available for the whole month (kw-hr./month) by the days in the month (30):

$$\frac{200 \text{ Kw-hr./month}}{30 \text{ days}} = 6670 \text{ watt-hr./day}$$

This is an excellent planning figure to design around when you are trying to estimate the number of watts consumed by devices, appliances, etc. to be used each day. In this example, with a windplant of the given size and average winds of 10 mph, you have 6670 watt-hrs. per day available to you.

2. The next step is to find the capacity in kw-hr. of the battery system in use. Assume a battery capacity of 270 amp/hrs. at 115 volts. Watts in the system is found by multiplying amps by volts. Therefore, 115 volts x 270 amp/hrs. = 31,050 watt-hours or 31.050 Kw-hrs.

3. Find the number of days you could expect to operate, assuming no wind, with 6670 watts/day load, from a 31,050 watt-hour storage system.

$$\frac{31,050 \text{ watts}}{6670 \text{ watts}} = 4.66 \text{ days}$$

Roughly, you could operate for at least four days directly from the batteries, with no input from the wind generator or the stand-by unit.

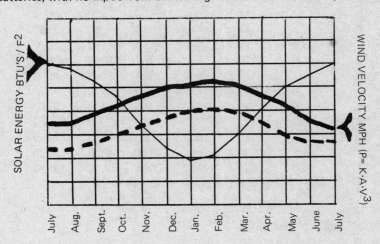

TABLE 6. COORDINATED WIND ELECTRIC DEMAND

Peak electrical demand could be coordinated with the peak energy period. This situation does not exist everywhere, but where it does it should be utilized.

Power company output is typically 60 cycles per second alternating current (A.C.). A battery storage system is direct current(D.C.). All resistance and heating devices (e.g., light bulbs, toasters) and universal motors (having brushes) can run on A.C. or D.C. The following require A.C. only; they will not operate on D.C:

1. Fluorescent lights (unless rewired),
2. Devices with transformers (e.g., televisions, radios, tape decks),
3. Appliances with standard induction motors.

If D. C. power is inadvertently supplied to an A.C. appliance, there is a high probability that the appliance will be destroyed.

BATTERY AND STORAGE SECTION

The most difficult and important calculation to make is the sizing of your storage system. Although there are many sophisticated and perhaps exotic methods of energy storage presently in development which could be used in this application, the most reliable at present is the lead acid storage battery. This battery still represents the cheapest practical method of electrical energy storage available for the individual user of wind power. Because the wind itself is an intermittent source of energy, a battery storage system must be capable of storing power through long, windless periods with reasonable efficiency at moderate cost.

Batteries used in wind plants are designed for repeated cycling over a period of many years. Their construction allows them to go repeatedly from a fully discharged to a fully charged state without damage. Some designs can withstand approximately 2000 complete cycles. These batteries are commonly known as stationary or houselighting batteries and are available in sizes from 10 amp/hr. to 8000 amp/hr. The normal voltage of the system is determined by the number of cells in series (each cell is approximately 2 volts), but the amount of storage capacity is determined by the plate thickness and area.

These batteries have thicker plates than the standard automobile battery and employ separators made of glass fiber material. The structural integrity

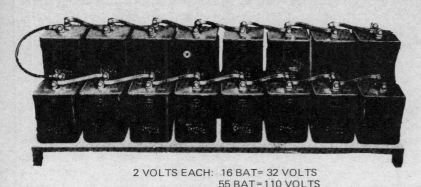

2 VOLTS EACH: 16 BAT= 32 VOLTS
55 BAT=110 VOLTS

CAR BATTERY — 6 VOLTS

FIG. 3 SIZE AND VOLTAGE DIFFERENCES BETWEEN CAR BATTERY
AND LIGHTING PLANT BATTERIES

of these cells is much greater, and large amounts of reserve space between the bottom of the plates and the case is common. This allows a large amount of area for material to collect without any damage from internal shock or shorting. Golfcart batteries have similar characteristics. Gould P B220, a 6 volt-220 amp cell, and Trojan P J217, a 6 volt-217 amp cell, can be used with reasonable success (see page 147).

A battery on charge is not a fixed or static potential, and is subject to change in its voltage and current output characteristics with changes in light, temperature elevation, etc. Simple testing methods compensate and adjust for these changes.

One important thing to remember is that when a battery is charging and discharging, there is a change not only in its amperage but also in its voltage level. Figure 4 shows that the battery voltage rises slowly until it reaches the 80-85 percent capacity-returned point, rises sharply between about 2.3 and 2.5 volts per cell (the gassing point where H_2 is created), and then flattens out at about 2.6 volts when the battery is fully charged. This sharp rise in voltage, during which only a small part of the charge is returned, is a characteristic of a lead acid cell and does create some design and operational problems. To return the last 15 percent of the charge in a reasonable time the charge voltage must increase by about 20%, or 0.5 volts. The high charge voltage required to complete the charge process may be unattainable because of the operating voltage of appliances. The current charge rate (amps) should be reduced at the gassing point 2.6 volts/cell to below the 20 hours rate. ($\frac{270 \text{ amp/hr}}{20 \text{ hr}}$ = 13.5 amps). This should be done by the electronic control and regulating system in your windplant.

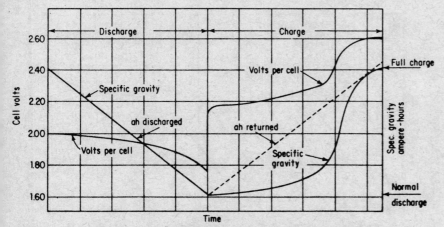

FIG. 4. TYPICAL VOLTAGE AND GRAVITY CHARACTERISTICS DURING A
CONSTANT RATE DISCHARGE AND RECHARGE.

BATTERY CAPACITY

The nominal rated capacity of most lead acid batteries is taken at 8, 10

or 20 hour discharge rates down to a cell voltage of 1.85 volts per cell. At higher discharge rates the capacity is reduced, while at lower discharge rates the available capacity is increased. Figure No. 5 illustrates the change in capacity with the rate of discharge. At the 12 hour rate, for example, a nominal 500 amp hour capacity battery has an available capacity of 550 amp hour, or 110%, down to a discharge voltage of 1.85 volts per cell. The capacity taken out of the battery is approximately 1040 amp hr. and about 83% of the total available capacity. This limitation on the extent of discharge is to ensure that the battery is not worked excessively and to provide a reserve capacity, which will greatly extend the life of the battery.

One of the key factors in guaranteeing the life of the battery system is the regularity of maintenance. It is vital that the electrolyte level in all batteries is maintained. This is done by adding distilled water to each battery when needed. All batteries must have tight connections to minimize (1) corrosion, (2) unequal voltage per cell, and (3) reduced capacity. The battery set must be an originally matched set. Do not mix batteries of different ages. Batteries should be located a few inches off the floor and away from the wall to permit good air circulation. For more exact information on batteries write the battery companies in your area.

A new battery does not become fully active until used the equivalent of about 30 complete cycles, because the plates are somewhat hard and do not absorb and deliver the full current. On a battery of 400 amp/hr. capacity this break-in period requires a few months with an average load.

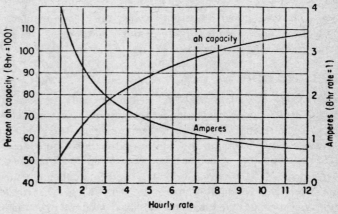

FIG. 5. CAPACITY-RATE CURVE BASED ON 8-HR. RATE

An important component in your overall electrical system is the wire. The correct size of wire for a given current must be accurately calculated. Keep in mind that D.C. travels through the wire and A.C. travels on the surface. This is important to know because fuses, circuit breakers, and switches for high current D.C. are different than for A.C. (D.C. outlets will have a polarity. Check when the system is wired to be sure they are correct.)

GENERATOR OR ALTERNATOR

The combined data of electrical needs and the size of the storage system necessary (which is based on the average wind speed) will determine the generator size necessary for your system. The last critical matching problem will be that of the blades to the generator.

Since propeller speed seldom exceeds 300-400 rpm, especially in the larger diameters, this calls for a low-speed generator designed to match the torque and horsepower output parameters of the prop in use. Dunlite compromises a bit by using an alternator which develops maximum output at 750 rpm, corresponding to a prop speed of only 150 rpm. Hence a 5:1 step-up gear is used. Elektro on the other hand utilizes all direct drive with the exception of the larger 6000 watt unit, which is geared. All major manufacturers are now using a multi-poled alternator which produces alternating current. An alternator is used for two reasons: (1) it operates at a lower cost than a comparable wattage D.C. generator, and (2) perhaps the most favorable factor, the high current is taken directly from the stator coils (see Fig. 6) and not through brushes as in a D.C. generator. Eventually the brushes wear down and need replacement.

One of the major cost factors in a modern wind system is the low speed continuous duty alternator or generator. Conventional alternators generally are designed for high rotational speeds (1800 to 3600 rpm), being driven by a gasoline engine, and therefore cannot be used for wind driven systems. Wind driven generators require a low cut-in speed. By using direct drive or a very low gearing ratio such as in the Dunlite unit, losses are kept to a

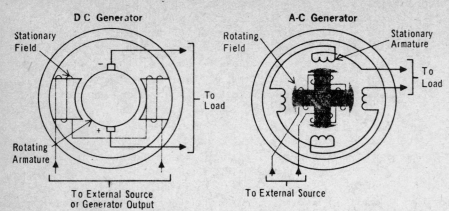

D-C Generator

Stationary Field

Rotating Armature

To Load

To External Source or Generator Output

A-C Generator

Rotating Field

Stationary Armature

To Load

To External Source

The d-c generator has a rotating armature and a stationary field. Voltage for its field can be obtained either from an external source or from the generator's own output.

The a-c generator has a stationary armature and a rotating field. Voltage for its field must come from an external source, called an exciter.

FIG. 6. AN AC GENERATOR AND A DC GENERATOR

minimum and overall machine efficiency remains relatively high. The design of direct drive generators is materials efficient and the low rotational speed keeps wear to a minimum.

Special electronic regulators are designed as part of generating systems to control the amount of current delivered to the batteries at a safe charging rate. Once the batteries are fully charged, the alternator will put out only a trickle charge, as is the case for the Dunlite unit. In the Elektro design, the entire windplant shuts down when the voltage regulator registers the battery cell voltage is above a predetermined level. This prevents any chance of overcharging the battery.

BLADES AND HUB

The efficiency of your entire wind system depends on what type of prop (blade profile) you use. All modern wind driven systems use two or three long slender blades with an airfoil section designed to produce maximum lift with minimal drag in the rpm range for which it operates. These efficient blades operate at a high tip speed ratio, which is the ratio of propeller tip speed (U) to wind velocity (V). The outstanding characteristic of the propeller-windmill is that at a given wind speed it rotates 5 to 10 times faster than a conventional multi-bladed windmill of the same size. The propeller type excels in lightness of construction, reduction of gearing difficulty, and is the only type of windmill through which direct generator drive can be accomplished. A higher tip speed ratio means higher rpm for a given wind speed, and higher rpm generally means more power output.

It is critical that the balance of the blade and hub system and its tracking are precisely adjusted. Static balancing insures equal mass distribution at all points in the rotation system. The procedure is fairly simple for two blade systems but becomes increasingly complex as more blades are added. Three dimensional configurations, such as Savonius rotors, are even more difficult. The figures below may be of some help.

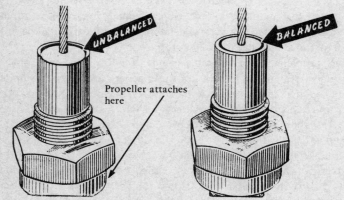

UNBALANCED

BALANCED

Propeller attaches here

FIG. 7. A METHOD DEVELOPED BY THE ARMY FOR BLADE BALANCING

Dynamic balance is very difficult to achieve. The main concern is to prevent vibrations. Therefore all blades must be identical in their lift or reaction components. Tracking means that the blades follow each other in exactly the same plane of rotation. The blade tips should track within 1/8", if possible, and should be no closer than one foot to the tower. The better the tracking the less the vibration.

Fig. 8 shows the relative tip speed ratio and power curves for various propeller types and VERTICAL AXIS rotors (SAVONIUS). The Figure shows the advantages of the propeller type windplant (curves 3,4,5,6,8 & 11) over the Savonius type plant (curves 1,7,9 & 10). For more information on the Savonius type rotor windmill see the discussion and figures on page 129 in the Energy Primer.

One has the choice of whether to use two or three blades. It is generally a decision based on economics and the relative output of the generator to be used. Although a two-bladed prop will run at a slightly higher aerodynamic efficiency than a similarly designed 3-bladed prop, it is seldom used in commercial plants with generator ratings above 2000 watts output. This is because the three-bladed prop provides the extra starting torque necessary to overcome difficulties found in lower winds, fluxing and eddying. They run slightly smoother when orienting to changes in wind direction. In comparison, it was found early in wind-generator design that the two-bladed system is "choppy" when the tail vane shifts during wind direction changes. When the blades are straight up and down they have no centrifugal resistance to the tail movements. When the blades are parallel to the ground the resistance is maximum. However, it is int—eresting to note that the Elektro two-bladed wind plants do not exhibit this trait since the variable pitch weight arms are placed at 90 degrees to the blade element and, in terms of balance, the assembly "looks" like a 4-bladed prop during rotation. Hence wind direction changes are very smooth.

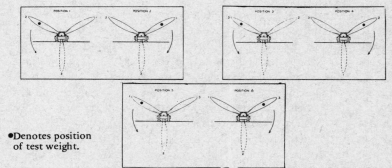

POSITION 1   POSITION 2   POSITION 3   POSITION 4

POSITION 5   POSITION 6

• Denotes position of test weight.

KNIFE EDGE BALANCING METHOD REQUIRES A PAIR OF PERFECTLY ALIGNED AND LEVELED EDGES OF HARD STEEL, A SPINDLE AND A HOIST

Once maximum power is developed at the rated windspeed of the generator (typically in the 16 to 25 mph range), excess power developed by the prop is potentially destructive and must be controlled. There are as many methods of overspeed control as there have been manufacturers of windmills, and each system has its own merits and disadvantages. The airbrake or airspoiler was used by several manufacturers during the 1920's and 30's on small diameter machines and is still used on the 200 watt unit built by Dynatechnology. Two small sheet metal vanes, resembling barrel slats, were placed at 90 degrees to each prop about the center axis of the hub. Springs held the vanes in normal position until centrifugal force pulled the vanes outward, diverting air away from the prop and thus decreasing rpm. This system was successful for small diameter units. Airbrakes are undesirable for larger scale operations since they throw heavy loads onto the entire structure.

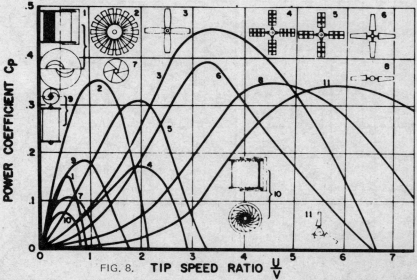

FIG. 8.   **TIP SPEED RATIO** $\frac{U}{V}$

POWER COEFFICIENT C_P

FIG. 9   "NORMAL"   "GOVERNING"

Quirk's, built in Australia, used a system in its early years which was similar in operation to the prairie windmills. The generator and prop assembly was mounted off axis from the supporting tower. The thrust developed by the prop would push the generator assembly around the tower axis, against a tail hung by gravity, at a design windspeed. Thus, the prop would turn partially sideways from the oncoming wind, slowing the prop down. A popular method among the home-built units in the 1920's was the use of a pilot vane set parallel to the prop. The arm of the vane was equal in distance to the radius of the propeller, and the area exposed to oncoming air would help to push the prop and generator assembly out of the wind. The disadvantage of these types of swinging systems is that they induce gyroscopic vibration and eliminate the power supply at the very time when the best winds are available. However, well designed systems are able to operate directly into the airstream and then be activated at a predetermined windspeed. More sophisticated designs utilize a feathering principle, which regulates the propeller rpm by changing the angle of attack (pitch) at high wind speeds. Pitch regulation is *attractive* since it holds the energy absorption nearly constant at all wind velocities, and allows continuous power output in high winds without inducing stress. Both Elektro and Quirk's utilize centrifugal force to act upon a set of spring-loaded weights to change pitch. Noting the Elektro diagram (Fig.10) it can be seen that the weights work on a direct line of centrifugal force and hence the actual density of the weight element itself is about 1 lb. For the two-bladed units, the assembly is designed to begin changing pitch (feathering) at about 400 rpm propeller speed. At this speed there is about 25 lbs. of centrifugal force on the weight arm itself. Quirk's utilizes a component of centrifugal force to act upon its weights, which are placed out from the prop shaft itself.

FIG. 10. CONTROL-PROPELLER HUB OF HEAT-TREATED NON-CORROSIVE LIGHT METAL WITH REGULATING MEMBERS, ADJUSTING MOVEMENT ON TAPEROLLER AND NEEDLE BEARINGS, EASY TO LUBRICATE, PROTECTED FROM WEATHER, NO RISK OF ICE-BUILDING. (ELEKTRO)

This positioning, and the lower prop speed of 150 rpm, results in the necessity of a 4 lb. weight at the end of each arm. Both systems are very successful in the control of any overloading, and can maintain safe rpm into winds of hurricane force. However, even though these units are stressed for 80 mph winds, the manufacturer still recommends that the propeller be stopped manually and/or rotated sideways to the wind. Most models have a brake control at the bottom of the tower for this purpose. Special wind-plants by these companies are available to operate unattended in winds up to 140 mph. Quirk's uses the conventional windplant, fitted with smaller diameter propellers (generally 10 ft. diameter). The latter, being shorter in arm length and with added internal stiffeners, allows the unit to operate head-on into high winds. Elektro prefers to shut down the plant completely, and has an optional high speed control package which can be added to the standard plants.

Hopefully the multitude of systems to come will be designed around some basic criteria like "Safety, Reliability, and Cost," in that order. These are important considerations particularly for those wishing to build their own design. A real understanding of the force and magnitude of the wind serves to reinforce the need for these criteria. Many a wind system has been literally swept away.

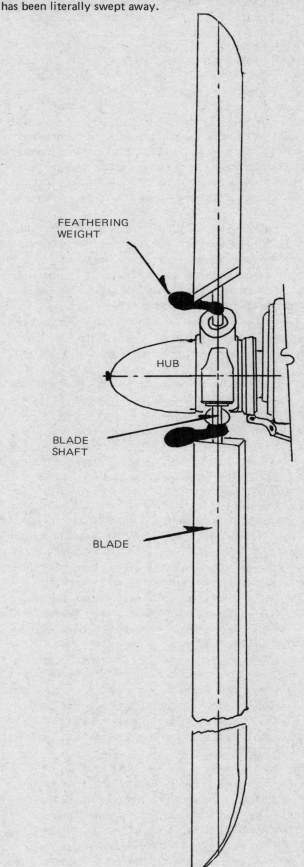

FEATHERING WEIGHT

HUB

BLADE SHAFT

BLADE

FIG. 11 QUIRK'S FEATHERING SYSTEM

One marvelous feeling that comes to you from taking these responsibilities, that is realizing the ability-to-respond, is the gathering of the sense of freedom. The individual benefits of these renewable energy systems represents only one part of the picture. Along with emoting their sense of ecological satisfaction, they provide a positive direction towards self-determination. A rare thing these days.

THOUGHTS ON
WIND DRIVEN GENERATORS

Donald Marier
Editor ASE Newsletter

The basic principle of a generator is that electric power is produced when a conductor moves through a magnetic field (Fig. 6). By increasing the magnetic field strength of a generator, the power output is increased proportionally. Voltage output can be stepped up by increasing the number of armature windings or by increasing the speed of the generator. For example, by doubling the number of turns in an armature, the generator will put out twice the voltage at the same speed as before, since the number of wire conductors crossing the magnetic field per unit time has been doubled. The current output is limited by the size of the wire used. See page 126.

Generators of vastly different physical size can have the same power rating. This is why the speed at which the generator puts out its rated power should be specified.

There are great differences in quality and cost between a generator rated at 3000 watts at 200 rpm and one rated at 3000 watts at 2000 rpm.

The blades (propeller) of a wind plant generally do not turn faster than about 200 rpm. (The speed of the blade tips is between 4 and 8 times the speed of wind.) Most electric motors and generators are built for speeds of from 100 to 4000 rpm, with 1800 and 3600 rpm being standard speeds for A.C. motors. A good share of wind designs in the past have used gears for matching the low speed of the blade to a relatively high speed generator. Besides the wear the gears are subject to, high speed generators can break down faster than direct slow speed generators.

In direct drive systems the gearing (so to speak) is electrical. The generator is designed to give its maximum output at the same speed as the blade system gives its maximum output. This results in a long-lived, reliable system, but uses more copper and steel.

The direct drive system can be used with either a D.C. generator or with an alternator. Old Jacobs' wind generators were a sturdy, direct drive design. Unfortunately, most were the 32 volt design which is not compatible with modern 100 volt equipment. (I am presently re-building one for 110 volt operation...) See page 153.

The generator, along with the storage system, is the most important part of the wind plant. From the day it is put up, the generator will be turning more or less continuously for years.

Most often, people have used car alternators in their homebuilt windplants. But car alternators are not well suited for wind generators as they are high speed devices and are very inefficient. (A D.C. car generator takes much less horsepower from the car's engine than an alternator.) When alternators were first used widely in cars, the industry promoted the myth that they charge better at low engine speeds, and that alternators had some inherent characteristic which made this possible. Actually the industry converted to alternators because of economics. They found that alternators were a higher speed device which needed less construction materials and had a shorter life. A D.C. generator could not function at the speeds at which alternators turn. The "trick" of the car alternator is that the pulley ratio is higher than with generators, allowing the alternator to be operated at top engine speeds . . . (10,000 ALT. rpm).

To show the difficulties of making a direct drive unit out of a car alternator, consider the following example. Assume the alternator will generate 1000 watts at 3000 rpm and that it is wound with 12 turns per coil of number 16 wire. To rewind the alternator for the speed range of the blades, cut the speed by a factor of 16 to 187 rpm. This would mean using 128 turns of number 28 wire per winding and the power output would be 62 watts. Obviously the only way to use a high speed alternator is with a gear or belt design. The disadvantage of this set-up is that its lifetime will not be very long.

Many people ask whether a wind system would be compatible with their present appliances, motors and tools. First it should be pointed out that when you change to a wind system, you will be changing the way you live. Many of the solar house designs to date have tried to duplicate exactly the characteristics of fossil fuel houses so the "public would accept the design." The result in the past was that these designs were often overly complicated and expensive . . .

The same applies to wind systems. There are two options—to use the D.C. voltage from the batteries directly or to convert the D.C. to A.C. with a rotary or electronic inverter. The D.C. system is simpler and cheaper in that no inverter is needed. Any heating element device such as light bulbs and irons will work equally well on A.C. or D.C. Most power tools can run on A.C. or D.C.—that is, tools using universal motors (they have brushes) and using A.C./D.C. switches. Electronic devices such as stereos can be run off of small A.C. inverters. The main problem is to find a refrigerator or freezer which will run on D.C. Refrigerators made in the last 15 or 20 years have sealed units. Before that, they had belt drive compressors allowing the use of any type motor. Commercial units still can be obtained with belt drive compressors. Also the recreational vehicle industry is now producing refrigerators which will run on A.C., D.C., or gas. They are expensive, but a used market should be developing soon. D.C. motors are not easily available, but car generators can be rewound for 110 or 32 volt operation and used where a 1/4 or 1/3 hp motor is needed. *If* you can't purchase one locally see page 147.

NOTES:

Winnie Redrocker has some information on high speed alternator use in ASE No. 15.

Martin Jopp of Princeton, Minnesota, has 55 years experience with motors and generators. He used a set of car batteries for his windplant for two years while waiting to find a set of used lighting plant batteries for storage. (See "Some Notes on Windmills in ASE No. 12, October 1973.)

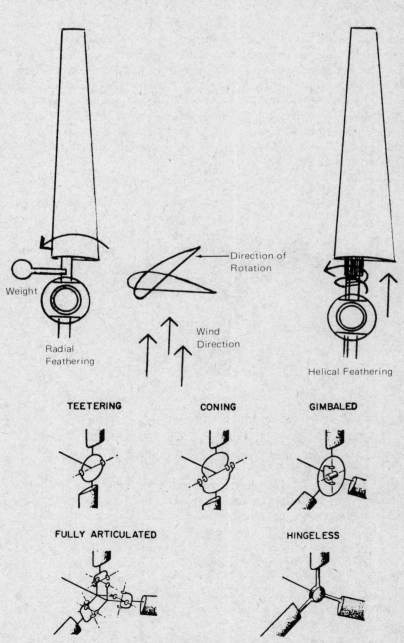

FIG. 12 TYPICAL ROTOR HUB CONFIGURATIONS
& FEATHERING SYSTEMS

SAVONIUS ROTOR

A detailed look at Fig. 15, below, and Fig. 8 on page 126 will illustrate the relationship of SAVONIUS type mills to the propeller type. The other graphs show a comparison of regular and modified SAVONIUS rotors. The changes in efficiency (coefficient of lift C_L over coefficient of drag C_D) by modifying the blade shape, are shown in graph No. 1.

This information plus the inherent problems with Bearings and Balancing, put the SAVONIUS rotor in a better perspective from a design and life-of-the-system standpoint. The life of the system is not usually viewed as an economic problem, but we find in the long run it is an important parameter. It becomes obvious that dependable, long term economic electrical power from the wind is best provided by the propeller type designs. The best application of the SAVONIUS rotor is for pumping and grinding (slow speed, medium torque uses).

The figures below show the C_L and C_D (coefficient of lift and coefficiency of drag) at different U/V "tip speed ratios:" U/V is the ratio of the speed of the outermost portion of the rotor to the wind speed. Coefficient of lift is the air pressure on the windward side of the rotor. Coefficient of drag is the air pressure on the leeward side of the rotor which slows, "drags" its movement. The rotor must first overcome the drag before it can begin to rotate.

The small arrows directed at the rotor profiles indicate the position of static zero or null starting torque. In other words the wind speed is not sufficient to turn that particular rotor and overcome the coefficient of drag.

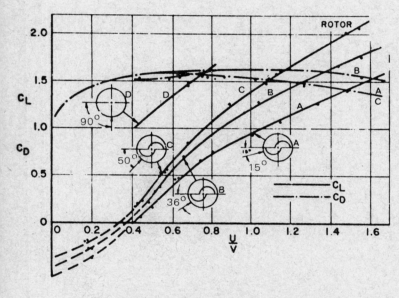

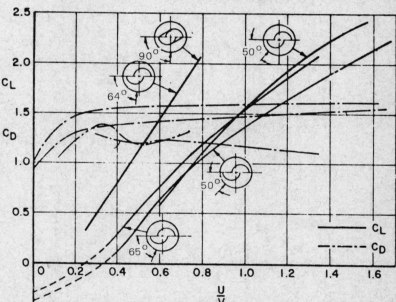

FIG. 14 AERODYNAMIC PERFORMANCE OF A COLLECTION OF BASIC AND MODIFIED SAVONIUS WING ROTORS THROUGH POWER EXTRACTING (BRAKED) U/V RANGE UP TO U/V MAX. IN FREE AUTOROTATION.

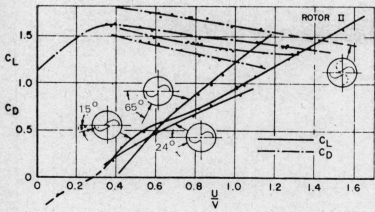

FIG. 13 AERODYNAMIC PERFORMANCE OF A COLLECTION OF BASIC AND MODIFIED SAVONIUS WING ROTORS THROUGH POWER EXTRACTING (BRAKED) U/V RANGE UP TO U/V MAX. IN FREE AUTOROTATION

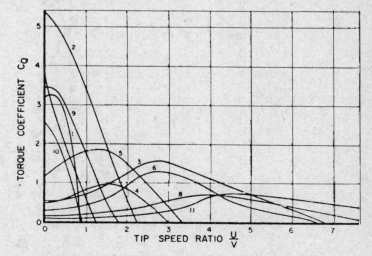

FIG. 15 PERFORMANCE SUMMARY OF 11 TYPES OF WINDMILLS.

RECYCLED WIND GENERATOR

The basic design (below) shows the components necessary for building a small, working wind generator from purchased ("off the shelf") and/or recycled materials.

FEATHERING: The system is modified from a windplant built by Paris-Dunn Co. of Iowa in the 1930's and 40's. Its main advantage is that it eliminates most of the typical speed-governing problems (i.e. loading and vibration) using a simple feathering device. It uses a spring and hinge that allow the propeller and generator to feather when wind speeds become so high (see photos on next page).

WIND PLANT MATERIALS: The plant is constructed mostly with nuts and bolts, so only some drilling and cutting (but no welding) is required. A wide choice of materials is possible, depending on your imagination and what is available. There are many variations and "tradeoffs" possible in the design.

BLADES: The Clark "Y" airfoil is a blade which is easy to carve, and fairly efficient (a tip-speed ratio of 8). It could be carved from hardwood, and the leading edges covered with copper foil for longer life. Care should be taken not to carve the trailing edge so thin that it will split. Construction instructions can be found in ASE Newsletter No. 14, pp. 10-13. (Note that the LeJay Manual has a layout for a Clark "Y" with a tip-speed ratio of 3 or 4.) Cut the prop from 2"x6"x 8' hardwood stock. Other high tip-speed ratio airfoils may also be used.

GENERATOR: A good generator to convert to windplant use is a Delco unit that comes from International Harvester.

2900 Z Stock	Rewind
600 Watts	300 Watts
at 1925 rpm	at 962.5 rpm
18 slots in Armature	18 slots
36 bar commutator	36 bars
number 12 wire	number 15 wire
8 turns/slot in armature	16 turns/slot in armature

The way one rewinds or gears the generator determines the amperage and voltage output. If you rewind the armature with twice as many turns per slot (for example, from 8 turns of No. 12 wire per slot (stock) to 16 turns of No. 15 wire per slot), you achieve the necessary voltage to comply with battery and appliance requirements (12V). This output can then be attained at a slower speed (962 rpm), which the prop will reach in approximately a 25 mph wind. Note that the wire size has been reduced from 50 amps to 25 amps. 12 volts x 25 amps = 300 watts.

TOWER: The tower could be a standard windmill tower, or Hans Meyer's Octahedron Tower, which is very nice and is materials efficient.

Consult the remaining pages in this section for additional information. Further information on small wind plants will be available from future Alternative Sources of Energy and New Alchemy Institute Newsletters.

Fig. 1 Hub   Fig. 2 Frame
Fig. 3$_A$ Slipring Assembly   Fig. 3$_B$ Sliprir
Fig. 4 Turntable   Fig. 5 Rear View of Ge
Fig. 6   Manual Pull-Out

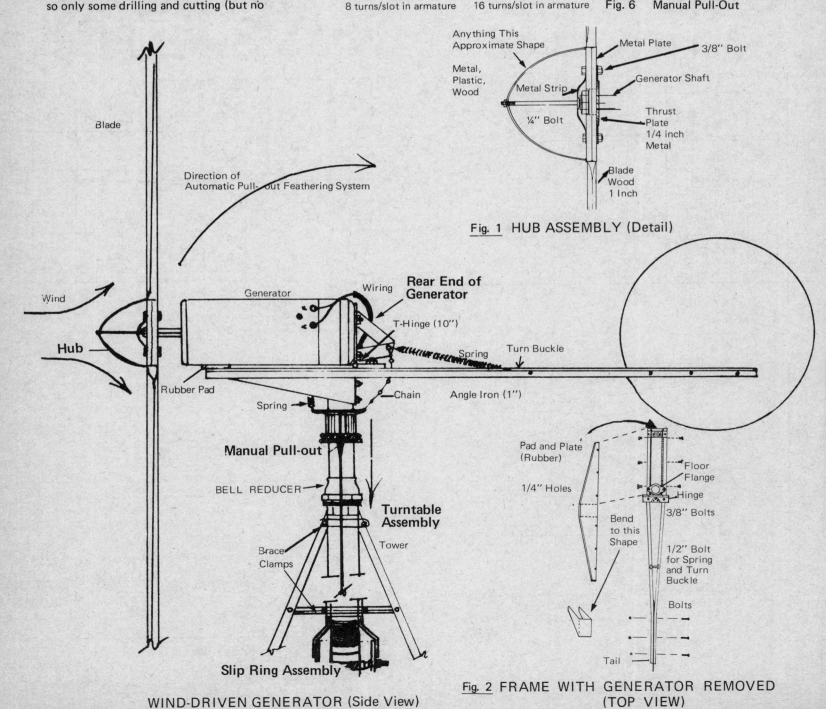

Fig. 1 HUB ASSEMBLY (Detail)

WIND-DRIVEN GENERATOR (Side View)

Fig. 2 FRAME WITH GENERATOR REMOVED (TOP VIEW)

Fig. 3_A SLIP RING ASSEMBLY AND HOUSING (DETAIL)

Spring Strip
Small Bolt
Brush
Brush Assembly Detail
Terminal 1/4" Bolt
3" P.V.C. Housing
Female Fitting
Shim
a
Bell 2" to 3"
Brass or Wire Slip Ring Assembly
b
3" P.V.C. Pipe
c
Brush Assembly
1—1/2" P.V.C. Pipe
P.V.C. Cap

Fig. 3_A SLIP RING ASSEMBLY AND HOUSING (DETAIL)

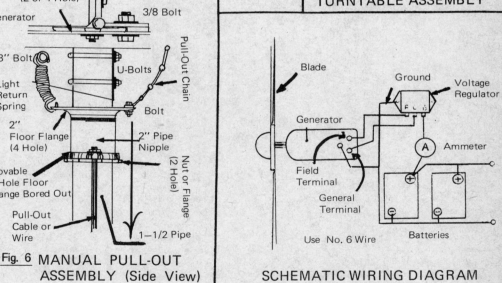

Connecting Lead Wires
PVC (Plastic) Pipe and Fitting
Pipe Wall
Hole for Insert
Thin Brass
B Type
c
Brass Insert Lock
Hole
Wire Wound Tightly
C Type
Hole

Fig. 3_B SLIP RINGS *

*Slip rings B or C type may be used.

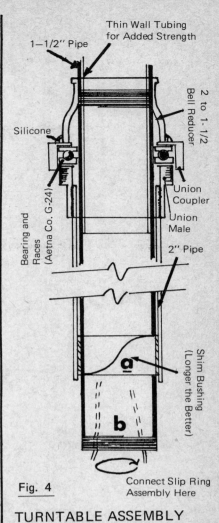

Thin Wall Tubing for Added Strength
1—1/2" Pipe
2 to 1-1/2 Bell Reducer
Silicone
Union Coupler
Bearing and Races (Aetna Co. G-24)
Union Male
2" Pipe
a
Shim Bushing (Longer the Better)
b
Connect Slip Ring Assembly Here

Fig. 4 TURNTABLE ASSEMBLY

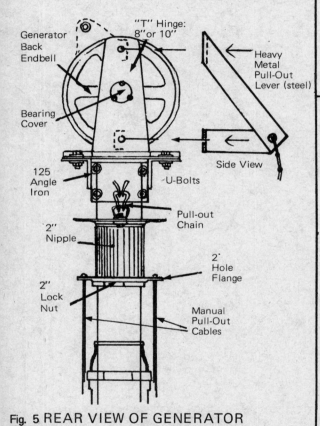

Generator Back Endbell
"T" Hinge: 8" or 10"
Heavy Metal Pull-Out Lever (steel)
Bearing Cover
Side View
125 Angle Iron
U-Bolts
2" Nipple
Pull-out Chain
2" Hole Flange
2" Lock Nut
Manual Pull-Out Cables

Fig. 5 REAR VIEW OF GENERATOR AND PULL-OUT SYSTEM

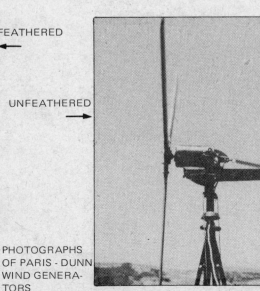

1—1/2 Floor Flange (2 or 4 Hole)
T Hinge 10"
Space if Needed
Generator
3/8 Bolt
3/8" Bolt
U-Bolts
Light Return Spring
Pull-Out Chain
2" Floor Flange (4 Hole)
Bolt
Movable 2 Hole Floor Flange Bored Out
2" Pipe Nipple
Nut or Flange (2 Hole)
Pull-Out Cable or Wire
1—1/2 Pipe

Fig. 6 MANUAL PULL-OUT ASSEMBLY (Side View)

Blade
Ground
Voltage Regulator
Generator
Field Terminal
General Terminal
Ammeter
Batteries
Use No. 6 Wire

SCHEMATIC WIRING DIAGRAM

FEATHERED

UNFEATHERED

PHOTOGRAPHS OF PARIS - DUNN WIND GENERATORS

COMPOSITE OF KILOWATT HOUR RATINGS FOR VARIOUS APPLIANCES
(PRE—ENERGY CRISIS ESTIMATES)

Name	Watts	Hrs/Mo	KWHRS/Mo.
Air conditioner, central			620*
Air conditioner, window	1566	74	116*
Battery charger			1*
Blanket	190	80	15
Blanket	50-200		15
Blender	350	3	1
Bottle sterilizer	500		15
Bottle warmer	500	6	3
Broiler	1436	6	8.5
Clock	1-10		1-4*
Clothes drier	4600	20	92*†
Clothes drier, electric heat	4856	18	86*†
Clothes drier, gas heat	325	18	6*†
Clothes washer			8.5*
Clothes washer, automatic	250	12	3*
Clothes washer, conventional	200	12	2*†
Clothes washer, automatic	512	17.3	9*
Clothes washer, ringer	275	15	4*†
Clippers	40-60		½
Coffee maker	800	15	12
Coffee maker, twice a day			8
Coffee percolator	300-600		3-10
Coffee pot	894	10	9
Cooling, attic fan	1/6-3/4HP		60-90*†
Cooling, refrigeration	3/4-1½ ton		200-500*
Corn popper	460-650		1
Curling iron	10-20		½
Dehumidifier	300-500		50*
Dishwasher	1200	30	36*
Dishwasher	1200	25	30*
Disposal	375	2	1*
Disposal	445	6	3*
Drill, electric, ¼"	250	2	5
Electric baseboard heat	10,000	160	1600
Electrocuter, insect	5-250		1*
Electronic oven	3000-7000		100*
Fan, attic	370	65	24*†
Fan, kitchen	250	30	8*†
Fan, 8"-16"	35-210		4-10*†
Food blender	200-300		½
Food warming tray	350	20	7
Footwarmer	50-100		1
Floor polisher	200-400		1
Freezer, food, 5-30 cu. ft.	300-800		30-125*
Freezer, ice cream	50-300		½
Freezer	350	90	32*
Freezer, 15 cu.ft.	440	330	145*
Freezer, 14 cu.ft.			140*
Freezer, frost free	440	180	57*
Fryer, cooker	1000-1500		5
Fryer, deep fat	1500	4	6
Frying pan	1196	12	15
Furnace, electric control	10-30		10*
Furnace, oil burner	100-300		25-40*
Furnace, blower	500-700		25-100*†
Furnace, stoker	250-600		3-60*†
Furnace, fan			32*†
Garbage disposal equipment			
	1/4-1/3 HP		½*
Griddle	450-1000		5
Grill	650-1300		5
Hair drier	200-1200		½-6*
Hair drier	400	5	2*
Heat lamp	125-250		2
Heater, aux.	1320	30	40
Heater, portable	660-2000		15-30
Heating pad	25-150		1
Heating pad	65	10	1
Heat lamp	250	10	2
Hi Fi Stereo			9*
Hot plate	500-1650		7-30
House heating	8000-15,000		1000-2500
Humidifier	500		5-15*
Iron	1100	12	13
Iron			12
Iron, 16 hrs/month			13
Ironer	1500	12	18
Knife sharpener	125		¼*

Name	Watts	Hrs/Mo	KWHRS/Mo.
Lawnmower	1000	8	8*†
Lighting	5-300		10-40
Lights, 6 room house in winter			60
Light bulb, 75	75	120	9
Light bulb, 40	40	120	4.8
Mixer	125	6	1
Mixer, food	50-200		1
Movie projector	300-1000		
Oil burner	500	100	50*
Oil burner			50*
Oil burner, 1/8 HP	250	64	16*
Pasteurizer, ½ gal.	1500		10-40
Polisher	350	6	2
Post light, dusk to dawn			35
Power tools			3
Projector	500	4	2*
Pump, water	450	44	20*†
Pump, well			20*†
Radio			8
Radio, console	100-300		5-15*
Radio, table	40-100		5-10*
Range	8500-1600		100-150
Range, 4 person family			100
Record player	75-100		1-5
Record player, transistor	60	50	3*
Record player, tube	150	50	7.5*
Recorder, tape	100	10	1*
Refrigerator	200-300		25-30*
Refrigerator, conventional			83*
Refrigerator-freezer	200	150	30*
Refrigerator-freezer 14 cu.ft.	326	290	95*
Refrigerator-freezer, frost free	360	500	180*
Roaster	1320	30	40
Rotisserie	1400	30	42*
Sauce pan	300-1400		2-10
Sewing machine	30-100		½-2
Sewing machine	100	10	1
Shaver	12		1/10
Skillet	1000-1350		5-20
Skil Saw	1000	6	6
Sunlamp	400	10	4
Sunlamp	279	5.4	1.5
Television	200-315		15-30*
TV, BW	200	120	24*
TV, BW	237	110	25*
TV, color	350	120	42*
TV, color			100*
Toaster	1150	4	5
Typewriter	30	15	.5*
Vacuum cleaner	600	10	6
Vacuum cleaner, 1 hr/wk			4
Vaporizer	200-500		2-5
Waffle iron	550-1300		1-2
Washing machine, 12 hrs/mo			9*
Washer, automatic	300-700		3-8*
Washer, conventional	100-400		2-4*
Water heater	4474	89	400
Water heater	1200-7000		200-300
Water pump (shallow)	½HP		5-20*†
Water pump (deep)	1/3-1 HP		10-60*†

AT THE BARN

Name	Capacity HP or watts	Est. KWHR
Barn cleaner	2-5	120/yr.*
Clipping	fractional	1/10 per hr.
Corn, ear crushing	1-5	5 per ton*
Corn, ear shelling	¼-2	1 per ton*†
Electric fence	7-10 watts	7 per mo.*†
Ensilage blowing	3-5	½ per ton
Feed grinding	1-7½	½-1½ per 100 lbs.*†
Feed mixing	½-1	1 per ton*†
Grain cleaning	¼-½	1 per too bu*†
Grain drying	1-7½	5-7 per ton*†
Grain elevating	¼-5	4 per 1000 bu*†
Hay curing	3-7½	60 per ton*
Hay hoisting	½-1	1/3 per ton*†
Milking, portable	¼-½	1½ per cow/mo.*†
Milking, pipeline	½-3	2½ per cow/mo.*†

Name	Capacity	Est. KWHR
Sheep shearing	fractional	1½ per 100 sheep
Silo unloader	2-5	4-8 per ton*
Silage conveyer	1-3	1-4 per ton*
Stock tank heater	200-1500 watts	varies widely
Yard lights	100-500 watts	10 per mo.
Ventilation	1/6-1/3	2-6 per day*† per 20 cows

IN THE MILKHOUSE

Name	Capacity	Est. KWHR
Milk cooling	½-5	1 per 100 lbs. milk*
Space heater	1000-3000	800 per year
Ventilating fan	fractional	10-25 per mo.*†
Water heater	1000-5000	1 per 4 gal

FOR POULTRY

Name	Capacity	Est. KWHR
Automatic feeder	¼-½	10-30 KWHR/mo*†
Brooder	200-1000 watts	½-1½ per chick per season
Burglar alarm	10-60 watts	2 per mo.*
Debeaker	200-500 watts	1 per 3 hrs.
Egg cleaning or washing	fractional	1 per 2000 eggs*†
Egg cooling	1/6-1	1¼ per case*
Night lighting	40-60 watts	10 per mo. per 100 birds
Ventilating fan	50-300 watts	1-1½ per day*† per 1000 birds
Water warming	50-700 watts	varies widely

FOR HOGS

Name	Capacity	Est. KWHR
Brooding	100-300 watts	35 per brooding period/litter
Ventilating fan	50-300 watts	¼-1½ per day*†
Water warming	50-1000 watts	30 per brooding period/litter

FARM SHOP

Name	Capacity	Est. KWHR
Air compressor	¼-½	1 per 3 hr.*
Arc welding	37½ amp	100 per year.*
Battery charging	600-750 watts	2 per battery charge
Concrete mixing	¼-2	1 per cu. yd.*†
Drill press	1/6-1	½ per hr.*†
Fan, 10"	35-55 watts	1 per 20 hr.*†
Grinding, emergy wheel	1/4-1/3	1 per 3 hr.*†
Heater, portable	1000-3000 watts	10 per mo.
Heater, engine	100-300 watts	1 per 5 hr.
Lighting	50-250 watts	4 per mo.
Lathe, metal	¼-1	1 per 3 hr.
Lathe, wood	¼-1	1 per 3 hr.
Sawing, circular 8"-10"	1/3-1/2	1/2 per hr.
Sawing, jig	1/4-1/3	1 per 3 hr.
Soldering, iron	60-500 watts	1 per 5 hr.

MISCELLANEOUS

Name	Capacity	Est. KWHR
Farm chore motors	½-5	1 per HP per hr.
Insect trap	25-40 watt	1/3 per night
Irrigating	1 HP up	1 per HP per hr.
Snow melting, sidewalk and steps, heating—cable imbedded in concrete	25 watts per sq. ft.	2.5 per 100 sq. ft. per hr.
Soil heating, hotbed	400 watts	1 per day per season
Wood sawing	1-5	2 per cord

Symbol Explanation

*AC power required

†Normally AC, but convertible to DC

Notes: Lighting in this table is assumed to be incandescent—if flourescent, the wattage bulbs consume the same power but deliver 3 times as much light—flourescent bulbs also require AC, but can be converted to DC.

These figures can be cut by 50% with conservation of electricity.

1) Sources for this table represent a conglomerate of several separate tables taken from:

a) Northern States Power Co., Mpls, Mn

b) University of Minnesota, Agricultural Extension Service

c) Seattle City Light, Seattle, Washington

d) Energy Conservation Techniques, National Bureau of Standards

e) Garden Way Labs

f) Henry Clews

g) Real Gas and Electric

WIND BOOK REVIEWS ———————— BASIC WIND ENERGY———

POWER FROM THE WIND

To the lay reader or the scientist, the value of Putnam's book is its logic and elucidation. He writes with candor of the conception, design, financing, erection, viability and shortcomings of the first large scale experimental wind generator.

The actualization of wind-turbines as a power source is exciting and here, supplemented with excellent drawings, charts and graphs, is an informative, highly readable basis for expanded research.

The surprise of this text is that it is such enjoyable reading. Beyond the mysteries of science, Putnam reflects the enthusiasm of all who collaborated on the project and he relates technical data so lucidly it is easily understood.

—Connie Meade

Power From the Wind
P.C. Putnam
1948; 223 pp

$9.95

from:
Van Nostrand Reinhold Co.
450 West 33rd Street
New York, NY 10001

or WHOLE EARTH TRUCK STORE

FEDERAL WIND ENERGY PROGRAM

The current federal wind program seems like it is one of the few federal programs that is actually addressing the needs of a small American industry. An approach that has been diverse, which is healthy, and imaginative means that it may receive full support of both the public and the wind energy industry. Write for it and find out what your government is doing.

— J.R.B.

Federal Wind Energy Program
(Summary Report ERDA 77-32)
Energy Research and Development Administration
1976

Free   from:
Wind Systems Branch
Division of Solar Energy
ERDA
Washington, D.C. 20545

WIND POWER

At any price this is the best, most concise description of wind energy on an individual user scale available. Some of the language used is difficult at times, but there is a glossary to help, plus a list of references, and a list of consultants for system planning and engineering. The book covers a very important item, which is a back up system for short periods of no wind. There are many places across the country where a back up system is needed. A major part also discusses assessing one's electric needs called the "Demand Analysis," including sample calculations and cost accounting.

— J.R.B.

Wind Power
Charles D. Syverson, P.E.
John G. Symons, Jr. Ph.D.
1973; 19pp

$3.00

from: J.G. Symons, Jr., PH.D., Box 233, Mankato MN
or WHOLE EARTH TRUCK STORE

WIND POWER FOR FARMS, HOMES AND SMALL INDUSTRY

"Everything you wanted to know about windmills . . ." is covered in this excellent primer on the selection and siting of small windmachines. Written for the layman, it proceeds through history, aerodynamic theory, wind behavior-site selection, power and energy requirements, wind system components, owning a windmill, windmill economics and legal hurdles. Examples with illustrations and graphics are presented throughout. A glossary and wind power data for 750 U.S. and southern Canada stations end the compendium. Kudos to ERDA's Lou Divone (Wind Energy Branch Manager), the authors, their consultants and reviewers for this modern classic, a vital addition to the library of all who are contemplating small wind systems.
—Lee Johnson
RAIN

Wind Power for Farms, Houses
and Small Industry
(ERDA Contract E 04-3-1270)
Jack Park and Dick Schwind
In Print

from:
Superintendent of Documents
U.S. Government Printing Office
Washington, D.C. 20402

LARGE SCALE SOLAR AND WIND ELECTRIC SYSTEMS FOR WASHINGTON STATE

This is an exploration of the cost and feasibility of using solar and wind electric generation to meet the agricultural energy demand in the northwest. Five methods of using solar energy to generate electricity are compared. Methods for computing cost per kilowatt and yearly power output of solar and wind energy conversion systems are explained.

—Richard Held

Large Scale Solar and Wind Electric
Systems for Washington State
Ecotope Group   from:
1976; 83 pp   The Ecotope Group
$10.00   747 - 16th Ave. E.
                    Seattle, WA. 98112

WIND POWER POTENTIAL IN SELECTED AREAS OF OREGON

If you're planning to build a windcharger in Oregon this book will be of limited use. It lists sources and individuals which would supplement anyone's data-lust. Commissioned by several of the Peoples Utility Districts in Oregon, Report No. PUD 73-1 is concerned with monolithic windplants able to kick out a thousand-plus kws.

Oregon State U's Department of Atmospheric Sciences apparently would be a good place to consult for wind statistics in the Oregon Coast. They monitor stations from Astoria to Yaquina and from Mt. Hebo to Winchester Bay.

Wind Power Potential in Selected   —Gar Smith
Areas of Oregon
E. Wendall Hewson, et al   from:
1973; Report No. PUD 73-1   Oregon State University
free   Corvallis, Oregon 97331

GEOTHERMAL AND WIND POWER ... ALTERNATIVE ENERGY SOURCES FOR ALASKA

Recommended plans, policies and programs developed from the proceedings of the Alaska Geothermal and Wind Resources Planning Conference held in Anchorage, Alaska, July 8-9, 1975.

The wind section of this publication is presented in a thorough manner, and is organized so clearly that its form might be used for studies in other areas of the country. The chapter covering wind potential, efficiency, sots, and applications, comes directly to the point of cost effectiveness in refreshingly plain English.
— J.R.B.

Geothermal and Wind Power ...
Alternative Energy Sources for Alaska
Robert B. Forbes (ed.)
1975   from:
           Geophysical Institute
           University of Alaska
           Anchorage, Alaska

THE GENERATION OF ELECTRICITY BY WIND POWER

Almost everyone working in wind energy research has probably read this book, and most of the articles in magazines and journals usually get their grounding from it. It is highly recommended reading. The best bet would be to try to get a copy through your local library and xerox it. Golding's book is considered a classic and is a thorough survey of wind energy. He covers taking wind data with no use of instruments, the history of windmills and how to do surveys of local sites. He also discusses different types of mills, blades, and generators, and the *economics* of wind generators, and compares them to other energy sources. He also has a good sense of the need for international cooperation and communication about these systems. One very nice thing about it is its 1955 technical jargon, which isn't too complex (about the level of high school or 1st year of college.)

— J.R.B.

The Generation of Electricity by Wind Power
E. W. Golding
1955; 318 pp.
           from:
           Philosophical Library, Inc.
           15 East 40th Street
           New York, NY 10016

ELECTRIC POWER FROM THE WIND

The best and cheapest source of information on wind energy. Check other sources for more information on batteries, inverters and their problems, etc.—but a very good place to start reading.
— J.R.B.

Electric Power From the Wind
Henry Clews
1973; 29 pp.

$2.00

from: or:
Solar Wind Co.   Enertech Corp.
P.O. Box 7   P.O. Box 420
East Holden, MA 04429   Norwich, VT

WIND MACHINES

WIND POWERED MACHINES

In the book are presented the basic problems connected with the selection of layouts, calculations and parameters of wind machines, their energy-producing character, technical and economic indices. Methods of optimal matching of wind engines with working machines, calculations for strength, and construction and automation of wind machines are analyzed in detail. A description is given of the set up of domestic and foreign wind installations for various purposes. Discussion of wind as a source, the theories of wind machines and aerodynamics are presented. The uses of machines are presented according to zone.

The book is intended for engineers, designers, workers and mechanics connected with utilization of wind energy and agriculture.

—J.R.B.

Wind-Powered Machines
Y. Shefter
1972; 288 pp
$8.95

from:
NTIS
P.O. Box 1553
Springfield, Virginia 22151

WIND-CATCHERS

A wind history book.
—J.R.B.

Wind-Catchers
Volta Torrey
$12.95

from:
The Stephen Greene Press
Brattleboro, VT 05301

or THE WHOLE EARTH TRUCK STORE

Windmills compete at the 1893 Columbian Exposition in Chicago.

Wind-Catchers provides the essential background of windmill development abroad, but the meat of the book is the story of windmill progress in America and its projection into the future of windmills as an energy source.

WINDMILLS

An absolutely beautiful book of the same genre as the expensive coffeetable books. The beauty of it is that it is more than just excellent photographs.
—J.R.B.

Windmills
Suzanne Beedel
1975; 143 pp.
$12.00

from:
Charles Scribner's Sons
Vreeland Avenue
Totown, New Jersey 07512

or WHOLE EARTH TRUCK STORE

HOMEMADE WINDMILLS OF NEBRASKA

Originally published in the 1890's, this booklet is packed with ideas for building practical low-cost and elegantly simple windmills. The idea of using wind power to cut wood has been proposed by many wind enthusiasts. Mr. A. G. Tingley of Verdon, Nebraska devised and built his Battle-ax windmill to do just that, saw wood. There are excellent ink drawings in some detail of many wooden windmills. They all seem to work!
—J.R.B.

Homemade Windmills of Nebraska
$4.00

from:
Farallones Institute
15290 Coleman Valley Road
Occidental, CA 95465

or WHOLE EARTH TRUCK STORE

FOOD FROM WINDMILLS

An incredible book showing the integration of energy and agriculture, a link that is all too often overlooked. This book gives so much information that the Farallones Institute has used it as their text to build their sail windmill.
—J.R.B.

Food From Windmills
Peter Fraenkel
1975; 75 pp.
$7.70 surface mail
$11.20 air mail

from:
Intermediate Technology Publications Ltd.
9 King Street
London, England, WCZE 8HM

or WHOLE EARTH TRUCK STORE

WINDMILLS AND MILL WRIGHTING

"Millwright; a builder and repairer of wind, water, and other mills." In this all-inclusive lucid text on windmills there is a glossary of windmill terms, an incredible folding plan of a windmill and its working parts, and many photographs and line drawings of windmills. It also describes vividly the life of the miller during high winds, like the wooden break catching fire from too high wind speeds. This is the finest book on the subject and is recommended for anyone working on or thinking about wind energy.
—J.R.B.

Windmills & Millwrighting
Stanley Freese
1957; 200 pp
$7.95

from:
Great Albion Books
Pierce Book Co., Inc.
Cranbury, New Jersey 08512

or WHOLE EARTH TRUCK STORE

WIND MACHINES

Wind Machines is a condensed and informative work. Attention is centered on large-scale systems both past and present. The graphics are excellent and the comparative wind generator designs section is concise and very articulate.

— J.R.B.

Wind Machines
(GPO 038-000-00272-4)
Frank R. Eldridge
1975

•$2.25

from:
Superintendent of Documents
Government Printing Office
Washington, D.C. 20402

or WHOLE EARTH TRUCK STORE

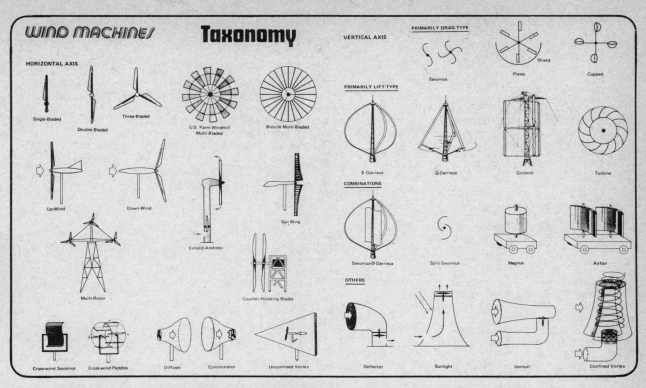

WIND MECHANICS

WIND FORCES IN ENGINEERING

Freedom equals responsibility. There are, in every country, numerous codes relating to wind loads on structures. The intention of this book is to give the designers a method of calculating wind loads from basic experiments and principles. It is difficult to compare codified rules and is unnecessary. A designer should, if he considers a code applicable:

1. Consult the ruling Code of Practice, and calculate his design loads.
2. Work out the wind loads from basic data in this book and associated references.
3. Select the greater of the two loadings.

No code covers the complete range of wind and structures. Most codes are many years old, and only provide an approximate answer; the best codes acknowledge this.

— J.R.B.

Also suggested reading for Towers: *On the Estimation of Wind Loads for Buildings and Structural Design.* Proc. Inst. Civ. Engrs. 25,1963 by C. Scruton and C. W. Newberry

Wind Forces in Engineering
Peter Sachs
1973; 390 pp

$36.00

from:
Pergamon Press, Inc.
Maxwell House, Fairview Park
Elmsford, New York 10523

APPLIED AERODYNAMICS OF WIND POWER MACHINES

Basic theory of the aerodynamics of various wind energy systems.

— J.R.B.

Applied Aerodynamics of
Wind Power Machines
(PB 238-595)
E. Wilson and P.B.S. Lissaman
1974

$5.25

from:
NTIS
Springfield, Virginia

THEORY OF WING SECTIONS

The wing sections or their deviations described in this book continue to be the most commonly used ones for airplanes designed for both sub and supersonic speeds, and for application to helicopter rotor blades, propeller blades, and high performance fans.

This book is intended to serve as a reference for engineers and is a resource to anyone studying at high technical levels. It includes a summary of air foil data. A knowledge of integral calculus and elementary mechanics is helpful, but if you take time to read it very carefully it is a valuable book. It does contain all aerodynamic information you'd need for designing windmill blades. It starts off hot and heavy but give it and yourself a chance to "sync in."

— J.R.B.

Theory of Wing Sections
Ira H. Abbott & A. E. Von Doenhoff
1959; 693 pp

$5.00

from:
Dover Publishing Inc.
180 Varick Street
New York, NY 10014

INTRODUCTION TO THE THEORY OF FLOW MACHINES

Introduction to the Theory of Flow Machines
A. Betz
1966

$27.00 photo-copy

from: Pergamon Press, Inc.
Maxwell House, Fairview Park,
Elmsford, NY 10523

WINDMILL TOWER DESIGN

Discussion of reinforced concrete and block-masonry towers.

— J.R.B.

Windmill Tower Design
F. Montesano and A. Fernandez
1973; 92 pp   Brace Research Institute
                     (see pg. 150)

STANDARD HANDBOOK FOR MECHANICAL ENGINEERS

Safety, Reliability, & Cost—Wind Energy is more complex and hazardous, in a technical manner, than other systems like solar heat and methane. SHME helps immensely in answering the difficult technical questions. It's a must, particularly if you are trying to build your own plant.

— J.R.B.

Marks Standard Handbook
for Mechanical Engineers
Ed. Theodore Baumeister
1963, new edition 1974;
1200 pp, with 6 pp on
wind power by E. M. Fales
with formulas and tables

$31.00

from: McGraw-Hill
1221 Avenue of the Americas
New York, New York 10020

or WHOLE EARTH
TRUCK STORE

LeJAY MANUAL

This marvelous manual from 1945 is still around, and loaded with useful information for the small scale home builder. The end of the book has a glossary, a discussion of wind electric plants and useful knowledge of electricity and batteries. There are 50 plans for devices from bug catchers to electric scooters. A definite bargain for the price. Might be considered the wind plant hotrodders manual.

— J.R.B.

LeJay Manual, 1945, 44pp $1.70
from: LeJay Mfg. Co., BellePlaine MN 56011

OCTAHEDRON TOWER

Hans Meyer, Windworks, Rt. 3, Box 329, Mukwonago, Wisconsin 53149

WIND SURVEYS

The National Climatic Center, NOAA, collects wind characteristics data from about 600 weather stations throughout the U.S. from which average wind speeds and directions, and monthly and annual average wind energy and power distributions can be obtained.

Choosing a site for a wind machine is critical. Local conditions and patterns must be considered and a detailed wind survey is essential to the performance and determination of tower height of a wind machine.

In the equipment section there is a list of firms providing this service and/or instrumentation.

— J.R.B.

**Wind Power Climatology
of the United States
(SAND74-0348)**
J.W. Reed
1975
$7.00

from:
NTIS
Department of Commerce
Springfield, VA 22151

━━ WIND AND CLIMATE ━━

CLIMATIC ATLAS OF THE U.S.

Very good and necessary information, expands and makes you aware, in another sense, of your environment. Is not very localized in terms of a particular plot of land, but a good place to start for a general look at the climate throughout the U.S. Two hundred seventy one climatic maps, 15 tables, covering mean temperature, snowfall, precipitation, wind speed, solar radiation, etc.

— J.R.B.

Climatic Atlas of the U.S.
Out of print

Individual pages (regions) available from:
Superintendent of Documents
U.S. Government Printing Office
Washington, D.C. 20402

WEATHER

The book is possibly the cheapest, best, most lucid, communicative little book about weather there is.

— J.R.B.

Of all aspects of the natural world, weather is outstanding in its beauty, its majesty, its terrors, and its continual direct effect on us all. Because weather involves, for the most part, massive movements of invisible air and is concerned with the temperature and pressure changes of this almost intangible substance, most of us have only a limited understanding of what weather is all about. This book will help you to understand it and also to understand, in some degree, how weather changes are predicted.

Weather
Paul E. Lehr
1965; 160 pp
$1.25

from:
Golden Press
Western-Publishing Co.
1220 Mound Avenue
Racine, Wisconsin 53404

or WHOLE EARTH TRUCK STORE

THE WINDS, ORIGINS & BEHAVIOR OF ATMOSPHERIC MOTION

This book says the same thing as *Sensitive Chaos* (see pg. 114   ENERGY PRIMER), but puts it in mathematically useable form. What the book has to say is, for the most part, basic and readable.

The study of flows of the earth's fluids forms a great part of the science called geophysics; the atmosphere and its motions. The book covers basic and technical information about the wind, its character, vortices, turbulence, force, patterns and measurement.

— J.R.B.

**The Winds, Origins & Behavior of
Atmospheric Motion**
George Hidy
1967; 174 pp
$3.95

from:
Van Nostrand Reinhold Co.
450 West 33rd Street
New York, NY 10001

CLOUDS ASSOCIATED WITH THE SIERRA WAVE IN THE
LEE OF THE SIERRA NEVADA RANGE IN CALIFORNIA.
The lowest cloud reaches an alt. of 6 km.   The middle one is
a lenticular cloud located along the crest of the lee wave at 10
km.   The upper one is another lee wave cloud at 13 km.

THE SELECTION & CHARACTERISTICS OF WIND POWER SITES

All ERA reports, including this one which costs $24.00, are very expensive, but the information is good. The report describes in very technical detail the development and successive stages, of a wind survey covering the British Isles area suitable for large-scale generation of wind power. Methods are shown for making a wind velocity survey and in employing the results to form a reasonably reliable estimate of wind potential. Also discussed is the relationship between *estimated* and *actual* output.

— J.R.B.

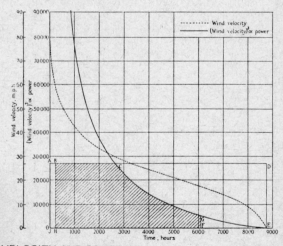

VELOCITY AND POWER-DURATION CURVES TYPICAL
OF AN EXCELLENT SITE

Method of Selection
1. Meteorological Records
2. Study of Maps
3. Inspection of Sites
4. Technique of Site Selection

**The Selection & Characteristics of Wind
Power Sites**
The Electric Research Association
E. W. Golding & A. H. Shodhart C/T 108
1952; 32 pp
$24.00

from:
E.R.A.
Cleeve Road, Leatherhead
Surrey KT22 7SA
England

UNDERSTANDING OUR ATMOSPHERIC ENVIRONMENT

This book's emphasis is on understanding, that is, on providing one with the physical explanation of atmospheric phenomena. As such, its purpose is twofold: to give you a deeper appreciation of this part of the universe, which we experience so intimately and continuously, and to impart an awareness of some of the laws of physics and the way in which they are applied.

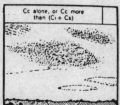

EXAMPLES OF THE C_h FAMILY OF CLOUD TYPES

Understanding Our Atmospheric Environment
Neiburger, Edinger, Bonner
1973; 193 pp
$10.95

from:
W. H. Freeman & Company
660 Market Street
San Francisco CA 94104

PLANS AND HOW-TO INFORMATION

WINDY TEN

Sells plans for an incredible replica of a Dutch windmill which also produces electricity. A sight to behold!

Windy Ten
Box 111
Shelby Michigan 4

SIMPLIFIED WIND POWER SYSTEMS FOR EXPERIMENTERS

The section on simplified aero dynamics of wind mill blade design is the main theme throughout the book. It offers information not found easily in other places, and has good material on hybrid designs. The author is an aerospace engineer, with a good feel for what he is talking about.

— J.R.B.

Simplified Wind Power Systems for
Experimenters
Jack Park, 73 pg.
$6.00

from:
Jack Park
Box 445
Brownsville, CA 95919

or WHOLE EARTH
TRUCK STORE

PRINCETON SAILWING

The Princeton Windmill Program has developed a sailwing wind turbine suitable for generating electricity, and recently licensed it to an aircraft company for development. A booklet on this program is available ($1.00) from:

Forestal Campus Library
Princeton University
Princeton, NJ 08540

Plans for the Princeton Sailwing are available from:

Flanagan's Plans
2032 23rd Street
Astoria, NY 11105

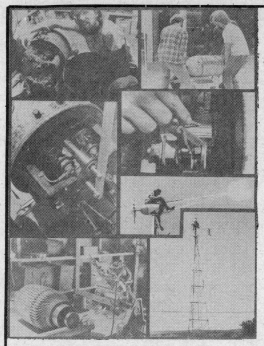

THE HOME BUILT WIND-GENERATED ELECTRICITY HANDBOOK

This second effort is the much needed extension of the basic knowledge imparted in his first book. The HomeBuilt Wind-Generated Electricity Handbook, however, concerns itself primarily with the uncovering, restoration, and insulation of previously discarded wind systems which were in the Midwest in abundance until the late 1950's. The title is misleading, but it is a good place to start to understand what has already gone on, learning from the past experiences of others.

— J.R.B.

VITA REPORTS

Low Cost Windmill for Developing Nations
Dr. Hartmut Bossel
1972; 40 pp; $1.00

Helical Sail Design
report number 1131.1
$0.75

Savonius Rotor Plans
report number 1132.1
$0.75

Vita reports are available from:
Volunteers in Technical Assistance
3706 Rhode Island Avenue
Mt. Rainier, Maryland 20822

WINDWORKS
Box 329, Route 3
Mukwanago, WI 53149

Windworks is an engineering firm active in the field of wind energy, power conditioning and load management, advanced structural systems, and publication of related educational information. Windworks has prepared an educational chart depicting the chronology of wind power development from the earliest known wind machine to the 125-foot diameter machine built by NASA. $3.25 post-paid from Windworks.

WIND AND WIND-SPINNERS

Somebody at last has done a good job of spelling out what is involved in making a wind electric system yourself. The design is for the well-known Savonius Rotor ("S" rotor), which is not a particularly efficient type, but is easy to build. The author is honest about it and gives a commendable amount of numbers and tables necessary to figure out expected performance for this and other machines. He also discusses the toughest riddle of the S rotor: how to get it up on a tower where the wind is and yet keep things simple. Altogether it seems to be very complete information and you could actually make a workable machine from these instructions.

— J. Baldwin

The HomeBuilt, Wind-Generated, Electricity Handbook
Michael Hackleman
1975; 194 pp.
$7.50

Wind and Windspinners
Michael Hackleman
1974; 139 pp.
$7.50

Both books from: Earthmind
5246 Boyer Rd.
Mariposa, CA 95338

The Sail Windmill was designed for mechanical power applications. The low speed, high torque output make it well suited for small manufacturing or agricultural situations. The sails are manually reefed in winds exceeding 20 mph. Construction plans are available from Windworks for $25.

ELECTRICITY AND ENERGY STORAGE

Wind Energy to electricity requires a system that levels out the energy flux, (i.e. changes in wind speed) or stores the energy for later use. Conventional and non-conventional forms of energy storage can be applied to wind electric systems. The books that follow are concerned with both.

BATTERIES AND ENERGY SYSTEMS

The size and characteristics of a battery system is the most important and difficult part of the wind plant to assess. I'd say this book covers the subject most completely.

—J.R.B.

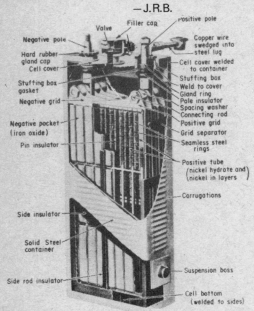

GENERAL CONSTRUCTION OF THE EDISON CELL.

Batteries and Energy Systems
C. L. Mantell Ph.D.
1970; 221 pp

$15.00

from:
McGraw-Hill Book Co.
1221 Avenue of the Americas
New York, NY 10020

INFORMATION ON OTHER STORAGE SYSTEMS

There is a good discussion of other storage systems in RADICAL AGRICULTURE (see pg. 166 EP).

HYDROGEN

Hydrogen is inexpensive, about half the current price of gasoline for equivalent usable energy. It's clean; water is the only emission. It's plentiful, even eternal, the most abundant element in the universe and on earth. It's safe; N.A.S.A. tests rate it safer than gasoline. It's adaptable; to almost any internal combustion application. An entire economy could be based upon hydrogen as a universal fuel.
—Carl A. MacCarley
UCLA Hydrogen Research

Electrochemisty of Cleaner Environments
J. O. M. Brockris, Ed.
"Hydrogen Economy"
D. P. Gregory and G. M. Long
1972

$22.50

From:
Plenum Press
New York, New York

WIND ENERGY STORAGE

The most economical form of storing wind energy today for small scale (.2 KW to 25 KW) applications is Lead Acid Batteries called "plant lighting batteries." These are different from the common auto battery (see pg.124 Energy Primer). There are many other forms of storing energy from the wind.

Compressed Air could be useful if one's needs were for mechanical to mechanical energy. A system more complex than elec. to mech. to mech. becomes very inefficient.

Heat Storage, by means of electrical resistance heating, could be useful in some situations and is very simple.

Water Storage (pumping into reservoirs) is a very well established technique of hydro-plant storage and would find easy application in wind systems.

Hydrogen, generated by the electrolysis of water, is a storage system proposed by many wind experts. Although the technology for widespread use has not yet been developed, it seems to have some really viable applications.

Flywheels are the simplest most exotic form of wind storage proposed. There are many considerations both economic and practical that we know very little about, it is very high materials technology.

—J.R.B.

STORAGE BATTERIES

This highly technical text is the COMPLETE WORD on batteries. Though not directly aimed at wind generator storage problems, you can find the information you need by combining information from various chapters. Considering the subject matter, it's surprisingly easy to understand. One of the oldie-but-goodies.

—J.R.B.

Storage Batteries
George Wood Vinal
1924, 1955; 446 pp.

$19.25

from:
John Wiley & Sons, Inc.
One Wiley Dr.
Somerset, NJ 08873

BATTERY STORAGE

Battery companies are the best sources of information on batteries.
—J.R.B.

ELECTROCHEMICAL SCIENCE

Want to understand batteries? This book is the best I've seen for it tries to explain, in a simple and largely qualitative way, the most important ideas of electrochemical science. It is not a specialized discipline. On the contrary, it is an interdisciplinary area which studies the behavior of electrified interfaces wherever they arise—in chemistry, materials, science, energetics, biology, engineering and so on.   —J.R.B.

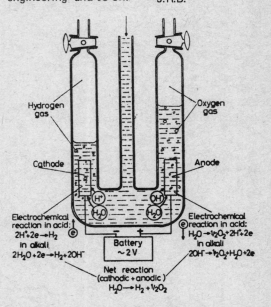

"This book is small, but the message it is written to convey is large: an environment unpolluted by power generation, in which we can live in the future, is possible through an electrically based technology."   — J.R.B.

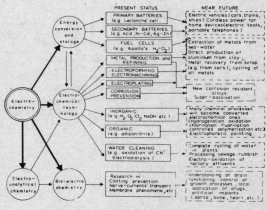

THE ELECTROCHEMICAL FUTURE

Electrochemical Science
J. O'M. Bockris & D. M. Drazic
1972; 300 pp

$20.50

from:
Barnes & Noble Books
10 E. 53rd Street
New York, NY 10022

COMPRESSED AIR

Hans Meyer
Windworks
Box 329, Rt. 3
Mukwonago, WI 53149

International Technical
Box 340
Warrenton, VA 22186

Alternative Sources of Energy
Rt. 2, Box 90A
Milaca, MN 56353
Issue No. 14; May 1974

FLYWHEELS

Science Magazine
"Energy Storage"
5/17/74 & 5/24/74

Wind Energy Conversion Systems,
Workshop, June 1973; $6.50
U.S. Dept. of Commerce
P.O. Box 1553
Springfield, VA, 22151

Popular Science magazine

Scientific American magazine

HEAT STORAGE

A partial assessment of the current state state-of-the-art and problems of wind energy—a good solution if you can use it.

—J.R.B.

Wind Power
1973; 24 pp.

free from:
Energy Unlimited
Palmer, MA 01069

BASIC ELECTRICITY

Basic Electricity was written for the Navy by the Navy to make available the fundamental information about electricity. The text is basic and not too technical, but requires the understanding of *basic* algebra—there are hundreds of illustrations that make it easy to read the pictures and find most of the information you need. It also covers the components of D.C. generators, and A.C. alternators, and motors.

— J.R.B.

Basic Electricity
Bureau of Naval Personnel
1970; 490 pp
$3.50
from:
Dover Publishing Inc.
180 Varick Street
New York, N.Y. 10014

or WHOLE EARTH TRUCK STORE

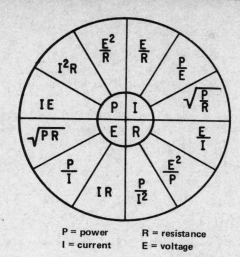

P = power   R = resistance
I = current   E = voltage

SUMMARY OF BASIC FORMULAS
Adjacent to each quantity are three segments. Note that in each segment, the basic quantity is expressed in terms of two other basic quantities, and no two segments are alike.

ELECTRIC MOTOR TEST & REPAIR

The book is <u>not</u> a college-level theory-ridden textbook. It is a workshop handbook, written in a highly useable form. What the book undertakes is covered in a logical and complete manner. For those who wish to rebuild motors, generators, etc., it is a welcome addition to your book shelf. It is the most practical book for learning the information on testing and rewinding small horsepower motors of every type, and most large basic motors AC & DC. Basic motors and generators are very similar.

Electric Motor Test & Repair   — J.R.B.
Jack Beater
1966; 160 pp
$6.95
from:
TAB Books
Blue Ridge Summit, PA 17214

or WHOLE EARTH TRUCK STORE

ELECTRIC GENERATING SYSTEMS

The purpose of this text is to discuss the application of engine generating systems to standby, portable, marine and mobile power needs. The book attempts to bridge the gap between books on engines, electricity, generators, and instruments, with reference particularly to ONAN generators. It has a glossary and index. An excellent book but mostly concerned with generators currently available "off the shelf." All of these are high RPM units when compared to the wind generator range of 0 to 200 RPM. Overall a very useful book.
— J.R.B.

Electric Generating Systems
Loren J. Mages
1970; 374 pp   from:
$5.95   Theodore Audel & Company
                       4300 West 62nd Street
                       Indianapolis, Indiana 46268

or WHOLE EARTH TRUCK STORE

BASIC ELECTRICITY

The most important thing, it appears, is to communicate, especially today, and in order to commune-icate, both parties must understand the language and use it. *Basic Electricity* is just what it says, Basic. However, it goes beyond this, but only through understanding and communicating in a very special manner, Logic. "People should do as well!"

— J.R.B.

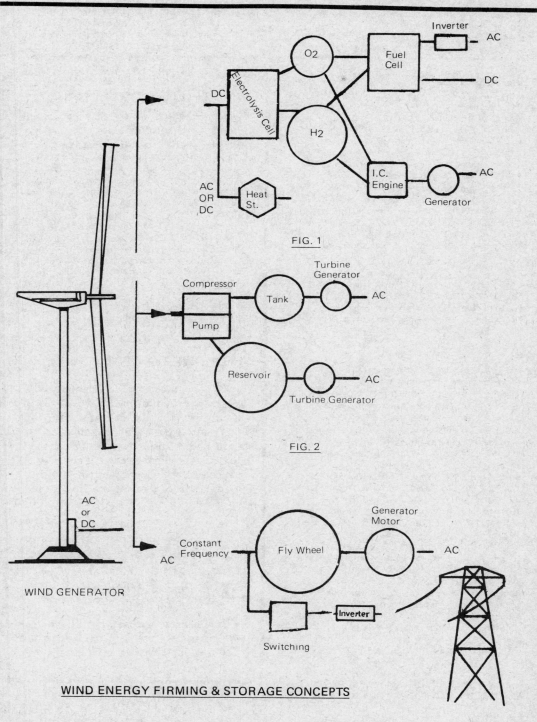

FIG. 1

FIG. 2

WIND GENERATOR

AC or DC

Constant Frequency

Fly Wheel

Generator Motor

AC

Inverter

Switching

WIND ENERGY FIRMING & STORAGE CONCEPTS

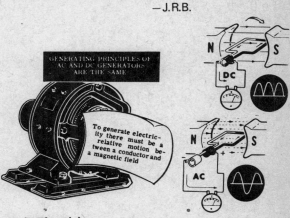

Basic Electricity
Vol. 1-5, 6 & 7
Van Valkenburgh, Hooger & Neville, Inc.
1953; 700 pp   from:
$13.75   Hayden Book Co.
                      50 Essex Street
                      Rochelle Park, N.J. 07662

or WHOLE EARTH TRUCK STORE

WIND PERIODICALS AND BIBLIOGRAPHIES

Except as noted, the material on this page can be ordered from:

NTIS
P.O. Box 1553
Springfield, VA 22151

Canadian National Aeronautical Establishment Report Low Speed Aerodynamics

A Wind Tunnel Investigation of a 14 ft. Diameter Vertical Axis Windmill   LTR-LA-105
P. South & R. S. Rangi
1972; 17 pp; cost unknown

U.S. Federal Power Commission. Fitting Wind Power to the Utility Network; Diversity, Storage, Firm Capacity, Secondary Energy
Percy H. Thomas, retired.
Washington, 1954. 24 p. TJ825.U515

U.S. Federal Power Commission. The Wind Power Aerogenerator, Twin-Wheel Type; A Study
Percy H. Thomas. Washington, Office of the Chief Engineer, Federal Power Commission, 1946. 77 p. TJ825.U52

The U.S. Energy Crisis: Some Proposed Gentle Solutions, Presented Before the ASME and IEEE
W. E. Heronemus
reprinted in Congressional record (daily ed)
Feb. 9, 1972, 92d Cong., 2d sess., v. 118:
E1043-1048.

A New Propeller-Type, High-Speed Windmill for Electric Generation
E. N. Fales
ASME Transactions, Vol 19, Paper AER-50-6, 1928.

Ramakumar, R. and others. Wind Energy Storage and Conversion System for Use in Underdeveloped Countries
from:
Intersociety Energy Conversion Engineering Conference 4th, Washington, D.C. 1969. Proceedings. New York, American Institute of Chemical Engineers, 1969. p. 606-613. TK2896.155 1969

The Windmill: Its Efficiency and Economic Use
Edward Charles Murphy
from:
Washington, U.S. Gov't Print. Off.
1901.2 v (U.S. Geological Survey. Water-supply and irrigation paper no. 41-42) TJ825.M96

U.S. Congress. House Committee on Interior and Insular Affairs
Production of power by means of wind-driven generator. Hearings, Eighty-second Congress, first session, on H.R. 4286. September 19, 1951. Washington, U.S. Gov't Print. Off., 1952. 41 pp. HD171.A18A32 82d, 1st no. 16

U.S. Federal Power Commission
Aerodynamics of the wind turbine, adapted for use of power engineers, prepared by Percy H. Thomas. Washington, Office of the Chief Engineer, Federal Power Commission, 1949.80 p. TK1541.U54

Windmills in the Light of Modern Research
A.Betz
U.S. National Advisory Committee for Aeronautics, Aug. 1928. 29 p. (Technical Memorandum no. 474) TL507.U57 no. 474

International Symposium on Molinology, 2nd, Denmark, 1969
Transactions of the 2. International symposium on Molinology, Denmark, May 1969. By Shmuel Avitsur and others; edited by Anders Jespersen. Lyngby, Danske Mollers Venner, Brede Baerk, 1971. 590 p. TJ823.I57 1969

U.S. Federal Power Commission. Electric Power From the Wind; a Survey
Percy H. Thomas. Washington, Office of the Chief Engineer, Federal Power Commission, 1945. 57 p. TK2081.U54

Studies on the Utilization of Wind Power in India
P. Nilakantan and R. Varadarajan
(Bangalore National Aeronautical Laboratory, 1962.) Technical Note, no. TN-WP-11-62) TL 504.B35 no. TN-WP-11-62, etc.

U.S. War Production Board. Office of Production Research and Development
Final report on the wind turbine. Research conducted by New York University, College of Engineering, Engineering Research Division. Washington, 1946, 278 p. (Its W.P.B. 144) PB 25370

GOVERNMENT REPORTS

Windmill Generator for the Bumblebee Buoy
D. M. Brown
1972; 19 pp; $3.00

Influence of Wind Frequency on Rotational Speed Adjustments of Windmill Generators
Ulrich Hutter
1973;. 16 pp; $3.00

The Development of Wind Power Installations for Electrical Power Generation in Germany
U. Hutter
1973; 29 pp; $3.00

The Influence of Aerodynamics in Wind Power Development
E. W. Golding
1961; 48 pp; $4.00

Wind Energy Conversion Systems, Workshop Proceedings, Held at Washington, D.C. on 11-13 June 1973
Joseph M. Savino
1973; 267 pp; $6.50

This is How You Can Heat Your Home With a Little Windmill
Leo Kanner Associates
1974; 13 pp; $3.00

Exploitation of Wind Energy
H. Christaller
1974; 12 pp; $3.00

Wind Energy Conversion Systems
J. M. Savino
1973; 270 pp; $15.50

H2-O2 Combustion Powered Steam-MHD Central Power Systems
G.R. Seikel, J.M. Smith, & L.D. Nichols
1974; 14 pp; $3.00

Status of Wind-Energy Conversion
R. L. Thomas and J. M. Savino
1973; 9 pp; $4.00

Utilization of Wind Power by Means of Elevated Wind Power Plants
F. Kelinhenz
1973; 30 pp; $3.00

Analysis of the Possible Use of Wind Power in Sweden. Part 1: Wind Power Resources, Theory of Wind-Power Machines, Preliminary Model 1 and 10 Mw Wind Generators
B. Soedergard
1974; 55 pp; $5.75

BIBLIOGRAPHIES

Burke, B. and R.N. Meroney. 1977. ENERGY FROM THE WIND: AN ANNOTATED BIBLIOGRAPHY, Engineering Sciences Library, Engineering Research Center, Colorado State University, Foothills Campus, Fort Collins, CO 80523. $10. Over 1100 references to books, conferences, journal articles, and technical reports, over 800 between 1973 and 1977. Most comprehensive to date.

Technology Applications Center, WIND ENERGY UTILIZATION BIBLIOGRAPHY, Order No. TAC-W-75-700, Energy Information Office, Technology Applications Center, University of New Mexico, Albuquerque, NM 87131, $10. A bibliography with abstracts of approximately 500 references regarding wind energy organized by subject area.

Van Steyn. 1975. WIND ENERGY: A BIBLIO-GRAPHY WITH ABSTRACTS AND KEY WORDS. Library Administration, University of Technology, Postbox 513, Eindhoven, Netherlands. $12. Excellent 2-volume work stressing non-U.S. wind experiments in the medium and large scale wind-turbine range.

RECOMMENDED READING

Chandery, M.J.. INITIAL WIND ENERGY DATA ASSESSMENT STUDY, Contract No. NSF-AG-517 NSFRAN-75-020 PB 244 132, NOAA/National Climatic Center, May 1975 132 pp.

Coonley, D. DESIGN WITH WIND, MIT Thesis Paper, May 1974, 148 pp. Explores the creative use of wind in the design of buildings.

Gilmore, E., and V. Nelson. POTENTIAL FOR WIND GENERATED POWER IN TEXAS, Department of Physics, West Texas State University and Department of Physical Science, Amarillo College, Report NT/8 for the Governor's Energy Advisory Council, October 1974, 159 pp. Estimations of power, possible uses, estimation of cost and storage are discussed for wind systems in Texas.

Mitre Corporation, WIND ENERGY WORKSHOP NUMBER TWO, ERDA/NSF, September 1975, Stock No. 038-000-00258-9, Superintendent of Documents, Government Printing Office, Washington, D.C. 20402, $10.00. An overview of projects and technical papers in the Federal Wind Energy Program as well as activities in a number of foreign countries, as presented on June 9-11, 1975 in Washington, D.C.

Rogers, S., et al., AN EVALUATION OF THE POTENTIAL ENVIRONMENTAL EFFECTS OF WIND ENERGY SYSTEMS DEVELOPMENT, FINAL REPORT, Contract No. NSF AER 75-07378, ERDA/NSF/07378-75-1, Battelle Memorial Institute, Columbus Laboratories, August 1976.

See also the Conference Proceedings section of the Solar Book Reviews, page 69.

WIND HARDWARE — WIND ELECTRIC — MANUFACTURERS AND DISTRIBUTORS

The following companies are manufacturers and/or rebuilders of wind energy systems. These companies have either developed new concepts and designs or they have purchased patents or ideas from earlier manufacturered systems.

HELION, INC.
Box 445
Brownsville, CA 95919

12/16 Construction plans.
Jack Park originally designed, built and conceived this wind-driven generator (see photo). Plans for construction include: 2000 watt generator, the 12/16 aluminum blades, a 12 or 16 foot diameter belt drive transmission, overspeed governor, and your choice of generator. Plans for tower are not included. Price $10 postpaid, book rate. California residents add State Tax.

KEDCO, INC.
9015 Aviation Blvd.
Inglewood, CA 90301

Kedco licenses and manufactures a 1200 and 1600 watt wind-driven generator originally developed by Jack Park of Helion. Kedco offers four wind generators for your application. Models 1200 and 1600 are rated at 1200 watts maximum for battery-charging applications, while Models 1210 and 1610 are supplied with 2000 watt DC generators of permanent magnet design for synchronous inverter operation, as well as wind furnace and other applications. A wind furnace is possibly the most efficient use of wind generated electricity. Electricity is converted directly into heat which can be stored in rocks, water, or salt solutions, for use in agricultural, industrial, and residential heating. The fundamental difference between the 1200 and 1600 series is the blade diameter. The increase in diameter from 12 to 16 feet nearly doubles the energy yield (kwh) you can expect from your wind generator, without doubling the price. Kedco wind generators are designed to produce energy at the lowest possible cost. Obviously, larger, more expensive wind generators will convert wind energy at an even lower unit cost ($/kwh), which is why the Kedco 1200/1600 family includes larger diameter blades. The 1600 offers the lowest rated windspeed available today.

WHAT DOES THE OUTPUT RATING OF A WIND GENERATOR REALLY MEAN?

Example: In round figures a 10 KW (10,000 watt) rated machine, in a 10 mph average wind, produces about 500 KWH per month. If a house uses 250 KWH/month, then a 10 KW machine, in the right location (10 mph average wind) might produce enough energy for 2 houses.

Price List:

Model 1200	$2,295.00
Model 1600	$2,895.00
Model 1210	$2,595.00
Model 1610	$3,195.00

- Automatic blade feathering by mechanical control.
- Ground shut-off/reset cables
- Aluminum blades—aerospace construction—all parts plated, painted or anodized.
- High efficiency transmission
- Automatic vibration sensing shut-off.
- Warranty—one year parts and labor for the Kedco Wind Generator. Alternator warranty per manufacturer specification.

TOPANGA POWER
Box 712
Topanga, CA 90290

"The Topanga Power generator kit is an outgrowth of original prototype work done by Helion. Versions of this machine have been spinning on towers for over two years now and have proven themselves under a variety of conditions. Topanga's kit is designed to be simple and easy to assemble. Once in operation it will perform as well as a factory-built unit. The kit is provided in a semi-finished state; all welding is completed and all power train alignments are accomplished, blade ribs are formed and installed on spars, feathering linkage is sized, counter weight arm is cast and drilled and cowling mounts are provided."

The Topanga Power generator is basically identical to the Kedco except for the following features. Instead of a belt drive system, there is a gear-box of an 8.6:1 ratio. The gear box directly drives the alternator through a flex coupling. The feathering system is also a little different with some interesting features not presently in the Kedco models. There is also built-in vibration protection which shuts the mill down completely. The unit was designed for the kit to use as much off-the-shelf materials as possible. The alternator can either be a Motorole or Lece NeVille. The gear box is also off the shelf, and bearings are all standard so that, in a remote situation, your nearest truck stop may in fact have the part you need. $1400 (fob)

SPECIFICATIONS:

The Topanga Power generator weighs 200 lbs. The blades are designed for a speed of 300 rpm (25 mph). Beyond 25 mph the blades will begin to feather automatically. The machine can be supplied with 12 or 16 foot diameter blades. Due to alternator limitations, the maximum power output will be the same with either set of blades, but the advantage of the longer blades will be that full power will come at a much lower windspeed. The long blades deliver full power at 18 mph, the short blades at 25 mph. Either 12 or 24 volt systems can be provided: 12 volt max = 1300 watts, 24 volt = 1600 watts.

THE AERO-POWER WIND GENERATOR
2398 Fourth Street
Berkeley, CA 94710

The Aero-Power wind generator consists of an alternator driven by a variable pitch propeller through a gearbox in order to achieve generating speed. At high wind speeds the blades automatically feather, preventing damage to the unit.

BLADES

Number	3
Tip Speed Ratio	8 to 1
Diameter	8' 6"
Material	Aircraft Sitka Spruce
GOVERNOR	Feathering
GEAR RATIO	2.5 to 1
GEARTYPE	Hardened Steel
	Helical Cut

ALTERNATOR

Voltage	14.5 volts
Amperes	75 (Max)
Power	1000 watts (Max)
Poles	6
Phase	3, Rect. to DC at Control Box
Diameter	7"
Weight	90 lbs.

OUTPUT

Wind Speed (mph)	Watts	Amps
6	0	0
7	42	3
10	140	10
15	350	25
20	700	50
25	1050	75

SENCENBAUGH WIND ELECTRIC

Design Features of MODEL 500-14:

- Slow speed, heavy duty 6 pole alternator with brushless stator, designed for a rugged long life.
- Direct drive. Propellers coupled directly to alternator input shaft.
- All current transferred from alternator through heavy duty slip rings with double collector brushes on each high current ring. Commutator enclosed within cast aluminum weatherproof housing with dust seal.
- Castings are of corrosion resistant aluminum. Alternator sealed within weatherproof, water-tight housing. Model 500 series uses same basic castings as the larger 1000 watt series.
- Generator bearings are double sealed against dirt and moisture.
- Available in special Marine version, with anodized castings, and stainless steel hardware.

Design features of the MODEL 1000-14:

- Use of large 12 foot diameter propellers allows plant to develop its rated output in winds of only 22-23 mph.
- Large heavy duty input shaft and bearings (1.375") carry propeller loads for smooth efficient operation.
- Same slow speed alternator design, double brush slipring assembly, and corrosion resistant aluminum castings as in the 500 watt series.
- Efficient, over designed helical gear transmission provides a positive, smooth and quiet transmission. Assures long life and low maintenance.
- Input bearings are double sealed against dirt and moisture. Bearings are prelubricated and sealed.
- Available in Special Marine version with anodized castings and stainless steel hardware.

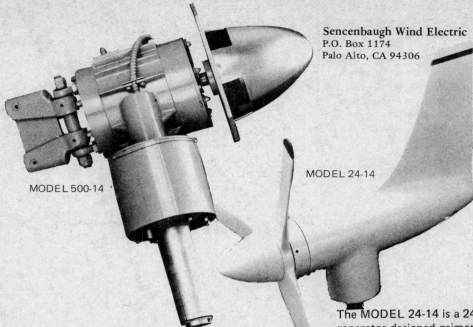

Sencenbaugh Wind Electric
P.O. Box 1174
Palo Alto, CA 94306

MODEL 500-14

MODEL 24-14

The MODEL 500-14 is a 500 watt direct drive wind-driven generator designed for battery charging applications. Its direct drive feature and small propeller lend it to an extremely rugged and low maintenance machine.

All related equipment, including towers, controls and storage systems, are also available, including a line of meterological instruments. Write for more information.

The MODEL 24-14 is a 24 watt wind driven generator designed primarily to trickle charge 12 volt battery storage systems. It's all aluminum and stainless steel construction makes it ideal for use in marine environments. This small windplant finds itself useful on sailboats, sound and light buoys or in any location where batteries need be maintained by trickle charging. Continuous duty output is 24 watts (1.7 amps at 14.4 VDC) in a 21 mph wind. Charging starts at 8-9 mph. This plant produces over 5 times the output as imported machines of the same price.

AERO POWER

AERO POWER
WIND
SYSTEM
(page 141)

PINSON ENERGY CORPORATION

Box 7
Marstons Mills, MA 02648

Pinson is now in production of a 4 KW cyclo-turbine vertical axis machine. They have built and tested several prototypes and their first run is expected to have 20 of these machines available.

The Cycloturbine is a vertical axis, straight-bladed wind turbine with cyclically pitched blades. Originally conceived and built at MIT with a National Science Foundation grant, it is similar in aerodynamic operation to a design patented in 1931 by G.J.M. Darrieus. The Cycloturbine differs initially from this classic "eggbeater" rotor. In that its blades do not remain at a fixed, "flat" angle, but follow a preset "schedule" of angle change (sometimes called "blade rock"), allowing more favorable use of aerodynamic forces on the blades. The amount and timing of pitch change are determined by a cam device mounted atop the main shaft, actuating the blades via pull rods. A tailvane affixed to the cam determines the correct orientation relative to the wind.

SPECIFICATIONS:

Length of blades	8 feet (2.4 M)
Chord of blades	11¾ inches (29.4 cm)
Type of airfoil	NACA 0015 modified
Diameter of rotor	12 feet (3.6 M)
Weight of rotor	111 lbs. (50.5 kg)
Type of overspeed control	Centrifugal
Type of driven device	4 KW Winco alternator
Starting windspeed	5 mph (2.2 M/s)
Governing windspeed	30 mph (13.5 M/s)
Governing RPM	200
Electrical Output (w/4KW Winco alternator)	2 KW @ 24 mph (11 M/s)
	4 KW @ 30 mph (13.5 M/s)

PRICE LIST (1977):
8 x 12' Rotor with aerodynamic controls, cam mechanism and:

parking brake	$3,000
Stub Tower Adapter	200
Power Transmission	200
4 KW Winco Alternator (4KS4PC3)	420
	$3,820

OCTAHEDRON TOWERS

30' Height	$800
40' Height	900
50' Height	1,100

Total System cost is application specific. As stated, the Cycloturbine can accept many driven components. The price breakdown is an example using a 4KW Winco Alternator without the electronic controls needed for many applications.

GRUMMAN ENERGY SYSTEMS,
A Division of Grumman Corporation
4175 Veterans Memorial Highway
Ronkonkoma, NY 11779

Grumman Energy Systems to date has produced 5 of their Windstream 25's, which have been sold to various groups for further wind energy investigation. Our information indicates that in 1977 there will be 12 more of these machines available for sale.

WINDSTREAM 25

(1) 25-ft., bladed all-aluminum variable pitch rotor with centrifugally actuated rotor tip speed brakes; all-aluminum nacelle; 15 kw brushless alternator; three phase rectifier providing 110V or 220VDC nominal;
(2) control panel;
(3) schematics and specifications of electrical power system for operation of the Windstream 25 either as (a) battery charging system or (b) for operation with a Gemini Synchronous Inverter (60 H 110 VAC);
(4) lightning arrester assembly for Grumman tower;
(5) maintenance manual.
$19,900 (1977) (F.O.B. Houston, Texas)

TOWER (recommended)

Centrifugally cast, steam cured prestressed steel reinforced freestanding concrete tower. Mounting points are provided for Servicing Package and/or the Lightning Protection Servicing System (FOB Cape Charles, Virginia)
        Standard — 45 ft.
Notes: (1) Length is overall tower length. Rotor is approximately 3 feet above tower interface resulting in hub-to-ground height of 40 ft.
        (2) Limited warranty in effect only if Grumman-supplied tower is used.
$1,965 (1977)

SERVICING PACKAGE (optional)

The Servicing Package consists of a Servicing Platform and tower steps. The all-aluminum Service Platform is compatible with the mounting points on the Grumman Tower. The platform can be permanently mounted to the tower or can be removed.

Steps compatible with the tower to permit ascent/descent of the tower from the ground to the Service Platform.
$2,750 (1977)

PERFORMANCE:
— Rated Power — 15 KW at windspeeds of 26 mph
— Annual Energy Output — approximately 37,500 KWH in annual mean windspeeds of 15 mph
— Alternator Cut-in Speed — 8 mph

ELECTRIC OUTPUT VERSATILITY:
— Variable voltage, variable frequency AC (110v, 220v, 440v)
— Low ripple, regulated DC (110v, 220v), for direct use of battery charging
— Utility grid tie-in via 15 KW Gemini Synchronous Inverter
— 60 Hz, 110v/220v single or three phase AC via static inverter

DESIGN:
— Electro/Mechanical Pitch Control Mechanism — controls output and feathering
— Down Wind Rotor — passively controls azimuth
— Modified NASA Airfoil — provides high torque at low RPM

CONSTRUCTION:
— Rotor and Nacelle — all metal for durability
— Tower — centrifugally cast concrete for strength and aesthetics

RELIABILITY:
    Brushless, self exciting field alternator

MAINTENANCE:
    Sealed ball bearings reduce lubricating requirement

SAFETY:
— Automatic shutdown at 60 mph
— Redundant aerodynamic speed brakes
— Structurally designed for 130 mph winds.

PHYSICAL DATA:
— Rotor: 25 ft. Diameter, weight 750 lbs.
— Design Rotor RPM: 125
— Nacelle Weight: 1,250 lbs.
— Tower Weight: 9000 lbs.

AMERICAN WIND TURBINE, INC.
1016 East Airport Road
Stillwater, OK 74074

American Wind Turbine, Inc., formerly known as AeroTech, is the first new wind company to go public. They have offered five million shares of common stock, priced at $0.10 per share. Specifications on their SST Wind Turbine (Super Speed Turbine) are as follows:

SST WIND TURBINE SPECIFICATIONS

	15 mph rating	Price (1977)
8-ft. diameter	.375 hp/225 watts	$ 250.00
12-ft. diameter	.75 hp 450 watts	525.00
16-ft. diameter	1.5 hp 900 watts	850.00

WIND-POWERED ELECTRIC DEEP WELL WATER PUMPING SYSTEM (12-ft. and 16-ft. TURBINES)

12-ft. turbine	525.00
belt	48.00
tower	650.00
mounting head and alternator	500.00
submersible pump	250.00
electric control box	125.00
Total	$2,098.00
16-ft. turbine	850.00
belt	64.00
tower	750.00
mounting head and alternator	825.00
submersible pump	475.00
electric control box	150.00
Total (1977)	$3,114.00

Other things offered by American Wind Turbine include towers, wind electric heating systems for houses, barns, grain drying, etc., as well as wind powered grinding systems, air compressors, and motors for shop, and a 2KW 110 Volt system as follows (1977 prices):

wind turbine	$850.00
belt	64.00
tower	750.00
mounting kit	125.00
alternator and controls	625.00
inverter system (2 KW)	850.00
battery bank	515.00
vented battery case	300.00
Total	$4,079.00
optional charger	160.00

WINCO DYNATECH, INC.
P.O. Box 3263
Sioux City, Iowa 51102

(As shown $395.00)

Capacity (Watts)	200
Approximate maximum amps	14
Approximate maximum volts	15
Generator Speed Range (RPM)*	270/900
Governor Type	22" air brake

NORTH WIND POWER COMPANY, INC.

P.O. Box 315
Warren, VT 05674

North Wind, as of 1977, has installed over 50 wind systems. That speaks for itself. North Wind offers reconditioned 32 and 110 Volt Jacobs Wind Electric plants from 1500 to 3000 watts, and also the following products and services:

— Solid state voltage controls and loading switches
— Rebuilding and rewinding service for Jacobs Wind Electric plants
— Aeropower 12 Volt 1000 watt Model "A" wind generator
— Environmental design, structural design and construction services
— Active and passive solar systems for space and hot water heating
— Gemini Synchronous Inverter Systems
— Their own "Tower of Power" and Rohn Towers
— High efficiency wood stoves and furnaces
— Exide, Mule and Surrette heavy duty storage batteries
— Soleq and Carter DC to AC inverters
— Wind data measuring and recording equipment
— Site analysis service for wind, sun and water
— Complete installation services

THE NORTH WIND EAGLES are completely reconditioned and rewound versions of the original Jacobs Wind Electric Plants. Over 50 million dollars worth of these plants were sold by the Jacobs brothers between 1931-1956. The North Wind Eagles are the only American-made medium capacity, direct-drive wind generators available today. There are no efficiency losses or complicated maintenance problems due to expensive gear boxes. North Wind Power Company has taken advantage of the latest aerodynamic, technological and electronic advancements to increase the performance of the original Jacobs plants with the addition of solid state controls, longer-life alloys, improved airfoils, and a custom 4 KW Gemini Synchronous Inverter especially designed for compatibility with the North Wind Eagles. Each North Wind Eagle is completely guaranteed from one to three years. Replacement parts and service are immediately available from your local distributor.

JACOBS-NORTH WIND
NORTH WIND EAGLE MODELS:

	1977 $
Eagle II	
2 KW 32 Volts with flyball governor and props	$2200
2 KW 110 Volts completely rewound with new blade-actuated governor and props	$3000
Eagle II - 110	
2 KW 110 Volts completely rewound with new blade-actuated governor and props	$3500
Eagle III	
3 KW 32 Volts with flyball governor and props	$3200
3 KW 32 Volts with new blade-actuated governor and props	$4000
Eagle III - 110	
3 KW 110 Volts with new blade-actuated governor and props	$4600

WIND ENERGY SYSTEMS

1. In areas where commercial power is available:

Eagle II - 110 Volts	$3500
40' Galvanized Guyed Tower	625
4 KW Gemini Synchronous Inverter	800
	$4925

Rated wind speed = 22 mph
Cut in speed = 8 mph
KW HR. output/month in average winds of:
   10 mph - 150-300 kw hr
   12 mph - 250-400 kw hr
   15 mph - 350-600 kw hr

2. In areas where commercial power is available:

Eagle III - 110 Volts	$4600
40' Galvanized Guyed Tower	625
4 KW Gemini Synchronous Inverter	800
	$6025

Rated wind speed = 24 mph
Cut in speed - 8.5 mph
KW HR. output/month in average winds of:
   10 mph - 200-400 kw hr
   12 mph - 300-600 kw hr
   15 mph - 400-800 kw hr

3. Remote DC System:

Eagle II - 32 Volts	$2200
358 Amp Hr Battery Bank	1025
40' Galvanized Guyed Tower	625
Solid State Controls	250
	$3600

Rated wind speed = 22 mph
Cut in speed = 8 mph
KW HR. output/month in average winds of:
   10 mph - 150-300 kw hr
   12 mph - 250-400 kw hr
   15 mph - 350-600 kw hr

4. Remote AC/DC System:

Eagle III - 32 Volts	$3200
358 Amp Hr Battery Bank	1025
40' Galvanized Guyed Tower	625
Solid State Controls	250
6 KW DC/AC Inverter	1500
	$6600

Rated wind speed = 22 mph
Cut in speed = 8 mph
KW HR. output/month in average winds of:
   10 mph - 200-400 kw hr
   12 mph - 300-600 kw hr
   15 mph - 400-800 kw hr

WINDPOWER SYSTEMS, INC.

P.O. Box 17323
San Diego, CA 92117

The machine you see pictured here was originally modeled as the RD-7000, with an output ideally of 7 kw. A decision was later made to reduce the size of the generator so that the maximum output range would be 4-5 kw. The decision was based on the fact that, with a machine of this size (300 square feet of swept rotor area), using a generator of this capacity will decrease the cost and result in very nearly the same energy output over a given period of time at most sites. This is due to the generally infrequent occurrence of winds in excess of 20 mph. Hence the RD-7000 has become the RD-4000. *Besides the rotor configuration and taking energy off the rim of the rotor*, it has a unique and reliable overspeed control referred to as the "tilt mode." That is, the system protects the rotor and supporting structure from overload imposed by wind above 25 mph while maintaining power output. This is done by allowing a turbine unit and generator to tilt backwards. For example, in a steady 35 mph wind, the rotors would be tilted back 45 degrees, and both the horizontal thrust on the rotors and the rotor speed would be nearly the same as at 25 mph and full output.

Windpower Systems is now directing most of its efforts to navigational aids and remote power situations as well as in a project in Santa Nella, California. This project is at a new, $3 million "Anderson Pea Soup" motel-restaurant complex. A seven-story windmill of the "Dutch" type, four-bladed, has been erected there. The diameter of the blades is 46 feet, producing a maximum of 10 kw in a 20 mph wind. Its step-up ratio is 50:1. It is then coupled to the generator, which is in turn connected to a Gemini Synchronous Inverter. Safety was an important consideration in the system design; it has several fail-safe mechanisms and was designed to withstand windspeeds of up to 140 mph with the blades stopped.

FOREIGN MANUFACTURERS

INDUSTRIAL INST. LTD.

Stanley Rd., Bromley BR2 9JF, Kent, England
Mr. Kirylok, .4 KW to 4 KW; 1 KW machine available.
Main business is inverters.

USSR INSTITUTE FOR FARM ELEC-TRIFICATION I-U VESHNIAKOVSKI

Prc. Dom Moscow J-456: Numerous machines from
½ to 25 KW.

DUNLITE ELECTRICAL CO.

Division of Pye Ind.
21 Frome St., Adelaide 5000
Australia

Three-blade wind-driven power plant,
with automatic, variable-pitch propeller.

WIND DRIVEN GENERATORS

DUNLITE Model "L," 1000 Watt, 32 or 36 Volt
sliping Windplant, metal 3-bladed full-feathering
propeller, Diotran voltage regulator and control
panel. Ideal for charging electric battery powered
vehicles!

DUNLITE Model "M", 2000 watt, Brushless wind-plant, available in 24, 32, 48, and 115 volts.   Dio-tran voltage regulator and control panel. Complete
system $6,350.00.

AEROWATT S.A.

37, Rue Chanzy, 75-Paris 11e
France

Aerowatt's machines are very expensive but
are exceptional in design. The generator is a
3 phase A.C. permanent magnet type which
starts charging in 6 mph winds and reaches
full output at 15 mph.

TABLE I AEROWATT WIND-BLOWN GENERATORS MAIN CHARACTERISTICS minus storage costs

	$1935. 24 FP 7	$3105. 150FRP7	$4565. 300FP7	$8375. 1100FP7	$18860 4100FP7
Wind speed at start of charge m/s (Miles/Hr.)	4	(6.7)3	3	3	4
Nominal wind speed m/s (Miles/Hr.)	7	(15.7)7	7	7	7.5
Nominal power watts	28	130	350	1125	4100
Nominal output voltage-volts	24	24	110	110 220/380	110 220/380

ELEKTRO G. m. b. H.

St. Gallerstrasse 27
Winterthur, Schweiz

Brushless 3 phase alternator-type wind generator with rectified D.C.
manufactured by ELEKTRO G.m.b.H. of Winterthur, Switzerland.

Model	Tropic Encased Unit	Rated Output In Watts	Rated Windspeed MPH	Voltages Available
W 50	Yes	50	39	6, 12, 24
W250	Yes	250	40	12, 24, 36
WV05	Yes	750	20	(12) 24, 36, 48, 65
WV15G	*	1200	23	(12), 24, 48, 115
WV25G	*	1800	22	(24), 36, 48, 65, 115
WV25/3G	*	2500	23	(24), 36, 48, 65, 115
WV35G	Yes	4000	24	48, 65, 115
WVG50G		6000	26	65, 115

*Tropic encased model available at extra cost.

Model	Propeller Diam. Ft.	No. Blades	Relative Cost of System(DC)
WV15G	9ft. 10in.	2	$5800
WV25G	11ft.6in.	2	5800
WV25/3G	12ft.6in.	3	6500
WV35G	14ft.5in.	3	8400
WVG50G	16ft.5in.	3	9500

$1500 for inverter (DC/AC)

DOMENICO SPERANDIO & AGER

Via Cimarosa 13-21
58022 Folloncia (GR) Italy

GARBINO 12V	250W
TURBINE 24V	500W
MONSONE 48V	1000W

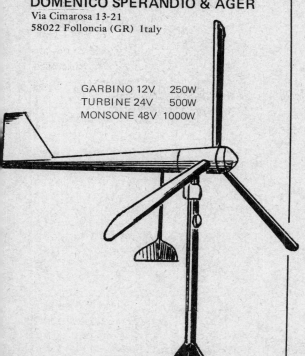

'LUBING'

Maschinenfabrik, Ludwig Bening
2847 Barnsdorf, P.O. Box 171, Germany

Lubing is a geared 1 to
5.5 step up alternator,
brushless type, 3 phase
24 volt D.C. or 12 volts
on request, very beauti-fully packaged system.

ENAG s.a.

Rue de Pont-l'Abbe
Quimper (Finistere)
France

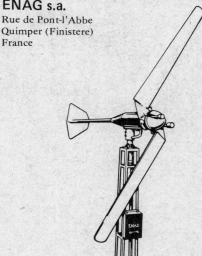

Two-Bladed Eolienne 24/30 Volts - 400 Watts $1,587

Three-Bladed Super Enag Eolienne 24/30 Volts - 1200 W
Complete with propeller and instrument panel $2,287

Three-Bladed Super Enag Eolienne 110 Volts - 2500 W
$4,164

DISTRIBUTORS of wind energy equipment are companies that sell and service the manufactured and/or rebuilt products of domestic and/or foreign wind energy conversion systems.

The following companies are listed with their addresses followed by their basic product lines. We suggest that, for further information, you contact these companies regarding current prices and other specifications. Some companies are listed who are also manufacturers or re-manufacturers. Most of these have been reviewed in the preceeding section.

Aeropower
2398 Fourth Street
Berkeley, CA 94710
Dempster

Boston Wind Power Co.
2 Mason Court
Charles Town, MA 92129
reconditioned *Jacobs*

Energy Alternatives
P.O. Box 223
Leverett, MA 01054
Dunlite/Quirks, Wincharger, Elektro, Jacobs, Lubing

Sencenbaugh Electric
Box 11174
Palo Alto, CA 94306

Steve Blake
Route 1, Box 93A
Oskaloosa, Kansas 66066
reconditioned *Jacobs*

SunWind Ltd.
P.O. Box 880
Sebastopol, CA 95472
Wind powered car

North Wind Power Company, Inc.
Box 315
Warren, VT 05674
Jacobs, Aeropower

Penwalt Automatic Power
213 Hutcheson Street
Houston, TX 77003
Aerowatt, Hutter Machine

Real Gas & Electric Company
Box A
Guerneville, CA 95446
Elektro, Dunlite/Quirks, Wincharger

Windlite Alaska
Box 43
Anchorage, AK 99510
Aerowatt, Sencenbaugh

Budgen & Associates
72 Broadview Avenue
Point Claire 710
Quebec, Canada
Elektro, Lubing

Merritt Windmill Inc.
P.O. Box 1374
B.C. Canada VOK 2BO

ADDITIONAL DISTRIBUTORS

Apollo Company
P.O. Box 5609
Inglewood, CA 90393
(213) 672-3312

Independent Energy Systems
6043 Sterrettania Road
Fairview, PA 16415
(814) 868-6211

Independent Energy Company
75 Minuteman Drive
Concord, MA 01742

KMP Parish Windmills
Box 441
Earth, Texas 79031
(806) 257-3411

Life Size Aero Design
342 Franklin Street
Alburtis, PA 18011

Prarie Sun & Wind Co.
4408 62nd Street
Lubbock, Texas 79409

Sigma Engineering
Box 5285
Lubbock, Texas 79417
(806) 762-5690

Windependence Electric
P.O. Box M1188
Ann Arbor, Michigan 48106
(313) 769-8469

ENERTECH CORPORATION
P.O. Box 420
Norwich, VT

Enertech specializes in the design and distribution of wind-powered generating systems and wind measuring equipment. Their standard systems are designed around some of the world's most reliable and popular windplants— Elektro, Dunlite and Wincharger. Components include Rohn and Dunlite towers; Exide, Mule and Surrette industrial batteries; Nova, Soleq and Topaz inverters; Onan and Winco generators. Enertech has prepared their own manual entitled "Planning a Wind Energy Conversion System," written specially for homeowners, architects, civil engineers and students interested in the basic principles and detailed workings of any wind energy conversion system which is available through the company.

REDE CORPORATION
P.O. Box 212
Providence, RI 02901

REDE Corporation will be handling the limited production of the Dominion Aluminum Fabricating Ltd's (DAF Darrieus Type wind generator).

DAF notes that the 15-foot wind turbine is available as follows:
  4000 watt/115 Volt
  3500 watt/24 Volt
The 20-foot diameter turbine is available as:
  8000 watt/115 Volt
  The 20-foot model achieves 8000 watts in an 18 mph wind.
  Estimated costs:
15-foot	$4500.00
20-foot	$6700.00

MISCELLANEOUS HARDWARE

ANEMOMETERS

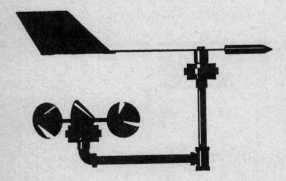

Reviewed below are anemometers from three different companies: Inexpensive (Dwyer), moderate (Taylor), and finest available (Bendix). The remaining companies are listed as additional sources of wind speed measuring devices.

Dwyer Instruments, Inc.
P.O. Box 373
Michigan City, Indiana 46360

Dwyer makes 2 interesting weather instruments. One is a hand held wind meter which you can align with the wind and read the speed in either of 2 scales, 2-10 MPH or 4-66 MPH. The other can be mounted permanently with the indicator placed indoors. Its cost is $24.95.

Taylor Instruments
Arden, North Carolina 28704

Full range of weather measuring instruments. Least expensive anemometer $85.00 with 60' of lead in wire, also rain gauges. The instruments are designed with the non-professional in mind and are attractive.

Bendix Environmental Science Division
1400 Taylor Avenue
Baltimore, Maryland 21204

The Bendix Environmental Science Division is an advanced development, engineering and manufacturing organization providing components, products and systems. The products are high quality, competitively priced and reliable. Their product line is constantly being expanded and new catalog sheets are provided to everyone on their mailing list.

The Bendix Aerovane® Wind Data System is unexcelled in performance, reliability, and service life. You want it they've got it. Its cost is $1220.00.

Davis Instrument Mfg. Co., Inc.
513 East 36th Street
Baltimore, Maryland 21218

Aircraft Components
North Shore Drive
Benton Harbor, Michigan 49022

Danforth
Division of the Eastern Co.
Portland, Maine 04103

Meteorology Research Inc.
Box 637
Altadena, California 91001

Robert E. White Instruments, Inc.
33 Commercial Wharf
Boston, Massachusetts 02110

Texas Electronics Inc.
5529 Redfield Street
P.O. Box 7151 Inwood Station
Dallas, Texas 75209

Maximum, Inc.
8 Sterling Drive
Dover, Massachusetts 02030

Climet Instruments Co.
1620 West Colton Avenue
P.O. Box 1165
Redlands, California 92373

Kahl Scientific Instrument Corporation
Box 1166
El Cajon, San Diego, California 92022

ANEMOMETERS (con't)

North Wind Power Co., Inc.
Box 315
Warren, VT 05674

North Wind Power Co. sells the Stewart 1/60th Mile Anemometer/Odometer. The device consists of a simple, single contact electrical switch mechanism that provides an on-off switch operation for each 1/60th mile of wind, or 60 contacts per mile of wind. The velocity of wind passing the instrument is indicated by the number of switch operations (1/60th of a mile) per minute of time (1/60th of an hour) directly in miles per hour. The switch operates on low voltage, low current AC or DC circuit using two wires. The anemometer is guaranteed to operate 500,000 miles or 5 years average service and to withstand a maximum wind velocity of 100 mph.

The Electronic Odometer is a 6 digit non-reset electromagnetic counter controlled by a sensitive electric circuit designed for remote wire connection to a contacting anemometer. The odometer serves as a totalizer and also as an audible signal of anemometer switch operation. The wind mileage passing the anemometer is registered continuously on the counter. The total mileage and the average wind speed are readily computed for any desired time period. Complete cost is $87.50.

Sign X Laboratories, Inc.
Essex, CT 06426

Analyzer Model 700 Series, $695. Utilizing low power-consuming, integrated circuits and transistors and non-volatile storage resistors, the wind spectrum analyzer records wind speed distribution. Using any contact anemometer, the wind speed is averaged for a specific period of time.

BLADES

Aeropower
2398 Fourth Street
Berkeley, CA 94710

Senich Corp.
Box 1168
Lancaster, PA 17609

North Wind Power Company, Inc.
Box 315
Warren, VT 05674

TOWERS

Advance Industries
2301 Bridgeport Drive
Sioux City, Iowa 51102

Eldon Arms
Box 7
Woodman, Wisconsin 53827

North Wind Power
Box 315
Warren, VT 05674

Has towers available in three types from 20 to 100 feet.

Additional information on two unique towers is available. The Octahedron tower from Windworks made of 1 1/8" electrical conduit and a concrete or concrete block tower design from Brace Research Institute (see pg. 135).

THOMAS REGISTER 1974 has five pages of tower manufacturers. Check your local library.

DC—MOTORS

The best source for fractional to five horsepower direct current motors. Five horsepower to five hundred horsepower available on order.
— J.R.B.

Reliance Electric Co.
24701 Euclid Ave.
Cleveland, Ohio 44117

GENERATORS/ALTERNATORS

Kato
3201 3rd Avenue N.
Menkato, Minnesota 56001

Onan
1400 73rd Avenue NE
Minneapolis, Minnesota 55432

Winco of Dyna Tech.
2201 E. 7th Street
Sioux City, Iowa 31102

Kohler
421 High Street
Kohler, Wisconsin 53044

Howelite
Rendale and Nelson Streets
Port Chester, New York 10573

McCulloch
989 S. Brooklyn Avenue
Wellsville, New York 14895

Sears and Roebuck

Winpower
1225 1st Avenue East
Newton, Iowa 50208

Ideal Electric
615 1st Street
Mansfield, Ohio 44903

Empire Electric Company
5200-02 First Avenue
Brooklyn, New York 11232

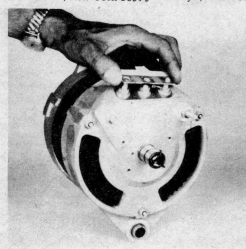

Leece-Neville Co.
Cleveland, OH 44103

Leece-Neville has alternators with slow speed characteristics that allow lower rotational speed of the alternator without sacrificing any amperage output at idle or top speed. They are self-current limiting units with a fully adjustable solid state integral regulator. These are excellent for wind systems of small wattage. Full information is available from the company, including maintenance instructions.

BATTERIES

Bright Star
602 Getty Avenue
Clifton, NJ 07015

Burgess Div. of Clevite Corp., Gould
P.O. Box 3140
St. Paul, MN 55101

Delco-Remy
Division of GM
P.O. Box 2439
Anderson, IN 46011

Eggle-Pichen Industries
Box 47
Joplin, Missouri 64801

ESB Inc.
Willard
Box 6949
Cleveland, OH 44101

Exide, 5 Penn Ctr., Plz.
Phila. PA 19103

Ever Ready
Union Carbide Corp.
270 Park Avenue
New York, NY 10017

Globe-Union
5757 N. Greenbay Ave.
Milwaukee, WI 53201

Gulton
212 T Dorham Avenue
Metuchen, NJ 08840

Keystone Battery Company
16 Hamilton Street
Saugus, Massachusetts 01906

Marathon Battery Company
8301 Imperial Drive
Waco, Texas 76710

RCA
415 S. 5th Street
Harrison, NJ 07029

Surrette Storage Battery Co., Inc.
Box 711
Salem, Massachusetts 01970

Batteries Mfg. Co.
14694 Dequindu
Detroit, Michigan 48212

C & D Batteries Eltuce Corp.
Washington & Chewy Street
Conshohocken, PA 19428

Gould Inc.
485 Calhoun Street
Trenton, NJ 08618

Mule Battery Company
325-T Valley Street
Providence R.I. 02908

Delatron Systems Corporation
20370 Rand Road
Palatine, Illinois 60067

STATIC INVERTERS

Emhiser Rand Industries
7721 Convoy Court
San Diego, CA 92111

300 and 500 watt transistorized square wave inverters with 6, 12, 24 and 36 or 48 volt inputs. Efficiency up to 80%, frequency regulation +1%.

Price $1.50 to $1.25/watt

Load demand at no extra cost

Dynamote Corporation
1130 Northwest 85th
Seattle, Washington 98117
(206) 784-1900

Manufacture SCR static inverters all with demand load provision. Primarily for RV/mobile communications market. Voltage input ranges of 12, 24, 32 and 36 volts with outputs of 120 Va to 1000 Va respectively.

Topaz Electronics
3855 Ruffin Road
San Diego, CA 92123
(714) 279-0831

Topaz Electronics is known throughout the computer industry for their work with uninterruptible power systems. They manufacture a complete line of sine-wave (Quasi-Square) inverters in ranges from 250 to 10,000 watt output. SCR commutating output with ferro-resonant filter and capacitor shaping network.

Frequency 50 - 60 hz +1% of fixed frequency
75% efficiency typical at full load
Net price $3 to $1.37/watt depending on size of inverter.

Delivery time: in stock to 6 weeks

Demand load at extra cost.

Gulton
13041 Genise Avenue
Hawthorne, CA 90250

Lorain
1122 F. Street
Lorain, Ohio 44052

Creative Electronics
3707 W. Touhy Avenue
Chicago, Illinois 60645

Interelectronics
100 U.S. Highway 303
Congers, N.Y. 10920

Globe
5757 N. Green Bay Avenue
Milwaukee, Wisconsin 53201

Wilmore
Box 2973
W. Dunham Station
Dunham, North Carolina 27705

Basku
603-5th
Highland, Illinois 60645

Systron-Donner Corp.
889 Galindo
Concord, CA 94518

GEMINI SYNCHRONOUS INVERTER

Windworks
Box 329
Route 3
Mukwonago, WI 53149

Synchronous Inversion is a process whereby intermittent and/or variable power sources can be used to supplement a primary source. Basically, this device is used in conjunction with a wind machine to feed power into an existing grid as a means of storing, the concept being that, while the wind power is available, it may either be used or sent back into the utility grid, ideally running an electric meter in reverse.

This is a very meaningful development in that it is the only option for small wind systems, other than battery storage, that appears to be cost effective.
8 kW capacity, $1,450.
4 kW capacity, $780.

ROTARY INVERTERS

Electro Sales
100 Fellsway West
Somerville, Massachusetts 02149

Ganter Motor Co.
2750 W. George Street
Chicago, Illinois 60618

Surplus Stores

Northwestern Electric
1752 N. Springfield Avenue
Chicago, Illinois 60647

WIND WATER PUMPERS

The American multi-vane fan type "farm windmill" has spread throughout the world. Mechanical wind power for pumping water has been the key to cattle production in the U.S. and the interior of Australia. The U.S. railroads were also dependent on water pumping windmills for water for steam locomotives. There are hundreds of thousands being used today, many of them have been running for over forty years. With minor maintenance and lubrication they continue to be put into this kind of service.

A complete wind water pumping system consists of the blade and rotor assembly; a gearbox to convert the rotary motion of the wind into the up and down motion of the pump; the pump itself; and a tower. The following companies are presently manufacturing such systems. They are sold both by the manufacturer and by local outlets.

AERMOTOR

Aermotor Water Systems
Broken Arrow, Oklahoma 74012

BAKER

Heller-Aller Co.
Perry & Oakwood
Napoleon, Ohio 43545

distributed by:
O'Brock Windmill Sales
Rt. 1 - 12st St.
North Benton, Ohio 44449

DEMPSTER

Dempster Industries, Inc.
P.O. Box 848
Beatrice, Nebraska 68310

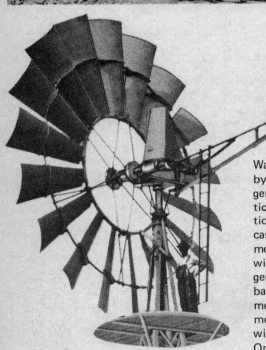

Water pumping can also be accomplished by an electric pump powered by a wind generator. Such a system would be particularly useful for a deep-well installation. It should also be noted that in most cases it is more efficient, although far more expensive, to pump water with a wind generator than a windmill. A wind generator pumping water without batteries will go through two conversions: mechanical to electrical and electrical to mechanical (pump). The entire process will be about $.40 \times .90 = 36\%$ efficient. On the other hand a windmill pumping water, which goes through one less conversion than a wind generator, is only about 30% efficient.

Diameter of Cylinder (Inches)	(15 MPH) Capacity per Hour, Gallons		Total Elevation in Feet SIZE OF AERMOTOR					
	6 Ft	8-16 Ft	6 Ft	8 Ft	10 Ft	12 Ft	14 Ft	16 Ft
1¾	105	150	130	185	280	420	600	1,000
1⅞	125	180	120	175	260	390	560	920
2	130	190	95	140	215	320	460	750
2¼	180	260	77	112	170	250	360	590
2½	225	325	65	94	140	210	300	490
2¾	265	385	56	80	120	180	260	425
3	320	470	47	68	100	155	220	360
3¼		550			88	130	185	305
3½	440	640	35	50	76	115	160	265
3¾		730			65	98	143	230
4	570	830	27	39	58	86	125	200
4¼		940			51	76	110	180
4½	725	1,050	21	30	46	68	98	160
4¾		1,170				61	88	140
5	900	1,300	17	25	37	55	80	130
5¾		1,700				40	60	100
6		1,875		17	25	38	55	85
7		2,550			19	28	41	65
8		3,300			14	22	31	50

AERMOTOR PUMPING CAPACITY

WADLER MANUFACTURING COMPANY
Galena, Kansas

The Wadler Pond Agitator is a modified Savonius wind turbine designed to convert power of the wind into mechanical motion for pond agitation. This simple device will inhibit iceover of ponds and lagoons in colder climates, providing open water for wildlife and farm animals.

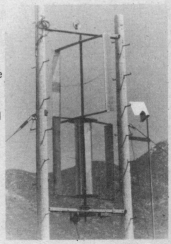

WIND RESEARCH GROUPS

FEDERAL GOVERNMENT

The Government's ability to clearly define the varying scales of wind energy applications from small to large has developed a successful network of communications among all levels. Hence, small companies, intermediate companies, and large corporations are sharing a growing amount of information and experience. The projects listed below have been pulled from the January 1, 1977, Summary Report, Federal Wind Energy Program. It is felt that these groups and organizations have the best vision of the specific wind project or problem described. However, this is not to be construed as the only meaningful research being funded by the Government. It is our suggestion that you contact either the Wind Energy Society of America or the American Wind Energy Association regarding other concepts and projects that may not appear here (see page 151).

Three new projects on advanced concepts were funded after evaluation of over 50 proposals received in response to a competition. The concepts include the Madaras (spinning cylinder) rotor, the Grumman tornado-type vortex augmenter, and a concept which seeks to tap the energy in humid air.

— J.R.B.

INVESTIGATION OF THE FEASIBILITY OF USING WINDPOWER FOR SPACE HEATING IN COLDER CLIMATES
William E. Heronemus, Principal Investigator
University of Massachusetts (Amherst)
Civil Engineering Department
Amherst, MA 01002

Grant No. E(49-18)-2365
Amount: $149,975.
Effective Date: May 1, 1976
Duration: 12 months

OUTPUT:
The project will provide an initial verification of the cost and practicality of heating homes by wind power.

BACKGROUND:
The use of wind power for space heating may be attractive since it minimizes the energy storage and interface problems associated with wind power in other applications. This project continues an effort to investigate the economic feasibility of heating buildings in cold, windy, cloudy portions of the country using wind energy. The solar habitat system at the University of Massachusetts has been equipped with a 32-foot wind turbine which generates power for direct resistance heating.

OBJECTIVES AND APPROACH:
Several alternate configurations will be studied and compared, including: (1) thermal water heat storage; (2) combined wind and flat-plate solar; and (3) direct conversion from wind to heat energy by mechanical means. The structural mechanics, aerodynamics and mechanical design of wind systems will be studied to improve the systems.

VERTICAL-AXIS WIND TURBINE RESEARCH PROGRAM
Richard H. Braasch, Project Manager
Sandia Laboratories
Aerodynamics Projects Department
Albuquerque, New Mexico 87115

Contract No. AT (29-1)-789
Amount: $1,500,000
Effective Date: October 1, 1976
Duration: 12 months

OUTPUT:
Engineering and economic data for the Darrieus concept will be gathered to allow assessment of its potential as an alternative to conventional (horizontal-axis, propeller-type) systems.

BACKGROUND:
The Darrieus vertical-axis (eggbeater-type) wind turbine is one of the most promising advanced systems in terms of its potential for early commercialization and increased energy output per unit cost. Testing of a 5-meter Darrieus prototype was carried out during FY 1976. In performing its research, Sandia is utilizing analytic tools which have been developed and refined during prior years of the contract.

OBJECTIVES AND APPROACH:
The effort to develop the Darrieus wind turbine is organized under three major task areas: (1) collection of performance data by continued operation of the prototype 5-meter Darrieus system in phase with the local utility; (2) fabrication and testing of a 17-meter device to obtain dynamic, performance and cost data; and (3) performance of system trade-off studies to allow performance and cost optimization of the Darrieus design.

MOD-1 1.5 MEGAWATT WIND SYSTEM
William H. Robbins, Project Manager
NASA-Lewis Research Center
Cleveland, OH 44135

Interagency Agreement No. E(49-26)-1010
Amount: $4,300,000
Effective Data: October 1, 1976
Duration: 12 months

OUTPUT:
Two MOD-1 experimental wind turbine generators will be designed, fabricated and installed at two of the 17 "candidate" utility sites during FY 1977 and 1978. Field-testing of the turbines will provide engineering and performance data for use in refining the design features of future systems and will contribute valuable information to wind energy applications and multi-unit systems studies. This system will form a basis for addressing the technical, economic and operational questions associated with wind power for utility use.

BACKGROUND:
General Electric has been selected as the primary contractor to develop a 1.5 megawatt horizontal-axis, propeller-type experimental wind turbine generator with a composite rotor 200 feet in diameter. The system will be optimized for 18 mph average wind sites.

OBJECTIVES AND APPROACH:
(1) Determine the economic characteristics of a utility-operated megawatt-scale wind turbine; (2) Involve industry in the design, fabrication and installation of large wind systems; and (3) Involve potential users of wind systems (such as utilities) so that institutional, operational and technical interface requirements can be clearly defined.

ROCKY FLATS TEST CENTER, WIND ENERGY SYSTEMS FOR HOME AND RURAL USE

ROCKWELL INTERNATIONAL
Rocky Flats Plant
P.O. Box 464
Golden, CO 80401

The Wind Energy Conversion Branch of the Energy Research and Development Administration's Division of Solar Energy has selected the Atomics International Division of Rockwell International to provide support at ERDA's Rocky Flats Plant, Colorado, in the technical management of small wind energy systems development. This effort is part of a large national program, the purpose of which is to develop wind energy to the point where it is a viable, supplemental source of energy for the country. The objective of the program involving small systems (i.e. systems having an output less than about 100 kilowatts) is to advance their commercialization and encourage their widespread use, particularly in the agricultural industries.

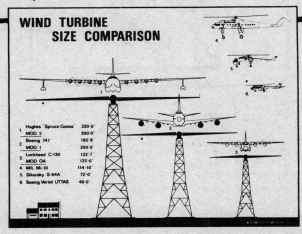

WIND TURBINE SIZE COMPARISON

Hughes "Spruce Goose"	320·0"
MOD 2	300·0"
Boeing 747	195·8"
MOD 1	200·0"
Lockheed C-130	132·7"
MOD OA	125·0"
MIL Mi-10	114·10"
Sikorsky S-64A	72·0"
Boeing Vertol UTTAS	49·0"

from: *Federal Wind Energy Program Summary Report*
(see page 133)

As a part of this objective, ERDA's Rocky Flats Test Center (RFTC) has been established to examine the performance of small wind energy systems. Twenty or more machines of varying sizes and types may be tested concurrently at the RFTC. Performance and reliability data of interest to the public as well as the manufacturers will be gathered and recorded continuously over a period of months or even years, if needed. Engineering data, of interest to manufacturers, will be provided by an intensive, high speed data collection system. It is hoped that the free availability of such test information will encourage the manufacturers to find ways to improve their product while keeping costs competitive. In turn, it is the intent to stimulate public interest in wind energy by publishing reports on the various wind energy systems and by encouraging visits to the test center by interested parties.

The RFTC, now under construction, should be available to the public by early spring 1977. It is expected that eight or ten different wind turbine generators will be operating at the site by that time. Machines with electrical outputs from 1-15 kW will be under test during the initial period. Machines with intermediate output of about 30-40kW are expected at the site later in 1977. Individuals or groups interested in a conducted tour of the RFTC should contact Walt Nelson of Rockwell International, Rocky Flats Plant, telephone (303) 497-2986.

UNIVERSITIES (partial list)

University of Massachusetts
Amherst, MA 01002

The projects at U of M are headed by William E. Heronemus whose main concept is off-shore wind power systems, the idea of anchoring towers off the New England coast, tapping the wind energy, then electrolizing seawater to produce hydrogen.

Cambridge University

A list of all manufacturers of windmills for both water pumping and electricity production by Gerry E. Smith $1. From University of Cambridge, Department of Architecture, Technical Research Division, 1 Scroop Terrace, Cambridge CB2 1PX, England (see also Architecture Section of ENERGY PRIMER).

Brown University
Prospect Street
Providence, Rhode Island 02912

One of the many groups doing research on the vertical axis darrieus hoop.

University of Hamburg
School of Naval Architecture
University of Hamburg
Hamburg, W. Germany

The energy crisis has brought up very much interest in wind-propulsion for merchant ships. During the last years there was done a lot of research work in the Institute concerning wind forces on ships and especially the properties of a new sailing ship concept based on modern aerodynamics and structural technology. Mr. Wilhelm Prolss is the inventor of this new propulsion system called "Dynaship."

Oklahoma State University

Since 1961, an interdisciplinary team of Engineers at Oklahoma State University has been developing a family of energy systems which could operate from intermittent sources of energy, such as solar radiation and wind energy, and utilize a mechanism of energy storage to insure that the input and output energy rates can be made relatively independent of one another. The research effort anticipated the ultimate end of inexpensive and domestically available hydrocarbon fuels, and sought to develop power systems which utilize sources of energy that are essentially non-polluting, non-depleting, and widely distributed over the surface of the earth.

Energy for the Future—The Oklahoma State University Effort
1974; 180 pp.

$5.00

A book from Engineering Energy Lab,
College of Engineering,
Oklahoma State University
Stillwater, Oklahoma 74074

Brace Research Institute
MacDonald Campus, McGill University
Ste. Anne de Bellevue 800
Quebec, CANADA HOA 1CO

Brace Research Institute investigates various solar/wind and fuel technologies, especially with reference to application in Third World countries.
One of their publications, "How to Construct a Cheap Wind Machine for Pumping Water" is a short 12-page booklet giving details on how to build a small water pumping windmill from mostly wooden materials. It seems especially valuable for use in arid areas for increasing food productivity.

How to Construct a Cheap Wind Machine for Pumping Water
A. Bodek
1973; revised 1975
Order No. L5

$1.50   from:
Brace Research Institute
MacDonald Campus, McGill University
Ste. Anne de Bellevue 800
Quebec, Canada HOA 1CO

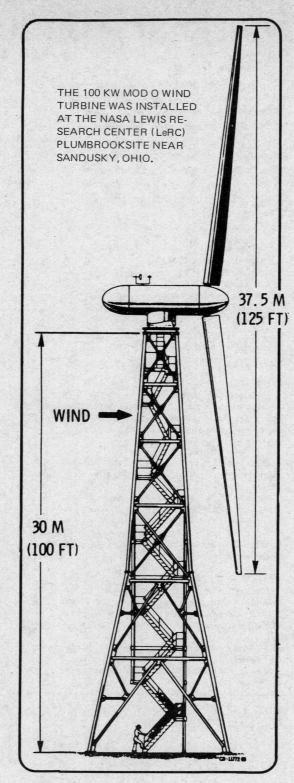

THE 100 KW MOD O WIND TURBINE WAS INSTALLED AT THE NASA LEWIS RESEARCH CENTER (LeRC) PLUMBROOKSITE NEAR SANDUSKY, OHIO.

37.5 M (125 FT)

WIND →

30 M (100 FT)

Montana Project
Dr. Ralph Powe
Montana State University,   Bozeman, Montana

Montana State University conceived a sail-powered car running on a 5 mile circular railroad-like track, driving generators to feed the system or to connect directly to high lines.

ELECTRICAL RESEARCH ASSOC.
Cleeve Road
Leatherhead Surrey KT22 7SA
England

Electrical Research Association has been conducting tests and publishing reports from 1933, the bulk of which were published from 1949 to 1960. Of the reports available and listed below a few were found to be highly valuable. From just a historical standpoint Report No. IB/T22 is excellent reading, originally published in *Engineering*, May 30, 1958. Another report found very useful was ERA Report No. C/T122, The Automatic Operation of a Medium-Sized Wind-Driven Generator Running in Isolation. This is a report on the Allgaier 7.5 KW Wind driven generator, which covers the installation, auxiliary supplies, operation and performance of this unit. Another useful report is C/T108, The Selection and Characteristics of Wind-Power Sites. In general, however, the amount of information derived from these publications is sparse for the relative cost but you are not going to find this published anywhere else. These publications and many others are available from the Publications Sales Department.

1. C6T101	LARGE-SCALE GENERATION OF ELECTRICITY BY WIND POWER—PRELIMINARY REPORT. E.W. Golding. 1949. 15 pp.	6.00
2. IB/T4	THE AERODYNAMICS OF WINDMILLS USED FOR THE GENERATION OF ELECTRICITY. L.H.G. Sterne and G.C. Rose. 1951. 12 pp.	6.00
3. C/T104	WIND- AND GUST-MEASURING INSTRUMENTS DEVELOPED FOR A WIND-POWER SURVEY. H.H. Rosenbrock and J.R. Tagg. 1951. 10 pp.	3.00
4. G/T105	AN EXTENSION OF THE MOMENTUM THEORY OF WIND TURBINES. H.H. Rosenbrock. 1951. 10 pp. 2 illustration sheets.	6.00
5. C/T106	THE DESIGN AND DEVELOPMENT OF THREE NEW TYPES OF GUST ANEMOMETER. H.H. Rosenbrock. 1951. 37 pp.	12.00
6. C/T108	THE SELECTION AND CHARACTERISTICS OF WIND-POWER SITES. E.W. Golding and A.H. Stodhart. 1952. 32 pp.	12.00
7. C/T110	THE UTILIZATION OF WIND POWER IN DESERT AREAS. E.W. Golding. 1953. 11 pp.	6.00
8. C/T111	THE ECONOMIC VALUE OF HYDROGEN PRODUCED BY WIND POWER. A.H. Stodhart. 1954. 8 pp.	3.00
9. C/T112	THE USE OF WIND POWER IN DENMARK. E.W. Golding and A.H. Stodhart. 1954. 16 pp. 1 illustration sheet.	6.00
10. C/T113	VIBRATION AND STABILITY PROBLEMS IN LARGE WIND TURBINES HAVING HINGED BLADES. H.H. Rosenbrock. 1955. 53 pp.	15.00
11. C/T114	AN EXPERIMENTAL STUDY OF WIND STRUCTURE (WITH REFERENCE TO THE DESIGN AND OPERATION OF WIND-DRIVEN GENERATORS). M.P. Wax. 1956. 24 pp.	9.00

Prices quoted are for members. For non-members, the price is double.

AMERICAN WIND ENERGY ASSOC.
54468 CR 31
Bristol, Indiana 46507

The primary purpose of the American Wind Energy Association is to promote the use of wind as a renewable energy source. A broad membership base includes *designers, distributors, manufacturers, wind machine owners,* and *interested individuals.* The association provides regular publications in the form of a newsletter and a technical journal; two major conferences each year with guest speakers, workshops, and displays; special reports and consulting services.

A one year membership in the Association is $25. Members receive the AWEA newsletter, a one year subscription to the Wind Power Digest and reduced rates for Association conferences and publications. Corporate memberships are $100.00 Because the AWEA is a non-profit organization, memberships are tax deductible.

BOSTON WIND, INC.
574 Boston Avenue
Medford, MA 02155

A non-profit organization, this is the first alternative energy center in the Boston area. They offer wind and solar workshops, disseminate information and conduct research. Their objectives are to provide an Alternative Energy Educational Center at which individuals may be trained to install, design, or build practical energy systems.

MAX'S POT
Maximum Potential Building Systems
6438 Bee Cave Rd.
Austin, Texas 78746

The following photos and drawings are from the work done by this group. Max's Pot is active in many areas of alternate energy. See page 204 and 246.

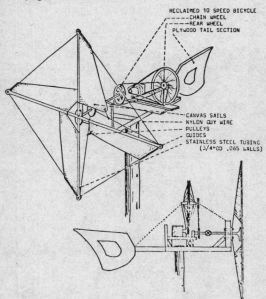

RECLAIMED 10 SPEED BICYCLE
CHAIN WHEEL
REAR WHEEL
PLYWOOD TAIL SECTION

CANVAS SAILS
NYLON GUY WIRE
PULLEYS
GUIDES
STAINLESS STEEL TUBING
(3/4"OD .065 WALLS)

WIND ENERGY SOCIETY OF AMERICA AMERICA
1700 East Walnut
Pasadena, CA 91106

A non-profit corporation concerned with information dissemination on wind energy, directed toward the academic community.

NEW ALCHEMY INSTITUTE
Box 432
Woods Hole, MA 02543

Since 1971 the New Alchemy Institute has been engaged in research on renewable food and energy systems (see pgs. 240, 244). Their wind energy research has focused on two projects: a sailwing water-pumper and a unique hydraulic wind generator.

Hydrowind Wind Generator: The Hydrowind system of the New Alchemy Institute is located on Prince Edward Island, Canada. In this system, four twenty-foot rotors power hydraulic motors at the top of the towers and these (by means of a hydraulic circuit) drive one large hydraulic motor which in turn drives a 25 Kw generator. The hydraulic drive train allows the generator (the most problematic part of windgenerators) to be installed on the ground where it is easy to work on it. This could be particularly useful on larger windgenerators. Another unique feature of the Hydrowind system is the hydraulic pitch control system. An anemometer measures the rise and fall of the windspeed and the blade pitch or angle is adjusted to be in the most efficient position for that windspeed. Beyond 25 mph the blades begin to feather so as not to be damaged, but they still maintain the full output of the system. Most conventional windgenerators use certrifugal forces and spring-loaded weights to feather their blades at high wind speeds. The pitch of the blades is not adjusted for different wind speeds and hence they are less efficient than the Hydrowind rotors. The New Alchemists have plans to develop larger Hydrowind systems in the future, and to integrate the Hydrowind plants into existing electric grid systems of Prince Edward Island. For more on the Hydrowind see Journal No. 4 of the New Alchemy Institute.

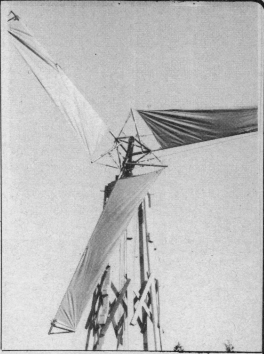

Mandurai-Type Sailwing: An eight meter diameter prototype sail wing windmill was erected Feb-March 1973 on a small farm owned by T.O. Heineman in a dry hill region of Mandurai, Tamilnadu, India. It lifted 300 pounds to a height of twenty feet in one minute in a slight breeze. This was accompanied by a rubber rope passing over a six-inch pulley on the horizontal drive shaft.

This original sail wing windmill was made of a one meter diameter bullock cart wheel to which three bamboo poles are lashed in a triangular pattern with overlapping ends. Each bamboo pole forms the leading edge of a wing, and a nylon cord stretched from the outer tip of the pole to the rim of the wheel forms the trailing edge. A stable and light weight airfoil results from stretching a long narrow triangular cloth sail "sock" over that bamboo-nylon frame. This wing configuration, a hybrid of low-speed eight bladed jib-sail wings and high speed two-bladed aerodynamic sail wings, produces high starting torque at low wind speeds. The bullock cart wheel is attached at the hub to the end of an automobile axle shaft which rotates in two sets of ball bearings. The shaft and bearing assembly is mounted horizontally on top of a ball bearing turntable.

A variation of this basic design, for materials common to North American conditions, is described in Journal No. 2 of the New Alchemy Institute.

— R.M.

The cost of a wind system that provides energy at our present rate of consumption is prohibitively expensive for a single family alone. It is as much a waste of materials as our present rate of consumption is a waste of energy. There is a definite *economy of scale* with wind systems. That is, the larger the system (up to a certain point), the more favorable the ratio of the cost of materials to the energy received. Sharing these costs and benefits with other families or small groups is economic, efficient, and energy conserving.

Wind has both a potential and kinetic form of energy. It is the kinetic form which we harness and the security in the potential energy which we seek. The Wind section, along with other sections in this book, offers some ideas, concepts and methods which contribute to a standard of living that provides an opportunity to further understand the human situation, without destroying the environment. The movement of the universe is creation.

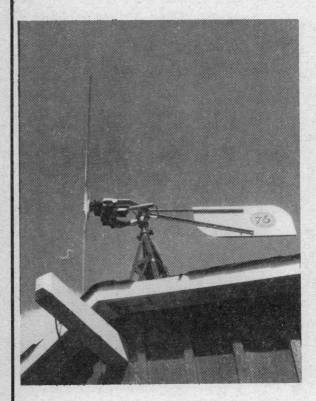

CLOCKWISE FROM UPPER RIGHT: PARRIS-DUNN — 1936 JACOBS (RECONDITIONED), RURAL-LITE, SMALLER RURAL-LITE GOVERNOR, DEKO ELECTRIC AND AIR-ELECTRIC 3000 WATT (CENTER).

VARIOUS <u>JACOBS</u>, 1930-1950 (UPPER HALF PAGE)
AND <u>WINCHARGERS</u> (LOWER HALF PAGE).

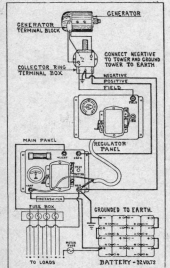

BIOFUELS

BioFuels are, as the name implies, renewable energy sources from living things. They are to be distinguished from *fossil fuels* which are also of biological origin, but which are non-renewable. All biofuels are ultimately derived from plants, which capture the sun's energy, convert it to chemical energy by photosynthesis, and in the process of being eaten or decayed, pass this energy onto the rest of the living world. In this sense, then, all forms of life, their byproducts and wastes, are storehouses of solar energy ready to be eaten, burned or converted into various organic fuels.

The following *BioFuels Section* is divided into six parts: *Biomass Energy, Agriculture, Aquaculture, Methane Systems, Alcohol* and *Wood*. For various reasons we have given special attention to some of these subjects and played down others. For example, *biomass energy* is a complex subject involving the dynamics of biology, photosynthesis, ecosystems, bioconversion process, etc. All we can do in our limited space is skim the top, oversimplify concepts and hopefully give an overview for the rest of the BioFuels section.

With respect to *agriculture*, so much has been written about the why and how of composting, organic farming/gardening, natural pest control, etc., that we have only included a partially annotated bibliography, a few overview comments and some pertinent book reviews as a *primer* to renewable strategies for agriculture . . . from the backyard to the urban lot to the farm.

The same can be said for *methane systems*, which have enjoyed recent popularity in a variety of places from industry to feedlots to municipal waste plants to the backyard. A wide range of information about digester designs, methane gas and sludge utilization is now available in both popular and technical books and articles. And here, too, rather than describe the nitty-gritty of "how-to-do-it" (requiring more space than we've got) we have simply tried to cull out the information that has helped us to understand the principles, problems and designs involved.

On the other hand, *aquaculture* is not so familiar. Although fish farming is big business in some places, there is really very little information about ways to manage fish ponds and small fish farms using simple materials, ecological techniques and local resources (including solar and wind power). But as food scarcities increase in the years to come, aquaculture will likely become more popular as an alternative source of animal protein and market item for local food economies. Dominick Mendola has outlined the background and basic ecological strategies for small aquaculture operations. With the help of Bill McLarney he has also annotated much of the pertinent information available in the field of aquaculture to help get things started.

As far as *alcohol* is concerned, we have real mixed feelings about its production and use as a practical biofuel for individuals and small groups.

BIOMASS

BIOMASS ENERGY
Richard Merrill

Every day, over 200 times more energy from the sun falls on our planet than is used by the U.S. in one year. About half of this energy is reflected back into space (Fig. 1). Most of the sunlight that finally penetrates the atmosphere ends up charging the great heat, wind and water systems of the biosphere. The rest (only 1/10 of 1%) is captured by green plants, algae and a few kinds of bacteria and converted by photosynthesis into the chemical energy of protoplasm. The plants are then eaten by other creatures who incorporate this chemical energy into their bodies.

All plant matter is called *biomass*, and the energy that is released from biomass when it is eaten, burned or converted into fuels is called *biomass energy*. Microbes, plants, trees, animals, vegetable oils, animal fats, manure garbage, even fossil fuels, all represent forms of biomass energy that can be produced, cultivated or converted in a variety of ways for human needs.

Compared to solar, wind and water devices, plants are very inefficient at converting solar energy into useful forms of energy. But only plants can

Interest in alcohol as a fuel source seems to crop up periodically during hard times . . . like today for example. But the legal restrictions controlling the production of ethyl alcohol and the technical problems associated with producing methyl alcohol seem overwhelming in terms of small scale operations. We've presented a brief overview of alcohol as a renewable energy resource in the hopes that the ethanol laws will become more flexible and methanol distillation technology more accessible and practical.

One spin-off from the energy "crisis" has been a renewed interest in *wood* as a fuel. From suburban woodlots to National Forests, scars of recent harvests are now a common sight. And in the marketplace firewood is nearly as expensive as coal. Because wood is such an exploitable biofuel, we felt that reasonable space ought to be devoted to a discussion of its efficient use.

A final thought about BioFuels. The harnessing of solar, wind and water power is basically a mechanical problem of capture, conversion and storage. The harnessing of BioFuel energy, on the other hand, is also a *biological problem* that must be sensitive to the chemistry, nutrition and ecology of living systems. It's hard to find people who are skilled or inspired in both disciplines of engineering and biology . . . of mechanical forces and life forces. This is why the building of a digester, a still, an efficient greenhouse, a fish pond or even an ecologically designed truck garden requires a special integration and "bringing together" of varied talents, skills and people. And this is why the utilization of BioFuels should be a top priority in organizing local efforts and designing for local needs.

ABOUT REFERENCES LISTED IN BIOFUELS

Many of the references listed in the BioFuels section were written during a simpler, more humanistic and more decentralized time, i.e., prior to pesticides, rural electrification, over-sized waste treatment plants, etc. Other information comes from periods of war, depression and other times of scarcity when there was a strong premium placed on self-sufficiency and survival technologies. Many of these references are out of print or hard to get.

Today, impending hard times have precipitated a rash of new "how-to and why" books, doomsday survival manuals, homecraft guides, expensive new-age textbooks and esoteric symposia from the "concerned" technical bureaucracy. Everywhere there is information, from the past *and* the present. We have found that if you try hard enough, virtually all of it is available in one way or another: libraries, cooperative purchases, copy shops, out-of-print bookstores and a publishing industry that is generally responsive to demands for republication. We needn't get hung up on books . . . simply use them as tools. Note:

WORDS GUIDE
EXAMPLES MOVE   —R.M.
BUT THERE IS NO SUBSTITUTE FOR DOING IT

convert solar energy into chemical energy; only plants can produce the fuel that sustains life. We can build fancy solar collectors and wind generators, but with all of our science and sophisticated gadgets, we still don't know how to construct a plant, or even how to produce a practical method of artificial photosynthesis. Biomass is the oldest and most fundamental source of renewable energy. And all we can do is grow with it, we can't build it.

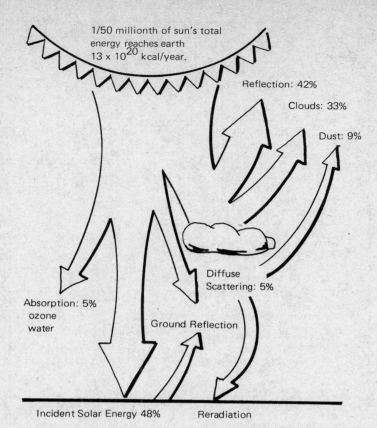

FIG. 1 AVERAGE DISPOSITION OF SOLAR ENERGY IN THE BIOSPHERE. EVERY DAY NEARLY 100 TIMES MORE ENERGY FALLS ON THE EARTH'S SURFACE THAN IS USED BY THE U.S. IN ONE YEAR.

Photosynthesis is more than just the basic process by which solar energy is converted into chemical (biomass) energy. It has also been one of the driving forces in the creation of the biosphere itself. Theory has it that for millennia following the cooling of the earth, organic molecules evolved in shallow seas from the ingredients of the primitive atmosphere (methane, hydrogen, ammonia and water). In time, some of these early compounds were able to reproduce and to subsist by "eating" other organic molecules. Other forms of proto-life survived, not by eating each other (which had its obvious limitations), but by evolving methods of using outside solar energy to synthesize their food from carbon dioxide and simple materials, i.e. by "photo"-synthesizing what they needed to maintain and propagate themselves. This was perhaps the most important step in organic evolution; for the first time one biological group could pass the vast and renewable resources of solar energy onto the rest of the living world. Equally im-

I. BASIC PHOTOSYNTHESIS FORMULA:

$$CO_2 + H_2 \xrightarrow{\frac{solar\ energy}{chlorophyll}} (CH_2O) + 2x + H_2O$$

carbon dioxide   hydrogen supply   biomass energy (carbohydrate)   element

II.   GREEN ALGAE AND HIGHER PLANTS

$$CO_2 + H_2O \longrightarrow (CH_2O) + O_2$$
                        oxygen

III.   SULFUR BACTERIA

$$CO_2 + H_2S \longrightarrow (CH_2O) + S + H_2O$$
    hydrogen   sulfur
    sulfide

IV.   SOME ALGAE AND BACTERIA

$$CO_2 + H_2 \longrightarrow (CH_2O) + H_2O$$
    hydrogen

V.   PURPLE BACTERIA

$$CO_2 + CHO + H_2O \longrightarrow (CH_2O) + H_2 + H_2O$$
    organic
    material

TABLE I GROSS EQUATIONS FOR DIFFERENT KINDS OF PHOTOSYNTHESIS. GREEN PLANTS ARE BY FAR THE MOST IMPORTANT CONTRIBUTOR TO THE ENERGY BUDGET OF THE BIOSPHERE.

portant was the fact that photosynthesis led eventually to the accumulation of oxygen and ultra-violet absorbing ozone in the atmosphere. This radically altered conditions for life on earth and probably triggered the explosive evolution of animal life and the eventual colonization of the land. Today the process of photosynthesis provides us with all of our oxygen, all of our food and most of our energy. Its importance cannot be overemphasized, although it is often forgotten.

Actually, organisms have evolved several ways to photosynthesize (Table I). But because water is so abundant, *green plant* photosynthesis is far and away the most important contributor to the energy budget of the biosphere. Likewise, other forms of life that don't get their energy directly from the sun have evolved a variety of methods for getting the energy and materials they need to survive (Table II).

I. LIFE FORMS BY THE WAY THEY OBTAIN ENERGY
   A. Phototrophs (energy from sunlight)
      Algae, Higher Plants, some Bacteria (green, purple)
   B. Chemotrophs (energy from chemicals)
      1. LITHOTROPHS (inorganic chemicals)
        some bacteria
      2. ORGANOTROPHS (organic chemicals)
        Animals, Fungi, Most Bacteria

OR

II. LIFE FORMS BY THE WAY THEY OBTAIN ENERGY AND CARBON
   A. Autotrophs (energy from non-living sources, carbon from CO_2)
      1. PHOTOSYNTHETIC (energy from light) . . . "A" above
      2. CHEMOSYNTHETIC (energy from inorganic chemicals) . . . "B1" above
   B. Heterotrophs (energy and carbon from organic chemicals) . . . "B2" above

---

TABLE II THE CLASSIFICATION OF LIFE FORMS ACCORDING TO THE WAY THEY OBTAIN ENERGY AND CARBON. ONLY TWO SOURCES OF ENERGY ARE AVAILABLE: SOLAR ENERGY AND BIOMASS ENERGY. LIKEWISE ONLY TWO SOURCES OF CARBON ARE AVAILABLE: CARBON DIOXIDE AND, AGAIN, BIOMASS.

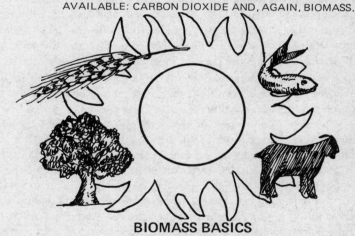

BIOMASS BASICS

To better understand the fundamentals of biomass production we need to define a few terms. The easiest way to do this is to trace the patterns of energy and materials as they pass through a simple community of plants and animals. As an example, we can take a hypothetical 1 acre pasture of grass and foraging cows.

The amount of solar energy falling on the field will vary greatly according to latitude and climate. For a typical mid-latitude area the average amount of incident solar energy is between 1 and 4 billion kilocalories per acre per year (kcal/acre/yr).** For our purposes we can take the lower value. Of this amount about 99% (990 million kcal/yr) is lost immediately from the grass plants by reflection and evaporation. The rest (10 million kcal/yr) is converted into plant tissue and is called the *gross primary productivity (GPP)*. The GPP does not represent the accumulated biomass since the plants must use some of their own energy to maintain themselves. In our pasture example, about 75% of the GPP (7.5 million kcal/yr) is actually converted into grass biomass and becomes available as food for the cows.[1,2] This is called the *net primary productivity (NPP)*. The NPP

**A kilocalorie, or 1000 calories, is the amount of heat needed to raise one kilogram of water (about a quart) one degree centigrade. It is equal to 3.96 BTU.*

is the measure of accumulated biomass available for food, fuel or conversion. The rest of the GPP not converted to plant biomass is radiated as unusable heat by respiration.

10 million kcal/yr	=	7.5 million kcal/yr	+	2.5 million kcal/yr
Gross Primary Productivity = (solar energy assimilated by plants)		Net Primary Productivity (biomass, yield or food energy)	+	Respiration (heat energy)

Now assume that the cows only eat about 1/2 of the available grass in the pasture.[2] This means that the amount of biomass eaten by the cows represents (1% x 75% x 50%) = .38% or less than 1/2 of one percent of the solar energy falling on the field. In our example, this amounts to 3.8 million kcal/yr and is called the *gross herbivore productivity*. In general about 90% of this food energy is lost as cow respiration and manure, leaving only about 10% as cow meat (380,000 kcal = 180 lbs of beef protein). This biomass is called the *net herbivore productivity*.

We can expand our simple community by assuming that if the cows were left to fend for themselves some would die naturally and decay in the pasture while others would fall prey to local predators. The predators, in turn, would lose about 90% of the cow flesh energy to respiration and wastes, converting only about 10% into their bodies as *net carnivore productivity*.

The passing of energy and nutrients from their source in plants through a series of organisms is called a *food chain*. So far we have been talking about a simple grass-cow-predator food chain. We can also refer to all animals feeding in the same level of the food chain as being in the same *trophic level*. In the pasture, grass-eaters like cows and gophers would be in one trophic level and cow-eaters like wolves and people would be in the next "higher" trophic level. We can now make some generalizations about biomass production:

Solar Efficiency of Biomass Production

Compared to mechanical systems, plants are very inefficient at converting solar energy into available (chemical) energy (Table IV). We can determine photosynthetic efficiency by comparing the chemical energy stored in plant tissue (either gross or net productivity) with the solar energy received by the plant. For our pasture/cow example we assumed an efficiency of 1% which is within the range of most crops and plants (Table V).

There are two major reasons why plants are so inefficient at converting solar energy. First, plants tend to put their maximum growth into short periods during favorable times of the year. If efficiencies are measured over the entire growing season, values will be much lower than if they are measured during peak growing periods (Table V).

Secondly, plants are simply unable to use most of the sunlight available to them. On land, from 70-80% of the incident light is reflected or absorbed by physical things other than plants.[3] We can get an idea of what happens to the remaining light energy from an elegant study done on an acre of corn during a 100 day growing season.[4] The study showed that 44.4% of the light received by plants was used to evaporate the 15 inches of rainfall received during the season: 54% was converted directly to heat and lost by convection and radiation, and the minute quantity remaining (1.6%) was actually converted into the tissues of the corn plants. About 33% of this gross productivity was used in respiration leaving 1.2% of the available light energy as corn biomass (Table III).

	Glucose (lbs)	kcal. (million)	Solar Efficiency
INCIDENT SOLAR ENERGY		2.043	
PRODUCTIVITY			
Net (N)	3040	25.3	1.2%
Respiration	930	7.7	0.4%
Gross (G)	3970	33	1.6%
Production efficiency = N/G			76.6%

TABLE III ENERGY BUDGET OF AN ACRE OF CORN DURING ONE GROWING SEASON (100 DAYS). 76.6% OF THE SOLAR ENERGY ASSIMILATED IS PUT INTO BIOMASS. FROM REF. 4

The fact that a portion of plant productivity is used for plant maintenance suggests another measure of solar efficiency for plants: the *production efficiency*. This is defined as the rate of energy lost during respiration to the net productivity.

		% Efficiency		
		Of Process	To Heat	To Electricity
I.	**BASIC PHYSICAL CONVERSIONS**			
	A. STEAM → MECHANICAL ENERGY	10-30		
	B. MECHANICAL → ELECTRICAL			80
	C. STEAM → ELECTRICAL			A.xB.= 8-25
II.	**SOLAR - MECHANICAL CONVERSIONS**			
	A. LOW TEMPERATURE SOLAR			
	1. Solar energy → hot water		20	
	B. HIGH TEMPERATURE SOLAR			
	1. Solar heaters, cookers, reflectors		50-80	
	2. Solar reflector → steam		40-60	
	3. "I-C" above			8-25
	4. Solar → steam → electricity			3-15
	C. SOLAR → ELECTRICITY (PHOTOCELLS)			
	1. Cadmium sulfide			5
	2. Silicon			12
	D. WIND			
	1. Wind → mechanical	44		
	2. "I-B" above			80
	3. Wind → mechanical → electrical			35
III.	**SOLAR - BIOLOGICAL CONVERSIONS**			
	A. FOOD CHAINS			
	1. Solar energy → plant chemical energy	0.3-3.0		
	2. Plant energy → herbivore energy	5-10		
	3. Herbivore → carnivore energy	5-15		
	B. WOOD			
	1. Solar energy → forest wood	0.5-3.0		
	2. Wood → heat (steam)		60-80	
	3. "I-C" above			8-25
	4. Solar → steam → electrical			.04-.8
	C. BIOGAS (DIGESTION)			
	1. Solar → plant	.3-3.0		
	2. Biomass → biogas*	40-70		
	3. Biogas → heat		75	
	4. Biogas → heat → mechanical	25-40		
	5. "I-B" above			80
	6. Organic waste → electricity (via biogas)			.02-.5
	D. ALCOHOL (DISTILLATION)			
	1. Fruits, grains → ethanol	75		
	2. Wood → ethanol	65		
	3. Biomass waste → methanol	55		

TABLE IV EFFICIENCIES OF SOLAR ENERGY CONVERSION SYSTEMS: COMPARING THE BIOLOGICAL WITH THE MECHANICAL

*Not including process heat

	% of Gross Productivity	% of Net Productivity
EXPERIMENTAL		
LABORATORY		
Algae (Chlorella)	20-35	
Dim light experiments	15-20	
FIELD		
Chlorella silt ponds	3.0	
Sewage ponds	2.8	
CULTIVATED CROPS		
PEAK OF SEASON		
Sugar beets, Europe	7.7	5.4
Sugar cane, Hawaii	7.6	4.8
Irrigated corn, Israel	6.8	3.2
DURING SEASON		
Sugar beets, Europe	2.2	
Rice, Japan		2.2
Sugar cane, Java		1.9
Corn, U.S.	1.6	1.3
Water hyacinth	1.5	
Tropical forest plantation	0.7	
ECOSYSTEMS		
Annual desert plants (peak)	6-7	
Tropical rain forest	3.5	
Freshwater springs, Florida	2.7	
Polluted bay, Texas	2.5	
Coral reef	2.4	
Beech forest, Europe	2.2	1.5
Scots pine, Europe		2.4
Oak forest, U.S.	2.0	.91
Perennial herb, grass		1.0
Cattail marsh	0.6	
Lake, Wisconsin	0.4	
Broomsedge community	0.3	
BIOMES		
Open ocean	0.09	
Arctic tundra	0.08	
Desert	0.05	
BIOSPHERE		
Land		0.4
Sea		0.2

TABLE V PHOTOSYNTHETIC EFFICIENCY OF VARIOUS PLANTS, CROPS, AND ECOSYSTEMS. (SEE TEXT FOR EXPLANATION)

Production efficiencies range between 85% (natural grasslands) to 29% (tropical rainforests). Generally speaking plant communities tend to fall into three groups (Fig. 2): (1) Those with high net productivity and low respiration. Included here are plants that have been selected for putting weight on fast, e.g., crops, weeds, young plants and vegetation in the early stages of ecological succession. (2) Those with low net productivity and low respiration. These are plant communities in marginal areas where water (desert), light (arctic) or nutrients (mid-ocean) are limited. (3) Those plants with low productivity and high respiration. These include mature ecosystems (old forests, coral reefs), polluted waters and other situations where the energy cost of maintenance is high. Note here for example that, beyond a certain age, the older a woodlot gets, the less efficient it is in converting solar energy to wood.

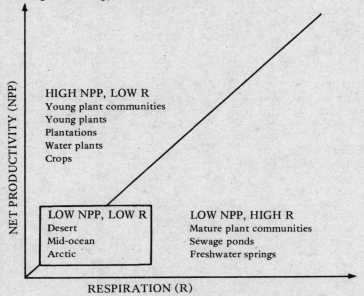

FIG. 2   THE RELATIONSHIP BETWEEN ENERGY GAINED IN NET PRIMARY PRODUCTIVITY AND ENERGY LOST TO RESPIRATION FOR VARIOUS PLANT COMMUNITIES.

Whereas plants convert about 75% of their gross productivity into biomass, animals, being mobile and feeding sporadically, are far less efficient. As a general rule about 90% of the available energy is lost at each exchange point between animals in the food chain.[5,6] In other words, the more removed an animal is from a plant diet, the more plant biomass it takes to maintain it. At the present time, over 80% of all U.S. grains are fed to livestock. Much of this valuable food energy is lost in cow respiration and feces (some of it could be retrieved by treating the manure as a resource rather than a waste). Obviously there is a great deal to be said for eating more vegetable proteins in one's diet and less meat, i.e. for eating closer to the bottom of the food chain.[7]

Energy Subsidies in Biomass Production

High photosynthetic and production efficiencies in crops are maintained with large outside energy inputs (fertilizers, supplemental foods, irrigation,

cultivation). Anything that reduces the energy cost of internal self-maintenance of a biological system is called an *energy subsidy*.[8] For example, modern agriculture is only productive because of a large energy subsidy of fossil fuels (see Agriculture Section). In terms of producing from its own internal energy supplies, modern agriculture is not efficient at all.

The Distribution of Biomass in Biological Systems

Energy flows through a biological community only once and is not recycled but is transformed into heat and ultimately lost to the community. Only the continual input of new solar energy keeps the community running. On the other hand, all matter is ultimately recycled through decomposers (mostly microbes) which break down complex organic molecules into simple materials to be used again by the plants. Thus in all biological systems . . . pastures, fish ponds, truck gardens, woodlots or the entire biosphere . . . *matter cycles and energy flows* (Figs. 3 and 4).

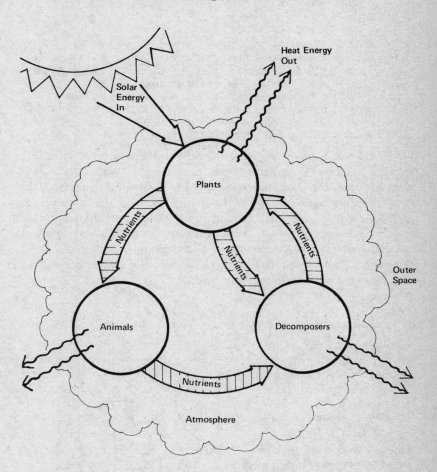

FIG. 3   RECYCLING OF NUTRIENTS AND FLOW OF ENERGY IN THE BIOSPHERE. VIRTUALLY ALL MATERIALS ARE RECYCLED IN THE BIOSPHERE, BUT ONLY A SMALL PORTION OF INCOMING SOLAR ENERGY IS STORED AT ANY ONE TIME. EVENTUALLY ALL BIOMASS ENERGY DISSIPATES INTO OUTER SPACE AS HEAT.

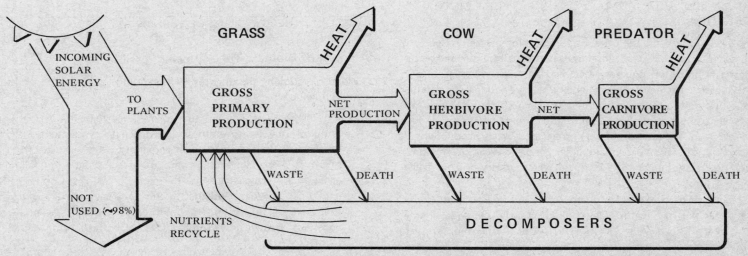

FIG. 4   RECYCLING OF NUTRIENTS AND FLOW OF ENERGY IN AN ECOSYSTEM FOOD CHAIN (GRASS, COW, PREDATOR). SEE TEXT.

For the efficient conversion of matter and energy back into the life cycle, ecosystems are organized into a biological hierarchy that can be shown graphically as *ecological or trophic pyramids*. The base of the pyramids are the abundant biomass of plants. Since energy and biomass are lost as they pass along the food chain, animals near the base of the pyramid (plant eaters) are more numerous (in terms of biomass, energy content and numbers) than the animals above which eat them (Fig. 5).

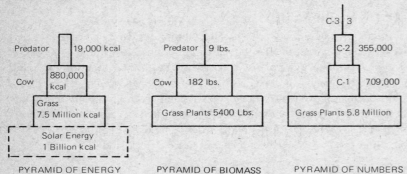

FIG. 5   THREE KINDS OF ECOLOGICAL PYRAMIDS BASED ON THE FOOD CHAIN OF A 1 ACRE FIELD OF GRASS OVER 1 YEAR. PYRAMIDS OF ENERGY AND BIOMASS ARE FROM GRASS/COW ECOSYSTEM DESCRIBED IN TEXT. PYRAMID OF NUMBERS IS FROM DATA PRESENTED BY ODUM 1971 (REF. 8), WHERE: C-1=PRIMARY CONSUMERS (HERBIVOROUS INVERTEBRATES); C-2=SECONDARY CONSUMERS (CARNIVOROUS SPIDERS AND INSECTS); C-3= TERTIARY CONSUMERS (MOLES AND BIRDS).

This concept helps us to understand the distribution of biomass in a biological system. For example, under normal conditions, we know that there will be a great deal more insect pests in a garden than there will be insect predators feeding on them. If a pesticide kills all insects equally (as most do), it will have the effect of actually increasing the number of pests . . . unless, of course, more pesticides are used. Knowing about ecological pyramids is also important in the biological design of a poly-culture pond (see Aquaculture Section), the pasturing of livestock and the management of other cultivated ecosystems.

BIOMASS PRODUCTION AND CONVERSION

Productivity Units

Primary productivity is the rate at which biomass is "fixed" by photosynthesis per unit area per unit time. It is usually expressed as dry organic matter in grams per square meter per day or per year ($g/m^2/day$; $g/m^2/yr$); kilograms per hectare per year (kg/ha/yr); pounds per acre per year (lb/acre/yr); metric tons per hectare per year (mT/ha/yr); short tons per acre per year (sT/acre/yr) or in terms more familiar and practical such as bushels per acre per season, lugs per tree, etc. Often productivity is measured in grams of carbon per square meter per day or year ($g\text{-}C/m^2/day$). As a general rule of thumb, about 45% of the weight of dry organic matter is carbon.

Biomass can also be converted to energy units and expressed as calories per square meter per day ($cal/m^2/day$), or kilocalories per square meter per day or year ($kcal/m^2/day$; $kcal/m^2/yr$). Converting biomass into energy units comes in handy when you are considering the energy budgets of natural ecosystems and agriculture, or the efficiencies of

bio-conversion processes. For "ball-park" estimates, 1 gram dry weight of organic matter is equivalent to 4-5 kcal of energy (1400-1750 kcal/lb or 5500-6900 Btu/lb). For more precise biomass to energy conversions see Table VI.

	SOLID		LIQUID	GAS
	kcal/g	BTU/lb	BTU/gal	BTU/ft³
FOSSIL FUEL				
COAL				
bituminous	9.2	13,100		
anthracite	8.9	12,700		
lignite	4.7	6,700		
COAL COKE	9.1	13,000		
CRUDE OIL		18,600	138,000	
FUEL OIL		18,800	148,600	
KEROSENE		19,810	135,100	
GASOLINE		20,250	124,000	
LP GAS		21,700	95,000	
COAL GAS				450-500
NATURAL GAS (Methane)		21,500	75,250	1050
PROPANE		21,650	92,000	2200-2600
BUTANE		21,250	102,000	2900-3400
BIOFUELS				
CARBOHYDRATES				
sugar	3.7-4.0	5300		
starch/cellulose	4.2	5800		
lignin	6.0	8300		
PROTEIN				
grain/legume	5.7-6.0	8050		
vegetable/fruit	5.0-5.2	7025		
animal/dairy	5.6-5.9	7850		
FATS				
animal	9.5	13,100		
vegetable oil	9.3	12,800		
MICRO-ALGAE	5.0-6.5	9500		
WOOD				
oak, beech	4.1	5650		
pine	4.5	6200		
all woods	4.2	5790		
BRIQUETS	8.1	11,500		
ALCOHOL				
methanol		8600	67,000	
ethanol		12,000	95,000	
BIOGAS (60% CH₄)				600-650
METHANE		21,500	75,250	1050
MISC. "WASTES"				
municipal organic refuse	2.8-3.5	4000-5000		
raw sludge	2.7-5.3	3700-7300		
digested sludge	2.7-5.0	3800-6900		
paper	5.5	7600		
glass	5.6	7700		
leaves	5.2	7100		
dry plant biomass	5.6	8000		
MISC. ANIMALS				
insect	5.4			
earthworm	4.6			
mammal	5.2			

TABLE VI   ENERGY VALUES OF VARIOUS FOSSIL FUELS AND BIOFUELS (ref. 9-12).

Amount of Biomass Theoretically Possible

We can estimate the theoretical maximum productivity of biomass in the following way[13]:

Assume that: a) The amount of available light imposes the upper limit to total photosynthesis and hence to productivity. b) Photosynthesis is limited to the visible region of the light spectrum (.4-.7 microns)** c) The average amount of solar radiation falling on a given area of average air mass during the growing season (June-Sept.) is 500 $cal/cm^2/day$, and that of this, 222 $cal/cm^2/day$ is attributed to the visible spectrum. Maximum productivity: If we make allowances for inactive absorption (10%), plant respiration (33%), inor-

**1 micron (u)=one-millionth of a meter.

ganic materials in the plant (8%) etc., the theoretical maximum net productivity is around 77 grams/m^2/day = 687 lbs./acre/day = 34sT/acre/100 days = 125 sT/acre/hr. As we shall see, even under ideal conditions (controlled laboratory pilot experiments and farming situations where high-yield crops are grown year around), the best we can hope for is around 25-35% of this theoretical maximum.

Net Productivity of the Biosphere

According to Fogg[19] if one year's yield of the earth's photosynthesis were amassed in the form of sugar cane it would form a heap over 2 miles high and with a base 43 miles square! A more precise estimate of our planet's productivity is around 175 billion dry tons per year (14). This figure is made by adding up the estimated productions of the major environments of the biosphere (Table VII). Note that 2/3 of the total productivity is generated on land . . . 1/4 by tropical forests alone. About 5% of the land productivity is due to agriculture, and about 0.5% of the total productivity is consumed as human food and livestock fodder. Interestingly enough, agricultural systems generally do not appear to be any more productive than many natural ecosystems.

Biome	Area 1 Million Square Miles	Average Net Primary Productivity Tons/Acre/Year	World Total Net Primary Productivity 1 Billion Dry Tons/Year
LAND			
Extreme (desert, ice)	9.3	.1	.1
Tundra	3.1	.6	1.2
Boreal forests	4.6	3.6	10.6
Temperate forests	6.9	5.8	25.8
Woodlands, shrublands	2.7	2.7	4.6
Grasslands:			
Temperate	3.5	2.2	5.0
Savannah	5.8	3.1	11.6
Desert scrub	6.9	.3	1.4
Tropical forests	7.7	8.9	44.1
TOTAL	50.5	27.3	104.4
AVERAGE		3.21	
Cultivated land:			
Mechanized	1.5	5.3*	5.1
Unmechanized	3.8	1.3*	3.3
TOTAL	5.3	6.6	8.4
AVERAGE		2.46	
FRESHWATER			
Lakes and streams	0.8	2.2	1.1
Swamps and marshes	0.8	8.9	4.4
TOTAL	1.6	11.1	5.5
AVERAGE		5.37	
MARINE			
Open ocean	125.9	.4	44.8
Coastal areas	13.1	.9	7.5
Upwelling areas	0.2	2.7	.3
Estuaries and reefs	0.8	8.9	4.6
TOTAL	140.0	12.9	57.2
AVERAGE		0.63	
BIOSPHERE TOTAL	197 Million Square Miles	1.39 Tons/Acre/Year	175.5 Billion Dry Tons/Year

TABLE VII GROSS ESTIMATES OF THE BIOMASS PRODUCTIVITY OF THE BIOSPHERE. Data culled from various sources (ref. 8, 14-18). * = including non-edible portions.

The energy equivalent of 175 billion dry tons of biomass is about 550 quadrillion (5.5x10^{17}) kcal. If we assume an annual insolation for our planet of 6x10^{20} kcal, we can then conclude that every year the biosphere fixes a little less than 1/10 of one percent of the solar energy falling on the earth's surface. Looked at another way, we can compare the average net productivity values of the major environments of the biosphere (Table VII) and note what a small fraction of the theoretical productivity (125sT/acre/hr) they represent. Again we see how inefficient plants are at converting solar energy to biomass, but emphasize how indispensable this "inefficient" process is to human survival.

Biomass Conversion

Solar energy remains trapped in plants until it is released at the time of being eaten, burned or decayed. We can distinguish between using green plants or organic wastes as raw materials for biomass conversions (Fig. 6). In the case of the former, growing crops means that nutrients must be recycled in order to maintain the fertility (productivity) of the land or waters. If nutrients can't be recycled they've got to be brought in from outside...often at great expense and energy. This is one reason for linking waste producing operations (e.g., livestock, sewage plants, methane digesters) with agriculture, algae ponds and other crops...to close the cycle.

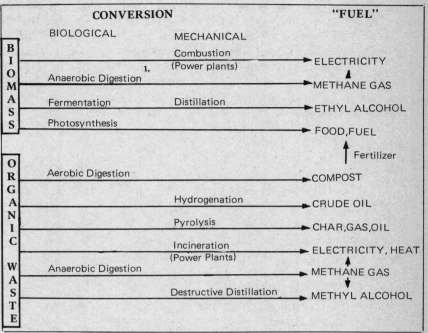

FIG. 6 BIOMASS CONVERSION PROCESSES

[1.]Anaerobic digestion also produces a liquid fertilizer, "sludge."

Although there are a variety of ways to convert plants directly into organic fuels, impending food shortages suggest that alternative fuel production may not be the wisest use of land and crops. Organic wastes, on the other hand, are produced all around us. Although they are still generally considered as a problem rather than a resource, there is little doubt that organic wastes will have to be used more and more as a raw material for conversion and recycling. Each year the U.S. produces over 870 million dry tons of discarded organic matter. Conservatively, about 16% of this (136 million tons) can be easily collected with even today's priorities (Table VIII). Virtually all of this "waste" can be converted in a variety of ways to fuels and fertilizers:

HIGH—TECH CONVERSIONS (Requiring high temperatures, pressures or corrosive chemicals, i.e., causing a relatively high impact on the local environment):

a) <u>Pyrolysis:</u> The heating of organic wastes (200-900°C) in the absence of air at atmospheric pressure to produce gas (500BTU/ft^3; oil (120,000BTU/gal.) and char (700-1200BTU/lb) all of which are combustible. The amount of energy recovered from the byproducts of pyrolysis is equal to 75-98% of the energy subsidy, i.e., there is a new energy loss for the process.[21]

b) <u>Hydrogasification:</u> Direct gasification of organic wastes (especially manure) with hydrogen at high temperatures (500-600°C) and pressures (1000 psi) to produce substitute pipeline gas (methane, ethane). About 6000 ft^3 of gas is produced (6 million BTU) per ton of dry organic matter. The hydrogen is generated from the residual char from the hydrogasification.

c) <u>Hydrogenation:</u> Conversion of organic wastes to oil by treatment with carbon monoxide and steam under pressure (2000-6000psi) at high temperatures (300-400°C). The net yield of low sulfur oil (16,000BTU/lb) is about 1.3 barrels per ton of dry organic waste.

d) <u>Destructive Distillation</u>: The heating of high-cellulose organic wastes (wood and refuse) and the distillation of liquid residues to produce methyl alcohol.

e) <u>Acid Hydrolysis</u>: The treatment of wood wastes with heat and acid to produce sugars for fermentation/distillation to ethyl alcohol.

LOW—TECH CONVERSIONS (i.e., processes with a relatively low impact on the local environment)

Organic wastes can be used directly as a livestock fodder, as a fertilizer (e.g., sewage sludge in algae ponds) or as a fuel for combustion. They can also be converted to other useful forms by COMPOSTING (decaying organic materials in carefully constructed piles to produce a soil conditioner and fertilizer); ANAEROBIC DIGESTION (decaying organic materials in air-tight containers to produce methane gas and a liquid fertilizer) and FERMENTATION/DISTILLATION to ethyl alcohol.

Source	Waste Generated	Readily Collectable
Agriculture		
Crops and food waste	390	22.6
Manure:		
Cattle	172.0	
Hogs	11.0	
Sheep	1.6	
Poultry	9.5	
	194.1	26.0
Urban		
Refuse	129	71.0
Municipal sewage solids	12	1.5
Industrial wastes	44	5.2
Logging and wood manufacturing	55	5.0
Miscellaneous	50	5.0
Total	874.1	136.3

TABLE VIII AMOUNTS (1 MILLION TONS) OF DRY, ASH-FREE ORGANIC WASTES PRODUCED IN THE UNITED STATES IN 1971. ADAPTED FROM REF. 21

Energy Plantations & High-Yield Biomass

One spin off from the "energy crisis" has been an interest in growing high-yielding plants in "energy plantations" and burning the biomass in conventional power plants to generate electricity. For example one proposal[22] describes how 100-650 square miles of rapidly growing trees could fuel a 1 million kilowatt electrical power plant. Another idea[23] calls for the growing of sea kelp in offshore waters of California and Peru in areas extensive enough to produce 1.8 billion dry tons of marine algae per year. This biomass would then be digested and converted to 12 trillion BTU of methane energy annually (17% of current U.S. demands).

According to proponents, the energy plantation has a few advantages over other sources of fuel[24,25]: 1) A favorable impact on our balance of payments and foreign relations (i.e., produced domestically). 2) A favorable impact on the environment (unlike strip mining and nuclear power plants). 3) Since biomass is low in sulfur it has an advantage over coal as a fuel. 4) Energy plantations generate a renewable supply of energy.

There are, however, major problems with the energy plantation concept: 1) It means that large land areas will be converted into single crop stands (monocultures) which are notoriously susceptible to disease and pest outbreaks. The ecological instability inherent in high-yield monocultures is rarely considered as part of the "cost" of large scale farming operations...whether they are producing for food or fuel. 2) It implies that large areas of land or offshore waters will be controlled by the limited interests of energy companies. 3) By centralizing energy production, the energy plantation perpetuates the need for elaborate transmission grids and tends to ignore the great potential of decentralized power generation, i.e., producing the energy where it is needed. 4) And finally, the idea of producing crops for anything but food is going to become increasingly difficult to justify in the years to come.

One interesting aspect of the energy plantation studies has been an evaluation of some high yielding plants that could conceivably be used as part of a local food/energy economy. According to Alich and Inman[25] the following land plants appear to have the most potential for biomass production (we have noted some additional advantages):

FORAGE GRASSES: Sudangrass (*Sorghum vulgare*); napiergrass (*Pennisetum purpureum*); silage corn (*Zea mays*); and forage sorghum (*Sorghum vulgare*). Sorghum is especially appealing since it is a perennial and can produce a ratoon crop (shoots from the roots of harvested plants). It is also generally more heat and drought resistant than corn and requires less water for normal growth.

KENAF (*Hibiscus cannabinus*): An annual plant of the Malvacea family with a stalk of 50% cellulose. Kenaf yields especially well when irrigated.

EUCALYPTUS: Native of Australia. One of the most productive trees in the world, *Eucalyptus* thrives under a variety of conditions, its flowers serve as an important source of nectar for honey bees, it sprouts profusely from cut stumps and it is very pest resistant.

SHORT ROTATION HARDWOODS: Several species of trees can be harvested at 1-3 year intervals in closely-spaced plantations. In this way the entire above ground portion of the tree can be used. In the case of many of the hardwoods, the young plants are able to regenerate themselves rapidly from the cut stump by sprouting (sycamore, poplar, sweetgum, ash) or by root sprouts (alder).

SUNFLOWER (*Helianthus annus*): There are several advantages to sunflowers: they are high-yielding, hardy and produce an edible seed that can also be extracted for its oil. The flowers seem to be a good refuge and food source (nectar and pollen) for a variety of beneficial insects (predators and parasites of crop pests). The stalks can be burned or used as silage. The Russians have succeeded in crossing the sunflower with the Jerusalem artichoke (*Helianthus tuberosus*) which produces large underground tubers that are good to eat.

In addition:

COMFREY (*Symphytum asperrimum*): A high-yielding perennial plant of the Borage family. Comfrey has one of the highest protein contents (21%) of any plant known. It is also one of the few plants that concentrates vitamin B-12. Its flowers are an important source of nectar, and its leaves have well known medicinal properties.

It is important to remember that the value of a plant is usually measured as a trade off between quality and quantity. For example, modern high yield grains are often low in protein content, susceptible to disease or reduced in some other *QUALITATIVE* way. The crops discussed above are no exception and they should always be grown with this trade off firmly in mind.

A few additional points: Solar–Physical conversions (pyrolysis, solar cells, solar boilers, wind generators etc.) are several times more efficient at converting solar energy to organic fuels or electricity than are conversions which include photosynthesis as a part of the process. However, photosynthesis is, and is likely to remain, the <u>only</u> practical process by which solar energy can be converted into chemical energy. Only plants can produce carbon fuels without an expensive electrochemical process as part of the system.

The capacity and efficiency of biomass systems are limited, but within those limitations they are the wisest source of renewable chemical energy.

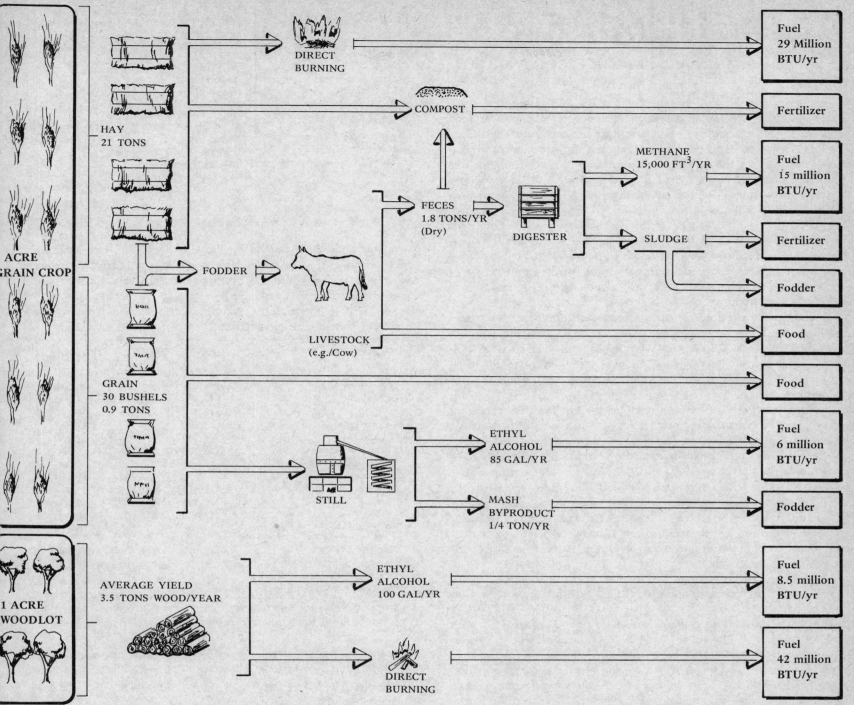

END PRODUCTS AND ROUGH ENERGY EQUIVALENTS OF VARIOUS BIOMASS CONVERSIONS FROM A 1 ACRE GRAIN FIELD AND WOODLOT.

References Cited:

[1] Golley, F.B. 1960. ENERGY DYNAMICS OF A FOOD CHAIN OF AN OLD-FIELD COMMUNITY. Ecol. Monogr. 30: 187-206.

[2] Love, M.L. 1970. THE RANGELANDS OF THE WESTERN UNITED STATES. "MAN AND THE ECOSPHERE" Sci. Amer., Feb., 1970.

[3] Rabinowitch, E. and Govindjee. 1969. PHOTOSYNTHESIS John Wiley, New York.

[4] Transeau, E.N. 1926. THE ACCUMULATION OF ENERGY BY PLANTS Ohio Journal of Science, 26:1-10.

[5] Slobodkin, L.B. 1962. ENERGY IN ANIMAL ECOLOGY. In: "Advances in Ecological Research", J.B. Cragg (ed.), Academic Press, New York.

[6] Kozlovsky, D.G. 1968. A CRITICAL EVALUATION OF THE TROPHIC LEVEL CONCEPT. I: ECOLOGICAL EFFICIENCIES. Ecology, 49:48-59.

[7] Lappe, F.M. 1971. DIET FOR A SMALL PLANET. Ballantine Books, Inc. New York.

[8] Odum, E.P. 1971. FUNDAMENTALS OF ECOLOGY. W.B. Saunders Co., Phil.

[9] Cohen, L.J. and J.F. Fernandes. 1968. THE HEAT VALUE OF REFUSE. Mech. Engin., Sept. 1968: 47-51.

[10] Fair, G.M. & E.W. Moore. 1932. HEAT AND ENERGY RELATIONS IN THE DIGESTION OF SEWAGE SOLIDS. I. THE FUEL VALUE OF SEWAGE SOLIDS. Sewage Works Journal, 4:242.

[11] Golley, F.B. 1961. ENERGY VALUES OF ECOLOGICAL MATERIALS, Ecology 43: 581-584.

[12] Merrill, A.L. and B.K. Watt. 1955. ENERGY VALUE OF FOODS---BASIS AND DERIVATION. USDA Handbook No. 74.

[13] Loomis, R.S. and W.A. Williams. 1963. MAXIMUM CROP PRODUCTIVITY: AN ESTIMATE. Crop Science, 3:67-72.

[14] Whittaker, R.H. 1970. COMMUNITIES AND ECOSYSTEMS. The Macmillan Co., Toronto.

[15] Rodin, L.E. and N.I. Bazilevich. 1965. PRODUCTION IN TERRESTRIAL VEGETATION. (ed., G.E. Fogg). Oliver and Boyd, London

[16] Ryther, J.H. 1963. GEOGRAPHIC VARIATION IN PRODUCTIVITY In: "The Sea", vol 2, M.M. Hill (ed.), Interscience, New York.

[17] Vallentyne, J.R. 1965. NET PRIMARY PRODUCTIVITY AND PHOTOSYNTHETIC EFFICIENCY IN THE BIOSPHERE. In: "Primary Productivity in Aquatic Environments". C.R. Goldman (ed.), Univ. Calif. Press, Berkeley.

[18] Westlake, D.F. 1963. COMPARISONS OF PLANT PRODUCTIVITY. Biol. Rev. 38: 385-425.

[19] Fogg, G.E. 1968. PHOTOSYNTHESIS. American Elsevier Publ. Co., Inc., N.Y. 116 pp.

[20] Cremtz, W.L. 1973. ENERGY POTENTIAL FROM ORGANIC WASTES: A REVIEW OF THE QUANTITIES AND SOURCES. U.S. Dept. Inter., Bureau of Mines, Info. Circular 8549

[21] Anderson, L.L. 1972. OIL FROM AGRICULTURAL WASTES. In: "Proceedings International Biomass Energy Conference." (SEE REVIEWS).

[22] Szego, G.C. and C.C. Kemp. 1973. ENERGY FORESTS AND FUEL PLANTATIONS. Chem. Tech., May 1973.

[23] Szetela, E.J. et al 1974. EVALUATION OF A MARINE ENERGY FARM CONCEPT. United Aircraft Research Labs. 400 Main St., Hartford, Conn., 06108.

[24] Szego, G.C. and C.C. Kemp. 1974. THE ENERGY PLANTATION. Inter-Technology Corp., P.O. Box 340, Warrenton, Va. 22186.

[25] Alich, J.A. and R.E. Inman. 1974. EFFECTIVE UTILIZATION OF SOLAR ENERGY TO PRODUCE CLEAN FUEL. Stanford Research Institute, National Science Foundation, "Research Applied to National Needs". NTIS, PB-233 956/2WE, $5.00.

AGRICULTURE

ENERGY AND AGRICULTURE

Richard Merrill

Agriculture is the means by which solar energy becomes our food energy. For thousands of years before farm machinery, pesticides, chemical fertilizers, etc., a great deal of human labor was required to grow food. But even in these "primitive" times more energy was available from the food than was required to grow it. Today, the tools of agriculture depend almost entirely on the high-energy inputs of fossil fuels (natural gas, petroleum, diesel, gasoline, LP gas, aviation fuel, etc.), and, the need for fossil fuels in agriculture is increasing every year. This is due not only to the demands of a rising population, but also to the demands of more sophisticated applications of power and chemicals to food production. In fact we are rapidly approaching a time (some argue that we are already there) when more energy will be required to produce our food than is obtained from it.

What is not often appreciated is that an agriculture nurtured by a single non-renewable resource is not only extremely vulnerable, but also, eventually, non-renewable itself: "What happens to food costs as fossil fuels become scarcer?," and, indeed: "What happens to agriculture when the oil wells and gas fields run dry?" These are important questions for our times.

Looking back we can now see that for the last few generations we have simply been exchanging finite reserves of fossil fuels for our supplies of food and fibre. Obviously this trade-off can't continue indefinitely. Soon we will have little choice but to adopt farm technologies that use a diverse base of alternative energy supplies. These include organic wastes to supplement chemical fertilizers, renewable forms of energy (solar, wind, and organic fuels) to help supply rural power needs, and integrated pest control programs to reduce the use of pesticides. Without the use of solar/wind/biological energies as back-up systems in the production of our food, modern agriculture could very well become self-defeating rather than self-sustaining.

Agriculture and Our Changing Diets

During the last two generations, the yields of most major U.S. crops have increased from 200-400% (Table I). The reason for this abundance is simple: intensive oil energy has replaced human labor as the principle input of crop production. Today virtually every phase of agriculture depends on fossil fuels in one form or another[1]. In fact agriculture consumes more petroleum than any other industry[2].

	1930	1940	1950	1960	1970
GRAINS:					
Rice	1.05	1.15	1.19	1.71	2.31
Corn	.57	.81	1.07	1.53	2.00
Sorghum	.30	.38	.63	1.11	1.42
Barley	.57	.55	.65	.74	1.02
Wheat	.42	.46	.50	.78	.93
Rye	.35	.35	.34	.55	.73
SUGAR BEETS	11.9	13.4	14.6	17.2	18.8
TOBACCO	.39	.52	.63	.85	1.06
SOYBEANS	.39	.49	.65	.71	.80

TABLE I. AVERAGE YIELD OF U.S. MAJOR CROPS. VALUES GIVEN IN TONS/ACRE (USDA Agricultural Statistics, 1971)

Some recent studies have examined the energy budget of various U.S. crops in an attempt to show patterns of energy consumption in agriculture. One way to do this is to add up the energy inputs involved in crop production (Fig. 1), and then to compare this total with the amount of energy provided by the yield of the crop. The ratio of yield energy to input energy (call it the "energy efficiency") is then used to reveal trends in the way energy is consumed in agriculture, or to compare the energy efficiencies of different crops.

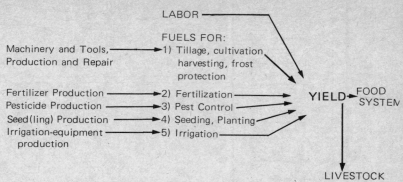

FIG. 1   MAJOR INPUTS OF ENERGY INTO AGRICULTURE.

Pimentel[3] measured the energy budget of the U.S. principal crop . . . corn. He showed that the energy efficiency has started to decline in recent years (Table II shows 3 of the 6 years measured). This alone has a profound effect on other food industries since corn supplies most of the livestock feed in this country. As the efficiency continues to decline, the price of meat will continue to rise.

Input	1945	1959	1970
Labor	12,500	7,600	4,900
Machinery	180,000	350,000	420,000
Gasoline	543,400	724,500	797,000
Nitrogen	58,800	344,400	940,800
Phosphorus	10,600	24,300	47,100
Potassium	5,200	36,500	63,000
Seeds for planting	34,000	60,400	68,000
Irrigation	19,000	31,000	34,000
Insecticides	0	7,700	11,000
Herbicides	0	2,800	11,000
Drying	10,000	100,000	120,000
Electricity	32,000	140,000	310,000
Transportation	20,000	60,000	70,000
TOTAL INPUTS	925,500	1,889,200	2,896,800
CORN YIELD (OUTPUT)	3,427,200	5,443,200	8,164,800
KCAL RETURN/INPUT KCAL	3.7	2.9	2.8

TABLE II. ENERGY EFFICIENCIES OF U.S. CORN PRODUCTION PER ACRE. Data are in Kilocalories. After Pimentel et. al, 1973 (Ref. 3).

Another study[4] examined the energy budget of all major field crops, vegetables, fruits and livestock produced in California. Because California is by far the largest producer of farm products in the country, the study has broader significance.

There were two major results. First, there was an accounting of the energy consumed by the different fossil fuels and agricultural inputs (Table III). Natural gas accounted for 53% of all the energy consumed, followed by diesel fuel (18%). The production, distribution and application of fertilizer accounted for nearly 15% of the total energy inputs.

Fertilizer production is so heavily geared to fossil fuel inputs that prices can only rise with increasing prices of petroleum products. It takes 8 million kilocalories of energy to make one ton of ammonia fertilizer[10]. The cost of natural gas as a raw material (source of hydrogen) and fuel (to fix atmospheric nitrogen) accounts for 60% of the manufacturing costs of ammonia, and ammonia now supplies 90% of all fertilizer nitrogen.

A second result was a list of the energy efficiencies of different crops (Table IV). In terms of energy, grains are among the most efficient of crops to produce, processed raw foods are the least efficient and raw fruits and vegetables are intermediate. In general, raw fruits and vegetables seem to require about as much energy to grow as they provide. Similar results have been obtained for crops grown in England[5].

So despite the high yields of modern farm technology, there does not always appear to be an obvious net return of energy to society. The benefits of solar energy fixed in our foods are increasingly being offset by the subsidy of fossil fuel energy needed to produce them. In fact, as far as energy is concerned, our agriculture is far less efficient than other forms of agriculture using more labor and less technology[6-9].

	Energy Source (In Millions of Units)						Total
Category	Natural Gas Therms	Electricity KWH	Diesel Fuel Gallon	Gasoline Gallon	LP Gas Propane Butane Gallon	Aviation Fuel Gallon	Million BBLS Crude Oil
Field crops	364.784	464.681	96.400	19.477	2.381	--	9.34
Vegetables	165.999	358.193	38.792	25.031	4.441	--	4.62
Fruits and nuts	127.168	410.773	26.158	12.602	3.296	--	3.39
Livestock	107.111	1,460.966	46.443	7.813	12.261	--	4.19
Irrigation	40.618	7,177.441	6.531	.487	4.521	--	5.16
Fertilizers	305.748	579.362	6.738	3.529	1.114	--	5.87
Frost protection	--	40.501	60.003	6.854	.904	--	1.63
Greenhouses	102.700	83.427	--	--	--	--	1.82
Agr. aircraft	--	--	1.072	1.607	--	8.994	0.25
Vehicles (farm use)	--	--	10.447	117.798	--	--	2.77
Others	--	--	--	--	23.711	--	0.39
TOTAL	1,214.128	10,575.344	292.584	195.198	52.629	8.994	
Equivalent (Million bbls crude oil)	20.93	6.21	7.06	4.17	0.86	0.19	39.43

TABLE III.   FOSSIL FUEL REQUIREMENTS FOR DIFFERENT ASPECTS OF AGRICULTURE. Data are for California (1971), from Ref. 4.

Commodity	Crop Caloric Content (A) 1,000 kcal/ton	Fuel and Electrical Energy Input (B) 1,000 kcal/ton	A/B* (Efficiency)
Field Crops			
Barley	3,166.1	479.0	6.6
Beans (dry)	3,084.4	2,683.1	1.2
Corn	3,338.4	1,027.3	3.3
Rice	3,293.1	1,289.3	2.6
Sorghum, Grain	3,011.8	1,188.8	2.6
Sugar	3,492.6	6,654.2	.5
Wheat Flour	3,020.9	563.3	5.4
		Average	3.2
Raw Vegetables and Fruits			
Beans, Green	1,115.8	2,048.0	.5
Broccoli	290.3	1,178.6	.2
Carrots	381.0	359.8	1.1
Cauliflower	244.9	986.4	.2
Celery	154.2	351.5	.4
Lettuce	163.3	484.3	.3
Melons	235.9	636.6	.4
Onions	344.7	390.3	.9
Potatoes	689.5	325.4	2.1
Strawberries	335.6	727.6	.5
Tomatoes	199.6	262.2	.8
Apples	508.0	401.1	1.3
Apricots	462.6	840.4	.6
Grapefruit	371.9	1,165.5	.3
Grapes	607.8	576.9	1.1
Oranges	462.6	1,089.5	.4
Peaches	344.7	471.6	.7
Pears	553.3	964.2	.6
Plums	598.7	1,650.9	.4
		Average	.7
Canned Vegetables and Fruits			
Beans, Green	870.9	3,021.5	.3
Tomatoes	190.5	1,138.9	.2
Apples	371.9	1,397.8	.3
Grapefruit	272.2	1,797.7	.2
Grapes	462.7	1,115.3	.4
Pears	417.3	1,734.1	.7
		Average	.3
Frozen Vegetables and Fruits			
Beans, Green	925.3	2,856.1	.3
Broccoli	254.0	1,911.2	.1
Cauliflower	199.6	1,619.8	.1
		Average	.2
Dried Fruits and Nuts			
Almonds	5,424.9	7,086.7	.7
Prunes	3,120.7	4,447.1	.7
Walnuts	5,697.1	10,745.6	.5
		Average	.6

*A/B = RATIO OF: CALORIC CONTENT/FUEL AND ELECTRICAL ENERGY. (The larger the number, the greater the efficiency.)

TABLE IV.   THE ENERGY EFFICIENCIES OF DIFFERENT CALIFORNIA CROPS (1972). From Cervinka et al. (Ref. 4).

There are strong implications to the fact that the principal raw material of agriculture is a dwindling resource. As oil prices go up, food prices go up. This applies especially to farm products that require heavy energy inputs, like processed foods and animal protein. In fact meat may become so expensive in the future that it will be replaced by vegetable protein in the diets of many people; for many it already has. As pointed out in the *biomass* section, from 80-90% of the food-energy eaten by animals is lost as metabolic heat. This is why it takes so much more energy to produce animal protein than it does plant protein** (Fig. 2).

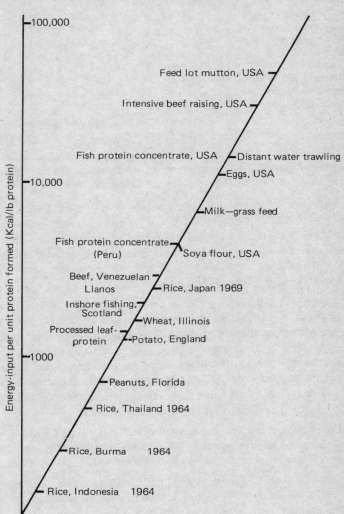

FIG. 2   TYPICAL ENERGY SUBSIDIES IN MODERN PROTEIN PRODUCTION. From Slesser (1973). Ref. 12.

It should not be inferred that because of the "energy crisis" we are all likely to become vegetarians. One of the hardest things for people to change is their diet. Furthermore, sources of good vegetable protein like soybeans and hard grains are also becoming hard to get as the United States implements its basic 1970's foreign policy of using domestic crops to reduce the balance of payments deficit and to barter for oil and natural gas on the foreign market. Inevitably, scarcities of fossil fuels (whether real or political) will lower the quality of food for most people, especially the poor...unless, of course, we begin producing some of our own food at the local level and with local resources.

The Food System and Our Changing Living Habits

Agriculture is just the starting point of a complex food industry that involves production, transportation, marketing, plus domestic storage and cooking. In 1963, food-related activities consumed about 12% of the total U.S. energy budget[13], or the equivalent of about 240 gallons of gasoline per person (Table V). Assuming that a well fed person eats about 1 million kilocalories of food energy per year (29 gallons of gasoline), we can see that the energy needed to put food on our table is nearly 8 times more

---

**In the U.S. it takes about 6500 kcal. to produce one pound of beef or about 38,000 kcal per pound of protein[11]. In contrast, one pound of corn (fed to the cattle) requires only 500-650 kcal to produce[3,4].

than the energy we get from our food. Or put another way, one actually eats 8 times more energy per day than is contained in the food eaten. If we considered all the energy lost in converting plants to animals, the figure would become even higher. *The further removed we are from our food sources, the more energy we consume when we eat.*

SOURCE	MILLION BTU'S	HEAT EQUIVALENT GALLON GASOLINE	% OF TOTAL
Agriculture	5.8	43.0	18
Food Processing	10.6	78.5	33
Transportation	0.9	6.7	3
Wholesale & Retail Trade	5.2	38.5	16
Domestic	9.9	73.3	30
Cooking	3.91		
Refrigeration, Freezing	4.61		
Appliance Production	.66		
Travel for Food	.72		
TOTAL	32.4 million BTU's per person per year	240 gal. gasoline	100.0%

TABLE V. ENERGY REQUIREMENTS OF FOOD-RELATED ACTIVITIES IN THE UNITED STATES (1963). Adapted from Ref. 13.

A more detailed breakdown of the energy used in the U.S. food system is shown in Table VI. Both Tables V and VI reveal something interesting. About 1/3 of the energy that goes into providing us with food is consumed domestically. This suggests that a great deal can be done to change our habits and activities in order to deal with the increasing costs and declining quality of our food. A few suggestions:

1. Start a food garden; your own in your backyard, or a community project. In many U.S. cities, municipal and industrial land is being loaned for community gardens. When you grow your own food, use methods that don't require large inputs of fossil fuel energy (e.g., pesticides, chemical fertilizers, elaborate tools etc.) Over half of the pesticides used in the U.S. each year are used in and around cities[16], and each year people use more chemical fertilizers on their urban lawns than are used in the entire country of India.

2. Change your diet to more natural and raw foods, and less meat. This would reverse a trend in our diets that has been interrupted only a few times in the last 50 years (Table VII).

3. Organize food cooperatives and local food economies; develop urban/rural food alliances. Help to get food directly from nearby producer to consumer with as few intermediate steps as possible.

4. Support small independent and neighborhood food stores near to home. Support small farmers in their attempt to grow for markets in your area.

More and more pressure is being put on farmers to produce food for the demands of both a rising population and expanding global markets. The existence of energy-consuming factory farms will probably be a part of our food system for some time to come whether they are fueled by oil or nuclear power. But it needn't be an either/or situation. Everywhere

people want more control over their lives, institutions and resources. An agriculture using the resources and wastes of the surrounding region and producing for local markets is one of the most important ways that a society can provide for itself.

COMPONENT	1947	1958	1970	(1970) % of Total
On Farm				
Fuel (direct use)	136.0	179.0	232.0	
Electricity	32.0	44.0	63.8	
Fertilizer	19.5	32.2	94.0	
Agricultural steel	2.0	2.0	2.0	
Farm machinery	34.7	50.2	80.0	
Tractors	25.0	16.4	19.3	
Irrigation	22.8	32.5	35.0	
Subtotal	272.0	356.3	526.1	24.2
Processing Industry				
Food processing industry	177.5	212.6	308.0	
Food processing machinery	5.7	4.9	6.0	
Paper packaging	14.8	26.0	38.0	
Glass containers	25.7	30.2	47.0	
Steel cans and aluminum	55.8	85.4	122.0	
Transport (fuel)	86.1	140.2	246.9	
Trucks and trailers (manufacture)	42.0	43.0	74.0	
Subtotal	407.6	542.3	841.9	38.8
Commercial and Home				
Commercial refrigeration and cooking	141.0	176.0	263.0	
Refrigeration machinery (home and commercial)	24.0	29.4	61.0	
Home refrigeration and cooking	184.0	257.0	480.0	
Subtotal	349.0	462.4	804.0	37.0
Grand Total	1028.6	1361.0	2172.0	100.0

TABLE VI. ENERGY USE IN THE U.S. FOOD SYSTEM. From Steinhart (1974), Ref. 14. Values: 10^{12} kcal.

	FRUITS		VEGETABLES			MEATS
	FRESH	PROCESSED	FRESH	PROCESSED Canned	Frozen	
1920	130	17	196	18	—	160
1925	121	18	190	25	—	163
1930	119	19	197	28	—	153
1935	134	22	198	26	—	139
1940	138	34	198	34	.6	166
1945	133	34	217	43	1.9	178
1950	107	43	170	41	3.2	177
1955	95	49	155	42	5.9	192
1960	90	50	150	43	7.0	195
1965	80	48	141	47	8.0	204
1970	80	55	141	51	9.6	230

TABLE VII. PER CAPITA CONSUMPTION OF FRESH AND PROCESSED FOODS IN THE U.S. (Retail-weight equivalent in pounds). U.S.D.A. Statistics (1971)

References Cited

[1] Perelman, M. 1972. FARMING WITH PETROLEUM. Environment 14:8-13.

[2] Committee on Agriculture, House of Representatives. 1971. FOOD COSTS... FARM PRICES: A COMPILATION OF INFORMATION RELATING TO AGRICULTURE. 92nd Congress, 1st Session, Washington D.C.

[3] Pimentel, D., et al. 1973. FOOD PRODUCTION AND THE ENERGY CRISIS. Science, 182:443-449.

[4] Cervinka, V., W.J. Chancellor, R.J. Coffelt, R.G. Curley and J.B. Dobie. 1974. ENERGY REQUIREMENTS FOR AGRICULTURE IN CALIFORNIA, California Department of Food and Agriculture, Sacramento, Calif. (with University of California, Davis.)

[5] Blaxter, K. 1974. POWER AND AGRICULTURAL REVOLUTION. New Scientist 61 (885): 400-403.

[6] Black, J.N. 1971. ENERGY RELATIONS IN CROP PRODUCTION—A PRELIMINARY SURVEY. Annals Appl. Biol. 67(2): 272-278.

[7] Rappaport, R. 1971. THE FLOW OF ENERGY IN AN AGRICULTURAL SOCIETY. Sci. Amer. 225: 117-132.

[8] Cottrell, F. 1955. ENERGY AND SOCIETY. McGraw-Hill, New York.

[9] Heichel, G.H. 1973. COMPARATIVE EFFICIENCY OF ENERGY USE IN CROP PRODUCTION. Conn. Agric. Exp. Sta., Bulletin 739, New Haven.

[10] Muller, R.G. 1971. AMMONIA, PROCESS ECONOMICS. Program Report, no. 44, Stanford Research Institute.

[11] Slesser, M. 1973. ENERGY ANALYSIS IN POLICY MAKING. New Scientist, 60 (870): 328:330, 1 Nov., 1973.

[12] Slesser, M. 1973. HOW MANY CAN WE FEED? Ecologist, 3(6): 216-220.

[13] Hirst, Eric. 1973. ENERGY USE FOR FOOD IN THE UNITED STATES, Natural Science Foundation Environmental Program, Oak Ridge, Tenn. Gov. Printing Office.

[14] Steinhart, J.S. and C.E. Steinhart. 1974. ENERGY USE IN THE U. S. FOOD SYSTEM. Science, 184: 307-316.

[15] Merrill, R. 1976. TOWARDS A SELF-SUSTAINING AGRICULTURE. In: "Radical Agriculture", Harper and Row, New York.

[16] Environmental Protection Agency. 1972. THE USE OF PESTICIDES IN SUBURBAN HOMES AND GARDENS, AND THEIR IMPACT ON THE AQUATIC ENVIRONMENT. EPA/ Gov. Print. Office, Washington D.C.

AGRICULTURE BOOK REVIEWS AND INFORMATION

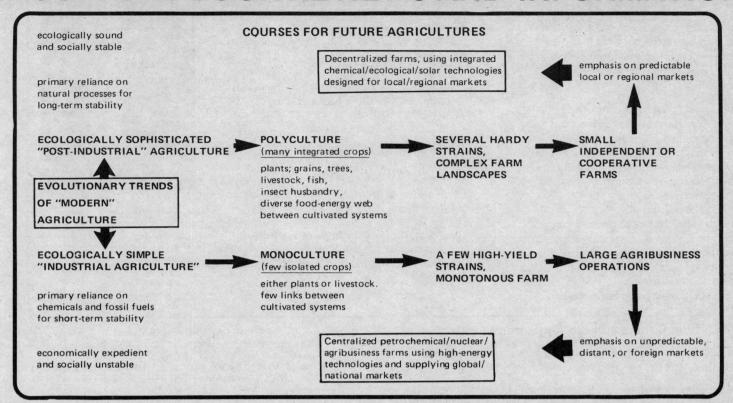

COURSES FOR FUTURE AGRICULTURES

ecologically sound
and socially stable

Decentralized farms, using integrated
chemical/ecological/solar technologies
designed for local/regional markets

emphasis on predictable
local or regional markets

primary reliance on
natural processes for
long-term stability

ECOLOGICALLY SOPHISTICATED
"POST-INDUSTRIAL" AGRICULTURE

POLYCULTURE
(many integrated crops)

SEVERAL HARDY
STRAINS,
COMPLEX FARM
LANDSCAPES

SMALL
INDEPENDENT OR
COOPERATIVE
FARMS

EVOLUTIONARY TRENDS
OF "MODERN"
AGRICULTURE

plants; grains, trees,
livestock, fish,
insect husbandry,
diverse food-energy web
between cultivated systems

ECOLOGICALLY SIMPLE
"INDUSTRIAL AGRICULTURE"

MONOCULTURE
(few isolated crops)

A FEW HIGH-YIELD
STRAINS,
MONOTONOUS FARM

LARGE AGRIBUSINESS
OPERATIONS

primary reliance on
chemicals and fossil fuels
for short-term stability

either plants or livestock.
few links between
cultivated systems

economically expedient
and socially unstable

Centralized petrochemical/nuclear/
agribusiness farms using high-energy
technologies and supplying global/
national markets

emphasis on unpredictable,
distant, or foreign markets

There are hundreds of good books on all aspects of gardening and agriculture. We have chosen those books that have been of special help to us. Whenever possible, we have grouped or keyed the references according to whether they are advanced, technical, and esoteric (A); popularized material with technical support (B); or basic and elementary (C). We have *listed important books that are out of print (OOP) in the hope that pressure will be brought to bear to bring them back. Other reference keys: Government Printing Office (GPO); National Technical Information Service, Springfield, Va. (NTIS).*

R.M.

——————— ECO-AGRICULTURE, FOOD AND PEOPLE ———————

There is a growing feeling in our culture today that the era of cheap abundant food is over, and that the cornucopia has been a short-term marvel with long-term costs to society. These costs include the loss of food quality, the destruction of our rural culture and environment, the rise of centralized food monopolies, and the consequences of a vast migration of people from farms to cities. Since 1948 over 25 million people have been relocated to urban centers by high technology and agribusiness economy. In less than two generations there has been a revolutionary change in the means of food production and in the patterns of human settlement and food distribution in this country.

The abandonment of farmlands and the separation of people from their land and food resources have become symbols of our social "progress." According to this view the success of our society can be measured by the degree to which our rural culture becomes a labor force for the urban machine and ceases to be a steward of the rural envrionment. But it is by no means obvious that the emigration of rural communities and the industrialization of agriculture have produced a just, stable, and fulfilling society. In fact, there is much to indicate that just the opposite has been the case, and that we have become affluent at the expense of agriculture, not because of it.

Since World War II we have so altered our rural environment and have become so totally dependent upon a single chemical strategy for food production that we face a future in which a major human concern will be the increasing hazards of the fuels and chemicals needed to continue the food supply. We brag of being a nation in which food is relatively cheap and agriculture is efficient; yet we ignore the fact that most measures of food prices and farm efficiencies fail to account for the endangerment of many valuable resources such as soil fertility, water, wildlife, public health, and a viable rural economy.

As a result of these trends there is a growing debate about the limits and hazards of modern farm technology and of the pollution and health problems caused by salt fertilizers, livestock chemicals, pesticides, etc. There is also much to indicate that the production potential of agriculture has already been reached and that higher yields will not warrant the costs of more energy and new technologies . . . especially on "left-over" marginal land.

The recent "energy crisis," however, has produced deeper insights into the major weakness of modern agriculture. Today's farms require massive inputs of fossil fuel energy to maintain them in a stable state. In fact, during the last few decades we have simply been exchanging finite reserves of fossil fuels for our food and fiber supply. Obviously this trade-off cannot continue indefinitely; if agriculture is energy-intensive, then fuel shortages must inevitably lead to food shortages. So in the very near future, we will have no choice but to adopt agricultural techniques that utilize *renewable* energy supplies. These include recycling of organic wastes to supplement synthetic fertilizers, using renewable forms of energy (solar, wind, and biofuel) to help supply rural power needs; and applying ecologically diverse cropping patterns and integrated pest management programs to reduce the use of pesticides.

It is not surprising, then, that rising energy prices have forced many farmers to reexamine some of the older, less energy-intensive, methods of soil husbandry (green manuring, and strip cropping and some of the newer methods of composting, integrated pest control, mixed cropping patterns renewable energy resources (solar-wind power and organic fuels), and the reintegration of livestock with agronomy. Without a broad approach to these alternatives, the future of modern agriculture could very well become self-defeating rather than self-sustaining.

The nub of the problem is this: for decades we have assumed that the only purpose of agriculture is to produce food. Over the years all sorts of propaganda . . . well-meaning and otherwise . . . have locked us into this illusion. As a result we tend to forget that agriculture is a dynamic biological process, not a factory. Its historic role has been to maintain productive land in order to sustain its people. In addition, a thriving rural culture has always played a vital part in providing food and fiber and absorbing dispossessed people during wars and economic depressions. In a healthy society, agriculture not only provides food; it also provides a reliable buffer during social crisis and a culture of land stewards for posterity. The practical use of renewable energy systems in agriculture can help us to put back the biology and culture in agriculture.

— Richard Merrill

RADICAL AGRICULTURE

... This anthology is the best introduction we've seen to the broad range of fundamental problems now facing the nation's food-production system. Highly recommended.
— Southwest Research and
Information Center, Albuquerque

If it is widely read, RADICAL AGRICULTURE could be an important book that has a far-reaching impact through its powerful synthesis of ideas from such diverse areas as soil science, philosophy and solar engineering. Ideally it should be stacked ... at supermarket check-out counters.
— Sam Love
The Washington Post

The newest and probably one of the best collections of essays as yet assembled on our changing agricultural scene.
— Maine Organic Farmers Assoc.

We are now able to say that the collective outlook represented by (such environmental pioneers as Rachel Carson, E.F. Schumacher, Howard Odum and Wendell Berry) has grown up and filled out to make a cultural plateau. There is unity, coherence, and vision in this outlook ... An impressive example of the comprehensive maturity of thinking in this cultural plateau is the just published RADICAL AGRICULTURE.
— Manas Magazine

I read many books, articles and papers about agriculture. If one half of what this book says is true—and I have no reason to doubt any of it—it amounts to the most serious indictment of an inefficient and wicked system that I have ever seen, and yet, in the alternatives it proposes, it offers the hope of the kind of reform that would bring America back to its philosophical origins, and that is bright optimism indeed.
— Michael Allaby
The Ecologist

Radical Agriculture
Richard Merrill, Editor
1976; 459 pp.

$6.95

from:
Harper and Row
General Books
Keystone Industrial Park
Scranton, PA 18512

Overviews: Rural Culture and Bio-Agricultural Systems

Allaby, M. and F. Allen. 1974. ROBOTS BEHIND THE PLOW. Rodale Press, Emmaus, Pa. A critical assessment of large-scale farming and a discussion of alternatives. A really fine book.

Commissie Onderzoek Biologische Landbouwmethoden. 1973. ALTERNATIVE INVENTARISATIE EVALUATIE EN AANBEVELINGEN VOOR ONDERZOEK LANDBOUW. (Alternative Manners of Farming: Survey, Assessment and Recommendations for Research). Centrum voor Landbouwpublikaties en Landbouwdocumentatie, Prinses Markjkeweg 17, Wageningen, The Netherlands. 159 pp. Important Dutch study outlining the economic advantages of ecological agriculture in North Europe. Being translated.

Council for Agricultural Science and Technology. 1972. DIRECTORY OF ENVIRONMENTAL SCIENTISTS IN AGRICULTURE. CAB Black, Dept. Agronomy, Iowa State University, Ames, Iowa 50010.

FARMING: SOURCES FOR A SOCIAL AND ECOLOGICALLY ACCOUNTABLE AGRICULTURE. Alternative Agricultural Resources Project (AARP), 870 Linden Lane, Davis, CA 95616. 68 pp. $2.00. Very good annotated reference book for post-industrial growers and teachers of the new agriculture.

Goldstein, J. 1973. THE NEW FOOD CHAIN: AN ORGANIC LINK BETWEEN FARM AND CITY. Rodale Press, Emmaus, Pa. A series of articles with a vision of cities and farms complementing each other instead of competing.

Horsfall, J.G. and C.R. Frink. 1975. PERSPECTIVE ON AGRICULTURE'S FUTURE: RISING COSTS ... RISING DOUBTS. Limits to Growth Symposium, Dallas, TX, October 21, 1975. The limits to continued growth in agriculture ... a landmark paper.

Koepf, H.H. et al. 1976. BIO-DYNAMIC AGRICULTURE: AN INTRODUCTION. Anthroposophic Press, Spring Valley, New York, and distributed by Bio-Dynamic Literature, Box 253, Wyoming, Rhode Island 02898. Most comprehensive English book on North American and European bio-dynamic farming techniques and actual farm descriptions ... the farm as an ecosystem.

LaVeen, P. 1972. PUBLIC POLICY AND THE FUTURE OF THE FAMILY FARM: A LAYMAN'S GUIDE TO THE ECONOMICS OF AGRARIAN REFORM. Department of Agricultural Economics, Univ. of Calif., Berkeley.

Lockeretz, W. et al. 1976. ORGANIC AND CONVENTIONAL CROP PRODUCTION IN THE CORN BELT: A COMPARISON OF ECONOMIC PERFORMANCE AND ENERGY USE FOR SELECTED FARMS. Center for the Biology of Natural Systems, Washington University, St. Louis MO. Two year study of 14 pairs of organic and conventional crop-livestock farms in the Corn Belt. First practical analysis of growing interdependence between economics and energetics in agriculture.

McWilliams, C. 1970. FACTORIES IN THE FIELD. Perregrine Press, Santa Barbara, California. The history of California agriculture and farm labor. A major indictment and devastating description of the most exploited labor class in the country.

Merrill, R. 1976. RADICAL AGRICULTURE. Harper and Row, NY (see Review).

Messerschmitt, H. et al. 1974. L'ENCYCLOPEDIE PERMENANTE DE L'AGRICULTURE BIOLOGIQUE (A Permanent Encyclopedia of Biological Agriculture). Editions Debard, Paris. A 2 volume, beautifully illustrated anthology. Loads of information, stressing balance between ecology, technology and economics in agriculture. In French.

POWER AND LAND IN CALIFORNIA. 1971. Ralph Nader Task Force Report. Center for the Study of Responsive Law, Washington D.C.

Rhodes, J.V. 1972. WHO WILL CONTROL U.S. AGRICULTURE? College of Agriculture, University of Illinois at Urbana-Champaign, Cooperative Extension Service, Special Publication No. 27.

Spedding, C.R. 1975. THE BIOLOGY OF AGRICULTURAL SYSTEMS. Academic Press, New York. 261 pp. How to identify more relevant integrated research programs in agriculture by using a systems approach.

Wolf, W. (ed.) 1975. ORGANIC FARMING: YEARBOOK OF AGRICULTURE. Rodale Press, Emmaus, PA. Summarizes practical information for the ecological farmer on soil fertility, pest control and markets. A nice response to the USDA Yearbook of Agriculture.

Walters, C. 1975. THE CASE FOR ECO-AGRICULTURE. Acres USA, 10227 East 61st Street, Raytown, MO 64133. 122 pp. Parity economics and ecological agriculture.

Williams, W.A. 1970. ROOTS OF THE MODERN AMERICAN EMPIRE. Random House, N.Y. American history as reflected in its changing agriculture.

Groups and Organizations

BROOKLYN BOTANIC GARDEN. 1000 Washington Ave., Brooklyn, N.Y. 11225. Extremely useful Garden and Horticultural Handbooks, priced from $1.25 to $1.60 on everything from soils and mulches to roses and dye plants.

CENTER FOR FARM AND FOOD RESEARCH, Inc., Box 166, Cornwall Bridge, CT 06754. Has written a FOOD POLICY STATEMENT for preserving farm-lands, for improving the American diet and for providing choice in the food marketplace. Organizes conferences, workshops and a food coalition; plus planning a monitoring program for pesticide use in Connecticut.

CENTER FOR RURAL AFFAIRS. Box 405, Walthill, NB 68067. Group is active in organizing legislation for the banning of corporate farming in the Midwest.

CENTER FOR STUDIES IN FOOD SELF-SUFFICIENCY. Vermont Institute of Community Involvement. 90 Main Street, Burlington, VT 05401. Group involved with techniques of assessing food needs at regional level. Excellent reports: LAND, BREAD AND HISTORY (Potential for Food Self-Sufficiency in Vermont) $2.50; and ENERGY UTILIZATION IN VERMONT AGRICULTURE, $1.50 (one of the few reports to demonstrate the energy efficiency of small farm operations).

COMMUNITY ENVIRONMENTAL COUNCIL. 109 E. De La Guerra St., Santa Barbara, CA 93101. One of the most progressive groups in the country in the areas of urban agriculture (AGRICULTURE IN THE CITY, $2.75), urban gardens and composting, recycling operations and urban planning around themes of appropriate growth and ways to make it possible.

GARDEN WAY PUBLISHING COMPANY, Charlotte, VT 05445, (802) 425-2171. Garden Way has been around for some time. Their publications are some of the most practical survival and how-to manuals around. At the present they are expanding into small scale machinery and alternative energy devices. Great people to deal with. Write and ask for their publication list.

INTERNATIONAL FEDERATION OF ORGANIC AGRICULTURE MOVEMENTS (IFOAM). c/o Research Institute for Biological Husbandry, Bootmingerstrasse 31, Postfach, CH-4104 OBERWIL/BL Switzerland. International networking quarterly and research organization. Valuable newsletter about worldwide events and results of the scanty research being done on the relationship between soil fertility and nutritional quality of food crops.

MIDWEST ORGANIC PRODUCERS ASSOCIATION, 606 S. 40th St., Omaha, NB 68105. Organization of midwest state organic growers groups. MOPA also conducts organic farming workshops, encourages research at agricultural colleges and publishes a newsletter

NATIONAL LAND FOR PEOPLE. 1759 Fulton, Rm 7, Fresno, CA 93721. Involved in long-term litigation with Bureau of Reclamation over enforcement of Reclamation Act and rights of small farmers to own land. Research into land ownership patterns in California opened a big crack in the wall of land monopoly. Films, newsletter and pamphlets very informative.

NATIONAL SHARECROPPER'S FUND. 2128 Commonwealth Ave., Charlotte, NC 28205. Through the Rural Advancement Fund of NSF the organization aids in the organization of farmers' marketing cooperatives in the south. Experimental Farm and training center is also active.

NATURAL ORGANIC FARMERS ASSOCIATION (NOFA) RFD 1, Plainfield, VT 05667. Membership is $10.00/year. They have a newsletter and advice on problems of ecological farming, direct marketing, local feed courses, etc.

NEBRASKA CENTER FOR RURAL AFFAIRS. Nebraska Low Energy Project, Box 405 Walthill, NB 68067. Working with integration of solar energy into agriculture. "Low Energy Agriculture" Newsletter.

NORTHERN CALIFORNIA LAND TRUST. 330 Ellis St., No. 504, San Francisco, CA 94102. Non-profit group seeking tax-deductible gifts of land or of money to buy land. This land is then held in perpetual trust, leasing it for life to low-income farming families or groups.

RODALE PRESS INC., Organic Park, Emmaus, PA 18049. The keeper of the "organic" movement during the skeptical years, the Rodale Press is now the major publisher (along with Garden Way) of popular books on food and energy self-reliance. Write for a list of their many books and recent research results.

SMALL FARM RESEARCH ASSOCIATION, Harborside, Maine 04642. Publish top quality pamphlets of interest to biological agriculture: The Use of Ground Rock Fertilizers in Agriculture, European Biological Agriculture 1976, and Annotated Bibliography of Biological Agriculture. "The credibility of biological agriculture suffers unfairly from misrepresentation and overstatement rather than from any functional weakness in its practices. This misrepresentation is sometimes the fault of overzealous amateurs but more often than not is compounded by those who wish for commercial reasons to sensationalize the latest 'miracle' product. It is our hope to counteract such hocus-pocus by publishing solid, concise, and dependable information which will enable those wishing to learn about biological agriculture to make decisions based on fact rather than on fiction."

SOIL ASSOCIATION. Walnut Tree Manor, Haughley, Stowmart, Suffolk, England IP 14 235. Founded in 1946 to promote organic husbandry. The "Haughley Experiments" (now run by Pye Institute) are the longest running organic farming experiments around. They publish a bi-monthly semi-technical magazine of highest quality sent free to members ($12/year).

SOIL ASSOCIATION OF MINNESOTA, INC. North Central group reporting about regional activities of organic growers and groups. $12.50/year membership and Newsletter.

Food and Nutrition

CENTER FOR SCIENCE IN THE PUBLIC INTEREST (CSPI). 1757 S. Street NW, Washington, D.C. 20009. Miscellaneous publications of their Nutrition Action Project: "Nutrition Action Newsletter," $10/year; "Food for People Not Profit" (Ballantine), $1.95; "Nutrition Scoreboard," $2.50.

Food Heritage. 1968. COMPOSITION AND FACTS ABOUT FOODS AND THEIR RELATIONSHIP TO THE HUMAN BODY. Health Research, 70 Lafayette St., Mokelumne Hill, CA 95245. 135 pp. Breaks down USDA Handbook No. 8 by foods richest in given nutrients instead of nutrient contents of different foods. Still the same data, though. Other good information on amino acids, vitamins and food combinations.

FROM THE GROUND UP: BUILDING A GRASS ROOTS FOOD POLICY. 1976. Center for Science in the Public Interest. 1757 S. St., N.W., Washington, D.C. 20009. Returning the control of our food system to people.

Hall, R.H. 1974. FOOD FOR NOUGHT: THE DECLINE IN NUTRITION. Vintage Books, New York. 308 pp. $3.95. Probably the most thorough critique of the role modern food industries play in the declining quality of our food . . . "(food) technology left unmonitored develops to suit technological, not biological ends."

Hightower, J. 1975. EAT YOUR HEART OUT. Crown Publisher, New York. Brief economic history and critique of our food distribution system.

Katz, D. and M.T. Goodwin. 1976. FOOD: WHERE NUTRITION, POLITICS AND CULTURE MEET . . . AN ACTIVITIES GUIDE FOR TEACHERS. Center for Science in the Public Interest. 1757 S. St., N.W., Washington, D.C. 20009.

Mitchell, D. 1975. THE POLITICS OF FOOD. James Lorimer & Co., Toronto, Canada.

NATIONAL FOOD SITUATION. Quarterly from: Economic Research Service, USDA, Washington, D.C. Free. Includes reports on food prices, predictions of future supplies, price trends and nutritional values.

New Alchemy Institute-West. 1977. Effects of Fertilizers and Soil Quality on the Nutritional Value of Food Crops: An Annotated Bibliography. NAI-West, Box 2206, Santa Cruz, CA 95063. $3. Scores of annotated references from technical journals, books and popular articles dealing with the entire range of opinion about the relative effects of climate, soil quality, and fertilizers on the nutritional value of crops.

Lappé, F.F. 1971. DIET FOR A SMALL PLANET. Ballantine, N.Y. A classic description of how (and why) to combine plant foods (legumes and grains) in cooking to obtain an adequate protein diet. Companion book: RECIPES FOR A SMALL PLANET by Ellen Buchman Ewald.

Schupan, W. 1965. NUTRITIONAL VALUES IN CROPS AND PLANTS. Faber and Faber, London. A summary of years of research.

Science Magazine. 1975. FOOD. American Association for the Advancement of Science, 1515 Massachusetts Ave., Washington, D.C. May 9, 1975 issue devoted entirely to food, nutrition and agricultural research.

Periodicals

ACRES, USA monthly newspaper, 10227 East 61st St., Raytown, MO 64133. $5.00/year. One of the few practical organic farming periodicals in this country. It is geared to midwest conditions, but has valuable information for everyone. Articles, organic merchants, and most important, a political base for what the movement is all about. Highly recommended.

AGRICULTURAL SCIENCE REVIEW. From: Government Printing Office, Washington, D.C. 20402. Quarterly, $2/year. Published by Cooperative State Research Service, USDA. Critical review journal of current state of agricultural research.

AGROBIOLOGIA. Centro Pedagogico y Cultural de Portales, Casilla 544, Cochabambra, Bolivia. In Spanish, perhaps the only periodical in South America devoted to the small diversified family farm and its role in the South American culture.

AGRO-ECOSYSTEMS. Elsevier Science Publishing Co., P.O. Box 211, Amsterdam, The Netherlands. $37.50/year. Something for your library to subscribe to. Quarterly journal of technical articles concerned with "ecological interactions between agricultural and managed forest ecosystems."

CALIFORNIA AND NORTHWEST ORGANIC JOURNAL. P.O. Box 540-H, Dept. A, Halcyon, CA 93420. bi-monthly, $4.20/year. Official journal of California Organic Growers.

COMPOST SCIENCE. Rodale Press, Emmaus, PA 18049. $6/year. Bi-monthly semi-technical articles on agricultural and municipal composting. A very important periodical . . . we need a format and energy like this for biological agriculture in general.

COUNTRY JOURNAL. 139 Main St., Brattleboro, VT 05301. $9.50/year, monthly. Practical survival skills directed to New England conditions. Fair articles on wood and other renewable energy systems.

COUNTRYSIDE. 312 Portland Rd., Waterloo, WI 53594. $9/year. Published since 1917, COUNTRYSIDE has a new format . . . a monthly magazine directed to rural folks, testimonials and articles about how-to in old and new ways . . . food and energy.

HARROWSMITH. Camden East, Ontario, Canada KOK 1JO. $6/year. Canadian back-to-the-land magazine; geared to norhtern living.

JOURNAL OF ENVIRONMENTAL QUALITY. 677 South Segoe Rd., Madison, WI 53711. Quarterly. Published by Crop Science and Soil Science Societies of America and the American Society of Agronomy. Technical reports include organic fertilizer utilization and soil dynamics.

MAINE ORGANIC FARMER AND GARDENER. P.O. Box 187, Hallowell, ME 04347; Bi-monthly, $2.50/year. Newspaper of the Maine Organic Farmers and Gardeners Association (MOGFA). Excellent compliment to ACRES, covering the theory and practice of biological agriculture, especially in Northeast Region.

MOTHER EARTH NEWS, P.O. Box 70, Hendersonville, NC 28739. $10/yr, bimonthly. The original back-to-the-land magazine, nice inspiration to city-dwellers and others just starting.

ORGANIC GARDENING AND FARMING MAGAZINE, Rodale Press, Emmaus, PA 18049 (monthly, $5.85/year). "OG and F" has been around since 1942. The magazine is only part of the Rodale Press which publishes numerous books and articles on food, health, gardening and other techniques for self-reliance.

SMALL FARMER'S JOURNAL. SFJ Inc., P.O. Box 197, Junction City, OR 97448; Quarterly, $8.50/year. Geared to the forgotten in-between scale of operation, the family farm. Journal emphasizes the use of horses and mules for motive power. "Practical Horse-Farming."

SCIENTIFIC HORTICULTURE. Journal of the Horticultural Education Association. Birbeck College, Malet St., London WC1E 7AX. £3.50/year once per year. Two to three hundred page journal on technical/ecological aspects of gardening and small farming methods.

TILTH NEWSLETTER. Rt. 2, Box 190-A, Arlington, WA 98223. $5/year, irregular. Outgrowth of the 1974 Northwest Alternative Agriculture Conference. Good articles and access information on agricultural technologies, organic farming, local food economies and marketing, etc., for the Northwest, but with applications everywhere. Loosely associated with Ecotope and Alternative Market Newsletter. Evolving (hopefully) into a regional research journal for radical agriculture.

━━━━━ URBAN AGRICULTURE ━━━━━

THE CULTURE OF URBAN AGRICULTURE

More and more these days one sees food gardens growing in and around cities . . . on rooftops, in backyards, vacant lots, parks, industrial greenbelts, etc. In fact, during the last few years there has been a dramatic increase in both home and community gardens throughout the country. To date, for example, there are major community gardening projects in the urban centers of Ann Arbor, Boston, Cleveland, Chicago, Los Angeles, New York, Portland, San Francisco, Seattle, Syracuse and Washington D.C., plus scores of suburban locations across the land.

There are a variety of traditional motives associated with community gardening. These include recreational exercise, fellowship with neighbors,

therapy, hobby and meditation (for many, gardening and not wilderness is a focus for a new commitment to the environment), creation of esthetic open space, etc. But perhaps the most compelling reason for the rising interest in urban food-gardening is the economic inflation of recent years that has caused many people to consider growing some of their own food in the cities. A recent Gallup Poll revealed that nearly 40% of the households (27 million) grow some of their own food, that nearly half of the nation's non-gardeners would have vegetable gardens if they could save money on their food bill, and that each year the number of vegetable gardeners in the U.S. increases by about 3 million.

There is nothing unique about this trend. Urban food-gardens seem to "crop" up every time there is a war ("Victory" gardens) or economic slump ("inflation" gardens). In fact public interest in food gardening has become an integral part of our recent history and its war-prosperity-depression cycles. When food is scarce and prices are high people tend to return to roots.

Nevertheless, a number of things suggest that today's gardening popularity is more than a casual interest or temporary response to hard economic times . . . that it is more of a transition than a trend. For one thing it is getting harder for city people to buy fresh food as urban and rural cultures grow further apart each year. It is likely that in the near future those that want fresh foods will simply have to grow it. There is also the fact that as our food system becomes more centralized, processed and dependent upon limited fossil fuels, the food we eat can only become lower in quality and higher in price. As noted in the previous article ENERGY AND AGRICULTURE, the further we get from our food source, the more energy we consume when we eat. Urban agriculture is a means of conserving energy by stimulating the local production and distribution of food stuffs where they are used.

There are other aspects of a transition in the sense that more and more community gardens are becoming integrated with a localized urban food economy that is unique to our recent history. This emerging food system includes food coops, farmers markets, alternative warehouses, regional trucking groups, and other forms of direct sales and regional distribution networks, all working to decentralize our inefficient food system, conserve energy and make available locally-produced, quality foods at reasonable prices. For many, urban agriculture is becoming a vital political reality with strong economic and environmental implications.

Finally, there is a growing interest in the application of urban agriculture to new ideas of urban and regional planning. Each year about 350,000 acres of farmland . . . roughly half the land area of Rhode Island . . . is lost to urban development, and each year an additional 1.9 million acres . . . about the size of Delaware . . . is removed from the "rural" or food production category and used for highways, recreation and preservation of wilderness. Obviously this pinch can't continue for long. Halting unnecessary urban growth by designating fertile urban areas for food production is a unique way to provide needed open space in cities and to help recycle a vast amount of municipal waste while providing jobs and food for the local population. Urban agri-culture is truly an emerging culture . . . new and unique to our times.

— Richard Merrill

Basic Information

SEE LISTS OF: "Agricultural Groups" and "Agricultural Periodicals" (pp.166•7) which describe activities useful to URBAN AGRICULTURE.

AGRICULTURE IN THE CITY. 1976. Community Environmental Council, Inc., 109 E. De La Guerra St., Santa Barbara, CA 93101. $3.00. Description of El Mirasol Polyculture Urban Farm Project. Techniques of Raised-Bed horticulture, natural pest control, small livestock husbandry and solar energy technologies as they apply to city conditions.

Assembly Committee on Environmental Quality. 1972. URBAN AND SUBURBAN USE OF PESTICIDES IN CALIFORNIA. California State Assembly, Sacramento, California No. 256.

COMMUNITY GARDEN NEWS. Emphasis on vacant lot gardens, school gardens, company/employee gardens, housing development gardens, gardens for the handicapped, and how cities can get programs under way. Write: Jean M. Davis, Suite G17, American City Bldg., Columbia, MD 21044.

COMMUNITY GARDENS. 1975. Hunger Action Center, Evergreen State College, Olympia, WA 98505. "Fostering greater individual and community control of food production and distribution." Community Garden Handbook ($1.00) and Farmers Market Organizers Handbook ($1.00).

Drake, S. and R. L. Lawrence. 1976. RECREATIONAL COMMUNITY GARDENING. Office of Information, Bureau of Outdoor Recreation, 18th & C, N.W., Washington D.C. 20240. 72 pp. Free. Based on 1975 nationwide series of Community Gardening Seminars, this book outlines criteria for establishing community gardens based on experiences of local programs.

Environmental Protection Agency. 1972. THE USE OF PESTICIDES IN SUBURBAN HOMES AND GARDENS AND THEIR IMPACT ON THE AQUATIC ENVIRONMENT. Pesticide Study Series No. 2 GPO.

Gardens For All, Inc. 1974. COMMUNITY GARDENING PROCEDURAL MANUAL. P.O. Box 164, Charlotte, Vt. 05445. Considerable experience and information about organizing community gardens. Write 163 Church St., Burlington VT 05401.

Cold, S.M. 1974. HUMAN RESPONSE TO PLANTS IN CITIES. California Horticultural Journal, Vol. 35, No. 4.

Goldstein, J. 1973. THE NEW FOOD CHAIN: AN ORGANIC LINK BETWEEN FARM AND CITY. Rodale Press, Emmaus, PA. A series of articles with a vision of cities and farms complementing each other instead of competing.

INSTITUTE FOR LOCAL SELF RELIANCE. 1717 18th St., N.W., Washington D.C. 20009. Working with the organization, economics and techniques of inner city food production from roof tops to vacant lots. Excellent bi-monthly newsletter includes:

VOL. 2 (June 1976). Getting the Lead Out: problems of contamination of city-grown food leads to the question . . . if plants are polluted aren't we? Urban gardens as necessary biological indicators . . . not dangerous carriers of heavy metals.

VOL. 4 (Nov. 1976). Organic Hydroponics: A Simple Solution: growing plants without soil is hard for hard-line growers to accept. For those in the inner city, it may be the only way to practically liberate roof tops for food production. ILSR is investigating low-chemical, light weight ways of doing hydroponics: fish-emulsion, blood meal, blended egg shells, compost teas, perlite, vermiculite, etc.

VOL. 5 (January 1977). How Does Your Garden Grow? estimating space needed for significant urban gardens. In single block area of very populated areas (Adams-Morgan), 82,000 ft^2 of rooftops and 44,000 ft^2 of backyards could provide 33% of vegetable needs for 468 people (per block). Thus, any attempt to grow food in inner city must be serious well coordinated effort using new and old techniques like local waste composting, solar greenhouses, perennial crops, roof-top hydroponics, canning and drying, etc.

Morris, D. and G. Friend. 1975. ENERGY, AGRICULTURE AND NEIGHBORHOOD FOOD SYSTEMS. Institute for Local Self-Reliance, 1717 18th St., N.W., Washington DC 20009.

PROFILES OF CALIFORNIA COMMUNITY GARDEN PROJECTS. 1975. Bureau of Outdoor Recreation, 450 Golden Gate Ave., San Francisco, CA 94102. Free.

The Wedell Group. 1976. INTENSIVE SMALL FARMS AND THE URBAN FRINGE. Landal Institute., Small Farm Research, 2300 Bridgeway, Sausalito, CA 94965. A land use study to determine the potential of agriculture in the urban fringe and if family-sized ecological farming can economically thrive in the urban/rural impact zones around the San Francisco Bay Area. A model of regional land use planning in urban areas.

TILTH NEWSLETTER. Rt. 2, Box 190-A, Arlington, WA 98223. 4 issues/yr, $5. Spinoff of 1974 Northwest Alternative Agricultural Conference. Good discussions of movement in Pacific Northwest.

URBAN HOMESTEADING: PROCESS AND POTENTIAL. National Urban Coalition. 2100 M St., N.W., Washington D.C. 20037. $2.50.

Food Production

REFER ALSO TO: "Crop Production," section (p. 174) and "Natural Gardening and Farming" section (p. 169).

Easey, Ben. 1974. PRACTICAL ORGANIC GARDENING. Faber and Faber Ltd., 3 Queen Square, London, WC1N 3AU. $4.50. British garden wisdom at its best.

Fryer, L.D. Simmons. 1975. ECOLOGICAL GARDENING FOR HOME FOODS. Mason-Charter, NY. Useful How-To book for urban gardener.

Jeavons, John. 1974. HOW TO GROW MORE VEGETABLES . . . Ecology Action of the Midpeninsula, 2225 El Camino Real, Palo Alto, CA 94306. The best introduction to the techniques of Bio-Dynamic/French Intensive market gardening . . . An excellent method for growing a great deal of food in small plots while conserving resources, energy and water.

Mittleider, J. 1970. FOOD FOR EVERYONE. Color Press, College Place, Washington 99324. A description of the "Mittleider Method" of intensive truck farming Large doses of chemical fertilizers plus artificial soils = big yields. May be of some value to roof top gardeners and other situations where soil and natural fertilizers are in short supply.

Newcomb, D. 1974. THE POSTAGE STAMP GARDEN BOOK. Hawthorn Books, Ltd., N.Y. Container and raised bed techniques for city people in small places.

Newcomb, D. 1976. THE APARTMENT FARMER: HASSLE-FREE WAY TO GROW VEGETABLES INDOORS, ON BALCONIES, PATIOS, ROOFS AND IN SMALL YARDS. J.P. Tarcher, Los Angeles.

Olkowski, H. and W. Olkowski. 1975. THE CITY PEOPLE'S BOOK OF RAISING FOOD. Rodale Press, Emmaus, PA 18049. Certainly one of the best books for urban growers. Also small livestock in crowded city areas are discussed.

Rateaver, B. and G. Rateaver. 1973. THE ORGANIC METHOD PRIMER. The Authors, Pauma Valley, CA 257. Discusses the various methods of small scale ecological food production. Very useful book especially discussions of "cover crops," "weeds," "tools," and "fertilizers."

Raymond, D. 1975. DOWN TO EARTH VEGETABLE GARDENING KNOW-HOW. Garden Way Publishing, Charlotte, VT 05445. Oriented to Eastern US, good practical advice and vivid graphics.

Schroeder, M. 1975. THE GREEN THUMBBOOK. (1000+ Sources for Seeds, Supplies and information). Valley Crafts, 168 Rainbow Lane, Cary, IL 60013. $2.95.

Thompson, H.C. and W. C. Kelly. 1957. VEGETABLE CROPS. McGraw-Hill Book Co., Hightstown, NJ 08520. $14.50. A classic Ag school textbook for decades . . . much useful information.

U.S. Department of Agriculture. COMMUNICATING HOME GARDEN INFORMATION 2 vol., Extension Service, USDA, Washington, D.C. 20250.

U.S. Soil Conservation Service. Information Division, USDA, Washington, D.C. 20250. Various primers for agriculture in the cities include: Know Your Soil (Home and Garden bulletin No. 267); Soil Conservation at Home; Mulches For Your Garden (Home and Garden Bulletin No. 185); Gardening On The Contour (Home and Garden Bulletin No. 179).

Wickers, David. 1976. THE COMPLETE URBAN FARMER. Julian Friedmann Publishers, London 173 pp; $6.50. Far from complete, but fair introduction to urban food production in small residential places (backyards, indoors and containers). Note conditions in England.

Marketing

Basic Information

ALTERNATIVE MARKET NEWS. c/o Earth Cyclers, Rt. 1, Edwal, WA 99008. Northwest regional newsletter . . . alternative food systems and economies.

McLeod, Darryl. 1976. URBAN-RURAL FOOD ALLIANCES: A PERSPECTIVE ON RECENT COMMUNITY FOOD ORGANIZING. In: "Radical Agriculture," R. Merrill, ed., Harper and Row, New York.

NATURAL ORGANIC FARMERS ASSOCIATION. RFD, Box 247, Plainfield, VT 05667. Major coordinating group for Northeast activities in local food marketing (food coops and farmer's markets). Valuable newsletter, $5/year.

Ridgeway, James. 1975. NEW FOOD SYSTEM IN VERMONT. The Elements, July 1975. Description of Northeast food distribution system involving food co-ops, warehouses, trucking and contract methods for organizing farm production on a regional scale.

Food Coops

Community Services Administration. 1200 19th St., N.W. Room 332, Washington D.C. 20506. NATIONAL CONSUMER DIRECTORY (listing of food coops); OPERATIONAL MANUAL . . . CO-OP STORES AND BUYING CLUBS (organization in low-income neighborhoods).

FOOD COOP PROJECT. 106 Girard S.E.; No. 110, Albuquerque, NM 87106. Useful publications include: The Food Coop Directory (1750 collective warehouses, processors, restaurants etc. $3.00); How to Start A Food Coop, $.40.

JAM TODAY: CALIFORNIA FOOD COOP NEWSLETTER. Jam Today, P.O. Box 1517, Palo Alto, CA 94301. bi-monthly, $2/year. Good newsletter for a regional food economy.

New England Federation of Co-ops (NEFCO). 1975. THE FOOD CO-OP HANDBOOK. Houghton Mifflin Co., 2 Park St., Boston, MA 02107. The most complete book on the subject, covering buying clubs, storefronts, financing, inventory, national food co-op directory, etc.

NORTH AMERICAN STUDENT CO-OP ORGANIZATION (NASCO). Box 1301, Ann Arbor, MI 48106. Clearinghouse of co-op information: Nasco News Bulletin, a Journal (New Harbinger), and many books and reprints (How To Start Your Own Food Co-Op: A Guide to Wholesale Buying by Gloria Stern).

NUTRITIONAL DEVELOPMENT SERVICES. Archdiocese of Philadelphia, 222 N. 17th St., Philadelphia, PA 19103. Distributes information about organizing Food co-ops.

ON THE MARKET. Citizen Action Press, 443 Russel Blvd., Davis, CA 95616. Monthly newsletter for farmers, food coops, truckers and warehouses of Northern California.

Ronco, W. 1974. FOOD CO-OPS: AN ALTERNATIVE TO SHOPPING IN SUPER-MARKETS. Beacon Press, Boston. A guide to organizing and developing large-scale food co-ops backed by a nice history of the movement.

THE COOPERATIVE LEAGUE. 1828 L St., N.W. Suite 1100, Washington, D.C. 20036. Tends to aim advice at those wishing to operate a co-op DEPOT (for dealing with food brokers), which requires 300-500 families, some $50,000 or so in capital and good paid management and labor force. Publications include: The Manual of Basic Co-op Management, A Primer of Bookkeeping for Cooperatives, A Brief Guide for Natural Food Co-Ops.

THE CULTIVATOR. Federation of Cooperatives, Inc., Box 107, 15 Central St., Hollowell ME 04347. Regional Co-op magazine for Maine and its local food economies.

Vellela, T. 1976. FOOD CO-OPS FOR SMALL GROUPS. Workman Publishing Co., 231 E. 51st St., New York, NY 10022. Good information for starting a food coop.

Wickstrom, L. 1975. FOOD CONSPIRACY COOKBOOK. 101 Productions, 834 Mission St., San Francisco, CA 94103. $4.95 Experiences of the Berkeley Food Co-op System . . . decentralizing local food economies for efficiency.

Farmers Markets

Bowler, D. 1976. FARMERS' MARKET ORGANIZER'S HANDBOOK. Hunger Action Center, Evergreen College, Olympia, WA 98505. $1.00 Experiences in the Northwest.

FARMERS' MARKET PACKET. 1976. Evanston Chamber of Commerce, 807 Davis St., Evanston, Ill. 60201. Experiences of citizens' group in Evanston.

ORGANIZING FARMERS' MARKETS. 1975. Natural Organic Farmer's Association, RFD 1, Box 247, Plainfield, VT 05667. $2.00 Experiences in the Northeast.

━━━━ NATURAL GARDENING AND FARMING ━━━━

Just as a grassland matures into a forest, a garden or field of crops matures in its own way through the years. When crops are first planted in a new area, especially if it has been fallow for a while, soil nutrients are often out of balance, weed seeds are plentiful, and local pests are apt to cause problems until you get to know their life histories. However, with each passing year, soils become richer and cleaner and the local environment becomes a familiar backdrop to the dynamics of the field or garden.

It's during the first year or two of a garden that people generally get discouraged. Weeds and bugs can take over quickly in an "immature" piece of land. Also, there is a tendency for people who are just starting out to celebrate the planting and the harvest, but to ignore the vital tasks of maintenance in between. A food-garden should be viewed as a dynamic habitat which, if given steady (not necessarily heavy) attention, will mature through the seasons. That is, it will return more and more food relative to the labor needed to produce it.

There are three parts to a self-renewing garden scene: soil fertilization, pest management, and crop propagation. Natural fertilizers (compost, green manures, rock minerals, etc.) are discussed in virtually every organic gardening book. Natural pest management is described to some degree in most of the books. But the practical aspects of plant propagation seem to be left to speciality books, most of which are out of print (see page 178 of the Primer). We are told how to compost and get rid of aphids, but not how to breed locally adapted varieties of crops. This seems to be the weak link in any really complete natural gardening book. This is why we have placed so much emphasis on the subject of propagation in this section.

Scores of books have been written about organic/bio-dynamic/ecological gardening and farming techniques, philosophies, and experiences. They all have one thing in common: they are general. No book can give you all the answers . . . unless of course you write it yourself for your own backyard, vacant lot or field. Ideally each region, community or city should have its own reference book for food production that describes local conditions of climate, pests, soils, adapted crop varieties, sources of organic materials, etc. In the meantime, maybe some of the books listed below will help things get started. Also refer to list in *Urban Agriculture* section.

— Richard Merrill

Gardening

Darlington, J. 1970. GROW YOUR OWN. Bookworks, Berkeley, Calif. 87 pp. (C) Very simple introduction to backyard organic gardening.

Foster, C.F. 1972. THE ORGANIC GARDENER. Vintage Books, N. Y. 234 pp. (C) Forty-three years of experience and advice. Good drawings, references, and some great ideas.

Heckel, A. (ed.). 1967. PFEIFFER GARDEN BOOK. Bio-Dynamic Farming and Gardening Assoc., Stroudberg, Pa. 200 pp. (C) Classic introduction to backyard bio-dynamic gardening.

Kramer, J. 1972. THE NATURAL WAY TO PEST-FREE GARDENING. Charles Scribner's Sons, N. Y. 118 pp. (C) Well-balanced between pest control and soil fertility. Talks more about the why and what than the how. Good list of organic merchants nationwide.

O'Brien, R. Dalziel. 1956. INTENSIVE GARDENING. Latimer Trend & Co., Plymouth. 183 pp. Details of labor intensive cultivation of organic vegetables, which focuses on no-dig methods of production.

Ogden, S. 1971. STEP-BY-STEP TO ORGANIC VEGETABLE GROWING. Rodale Press, Emmaus, Pa. 182 pp.(C) A very good "how-to" book of one organic grower's methods.

Philbrick, J. & H. 1971. GARDENING FOR HEALTH AND NUTRITION. Steiner Publications, Blauvelt, N. Y. 93 pp. (C) The most succinct introduction to bio-dynamic gardening.

Rodale, R. (ed.). 1971. THE BASIC BOOK OF ORGANIC GARDENING. Ballantine Books, N. Y. A more concise version of earlier books. A very good introduction.

Rodale, J.I. (ed.). 1961. HOW TO GROW VEGETABLES AND FRUITS BY THE ORGANIC METHOD. Rodale Press, Emmaus, Pa. A classic reference book of organic technique. Mostly testimonials that seem relevant to Eastern U.S. But for first approximations it is quite valuable.

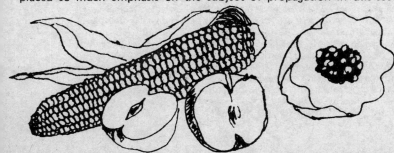

Rodale, J.I. (ed.). 1959. THE ENCYCLOPEDIA OF ORGANIC GARDENING. Rodale Press, Emmaus, Pa. A book can be judged by how often it's used. This book gets used an awful lot. Outdated in the area of natural pest control, but a valuable reference book.

Rodale, J.I. 1945. PAY DIRT. Rodale Press, Emmaus, Pa. 245 pp. Rodale's first and perhaps best book. Now in its 14th printing.

Sunset Books. 1967. SUNSET WESTERN GARDENING BOOK. Lane Books, Menlo Park, Calif. 449 pp. East or West, this is the most useful book we have found for everyday gardening advice. Beautifully organized, cross-referenced in encyclopedic fashion. The growing zone maps for the Western states ought to be expanded for the entire U.S. (Non-organic).

Sunset Books. 1971. GUIDE TO ORGANIC GARDENING. Lane Books, Menlo Park, Calif. 72 pp. An excellent PRACTICAL guide to methods and tools.

Tyler, H. 1972. ORGANIC GARDENING WITHOUT POISONS. Simon and Schuster, N. Y. 224 pp. Good photos, well rounded between soil fertility and pest control.

Farming

Allen, F. 1971. MAKING A GO OF ORGANIC FARMING: SOME COMMERCIAL CONSIDERATIONS. Rodale Press, Emmaus, Pa. 28 pp. $0.50.

Bromfield, L. 1947. MALABAR FARM. Ballantine Books, N. Y. 470 pp. (C) The first book to read if you want to run a farm without chemicals and with the cycles of life. A personal account.

Hainsworth, P. 1954. AGRICULTURE: A NEW APPROACH. Faber & Faber, Ltd., London. 248 pp. (B) One of the best sources for documenting the validity of organic farming techniques. Evidence culled from many places.

Howard, A. 1947. THE SOIL AND HEALTH: A STUDY OF ORGANIC AGRICULTURE. Devin-Adair (now Schocken Books, N. Y., 1972). 307 pp. (C) A classic overview by a pioneer in organic soil fertility. Good practical stuff from years of experience, but the other half of the problem (pests) is taken too lightly.

King, F.H. 1911. FARMERS OF 40 CENTURIES: PERMANENT AGRICULTURE IN CHINA, KOREA AND JAPAN. Rodale Press, Emmaus, Pa. 441 pp. (C) $7.95. Classic treatise of aboriginal farming in the Far East. Organic recycling at the epitome of cultural developmment. Valuable photos of early rural China.

Pfeiffer, E. 1940. BIO-DYNAMIC FARMING AND GARDENING. Anthroposophic Press, N. Y. 240 pp. (C) THE introduction to bio-dynamic farming techniques.

Steffen, R. et al. 1972. ORGANIC FARMING: METHODS AND MARKETS. Rodale Press, Emmaus, Pa. 124 pp. Testimonials of organic farmers, some techniques, marketing problems, directory of organic farmers in U.S. Superficial but handy.

U.S. Dept. Agriculture. 1938. SOILS AND MAN. USDA Yearbook for 1938. (GPO) (OOP) After the dustbowl of the mid-thirties, the USDA sent scientists and technicians into rural America. A survey in 1934 of 2 million acres was taken with regards to erosion, soil fertility depletion, etc. The results of the survey are published as the yearbook. Most of the book emphasizes organic materials and methods. This book is absolutely essential for the serious organic agriculturist. It is without equal in its depth of treatment and source of data. Write the USDA and demand that it be reprinted.

SOILS: BIOLOGY AND FERTILITY

SOIL FUNDAMENTALS

Soil is perhaps the most complex ecosystem in the biosphere. In many ways, less is known about it than the surface of the moon. Soil consists of both the living and the dead, the geological and the biological. Microbes, animals, minerals, water, and gases all interact in a dynamic balance of life and death. The soil isn't just a medium for holding plants upright. It's also a habitat for many living things, whose related activities feed the plants. And it is this biological energy in the soil that can be tapped to fertilize crops and give us food.

Some of the larger soil animals are familiar: insect larvae, earthworms, etc. are worm shaped or otherwise equipped for burrowing. Other animals like sow bugs and ground beetles live on top of the soil under refuges.

Far richer in species and numbers are the microbes that don't have to make their spaces, but which live in the tiny pores between bits of soil. Although they escape our eyes, soil microbes —algae, fungi, and bacteria— make up the bulk of living matter and form the energy foundation upon which all other members of the soil community ultimately depend. Most soil microbes live inside other animals or in the film of moisture that lines the hollows of the soil.

Soil is the product of two processes: the decomposition of rock (biological and by natural weathering) and the decay of plant and animal life. The combination of these processes determines the ability of soils to support plant life. As organic matter decays, it gives off acids that release nutrients held in a complex form in rock minerals (mostly clay). This is why the best natural fertilizer is a combination of manure or compost and finely ground rock mineral fertilizers (limestone, dolomite, rock phosphate, rock granite, etc.). As usual, the geological and the biological form the optimum symbiosis.

The formation of *humus* is the most important component of soil fertility. The process begins with large soil animals eating decaying organic debris that has been partly broken down by certain bacteria. Other kinds of bacteria continue the breakdown in the intestines of the soil animals. As the excretions of the large soil animals are eaten by smaller animals, they become progressively finer and richer in nutrients that plants can take up through their roots. This "soil manure" is called *humus*. Just as cows give manure from plants they eat, so soil animals give manure *(humus)* from the soil's organic matter that they eat.

Humus serves a variety of purposes in the soil: 1) It provides nutrients. 2) It helps to release nutrients from rock materials. 3) It holds water in the soil. 4) It stimulates beneficial soil fungi such as *mycorrhiza* (symbiotic root fungi) and predatory forms that prey on the nematodes. 5) Decaying organic matter releases carbon dioxide that can be used by plants during photosynthesis. 6) It provides trace minerals typically lacking in chemical fertilizers. Trace minerals (e.g., boron, manganese, copper, cobalt) are the building blocks of enzymes and vitamins that help plants to resist disease. 7) To some degree (and there is controversy here) humus provides antibiotics to plants that also help them to resist disease. All of these processes operating in concert from the one main source . . . humus . . . serve to promote a healthy soil and healthy plants.

Humus can be made by gathering up local organic wastes into "compost piles" and letting the materials decay into humus. By putting the pile together carefully and with attention to detail, a rich fertilizer and soil conditioner can be produced.

The increasing cost of chemical fertilizers has placed a premium on organic wastes in rural areas. Where livestock operations used to give manure away, now they are selling it. However, in the cities, organic residues are still considered a waste rather than a resource. Go to your local dump and watch it being thrown away . . . by the ton! About 30% of all the solid waste generated in the home is capable of being composted. Unfortunately most of this still goes down garbage disposals or into trash cans. This is too bad, because there is plenty of good information available about ways to fertilize with local organic wastes. It only takes a little time and change of habit.

—Richard Merrill

Basic Soil Books

Farb, Peter. 1959. LIVING EARTH. Harper & Row, N. Y. 178 pp. Simple introduction to the ecology of the soil.

Kohnke, H. 1966. SOIL SCIENCE SIMPLIFIED. Balt Publ. Lafayette, Indiana. 77 pp.

Kuhnelt, W. 1961. SOIL BIOLOGY. Rodale Press, Emmaus, Pa. 397 pp. Draws on work mostly from European soils, but still very good for general descriptions of soil properties and adaptations of soil animals. Good bibliography.

Ortloff, H. and H.B. Raymore. 1972. A BOOK ABOUT SOILS FOR THE HOME GARDENER. William Morrow & Co., Inc., N. Y. 189 pp.

Russell, E.J. 1957. THE WORLD OF SOIL. Fontana, c/o Watts, Franklin, Inc., N. Y. 237 pp. Clear and smooth descriptions by an expert.

Russell, E.J. 1950. LESSONS ON SOIL. Cambridge Univ. Press, Cambridge. 133 pp. Great learning guide . . . for schools, for anyone.

Schaller, F. 1968. SOIL ANIMALS. Univ. Mich. Press, Ann Arbor. 145 pp. Brief outline of soil ecosystems, emphasizes larger soil animals. Introductory.

Schatz, A. 1972. TEACHING SCIENCE WITH SOIL. Rodale Press, Emmaus, Pa. 133 pp.

ENERGY and CO_2 INPUT

ENERGY and CO_2 LOSS

SOIL FAUNA

DETRITIVORES → CARNIVORES (Predators, Parasites)

DETRITIS

PRIMARY CONSUMERS   SECONDARY CONSUMER → TERTIARY CONSUMER

MICROPHYTIC FEEDERS

FECES and DEAD BODIES

FECES and DEAD BODIES

SOIL HUMUS

SOIL MICROFLORA

From: THE NATURE AND PROPERTY OF SOILS, by N.C. Brady (Macmillan, N.Y.).

ULTIMATE DECOMPOSERS

Intermediate Soil Books

Albrecht, W. A. 1958. SOIL FERTILITY AND ANIMAL HEALTH. Fred Hahue Printing Co., Webster, Iowa. 232 pp. A vital book on soil fertility, livestock health, and human disease.

Balfour, E. B. 1950. THE LIVING SOIL. Devin-Adair, N. Y. 270 pp. One of the first to document organic technique with research evidence (albeit weak). A classic in soil biology/fertility relationships.

Berger, K. 1972. SUN, SOIL, AND SURVIVAL. Univ. Oklahoma Press, Norman, Oklahoma. 371 pp. An excellent intermediate level textbook, but lacks good information on soil biology.

Bridges, E.M. 1970. WORLD SOILS. University Press, Cambridge. 89 pp. Patterns of soils in the world. Superb color plates of soil profiles.

Buckman, H.O. and N.C. Brady. 1974. THE NATURE AND PROPERTIES OF SOILS. Macmillan Co., N. Y. 639 pp. Best college text on soils. It even has "humus" in the index.

Davies, Nancy et al. 1973. SOIL ECOLOGY. Prentice-Hall, Englewood Cliffs, New Jersey. 197 pp. Learning ecology through the soil. Textbookish, but simple . . . top.rate.

Donahue, R.L. et al. 1977. SOILS: AN INTRODUCTION TO SOILS AND PLANT GROWTH. Prentice-Hall, Inc., Englewood Cliffs, NJ 07632. On a par with Buckman and Brady, more practical, less soil formation theorizing.

Eyre, S.R. 1963. VEGETATION AND SOILS. Aldine Publishing Co., Chicago. 328 pp. How plants effect soils and human settlements and vice versa.

Garrett, S. 1963. SOIL FUNGI AND SOIL FERTILITY. The Macmillan Co., N.Y. 165 pp. Overview by an expert.

Jackson, R.M. and F. Raw. 1966. LIFE IN THE SOIL. St. Martin's Press, N. Y. 59 pp. Authors from Rothamsted discuss soil ecology and ways of studying it. Good stuff.

Parkinson, D. et al. 1971. METHODS FOR STUDYING THE ECOLOGY OF SOIL MICRO-ORGANISMS. IBP Handbook No. 19, Blackwell Scientific Publ., Oxford. 116 pp.

Rodale, R. (ed.). 1961. THE CHALLENGE OF EARTHWORM RESEARCH. Soil and Health Found., Emmaus, Pa. 102 pp. Modern research on earthworm ecology. Top bibliography.

Russell, E. and J. Russell. 1961. SOIL CONDITIONS AND PLANT GROWTH. John Wiley & Sons, Ltd. N.Y. 688 pp. First published in 1912, now nine editions later it is still a classic.

U.S. Dept. of Agriculture. 1957. SOILS. USDA Yearbook, 1957. Gov. Printing Office, Washington D.C.

Waksman, S.A. 1952. SOIL MICROBIOLOGY. John Wiley & Sons, Inc., N. Y. 356 pp. (OOP) Classic textbook still of MUCH value.

Technical Soil Books

Burges, A. 1958. MICRO-ORGANISMS IN THE SOIL. Hutchinson, London. More recent and ecologically oriented than Waksman's SOIL MICROBIOLOGY.

Doekson, J. and J. van der Drift (ed.). 1963. SOIL ORGANISMS. North Holland Publ. Co., Amsterdam. Current research on the biology and ecology of soil life and ecosystems.

Gray, T. and S. Williams. 1971. SOIL MICROORGANISMS. Hafner Publ. Co., N. Y. 240 pp.

Gray, T. and D. Parkinson (ed.). 1968. THE ECOLOGY OF SOIL BACTERIA. Univ. Toronto Press, Canada. Anthology of technical papers.

Harley, J. 1959. THE BIOLOGY OF MYCORRHIZA. Leonard Hill, London. 233 pp. A bit out of date, but lays out the biology of beneficial root fungi.

Krasil'nikov, N.A. 1958. SOIL MICRO-ORGANISMS AND HIGHER PLANTS. NTIS, Washington D. C. 474 pp. $4.75. English translation of Russian works in soil science; top rate for relations between soil microbes and higher plants.

McLaren, A. and G. Peterson. 1967. SOIL BIOCHEMISTRY. Edward Arnold, London. 509 pp.

Parkinson, D. and J. Waid (ed.). 1960. THE ECOLOGY OF SOIL FUNGI. Liverpool Univ. Press, England. 324 pp. Technical anthology. Good on mycorrhiza.

Pauli, F. 1967. SOIL FERTILITY: A BIODYNAMIC APPROACH. Adam Hilger, London. 204 pp. Semi-technical book for soil researchers. Humus can be studied in the context of science as a whole.

GREEN MANURE NOTES

SAMUEL AND LOUISE KAYMEN
Natural Organic Farmers Association
Plainfield, Vt. 05667

The term green manure refers to the practice of incorporating young green plants into the soil. These plants are usually grown with the single purpose of being used as a soil improver. This practice is usually applied to plots or fields that are in a planned system of rotation. The green manure crops can be annuals, biennials or even perennials planted at almost any time of the year depending on the rotation system and the crop chosen. On the basis of their time of seeding and their occupation of the land, green manure crops are used quite differently. One useful succession that works well in New England is oats and Canada field peas, early spring, then one or two crops of buckwheat between late spring and late summer, then winter rye and hairy vetch planted in the fall. The rye and vetch occupy the ground throughout the winter and are considered a cover crop, i.e. they prevent the topsoil from being washed away by rains. The rye and vetch are allowed to grow some in the spring and are dug into the soil when about 9-12 inches high.

Rye and vetch are planted on land that has a food crop all summer. Many people seed rye into the corn crops at the time of the last cultivation and use it to cover the soil all winter. Some use buckwheat during the summer in between the food crop rows. Some crops are planted in the spring with a spring grain crop, such as a slow growing clover with spring wheat. After the wheat is harvested, the clover may be incorporated into the soil or allowed to winter over and incorporated in the next spring. There are many specific reasons for using green manures:

1. *The addition of organic matter* as food for the life of the soil.

2. The ability of legume crops to *accumulate atmospheric nitrogen* is another important function.

3. Another reason to grow green manure crops is to *conserve water soluble elements* needed for plant growth. A crop of rye, clover or buckwheat could pick up the various elements and hold them in plant form until they are needed.

4. Many crops, especially deep-rooted ones such as alfalfa, sweet clover and ragweed, *concentrate needed plant nutrients* from the subsoil. Their subsequent decay releases those deep gathered elements in the topsoil. This permits the use of these elements by succeeding shallow-rooted crops.

5. The *solubility of* calcium, phosphorous, potassium, magnesium and other *elements is increased* through the effect of the organic and inorganic acids produced as a result of the decomposing organic matter of green manures.

6. Green manure crops *improve the subsoil* by penetrating deeply. As the roots decay, numerous channels are formed which facilitate the circulation of air and the movement of water up and down in the soil.

7. The *protection of the surface soil* is often considered one of the most important functions of the green manure crop. On sloping land the soil is protected from water erosion, and light soils are protected from wind erosion.

It is important to distinguish between green and "brown" manures. Green manures are young plants that are high in protein (nitrogen), sugar and moisture, and low in lignin and cellulose (carbon). When dug into the ground in the spring, green manure crops add nutrients accumulated by the crops during the winter period. "Brown" manures, on the other hand, are plants in a mature stage of growth; that is, they are low in protein, sugar and moisture, but high in lignin and cellulose (i.e. very close to seed production). Their purpose is to add humus-building substances and not nutrients to the soil. Because of their high carbon content, brown manures are very resistant to decay and will tie up nitrogen in the soil. Thus crops should not be planted immediately after incorporating brown manures into the soil.

Now let's assume you have a piece of abused, exploited and mismanaged earth. One way to get going is to make use of the weeds that grow there naturally, since their very function is to improve the soil. But, left to nature, it would take a long time. We can speed up the process by incorporating the weeds into the topsoil. Then allow the weeds to germinate and grow again. Always dig in the weed crop at the time when it is young, green and succulent so it will decompose very rapidly.

If your piece of earth has a little grass growing with the weeds then you *don't* have to "weed fallow" the first season. You can start immediately with the first green manure crop to use on poor land . . . buckwheat. This wonderful plant has the ability to make use of minerals in the soil better than most other plants. It accumulates calcium, and is an effective competitor with weeds. You should broadcast it at about 100 lbs per acre, or ¼ lb for 100 square feet. Sow buckwheat after July, for it likes to make its flowers in cool moist weather. When it is about 10% in bloom, dig it in.

After buckwheat the next green manure to be used is winter rye and hairy vetch. The vetch is a legume and fixes atmospheric nitrogen. It is also a vine and will climb up the ryegrass. The mixture should be Balbo rye (125 lbs/acre) and vetch (60 lbs/acre).

The final green manure to use, after your soil has received the benefits of the above, is sweetclover, either white or yellow (15-20 lbs/acre). Sweetclover can accumulate over 200 lbs of nitrogen per acre, the highest of any legume. That was its primary use in agriculture in the early part of this century, before cheap nitrogen fertilizer came along. With the price of fertilizer going up, sweetclover may again resume its role.

Organic Matter, Green Manures and Soil Fertility

Allison, F.E. 1973. SOIL ORGANIC MATTER AND ITS ROLE IN CROP PRODUC-TION. Elsevier Scientific Publishing Co., New York. $52.00 Author from the USDA. Good up-to-date overview of organic farming methods. A bit hard on the organic approach in general. Outrageous price.

Alther, R. and R.O. Raymond. 1974. IMPROVING GARDEN SOIL WITH GREEN MANURE: A GUIDE FOR THE HOME GARDENER. Garden Way Publishing Co., Charlotte, VT. 05445. 36 pp. $2. (C). Only book of its kind. Useful, but needs to be expanded into areas of green/brown manures; phytotoxicity of fresh green manures, specific techniques of incorporation WITHOUT rototillers, etc.

Azevedo, J. 1974. FARM ANIMAL MANURES: AN OVERVIEW OF THEIR ROLE IN THE AGRICULTURAL ENVIRONMENT. California Agricultural Experiment Station, Extension Service, Manual 44, from: Agricultural Publications, University of California, Berkeley, CA 94720. 109 pp., $4. (B). A definitive summary and complete bibliography.

Evans, H.J. 1975. ENHANCING BIOLOGICAL NITROGEN FIXATION. National Science Foundation, from: Division of Biological and Medical Sciences, NSF, Washington D.C. 20550. 52 pp. (A). Proceedings of technical conference on state-of-arts.

Food and Agriculture Organization of the United Nations. 1975. ORGANIC MATERIAL AS FERTILIZERS. from: UNIPUB, PO Box 433, Murray Hill Station, NY 10016, 395 pp, $33. Use of organic material in Third World Agriculture. Great reference sections. Ask library.

Kononova, M.M. 1961. SOIL ORGANIC MATTER: ITS NATURE, ITS ROLE IN SOIL FORMATION AND IN SOIL FERTILITY. (second English Ed.), Pergamon Press, New York. 544 pp. (A). A technical bible of Russian humus science.

Stewart, W.D. 1966. NITROGEN FIXATION IN PLANTS. University of London, Athlone Press, London.

Waksman, S.A. 1938. HUMUS. Williams & Williams Co., Baltimore. 526 pp. (OOP) (B). Definitive up to its time. No popular sequel yet.

Fertilizers

Blake, M. 1970. DOWN TO EARTH: REAL PRINCIPLES FOR FERTILIZER PRACTICE. Crosby Lockwood & Son, Ltd. London. 93 pp. (B) Chemical and organic fertilizers can be mixed and used in a truly beneficial way.

Coleman, Eliot, 1976. THE USE OF GROUND ROCK POWDERS IN AGRICULTURE. Small Farm Research Association, Harborside Maine 04642.

Kingman, A.R. 1972. A SURVEY OF SEAWEED RESEARCH. Dept. Horticulture, Clemson, University, Clemson, S.C. State-of-arts review discussing biological advantages of using seaweed in agriculture.

Ministry of Agriculture, Fisheries and Food. 1971. TRACE ELEMENTS IN SOILS AND CROPS. Proceeding of a Conference, Soil Scientists of the National Agricultural Advisory Service. Technical Bull. 21. Her Majesty's Stationery Office, London. 217 pp. (A)

Organic Gardening and Farming. 1973. ORGANIC FERTILIZERS: WHICH ONES AND HOW TO USE THEM. Rodale Press, Inc., Emmaus, Pa. 129 pp. (C) A good, simple, practical book.

Shaw, E.J. 1965. WESTERN FERTILIZER HANDBOOK. California Fertilizer Association, 719 K St., Sacramento, CA 95814. 200 pp., $2. Best guidebook for using chemical fertilizers, if you're into it.

Stephenson, W.A. 1968. SEAWEED IN AGRICULTURE AND HORTICULTURE. Faber & Faber, London. 231 pp. (B) Definitive!

Stiles, Walter. 1951. TRACE ELEMENTS IN PLANTS AND ANIMALS. (2nd ed.), Cambridge Univ. Press, Cambridge, England.

Taft, D.C. and C.D. Burris. 1974. PHOSPHATE CONSUMPTION AND SUPPLY. Garden Way Laboratories, Charlotte, Vt. 05445. 9 pp. (C) There is no known substitute for phosphate rock. It is probably one of the major limiting resources on this planet.

Mulch

A mulch is a layer of material, preferably organic, that is placed on the soil surface around plants to conserve moisture, hold down weeds, stabilize soil temperatures and build soil humus and soil fertility.

Brooklyn Botanic Garden. 1957. HANDBOOK ON MULCHING. Brooklyn Botanic Garden, 1000 Washington Ave., Brooklyn, N. Y. 79 pp. (C)

Campbell, S. 1973. THE MULCH BOOK: A GUIDE FOR THE FAMILY FOOD GARDENER. Garden Way Publishing Co., Charlotte, Vt. 136 pp. (C)

Rodale, R. et al (eds.). 1971. THE ORGANIC WAY TO MULCHING. Rodale Press, Inc., Emmaus, Pa. 192 pp. (C)

Rowe-Dutton, P. 1957. THE MULCHING OF VEGETABLES. Technical Communication No. 24, Commonwealth Bureau of Horticulture and Plantation Crops, Commonwealth Agricultural Bureaux, Farnham Royal, England. 169 pp. (B) A summary of over 350 references on mulching, mostly from Europe. A lot of words just to say that mulch works.

COMPOSTING AND COMPOST PRIVIES

Composting

Billington, F.H. 1955. COMPOST FOR GARDEN PLOT OR THOUSAND-ACRE FARM. Faber & Faber, London. (B) (OOP)

Breidenbach, A.W. 1971. COMPOSTING OF MUNICIPAL SOLID WASTES IN THE UNITED STATES. (EPA, GPO) (SW-47r). 101 pp. (B) Best popular overview of municipal composting. Conclusion: in the U.S. it is not yet economically practical. So what else is new?

Bruce, M. 1967. COMMON SENSE COMPOST MAKING BY THE QUICK RETURN METHOD. Faber & Faber, London. (C) Method of backyard composting designed during WWII for victory gardens in England. Great practical "how-to" book.

Campbell, S. 1974. LET IT ROT. Garden Way Publishing Co., Charlotte, VT. Probably the best popular introduction to composting.

COMPOST SCIENCE, (bimonthly magazine, $6.00/year.)Rodale Press, Emmaus, Pa, , 18049. Best journal in the country on organic waste recycling. Describes both experiences and research in a popular style with technical back-ups. Too bad there isn't a comparable journal for all natural farming.

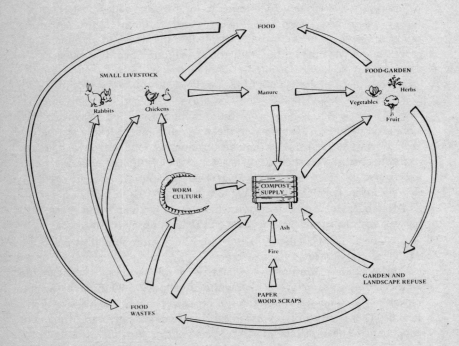

Dindal, D.L. 1972. ECOLOGY OF COMPOST: A PUBLIC INVOLVEMENT PROJECT. Office of Public Service, State University of New York, Syracuse. 12 pp. (C) Great little pamphlet ($0.10) for getting anyone started.

Gainesville Municipal Waste Conversion Authority, Inc. GAINESVILLE COMPOST PLANT: FINAL REPORT ON A SOLID WASTE MANAGEMENT DEMONSTRATION. NTIS, Pb-222 710. 237 pp. (B) Example of the struggles of a municipal composting operation.

Golueke, C. 1972. COMPOSTING: A STUDY OF THE PROCESSES AND ITS PRINCIPLES. Rodale Press, Inc., Emmaus, Pa. 110 pp. (B) Principles, technology, health aspects, home composting methods . . . it's all here in a truly fine book about composting.

Golueke, C.G. et al. 1954. A CRITICAL EVALUATION OF INOCULUMS IN COMPOSTING. Applied Microbiology, 2(1): 45-53.

Gotaas, H.B. 1956. COMPOSTING. SANITARY DISPOSAL AND RECLAMATION OF ORGANIC WASTES. World Health Organization. From: The American Public Health Association, 1740 Broadway, New York 10019. 205 pp. (B) Loaded with general information. Focuses on intermediate (villages, small towns) and large scale applications.

Hortenstine, C. and D.F. Rothwell. 1973. COMPOSTING MUNICIPAL REFUSE AS A SOIL AMENDMENT. EPA/NTIS, PB-222 422. 62 pp. (B)

Koepf, H.H. 1966. COMPOST: WHAT IT IS, HOW IT IS MADE, WHAT IT DOES. Bio-Dynamic Farming and Gardening Assoc., R.D. 1, Stroudberg, Pa. 18360. 18 pp., $0.55.

McGauhery, P.H. 1971. AMERICAN COMPOSTING CONCEPTS. EPA, GPO (SW-2r). 23 pp. (C). Brief popular overview.

Poincelot, R. 1972. THE BIOCHEMISTRY AND METHODOLOGY OF COMPOSTING. The Connecticut Agricultural Experimental Station, New Haven, Conn. 06504 (A) One of the most concise descriptions of the technical aspects of composting. Lots of references (169) and lots of facts.

RECLAMATION OF MUNICIPAL REFUSE BY COMPOSTING. Technical Bulletin No. 9. Series 37. Sanitary Engineering Research Project. Univ. Calif., Berkeley. June 1953. 89 pp. (A) Technical, but with some useful hints that can be culled from controlled experiments.

Rodale, J. (ed.). 1969. THE COMPLETE BOOK OF COMPOSTING. Rodale Press, Emmaus, Pa. 1007 pp. (C) Despite the title and length, the book is far from complete. Useful in places.

Schatz, A. and V. Schatz. 1971. TEACHING SCIENCE WITH GARBAGE. Rodale Press, Emmaus, Pa. (C) Simple and direct ways of teaching scientific principles and importance of waste recycling at the same time.

Snyder, W.C. et al. 1972. BIOLOGICAL CONSEQUENCES OF PLANT RESIDUE DECOMPOSITION IN SOIL. EPA-NTIS, PB-221 113. 136 pp. (A) The other side of the coin. Toxicity of decomposing plant wastes on seedlings and roots of field crops.

Tietjen, C. and S. Hart. 1969. COMPOST FOR AGRICULTURAL LAND? Journal Sanit. Engin. Div. of P.A.S.C.E. 95 (SA2): 269-287. (B)

Wadleigh, C.W. 1968. WASTES IN RELATION TO AGRICULTURE AND FORESTRY. U.S. Dept. Agric., Misc. Publ. No. 1065. 112 pp.

EXPERIENCES WITH COMPOST PRIVIES

MAX KROSCHEL
Farallones Institute
Occidental, California

"Dein eigenes scheisse stinkt nicht"
– Albert Einstein

A compost privy, or waterless toilet, is a small-scale, on-site waste recycling system which stabilizes human excrement (night soil) and household wastes. These wastes are deposited in a container, but instead of being filled over like the old privy, they are composted with air, removed from the container after a period of time and recycled on local land. The compost privy takes the place of the toilet, septic tank and garbage can.

There are several advantages to the compost privy: (1) The compost privy can be used where sewer hookups or septic tanks are unavailable or not practical. (2) The compost privy saves water normally flushed down the toilet. About half the annual domestic water consumption (7,000-10,000 gallons per person) is saved each year. (3) The use of a compost privy eliminates the largest pieces of waste travelling through pipes. This means that you can use 2" waste pipe instead of 4" in a house, thereby saving money on plumbing. Because there is no need for water traps (more money saved), water pipes can drain clear after use, thus reducing the possibility of freezing. (4) The compost privy returns valuable nutrients and humus to the soil. Between 1 and 2 cubic feet of humus, safe to use in a garden, are produced from each person's excreta in a year. (5) The compost privy allows you the use of the squatting position, which is the healthiest posture for defecation.

There are also some limitations to compost privies: (1) The compost privy does not receive "gray" water (water from sinks, showers, etc.). However, this can be diverted into the garden, recycled through a solar still, or emptied into a sump pit and leaching lines. (See bibliography.) (2) The idea of a compost privy conflicts with the vast majority of local health codes. (3) The compost privy requires considerable floor space, and may be hard to incorporate into existing structures.

Several manufactured composting toilets are available commercially, and owner built plans are also available from several sources (see below).

The basic components of a compost privy are: (1) a receiving chamber fitted with a toilet seat or squat plate and constructed so as to exclude disease carrying animals, (2) a screened air inlet pipe and exhaust outlet vent-stack to carry away odors, water vapor and carbon dioxide while bringing oxygen to the decomposing material, and (3) an access door for removing the compost. Operation requires priming the chamber with decomposable carbonaceous material and adding additional carbon matter after each use.

The consensus among investigators (see Bibliography) is that the heat of a compost pile, balanced in carbon and nitrogen materials, with proper moisture and adequate aeration, turned several times after each four or five days is sufficient to kill all known fecal-borne diseases caused by bacteria (e.g., typhoid, cholera); protozoans (e.g., amoebic dysentery); and intestinal parasites. Death is caused either by the heat of aerobic decay or antibiotic agents in the pile. Thermal death points are known for these organisms, and large compost piles attain and exceed these temperatures for extended periods of time. However there is doubt among health officials that small compost piles, commonly found in compost privies, heat up sufficiently and thoroughly enough for proper treatment. For this reason (and numerous other cultural and prejudicial ones) they have been reluctant to sanction the use of compost privies even on an experimental basis.

The Farallones Institute, in conjunction with state and local health officials, has been investigating compost privies for the past several years. The main concern of our research has been related to temperatures, but we have also monitored loading rates and experimented with various carbonaceous add-mixtures. Several samples of aged privy compost have been submitted for bacterial and parasitilogical examination. Although early samples showed "no enteric pathogenic organisms and a marked reduction in coliform bacteria," a recent sample showed a positive reading. A well decomposed 6 month sample was submitted to a normal fecal analysis at a local medical lab and the County Health Department. They reported finding viable eggs of hookworm *(Necator americanus)*, round worm *(Ascaris lumbricoides)*, whipworm *(Trichuris trichiura)* and unidentifiable protozoan cysts. Examination of the residents of the research community indicated that two persons were carrying these parasites. Both were former Peace Corps volunteers and had histories of parasite infestations while living in the tropics. No other persons were found to be infected including the two persons who regularly turned and maintained the privies, indicating that no cross contamination had occurred and that the privies were vector proof and being operated in a safe and sanitary manner.

The Farallones Institute has been a long-term proponent of waterless toilets. We are opposed to the mixing of 100 parts of clean potable water with one part of valuable fertilizer, sending it through miles of elaborate tubing and then spending money in a futile attempt to separate the two. The discovery of viable parasites in our privy compost does not spell doom for the compost privy idea. It merely points out flaws in our present design and management procedures and is a step along the way to the development of a workable composting toilet. It does, however, reinforce our urge of caution in using composted human excrement. We have always recommended against using it in the garden on vegetables (especially leafy greens and root crops) but rather urged that it be used around ornamentals, fruit trees or berry bushes. Hookworm larvae can migrate several inches through the soil, so a recommended Public Health measure for dealing with infected night soil is to bury it in 12" trenches and to leave the area undisturbed for one year. Trenches can be dug around young fruit trees that are still expanding their root zones, or in areas later to be planted in fruit trees, berries or other perennials.

The major problem of the Farallones privy has been heat loss to the cement block walls and concrete slabs of the chamber. We are modifying the design to use an insulated stud wall with a plastic vapor barrier and redwood or cedar sheathing to resist rot on the inside. We are experimenting with an insulating concrete using vermiculite as the aggregate (durability may be a problem). The interior space has been expanded to three chambers; one for collection below the privy seat and two for composting and storage.

During operation, the floor of the first compartment is primed with straw or other dry absorbent carbonaceous material. After each use a small amount of cut grass or chopped straw is added. Depending on the number of persons using the privy, it takes one to six months to accumulate about 2/3 cubic yard of material. This material is then used as a nitrogen source to build a compost pile in the second chamber. An equal amount of straw, cut grass, spoilt hay, fresh garden refuse and kitchen scraps is put in. Proper moisture content is important and water or urine should be added if necessary.

The pile should heat up in two or three days and can be monitored with a long probe thermometer. After a week the pile should be turned into the last bin to aerate it and to mix the outside layers into the center, so that all the material has a better chance to heat up and be pasteurized.

A cheaper, easier and safer privy is the drum privy. An open-top 55 gallon drum is fitted with perforated aeration pipes that thread into the 2" bung hole in the side. The interior is painted with epoxy paint or fiberglass resin to prevent corrosion. The drum can be fitted with a seat directly or jacked up under the floor of the bathroom or privy outbuilding. To mix and aerate the collected material, one simply clamps on the lid and rolls the drum around on its side a few times. When the drum is full there are two options for treatment. One is to compost the material into a large window. A large compost pile is made at least 6 feet high using garden and agricultural wastes and animal manures. After a day or two when the pile has attained a high enough temperature (over 150°F) it is opened up with a pitchfork, and the contents of the drum poured into the center and the pile closed up again, completely enclosing the privy material. It is then allowed to compost for 6 months. The second option is to pasteurize the drum and contents in a solar oven. After pasteurization, the contents can be composted with less stringent measures and used more liberally. The solar oven concept is still under development by the Farallones Institute and more information should be available soon from the Rural Center.

Compost Privies; Plans and Information

Berry, W. 1973. A COMPOSTING PRIVY. Organic Gardening and Farming Magazine. Dec. 1973, pp. 88-97. Some useful hints plus a discussion of health aspects by Clarence Golueke.

Clark, S. and S. Tibbetts. 1976. COMPOSTING TOILETS AND GREYWATER DISPOSAL: BUILDING YOUR OWN. The Alternative Waste Treatment Assoc., Star Route 3, Bath, ME 04530. $3.00 Cold weather info., several plans.

THE COMPOST TOILET NEWS. from: Alternative Waste Treatment Assoc., Star Route 3, Bath, ME 04530. $4/yr. Newsletter on compost privies and greywater systems. Designs and trouble shooting.

Farallones Institute. 1976. COMPOST PRIVY BULLETIN AND GREYWATER FACT SHEET. FI Rural Center, 15290 Coleman Valley Rd., Occidental CA 95465. $3.00. Includes plans for a compost privy under investigation.

Golueke, C. & H. Gotaas. PUBLIC HEALTH ASPECTS OF WASTE DISPOSAL BY COMPOSTING. American Journal of Public Health. 44(3): 339-348

Hills, Lawrence. 1972. THE CLIVUS TOILET: SANITATION WITHOUT POLLUTION. Compost Science, May-June 1972.

Kern, Ken. 1977. COMPOST TOILET DESIGN. Owner-Builder Access, Box 550, Oakhurst, CA 93644. Compost privy in slip-form concrete tower containing: septic tank, compost bins, shower, tub, sauna, water storage, etc.

Leich, H.H. 1975. THE SEWERLESS SOCIETY. Bulletin of the Atomic Scientists, Nov., 1975. Overview of Incinerating, composting, oil-flush, vacuum and aerobic tank toilets.

Lindstrom, C.R. 1969. EXAMINATION OF THE OPERATING CHARACTERISTICS OF A COMPOSTING INSTALLATION FOR ORGANIC HOUSEHOLD WASTES. from: Clivus Mulstrum USA, 14A Eliot St., Cambridge, MA 02138.

Lindstrom, R. 1965. A SIMPLE PROCESS FOR COMPOSTING SMALL QUANTITIES OF COMMUNITY WASTE. Compost Science, Spring 1965, p. 30.

Mann, H.T. and D. Williamson. 1976. WATER TREATMENT AND SANITATION: SIMPLE METHODS FOR RURAL AREAS. Intermediate Technology Publications Ltd., 9 King St., London WC2E 8HN. 90 pp.

Nichols, H.W. 1976. ANALYSIS OF BACTERIAL POPULATIONS IN THE FINAL END PRODUCT OF THE CLIVUS MULSTRUM. Center for the Biology of Natural Systems, Washington Univ., St. Louis, MO 63130.

Nishihara, S. 1935. DIGESTION OF HUMAN FECAL MATTER, WITH pH ADJUSTMENTS BY AIR CONTROL (EXPERIMENTS WITH AND WITHOUT GARBAGE). Sewage Works Journal, Vol. 7, No. 5.

RAIN Magazine. 1977. COMPOST PRIVY UPDATE; 55-GALLON DRUM COMPOSTING PRIVY and COMPOSTING DRUMS IN PARADISE VALLEY. RAIN Magazine, Vol. 3, No. 8, June, 1977. Nice review articles on current state-of-the-arts of compost privies.

Rodale, Robert. 1971. GOODBYE TO THE FLUSH TOILET. Compost Science, November-December 1971.

Scott, J.C. 1952. HEALTH AND AGRICULTURE IN CHINA. Faber & Faber, London. 279 pp. A fantastic account of composting experiments in China using night soil. A very valuable book for anyone interested in composting human wastes.

Shell, G.L. and J.L. Boyd. 1969. COMPOSTING DEWATERED SEWAGE SLUDGE. U.S. Dept. of Health, Education, and Welfare. (GPO) 28 pp. (C)

STOP THE FIVE GALLON FLUSH: A SURVEY OF ALTERNATIVE WASTE DISPOSAL SYSTEMS. Minimum Cost Housing Group, School of Architecture, McGill University, Montreal, Canada. 1976. 60 pp. $1.75. An excellent description of every alternative toilet imaginable. One of the classic AT booklets of the 70's.

Swanson, C. 1949. PREPARATION AND USE OF COMPOSTS, NIGHT SOIL, GREEN MANURES AND UNUSUAL FERTILIZING MATERIALS IN JAPAN. Agronomy Journal, 41(7): 275-282.

United Stand. 1975. UNITED STAND PRIVY BOOKLET. United Stand, Box 191, Potter Valley, CA 95469. $2.50 Construction drawings and explanation of several different homemade composting privies and greywater systems for rural application.

Voell, A.T. and R.A. Vance. 1974. HOME AEROBIC WASTEWATER TREATMENT SYSTEMS, EXPERIENCES IN A RURAL COUNTY. Ohio Home Sewage Conference Proceedings, Ohio State University, 1974.

Wagner, E.G. and J.N. Lanoix. 1958. EXCRETA DISPOSAL FOR RURAL AREAS AND SMALL COMMUNITIES. World Health Organization, Geneva. Many useful designs for privies including compost privies. Important review of health aspects of human wastes.

Warshall, P. 1976. SEPTIC TANK PRACTICES. Box 42, Elm Rd., Bolinas, CA 94924. 76 pp. This is the most complete and beautifully written book on home-site sewage treatment available today.

Wiley, B. and S. Westerberg. 1969. SURVIVAL OF PATHOGENS IN COMPOSTED SEWAGE. Applied Microbiology, Vol. 18, pg. 994.

Wiley, J.S. 1962. PATHOGEN SURVIVAL IN COMPOSTING MUNICIPAL WASTES. Journal, Water Poll. Cntrl. Fed., 34(1): 80-90.

HOMEBUILT COMPOST PRIVY

Source: Farallones Institute

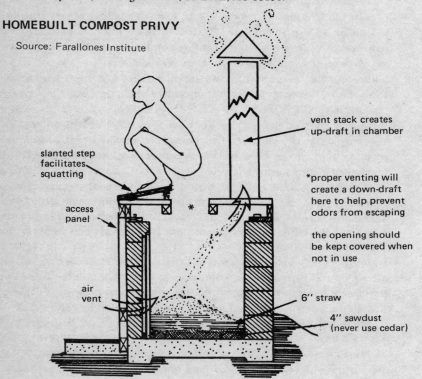

vent stack creates up-draft in chamber

slanted step facilitates squatting

access panel

air vent

*proper venting will create a down-draft here to help prevent odors from escaping

the opening should be kept covered when not in use

6" straw

4" sawdust (never use cedar)

Compost Privy Manufacturers

Biomat (same as Mull-Toa)
Biomat Enterprises
739 - 2nd Street
Coeur d'Alene, ID 83814

Bio-let Composting Toilet
Bio-Utilities, Inc.
Box 135
Narberth, PA 19072

Clivus — U.S.
Clivus Multrum, Inc.
14 Eliot Street
Cambridge, MA 02138

Ecol Sanitary Unit
Minimum Cost Housing Group
Brace Research Institute
McGill University
Montreal, CANADA

Ecolet
Recreation Ecology Conservation
of the United States, Inc.
9800 W. Blue Mound Road
Milwaukee, WI 53226

Kern Compost Privy
Ken Kern
P.O. Box 550
Oakhurst, CA 93664

Mullbank
Inventor AB
Prastgatan 42
931 00 Ostersund, SWEDEN

Multrum
Scan Plan
3 Sankt Kjelds Gade
DK-2100 Copenhagen
DENMARK

Mull-Toa
Hahs Kr. Nielsen
Sorkesalsveien 22
Oslo 3, NORWAY

Saniterm
AB Electrolux
Luxbacken 1,
112 62 Stockholm, SWEDEN

Toa-Throne
Enviroscope Inc.
Attention: Lars deJounge
P.O. Box 752
Corona del Mar, CA 92625

CROP PRODUCTION

Plant Basics

Bailey, L.H. 1915. FARM AND GARDEN RULE-BOOK. Macmillan Co., N.Y. (OOP) Easily one of the more valuable books one can own. Should be reprinted.

Elliott, J.H. 1958. TEACH YOURSELF BOTANY. The English Universities Press Ltd., London. 214 pp. (B) Combine this with Fogg's book and you've got what you need to know about plants.

Fogg, F.E. 1963. THE GROWTH OF PLANTS. Penguin Books. Baltimore, M.D. 288 pp. (B) Photosynthesis, plant biology/chemistry/ecology, and the rhythms of growth and flowering . . . it's all there. Great introduction.

Fogg, F.E. 1968. PHOTOSYNTHESIS. American Elsevier Publ. Co., N.Y. (B) Elegant and simple while staying deep into the subject.

Galston, A.W. 1961. THE LIFE OF GREEN PLANTS. Prentice-Hall, Englewood Cliffs, N.J. One of the best introductory books on plant science.

Harrington, H.D. and L.W. Durrell. 1957. HOW TO IDENTIFY PLANTS. The Swallow Press Inc., Chicago. 203 pp. (C) If you want to collect seeds or plants, take this book as a back up to your local plant key. Explains all the fancy words.

Hayward, H. 1967. THE STRUCTURE OF ECONOMIC PLANTS. Hafner Service. Detailed anatomy of 16 cultivated plants: corn, hemp, onion, alfalfa, wheat.

Janick, J. et al. 1969. PLANT SCIENCE: AN INTRODUCTION TO WORLD CROPS. W.H. Freeman, San Francisco. 269 pp. (B) Classic textbook of the foundations of world crop production. Illustrations unmatched.

Jaques, H.E. 1958. HOW TO KNOW THE ECONOMIC PLANTS. Wm. C. Brown Co., Publ. 174 pp. (C) One of the "Picture Key Series".

Magness, J. et al. 1971. FOOD AND FEED CROPS OF THE UNITED STATES. Bulletin 828, New Jersey Agric. Exper. Station, New Brunswick. 255 pp. (C) Handy reference for all major crops grown in the U.S.

Rickett, H.W. 1957. BOTANY FOR GARDENERS. Macmillan Co., N.Y. 236 pp. (C) Simple coverage of the basic principles of botany using cultivated plants.

Shurtleff, M.S. 1966. HOW TO CONTROL PLANT DISEASES IN HOME AND GARDEN. Iowa State Univ. Press, Ames. 649 pp. (B) Probably the best book to have, although no color plates.

Sprague, H.B. 1964. HUNGER SIGNS IN CROPS. McKay Co., Inc., N.Y. 390 pp. (B) Color photos, keys and text allow first approximations to nutritional deficiencies in common temperate crops.

United States Department of Agriculture. 1960. INDEX OF PLANT DISEASES IN THE UNITED STATES. Agric. Handbook No. 165, Gov. Print. Office, Washington D.C. 531 pp. Known diseases of cultivated plants and range of occurrence.

Truck Crops

Hills, L. 1971. GROW YOUR OWN FRUITS AND VEGETABLES. Faber & Faber, London.

John, G.F. 1968. BEST IDEAS FOR VEGETABLE GROWING. Rodale Books, Inc. Emmaus, Pa. (C) Compilation of testimonials by O. G. & F. editors.

Jones, H.A. and T.R. Joseph. 1928. TRUCK CROP PLANTS. McGraw Hill Co., N. Y. (OOP) A well-known old source of data on propagation, cultivation, manuring, storage, diseases, etc. Pre-chemical ideas.

Knott, J.E. 1957. HANDBOOK FOR VEGETABLE GROWERS. John Wiley & Sons, Inc. 245 pp. (C) $8.00. Mostly tables and charts of every aspect of vegetable biology. One of the most useful reference books you can own.

MacGillivray, J.H. 1952. VEGETABLE PRODUCTION. The Blakiston Co., N. Y. 397 pp. (B)

MacGillivray, J.H. 1968. HOME VEGETABLE GARDENING. California Agricultural Exper. Sta., Extension Serv., Circular 449. 31 pp. (C)

Riotte, L. 1974. THE COMPLETE GUIDE TO GROWING BERRIES AND GRAPES. Garden Way Publishing Co., Charlotte, Vt. 144 pp. (C)

Robbins, A.R. 1974. TWENTY-FIVE VEGETABLES ANYONE CAN GROW. Dover Publications, Inc., N. Y. 216 pp. (C) Reprint of 1942 book. Biology, horticulture, harvest, storage, uses of them all.

Rodale, J.I. et al (eds.). 1969. BEST IDEAS FOR ORGANIC VEGETABLE GROWING. Rodale Books, Inc. Emmaus, Pa. 197 pp. (C)

Shewell-Cooper, W.E. 1967. THE COMPLETE VEGETABLE GROWER. Faber & Faber, London. 288 pp. (C) Geared to English conditions, but one of the best introductions and reference books on vegetables.

Sunset Books. 1970. VEGETABLE GARDENING. Lane Books, Menlo Park, Calif. 72 pp. (C) Simple, practical and visual.

Wester, R. 1972. GROWING VEGETABLES IN THE HOME GARDEN. Home and Garden Bulletin No. 202, USDA, Gov. Printing Office, Washington D.C. 49 pp. (C)

Field Crops and Grains

Heath, M.E. et al. 1973. FORAGES: THE SCIENCE OF GRASSLAND AGRICULTURE. Iowa State University Press, Ames. 755 pp. (B) Comprehensive textbook and a lot of information.

Highstone, H.A. 1974. PRACTICAL HAY-MAKING ON A SMALL PLACE. Garden Way Publishing Co., Charlotte, VT 05445. 10 pp. $0.50.

Hills, L.D. 1975. COMFREY REPORT. Bargyla Rateaver, Pauma Valley, CA 92061. $5.50. Under proper conditions, Comfrey can produce more protein per acre than soybeans. One of the few plants to concentrate vitamin B12.

Hollins, A. 1968. MY FARMING SYSTEM. Journal of the Soil Association, Jan. 1968, from The Soil Association, Haughley, Suffolk, England. Description of a 15 species grass/herb pasture system. Old and mellow.

Kipps, M.S. 1970. PRODUCTION OF FIELD CROPS. McGraw-Hill, New York.

Langer, R.H. 1972. HOW GRASSES GROW. Edward Arnold, London (Studied in Biology Series, No. 34). 60 pp. $2.50. Excellent description.

Logsdon, E. 1976. SMALL-SCALE GRAIN RAISING. Rodale Press, Emmaus, PA 18049. Best book on the subject of raising, harvesting and storing small grains in small areas.

Martin, J.H. and W.H. Leonard. 1967. PRINCIPLES OF FIELD CROP PRODUCTION. Macmillan Co., N.Y. Highly recommended for grain crops.

Morrison, F.B. 1946. FEEDS AND FEEDING. The Morrison Publishing Co., Ithaca, NY 1050 pp. The classic text for those growing grasses, root crops and legumes for animal stock feed.

Staten, H.W. 1958. GRASSES AND GRASSLAND FARMING. Devin-Adair. NY One of the most important books on the subject for the farmer.

STOCK SEED FARMS. Route 1, Box 112, Murdock, NB 68407. Specializes in native American prairie grass seed and prairie wildflower seed. Handouts and booklets.

U.S. Dept. of Agriculture, 1948. GRASS. USDA Yearbook, 1948. Washington D.C., U.S. Gov. Printing Office.

Voisin, Andre. 1959. GRASS PRODUCTIVITY. Philosophical Library, New York. Classic description of rotational grazing system applicable to chickens and cattle alike. Hard to find but worth looking for.

Voisin, Andre. 1960. BETTER GRASSLAND SWARD. C. Lockwood, London.

NATURAL PEST MANAGEMENT

Pesticides are poisons that kill pests: insecticides (insects), herbicides (weeds), and fungicides (plant disease). Although little is known about the extent to which pesticides have been used, a great deal is known or suspected about their abuses. Suffice it to say that pesticides have been used so much on farmlands and around cities (50% are used in cities) that they must now be considered an integral part of our biological systems: present in our bodies, flowing with the rivers, drifting in the air and falling with the rain. After a generation of unrestrained use, pesticides have produced —and are still producing—serious side effects to public health, wildlife, soil life, and the balance of natural ecosystems. The FULL consequences of these effects are as yet unknown and remain for future generations to discover.

Besides their dangers, pesticides commonly fail to control pests. There are at least three reasons for this: 1) Pests often become resistant to pesticides. 2) Pesticides cause outbreaks of secondary pests. 3) For various ecological reasons, pesticides tend to free pests from control by their natural enemies. All these factors produce a series of ecological backlashes that cause farmers to use larger doses of more toxic pesticides at greater economic, environmental and energy costs.

The use of pesticides will probably remain an integral part of agriculture for years to come. But their *unrestrained use as an end in themselves* can only benefit the petrochemical companies, and not agriculture or the environment in which it operates.

There are two alternative strategies to the use of pesticides: 1) The use of safe pesticides in a controlled way. 2) The use of techniques that don't use pesticides at all. The prerequisite for better pesticide use is accurate pest monitoring. By keeping a regular account of the numbers of pests, pesticide application can be planned more rationally, giving adequate control with a few well-timed applications. Often a crop is vulnerable to attack for only a relatively short period. Massive applications at regular intervals is a consuming, not a rational, approach to pesticide use. Another way is the use of selective poisons which kill the target pest but are harmless to natural enemies. Finally, there is a whole host of non-persistent botanical pesticides derived from wild plants: Ryania, rotenone, pyrethrum and many more yet unknown. Most research is done to find more lethal chemicals, rather than less persistent but more specific ones.

Alternatives to pesticides are based on a more rigorous understanding of the term "pest control". A "pest" is not anything that crawls in the field.

A pest is any creature whose numbers jeopardize the crops being grown. "Control" doesn't mean total eradication (impossible anyway), but rather keeping a potential pest below a certain level. In order to work, alternatives to pesticides should do three things: 1) They should be capable of keeping pests at a harmless density. 2) They should not cause pests to develop resistances. 3) They should work with and not against the controls provided by natural enemies of pests.

There are several methods of pest control that fulfill these requirements to one degree or another. Sex-scent attractants, baits, light traps, sterile-male radiation, resistant crop varieties, and "soft" pesticides have all been used alone or as part of an integrated control program. But these tactics are based on one narrow strategy: *discouraging the pests*. A more permanent and stabilizing method is based on *encouraging the natural enemies* of pests, their predators, parasites and diseases. In other words: controlling pests biologically, rather than chemically. Listed below are some aspects of biological control that are currently being studied or practiced:

1) Finding natural enemies suitable for specific pests and environments. Most familiar are the predatory insects . . . the bugs that eat bugs. These include certain beetles, lacewings, mantises and dozens of other less familiar species. More numerous and effective in controlling pests are the parasitic wasps, such as trichogramma. These tiny insects deposit their eggs in the soft bodies of pests or their eggs. More recent discoveries include a few types of insect diseases. Bacteria such as milky spore disease on boll weevils and tobacco budworms, and *Bacillus thuringiensis* ("Biotrol", "Thuricide", "Parasporin") have been marketed and recommended for use against caterpillars. Also under consideration at the present are some fungi and virus diseases. All of these diseases are specific to certain insects and harmless to other animals.

2) Some private businesses have begun rearing large numbers of natural enemies under artificial conditions (insectaries). Several kinds of natural enemies have been raised successfully and sold to the public by the insectaries or by distributing companies. In a few cases farmers have formed cooperative insectaries to serve the needs of several farms.

3) New ecological techniques are also being developed for nurturing natural enemies that occur naturally in the fields. What is not generally appreciated is that methods of biological control are usually most effective in a cultivated habitat that itself is diverse. Planting several kinds of crops simultaneously or rotating crops in a specific way gives natural enemies

more options for survival and makes them more effective at controlling pests. This subject is discussed in more detail in the following section.

Controlling insect pests by using their natural enemies is not so much a technology as it is an understanding of ecology. All animals are kept from overrunning the countryside by their natural enemies. The process goes on all around us, although we are likely to take it for granted. Natural enemies are small, relatively rare and don't kill bugs as fast as chemicals. Instead, there is a lag time between peak numbers of pests and the time when their enemies can respond and bring their numbers down. The whole key to using predators, parasites and disease to combat pests is a fundamental understanding of ecology and some patience. No simple answers exist to problems of pest control, whether in the backyard garden or field. What is clearly needed is an approach that looks to each pest situation as a whole and then comes up with a program that integrates the different alternatives outlined above. This integrated approach is nothing more than reacting to a biological problem in a holistic way with basically biological tools.

— Richard Merrill

Basic Books

Brooklyn Botanic Garden Record. 1966. HANDBOOK ON BIOLOGICAL CONTROL OF PESTS. Brooklyn Botanic Garden, N. Y. 97 pp. One of the best places to start for learning about the subject. Many photos.

Cox, J. 1974. HOW TO USE BIOLOGICAL CONTROLS EFFECTIVELY. Organic Gardening and Farming Magazine. Feb. 1974. pp. 76-82. Twenty-five points by experts on how to select and use beneficial insects in the garden . . . on the farm.

Olkowski, H. 1971. COMMON SENSE PEST CONTROL. Consumers Cooperative of Berkeley. 4805 Central Ave., Berkeley, CA 52 pp. Probably the best place to start if you want to manage your household/garden with respect to "pests". Simple, Clear, Beautiful.

Olkowski, H. & W. Olkowski. 1974. INSECT POPULATION MANAGEMENT IN AGRO-ECOSYSTEMS. In: "Radical Agriculture", R. Merrill, (ed.), Harper and Row.

Philbrick, J. and H. Philbrick. 1963. THE BUG BOOK: HARMLESS INSECT CONTROLS. Garden Way Publ. Co., Charlotte, Vt. 143 pp.

Rodale, J.I. 1966. THE ORGANIC WAY TO PLANT PROTECTION. Rodale Press, Emmaus, Pa. 355 pp. Principles of non-chemical pest control. Troubleshooting text geared to 1) crop, and 2) pest, what it is and how to subdue it.

Van den Bosch, R. and P. Messenger. BIOLOGICAL CONTROL. Intext Educational Publishers, N. Y. 180 pp. Great non-technical introduction for layman.

Intermediate Books

Clausen, C.P. 1940. ENTOMOPHAGOUS INSECTS. McGraw-Hill, N. Y. 688 pp. Overview biology of insects that eat insects. An oldie but a goodie.

Clausen, C.P. 1956. BIOLOGICAL CONTROL OF INSECT PESTS IN THE CONTINENTAL UNITED STATES. USDA Tech. Bull. 1139 151 pp.

DeBach, P. 1974. BIOLOGICAL CONTROL BY NATURAL ENEMIES. Cambridge Univ. Press, London. 323 pp.

Hunter, B. 1971. (2nd ed.) GARDENING WITHOUT POISONS. Berkeley Publ. Co., N. Y. 352 pp. Factual book of non-chemical pest control. Good introduction.

National Research Council. 1969. INSECT-PEST MANAGEMENT AND CONTROL. Publ. 1695. Washington D.C. 420 pp. Good practical manual including info on surveying techniques and cultural control.

Swan, L. 1964. BENEFICIAL INSECTS. Harper and Row, N. Y. 429 pp. Simple coverage of bio-control and biologies of predators and parasites.

Technical Books

Burges, J.D. and N. Hussey. 1971. MICROBIAL CONTROL OF INSECTS AND MITES. Academic Press, N. Y. 861 pp. Technical anthology on the use of diseases to control pests.

DeBach, P. (ed.). 1964. BIOLOGICAL CONTROL OF INSECT PESTS AND WEEDS. Reinhold Publ. Co., N. Y. 844 pp. Great overview of history and basic concepts. Anthology by all the experts.

ECOLOGICAL ANIMAL CONTROL BY HABITAT MANAGEMENT. 1969-71 (3 vols.) Tall Timbers Research Station, Tallahassee, Fla. Technical, but valuable discussions of integrated control techniques and experiences.

Huffaker, C.B. 1971. BIOLOGICAL CONTROL. Plenum Press. N. Y. 511 pp. Easy up-to-date textbook coverage of subject by pioneer researcher.

Metcalf, R.L. and W. Luckmann. 1975. INTRODUCTION TO INSECT PEST MANAGEMENT. John Wiley and Sons, N.Y. A very comprehensive and important anthology of integrated pest management techniques.

WASP PARASITIZING APHID

National Academy of Sciences. 1969. PRINCIPLES OF PLANT AND ANIMAL PEST CONTROL. Vol. 3, "Insect Pest Management and Control". NAS, Washington D.C.

Rabb, R.L. and F. Gutherie (ed.). 1970. CONCEPTS OF PEST MANAGEMENT. North Carolina Univ. Press, Raleigh. 242 pp. Technical, but excellent information on the ecology behind integrated control.

Identifying Insects and Pest Problems

Borror, D.J. and R. White. 1970. A FIELD GUIDE TO THE INSECTS OF AMERICA NORTH OF MEXICO. Houghton Mifflin Co., Boston. (C) Best introductory guide to insect identity. Color plates and keys to get them to "family".

Borror, D.J. and D. DeLong. 1971. (3rd ed.) AN INTRODUCTION TO THE STUDY O INSECTS. Holt Rinehart and Winston, N. Y. 812 pp. One of the better textbooks to families of insects.

Davidson, R.H. and L.M. Peairs. 1966. (6th ed.) INSECT PESTS OF FARM, GARDEN AND ORCHARD. Wiley.

Herrick, G. 1925. MANUAL OF INJURIOUS INSECTS. Henry Holt & Co., N. Y. 489 pp. (B) (OOP) Valuable pre-pesticide information. Good life-history stuff on pests.

Metcalf, C.W. et al. 1962. DESTRUCTIVE AND USEFUL INSECTS; THEIR HABITS AND CONTROL. McGraw-Hill, N. Y. 981 pp. (B) The best book for identifying pest problems from the way the plant looks. GREAT troubleshooter.

Pictured Key Nature Series. HOW TO KNOW THE IMMATURE INSECTS: INSECTS; BUTTERFLIES; BEETLES; GRASSHOPPERS. William C. Brown and Co., Dubuque Iowa. Illustrated keys to insects you might find in the garden, on the farm.

Romoser, W.S. 1973. THE SCIENCE OF ENTOMOLOGY. Macmillan, N.Y. Text book discussion of the biological and behavioral systems of insects. Great illustrations and photos. Highly recommended.

Swan, L.A. and C.S. Papp. 1972. THE COMMON INSECTS OF NORTH AMERICA. Harper and Row, New York. 750 pp. $15.00.

U.S. Department of Agriculture. 1971. INSECTS AND DISEASES OF VEGETABLES IN THE HOME GARDEN. USDA Home and Garden Bulletin. No. 46. Gov. Printing Office, Washington D.C. 50 pp. (C) Good photos, drawings, simple descriptions . . . forget the chemical advice.

Westcott, C. 1973. THE GARDENER'S BUG BOOK. Doubleday and Co., Inc., Garden City, N.Y. 689 pp. (B) A "must" reference for identifying pests, their life histories, and what they eat . . . with chemical advice.

Zim, H. 1966. INSECT PESTS. Golden Press, N. Y. 160 pp. (C) Grade-school level, but can get anyone started. Top rate color drawings.

Resistant Varieties

Atema, H. 1974. A PRELIMINARY STUDY OF RESISTANCE IN 20 VARIETIES OF CABBAGES TO THE CABBAGE WORM BUTTERFLY. Journal No. 2, New Alcher Institute, Box 432, Woods Hole, Mass. (B)

Cox, J. 1973. INSECT RESISTANT VARIETIES: TOOLS FOR THE GARDENERS. Organic Gardening and Farming Magazine, May 1973. pp 85-90. (C)

Painter, R.H. 1968. INSECT RESISTANCE IN CROP PLANTS. University Press of Kansas, Lawrence. 520 pp. (B) On the method of finding resistant plants. Covers the field. Some useful varieties listed.

Radcliff, E.B. and R.K. Chapman. 1966. PLANT RESISTANCE TO INSECT ATTACK IN COMMERCIAL CABBAGE VARIETIES. Journal of Economic Entomology, 59 116-120. (A) An example of some handy stuff being tucked away in technical journals.

Pest Control Dealers, Insectaries and Consultants

The following companies supply biological control agents. Write the companies and ask for information about prices and availability, and the techniques of using the living things they sell. Don't buy anything until you are ready to use it.

ASSOCIATION OF APPLIED INSECT ECOLOGISTS. 10202 Cowan Heights Dr., Santa Ana, CA 92705. Undertake biological and integrated control programs on major California field, orchard and truck crops. Publish newsletter.

Beneficial Bio Systems. 1523 63rd St., Dept. CD4, Emeryville, CA 94608. Dealer in beneficial insects (predators, parasites) specific to flies and fly control.

Bio-Control Co., Rt. 2, Box 2397, Auburn, Calif. 95603. Ladybug beetles gathered in the Sierras from winter colonies; sometimes mantis cases.

Biotactics, 4436 Elora Court, Riverside, Calif. 92503. Phytoseiulus (predator of red spider mites).

Eastern Biological Con. Co., Rt. S Box 379, Jackson, New Jersey 08527.

Ecological Insect Services. 15075 W. California Ave., Kerman, Calif. 93630. Consultation and control for farms and gardens.

Fairfax Biological Laboratory, Clinton Corners, N.Y. 12514. "Doom", trade name for milky spore disease (a bacteria for controlling Japanese beetle).

Gothard, Inc., P.O. Box 370, Canutillo, Tex. 79835, Specializes in trichogramma wasps, mantis egg cases.

International Center for Biological Control. University of California, 1050 San Pablo Ave., Albany, CA 94706. Offers computerized information service concerning technical information on the natural enemies of agricultural pests.

International Mineral and Chemical Corp., Crop Aid Products, Dept. 5401, Old Orchard Road, Skokie, Ill. 60076. Thuricide, trade name for Bacillus thuringiensis.

Rincon-Vitova Insectaries, Inc., P.O. Box 95, Oak View, Calif. 93022. Lacewings; trichogramma; parasites for fly larvae and army worms; predatory mites for red spider mites; cryptolaemus for mealybugs; and many others. Best outfit in the country for farmers.

Thompson Hayward Chemical Co., Box 2382, Kansas City, Kansas. 66110. Biotrol, trade name for Bacillus thuringiensis.

Spalding Laboratories. Route 2, Box 737, Arroyo Grande, CA 93420. Dealer in "fly predators" (actually parasites).

MIXED CROPS AND COMPANION PLANTING

In wildlife communities there is a wide variety of plants growing together in the same area that form a natural plant association. These plants are able to co-exist because they complement each other's requirements, provide direct benefits to each other, or make demands on resources at different times. Some thrive in direct sunlight, others live in the shade provided by larger plants; some have shallow roots, while others penetrate deeply; some grow quickly and can be harvested early, making room for slower growing forms; some serve as food for natural enemies that in turn feed on pests of other plants. All available space is used at all times by an array of differently adapted plants.

The possible combinations of cultivated plants helping or harming one another are far too numerous to describe here in detail. A few obvious kinds of mixed cultures may be familiar:

Intercropping: These are the well-known space-saving combinations . . . fast and slow-growing plants, shallow and deep-rooted ones, tall and shade-loving crops, vines and supporting plants, etc. These combinations increase the dimensions of the garden by exploiting available space.

Rotation: Some plants may leave behind beneficial conditions for other plants that follow. Nitrogen-fixing legumes (peas, beans, vetch, clover, etc.) planted before a major crop are the most familiar example. Also heavy-feeding crops (potato, cabbage, corn) can be followed by light-feeding crops.

Companion crops: Crop patterns and local environments can be modified so as to favor beneficial insects already in the field, or to discourage the presence of pests. Changes might include the cultivation of plants that provide food for natural enemies or that discourage the invasion of pests, and the maintenance of uncultivated areas or permanent refuges. Some examples are listed below.

a. Flower crops: The nectar and pollen of many flowers provide food for adult beneficial insects. In orchards, for example, wildflowers can nurture populations of parasitic wasps and thereby reduce certain pests. Research in Russia has shown that, when the weed *Phacelia* was planted in orchards, a parasite of the tree's scale pest thrived in the orchard by subsisting on the nectar of the weed. Another Russian study showed that, when small plots of umbellifers (a family of plants) were planted near vegetable fields in a ratio of 1 flower plant to 400 crop plants, up to 94% of the cabbage cutworms were parasitized. Flowers of crop-plants such as the cabbage family, legumes and sunflowers can also serve as alternative food sources for beneficial insects.

b. Repellent crops: Most insects are selective as to the kinds of plants they eat. It is believed that insects are attracted to the odors of "secondary" substances in plants rather than to the food value of the plant itself. Experiments have shown that odors given off by aromatic plants interplanted with crops can interfere with the feeding behavior of pests by masking the attracting odor of the crops. Repellent crops so far described include various pungent vegetables *(Solanum, Allium)* and aromatic herbs *(Labiatae, Compositae and Umbelliferae).*

c. Trap crops: Some plants can be used to attract pests away from the main crop. With careful monitoring these "trap" crops can provide food in the form of large numbers of pests, which then attract their own natural enemies. For example, when alfalfa strips were interplanted with fields of cotton, the Lygus bug (a serious pest in California) migrated away from the cotton and into the alfalfa. With their concentrations of Lygus bugs, the alfalfa plots then provided a food source for several predatory insects in the area. Trap crops of alfalfa may also have applications in walnut and citrus orchards and bean fields. In the coastal climate of California, brussels sprouts, which attract large numbers of aphids, can serve as winter insectaries for parasitic wasps. When aphids attack other crops in the spring, the wasp populations, having fed on aphids during the lean winter, are large enough to respond quickly and control the aphids.

d. Hedgerows and Shelter Belts: For centuries hedgerows have been planted between field crops to slow down winds. Uncultivated land also affects the distribution and abundance of insects associated with nearby crops. Wild plant stands can provide alternative food and refuge for pests and their natural enemies alike. In England, where much farming is done near wild vegetation, pest problems are generally less severe than in the United States where monoculture farming persists.

Many other kinds of plant relationships can be cultivated to advantage. Some "component" species probably serve more than one beneficial function. Repellent herbs, for example, also produce food-rich flower heads, as do many trap crops. Garden models exist for a variety of mixed cropping schemes, and these undoubtedly could be tested and applied on a larger scale. Interest, however, will probably remain focused on monocultures until the "costs" of pesticides and poor farm management exceed the "costs" of ecological designs.

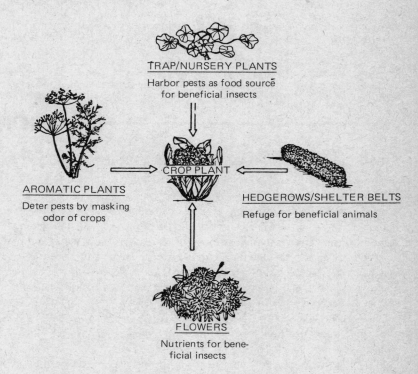

TRAP/NURSERY PLANTS
Harbor pests as food source for beneficial insects

AROMATIC PLANTS
Deter pests by masking odor of crops

CROP PLANT

HEDGEROWS/SHELTER BELTS
Refuge for beneficial animals

FLOWERS
Nutrients for beneficial insects

SOME POSSIBLE COMPONENTS AND PLANT INTERACTIONS OF A DIVERSE CROPPING SYSTEM. Based on a garden model of "companion planting" arrangements.

Interplanting the farm landscape with trees, hedgerows and other perennial stands, together with rotations, strip cropping and mixed stands will serve to promote stability and effective natural pest control. But diverse landscapes and mixed farming methods *per se* will not create stability. Sometimes diversity decreases pest damage, other times it may increase it. The web of possible plant/animal relationships is immeasurably complex, and each situation and crop ecosystem is unique. In other words, the right kind of diversity must be established, and we can only know that by practical experience in local areas and ecological studies of the agricultural environment.

—Richard Merrill

General Books/Articles

Allen, R. 1974. DOES DIVERSITY GROW CABBAGES? New Scientist, 29 Aug., 1974. (B)

Cox, J. 1972. HOW TO PLANT A COMPANION GARDEN. Organic Gardening and Farming Magazine, Feb., 1972. pp. 50-55;   (C)

Dethier, V. 1947. CHEMICAL INSECT ATTRACTANTS AND REPELLENTS. Blakiston Co., Phil. (B) (OOP) Discusses the mechanisms behind plants that attract and repel insects.

Dethier, V.G. 1970. CHEMICAL INTERACTIONS BETWEEN PLANTS AND INSECTS. In: "Chemical Ecology", E. Sondheimer and J.B. Simeone (ed.), Academic Press, London. (B) Semi-technical overview of the subject.

Merrill, R. 1972. ECOLOGICAL DESIGN IN THE ORGANIC GARDEN: COMPANION PLANTING AND NATURAL REPELLENTS. Organic Gardening and Farming Magazine: April 1972, pp. 48-53. (B)

Philbrick, H. and R.B. Gregg. 1966. COMPANION PLANTS AND HOW TO USE THEM. Devin-Adair Co., N.Y. 111 pp. (C) Admittedly only a primer, the book is still the most complete compilation of formal experiments and casual observations of beneficial plant combinations for the home gardener.

Root, R. 1973. ORGANIZATION OF A PLANT-ARTHROPOD ASSOCIATION IN SIMPLE AND DIVERSE HABITATS: THE FAUNA OF COLLARDS (Brassica oleracea). Ecological Monographs, 43: 95-124. (A)

Scott, G. 1969. PLANT SYMBIOSIS. William Clowes and Sons, Ltd., London. (B)

Trap Plants, Repellent Plants, Flowers, and Hedges as Companion Crops

Brooklyn Botanic Garden. 1958. HANDBOOK ON HERBS. Brooklyn Botanic Garden, Brooklyn, N. Y. 96 pp. (C) A lot of books describe herbs as food and medicine. This seems to be the most concise book for telling you how to grow and identify them.

Fraenkel, G.S. 1959. THE RAISON D'ETRE OF SECONDARY PLANT SUBSTANCES. Science 129: 1466-1470. (B) Why plants put out odors that attract pests.

Leius, K. 1967. INFLUENCE OF WILD FLOWERS ON PARASITISM OF TENT CATERPILLARS AND CODLING MOTH. Can. Ent. 99: 444-446. Eighteen times as many tent caterpillars (pupae) parasitized in orchards with rich undergrowths of wild flowers as in orchards with poor floral undergrowths.

Leius, K. 1960. ATTRACTIVENESS OF DIFFERENT FOOD AND FLOWERS TO THE ADULTS OF SOME HYMENOPTEROUS PARASITES. Can. Ent. 92: 369-76. (B)

Stern, V. 1969. INTERPLANTING ALFALFA IN COTTON TO CONTROL LYGUS BUG AND OTHER INSECT PESTS. In "Tall Timbers Conference on Ecological Animal Control by Habitat Management," Proceeding, No. 1; 55-69. Tall Timbers Research Station, Tallahassee, Fla.

Tahvanainen, J.O. and R. Root. 1972. THE INFLUENCE OF VEGETATIONAL DIVERSITY ON THE POPULATION ECOLOGY OF A SPECIALIZED HERBIVORE, PHYLLOTRETA CRUCIFERAE. Oecologia (Berl) 10: 321-346. (A) Adult flea beetles more abundant on collards grown in monocultures than on those grown adjacent to natural vegetation. Mixed crops per se tended to "confuse" pests.

van Embden, H.F. 1964. THE ROLE OF UNCULTIVATED LAND IN THE BIOLOGY OF CROP PEST AND BENEFICIAL INSECTS. Scientific Horticulture, Vol. 17.

Harmful Crops

Garb, S. 1961. DIFFERENTIAL GROWTH-INHIBITORS PRODUCED BY PLANTS. Botanical Reviews, 27:422-443. (B) Useful list of crops that inhibit each other.

Lawrence, T. and M.R. Kilcher. 1962. THE EFFECT OF FOURTEEN ROOT EXTRACTS UPON GERMINATION AND SEEDLING LENGTH OF FIFTEEN PLANT SPECIES. Canadian Journal of Plant Science, 42(2): 308-318. (A)

Rice, E.L. 1974. ALLELOPATHY. Academic Press, New York (A). Comprehensive overview of biochemical (harmful) interactions between plants. Examples in agriculture and horticulture.

Whittaker, R.H. and P.P. Feeny. 1971. ALLELOCHEMICALS: CHEMICAL INTERACTIONS BETWEEN SPECIES. Science, 171: 757-770. (A)

PROPAGATION

GENETIC EROSION AND MONOCULTURES

For millennia, people have domesticated wild species of plants and animals for food, selecting strains that were palatable and easy to grow. About 75 years ago, engineers began to develop controlled breeding programs by selecting crop varieties that were resistant to some of the diseases and pests that have plagued societies throughout history. These efforts continue today, but they have taken a back seat to the development of a few high-yielding and uniform crops which meet the demands of mechanical harvesters and a competitive market economy.

For several reasons the new genetic strategy has placed agriculture in its most vulnerable position ever, by forsaking biological quality for yield and appearance:

1) The genetic base for most major crops has become dangerously narrow. As farming practices rely more and more on a few productive varieties (Table I), the numerous strains once grown in and adapted to local communities are being abandoned. Changing land use patterns further reduce the diversity and distribution of genetic types, by destroying habitats of local wild plants. A genetic reserve is important because the evolutionary contest between diseases and cultivated plants is a continuous exchange of mutual adaptations; short-lived microbes mutate and recombine to new diseases while longer-lived crops struggle to adapt resistances. The development of resistant varieties is also a continuous process and needs a diverse genetic base from which to operate. Unfortunately, many of the old varieties of crops have been discarded irretrievably; new varieties represent only a small fraction of the gene pool once planted. This loss, although not well publicized and understood by the public, has serious implications for the availability of future food supplies.

2) Planting large areas in the same kind of crop encourages the spread of disease and pest outbreaks. When the monocultures are extended over broad geographic areas, as they are today in the United States, the potential for crop epidemics is compounded. This is precisely the condition that caused the great Irish potato famine of the 1840's and the U.S. corn leaf blight in the early 1970's when over 15% of the total U.S. corn crop was destroyed. These and other examples show that crop monocultures and genetic uniformity actually invite crop epidemic, increase pesticide use, and increase the potential for higher food costs and food shortages.

Unfortunately, the market economy rather than common sense determines whether a new crop variety is used. The farmer requires uniform crops for tending and mechanical harvesting. The middlemen require uniformity for processing and mass merchandising. The competitive market permits no alternatives. This dilemma was highlighted by the 1970 U.S. corn leaf blight and was described in a report issued by the National Academy of Sciences, which concluded that:

a) . . . most major crops are impressively uniform genetically and impressively vulnerable. b) This uniformity derives from powerful economic and legislative forces. c) . . . increasing vulnerability to epidemics is not likely to generate automatically self-correcting tendencies in the marketplace.

3) Most high-yield crops encourage the use of pesticides and synthetic fertilizers; in fact, they have actually been developed together with agricultural chemicals. This means that farmers are forced to use pesticides to protect their high yields and to stretch production potential. High-yield varieties are productive only because they are responsive to heavy doses of chemical fertilizers; they are ineffective without this input.

**   **   **   **

What is needed is a diversified breeding program which can continually breed new varieties of hardy, domestic stocks adapted to *local* conditions. In this way, a wide genetic reservoir for the selection process is maintained, and the patchwork of diverse crops establishes buffer areas against epidemics. In particular, as noted by Howard Odum: "We need to get livestock back onto the farm and develop varieties that can fend for themselves in reproduction, protect themselves from weather and disease, move with the food supply and develop their patterns of group behavior. We must conserve and develop further a wide variety of competitive plant species that can also provide a reasonable growth in unfertilized soils and resist pests and disease."

—Richard Merrill

CROP	NUMBER OF POPULAR COMMERCIAL, OR CERTIFIED VARIETIES	MAJOR VARIETIES NO.	% OF CROP ACREAGE
Beans, snap	70	3	76
Cotton	50	3	53
Corn[1]	197	6	71
Peanut	15	3	70
Peas	50	2	96
Potato	82	4	72
Rice	14	4	65
Soybeans	62	6	56
Sugar beet	16	2	42
Sweet potato	48	1	69
Wheat	269	9	50

TABLE I.   The extent to which a few crop varieties dominate American agriculture. From: National Academy of Sciences. GENETIC VULNERABILITY OF CROPS. [1] Released public inbreds only, expressed as percentage of seed requirements.

Agricultural Origins and Genetic Resources

Anderson, E. 1969. PLANTS, MAN AND LIFE. University of California Press, Berkeley and Los Angeles. 251 pp. Popular account of the origins and evolution of cultivated plants . . . showing them to be largely mongrel weeds . . . hybrids which sprang up on the village dump heaps of early man.

Brukill, I. H. 1953. HABITS OF MAN AND THE ORIGIN OF THE CULTIVATED PLANTS OF THE OLD WORLD. Proceedings of the Linnean Society of London, 614:12.

Frankel, O. and E. Bennett. 1970. GENETIC RESOURCES IN PLANTS . . . THEIR EXPLORATION AND PRESERVATION. (Inter. Biol. Progr., No. 11), Davis Co., Philadelphia, PA.

Harlan, J.R. 1971. AGRICULTURAL ORIGINS: CENTER AND NONCENTERS. Science, 174:468.

Issac, E. 1970. GEOGRAPHY OF DOMESTICATION. Prentice-Hall, Inc., Englewood Cliffs, N.J. 119 pp. (B) Probably the best short overview of the subject without the usual professional rhetoric. Lacks some of the more ecologically sophisticated ideas, however. (See Ucko and Dimbleby below.)

Klose, N. 1950. AMERICA'S CROP HERITAGE: THE HISTORY OF FOREIGN PLANT INTRODUCTION BY THE FEDERAL GOVERNMENT. Iowa State College Press, Ames, Iowa. 156 pp.

National Academy of Sciences. 1972. GENETIC VULNERABILITY OF MAJOR CROPS. NAS, Washington D.C.

Sauer, C. AGRICULTURAL ORIGINS AND DISPERSALS. MIT Press, Cambridge, Mass. 175 pp. (C)

Schwanitz, F. 1966. THE ORIGIN OF CULTIVATED PLANTS. Harvard University Press, Cambridge, Mass. 174 pp. (C)

Struever, S. (ed.) 1971. PREHISTORIC AGRICULTURE. American Museum of Natural History. The Natural History Press, Garden City, N.Y. (B). Anthology of current ideas on how-why-where the first crops were cultivated.

Ucko, P.J. and G.W. Dimbleby (eds.) 1969. THE DOMESTICATION AND EXPLOITATION OF PLANTS AND ANIMALS. Aldine Publ. Co., Chicago. 581 pp. (B). Proceedings of research seminar in archaeology, London University. Provocative chapters suggest that under conditions of abundant food supply in biologically diverse regions (river delta at ocean edge), semi-nomadic tribes may have discovered agriculture as a result of leisure pursuits rather than the pressures of dwindling food supplies.

Vavilov, N.J. 1949. THE ORIGIN, VARIATION, IMMUNITY AND BREEDING OF CULTIVATED PLANTS. Chronica Botanica 13(1-6). Waltham, Mass. (B) (OOP). A classic among classics.

General Horticulture Information Sources

BROOKLYN BOTANIC GARDEN. 1000 Washington Ave., Brooklyn, N.Y. 11225. Extremely useful Garden and Horticultural Handbooks, priced from $1.25 to $1.60 on everything from soils and mulches to roses and dye plants.

THE NEW YORK BOTANIC GARDEN. Bronx, New York 10458. The library offers bibliographies on different plant subjects ALL FREE except for: "The Medicinal and Food Plants of the North American Indians." ($1.25).

INFORMATION DIVISION, USDA Agricultural Research Service. Federal Center Building, Hyattsville, Md. 20782.

NORTH AMERICAN FRUIT EXPLORERS. Sec. Robert Kurle, 87th and Madison, Hinsdale, Ill. 60521. Publish North American Pomona quarterly, information and resources . . . advantages of personal exchange members and library.

SOIL CONSERVATION SOCIETY OF AMERICA, Plant Resources Division. Write to: D. E. Hutchinson, Chairman, Natural Vegetation Subcommittee, 5717 Baldwin Ave., Lincoln, Nebraska 68507.

UNITED STATES NATIONAL ARBORETUM. Washington, D.C. 20002. For plant identification problems or lists of specific kinds of plant sources (i.e. "Native Flower Sources").

THE AMERICAN HORTICULTURAL SOCIETY. 901 N. Washington St., Alexandria, Va. 22314. If you cannot find plants you are looking for the AHS can probably help. They also have an important sourcebook: DIRECTORY OF AMERICAN HORTICULTURE. ($5.00) that lists the major plant organizations and societies plus the agricultural extension services and experimental stations.

CALIFORNIA RARE FRUIT GROWERS. Star Route Box P, Bonsall, CA 92003. Seed exchange, lists of plant sources, lots of pratical information, exchanges between members, quarterly newsletter, dues $3.00.

HERB BUYER'S GUIDE. Richard Hettern, Pyramid Communications, Inc., 919 3rd Ave., New York, N. Y. 10022 ($1.25).

HERB GROWER MAGAZINE. Falls Village, Conn. 06031. "The Marketplace" quarterly $5.00.

HERB SOCIETY OF AMERICA. 300 Massachusetts Ave., Boston, Mass. 02115. Sells a list of herb sources.

PLANT BUYERS' GUIDE. Massachusetts Horticultural Society (Ahrno H. Nehrling, Director of Publications), Bellman Publishing Co., Inc. Boston, Mass.

USDA FOREST SERVICE. Washington D.C. "Forest Tree Seed Orchards: A Directory of Industry, State and Federal Tree Seed Orchards in the U.S."

General Propagation

Bailey, L. H. 1967. (18th printing from 1896). THE NURSERY MANUAL. The Macmillan Co., N. Y. 456 pp. (B) Dated, but still in print and still a good place to start.

Brooklyn Botanic Garden. 1957. HANDBOOK ON PROPAGATION. Brooklyn Botanic Garden, Brooklyn, New York 11225. $1.25. (C) Great simple introduction.

Free, Montague. 1957. PLANT PROPAGATION IN PICTURES. Doubleday & Co., N. Y. 249 pp. $6.95.

Hartmann, H.T. and D.E. Kester. 1968. PLANT PROPAGATION: PRINCIPLES AND PRACTICES. Prentice-Hall, Inc., Englewood Cliffs, N. J. 702 pp. (B) Comprehensive and slightly technical, this is probably the most thorough popular book on plant propagation in print. A book with which to end . . . not start.

Mahlstede, John P. & E.S. Haber. 1957. PLANT PROPAGATION. John Wiley & Sons, Inc., New York. 410 pp. $10.25. Very good.

Prockter, Noel J. 1950. SIMPLE PROPAGATION. A BOOK OF INSTRUCTIONS FOR PROPAGATION BY SEED, DIVISION, LAYERING, CUTTINGS, BUDDING AND GRAFTING. W.H. & L. Collingridge Ltd., London, Transatlantic Arts, Inc., Forest Hills, New York. 144 pp. (C) (OOP)

van der Pijl. 1972. PRINCIPLES OF DISPERSAL IN HIGHER PLANTS. 2nd ed. Springer-Verlag. $12.60. (A)

Wright, R.C.M. 1956. PLANT PROPAGATION AND GARDEN PRACTICE. Criterion Books, New York. 192 pp. (B)

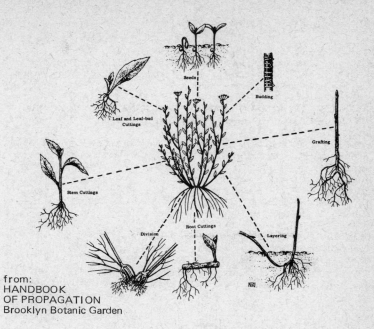

from:
HANDBOOK
OF PROPAGATION
Brooklyn Botanic Garden

The production of seeds and crop varieties adapted to local conditions is a vital part of regional self-reliance in food production. In the far west, where water is a premium and pests are rarely set back by winter kill, growers are interested in drought-hardy, pest-resistant crop varieties. In the South, with its long growing season, high temperatures and humidity, attention is given to breeding in insect and disease resistance and breeding out premature flowering. For northern conditions, plant breeders and seed selectors should work to develop early varieties that will germinate fast, then grow and produce a crop in fewer days, with less total heat and light than later-maturing varieties. Obviously there are many more specific requirements than this based on the thousands of micro-climates throughout the country.

There are several good, reliable seed companies in North America, but most of them pay little attention to the purchase or development of local or pest-resistant varieties (only a few seed companies actually produce their own seeds). In response to this increasing interest in regional food self-reliance and family food production, a new type of seed-supplier is surfacing in various parts of the country. For the most part these new seed people are interested in exchanging seeds, breeding and purchasing regionally adapted and open-pollinated varieties, and encouraging the home production of seeds. Listed below are some of these new cottage industry seed people.

Johnny's Selected Seeds
attn: Rob Johnston
Albion, ME 04910
Vegetable seeds for growers in the North. Testing European varieties. Good selection of open-pollinated corn and cold hardy soybeans. Promoting the production of home-grown seeds and the availability of old "heirloom" varieties now facing extinction.

Abundant Life Seeds
attn: Forest Glen Roth
Box 30018
Seattle, WA 98103
A policy of regional self-sufficiency in agriculture through the production and distribution of acclimated vegetable, flower and herb seeds. About 200 varieties of locally-grown, untreated seeds that do well along the North Pacific rim of the US are available.

J. L. Hudson World Seed Service
Box 1058
Redwood City, CA 94064
Emphasizes rare wild plants and herbs, including many imported medicinal varieties. Catalog includes hundreds of trees, shrubs, herbs and flowers listed and indexed according to their medicinal and edibility characteristics.

The Redwood City Seed Co.
attn: Craig Dremann
Box 361
Redwood City, CA 94064
Interested in swapping for seed from wildflowers, food, dye, medicinal and fiber plants as well as collecting seeds from native California plants (many of which are drought-hardy) and vegetable varieties adapted to Western regional conditions.

Stock Seed Farms (formerly Wilson Seed Farms)
Route 1, Box 112
Murdock, NE 68407
Specializes in seeds of native American prairie grasses such as Buffalo grass, Sideoats Grama, Western Wheat grass etc.

Seeds & Seed Production

Barton, Lela V. 1961. SEED PRESERVATION AND LONGEBITY. Leonard Hill Books Ltd., Interscience Publ. Inc., New York 216 pp. (B) (OOP).

Cox, Joseph and George E. Starr. 1927. SEED PRODUCTION AND MARKETING. John Wiley & Sons, Inc. (OOP) How to grow seeds; production of vegetables, field legumes, grasses. Should be reprinted.

Dremann, C. 1974. VEGETABLE SEED PRODUCTION IN THE SAN FRANCISCO BAY AREA OF CALIFORNIA AND OTHER WARM-WINTER AREAS OF THE UNITED STATES. from: Redwood City Seed Co., P.O. Box 361, Redwood City, CA 94064. $0.50.

Grotzke, H. FROM SEED TO SEED. from: Biodynamic Farming and Gardening Association, RD1, Stroudsberg, PA 18360. 28 pp. A refreshing view of seeds and seed production, especially differences between "selection" and "crossing" strategies.

Hawthorne, Leslie R. and Leonard H. Pollard. 1954. VEGETABLE AND SEED PRODUCTION. The Blakiston Co., Inc., New York. (OOP) Commercial seed harvesting and breeding of vegetables for commercial operations, but information can be gleaned for small timers. Should be printed!!

Hill, Lawrence D. 1975. SAVE YOUR OWN SEED. from: Henry Doubleday Research Association, Bocking, Braintree, Essex, England. 46 pp; $1.50. Practical descriptions of English experiences.

Johnston, R. 1976. GROWING GARDEN SEEDS: A MANUAL FOR GARDENERS AND SMALL FARMERS. from: Johnny's Selected Seeds, Albion, Maine 04910. $2.30. Excellent how-to pamphlet on growing vegetables for seed production; harvesting and storing seeds for future use. Tends to be oriented to harsh winter areas. Badly needs graphics. For milder climates use Craig Dremman's pamphlet.

Knott, James K. 1962 (Rev. ed.). HANDBOOK FOR VEGETABLE GROWERS. Chap. 6, "Seed Production," John Wiley & Sons, Inc., New York.

Kozlowski, T.T. (ed.). 1972. SEED BIOLOGY. Vol. 3: "Insects and Seed Collection, Storage, Testing and Certification." Academic Press, New York.

Martin, A.C. and W.D. Barkley. 1973. SEED IDENTIFICATION MANUAL. Univ. of Calif. Press, Los Angeles. 221 pp. Photos/drawings of seeds of cultivated and wild plants found in farmlands, and woodlands of North America.

Mayer, A.M. and A. Poljakoff-Mayber. 1963. THE GERMINATION OF SEEDS. Monograph Series on Pure and Applied Biology. Pergamon.

Roberts, E.H. (ed.) 1972. VIABILITY OF SEEDS. Syracuse Univ. Press, New York. 448 pp.

U.S. Dept. Agriculture. (1) TESTING AGRICULTURAL AND VEGETABLE SEEDS, handbook 30. (2) 1942. STORAGE OF VEGETABLE SEEDS, Leaflet 220. (3) 1936. SEEDS, USDA YEARBOOK, 1936. All from: Government Printing Office, Washington D.C.

VEGETABLE AND HERB SEED GROWING FOR THE GARDENER AND SMALL FARMER. from: Bullkill Creek Publishers, Hersey, MI 49639.

Plant Genetics and Breeding

Allard, R.W. 1960. PRINCIPLES OF PLANT BREEDING. Wiley, N. Y.

Brewbaker, J. 1964. AGRICULTURAL GENETICS. Prentice-Hall, Inc., Englewood, N.J. 156 pp. A succinct and well-illustrated introduction to the principles of genetics using domestic plants and animals as examples.

Briggs, F.N. and P.F. Knowles. 1967. INTRODUCTION TO PLANT BREEDING. Reinhold, N. Y.

Crane, M.B. and W.J. Lawrence. 1952. GENETICS OF GARDEN PLANTS. Macmillan. N. Y. (OOP)

Darlington, C.D. 1973. CHROMOSOME BOTANY AND THE ORIGINS OF CULTI—VATED PLANTS. Hafner Publ. Co., New York. 231 pp. (B)

Lawrence, W.J. 1945. PRACTICAL PLANT BREEDING. Allen and Unwin. London.

Lawrence, W.J. 1968. PLANT BREEDING. Edward Arnold, London. Excellent, concise introduction.

Nelson, R.R. 1973. BREEDING PLANTS FOR DISEASE RESISTANCE . . . CONCEPTS AND APPLICATIONS. Penn. State University Press, University Park. State-of-the-arts summary with chapters on major crops.

Williams, W. 1964. GENETIC PRINCIPLES AND PLANT BREEDING. Blackwell Sci. Publ., Oxford, England. 504 pp. (A)

Pollination Ecology

Faegri, K. and L. van der Pijl. 1971. THE PRINCIPLES OF POLLINATION ECOLOGY. Pergamon Press, Oxford. Definitive work.

Free, J.B. 1970. INSECT POLLINATION OF CROPS. Academic Press, New York. 544 pp. (A)

Grout, R.A. 1953. PLANNED POLLINATION: AN AGRICULTURAL PRACTICE. Dadant & Sons, Inc., Hamilton, Ill. 23 pp (C)

Lovell, J.H. 1918. THE FLOWER AND THE BEE: PLANT LIFE AND POLLINATION. Charles Scribner's Sons, New York. 282 pp. (B) (OOP)

Meeuse, J.D. 1961. THE STORY OF POLLINATION. The Ronald Press,Co., New York. 243 pp. (B)

Mittler, T.E. (ed.). 1960. PROCEEDINGS OF THE FIRST INTERNATIONAL SYMPOSIUM ON POLLINATION. Copenhagen. Swedish Seed Growers Assoc., Stockholm. Lindhska Press, Orebro, Sweden. (A)

Proctor, M. and P. Yeo. 1973. THE POLLINATION OF FLOWERS. William Collins Sons and Co., Ltd. London.

GREENHOUSES

Greenhouses are controlled environments for growing useful plants and other food. The first greenhouses were built over a century ago; they were probably pit-structures facing south and using the surrounding earth for insulation and heat storage . . . similar to today's "solar-heated" greenhouses. Since then greenhouses have been used to germinate seeds, produce food-crops and ornamentals, force plants for early transplanting and, more recently, to cultivate (sub)-tropical fish and plant systems for year-around production. Greenhouses can be built from easily available materials; they can be small and simple or they can be elaborate mini-ecosystems in which plants, fish and wastewater interact to provide food and to filter wastes. Greenhouses expand the options for people interested in extending the growing season, producing warm-climate food-stuffs or providing a nursery for enhancing their local landscape.

By their very nature greenhouses accumulate a great deal of heat during the day; in fact, daytime heating is a major "problem" with most greenhouses. During the night most of this excess heat is lost through the translucent skin, and the greenhouse must be heated with auxiliary fossil fuels. As a result there is growing interest in designing greenhouses that use solar energy for heating. Large commercial greenhouses, because of obvious scale problems, are tending toward the use of active solar systems using detached collectors and elaborate storage devices (see Reviews, pg. 94). Small greenhouses, however, are more suited to passive solar heating. Here, the excess solar heat gained by the greenhouse during the day is stored in heat-absorbing "sinks" (e.g. rocks, or water). This heat, then, is released at night into the greenhouse by natural heat flows from the sink to the greenhouse space. This is accomplished in one or more of the following ways: 1) orientation of the greenhouse along an east-west axis so that the largest transparent surface is facing within 20° of true south (in the northern hemisphere), 2) insulation of the north wall, 3) placement of adequate ventilation at bottom and top to stimulate natural convection flow, 4) weatherstripping of all doors and vents, 5) use of heat sinks inside the greenhouse for absorbing heat, 6) insulation of south glazing area at night.

Solar-heated greenhouses may serve another important function . . . heating the home. By attaching a solar-heated greenhouse to a house or shelter, excess heat generated in the greenhouse can be passed into the house or stored in its walls. In addition the attached solar-heated greenhouse can provide food and serve as a repository for kitchen/human wastes (compost) while providing oxygen and water vapor (if needed). Attached solar-heated greenhouses are the cheapest and most practical way to retrofit solar-heating to a home while providing food, purification and valuable esthetic quality to the living space.

Listed below are references we have found useful in understanding greenhouse design and husbandry. FS-SHG refers to FREE-STANDING SOLAR-HEATED GREENHOUSES and A-SHG refers to ATTACHED SOLAR-HEATED GREENHOUSES.

—Richard Merrill

From: YOUR HOMEMADE GREENHOUSE by J. Kramer (Cornerstone, N.Y.).

Conventional Greenhouse: Design and Construction

Augsburger, N.D. et al. THE GREENHOUSE CLIMATE CONTROL HANDBOOK: PRINCIPLES AND DESIGN PROCEDURES. Acme Engineering and Manufacturing Corp., Muskogee, Oklahoma. 32 pp; $2.00 Excellent manual for describing basic principles and design procedures for selecting standard climate control systems for standard greenhouses. Many design implications for SHG's as well.

Bitterman, M. and D. Dykyjova. 1973. OPTIMAL SHAPE OF GREENHOUSE ROOFS DEDUCED FROM THE SOLAR SHAPE OF TREE CROWNS AND OTHER PLANT SURFACES. Paper V22 from the International Congress: "The Sun in the Service of Mankind." UNESCO, Paris.

Brann, D. R. 1972. HOW TO BUILD A WALK-IN OR WINDOW GREENHOUSE. Directions Simplified, Inc. Briarcliff Manor, NY.

Bugbee, B. 1975. GREENHOUSE DESIGN. Alternative Sources of Energy Magazine, No. 18, July 1975, Milaca, MN 56353. Fine brief overview of basic design components of small greenhouses and solar components of SHG's.

THE GREENHOUSE CLIMATE CONTROL BOOK. Acme Engineering and Mfg. Corp., Muskogee, Oklahoma.

Holloway, D.G. 1968. THE PHYSICAL PROPERTIES OF GLASS. Springer-Verlag, NY. Useful technical discussion.

Hudson, J.P. 1957. CONTROL OF THE PLANT ENVIRONMENT. Butterworth Scientific Publications, London.

Kramer, J. 1975. YOUR HOMEMADE GREENHOUSE AND HOW TO BUILD IT. Cornerstone Library, New York, NY 96 pp; $2.95. Poor construction details, but, if you have some building experience, excellent source of ideas from good working drawings and photos. Frame, geodesic, arched, porch, lean-to, loft, atrium and window greenhouses.

Lawrence, William. 1950. SCIENCE AND THE GLASSHOUSE. Oliver and Boyd, London. A very important book for design criteria in greenhouse construction.

Ministry of Agriculture, Fisheries and Food. 1964. COMMERCIAL GLASSHOUSES: SITING, TYPES, CONSTRUCTION AND HEATING. Her Majesty's Stationery Office, Bulletin 115, London. 112 pp. An excellent introduction to standard commercial-scale greenhouse design.

Neal, C.D. 1975. BUILD YOUR OWN GREENHOUSE: HOW TO CONSTRUCT, EQUIP AND MAINTAIN IT. Chilton Book Co., Radnor, PA 130 pp; $9.95 Best popular book for construction details (foundation, plumbing, ventilation etc.) of small greenhouses and coldframes.

Nicholls, R. 1975. THE HANDMADE GREENHOUSE FROM WINDOWSILL TO BACKYARD. Running Press, Philadelphia PA 19103. 127 pp; $4.95. Very simple descriptions of tools, site location, how to read plans, plus line drawing construction of window greenhouses, cold frames, dome and free-standing greenhouses. Bibliography and greenhouse resources and manufacturers. Good place to start if you know nothing about building.

Preston, F.G. 1958. A COMPLETE GUIDE TO THE CONSTRUCTION AND MANAGEMENT OF GREENHOUSES OF ALL KINDS. Taplinger Publishing Co., NY.

Roberts, W.J. HEATING AND VENTILATING GREENHOUSES. Cooperative Extension Service, Cook College, Rutgers University, New Brunswick NJ. Nice booklet for calculating heating and venting requirements for standard greenhouses.

Seeman, J. 1974. CLIMATE UNDER GLASS. World Meteorological Organization, Technical Note No. 131, WMO, Geneva, Switzerland. Technical analysis of the effects of climate on greenhouse design. Radiation and heat balance; heat transformation; air, soil and plant temperature, humidity, CO_2, light, shading and ventilation. Highly recommended for the technical at heart. Extensive bibliography of foreign literature.

Simons, A. 1957. ALL ABOUT GREENHOUSES. John Gifford Ltd., London. Construction and design of standard greenhouses.

Van Wijk, W.R. 1963. PHYSICS OF PLANT ENVIRONMENT. Interscience Publishers, John Wiley and Sons, Inc., New York. 382 pp. A highly technical, but invaluable sourcebook for SHG designers. Chapters include: "Thermal Properties of Soils;" "Temperature Variation in Soils;" "Turbulent Transfer in Air;" "The Glasshouse Climate." Incredible bibliography. Pursue this one!

White, J. W. and R. A. Aldrich. 1973. THE DESIGN AND EVALUATION OF RIGID PLASTIC GREENHOUSES. Transactions of American Society of Agricultural Engineers, Vol. 16, No. 5, pp. 984. ASAE, St. Joseph, Michigan.

Conventional Greenhouse: Management and Horticulture

Abraham, G. and K. Abraham. 1975. ORGANIC GARDENING UNDER GLASS. Rodale Press, Emmaus, PA 18049. 308 pp; $8.95. This is definitely one of the better books on sensible greenhouse management. Chapters include: design, soils; internal climate control; pest management; propagation; fruits; herbs and flowers. Pest control chapter is weak on techniques of bio-control application, and a chapter is needed on vegetable production. Otherwise well worth the price.

Brooklyn Botanic Garden. 1976 (rev. ed.). GREENHOUSE HANDBOOK FOR THE AMATEUR. BBG, 1000 Washington Ave., Brooklyn, NY. 97 pp; $1.50 Well illustrated, brief descriptions of design and management of small greenhouses, plus culture of many flower crops. Fine place to start with little money.

Chabot, E. 1955. THE NEW GREENHOUSE GARDENING FOR EVERYONE. M. Barrows & Co., Inc. NY.

Dulles, M. 1956. GREENHOUSE GARDENING AROUND THE YEAR. Macmillan Co., NY. 195 pp.

Eaton, J.A. 1973. GARDENING UNDER GLASS. Macmillan, NY.

Hussey, N.W. and L. Bravenboer. 1971. CONTROL OF PESTS IN GLASSHOUSE CULTURE BY THE INTRODUCTION OF NATURAL ENEMIES. In: "Biological Control," C. B. Huffaker (ed), Plenum Press, NY.

Large, J. 1972. GROWER MANUAL FOR GLASSHOUSE LETTUCE. Grower Books, London. Fine manual for growing lettuce under glass with implications for other crops.

Laurie, A. et al. 1969. COMMERCIAL FLOWER FORCING (7th ed) McGraw-Hill Book Co., New York, NY. Excellent source of info for all major greenhouse flower and plant crops, including pot plant rotation.

McDonald, E. 1971. HANDBOOK FOR GREENHOUSE GARDENERS. Lord and Burnham, Irvington-on-Hudson, NY. Nice description by major greenhouse company.

Markel, J.L. and M. Noble. 1956. GARDENING IN A SMALL GREENHOUSE. D. Van Nostrand Co., NY.

Nelson, K.S. 1973. GREENHOUSE MANAGEMENT FOR FLOWER AND PLANT PRODUCTION. The Interstate Printers and Publishers, Inc., Danville, IL. Primarily for commercial greenhouse foremen, but discussions of soil/air environments as well as marketing info may be valuable for large-scale SHG's.

Tauber, M. J. and R. G. Helgesen. 1974. BIOLOGICAL CONTROL OF WHITE-FLIES IN GREENHOUSE CROPS. New York's Food and Life Sciences, Vol. 7: 13-16.

Veen, R. and G. Meijer, 1959. LIGHT AND PLANT GROWTH. Macmillan Co., NY. Somewhat dated, but good basic information.

Walls, Ian. 1970. GREENHOUSE GARDENING. Ward Lick, Ltd., London. Describes history of greenhouse design in England. Especially valuable for discussion of optical filter experiments (e.g., blue light) in greenhouses for enhancing photosynthesis.

Wittwer, S.H. and S. Honma. 1969. GREENHOUSE TOMATOES . . . GUIDELINES FOR SUCCESSFUL PRODUCTION. Michigan State University Press, East Lansing. MI.

Wyatt, I.J. 1974. PROGRESS TOWARDS BIOLOGICAL CONTROL UNDER GLASS. In: "Biology in Pest and Disease Control." D.P. Jones and M.E. Solomon (eds), Blackwell, Oxford.

A-FRAME GREENHOUSE

From: YOUR HOMEMADE GREENHOUSE by J. Kramer (Cornerstone, N.Y.).

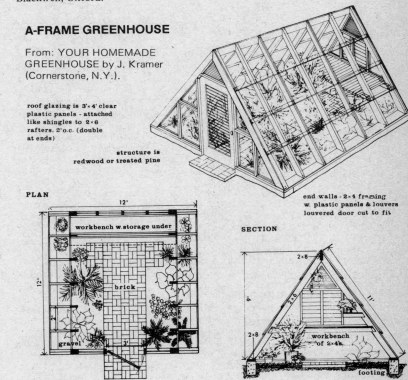

roof glazing is 3'x 4' clear plastic panels - attached like shingles to 2×6 rafters. 2'o.c. (double at ends)

structure is redwood or treated pine

PLAN

workbench w. storage under

brick

gravel

end walls - 2×4 framing w. plastic panels & louvers louvered door cut to fit

SECTION

workbench of 2×4k

footing

Solar-Heated Greenhouses: Views, Design, Construction, Horticulture

Alward, R. 1976. THE GREENHOUSE AS A SOURCE OF FOOD AND WINTER HEAT. In: "Low-Cost, Energy-Efficient Shelter for the Owner and Builder." E. Eccli (ed). Rodale Press, Emmaus, PA. Design constraints and construction of A-SHG. Brief and superficial.

Anderson, B. and M. Riordan. 1976. THE SOLAR HOME BOOK. Cheshire Books, Harrisville, New Hampshire. Fine source for explanation of heat loss/gain calculations, properties of building materials, solar climate data, etc.

A REVOLUTION IN GREENHOUSES. Sunset Magazine, March 1977. Sunset Publications, Menlo Park, CA. Photo description story of SHG's at Solar Sustenance Project; Ecotope, Helion "Solar Room"; and 2 private Arizona SHG's.

Balcomb, J.D. et al. 1975. SIMULATION ANALYSIS OF PASSIVE SOLAR HEATED BUILDINGS . . . PRELIMINARY RESULTS. from: Los Alamos Scientific Laboratory, Los Alamos, NM 87544. Easily understood computer simulation analysis of the solar gains through windows, walls, modified walls, skylights and clerestories. Concludes that passive systems are as effective as active ones but that placement and type of thermal storage is crucial. Landmark paper for anyone seriously interested in designing SHG's.

Calthorpe, P. 1977. A ONE HUNDRED PERCENT SOLAR HEATED GREEN-HOUSE. 1977 Report, Farallones Institute Rural Center, 15290 Coleman Valley Rd., Occidental, CA 95465. Simple description, no design analysis . . . badly needed for this nicely constructed FS-SHG.

Fisher, R. and B. Yanda. 1976. THE FOOD AND HEAT PRODUCING SOLAR GREENHOUSE: DESIGN, CONSTRUCTION AND OPERATION. John Muir Publications, Box 613, Santa Fe, NM 87501. 161 pp; $6.00. Far and away the most complete book on the subject of small A-SHG's. See review pg. 95

Gillett, D.A. 1975. A PAPER ON SOLAR POWERED GREENHOUSES. Kalwall Corp., Solar Components Division, Manchester, NH 03103. A thermal comparison of three small SHG's with varying amounts of thermal mass supplied by Kalwall Solar Battery Collector (fibreglass) Tubes.

Heeschen, C. 1976. DESIGNING A SOLAR GREENHOUSE. The Maine Organic Farmer and Gardener, Vol. 111, No. 5, Sept/Oct 1976. Nice discussion of design components and comparison of three different SHG designs.

Hoff, E. et al. 1977. NOTI SOLAR GREENHOUSE: PERFORMANCE AND ANALYSIS. Center for Environmental Research, School of Architecture, University of Oregon, Eugene, OR 97403. 31 pp, $2.00. Specs of 12'X17', post and beam FS-SHG with 14 tons north wall rock storage, wood heat back-up and forced-convection of excess heat up through rock storage.

Kramer, J. 1975. SUNHEATED INDOOR/OUTDOOR ROOM FOR PEOPLE, FOR PLANTS. Charles Scribner's Sons, NY. 110 pp. Excellent drawings of cold and hot climate greenhouses as A-SHG.

Lawand, et al. PAPERS ON SOLAR GREENHOUSES; An Investigation of the Contribution of Solar Energy in Heating Greenhouses in Quebec; The Development and Testing of an Environmentally Designed Greenhouse for Colder Regions. from: Brace Research Institute, c/o MacDonald College, Ste. Anne de Bellevue, Quebec, Canada (latter also in Solar Energy, Vol. 17). Heat budget analysis of SHG during Quebec winter and the evolution of the tilted north wall greenhouse.

Lucas, R.F. and C.D. Baird. 1976. APPLICATION OF SOLAR HEATED WATER TO GREENHOUSES. Inst. Food and Agric. Sci., University of Florida, Gainsville. Shows calculations of seasonal heating load, collector size and water storage capacity for solar-heated commercial greenhouse.

Mears, D. et al. 1976. GREENHOUSE SOLAR HEATING: A PROGRESS REPORT. Dept. Agric. Engineering, Cook College, Rutgers Univ., New Brunswick, N.J. 08903. 74 pp. Potentials of solar-heat exchanges and insulating night curtains in commercial greenhouses. Valuable summary.

Moodie, P., K. Smith and H. Reichmuth. 1976. EVALUATION OF A SOLAR HEATED GREENHOUSE FOR THE ENVIRONMENTAL FARM PROJECT AT CANYON PARK JUNIOR HIGH SCHOOL IN BOTHELL, WASHINGTON. from: Ecotope Group, 747 16th Ave., E., Seattle, WA 98112. Design specs for FS-SHG (see "Research Groups" below).

Nash, R.T. and J.W. Williamson. 1974. TEMPERATURE STABILIZATION IN GREENHOUSES. Dept. Electrical Engineering. Vanderbilt University, Nashville, Tennessee.

PASSIVE SOLAR HEATING AND COOLING. Conference and Workshop Proceedings, May 18-19, 1976, Albuquerque, NM. from: Los Alamos Solar Energy Group, Los Alamos, NM. Landmark meeting with many implications for SHG design in the scores of papers.

PROCEEDINGS: SOLAR ENERGY . . . FUEL AND FOOD WORKSHOP. ERDA/USDA. from University of Arizona Tucson Environmental Lab, Tucson International Airport, Tucson, AZ 85706. $5.00. Definitive review of techniques for solar-heating commercial greenhouses; active, passive, retrofit and original design. Small A-SHG also discussed.

PROCEEDINGS OF A JOINT CONFERENCE: SHARING THE SUN. American Section of the International Solar Energy Society and the Solar Energy Society of Canada, 1976. ISES. 300 State Rd., 401, Cape Canaveral, FL 32920. Volume 7 contains several papers on SHG's.

PROCEEDINGS, INTERNATIONAL SOLAR ENERGY SOCIETY. 1975. Los Angeles (UCLA), July 28-August 1, 1975. Papers include: Kusianovich, J., Solar Heated Greenhouse with a One Year Payout; Nash, R.T., Temperature Stabilization in Greenhouses (analysis used in calculating solar energy requirements of 7'x12' greenhouse).

Ruttle, J. 1976. A NICE GREENHOUSE FOR NASTY WINTERS: ALL SOLAR. Organic Gardening and Farming Magazine, Sept. 1976. Description of FS-SHG featuring low-cost design and removable plastic film frames for insulating south face glazing.

SOLAR GREENHOUSE COMES OF AGE. 1976. Organic Farming and Gardening Magazine, Rodale Press, Emmaus, PA 18049. A series of nice articles in the September 1976 issue of OG&F.

Telkes, M. 1974. SOLAR ENERGY STORAGE. ASHRAE Journal, Sept. 1974, pp. 38-44. Discussion of eutectic salts for greenhouse solar heat storage.

Yanda, W. and S. Yanda. 1975. AN ATTACHED SOLAR GREENHOUSE. The Lightning Tree, Box 1837, Santa Fe, NM 87501. $1.50. Bilingual (Spanish) booklet describing construction of low-cost, 10x16 A-SHG for food and (up to 25%) space heat. Prototype of many such A-SHG's now working all over the Southwest, from good work of Solar Sustenance Project.

Solar-Heated Greenhouses: Aquaculture Systems

Angevine, R., E. Barnhart and J. Todd. 1974. A PROPOSED SOLAR HEAT AND WIND POWERED GREENHOUSE COMPLEX ADAPTED TO NORTHERN CLIMATES. Journal No. 2, New Alchemy Institute, Box 432, Woods Hole, MA 02543.

Dekorne, J. 1975. THE SURVIVAL GREENHOUSE (see below).

Deryckx, W. and B. Deryckx. 1976. TWO SOLAR AQUACULTURE/GREENHOUSE SYSTEMS FOR WESTERN WASHINGTON: A PRELIMINARY REPORT. 50 pp. $5.00. Good basic description of small fresh water fish culture systems and design considerations of parabolic and Rhombi-cube Octahedron FS-SHG (see "Research Groups" below).

Mackie, B. and M. 1977. SOLAR HEATED GREENHOUSE. Sun Experimental Farms, 835 Fleishauer Lane, McMinnville, OR 97128. 16 pp., $2. Plans for FS-SHG with oil drums and graywater fish tanks for thermal mass.

Solar-Heated Greenhouse: Pit Structures

Dekorne, J. 1975. THE SURVIVAL GREENHOUSE: AN ECOSYSTEM APPROACH TO HOME FOOD PRODUCTION. Walden Foundation, Box 5, El Rito, NM 87530. 165 pp. $7.50. Describes the philosophy, design and management of a pit FS-SHG that integrates hydroponic crops, small livestock, fish culture, bluegill, Tilapia, and wind power as a system. Mistakes are pointed out. Good description of basic greenhouse design principles.

Hinds H. and J. 1976. THE SUNPIT GREENHOUSE. In: "Cloudburst 2," Marks, V. (ed), Cloudburst Press, Mayne Island, BC V0N 2J0. General plans of a tested pit greenhouse adapted for cold temperate climate.

Kern, Ken. 1974. THE PIT GREENHOUSE. In: "The Owner-Built Homestead. Owner-Builder Publications, Box 550, Oakhurst, CA 93644. Kern presents some interesting facts about light, heat and solar design for greenhouses.

Taylor, K. 1974. COLD PIT GARDENING. Horticulture Magazine, MA. Hort. Soc., 300 Mass. Ave., Boston 02115.

Taylor, K. and E. Gray. 1969. WINTER FLOWERS IN GREENHOUSES AND SUN HEATED PITS. Charles Scribner's Sons, New York. Presently the Sunpit enthusiast's bible.

Solar-Heated Greenhouses: Research and Community Groups

AGRICULTURAL RESEARCH AND EDUCATION CENTER. 5007 60th St., East, Bradenton, FL 33507. Working with Dept. of Agricultural Engineering, Univ. Florida on solar methods of greenhouse heating and cooling; active air and water systems plus passive designs.

ARIZONA SUNWORKS, Star Route, Chino Valley, CA 86323. A-SHG, 10°F design temperature, misted plant system.

ATMOSPHERIC SCIENCES RESEARCH CENTER. State University of New York, Albany, New York.

BROTHER'S REDEVELOPMENT INCORPORATED. Box 2043, Evergreen, CO 80439. A group building small SHG's as part of low-income housing using renovation project.

CITIZENS FOR ENERGY CONSERVATION AND SOLAR DEVELOPMENT. Box 49173, Los Angeles, CA 90049. Designing and building FS-SHG for subtropical subclimate of Western US. Plans and brief research paper.

ECOTOPE GROUP, 747 16th Ave. East, Seattle, WA 98112. (See pg. 247) With Bear Creek Thunder (Box 799, Ashland, OR 97520) Ecotope Group has designed and built 3 significant free-standing solar heated greenhouses: 1) an aquaculture-food crop complex with a parabolic north wall, 2) a rhombo-cube octahedron aquaculture-food crop complex, and 3) a solar heated greenhouse with stacked polyethylene bags on the south wall, and a 20 ton concrete north wall. Has unique venting chimney stack.

ENVIRONMENTAL RESEARCH LABORATORY, University of Arizona, Tucson, AZ 85706. Developing integrated solar heated greenhouse systems including: 1) shrimp-hyacinth ponds and vegetable growing in a polyethylene covered greenhouse, 2) large-scale desalinization-agriculture complex, and 3) controlled environment agricultural facility.

FARALLONES INSTITUTE RURAL CENTER, 15290 Coleman Valley Rd., Occidental, California. (see pg. 245). FS-SHG for propagation and vegetable production. Water storage, hand vent.

NEW ALCHEMY INSTITUTE EAST, P.O. Box 432, Woods Hole, MA 02543. (see pg. 244). One of the leading groups in the country working with food production (especially aquaculture) in solar heated greenhouses. Projects include: 1) pit greenhouse aquaculture complex, 2) dome greenhouse fish ponds with unique biofilters, 3) closed system "Mini-Ark" with separate SHG structures for bio-filter, fish food culture, and fish stock rearing. Water circulated by sail-wing windmills, and 4) a self contained bio-shelter ("Ark") that functions as a food and energy producing, live-in research center. Has solar fish ponds, solar heating and hydrowind power.

SOLAR SUSTENANCE PROJECT. Rt. 1, Box 107AA, Santa Fe, NM 87501. One of the first groups in the country to actively design and build low-cost SHG's for food and space heat. Starting out with individual SHG for low-income people, group now has several community greenhouses under construction in urban areas of NM. See: THE FOOD AND HEAT PRODUCING SOLAR GREENHOUSE by director Bill Yanda (see pg. 95).

RURAL RESEARCH HOUSING UNIT, USDA. Clemson University, Clemson, SC. ERDA-funded project to develop, design and monitor various A-SHG as "Greenhouse Residences." Good cost and heat analyses . . . see PROCEEDINGS: FOOD AND FUEL WORKSHOP (pg. 94).

UNITED STATES DEPARTMENT OF AGRICULTURE, AGRICULTURAL RESEARCH SERVICE. Northeastern Region, Beltsville, MD 20705. Designing an active A-SHG with south-facing collectors on north inside wall of greenhouse to improve collector efficiency.

Solar-Heated Greenhouses: Manufacturers, Plans and Hardware

DALEN PRODUCTS, INC. 201 Sherlake Dr., Knoxville, TN 37922. Solar vent. A small thermal element opens (up to 9 lbs) at 62°F. At 75°F, it is fully open (7 1/2''): about $30.00 Counterbalance spring attachment ($5.00) will extend lift to 20 lbs.

ENERGY SYSTEMS. 77 La Paloma, White Rock, NM 87544. Detailed plans for an A-SHG using water-in-drums as heat sinks.

GARDEN WAY LABORATORIES. Box 66, Charlotte, VT 05445. 15 pp. Excellent detailed plans for a "Solar Room," an A-SHG with forced air convection (to and from house) and 55-gal water drum storage.

GLAZING. For various glazing products see the solar hardware matrix on pg 70 of EP.

HEAT MOTORS, INC. 635 West Grandview Ave., Sierra Madre, CA. "Automatic thermal activator" expands internal fluids in ram cylinder. For each 10°F, ram will move 50 lbs. Temperature range is adjustable. About $60. ($600.00 model moves 1 ton.)

SOLAR ROOM. P.O. Box 1377, Taos, NM 87551. Inflatable, double-wall, durable plastic film A-SHG, available in kit form and easily attached to houses and mobile homes. Can be stored in summer.

SOLAR SUNSTILL, INC. Setauket, NY 11733. Sells coatings for the control of condensation and light. 1) SUN CLEAR converts non-wettable surfaces of plastics and glass to wettable conditions. 2) VARISHADE, a coating that shades when dry and is transparent when wet. 3) THERMOSHADE, a temperature sensitive film transparent below a certain temperature, but becomes cloudy above 80-100°F.

SOLAR TECHNOLOGY CORPORATION. 2160 Clay St., Denver, CO 80211. Offers 3 A-SHG and FS-SHG models. Deluxe "Solar Garden" features: "enclosed thermal storage and thermostatically controlled internal circulation fan . . and timer controlled automatic watering system."

ZOMEWORKS. P.O. Box 712, Albuquerque, NM 87103. (1) BEADWALL®: Photocell system for blowing and evacuating styrofoam beads in between double glazing for night insulation. Initial electrostatic problems seemed to have been solved with glycerine coatings (see SOLAR ENERGY Vol. 15, No. 7, Nov. 1975). (2) NIGHTWALL®: Rigid styrofoam sheets adhered to perimeter of windows (south glazing) surface with magnetic strips. (3) SKYLID®: Device that uses counter-balanced weights and freon containers (totally mechanical—no electricity) to open vents or skylights.

AQUACULTURE

AQUACULTURE: BRINGING IT
HOME WITH THE NEW ALCHEMISTS

The art of culturing fish and shellfish for food has been a flourishing tradition in Asia for centuries. In many "polyculture" ponds several species of edible plants and animals are grown together with a degree of ecological sophistication unrivaled in animal husbandry. In the United States aquaculture has been limited to the growing of single cash crops (e.g., trout, catfish, oysters) in commercial ponds or coastal waters. There never has been much of an interest in an aquaculture that could work for small groups, farmers, or communities using local resources and ecological techniques in temperate North American climates.

However as our beef/oil economy continues to decay, more and more people are beginning to think about producing food at the local level for their families, collectives, neighborhoods, communities, etc. For some the old "Victory garden" will flourish on roof-tops and street corners, in backyards and vacant lots. For others the bounties of vegetables, herbs and fruits can be enhanced with animal protein supplied by a low energy, low cost, ecologically-sound aquaculture. In order to work, such a mini-fish farm must be suited to the needs and local resources of individuals and small groups with little in the way of money or land.

To this end the New Alchemy Institute (Box 432, Woods Hole, Massachusetts 02543) has been experimenting with various "backyard fish farms" for the past few years. So far the New Alchemists have been concentrating on five basic strategies in an attempt to develop a low-cost, indigenous aquaculture adapted to northern climates: (1) the use of ecological models from Asian polyculture ponds, (2) the use of inexpensive greenhouses over fish ponds to keep the waters warm, (3) the use of biological filters for the transformation of toxic substances into useful nutrients, (4) the use of solar/wind power to regulate the internal climate of the pond-greenhouses and to pump filtered water through the system, (5) the use of fertilizers and supplemental fish foods produced in local gardens and insect cultures. The New Alchemists have published numerous useful how-to-do-it pamphlets, books and articles on the philosophy/ecology/hardware of backyard fish farms (see Aquaculture, page 137). Because the following paper is written only as a perspective, we refer the reader to the New Alchemy Institute for the "nuts and bolts" stuff. However, as the New Alchemists themselves are the first to admit, very little is known about small aquaculture operations in temperate climates. We can put together information from esoteric articles, the experiences of a few aquaculturists, the records of perennial fish ponds in the tropics or even the pertinent fallout from commercial set-ups, but perhaps more than any other kind of food/energy system described in the *Primer*, aquaculture is still in our heads . . . it needs participants and the flow of our experiences.

—R. M.

AQUACULTURE
by Dominick Mendola

The purpose of aquaculture is to provide fish and shellfish protein for human diets. The need for a substantial increase in the world's supply of protein is obvious . . . as is the place of fish in human nutrition. What is not so obvious is why aquaculture is a beneficial means of food production, especially protein. The reason is because aquaculture is more akin to agriculture than to ocean fisheries since the means of production (water, land stock and equipment) are accessible to many people and resources can be controlled by individuals, small groups, or even entire communities.[1] Also, since many of the fish populations presently exploited by ocean fisheries have reached their maximum sustainable yields, aquaculture can help supply the additional needs of an increasing demand for fish products.[2]

There are definite advantages to culturing fish as a food source. For one thing fish are able to produce more protein per pound of food eaten than, say, a cow, chicken, or any other land animal (Fig. 1).

FIG. 1 POUNDS OF PROTEIN GIVEN TO VARIOUS LIVESTOCK TO PRODUCE 1 POUND OF PROTEIN FOR HUMAN CONSUMPTION. REDRAWN FROM FIGURE OF LAPPÉ (REF. 3) AND DATA OF BARDACH ET AL (REF. 1).

There are two basic reasons for this: (1) Fish and shellfish live in a medium that has about the same density as their bodies and as a result they don't have to spend much energy supporting themselves against gravity. Thus, aquatic animals have a reduced skeleton and a greater ratio of flesh to bones . . . a definite advantage for a food species. (2) Fish and shellfish are cold-blooded, and their body temperatures are essentially that of their surroundings. Because of this they don't have to spend much energy maintaining warm body temperatures as do warm-blooded birds and mammals. This "saved" energy can go into the production of animal protein.

There are other advantages. Fish live in a three dimensional world where nutrients necessary for growth are distributed throughout a volume. So for a given area, aquaculture can yield more food than land-based animal husbandry.[4] For example, in an area once used for cattle grazing in Tanganyika, beef was produced at a rate of 9.8 pounds per acre per year (lbs/acre/yr). When this was replaced with aquaculture (with artificial feeding) fish production provided meat at the rate of about 2200 lbs/acre/yr.[5] Table I shows yields of selected aquaculture systems compared to typical land-animal husbandries.

As Table I indicates, there are different kinds of aquacultures; fresh water, brackish and marine. There are also a wide variety of animals (fish and shellfish) and plants (seaweeds, watercress, water lettuce, etc.) that are widely grown. We shall emphasize fresh water pond aquaculture that uses indigenous resources (simple materials, locally-grown fish foods and solar/wind power supplies). There are several reasons for this approach: (1) The methods and materials needed for this form of aquaculture are available to more people than for mariculture (farming the sea) or commercial aquaculture. (2) Costs can be kept low and human labor can provide the energy for construction and maintenance. (3) Many existing bodies of water can be easily converted to this form of aquaculture. (4) With small, self-contained fish operations one can easily control most of the inputs affecting the health, productivity and stability of the system. For example, many of the pollutants (pesticides, heavy metals and petroleum products) increasingly found in fish products can be eliminated or at least reduced with small aquaculture systems. However, even with local freshwater supplies one must consider the history of the use of toxic chemicals in the area. In fact it is always best to have water checked by a water quality laboratory since many toxins, especially pesticides, are slowly but surely finding their way into almost every supply of surface and ground waters in the U.S.[8,9,10,11] *

Aquaculture	Place	Culture/Feeding	Annual Yield* (lb/acre/yr)
FISH			
Largemouth bass, bluegill	U.S.	Polyculture/natural	225-400 (average)
Channel catfish	U.S.	Commercial ponds/artificial	2,000-3,000 (maximum)
Channel catfish	U.S.	Experimental, sewage ponds	3,600
Chinese carps	S.E. Asia	Polyculture/heavy fertilizer	6,300-8,000 (maximum) 2,700-3,600 (average)
Common carp	Poland	Sewage ponds	1,200
Common carp	Japan	Intensive ponds/heavy	4,500
Estuarine (Mullet)	U.S.	Tidal ponds/natural	185 (average)
Rainbow trout	U.S.	Flowing water ponds/heavy	60,000 (maximum)
Tilapia	Tropical world	Ponds/fertilization, feeding	2,000-5,500 (large fish) 18,000 (small fish)
Walking catfish (Clarias)	Thailand	Commercial/heavy feeding	88,000 (maximum)
SHELLFISH			
Sea mussel	Spain	Floating rafts	540,000 (with shell) 270,000 (meat)
Freshwater mussel	Alabama	Polyculture, experimental ponds	1,131 (with shell) 413 (meat)
Oyster	U.S.	Bottom culture, mechanical pest control	4,500 (maximum)
Oyster	U.S.	Bottom culture, chemical pest control	45,000 (maximum)
Oyster	Japan	Floating lines, no pest control	50,000 (average)
LAND HUSBANDRY			
Pasture beef	U.S.	Animals range free	50-200
Hogs	Malaysia	Natural forage	2,300-11,000
Chickens	U.S.	Cages/artificial feed	160,000

TABLE I YIELDS OF SOME DOMESTIC FISH, SHELLFISH AND LAND ANIMALS. FROM REF. 4, 6, AND 7.

*In terms of energy production fish have a caloric value of about 1 kcal/gram of wet weight; e.g., the "Walking Catfish" produces about 4 million kcal/acre/year. For comparison, beef production yields only about 162,000 kcal/acre/year. (Ref. 10).

Many of the aquatic species that can be cultured for food are from tropical countries where aquaculture has been a way of life for centuries. Some of these animals can be imported or bought from distributors on this continent (see Reviews), and will thrive under artificially heated conditions. However, with a few notable exceptions (e.g., *Tilapia*) it is probably best to rely on temperate climate species, especially those found in the local area. (See Aquaculture appendix page 192, for list of some commonly available species together with their general habits.)

Ecological Food Production and Aquaculture

In nature it is common to see many kinds of plants and animals living together all with different habits. Generally speaking, in natural communities with a high degree of ecological diversity (i.e. with many kinds of plants and animals exchanging energy among themselves) like tropical rainforests or coral reefs, one usually finds a greater degree of stability than "simple" communities like pine forests or corn fields. With high ecological diversity there is less chance that disease or sudden changes from outside will destroy the integrity of the community or kill off individual species.[12,13]

In aquaculture, ecological diversity is called *polyculture*, where a variety of plants and animals, picked for their mutually beneficial interactions, are grown together in the same system (pond, tank or combina-

*A few filter materials (e.g. polyurethane foam and activated carbon) have been used successfully to remove toxins from water solutions. However such filters are impractical in all but the smaller or commercial systems since they must be changed often or cleaned with organic solvents. Also, to be effective, the filters require a very slow flow rate.

tion). In ideal polyculture, large plants provide food for some fish, while microscopic floating plants (phytoplankton) serve as food for others. These plant-eating fish in turn serve as food for flesh-eating fish. All waste products are then eaten as detritus by certain other fish and shellfish. In this way, all available food niches are filled so that energy flow and nutrient recycling can proceed at a maximum rate. There are of course many possible combinations of fish and shellfish that are ecologically impossible as for example when a particularly voracious fish eats everything else in the pond. In other words, a lot of aquatic organisms living together *per se* do not produce ecological diversity. Only certain combinations can do that.

Once a proper polyculture system is found, it usually turns out to be synergistic, that is the growth of each individual species is enhanced beyond the point where it would be if it were raised separately in a monoculture. Put another way, polyculture assures that all food materials added to the system are completely used for growth. And herein lies the great advantage of polyculture over monoculture in aquatic food production systems.

Polyculture and the Sacred Carp

The polyculture model comes from the Chinese and Asian peoples who have been practicing the art of aquaculture for thousands of years. Today, in China, freshwater aquaculture provides at least 1.5 million tons of protein food every year.[14] As noted above, the Asian polyculture system is based on the belief that the pond is an ecosystem and that it should contain a variety of edible species for the maximum use of food and habitats.

The mainstay of the Asian polyculture pond is the Chinese Carp, of which there are several species (Fig. 2).

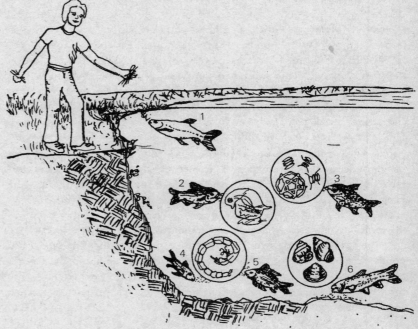

FIG. 2   FEEDING HABITS OF PRINCIPAL CHINESE CARP SPECIES. (1) GRASS CARP (CTENOPHARYNGODON IDELLUS) FEEDING ON LARGE FLOATING PLANTS. (2) BIG HEAD (ARISTICHTYS NOBILIS) FEEDING ON MICROSCOPIC ANIMALS (ZOOPLANKTON) IN MID-WATER. (3) SILVER CARP (HYPOPHTHALMICHTYS MOLITRIX) FEEDING ON MICROSCOPIC PLANTS IN MIDWATER. (4) MUD CARP (CIRRHINUS MOLITORELLA), AND (5) COMMON CARP (CYPRINUS CARPIO) FEEDING ON BOTTOM ANIMALS, DETRITUS AND CARP FECES. (6) BLACK CARP (MYLOPHARYNGODON PICEUS) FEEDING ON MOLLUSKS. FIGURE REDRAWN FROM BARDACH ET AL (REF. 1)

Typical yields from Chinese polyculture ponds average about 2400 lbs/acre/year, with some of the better ponds producing 4800-6400 lbs/acre/year.** Yields are increased by feeding manures, vegetable wastes and other organic materials to the ponds. These fertilize the water and increase the primary productivity (yields of plants). This excess food energy is then passed up the food chain to the carp.

Often the Asians situate their fish ponds at the bottom of a hill or sloping farmland, allowing natural drainage to carry the manures and agri-

**1 kilogram/hectare = 2.2 lbs/2.47 acres or about 1 lb/acre (actually 1 kg/ha = .89 lbs/acre).

cultural runoff into the pond (Fig. 3).

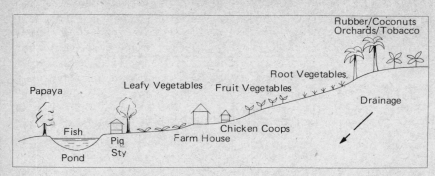

FIG. 3   AN INTEGRATED AQUACULTURE-AGRICULTURE SYSTEM USED IN SINGAPORE. REDRAWN FROM BARDACH ET AL (REF. 1) AND DESCRIPTION OF HO (REF. 6).

The Asians also use the pond water to irrigate (and fertilize) their crops. At regular intervals the ponds are drained and the accumulated bottom sludge is dug out and added to vegetable beds as compost.

Table II has been compiled to show various carp polyculture schemes that have been tried throughout the world. These examples give an idea of optimal conditions for the stocking, management and feeding of carp.

Simple Energy Budgets for Aquaculture

There doesn't seem to be any information or exact accounting of all energy inputs into freshwater fish ponds. (But see the materials list for the 3 backyard fish models built by the New Alchemists, in their Journal No. 2). For now we have to be content with budgets based on labor inputs for Asian pond operations.

Le Mare[7] described a very efficient 11 acre farm near Penang, Malaya on which a tenant farmer kept pigs and fish, and tended a small diverse garden and some rice fields. The pigs were bred, kept in a nursery, and then fattened in pens. Some pigs were kept in sties on the banks surrounding a series of 8 small ponds (about 2 acres total). Daily the farmer allowed running water to wash through the pig sties and carry the dung and excess pig food into the ponds by a series of drainage ditches.

The ponds were about 3 feet deep and were stocked with various Chinese carp (Common, Silver and Grass), in polyculture with *Tilapia mossambica*. All ponds were planted with "water lettuce" *(Ipomea repens)*, a floating aquatic plant suitable for pig food (21% protein—dry weight). Under these conditions the water lettuce grew so rapidly that about 1100 lbs/day (wet weight), *in excess of that eaten by the carp*, were harvested and added as a supplement to the pig food (broken rice, rolled and ground oats, groundnut cake, copra cake, brain, fish meal, and cod liver oil) in a ration of 1/3 water lettuce to 2/3 feed.

Production of fish during the first year was about 3200 lbs/acre. The farm also produced 700 pigs (about 2700 lbs. of pig meat). The records of production costs showed that the fish gave a higher proportional return than the pigs because little labor and materials were spent tending the fish, whereas a great deal of labor plus additional food was spent on the pigs. In other words, the *energy subsidy* of the fish culture was lower than for the pigs. Since the only inputs were human labor, we can easily estimate the time (energy) necessary to manage and harvest the 2 acres of ponds from what we know of fish raising (Table III, next page).

Species, Stocking Ratio		Treatment and Stocking Density	Yields and Survival
1. KWUNGTUNG PROVINCE, CHINA			
TENANT FARM, 1/5 ACRE			
a. Mud Carp	(Ref. 15)	Mud Carp—600 fingerlings (30 lbs.)	≈ 4000 lbs/acre/year
b. Grass Carp	(Ratio not given)	Grass Carp—200 fingerlings (14 lbs.)	
c. Big Head Carp		Big Head Carp—25 fingerlings (9 lbs.)	
d. Silver Carp		Silver Carp—25 fingerlings (9 lbs.)	
e. Black Carp		Black Carp—10 fingerlings (2 lbs.)	
f. Common Carp		Common Carp—25 fingerlings (2 lbs.)	
2. INDIA EXPERIMENTAL PONDS			
a. Grass Carp	(Ref. 16)	24,300-36,500 per acre	2590 lbs/acre/year
b. Silver Carp	(3:4:3)	Carp fry	80% survival
c. Common Carp		No feeding data	
3. ISRAEL, BRACKISH WATER PONDS			
a. Common Carp	(Ref. 17)	Variable stocking rates	(estimated) 6250-8000 lbs/acre/year
b. Silver Carp		Ponds fertilized	
c. Tilapia aurea Food habits unknown		Fish fed pelletized diet high in protein	
d. Tilapia nilotica Plankton feeder?			
Omnivore or feeds on high plants			
(exact habits unknown)			
e. Grey Mullet (Mugil spp.)			
Plankton, benthic algae			
4. ROMANIA, 300 ACRE POND			
a. Common Carp	(Ref. 18)	1700, 570, 40 respectively lbs/acre	1750 lbs/acre/year
b. Silver Carp	(3:1:1)	Yearling fish, supplemental feeding	56% Common, 34% Silver, 10% Grass
c. Grass Carp			Food Conversion Ratio 3:1
5. ROMANIA, 300 ACRE POND			
Same as above	(Ref. 18)	2350, 1300, 160 per acre respectively	2050 lbs/acre/year
	(14:8:1)	Supplemental feeding.	56% Common, 39% Silver, 5% grass
			Food Conversion 2.7:1
6. CHINA, 1½ ACRE POND			
a. Black Carp	(Ref. 15)	Black Carp—773 (844 g)	Total weight 11,800 lbs/year
b. Silver Carp		Silver Carp—2000 (23 g)	≈ 50% Blacks
c. Big Head		Big Head—400 (30g)	Net Production 10,000 lbs/year or
d. Bream (Parabramu pekinesis)		Bream—1214 (25 g)	6700 lbs/acre/year
Feeds on insects, worms, small fish		Grass Carp—110 (191g)	95.6% survival
e. Common Carp		Common Carp—1905 (26g)	
		Total 6402 (812 kg)	
		Fish (1800 lbs)	
4. BURMA, 3 REARING PONDS, 0.2 ACRES EACH			
a. Grass Carp	(Ref. 19)	1370-2035 2g fingerlings/acre	(After 6 months)
b. Silver Carp	(Variable)	Fed with mixture of rice bran, peanut cake, and chopped green vegetation. (Ratio 1:1:2)	Total production 1700, 1820, 1242 lbs/acre
			Average production: Silver —868 lbs/acre
		7—11 lbs/day (quantity doubled after 3 months)	Grass—720 lbs/acre
			Survival: Silver 95%, Grass 97%

TABLE II   POND STOCKING DENSITIES, MANAGEMENT TREATMENTS AND YIELDS OF VARIOUS CARP POLYCULTURE SCHEMES

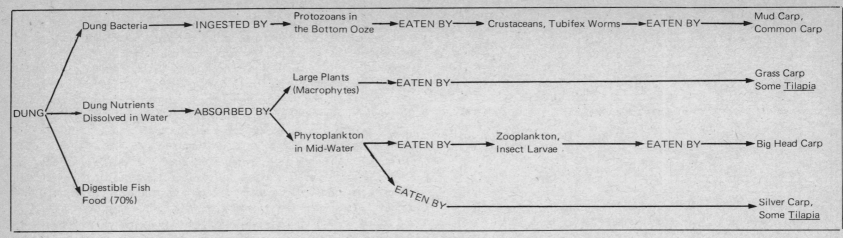

FIG. 4   THE CONTRIBUTION OF DUNG FERTILIZER TO THE DIFFERENT
FOOD CHAINS OF A POLYCULTURE FISH POND.

LABOR ENERGY INPUT

A. Assume that a person doing moderate work consumes about: 170 kcal/hr.

B. Assume that the farmer must spend about 500 hours per year tending and harvesting the fish from the 2 acres of ponds.

C. Therefore total labor energy expended is:
500 hrs/year x 170 kcal/hour . . . . E = 85,000 kcal/year

FISH ENERGY OUTPUT (PRODUCTION)

D. Annual fish production from the 2 acres of ponds is: 3600 kilograms (7900 lbs)

E. Assume energy value of fish flesh to be:
1 kcal/gram wet weight

F. Then energy value of total production is:
(3600 Kg) x (1 Kcal/gram) . . . . P = 3,600,000 Kcal/year

ENERGY BUDGET OF MALAYSIAN POLYCULTURE FISH POND

C/F . . . . . . 0.02 = 2% i.e. a labor energy subsidy of 2 Kcal is required for every 100 Kcal of fish produced

TABLE III·   ENERGY BUDGET FOR POLYCULTURE FISH PONDS IN MALAY-
SIA. NOTE: THE PIG MANURE AND SPILT PIG FOOD THAT FER-
TILIZE THE PONDS ARE CONSIDERED AS BY-PRODUCTS OF THE
PIG PRODUCTION, AND FOR OUR PURPOSES HERE ARE NOT FI-
GURED IN THE ENERGY BUDGET. BASIC DATA FROM REF. 7

It is interesting to note that the pig dung serves as food for the fish in at least four important ways (Fig. 4). This illustrates the importance of organic fertilizers to the balanced diversity of the fish pond (i.e., more than just one food chain is supplied with food).

Taiwan is another Asian country where fish is an important part of the diet (63% of the animal protein eaten). Aquaculture in Taiwan covers the range from intensive, coastal oyster farms to carp and *Tilapia* polycultures in inland ponds. The results of one study from Taiwan[20] are important because they suggest that under comparable conditions freshwater pond aquaculture has a higher return on energy/material investment than either pig farming or ocean fishing (Table IV).

Type of Husbandry	Kg*/Man Year Average	High	Kg/Hectare/Year** Average	High	U.S. $/Kg Average	Low
Brackish Water (Milk fish)	5,098	11,022	2112	2687	0.37	0.29
Freshwater Ponds (carp, _Tilapia_)	10,453	70,607	1537	2413	0.31	0.20
Shallow Sea (oysters)	45,575	--	1292	2096	0.16	0.10
Hog Farming	12,000	--	(see Table I)		0.43	--

* 1Kg = 2.2 lbs.
** 1 Kg/ha - .89 lbs/acre

TABLE IV   PRODUCTIVITY AND PRODUCTION COSTS FOR DIFFERENT
TYPES OF WATER AND LAND HUSBANDRY IN TAIWAN.
AFTER REF. 20

When compared in terms of protein production, the results stay relative-

ly the same. Shallow sea oyster farming is the most economical, followed by freshwater pond aquaculture (Table V). However, it is worth re-emphasizing the obvious: freshwater pond aquaculture is practical for many more people than is mariculture. And in this sense, it may be the most efficient means of animal protein production.

Product	Production Kg Protein/Man-Year Average	High	Cost U.S. $/Kg Average	Low	Protein %	Fat %
Brackish Water (Milk fish)	519	1,123	3.63	2.84	20	2
Freshwater (Carp)	1,148	7,759	2.80	1.81	22	9
Shallow Sea (Oyster)	4,552	--	1.60	--	10	4
Hog	757	--	6.81	--	8	41

TABLE V·   ESTIMATED COST OF PROTEIN PRODUCTION IN TAIWAN.
FROM REF. 20

Using the same logic we did for the Malaysian fish-pig farm we can now estimate the energy budget of the Taiwan polyculture system (Table VI).

LABOR ENERGY INPUT

A. Assume the labor energy of a man-year to be:
(8 hrs/day) (170 kcal/hr)(365 days/yr) . . . . . 496,000 kcal

FISH ENERGY OUTPUT (PRODUCTION

B. 23,000 lbs/man year* . . . . 10.5 million kcal

ENERGY BUDGET OF TAIWAN FISH FARMING

A/B = .05 = 5%

TABLE VI·   PARTIAL ENERGY BUDGET OF FRESH WATER POND AQUA-
CULTURE IN TAIWAN. AFTER REF. 20. BUDGET INCLUDES
MAINTENANCE ENERGY ONLY AND NOT SECONDARY INPUTS
LIKE: PRE-PLANNING, SITE CONSTRUCTION, PRODUCTION
ENERGY OF MATERIALS USED, FERTILIZERS, ETC.

Although these examples are for tropical countries, they do suggest that an aquaculture which is practiced with ecological techniques is an extremely efficient means of obtaining protein for human nutrition. Also, aquaculture can thrive on waste products and need not depend upon grains which themselves can be used for human food.

As a final point of interest we can compare the energy budgets (protein ratios) of various foods produced in the U.S. (Fig. 5).

This information suggests that fish provide more animal protein per energy input than any other kind of U.S. food (compare to Fig. 1), even though most of the fish produced by the U.S. is from offshore fishing.

The Small Fish Farmer in the U.S.

We have given examples of pond aquaculture systems operated in the tropics. Now what about comparable systems for temperate latitudes . . . specifically the U.S?

First of all it is important to understand that the single most important

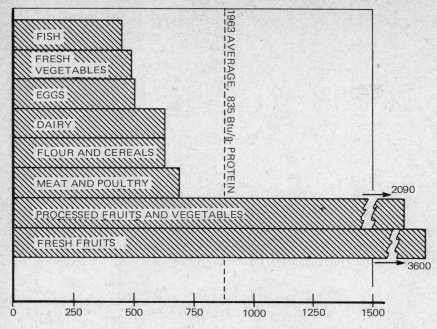

PROTEIN RATIO (Btu-primary energy/g - food protein

FIG. 5   PROTEIN RATIO OF VARIOUS FOOD GROUPS IN THE U.S.
          PROTEIN RATIO = RATIO OF ENERGY INPUTS TO ENERGY
          OUTPUTS (PROTEIN ENERGY). REDRAWN AFTER REF. 21.

factor with regards to fish production is water temperature. Being cold-blooded, fish and shellfish respond immediately to changes in temperature, generally growing better in warmer water. The tropics receive more solar energy and have a warmer average climate than the higher latitudes. Thus aquatic organisms grow faster and for longer periods (year around) in the tropics than they do in temperate climates. Generally speaking, within the temperature range in which a fish can thrive, an increase of about 10°C can cause growth and metabolism to double. This relationship is also reversible. Growth rates can be cut in half for each 10°C drop in average water temperature. Below a critical temperature, of course, growth ceases altogether (e.g., when fish are over-wintering), and if the water temperature continues to fall, a lower *lethal limit* is reached.

As a result of this temperature relationship, natural or managed aquatic systems in most of the U.S. aren't very productive. This is why the majority of commercial aquaculture operations (except trout which thrive in cold waters), are located in the southern Gulf states.

The small farmer not living on the Gulf of Mexico, in the southwestern U.S. or southern California has few choices if he wants to raise fish for food in farm ponds. He is bound to choose an endemic species that can survive over winter, until summer when growth can resume. Since the growing season is so short (3-6 months in most of the U.S.) productivity is low compared to the tropics.

In the U.S. the traditional regime is the Largemouth Bass-Bluegill community. The bass is a predator and feeds only on smaller fish (including their own offspring) and invertebrates it can find in the pond. The bluegill feeds mostly on bottom animals. Because they eat mostly animals and not plants, bass-bluegill cultures rarely produce more than 400 lbs/acre/yr.

The same situation exists for the other traditionally cultured fish in the U.S. (catfish, trout and perch), which are basically flesh eaters growing only during the warm part of the year. When raised in small farm ponds without supplemental feeding, yields are very low. It is only when these fish are raised commercially that the yields increase substantially. Commercial trout and catfish are big business in the U.S., and like the rest of agriculture they have already been locked into the agribusiness syndrome. Monoculture is practiced with heavy overstocking and feeding with high protein trout and "cat chows." The ponds must be artificially aerated to keep the fish from dying. Plants are kept out of the system lest they foul up the "efficient" workings, and if they should happen to become established . . . herbicides are often used![14]

But even with chemical aids and technical "advances," American catfish farmers seldom produce more than 3000 lbs/acre/yr. Trout have yielded up to 60,000 lbs/acre/yr (Idaho trout farms), but the fish are very expensive

due to the high cost of the operation.

There is another point to be made. The bass-bluegill system is traditional primarily because these species are a favorite of sport fishermen, and not because they are a food-species *per se*. Herein lies the difference between U.S. and Asian pond fish cultures. Whereas the Asians are serious about trying to get as much food out of their ponds as possible, Americans are content with an occasional fish dinner to augment their beef/pork diets. Also Americans are fairly squeamish about the fish they eat; very few seem willing to accept a more productive plant-eating fish like carp. In fact, most herbiverous fish have been dubbed "trash fish" in this country. Possibly these prejudices will give way in time as animal protein becomes more expensive. For example, per capita consumption of fish increased 15% between 1967-1970.[2] And since there will always be people who want to "grow their own" we just might see an increased interest in rural fish ponds.

According to Bill McLarney,[22] a 1 acre farm pond somewhere in rural America, with bass and bluegill, could produce 108 pounds of edible meat per year if the following management practices were followed: (1) Periodic harvesting of larger fish by partial draining and seining. This provides additional food for the remaining smaller fish. (2) Leaving about 25% of the small fish that remain after the last harvest (in mid-late fall), and a small breeding population of adult fish making the pond a perennial source of food. (3) Using proper skill in dressing the fish. McLarney quotes a 40% dressing loss for fish caught in Massachusetts ponds. With these precautions one could get over 100 pounds of edible fish meat per acre per year in a growing season of about 6 months.

Tending a large farm pond is only practical if you happen to live on a farm or other open space; but what about people with just a backyard. Are there fish rearing systems for them?

The Backyard Fish Farm

At the present time a small research group in Massachusetts, the New Alchemy Institute, is experimenting with different kinds of small scale aquaculture systems geared to individuals, families and small groups. The Alchemists have pioneered in methods of raising herbivorous fish in small ponds and tanks enclosed by greenhouses, and fueled with local resources.

The Alchemists' systems are modeled after Asian polyculture farms, but are adapted for northern climates. They have done this by: (1) Incorporating home-built solar heat collectors to add to the warming effect of the greenhouse and raise the water temperature of the ponds to 20°–30°C. At these temperatures *Tilapia* and carps undergo rapid growth and reach harvestable size in as little as 10 weeks if given proper food. (2) Stocking, fertilizing and feeding with organic wastes and natural foods grown either near or right in the cultures. (3) Using biological filters for removing toxic materials, and wind energy for moving the filtered water. The systems are designed to be ecologically efficient without requiring large sums of money for their construction, or much time for their maintenance.

To date the New Alchemists have tested three basic aquaculture models (Figs. 6, 7, 8). Figure 9 shows their next planned project . . . The Ark. Journal No. 2 of the New Alchemy Institute has a complete description of all four set-ups. We will describe briefly one of their systems (the Mini-Ark). It illustrates the basic strategy behind backyard fish farming: to mimic the fish culture after a natural pond ecosystem.

Figure 8 shows the "Mini-Ark." Designed as a precursor to the Ark (Fig. 9), it is a closed recirculating system in three stages. A water pumping windmill made from simple materials (see Wind Section page 95 of *Primer*) circulates the water by pumping it from the lower fish pool to the upper pool where the water then flows back down through the middle and lower pools by gravity.

The *upper pool* is a purifying filter consisting of three separate compartments: a biological filter, an earth filter and an algae culture. Basically the entire filter pool takes the place of the earth-bottom of a natural pond or lake. The biological filter is a bed of crushed clam (Quahog) shells which provides a large surface substrate for the growth of nitrifying bacteria. These microbes convert the toxic waste materials produced by the fish (mostly ammonia) into nitrates and nitrites (NO_3 and NO_2). If allowed to accumulate, these toxins would either kill the fish directly or stunt their growth. Also the nitrates and nitrites are the form of nitrogen most pre-

COMPOST INSULATOR/HEATER
For Backyard Fish Farm Pool/Dome System

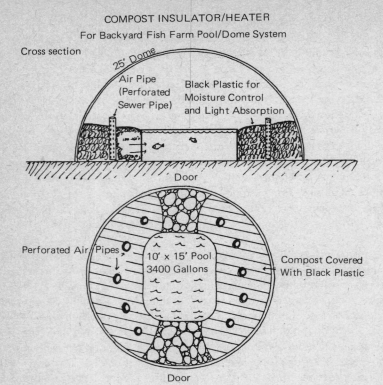

FIG. 6   A DOME-COVERED FISH POND FOR NORTHERN CLIMATES.
FROM THE JOURNAL NO. 2 OF THE NEW ALCHEMY INSTITUTE.

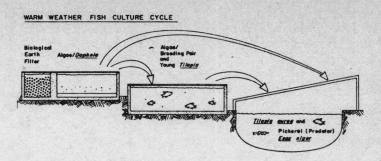

FIG. 7   A THREE-TIERED, FLAT TOP FISH RAISING COMPLEX.
FROM THE JOURNAL NO. 2 OF THE NEW ALCHEMY INSTITUTE.

FIG. 8   THE MINIATURE ARK: A WIND-POWERED, SOLAR-HEATED, COMBINED GREENHOUSE AND BACKYARD FISH FARM. FROM THE JOURNAL NO. 2 OF THE NEW ALCHEMY INSTITUTE.

ferred and acceptable to plants.** Thus potentially harmful wastes are recycled into useful nutrients for the plants, which in turn feed the fish, etc. etc. The clam-shell filter does two other things: it traps suspended organic matter which tends to use up available oxygen, and it buffers the pH of the culture water.

The earth filter section consists of small water plants growing in dirt. The earth does two things: first, it provides trace elements vital to growing plants and animals; second, it inoculates the system with nitrifying bacteria for the biological filter. The aquatic plants growing in the earth bed contribute oxygen to the filter thus ensuring the life of the (aerobic) nitrifying bacteria even when wind and water circulation are poor.

The third compartment of the upper pool becomes rich with microscopic algae (phytoplankton) as the nutrients from the biological filter flow into this space. This rich "soup" is allowed to flow into the *middle pool* where tiny crustaceans (*Daphnia*) eat the phytoplankton and increase in numbers. The *middle pool* is periodically drained into the *bottom pool* where the *Daphnia* serve as food for juvenile fish growing there.

The *bottom pool* (8500 gallons) is the main culture pool for the fish. Presently the Alchemists are stocking species of *Tilapia* in polyculture. These fish grow well in the warm water (25°-30°C) made possible by the

***Recent research [23] has indicated that when raising herbivorous fish like Tilapia or Carp it may not be wise to include large water plants in the system since they remove large amounts of nutrients (especially nitrates and phosphates) which would otherwise be used by the phytoplankton for primary productivity.*

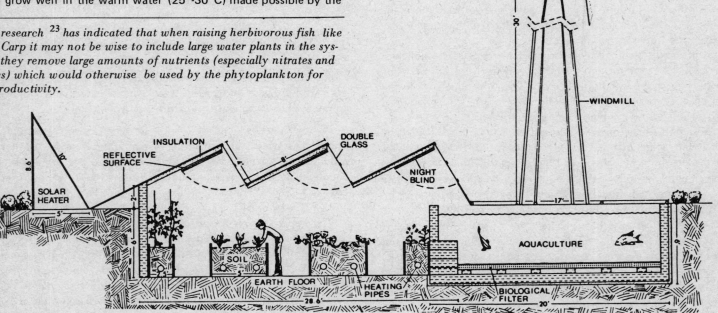

FIG. 9   THE ARK; A PROPOSED ADVANCED MODEL OF THE MINI-ARK.
FROM JOURNAL NO. 2 OF THE NEW ALCHEMY INSTITUTE.

greenhouse cover, good pond insulation and a solar collector. Since the Alchemists have been conducting their experiments as models for others to use, they have not been trying to produce especially high yields. However Bill McLarney feels that they might be able to harvest about 100 lbs/yr of fish from one of their dome ponds (3000 gallons). This is quite an improvement over natural non-managed systems when we consider that this pond is only 0.02 acres (remember the maximum 400 lbs/acre/yr for the bass-bluegill farm ponds).

There are many techniques used by the New Alchemists to increase fish productivity. They are listed here by way of summary:

1. Increase water temperature by using insulated pools, greenhouse structures with double insulating skins, solar water heaters, insulating night covers for the tanks and reflective panels which bring more heat into the pools.

2. Fertilize with manures to increase the primary productivity of the system.

3. Raise herbivorous fish which feed at the primary consumer level of the food chain and are thus more efficient at converting primary productivity (aquatic plants) to fish meat.

4. Polyculture . . . growing a variety of animal species that feed in different ecological niches and utilize different living habitats of the contained ecosystem (culture tank or pond).

5. Raise natural foods in or near the system for supplemental feeding (see Fig. 10). These foods may include midge fly larvae *(Chironomids)*, fly maggots, other insects caught by "bug lights" at night, earthworms, amphipods as well as plants such as garden weeds, vegetable wastes (especially carrot tops), grass clippings, etc. (See the bibliography at the end of this article for further information on the cultivation of natural animal foods for fish.) When animal protein is used as a supplement to the plant diets of herbivorous fish, especially in the early stages of growth, it enables the fish to grow faster and stay in better health than when they are fed strictly on plant materials. The New Alchemists have found that about 10% of the diet of *Tilapia* should be made up of animal foods. Animal protein also has a high conversion efficiency into fish flesh. In other words it takes less animal foods to produce a given amount of fish flesh than it does plant material (Table VII).

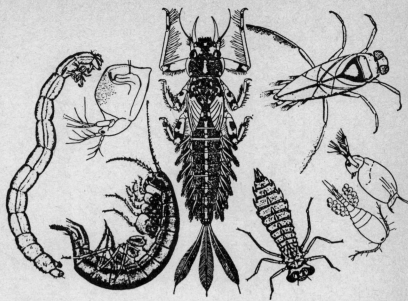

FIG. 10   AQUATIC INVERTEBRATES GROWN AS NATURAL FOODS FOR CULTURED FISH AND PRAWNS. CLOCKWISE FROM UPPER RIGHT: WATER BUG, CLADOCERON AND COPEPOD, DRAGONFLY NYMPH, GAMMARUS, CHIRONOMID MIDGE LARVAE, DAPHNIA, CENTER, MAGFLY NYMPH.

materials that have a long life and low energy cost.

8. Consideration of the cost-effectiveness of the system. Is it economical to build and maintain? The New Alchemists have taken pains to design for the homesteader or small group, but yet have not skimped on design. Their systems are designed to last and provide paybacks in terms of edible products for many years.

The Neighborhood Fish Farm

Another system worth mentioning is one that has been designed to serve the animal protein needs of a larger number of people, say a group of families or a community. It is derived from the greenhouse-polyculture schemes described above. It is a fairly large system and it emphasizes, in addition to fish production, shellfish culture and vegetable hydroponics.

The modular unit is shown in Fig. 12. It is a quonset-type greenhouse 100' by 30' that is equipped with double plastic walls, insulation on the north side and ends, and solar heat collectors alternated with clear double-wall sections on the south side. Inside are three large culture tanks (60' x 6' x 4') built of polyurethane foam, hand layed-up cement and fiberglass or wood (the tradeoffs on tank design haven't been completely explored as yet). Also smaller tanks for rearing natural foods and housing biological filters are included as partitions from the larger culture tanks. An area at one end of the greenhouse is set up for rearing juvenile animals and keeping breeding adults. Finally, associated with each main tank is a hydroponic growing compartment where the culture water from the main tanks (a "soup" of excellent fertilizer) is flushed through gravel beds planted with vegetable crops (Fig. 11).

Among the many kinds of animals grown is the giant Malaysian fresh-

Foods of Plant Origin		
Lupine seeds	3-5	
Soyabeans	3-5	
Maize	4-6	
Cereals	4-6	
All cereals	5	
Potatoes	20-30	
Guinea Grass	48·0	
Cottonseed	2·3	
Cottonseed cake	3·0	
Groundnut cake	2·7	
Ground maize	3·5	
Ground rice	4·5	
Oil palm cakes	6·0	
Mill sweepings	8·0	
Rice flour	8·0	
Manioc leaves	13·5	

Household scraps	1
Banana leaves	25·0

Foods of Animal Origin	
Gammarus	3·9-6·6
Chironomids	2·3-4·4
Housefly maggots	7·1
Fresh sea fish	6·9
Fish flour	1·5-3·0
Freshwater fish	2·9-6·0
Fresh meat	5-8
Liver, spleen and abbatoir offals	8
Prawns and shrimp	4-6
White cheese	10-15
Dried silkworm pupae	1-8

Mixtures	
Fresh sardine, mackerel scad, dried silkworm pupae	5·5
Liver of horse and pig, sardine, silkworm pupae	4·5
Silkworm pupae, silkworm feces, grass, soyabean cake, pig manure, night soil	4·1
Cortland Trout diet No. 6	7·1
Raw silkworms pupae, pressed barley, Lema and Gammarus	2·5

TABLE VII   CONVERSION RATES OF VARIOUS FOODS INTO FISH. CONVERSION RATE = WEIGHT OF FOOD/INCREASE IN WEIGHT OF FISH. THE LOWER THE NUMBER THE MORE EFFICIENT IS THE FOOD IN PRODUCING FISH FLESH. ADAPTED FROM HICKLING (REF. 15).

6. Periodic harvest of larger fish. This allows the remaining smaller fish to use the available food more efficiently.

7. Use of low-impact technology for building the equipment needed to run the fish farm. These include windmill pumps for circulating water, solar collectors for warming the pond water, wind generators and heating coils for heating the water during periods of cloudy weather, and use of building

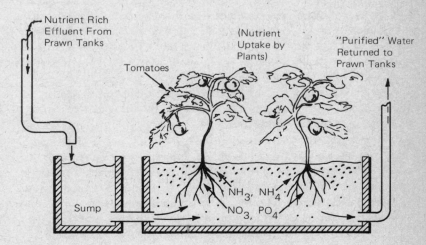

FIG. 11   HYDROPONIC VEGETABLE GROWING TANK. ILLUSTRATES PURIFYING ACTION OF PLANTS ON NUTRIENT LADEN CULTURE WATER.

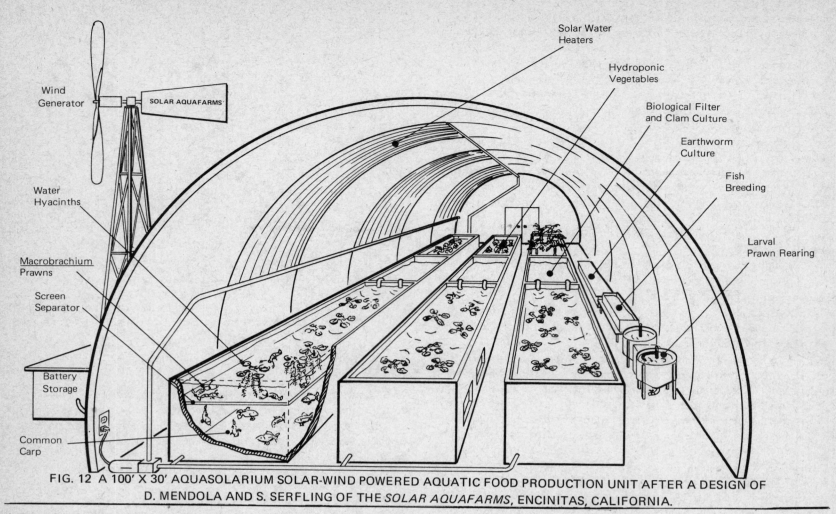

FIG. 12 A 100' X 30' AQUASOLARIUM SOLAR-WIND POWERED AQUATIC FOOD PRODUCTION UNIT AFTER A DESIGN OF D. MENDOLA AND S. SERFLING OF THE *SOLAR AQUAFARMS*, ENCINITAS, CALIFORNIA.

water prawn *(Macrobrachium rosenbergi)*. A relative of saltwater shrimp, this animal is well suited to freshwater tank culture. It is fast growing (egg to adult in 7-9 months at 27°C), hardy, omnivorous, disease resistant, easy to breed and tasty. The Solar Aquafarms, Encinitas, California, have spent the last 4 years working with this animal and heartily recommend it. It is ideal for culture as human food since it has a high protein content (22% wet weight) and a high yield of edible meat (about 50%).

The females bear from 10,000 to 60,000 eggs (15,000 average). About 50 to 80% of these hatch as planktonic larvae. After spending about 30 days in special rearing vessels, the larvae "settle out" as bottom-crawling, diminutive versions of the adult animal. Survival from eggs to this post-larval stage is usually about 50% with good culture conditions.

The main culture tanks are partitioned horizontally to keep prawns in the top half and common carp in the bottom half. The top half is planted with floating aquatic plants (water hyacinths) whose fibrous roots serve as a habitat for the juvenile prawns. The hyacinths also act as hiding places for the adults, since the males are very aggressive towards each other (i.e. territorial) and tend to fight if allowed to come into contact.

Other methods have also been proposed to house the adult prawns, but these are primarily designed for intensive, high-yield commercial operations.

Common carp are kept beneath the horizontal screen. These feed on the rain of organic wastes coming from above. The carp are bred in the laboratory and stocked as juveniles in the subspace. The fish wastes are taken up by the water plants and the hydroponic vegetables.

After the water is passed through the hydroponic beds it is pumped into the biological filter beds and then back into the main culture tanks. A scheme is also being worked out where a portion of the culture water including organic debris is passed into a bed of fresh water clams *(Corbicula)*. The bivalves filter out the detritus and use this food energy for their own growth. They multiply readily without any management, and can be periodically cropped to be fed back to the prawns or used for human food. Still another cycle involves some of the culture water passing into a tank where *Daphnia* are grown as food for the juvenile prawns.

The prawns must eat particulate materials. They will take almost any garden scraps chopped finely and are particularly fond of cooked rice and raw oats. Only supplemental amounts of these foods are needed, since a good portion of their food can be the clams, *Daphnia*, and plants cultured in the aquaculture system.

The energetics of the prawn-carp operation are still being worked out by the west coast New Alchemists who plan on writing a workbook on it next year (Fig. 13, next page). What is known already, however, is that about 900 pounds of prawns (450 pounds of meat) and 400-600 pounds of fish meat per year can be harvested from a system whose initial costs (materials plus maintenance) average $1000 per year amortized over 10 years. This is a reasonable expenditure for a system that yields over 1000 pounds of edible meat . . . not to mention the unknown quantity of clams and vegetables.

* *

We have seen examples of backyard, neighborhood and farm-scale fish farms tailored for the American environment and modeled after natural aquatic ecosystems. Obviously any aquaculture set-up requires a certain amount of biological skill to construct and maintain. However, with time, patience and proper attention, we believe these schemes, or ones like them, can be made practical and serve as integral parts of a local food producing operation.

OBTAINING FISH AND SHELLFISH FOR AQUACULTURE

Some freshwater animals suitable for aquaculture are legally controlled in many states. For example, in California, it is illegal to import grass carp and the freshwater clam *Corbicula* across state lines, or to release these species into natural waters. In closed aquaculture systems (those recirculating the same water) you have control over species escaping into local streams, etc., and possibly upsetting their ecological balance. In open systems (those using water from local sources and releasing culture water into them) there is always a possibility of contamination. Tropical or sub-tropical species like *Tilapia* or *Macrobrachium* would probably not survive a winter in streams and rivers in most of the United States. Others adapted to cool waters might thrive. In any event it is probably best to check with your local Fish and Game Office. They can tell you the legal implications of growing a particular fish or shellfish for food. As far as obtaining *Tilapia*,

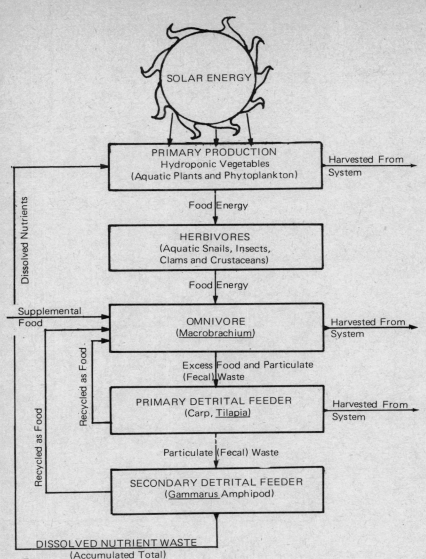

FIG. 13   ENERGY AND NUTRIENT FLOW SCHEME FOR A CLOSED-CYCLE POLYCULTURE SYSTEM DESIGNED FOR MAXIMIZING PRODUCTION OF FISH AND PRAWNS AND MINIMIZING ENERGY LOSSES

Inquiries should be made with local aquarium stores, or by contacting one of the distributors listed below.

Sources for fish stocks:

Perry Minnow Farm
Rt. 1, Box 128-C
Windsor, VA 23487
(Israeli Carp)

J. M. Malone and Son Enterprises
P.O. Box 158
Lanoke, Arkansas 72086
(Grass Carp, Tilapia)

Finally, Journal No. 2 of the New Alchemist includes further information on how and where to obtain stocks of *Tilapia*. Journal No. 2 and other pamphlets on aquaculture (see below) can be obtained from:

New Alchemy Institute
Box 432
Woods Hole, Massachusetts 02543

NEW ALCHEMY AQUACULTURE PUBLICATIONS

1. AQUACULTURE BIBLIOGRAPHY
William O. McLarney. Includes references on polyculture and pond construction. Available from New Alchemy, or Readers' Service, Organic Gardening and Farming magazine. Emmaus, Pennsylvania 18049. Price $1.00
2. THE BACKYARD FISH FARM WORKING MANUAL FOR 1973
W. O. McLarney, ed. This first "how-to-do-it" manual for backyard fish farmers has been updated and much improved in a comparable article in The Journal of the New Alchemists No. 2.
3. WALTON II: A COMPLEAT GUIDE TO BACKYARD FISH FARMING
W. O. McLarney and J. H. Todd, 1974. It's all we have learned about raising fish under intensive and ecological conditions. An extensive treatment intended to assist those wanting to start raising low-cost, high-quality fishes. Includes sources of fish. The Journal of The New Alchemists No. 2, 1974. $6.00.

4. AQUACULTURE SERIES FOR ORGANIC GARDENING AND FARMING MAGAZINE (EMMAUS, PA 18049).
August 1971, W. McLarney Aquaculture on the Organic Farm & Homestead
November 1971, W. McLarney, The Fish Pond Revisited
January 1972, J. Todd and W. McLarney The Backyard Fish Farm
February 1972, W. McLarney Why Not Carp?
April 1972, W. McLarney Pond Construction: First Step to Successful Aquaculture
5. AQUACULTURE: THE FARMING AND HUSBANDRY OF FRESHWATER AND MARINE ORGANISMS
J. Bardach, J. Ryther and W. O. McLarney. John Wiley and Sons, 1972, 868 pages, $37.50. This is the definitive English language text in the field. NAI's McLarney was the primary contributor to the book. It's very expensive, but if you are going to commit yourself to aquatic farming, you will need to read it. Ask your library to buy it. The cost is largely in the plates and illustrations which add a lot of value to the text. See Reviews.
6. STUDIES OF THE ECOLOGY OF THE CHARACID FISH, BRYCON GUATE-MALENSIS, IN THE RIO TIRIMBINA, COSTA RICA, WITH SPECIAL REFERENCE TO ITS SUITABILITY FOR CULTURE AS A FOOD FISH
W. O. McLarney. The Journal of The New Alchemists No. 1, $2.00, 1973. Yep, it's what it says it is!
7. THE ARK: A SOLAR HEATED GREENHOUSE AND AQUACULTURE COMPLEX ADAPTED TO NORTHERN CLIMATES
By several New Alchemists. The Journal of The New Alchemists No. 2, 1974, $6.00. Design and rationale for such a structure. All our aquaculture structures employ a variety of methods for trapping and storing the sun's heat. Discussion of these methods are in articles on aquaculture.
8. AQUACULTURE: TOWARD AN ECOLOGICAL APPROACH
W. O. McLarney, In: "Radical Agriculture," ed. Richard Merrill, Harper & Row

REFERENCES CITED

[1]Bardach, J. E., J. H. Ryther, and W. O. McLarney, 1972. AQUACULTURE: THE FARMING AND HUSBANDRY OF FRESHWATER AND MARINE ORGANISMS. See reviews.
[2]Whitaker, Donald R., 1972. AQUACULTURE VERSUS LATENT RESOURCES. 17th Annual Meeting, Atlantic Fisheries Technology Conference, Annapolis, Maryland, October 22-25, 1972.
[3]Lappé, Frances M., 1971. DIET FOR A SMALL PLANET. Ballantine Books, Inc. New York.
[4]McLarney, W. O., 1971. AQUACULTURE ON THE ORGANIC FARM AND HOMESTEAD. Organic Gardening and Farming Magazine, August 1971.
[5]Schuster, W. H., G. L. Kesteven, and G. E. D. Collins. 1954. FISH FARMING AND INLAND FISHERIES MANAGEMENT IN RURAL ECONOMY. F.A.O. Fisheries Study No. 3. Rome, Italy.
[6]Ho, R., 1961. MIXED FARMING AND MULTIPLE CROPPING IN MALAYA. In: Proceedings of the Symposium on Land-Use and Mineral Deposits in Hong Kong, Southern China and Southeast Asia. S. G. Davis (ed.), Hong Kong University Press.
[7]LeMare, D. W., 1952. PIG-REARING, FISH-FARMING AND VEGETABLE GROWING. Malayan Agricultural Journal, Vol. 35, No. 3, pp 156-166, Kuala Lumpur, Malaya.
[8]McLarney, William O., 1970. PESTICIDES AND AQUACULTURE. American Fish Farmer 1(10):6-7, 22-23.
[9]Woodwell, George M., P. P. Craigard, and H. A. Johnson. 1971. DDT IN THE BIOSPHERE: WHERE DOES IT GO? Science 174:1101-1107.
[10]Manigold, D. B. and J. A. Schulze, 1969. PESTICIDES IN WATER: PESTICIDES IN SELECTED WESTERN STREAMS. A PROGRESS REPORT. Pesticide Monitor Journal 3:124-135.
[11]Risebrough, R. W., 1969. CHLORINATED HYDROCARBONS IN MARINE ECOSYSTEMS. In: Chemical Fallout, M. W. Miller and G. C. Berg (eds.), Charles C. Thomas, Springfield, Illinois.
[12]Margalef, Ramon, 1968. PROSPECTIVES IN ECOLOGICAL THEORY. University of Chicago Press, Chicago, Illinois.
[13]Odum, Eugene P., 1971. FUNDAMENTALS OF ECOLOGY, 3rd edition, W. B. Saunders Company, Philadelphia, Pennsylvania.
[14]McLarney, W. O., 1974. AQUACULTURE: TOWARDS AN ECOLOGICAL APAPPROACH. In: "Radical Agriculture," R. Merrill (ed.), Harper & Row
[15]Hickling, C. F., 1971. FISH CULTURE. See reviews.
[16]Anon. 1968. F.A.O. AQUACULTURE BULLETIN. Vol. 1, No. 1, July 1968. F.A.O., Rome, Italy.
[17]Anon. 1971. F.A.O. AQUACULTURE BULLETIN, Vol. 4, No. 1, October 1971. F.A.O., Rome, Italy.
[18]Anon, 1967. Indo-Pacific Fisheries Council. Current Affairs Bulletin 50:25, 1967. F.A.O. Bangkok, Thailand.
[19]Anon, 1971. Indo-Pacific Fisheries Council. Occassional Paper 71/5, 1971. F.A.O. Bangkok, Thailand.
[20]Shang, Yung C., 1973. COMPARISON OF THE ECONOMIC POTENTIAL OF AQUACULTURE, LAND ANIMAL HUSBANDRY AND OCEAN FISHERIES: THE CASE OF TAIWAN. Aquaculture 2:187-195.
[21]Hirst, Eric, 1973. ENERGY USE FOR FOOD IN THE U.S. Oak Ridge National Laboratory Report No. ORNL-NSF-EP-57.
[22]McLarney, W. O., 1971. THE FARM POND REVISITED. Organic Gardening and Farming. November 1971.
[23]Rogers, H. H. and D. E. Davis, 1972. NUTRIENT REMOVAL BY WATER-HYACINTH. Weed Science 20(5):423-428.

POPULAR FISH AND SHELLFISH SPECIES SUITABLE FOR SMALL-SCALE FRESH WATER POND CULTURE IN THE U.S.

Species	Feeding and Habits	Area of Origin/Culture	Temperature Range (°C) Optimum	Minimum	Comment
CYPRINIDS					
Common carp _Cyprinis carpio_	Eats attached algae, bottom detritus and benthic animals.	China/worldwide	18°-28°	13°	Can be polycultured
Mud carp _Cirrhina molitorella_	Eats attached algae, bottom detritus and benthic animals.	China/worldwide	20°-28°	13°	Can be polycultured
Grass carp _Ctenopharyngodon idella_	Eats large plants, floating and attached. Also grass clippings, weeds.	Amur river, China	23°-29°	13°	Can be polycultured
Big Head carp _Aristichthys nobilis_	Eats microscopic animals (Zoo plankton) in mid-water.	China	20°-28°	13°	Can be polycultured
Silver carp _Hypopathalmichthys molitrix_	Eats microscopic plants (phytoplankton) in mid-water.	China	20°-28°	14°	Can be polycultured
Black carp _Mylopharyngodon piceus_	Eats bottom molluscs (snails, clams, mussels)	China	20°-28°	13°	Can be polycultured
CICHLIDS					
Tilapia heudeloti (or _microcephala_)	Eats large and maybe microscopic plants. Tolerates brackish water.	Coastal West Africa from Senegal to Congo	20°-30°	12°	All _Tilapia_ will breed in captivity.
Tilapia mossambica (Java Tilapia)	Eats mostly plankton, but also all plant material and some animal feed. Tolerates mild salinity.	East Africa/S.E. Asia, Japan, Latin America, U.S.	15°-30°	16°	Slightly aggressive to other species. Difficult with small carp in polyculture.
Tilapia nilotica (nile Tilapia)	Eats plankton and large plants.	Syria to East Africa/throughout world.	15°-30°	16°	Little known.
Tilapia zillii	Eats only plants, mostly large. Used in weed control.	Equatorial East Africa/throughout world	22°-26°	15°	Aggressive toward other species.
Tilapia aurea	Eats a variety; chopped plants, grain and animal foods.	West Africa	25°	13°	
SALMONID					
Rainbow trout _Salmo gairdneri_	Eats mostly aquatic insects. Needs high protein diet when fed artificially and lots of oxygen.	Pacific coast of North America	13°-20°	12°	Generally monoculture but possible for polyculture.
CENTRARCHIDS					
Bluegill _Lepomis macrochiris_	Eats mostly bottom animals, possibly algae	U.S. principal species stocked in farm ponds for angling.	20°-25°	10°	Eats young of other species. High reproduction causes stunting.
Redear Sunfish _Lepomis microlophus_	Eats bottom animals especially snails.	U.S. principal species stocked in farm ponds for angling.	20°-25°	10°	Eats young of other species. High reproduction causes stunting.
Largemouth bass _Micropterus salmoides_	Eats small fish and bottom animals. Stocked with sunfish in farm ponds, also bluegill (1 bass:10 bluegill) needs lots of oxygen.	U.S. lakes and ponds	20°-30°	10°	Obligate carnivore—eats other fish and invertebrates.
Crappies and _Pomoxis Spp._	Eats bottom animals.	U.S. lakes and ponds. Stocked with bass.	20°-30°	10°	Very low reproduction. Used only in fish-out ponds.
MISCELLANEOUS FISH					
Bigmouth Buffalo _Ictiobus cyprinellus_	Eats plankton, bottom animals and detritus.	Popular in U.S. in early 1900's. Renewed interest in hybrids.	20°-25°	13°	♀ Black X ♂ Bigmouth = fast growing hybrid.
Smallmouth Buffalo _Ictiobus bubalus_	Eats plankton, bottom animals and detritus.				Stocked in catfish ponds
Black Buffalo _Ictiobus niger_	Eats plankton, bottom animals and detritus.				Also used in rice field rotations (900 lbs/acre in 18 months).
Channel catfish _Ictalurus punctatus_	Eats other fish and bottom animals.	U.S. rivers, lakes and ponds. Widely cultured. Survives freezing.	22°-28°	15°	Mostly grown in commercial monocultures. Not practical for small scale rearing.
Brown bullhead _Ictalurus nebulosus_	Eats other fish and bottom animals.	U.S. rivers, lakes and ponds	22°-28°	15°	Smaller and hardier than Channel catfish, but more susceptible to disease. Reproduce easily without management, but stunting and over-population can occur.
Yellow bullhead _Ictalurus natalis_	Eats other fish and bottom animals.	U.S. rivers, lakes and ponds	22°-28°	15°	
Black bullhead _Ictalurus melas_	Eats other fish and bottom animals.	U.S. rivers, lakes and ponds	22°-28°	15°	
SHELLFISH					
Red crayfish _Procambris clarki_	Scavenger: table scraps, detritus, etc. but needs some plants for best growth.	Southern U.S. to 35°N, into Southern California. Found in streams, creeks, lakes and flood control channels.	10°-30°	13°	Due to rapid growth, more suitable for growth than _Astacus_ or _Pacifastacus_. Use in polyculture with carps, bass, bluegill, etc.
Bivalve molluscs _Lampsilis_ spp. _Corbicula_ spp.	Live in bottom mud/ooze. Filter out plankton and floating detritus.	_Lampsilis_ in Mississippi Valley; _Corbicula_ introduced from Asia, now found in waterways of East and West coasts.	10°-30°	5°	_Lampsilis_ needs fish as host for parasitic young—drawback, _Corbicula_ does not. Do poorly in turbid or low O_2 water. Very susceptible to copper and other heavy metals. Good polyculture animals; improves O_2 supply for fish.

REVIEWS: AQUACULTURE by Dominick Mendola

A GUIDE TO THE STUDY OF FRESH-WATER BIOLOGY

A hip-pocket guide to the identification of fresh-water animals. A recognized classic, the fifth edition adds fish to the algae, protozoans, rotifers, molluscs, and insects (immature ones too) described in earlier editions. Excellent drawings illustrate the key characteristics of each organism, while a key-guide fills in the gaps for identification. Also included: materials for collecting in the field and methods of determining water chemistry. A valuable book for those learning to identify food organisms of cultured fish . . . and, of course, the fish themselves.

A Guide to the Study of Fresh-Water Biology
James Needham and Paul Needham
1962, 5th ed. (1st ed., 1938); 108 pp

$2.95

from:
Holden-Day, Inc.
500 Sansome Street
San Francisco, California 94111

TEXTBOOK OF FISH CULTURE: BREEDING AND CULTIVATION OF FISH

More information is included in this book for the would-be organic fish farmer than any other single book I've read. A major sourcebook for the breeding and cultivation of freshwater food-fish. Packed with drawings, tables and photos of culture hardware and systems presently in use throughout the world. There is also information on pond construction, maintenance and improvements, natural methods of increasing pond production, enemies and diseases of fish, harvesting of fish etc. The high price and lack of an index are drawbacks, but this book is a must for the serious aquaculturist.

Textbook of Fish Culture:
Breeding and Cultivation of Fish
Marcel Huet
1970; 436 pp

$27.00 (Ł 12,50) plus shipping

from:
Fishing News (Books) Ltd.
23 Rosemount Avenue
West Byfleet, Surrey, England

or
110 Fleet Street
London EC4A2JL

FISH CULTURE

One of the finer books on the subject: well organized, full of illustrations and many references (312). Also: aquaculture history, pond construction and management, fertilizers, supplemental feeding and foods, brackish and sea water culture, stocking rates, yields, fish genetics, diseases etc. The photos (66) take you around the world in pond culture. The data on stocking rates are especially valuable. This book is cheaper than Huet or Bardach, et. al., but not as inclusive.

Fish Culture
C. F. Hickling
1971 (2nd ed); 317 pp

Ł 4,50 ($10.85 U.S.)

from:
Faber and Faber
24 Russell Square
London WC1 England

GROWTH AND ECOLOGY OF FISH POPULATIONS

Definitely for the serious practitioner only . . . not a book for the beginner, but for those who want a deeper understanding of the ecological principles behind rearing fish. Sections include: growth processes in fish, food, competition and niches, growth and maintenance of populations, predator-prey relationships among fish, the trophic environment and fish growth, etc. Contains information not found in other books reviewed here. Well illustrated in textbook fashion, a little heavy on the mathematics.

Growth and Ecology of Fish Populations
A. H. Weatherley
1972; 293 pp

$13.50

from:
Academic Press
111 5th Avenue
New York, N.Y. 10003

AQUACULTURE: THE FARMING AND HUSBANDRY OF FRESHWATER AND MARINE ORGANISMS

Hailed as a "bible" by American aquaculturists, this comprehensive volume covers both freshwater and marine aquacultural practices. There are sections on general principles and economics, plus sections on culture methods and techniques for every major species being cultured throughout the world. These include: Common, Chinese and Indian carp; pike; perch; bass; Tilapia; mullet; eels; salmon and trout; pompano; yellowtail; marine flatfish; freshwater crayfish; crabs; scallops; mussels; abalone; squid; shrimp, lobster and frogs; seaweeds and edible freshwater plants. There are many references, excellent illustrations and photos showing the systems and apparatus in use for the different culture practices. Although the emphasis is towards commercial aquaculture, there is much information for the organic fish farmer. Recent printing of the paperback edition will bring this valuable sourcebook into the reach of all those interested in aquaculture.

Aquaculture: The Farming and Husbandry of Freshwater and Marine Organisms
John E. Bardach, John H. Ryther and William O. McLarney
1972; 868 pp

$37.50 hardback

paperback edition late 1974

from:
Wiley-Interscience
605 3rd Avenue
New York, NY 10016

or WHOLE EARTH TRUCK STORE

ECOLOGY OF FRESH WATER

A concise guide to life in fresh waters. The author takes you on a field trip to a pond, a quiet canal and a running stream to sample the plants and animals in a manner that unfolds the beauty of the aquatic world. Chapter on aquatic plants has good drawings (usually hard to find), and chapter on energy transfer in aquatic ecosystems tells it like it is . . . and neatly too. The book introduces the basic concepts and terminology of aquatic ecology without getting esoteric . . . a real must for the beginner. Very well illustrated and referenced, the book fits in your hip pocket or satchel.

Ecology of Fresh Water
Alison Leadley Brown
1971; 129 pp

$4.00

from:
Harvard University Press
79 Garden Street
Cambridge, Massachusetts 02138

FISH AND INVERTEBRATE CULTURE: WATER MANAGEMENT IN CLOSED SYSTEMS

Definitely the first book to buy if you're ready to get *started* in fish farming. An excellent manual covering the necessities—biological, mechanical and chemical filtration, the carbon dioxide system, respiration, salts and elements, toxic metabolites, disease prevention by environmental control, laboratory tests, etc. A real "nuts and bolts" book for the culturist. Many fine drawings show water management hardware, water circulation and flow schemes, relevant biological and chemical cycles and water treatment procedures. The information is especially directed to the problems encountered in closed-system culture in tanks or small ponds, but the chemistry is also applicable to water management in larger systems.

Fish and Invertebrate Culture:
Water Management in Closed Systems
Stephen H. Spotte
1970; 145 pp

$9.50

from:
John Wiley & Sons, Inc.
1 Wiley Drive
Somerset, N.J. 08873

or 1530 South Redwood Road
Salt Lake City, Utah 84104

or WHOLE EARTH TRUCK STORE

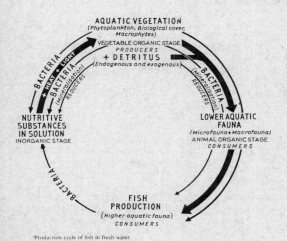

Production cycle of fish in fresh water.

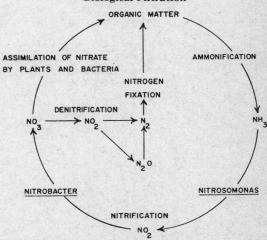

Figure 1. The nitrogen cycle.

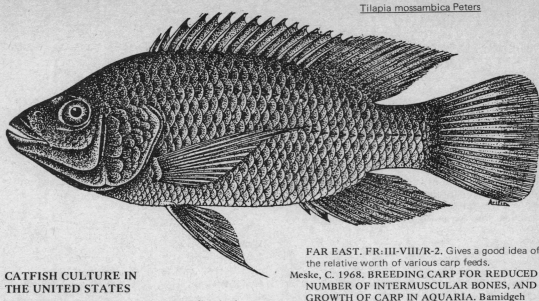

Tilapia mossambica Peters

BIBLIOGRAPHY: AQUACULTURE

BOOKS, SYMPOSIA AND GENERAL ARTICLES

Introductory Note: The literature on aquaculture is scattered among various obscure scientific and popular publications, many of them of rather poor quality. Literature dealing with the broad field of aquaculture is scarcer than writings on specific forms of the art. (Adapted from Aquaculture Bibliography of Bill McLarney, NAI-East).

Bardach, J.E. 1968. AQUACULTURE. Science, 161: 1098-1106. Brief world survey of aquaculture and its future.

Bardach, J. E. and J. H. Ryther. 1968. THE STATUS AND POTENTIAL OF AQUACULTURE. Vol. No. 1, Culture of Invertebrates and Algae, Vol. No. 2, Fish Culture. NTIS.

Borgstrom, G. (ed.) 1961. FISH AS FOOD. Vols. I & II. Academic Press, N.Y.

Eddy, Samuel. 1957. HOW TO KNOW THE FRESH-WATER FISHES. Wm. C. Brown Co., Dubuque, Iowa. Identification of the native fishes of North America.

FAO (Food and Agricultural Organization of the United Nations). 1966. FAO WORLD SYMPOSIUM ON WARM WATER POND FISH CULTURE. Hereafter—FAOWS. 123 papers by top aquaculturists from all over the world. Mostly in English.

Gerking, S. D. (ed.) 1967. THE BIOLOGICAL BASIS OF FRESHWATER FISH PRODUCTION. John Wiley & Sons, N.Y.

Hora, S. L. and T.V.R. Pillay. 1962. HANDBOOK ON FISH CULTURE IN THE INDO-PACIFIC REGION. FAO Fisheries Biology Technical Paper No. 14, 204 pp. Detailed treatment of Asian fish culture. Techniques of pond fertilization, polyculture stocking, etc.

Jones, W. 1970. COMMERCIAL FISH FARMING: HOW TO GET STARTED. American Fish Farmer 1(2):5-8.

Prowse, G. A. 1963. NEGLECTED ASPECTS OF FISH CULTURE. Indo-Pacific Fisheries Council, Current Affairs Bulletin. 36:1-9. Digestibility of algae, effects of different fertilizers on algae, pros and cons of natural aquatic plants as fish food, optimal size of ponds, genetic selection, etc.

Schaeperclaus, W. 1948. TEXTBOOK OF POND CULTURE. Fishery Leaflet No. 311, U.S. Dept. Inter., Fish and Wildlife Series, Wash. D.C.

U.S. Bureau of Sport Fisheries and Wildlife. 1970. REPORT TO THE FISH FARMERS, THE STATUS OF WARM-WATER FISH FARMING AND FISH FARM RESEARCH. Resource Publ. No. 83, U.S. Dept. Inter., Division of Fisheries Research, Wash. D.C.

MAGAZINES AND JOURNALS

AMERICAN FISH FARMER. P.O. Box 1900. Little Rock, Arkansas 72203 (monthly). Good coverage of aquaculture in this country and throughout the world.

AQUACULTURE. Elsevier Scientific Publishing Co., 52 Vanderbilt Ave., N.Y. Scientific articles on aquaculture . . . both freshwater and marine. Up to date research.

AUSTRALIAN FISHERIES. Fisheries Division of Australian Government, Public Services. Dept. of Primary Industry Canberra, A.C.T. 2600.

BAMIDGEH. (Bulletin of Fish Culture in Israel), Nir-David, D.N., Israel. (monthly). Scientific papers on fish culture in Israel and elsewhere (in English). Deals with carp, mullet and *Tilapia*. Israel has done quite a bit in fish culture in arid climates.

FAO FISH CULTURE BULLETIN, F.A.O. Rome, Italy. (Quarterly). Short reports on aquaculture development worldwide.

FAO PUBLICATIONS—REPRINTS. Available from: Unipub, Box 433, New York, NY 10016. Send for list of available titles under FAO headings.

FARM POND HARVEST. 372 South East Ave., Kankakee, Illinois 60901. (Quarterly). Mostly items on management of American farm ponds for sport and food fish production.

PROGRESSIVE FISH CULTURIST. Gov. Print. Office. (Monthly). Mostly hatchery culture of trout and salmon.

CATFISH CULTURE IN THE UNITED STATES

Introductory Note: The channel catfish is the most important aquacultural species in the United States. It may or may not be well suited to the homesteader, but they could conceivably become the first fish to be raised on a commercial scale, using organic methods, in this country.

— William McLarney

Allen, F. 1971. WEHAH FARM — RICE RAISERS THE RIGHT WAY. Organic Gardening and Farming 18(7):66-72. Contains a brief description of growing catfish in conjunction with rice.

AMERICAN FISH FARMER 2(3). This issue contains three good articles on the economics of catfish farming.

Brown, E. E., M. G. LaPlants, and L. H. Covey. 1969. A SYNOPSIS OF CATFISH FARMING. University of Georgia College of Agriculture Experiment Stations. Bulletin 69, 50 p.

CATFISH FARMERS OF AMERICA. The Catfish Farmer. The monthly official publication of the CFA. $6.00 per year from: Catfish Farmers of America, Tower Building, Little Rock, Arkansas 72201.

CATFISH FARMING. From: Agri-Books, Box 5001-AC, San Angleo, Texas 76901. $12.00.

Lee, J. S. 1971. CATFISH FARMING. Mississippi State University Curriculum Coordinating Unit for Vocational-Technical Education. State College, Mississippi 39761. 103 p. $2.00. Excellent detailed treatment of all phases of catfish culture.

Mahan, P. 1973. RAISING CATFISH IN A BARREL. Organic Gardening and Farming. Nov. 73:112-117.

Swingle, H. S. 1957. COMMERCIAL PRODUCTION OF RED CATS (SPECKLED BULLHEADS) IN PONDS. Proc. S. E. Ass. Game Fish Commissioners. 10:156-160. The only paper on the possibility of culturing bullhead catfish, which might, under some conditions, be more suitable to the home-steader than channel catfish.

Tiemeier, O. W. and C. W. Deyoe. 1967. PRODUCTION OF CHANNEL CATFISH. Kansas State University of Agricultural Experiment Station. Bulletin 508.

CARP CULTURE

Many good papers on carp culture also appear in Bamidgeh, the Bulletin of Fish Culture in Israel, and the serious reader should consult the annual indices for this journal.

William McLarney

Alikunhi, K. H. 1966. SYNOPSIS OF BIOLOGICAL DATA ON COMMON CARP, *Cyprinus carpio* (Linnaeus) 1958, Asia and the Far East. FAO Fisheries Synopsis 31.1.

Backiel, T. and K. Stegman. 1966. TEMPERATURE AND YIELD IN CARP PONDS. FAOWS, FR: V/E-2.

Ehrlich, S. 1964. STUDIES ON THE INFLUENCE OF NUTRIA ON CARP GROWTH. Hydrobiologia 23(½):196-210.

Kirpichnikov, V. S. (ed.). 1970. SELECTIVE BREED-IND OF CARP AND INTENSIFICATION OF FISH BREEDING IN PONDS. Bulletin of the State Scientific Research Institute of Lake and River Fisheries. Vol. 61. 249 pp. NTIS.

Ling, S. W. 1966. FEEDS AND FEEDING OF WARM WATER FISHES IN PONDS IN ASIA AND THE FAR EAST. FR:III-VIII/R-2. Gives a good idea of the relative worth of various carp feeds.

Meske, C. 1968. BREEDING CARP FOR REDUCED NUMBER OF INTERMUSCULAR BONES, AND GROWTH OF CARP IN AQUARIA. Bamidgeh 20(4):105-119. Describes a highly intensive system of growing carp in a closed recirculating system, with spectacular yields.

Nair, K. K. 1968. A PRELIMINARY BIBLIOGRAPHY OF THE GRASS CARP. FAO Fisheries Circular 302, 155 pp.

Nambiar, K.P.P. 1970. CARP CULTURE IN JAPAN—A GENERAL STUDY OF THE EXISTING PRACTICES. Indo-Pacific Fisheries Council, Occ. Pap. 1970/1, 41 p. Fairly complete treatment of Japanese practices.

Sarig, S. 1966. SYNOPSIS OF BIOLOGICAL DATA ON COMMON CARP. *Cyprinus carpio* (Linnaeus) 1785, Near East and Europe, FAO Fisheries Synopsis 31.2.

Stevenson, J. H. 1965. OBSERVATIONS ON GRASS CARP IN ARKANSAS. Progressive Fish Culturist 27(4):203-206.

TILAPIA CULTURE

Introductory Note: Despite the great importance of *Tilapia* in fish culture, there is not extensive literature on these fishes. Below are listed a few publications which may be helpful. Many good papers on *Tilapia* culture also appear in Bamidgeh, the Bulletin of Fish Culture in Israel, and the serious reader should consult the annual indices for this journal.

William McLarney

Avault, J. S., Jr. E. W. Shell, and R. O. Smitherman. 1966. PROCEDURES FOR OVERWINTERING TILAPIA. FAOWS FR: V/E-3.

Chimits, P. 1957. THE TILAPIAS AND THEIR CULTURE, A SECOND REVIEW AND BIBLIO-GRAPHY. Fisheries Bulletin FAO 10:1-24.

Hickling, C. F. 1963. THE CULTIVATION OF TILAPIA. Scientific American. 208:143-152.

Lasher, C. W. 1967. TILAPIA MOSSAMBICA AS A FISH FOR AQUATIC WEED CONTROL. Progressive Fish Culturist 29(1):48-50.

Maar, A., M.A.E. Mortimer and I. Van der Lingen. 1966. FISH CULTURE IN CENTRAL EAST AFRICA. FAO, Rome. 158 p. Good, simple description of African methods, many of them involving quite small ponds.

Myers, G. S. 1955. NOTES ON THE FRESHWATER FISH FAUNA OF MIDDLE CENTRAL AMERICA WITH SPECIAL REFERENCE TO POND CULTURE OF TILAPIA. FAO Fisheries Paper No. 2:1-4.

Shell, E. W. 1966. RELATIONSHIP BETWEEN RATE OF FEEDING, RATE OF GROWTH AND RATE OF CONVERSION IN FEEDING TRIALS WITH TWO SPECIES OF TILAPIA, T. MOSSAMBICA AND T. NILOTICA. FAOWS FR:III/E-9.

Swingle, H. S. 1966. BIOLOGICAL MEANS OF IN-CREASING PRODUCTIVITY IN PONDS. FAOWS, FR: V/R-1. Tilapia-catfish combinations for culture in the United States.

Swingle, H. S. 1960. COMPARATIVE EVALUATION OF TILAPIAS AS PONDFISHES IN ALABAMA. Transactions of the American Fishery Society 89(2):142-148.

Uchida, R. N. and J. E. King. 1962. TANK CULTURE OF TILAPIA. Bulletin of United States Fish and Wildlife Service: 199(62):21-52.

BASS AND SUNFISH IN FARM PONDS

Introductory Note: Most state conservation departments and many university agricultural extension services have booklets on farm ponds which may be useful, particularly to neophytes.

William McLarney

Davison, V. E. and J. A. Johnson. 1943. FISH FOR FOOD FROM FARM PONDS. U.S. Department of Agriculture, Farmers' Bulletin 1938. 22 pp.

Davison, V. E. 1947. FARM FISH PONDS FOR FOOD AND GOOD LAND USE. U.S. Department of Agriculture, Farmers' Bulletin 1938. 29 pp.

Etnier, D. A. 1971. FOOD OF THREE SPECIES OF SUNFISHES (Lepomis) AND THEIR HYBRIDS IN THREE MINNESOTA LAKES. Transactions of the American Fisheries Society 100(1):124-128. May be useful to anyone planning polyculture including sunfish.

Lewis, W. M. and R. Heidinger. 1971. AQUACULTURE POTENTIAL OF HYBRID SUNFISH. American Fish Farmer 2(5):14-16.

Ricker, W. E. 1948. HYBRID SUNFISH FOR STOCKING SMALL PONDS. Transactions of the American Fisheries Society 75:84-96.

Stockdale, T. M. 1960. FARM POND MANAGEMENT. Agricultural Extension Service, the Ohio State University. 24 pp.

TROUT

Bussey, G. 1971. HOW TO GROW TROUT IN YOUR BACK YARD (OR OTHER UNLIKELY PLACES). Available for $3.00 from Life Support Systems, Inc., Box 3296, Albuquerque, New Mexico 87110. Gene Bussey manufactures and sells closed recirculating systems which are claimed to make it possible to grow large numbers of trout—organically, if one wishes—at low cost on any scale from very small for individual use to a large commercial operation.

Lavrovky, V. V. 1966. RAISING OF RAINBOW TROUT (Salmo gairdneri Rich.) TOGETHER WITH CARP (Cyprinus carpio L.) AND OTHER FISHES. FAOWS, FR:VIII/E-3. That's right, they grow trout and carp together in Russia.

Scheffer, P. M. and L. D. Marriage. 1969. TROUT FARMING. U. S. Soil Conservation Service. Leaflet 552, GPO $0.10. Should be read by would-be beginners of trout culture.

Borell, A. E. and P. M. Scheffer. 1966. TROUT IN FARM AND RANCH PONDS. U.S. Department of Agriculture. Farmers' Bulletin. No. 2154.. 18 p. GPO $0.10.

AMERICAN CRAYFISH CULTURE

Introductory Note: Most of the available literature on crayfish culture deals with Louisiana species and methods. Techniques should be developed for crayfishes native to other parts of the country. Anyone seeking to do so should explore the purely biological literature on crayfish as well as the culture literature, paying particular attention to size, feeding habits, and reproductive habits of the species which are of interest.

William McLarney

Anonymous. 1970. CRAWFISH: A LOUISIANA AQUACULTURE CROP. American Fish Farmer 1 (9), pp. 12-15.

Fielding, J. R. 1966. NEW SYSTEMS AND NEW FISHES FOR CULTURE IN THE UNITED STATES. FAOWS FR: VIII/R-2.

Ham, B. Glenn. 1971. CRAWFISH CULTURE TECHNIQUES. American Fish Farmer 2 (5), pp. 5-6, 21 and 24.

Lacaze, Cecil. 1970. CRAWFISH FARMING. Fisheries Bulletin. No. 7. Louisiana Wild Life and Fisheries Commission, P.O. Box 44095, Capitol Station, Baton Rouge, Louisiana 70804.

POND POLYCULTURE

Introductory Note: Although polyculture is one of the oldest and most important methods of increasing fish pond productivity, there is little literature dealing with polyculture per se. What little there is deals mostly with Oriental systems, but it should be read by the serious aquaculturist, for these must serve as the models for analogous systems in North America.

William McLarney

Buck, D. Homer, Richard J. Baur and C. Russell Rose. 1973. AN EXPERIMENT IN THE MIXED CULTURE OF CHANNEL CATFISH AND LARGE-MOUTH BASS. The Progressive Fish Culturist. 35 (1):19-21.

Childers, W. F. and G. W. Bennett. 1967. EXPERIMENTAL VEGETATION CONTROL BY LARGE-MOUTH BASS—TILAPIA COMBINATION. Journal Wildlife Management 31:401-407.

Coche, A. G. 1967. FISH CULTURE IN RICE FIELDS, A WORLDWIDE SYNTHESIS. Hydrobiologia 30(1):1-44.

Prewitt, R. 1970. RAMBLING ALONG. American Fish Farmer 2(1):23-24. Brief discussion of the beginnings of polyculture in the U.S.

Sarig, S. 1955. CULTURE OF TILAPIA AS A SECONDARY FISH IN CARP PONDS. Bamidgeh, 7(3):41-45.

Tang, Y. A. 1970. EVALUATION OF BALANCE BETWEEN FISHES AND AVAILABLE FISH FOODS IN MULTISPECIES FISH CULTURE PONDS IN TAIWAN. Transactions of the American Fisheries Society 99(4):708-718. Technical discussion of Chinese carp polyculture in relation to feeding and fertilization.

Yashouv, A. 1969. MIXED FISH CULTURE IN PONDS AND THE ROLE OF TILAPIA IN IT. Bamidgeh 21(3):75-82.

Hashouv, A. 1958. ON THE POSSIBILITY OF MIXED CULTIVATION OF VARIOUS TILAPIA WITH CARP. Bamidgeh, 10(3):21-29.

FERTILIZATION

Ball, Robert C. 1949. EXPERIMENTAL USE OF FERTILIZER IN THE PRODUCTION OF FISH-FOOD ORGANISMS AND FISH. Michigan State College Agricultural Experiment Station Technical Bulletin 210, 28 pp.

McIntire, C. David and Carl E. Bond. 1962. EFFECTS OF ARTIFICIAL FERTILIZATION ON PLANKTON AND BENTHOS ABUNDANCE IN FOUR EXPERIMENTAL PONDS. Oregon Agricultural Experiment Station Technical Paper No. 1423, Corvallis, Oregon.

Swingle, H. S. and E. G. Smith. 1939. INCREASING FISH PRODUCTION IN PONDS. Transactions of 4th North American Wildlife Conference. American Wildlife Institute, Washington, D.C.

Swingle, H. S. 1947. EXPERIMENTS ON POND FERTILIZATION. Alabama Agricultural Experiment Station Bulletin 264, 34 pp. Auburn, Alabama.

Tanner, Howard A. 1960. SOME CONSEQUENCES OF ADDING FERTILIZER TO FIVE MICHIGAN TROUT LAKES. Transactions American Fishery Society 89(2):198-205.

Walny, Pawel, 1966. FERTILIZATION OF WARM-WATER FISH PONDS IN EUROPE. FAOWS, FR:II/R-7.

NATURAL FOODS

De Witt, John W. and Wendell Candland. 1970. THE WATER FLEA, DAPHINA, AS A COMMERCIAL FISH-FOOD ORGANISM FROM MUNICIPAL WASTE OXIDATION PONDS. Humbolt State College, Arcata, California.

McLarney, William O. 1974. AN IMPROVED METHOD FOR CULTURE OF MIDGE LARVAE (Chironomidae) FOR USE AS FISH FOOD. Journal of the New Alchemists 2:118-119.

Sorbeloos, P. 1973. HIGH DENSITY CULTURING OF THE BRINE SHRIMP, ARTEMIA SALINA. Aquaculture (1):385-391.

Spotte, Stephen. 1973. MAKING YOUR OWN FISH FOOD AND HOW TO RAISE EARTHWORMS AND BRINE SHRIMP FOR FISH FOOD. Chapter in: Marine Aquarium Keeping. Wiley-Interscience, New York.

U. S. Fish and Wildlife Service. 1962. FISH BAITS: THEIR COLLECTION, CARE, PREPARATION AND PROPAGATION. U.S.D.I. Fish and Wildlife Service, Leaflet Fl-28. Includes mealworms, blood-worms, earthworms, Hellgrammites and other aquatic forms, among others.

Yashouv, A. 1956. PROPAGATION OF CHIRONOMID LARVAE AS FOOD FOR FISH FRY. Bamidgeh 22(4):101-105.

Yount, James L. 1966. A METHOD FOR REARING LARGE NUMBERS OF POND MIDGE LARVAE—WITH ESTIMATES OF PRODUCTIVITY AND STANDING CROP. The American Midland Naturalist 76(1):230-238.

NUTRITION, GROWTH AND ENERGETICS

Davies, P.M.C. 1967. THE ENERGY RELATIONS OF CARASSIUS AURATUS (GOLDFISH)—III. GROWTH AND THE OVERALL BALANCE OF ENERGY. Comparative Biochemistry and Physiology 23:59-63.

Halver, John E. (ed.) 1972. FISH NUTRITION. Academic Press. New York.

Hastings, W. H. (e.a. 1966). WARMWATER FISH NUTRITION. U.S.D.I. Bureau of Sport Fisheries and Wildlife. Fish Farming Experimental Station, Stuttgart, Arkansas.

Mann, K. H. 1965. FISH IN THE RIVER THAMES. Journal Animal Ecology 34:253-75. Deals with metabolism and production of natural and culture situations.

Mann, Hans. 1961. METABOLISM AND GROWTH OF POND FISH. In: Borgotrom, G. (ed.) Fish as Food. Vol. I:82-89.

POND CONSTRUCTION

Delmendo, M.N. et al. 1970. CONSTRUCTION OF PONDS FOR AQUACULTURE. Indo-Pacific Fisheries Council. Symposia 18-27 Nov. 1970. Paper No. IPFC/C70/SYM 26. FAO Bangkok 2, Thailand. Deals with larger ponds—as for commercial culture. Some general guidelines, useful to smaller systems.

Dillon, O. W., Jr. 1970. POND CONSTRUCTION, WATER QUALITY AND QUANTITY. Paper presented to the California Catfish Conference, Sacramento, California. January 20-21, 1970. Probably available from U.S. Soil Conservation Service, Fort Worth, Texas.

Mitchell, T. E. and M. J. Usry. 1967. CATFISH FARMING—A PROFIT OPPORTUNITY FOR MISSISSIPPIANS. Mississippi Research and Development Center, 787 Lakeland Drive, Jackson, Mississippi. 83 pp. Deals specifically with catfish farming, but the information on pond construction is among the best available. Probably some charge for this publication.

Renfro, G. Jr. 1969. SEALING LEAKING PONDS AND RESERVOIRS. U.S. Soil Conservation Service. SCS-TP-150. 6 p.

Vanicek, C. David and A. Wendell Miller. 1973. WARMWATER FISH POND MANAGEMENT IN CALIFORNIA. U.S. Department of Agriculture Soil Conservation Service Report No. M7-N-23056. GPO.

Cultured "Mirror" Carp

METHANE

METHANE SYSTEMS: PRINCIPLES AND PRACTICE

Ken Smith

Basic Process of Methane Digestion

When organic material decays it yields useful by-products. The kind of by-product depends on the conditions under which decay takes place. Decay can be *aerobic* (with oxygen) or *anaerobic* (without oxygen). Any kind of organic matter can be broken down either way, but the end products will be quite different (Fig. 1).

It is possible to mimic and hasten the natural anaerobic process by putting organic wastes (manure and vegetable matter) into insulated, air-tight containers called *digesters*. The digester is fed with a mixture of water and wastes, called "slurry." Inside the digester, each daily load of fresh slurry displaces the previous day's load which bacteria and other microbes have already started to digest.

Each load of slurry moves in the digester to a point where the methane bacteria are active. At this point large bubbles force their way to the surface where gas accumulates. The gas is very similar to natural gas and can be burned directly for heat and light, stored for future use, or compressed to power heat engines.

Digestion gradually slows down toward the outlet end of the digester and the residue begins to stratify into distinct layers (Fig. 2):

Sand and Inorganic Materials at the bottom.

Sludge, the spent solids of the original manure reduced to about 40% of the volume it occupied in the raw state. Liquid or dry sludge makes an excellent fertilizer for crops and pond cultures.

Supernatant, the spent liquids of the original slurry. Note that the fertilizing value of the liquid is as great as sludge, since the dissolved solids remain.

Scum, a mixture of coarse fibrous material, released from the raw manure, gas, and liquid. It resembles, more or less, the "head" of beer. The accumulation and removal of scum is one of the most serious problems with digesters. In moderate amounts, scum can act as an insulation. But in large amounts it can virtually shut down a digester.

Biogas, the gas produced by digestion, known as marsh gas, sewage gas, dungas, or bio-gas, is about 70% methane (CH_4) and 29% carbon dioxide (CO_2) with insignificant traces of oxygen and sulfurated hydrogen (H_2S) which gives the gas a distinct odor. (Although it smells like rotten eggs, this odor has the advantage of making leaks easy to trace.)

Biochemistry of Digestion

BIOLOGICAL ACTIVITIES: The anaerobic digestion of organic matter is a complex and sensitive process. This microbial process involves three different groups of bacteria and takes place in three stages: *solubilization, acidogenesis* and *methanogenesis* (Fig. 3).

Solubilization: In this first stage, the complex proteins, carbohydrates, celluloses, fats and oils are dissolved by enzymes. This hydrolysis transforms the complex organics into simple amino acids, simple sugars, fatty acids and glycerol. The simple compounds are now in a soluble form and can pass through the cell walls of the acid-forming bacteria for fermentation.

Acidogenesis: The second stage is a fermentation process which consists of the oxidation and reduction of these simple compounds into simple or "short chain" organic acids. These "acid forming" bacteria are facultative anaerobes (living both in the presence and absence of air). They provide the food for the final stage of decomposition, stabilization and methane production.

Methanogenesis: In this final stage, the simple organic acids are consumed by a group of strictly anaerobic bacteria known as the "methane formers." The consumption of these acids by the methane formers results in the stabilization of the organic matter by converting it to methane and carbon dioxide gas—biogas. These methane bacteria are characterized by slow growth rates and extreme sensitivity to toxicity from air. Methane

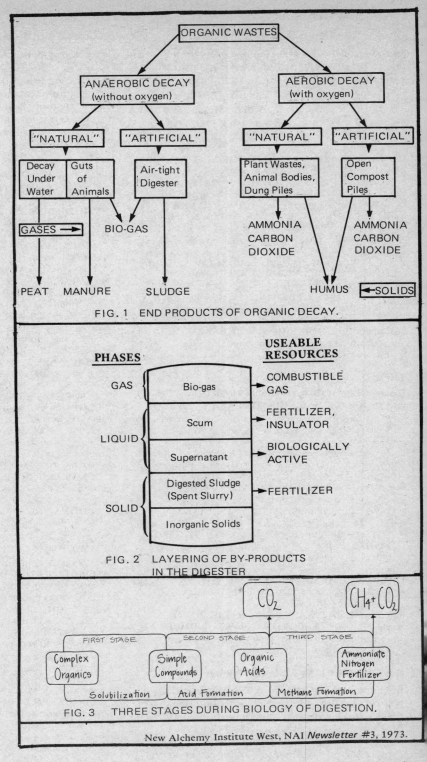

FIG. 1   END PRODUCTS OF ORGANIC DECAY.

FIG. 2   LAYERING OF BY-PRODUCTS IN THE DIGESTER

FIG. 3   THREE STAGES DURING BIOLOGY OF DIGESTION.

New Alchemy Institute West, NAI *Newsletter* #3, 1973.

formation is accomplished chemically by combining hydrogen and carbon dioxide gases produced in the acidogenic stage. Because of their slow rate of reproduction and their dependence on the acid formers, *the methane bacteria are the sensitive and rate-limiting factors in the process.*

SIMPLE DIGESTER MONITORING: A sensitivity and understanding of this simple biochemistry is essential to the successful operation of a digester. Some very simple and easily learned laboratory tests can greatly facilitate this understanding. The three most vital tests to understand are *pH, volatile acids,* and *methane gas content.* Instructions and detailed procedures for these tests are given in a booklet—*WPCF #18: Simplified Laboratory Procedures,* which is available from the following:

Water Pollution Control Federation
2626 Pennsylvania Ave. N.W.
Washington, D.C. 20037

Members — $4.00   Non-members — $8.00

The *pH* is the measure of acidity or alkalinity concentration in the digester slurry. The discharge or effluent from the digester is most commonly monitored for pH. The pH is measured on a scale of 1 to 14 (see Fig. 4). Seven is neutral. Effluent (digested discharge) from a healthy

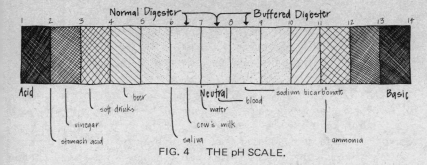

FIG. 4   THE pH SCALE.

digester will be in the range of 7.0 to 7.5. pH's as high as 8.5 have occurred in digesters where the substrate is raw plant or animal waste. When the pH dips below 6.5 in a working digester, there is usually something going "sour." This is often corrected by reduced loading of fresh material and sometimes by the addition of lime or anhydrous ammonia. The pH will generally be very low (6.0—6.5) during the initial start-up phase of a digester and may last as long as two weeks. This period of acid forming will also be characterized by production of high concentrations of carbon dioxide. Stabilization of the methane process is characterized by a gradual increase in the pH as well as an increased production of methane gas.

The pH is very easily and simply monitored by pH paper. The pH paper is dipped into the effluent or the digester. Upon removal it will show a color between red and blue (pH 1-14). Matching the paper color to a chart will give an indication of the exact pH within a range of 0.2 gradations.

Volatile Acids is a critical measure of digester performance. It clearly indicates the interface between the acid formers and the methane formers. An increase in volatile acids will precede a decrease in gas production, methane content and finally a low pH. The acid formers are very hearty bacteria being both facultative and anaerobic; they are fast growers which are not sensitive to oxygen, pH or toxicants. In contrast, the methane formers are only a few species of relatively slow growers which are very sensitive to oxygen, pH, changes in temperature and toxicants. The methane formers are dependent on volatile acids as a feedstock for methane production, but too many "free" volatile acids will lower the pH and cause the methane formers to die, leading to a "sour" digester. The test for this requires some special chemicals and laboratory glassware, but it is well worth the effort if one is serious about a continuous supply of combustible gas.

Methane Gas Content is an indicator of the quality of waste stabilization. The methane bacteria utilize volatile acids to produce methane gas by combining hydrogen and carbon dioxide. Total gas production from a digester is a combination of carbon dioxide and methane. A healthy digester will produce a gas containing between 60 to 70% methane. There is a simple test for determining the methane content of the biogas from a digestion system by testing the reaction of the biogas with a solution of sodium hydroxide. The carbon dioxide (but not the methane) will dissolve in the sodium hydroxide and in the process form sodium bicarbonate:

$$NaOH + CO_2 = NaHCO_3$$

This solution can then be measured to determine the amount of carbon dioxide gas. It is assumed that all that is left is methane gas. Partial proof of this is to ignite the left-over gas. Fig. 5 shows the equipment and procedure for testing for methane. It should be noted that this test can be off as much as 10% and is only a rough measure of the gas quality, but it will suffice for most small scale experiments and applications.

Digestible Properties of Organic Matter

When raw materials are digested in a container, only part of the waste is actually converted into methane and sludge. Some of it is indigestible to varying degrees, and accumulates in the digester or passes out with the effluent and scum. The "digestibility" and other basic properties of organic matter are usually expressed in the following terms:

MOISTURE: The weight of water lost upon drying at 220° F until no more weight is lost.

TOTAL SOLIDS (TS): The weight of dry material remaining after drying as above. TS weight is usually equivalent to "dry weight." (How-

ever, if you dry your material in the sun, assume that it will still contain around 30% moisture.) TS is composed of digestible organic or "Volatile Solids" (VS), and indigestible residues or "Fixed Solids" (FS).

Volatile solids (VS): The weight of organic solids burned off when dry material is "ignited" (heated to around 1000° F). This is a handy property of organic matter to know, since VS can be considered as the amount of solids actually converted by the bacteria.

Fixed Solids (FS): Weight remaining after ignition. This is biologically inert material.

LIGNIN: This volatile organic material is generally considered to be non-digestible, refractory material. In actual digestion not all of the VS are converted to biogas. Conversion of the VS is dependent on the ability of the methane formers to access and attack these VS. In actuality only 40 to 70 percent of the VS are destroyed through the methane fermentation process. The lignin content of various substrates varies from 0% in pure cellulose to almost 35% for wood. Lignin inhibits anaerobic digestion by entrapment of digestible materials. Wood is a prime example of a lignocellulosic compound.

Lignin is, however, combustible, as is evidenced by burning wood which results in an ash content of about 12%. Lignin is a primary component of the *scum* formation in digesters along with other light organic refractories such as animal hair and oils.

TOTAL DIGESTIBILITY: According to what we have noted so far, if there were 100 pounds of chicken manure, 72-80 pounds of this would be water, leaving only 20-28 pounds of TS. Of these TS only 75-80% are VS, or about 15-24 pounds VS are available for digestion. Of the 15-24 pounds of VS only 40-70% can actually be converted to gas by the methane bacteria. If each pound of converted VS yields 5 cubic feet of gas, then there will be about 30-84 ft^3 (24 lbs. VS × .7 = 16.8 × 5 ft^3 = 84) of gas available from the 100 pounds of chicken manure.

Resource Assessment

Broad experimentation and application of the anaerobic digestion process to various organic materials has been clearly demonstrated. Primary application has been in the treatment of municipal sewage. Laboratory experiments, pilot scale demonstrations and small scale applications have clearly proven that methane production is feasible and practical with organic matter ranging from paper and garbage to animal manures and crop residues. For purpose of discussion, these will be assessed in the following categories: *Urban Residues*, *Agricultural Residues* and *Biomass Plantations*.

URBAN RESIDUES: Residues from urban areas fall into two broad categories: (1) *sewage* and (2) *solid waste*.

Sewage: This waste stream contains both human and industrial waste. The primary resource is human excrement, especially feces. Many cities already employ digesters to treat this waste stream. Table I includes a breakdown of human wastes. These make up the major portion of gas-

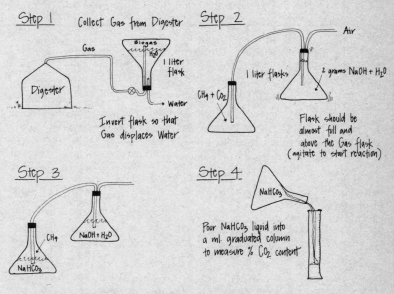

FIG. 5   SIMPLE TEST FOR METHANE CONTENT OF BIOGAS.

TABLE I MANURE PRODUCTION OF VARIOUS LIVESTOCK AND HUMANS. "Livestock Unit" = VS production relative to laying hens. New Alchemy Institute West, NAI *Newsletter* #3, 1973.

	Wet Raw Manure * lb/day	ton/yr	gal/day	Total Solids lb/day	ton/yr	Volatile Solids lb/day	Livestock Unit
Bovine							
Dairy Cow	1600 132	24	18	16.6	3.1	13.8	300-350
	1300 107	19.5	15	13.5	2.5	11.2	250-300
Dairy Heifer	1000 85	15.5	11.2	9.2	1.7	7.5	150-200
Beef Feeder	1000 60	11	7.5	6.9	1.3	5.9	150
Beef Stocker	500 45	8.2	5.2	5.8	1.0	4.8	120
Horse							
Large	1000 45	8.2	6.7	9.4	1.7	7.5	180-200
Medium	850 36	6.6	5.4	7.0		5.5	120-150
Pony	15.4			3.0		2.4	50-70
Swine							
Hog Breeder	500 25	4.6	3	2.2	0.4	1.6	40
Hog Feeder	200 13	2.4	2.2	1.2	0.22	1.0	25
	100 6.5	1.2	1.1	0.6	0.11	0.5	13
Wiener	15 1.0	0.2		0.1			
Sheep							
Feeder	100 4	0.7	0.8	1.0	0.18	0.8	20
Lamb	30 1.5	0.3		0.4		0.2	5
Fowl							
Geese, Turkey	15 .6	220 lb	.2 qt	.15	55 lb	.10	2.5
Ducks	6 .4	250 lb	.15 qt	.10	37 lb	.07	1.8
Broiler Chicken	4 .3	110 lb	.1 qt	.07	26 lb	.05	1.3
Laying Hen	4 .2	75 lb	.1 qt	.05	18 lb	.04	1.0

	Portion	Amount	% TS	TS/day	%VS	VS/day	Livestock Unit
Humans	Urine	2 pt., 2.2 lb	6%	.13	75%	.10	6
(150 lbs)	Feces	0.5 lb	27%	.14	92%	.13	
	Total	2.7 lb	11%	.27	84%	.25	

*Bulk density of raw manure = 34 ft³/ton or 60 lb/ft³, or 8 lb/gal with no flushing water.

producing resources in municipal sewage. About 0.9 cubic feet of biogas is produced per capita per day in municipal treatment plants.

Solid Waste: This is a very good resource for the production of methane gas. It is, however, almost undiscovered and certainly not used on any scale to date. It has been well documented by several studies and pilot scale projects as a highly potential resource for producing pipeline-quality natural gas. Per capita generation of solid waste is 5-6 lbs/day. Of this amount, three to four pounds is a potential feedstock for methane production (glass, metal, wood and plastics should be removed). Each pound of organic refuse will produce about five cubic feet of methane gas. Table II gives a representative sample of urban organic waste and its gas potential.

Combined Urban Waste: Municipal solid waste and sewage can be combined for gas production. If there are three pounds of digestible matter per capita per day and each pound of matter will yield five cubic feet of methane gas, then each person's garbage will potentially create 15 cubic feet of methane gas. In addition, each person's excrement will produce about one cubic foot of methane per day. The potential gas production for a city of 30,000 people is as follows:

$$16 \text{ ft}^3 \times 30,000 = 480,000 \text{ ft}^3/\text{day}$$

If half of this gas is required to maintain the facility, then 240,000 ft³/day

TABLE II SOLIDS REDUCTION AND GAS PRODUCTION EFFICIENCY OF DIGESTION MIXTURES CONTAINING SOLID WASTE COMPONENTS

Component	Proportion Of Comp. In Sludge Feed (%)	Solids Reduction Total (%)	Volatile (%)	Gas Production Per Volatile Solids Added (cu ft/lb)	Destroyed (cu ft/lb)	Component Efficiency (%)
Garbage	100	51.6	66.2	8.8	13.9	90.8
Kraft paper	60	68.2	76.9	9.2	12.1	94.6
Newspaper	30	47.5	58.0	7.5	13.0	44.3
Garden debris	50	54.3	66.2	7.8	11.9	73.1
Wood	60	—	—	4.3	7.7	4.4
Chicken manure	100	57.7	67.5	5.0	17.1	53.2
Steer manure	100	22.1	27.9	1.4	8.7	14.9
Sewage sludge	0	49.9	64.0	9.7	15.2	—

from: Kelin, 1972. *Anaerobic Digestion of Solid Wastes.* Compost Science, Jan/Feb 1972, pg. 6.

would be available for use as pipeline gas. Cooking requirements per capita range between 7 and 10 ft³/day; therefore, cooking requirements are between (7 ft³ × 30,000) and (10 ft³ × 30,000) = 210,000-300,000 ft³/day. This need could be more than met by the potential gas production of the city's inhabitants.

AGRICULTURAL RESIDUES: These resources are of two types: (1) *crops and crop residues* and (2) *manures.* The problem with utilizing the resources is both their availability and their concentration. For example, while feedlots and dairies offer both availability and concentrations of manures, other more dispersed and ecological types of agriculture that encourage grazing and small-scale applications make access to manure difficult. Likewise, the availability of crop residues at processing plants is much more reasonable than collection and harvesting of residuals left in the field.

Crops and Crop Residues: As early as 1930, Buswell was experimenting with agricultural residues from plant materials and their application to anaerobic digestion. His work at the University of Illinois proved that the cellulosic material from plants was a prime feedstock for digesters. Several designs for digesters to specifically use such crop residues as cornstalks were developed and patented by Buswell in the 1930's. This work was conclusive, and some applications were made for papermill pulp, strawboard residues and distilleries. In general, however, these resources have not had broad applications. This is probably because the materials are often coarse and fibrous and not as easy to digest as more liquid resources such as municipal sewage or manure. Such fibrous materials take extra preparation from grinding or chopping to make them suitable for digestion. A mixture of crop residue and manure will increase the solids content of the slurry and could increase gas production under conditions in which moisture content of manure is too high from wash-down water.

Manure: When you see a table that shows the amount of manure produced by different kinds of livestock, bear in mind that the amount on the table will probably not be the amount that is actually available from your animals. There are three major reasons for this: (1) Manure production varies with the size of the animal. (2) Not all of the manure produced is easily collected, and depends on the degree of livestock confinement. (3) The kind of manure that can be collected may be fresh excrement (feces *and* urine), excrement plus bedding, wet feces only, etc.

The Livestock Unit: Keeping in mind all the things that can affect the type and amount of manure that can be collected for digestion, we can assemble a general manure production table that shows *rough average amounts* of fresh manure, total solids and volatile solids produced by various livestock. The table enables us to get some idea of the relative production of readily available digestible material (VS) by different livestock (the "livestock unit"). Thus, on the average one large horse produces about 180 times as much digestible material per day as a laying chicken (see Table I).

BIOMASS PLANTATIONS: Conversion of solar energy into chemical storage through the production of rapidly growing plants has become a popular concept. The idea is not new; e.g., almost all old farms have a designated woodlot to continually supply the need for firewood. Methane production from high rate biomass has been very popular for some time. Focus has generally been on aquatic plants which can be grown in nutrient-rich waste streams. Most notable work has been done for algae, kelp and water hyacinths. Large-scale biomass farms have been developed and demonstrated on a pilot scale. Other small-scale units for farms have also been demonstrated.

Algae: Most of the work with these unicellular plants has been with fresh water species grown in shallow, artificial ponds (e.g., *Chlorella, Scenedesmus, Ulothrix,* and *Spirulina).* Algae have been grown in digester waste water as feed for animals; they can also be fed directly to digesters for gas production. The production of algae is dependent on sunlight, temperature and nutrients. In sunny Richmond, California, algae have grown at a maximum rate of 39 tons dry weight per acre per year . . . about 200 lbs. per day. If it is assumed that all of this can be harvested and digested at a conversion efficiency of 60%, then the 200 pounds of algae will produce 720 ft³ of methane per day (figures based on Oswald and Golueke [1960]: CH_4 = 65.5% and one pound dry algae produces 3.6 ft³ of CH_4).

Water Hyacinths: This hardy, floating aquatic weed reproduces rapidly from seed propagation, as well as by sprouting from a mother plant. It is noted for being a nuisance in many warm regions, and it thrives in polluted waterways. The National Aeronautics and Space Administration has been experimenting with water hyacinths as biological treatment systems for domestic waste streams. Growing in warm, sunny, waste water, water hyacinths have attained growth rates of 800-1,600 pounds per acre/*day*. This growth rate is by far the highest for any known plant. At these growth rates there is enough biomass produced to generate between 3,500 and 7,000 ft^3 of methane gas, as well as 0.5 tons of fertilizer per acre per day (reported by Wolverton and McDonald [1976]).

Kelp: Ocean farms growing large brown marine algae (kelp) using nutrients from 1,000 ft. "sea wells" have been proposed, and pilot scale farms have been tested at the *Ocean Farm Project.* This program is directed by Dr. Howard Wilcox of the Naval Undersea Center in San Diego, California. The kelp are cultivated on large floating rafts of plastic lines radiating from a central float platform. In the full scale, this central float platform would house digesters and other kelp processing activities. These farms are envisioned in 1,000-acre modules capable of producing about 160 million cubic feet of gas per year. This yield is equivalent to about 500 ft^3 of methane per acre/day.

Methane Digesters: Some Basic Questions

With this very general background we are ready to ask a fundamental question about the use of digesters: Are methane digesters practical? Do we have the resources (easily available organic wastes, time and money) to build one? Will a methane digester provide us with enough gas or liquid fertilizer to make it worth our while to build and *maintain* one?

Are methane digesters practical? The answer has to do with scale and needs. On a large-scale, that is, in centralized waste treatment facilities (municipal, community or village), livestock operations, and food processing plants there is no doubt that anaerobic digestion is a sound way of recycling organic wastes. Increasing work is being done on digesters at this level all the time. On an intermediate scale (ranches, stables, large kennels, small livestock farms, etc.) the big trade-off is in the cost of the operation. On a small scale there seems to be real doubt whether digesters are really practical for anything except isolated homesteads or as educational tools.

Certainly there would be argument on this from people experimenting with small scale digester systems. But the common complaint among people who have tried to set up small digesters is that the amount of gas available is usually too small to justify the time and expense of building one.

Let's look at a few facts. Suppose you had 60 laying hens and all of their manure available for digestion. From Table I we can see that 60 laying hens produce about 2½ lbs of VS per day. If we assume that 1 lb of VS can yield 5-7 ft^3 of methane gas, then 60 chickens will provide us with about 12-18 ft^3 of methane/day, enough to cook all the simple meals of a person for one day (Table III). Notice also that it takes the wastes from 8 adults to provide enough gas to cook the meals of one person. If you think it is

worth it ... do it. In cases where people are willing to reduce their standard of living drastically, and have the time to maintain a digester, small digesters could be an asset and help to provide a small portion of the energy needs and a valuable fertilizer. But just because it produces energy, a digester is not necessarily worth building. Do some serious figuring first. Use Tables I, II and III as guidelines and refer to the books in the review section for more detailed information.

Once the decision has been made to build a digester two additional questions need to be asked: 1) With the organic wastes and resources at hand, what kind of digester should be built and how big should it be? 2) What is the best way of using the gas and sludge produced? The first question is explained in detail in the many books described in the review section. With regards to the second question, remember that a digester is just the heart of a system that produces fuel gas *and* fertilizer. These by-products can be used in a variety of ways (Figs. 6 and 7). Perhaps the biggest advantage to digesters is not so much that they produce fuel and fertilizer. but that they serve so well as back-up components to an integrated system. Their biological requirements for heat mean that digesters can be fed by the excess energies of solar and wind devices while at the same time they can use their own waste products as fuel. Digesters are literally the guts of an integrated energy system

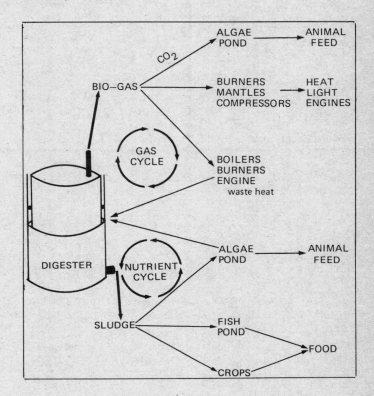

FIG. 6   THE POTENTIALS OF A DIGESTER SYSTEM.

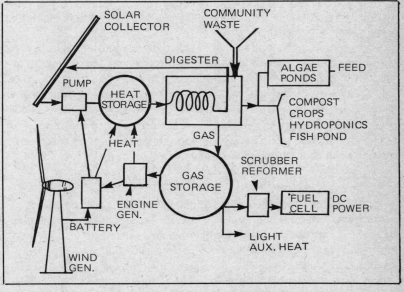

FIG. 7   THE POTENTIALS OF AN INTEGRATED RENEWABLE
ENERGY SYSTEM USING A DIGESTER.

TABLE III   USES FOR METHANE. Consumption of methane and biogas for different uses.

USE	FT^3	RATE
Lighting	2.5	per mantle per hour
Cooking	8 - 16	per hour per 2 - 4" burner
	12 - 15	per person per day
Incubator	.5 - .7	ft^3 per hour per ft^3 incubator
Gas Refrigerator	1.2	ft^3 per hour per ft^3 refrigerator
Gasoline Engine*		
CH$_4$	11	per brake horsepower per hour
Bio-Gas	16	per brake horsepower per hour
For Gasoline		
CH$_4$	135 - 160	per gallon
Bio-Gas	180 - 250	per gallon
For Diesel Oil		
CH$_4$	150 - 188	per gallon
Bio-Gas	200 - 278	per gallon
*25% efficiency		

Methane Digester Design

DIGESTER CLASSIFICATIONS: Digesters can be classified by the way they are loaded: (1) *batch* or (2) *continuous.* Further delineation of these types considers the actual hydraulic flow characteristics which can be related to digester performance and specific sizing. There are four principal digester types described by their hydraulic flow characteristics: *batch, plug flow, arbitrary flow* and *complete mix.* These types of flow are depicted in Fig. 8.

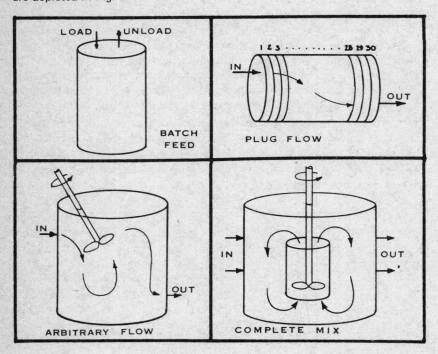

FIG. 8   DIGESTERS CLASSIFIED BY HYDRAULIC FLOW CHARACTERISTICS
(from: Anaerobic Digester Feasibility Study, Ecotope Group).

Batch Loading describes a process where materials neither enter nor leave the digester on a continuous basis. Usually a batch digester is loaded only one time—the material is left to digest and spend all of its gas, at which time the process is complete and the contents emptied for a new loading.

Plug Flow describes a process where material is discharged from the digester (usually no mixing) and theoretically remains in the digester for a time equal to the detention required for decomposition.

Arbitrary Flow is any degree of partial mixing between a plug flow and a completely mixed digester. This partial mixing is an attempt to insure that material passes through the proper sequence of events to be fully digested.

Complete Mixing occurs when material entering the digester is immediately dispersed throughout the tank. This further implies that "the particles leave the tank in proportion to their statistical population" (Metcalf and Eddy, 1972).

The hydraulic flow characteristics of these various designs have a relationship to the growth of the microorganisms. These flow characteristics play a major role in the sizing of a digestion reactor tank, since biological activity determines the "residence" or "detention" time of the substrate (organic material) within the tank. The residence time depends on the amount of biological activity and thus the rate of digestion. The other important characteristic associated with residence time is temperature, which will be discussed separately.

DESIGN CONSIDERATIONS: FLOW CHARACTERISTICS:

Batch Loaded digesters with no mixing apparatus are the easiest to construct and operate, but are also the least productive and reliable. Because they are opened and loaded/unloaded periodically, each new loading requires the biological cycle to begin again. This means that combustible gas is only produced in the latter half of the cycle, since the methane cycle must await food from the acid formers. One design, which smoothes out the combustible gas cycle, is to have a series of batch digesters working in different stages. The residence time for complete digestion of the substrate will vary between 60 and 90 days. Fig. 2 shows the probable stratification of a batch digester. Complete digestion is indicated by reduced and subsided gas production. Scum build-up in batch loaders is significant and can cause gas production to cease before the digestion process is completed by preventing the release of gas from the active layers of the digester.

Plug Flow Digesters, sometimes called "horizontal displacement digesters" require a residence time of between 30 and 40 days. It is difficult to have a true "plug flow" because there is naturally some dispersion along the length of the tank since vertical "plugs" (daily loadings) will not be totally segregated. In theory, the plug flow digester separates the three stages of digestion along the length of the digester tank. The plug flow system is analogous to an assembly line with different biological processes in different sections of the digester (Fig. 3). Because of this "assembly line" digestion the plug flow system is less efficient than mixed digesters since only about 50% of the tank is actually used for methane production, resulting in a long residence time. This may, however, be of benefit to small to medium-scale farm operations where a covered lagoon can be used as a digester and the cost of a large tank is not an issue. Another benefit of the plug flow system is that it can be designed and built using a minimum of mechanical and power-driven devices which are essential in mixed digesters. These systems do, however, require periodic shutdown for removal of scum formation, but their high surface to volume ratio lessens the problem to periods as long as a year.

Arbitrary Flow digesters can be of various designs. For example, a plug flow digester with a mixer installed. Other designs might include mixers that were agitated on an intermittent basis, e.g., a hand-operated or a windmill-driven mixer. These schemes lend themselves well to home and farm-scale applications, and they can be made and maintained at minimum cost with excellent results. Any amount of mixing will increase the biological activity and reduce the residence time which in turn will reduce the size requirements for the digester tank.

Complete Mix digesters were developed through the evolution of sewage treatment plant designs. Their improved hydraulic flow characteristics, reliability and efficiency have been proved in many decades of applications for the treatment of municipal solid wastes. As early as the 1930's these systems were widely accepted as the most reliable means of treatment for sewage sludge. Long-term maintenance of anaerobic treatment systems for stabilizing organic waste efficiently requires that the acid formers and the methane formers be kept in a state of dynamic equilibrium. This dynamic equilibrium can only be maintained through high-rate agitation which causes the entire contents of the digester to be vigorously mixed by either a power-driven mechanical device or by recirculation of compressed digester gas. This mixing must be done on a continuous basis. The complete mix digester can produce 98% of the gas available in a 15 to 20-day residence period and requires a tank of about 1/3 to 1/2 the size of a plug flow system. This vigorous mixing also prevents scum accumulation by keeping it mixed in the slurry so that it flows out with the effluent.

The disadvantages of a high-rate, complete mix digestion process are as follows:

(1)   High energy input for mixing: mechanical mixing requires 1 hp per 10,000 gallons and gas recirculation mixing requires 2 hp per 10,000 gallons.

(2)   High maintenance cost for equipment, including pumps, compressors, mixers and related equipment.

(3)   High capital investment for equipment.

Traditionally, where municipal treatment of human sewage was considered a public responsibility to the health of a community, cost—both energy and dollar—have been ignored in face of reliability and insurance of treatment for waste stabilization. The application of these systems at any but the largest of scales, however, is unlikely and at best questionable.

TEMPERATURE: Equally important and critical to residence time and digester size is the temperature of the digesting liquor. Three different groups of bacteria have been identified according to their temperature sensitivity.

(1)   Psychrophylic (40° - 50°F)

(2)   Mesophylic (50° - 110°F)

(3)   Thermophylic (110° - 140°F)

Almost no biological activity occurs below 40°F, and most natural anaerobic activity occurs in the mesophylic range, especially near 95°F (bovine and other ruminants are good examples of natural mesophylic digesters). Temperature conditions are so vital to the digestion process that digestion systems tend to be classified by *both* their flow and temperature characteristics; for example, most sewage treatment digesters are classified as "complete-mix, mesophylic systems."

Biological activity increases with temperature, but there is no indication that this will also cause more gas or degradation to occur. Increasing temperature only decreases the time required for complete digestion with a corresponding increase in gas production *for a given period of time.* Figure 9 graphically presents the digestion (residence) time relative to temperature.

New Alchemy Institute West, NAI *Newsletter* #3, 1973.

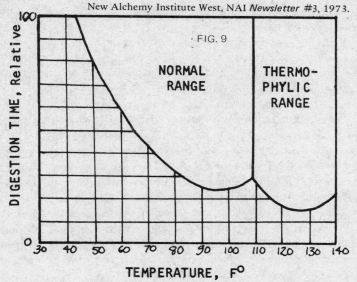

FIG. 9

NORMAL RANGE

THERMO-PHYLIC RANGE

Mesophylic range anaerobes seem to be the most hearty and least susceptible to going sour. The high-performance thermophylics are more sensitive, but are most likely to be used in situations of greatest control.

A temperature of 95°F has been most frequently used as an optimum for most digestion systems. There is, however, a renewed interest in the thermophylic range since this can act to reduce digester volume by 50%. In highly automated digestion systems with careful monitoring and sensitive controls, it is clearly an effective option. The residence times for such substrates as solid waste and bovine manure have been reduced to 8 to 10 days with 95% to 98% degradation.

In contrast to this direction, and equally valid for consideration for farm or home-scale digesters is the mid-range mesophylic option. If tank size is not a significant issue and if extensive heating and monitoring equipment is to be avoided, then temperatures of about 70°F might be an attractive option.

Another most important issue to consider with regard to temperature is the heat added versus the heat lost from the system. Heat added is usually provided by gas produced, and maximization of production is of most importance in determining the usefulness of a digester. Heat losses then are critical to digester design. First, there is the "skin" or surface loss from the tank itself. Second, there is the heat lost from the addition of cold new material and the discharge of warm effluent in continuous flow system.

Skin losses can be estimated using a basic formula for "conductive heat transfer" from a mechanical engineering handbook. The method used is to multiply the sum of the conductivities of the tank material and insulating jacket (U) by the surface area of the tank (A) and the average temperature difference (dT):

$$Q_{heat\ loss} = UAdT$$

Regardless of how much insulation is applied to a digester tank, there is the more critical issue of heat needed for daily exchanges through loading. The greater the volume of the daily load, the larger the quantity of gas required to maintain constant digester temperature. One means of reducing this loss is to incorporate a counter-flow heat exchanger in the loading and discharge areas of the digester. A counterflow heat exchanger requires that the fresh load (influent) and the discharge (effluent) streams pass beside each other for a sufficient distance to exchange heat. This can be accomplished by having the streams flow in trenches beside one another.

To calculate the heat requirements of a daily loading:

$$Q_{heat\ loss}\ Btu's = gallons \times 8.33 \times dT$$

where: dT = digester temperature - influent temperature
and: 1 Btu = 1 pound water raised 1 degree F
and: 1 gallon of slurry = 8.33 pounds

All the heat which can be extracted from the effluent through a heat exchanger can be subtracted from this heat requirement. To determine the length and surface area for a counterflow heat exchanger, it will be necessary to consult an engineering handbook. Crude heat exchangers can be expected to reclaim between 30% and 50% of the heat loss; more sophisticated design can reclaim 60% to 70%.

DIGESTER SIZING: Sizing digesters is dependent on several related and independent variables. These include (1) volume of resource, (2) solids content of resource, (3) water added, (4) pounds of VS/ft³ of digester volume, and (5) residence (detention) time.

Let's size a plug flow digester for one dairy cow. To do this, we need to know two things: the loading rate and the volume of manure available.

A dairy cow produces 18 gallons of wet manure per day; about 12% is Total Solids (Table I). To this some water must be added for ease of handling. The added water will affect the solids content, the loading rate and the residence time. As little water as possible should be added because water reduces the percent of digestible material (Volatile Solids) and increases the heating needs (cold water). If one part water is added to two parts fresh manure, the % of Total Solids (TS) is reduced by 33%, to 8% TS (.66 × 12% TS) and a new volume of 27 gallons (9 + 18 gallons) results. Put another way: 2 parts fresh manure (18 gallons) is composed of 88% (15.8 gallons) water and 12% (2.2 gallons) TS. Adding this to 1 part water (9 gallons) gives 27 gallons of slurry composed of 24.8 gallons water (9 + 15.8 gallons) and 2.2 gallons TS, or about 2.2 ÷ 27 = 8% TS . . . a reduction of 33% from the original 12% TS. With a normal residence time of 30 days (for plug flow), the digester volume required is 810 gallons (30 days × 27 gal/day) or 108 ft³ (810 gal. divided by 7.5 . . . the number of gallons of any liquid per ft³ of space). From Table I, 13.8 lbs. of Volatile Solids (VS) are produced daily by a dairy cow. Thus the loading rate is 13.8 lbs. VS ÷ 108 ft³ = 0.13 lbs. VS per ft³ of digester per day.

Ideally, the greater the loading rate, the better the performance of the digester, but the interaction of volume, ease of handling and residence time play a significant and independent role in this determination. Well-mixed, high-rate digesters are a benefit in this regard, since they have shorter residence times. For example, a complete-mix mesophylic system with 95°F operating temperature has a residence requirement of only 15-20 days. With the above example the digester volume is reduced to 540 gallons, and the loading rate is increased to 0.19 lbs. VS per ft³ (540 gal. ÷ 7.5 = 72 ft³; 13.8 lbs. VS ÷ 72 = 0.19 VS per ft³ of digester volume per day) or about 132 lbs per day total (1.81lbs/ft³ × 72 ft³). We have come full circle to the daily production of the cow (132 lbs per day):. . .a nice way to check these rather tedious calculations.

GAS AND FERTILIZER USE:

Gas Use: Biogas has a heating value of between 500 and 800 Btu/ft³ depending almost exclusively on the methane content of the gas. Natural gas, by comparison, has a heating value of about 1,000 Btu/ft³ and is almost all methane. Biogas can be easily burned without removing the carbon dioxide part of the gas. It is most commonly used as a natural gas substitute.

First off, methane is 120 octane indicating that it can be used most efficiently in an engine with a high compression ratio like a diesel (15:1) rather than a gasoline engine (8:1). (For a good description of using biogas to power diesel engines see John Fry's book, page 205.)

A first choice for using biogas is in a boiler, especially if there is some use for steam. Boilers can be as high as 85% efficient and, since biogas burns clean, this efficiency will not decrease over time. In contrast, coal or oil-fired boilers have to be cleaned of soot frequently.

Refrigeration is another high temperature application of biogas . . . for two reasons. First, it is very efficient, on the order of 80% to 90% conversion efficiency. Second, demand for energy is in phase with the natural cycle. This is to say that there is a high demand for gas for cooling when there is a high production of gas from the digester because of the naturally warm ambient temperatures in the summer. Absorption (gas) refrigerators are still available and can be found at camper and marine dealers or in remote location applications.

Another important use for biogas is as a cooking fuel. This high temperature application, although it is not as efficient—on the order of 40%-60%—matches an essential function with an acceptable form of energy. Biogas is easily adaptable to natural gas appliances, although it may require a slightly enlarged burner orifice and air mixture adjustment.

Another option for gas use is to run a stationary gasoline or diesel engine. Gasoline engines are on the order of 20%-25% efficient at doing work while diesels are 25%-30%. As mentioned previously, the high octane rating of methane requires a high compression ratio for efficient utilization. Diesels have these high compression ratios and can be run on a mixture of 5%-20% diesel fuel (for ignition) with the remainder being biogas. Gasoline engines can be bought which are designed to run on natural gas or biogas, or they can be modified by increasing the compression ratio (a local automotive speed shop can be a good resource for this modification).

The waste heat from internal combustion engines can be used to heat the digesters or for other heating needs. Water-cooled engines are the easiest to adapt to heat recovery. About 25%-30% of the heat can be recaptured from the engine coolant, and another 10%-15% can be saved by putting a heat exchanger in the exhaust of the engine. The total heat recovery then could be 45%. If the engine performs at 25% work efficiency, then the overall efficiency would be 70%, but the engine has additional losses if it is used to do work. Used to run a generator it will lose 30% to 50% to make the electricity. If there is no place to use the waste heat from the engine, then that energy is lost. Suppose that the digesters don't require as much heat as the engine is generating. The engine must still give off that heat for cooling purposes, so it is important to do careful sizing by evaluating the production of biogas and the demands for various forms of energy available from the system. Consumption rates of methane and biogas for different functions are listed in Table III.

Fertilizer: The big dispute and question here is the following: "Is anaerobically treated effluent a better fertilizer or soil conditioner than if the raw waste were simply spread on the land and allowed to decompose?"

During the digestion process the organic nitrogen, which is bound to the proteinaceous material of the substrate, is released and appears as ammoniated nitrogen. This appears to parallel the production of gas and the reduction of VS. Ammoniated nitrogen, as found in urea, will easily hydrolize and be lost as free ammonia. In the soil where plant growth is active, this ammoniated nitrogen is quickly consumed as a nutrient. In contrast, organically bound nitrogen must be decomposed by soil organisms before it is in an ammoniated form and can be used by plants.

It appears that there is some added value to the application of digested effluent to growing crops. Another benefit of the digestion process is that it will preserve the ammoniated nitrogen in storage lagoons. This ammoniated nitrogen remains in solution at a pH below 7.5. The pH of digested effluent is generally between 6.8 and 7.0. The spreading of raw, untreated matter can lead to pollution through run-off into streams from rainfall.

The benefit of digested over raw manure appears to depend on its handling and application. As greater emphasis is placed on waste management in agricultural areas to abate pollution, the prospect of utilizing digesters will have additional benefits to farmers. The combination of combustible gas, a treated fertilizer and a pollution abatement facility will be most appealing.

Methane Conversion Applications: Past, Present, Future

Any attempt to cover the applications of operating methane conversion systems of the past, present and future will at best be representative. The systems, especially those of the present, will always be changing, and any attempt at reporting the "state-of-the-art" will be dated as our consciousness is shocked by the obvious necessity to shift to renewable energies.

HISTORICAL PERSPECTIVE: Great names are associated with the discovery of biogas from anaerobic decomposition of organic material. "Swamp or marsh gas" was well known to even the Dark Ages and was reported as early as 1667 by Shirley. Such names as Volta (1778) and the great Pasteur (1884) have reported the properties of this biological activity.

Possibly the earliest applied use of methane conversion was at a leper colony in Bombay, India. In 1907 an internal combustion engine was operated off of "sludge gas" being generated from a demonstration sewage treatment facility there. Gas from the sewers of Exeter, England, was used for street illumination in 1896.

The first widespread application of anaerobic digestion was for treating municipal sewage sludge. These treatment plants were first installed in Birmingham, England (about 1927), Berlin, Germany (1928), Charlotte, North Carolina, U.S.A. (1928), Baltimore, Maryland, U.S.A., and many other cities by 1935.

Farm applications came later and were greatly influenced by the work of Buswell at the University of Illinois who pioneered experimental work on digestion of farm related residues. World War II and fuel shortages in Europe accelerated the development of farm-scale digesters for fuel production as opposed to sewage treatment.

Descriptive accounts and designs of many of the systems listed below are given in the recent publication by the Energy Research and Development Administration prepared by W. J. Jewell, *et al.* (see Book Reviews: *Bioconversion of Agriculture Waste for Pollution Control and Energy Conservation*). Table IV is a representative list of methane digesters which were built in the period between 1930 and 1960.

After 1960 the return of inexpensive fuel sources and the high maintenance requirements for digesters caused almost all of these plants to be abandoned. Of the fifteen plants operating in 1957 in Germany, only one remains today. This is a Schmidt-Eggersgluess type at a monastery in Bad Toelz, Germany.

Name	Location/Year	Application	Remarks
Buswell	Univ. of Illinois (1933)	Cornstalks	Arbitrary flow
Darmstadt	Czechoslovakia (1952)	Manure w/ bedding	"Canal process"
Weber	Germany	Manure and straw	for 26 cows
Ducellier & Inman	France (1952)	Manure w/ bedding	Silos used as digesters
Schmidt & Eggersgluess	Germany (1950)	Dairy farms	Several installations, complete-mix, mesophylic
Munich System	Grub, Germany (1956)	Manure	Pilot plant only, no full-scale built
Desai	New Delhi, India (1946)	Cow and vegetable	Pilot plant at Indian Ag. Exp. Station
P.C. Winters	South Africa (1957)	Manure and crop residue	Plug flow
John Fry	South Africa (1959)	Pig manure	Plug flow, 1000 animals

TABLE IV   REPRESENTATIVE METHANE DIGESTER OPERATIONS, CIRCA 1930-1960.

PRESENT U.S. WORK: Recent interest, both private and governmental, in solar energy as well as environmental quality has led to a renewed interest in methane digesters for sewage processing, agricultural applications and biomass plantations. The following descriptions, while not exhaustive, are representative of the activity in the development of methane digesters in the United States.

U.S. Energy Research and Development Administration (ERDA)- Division of Solar Energy has a special program called "Fuels from Biomass." A part of this program is a coordinating group of university experiments, government programs and private contractors working on "fuel gas production from animal residue." The group of ERDA contractors and their activities are listed below:

(1) DYNATECH R/D COMPANY, Cambridge, MA

Working on computer simulation of various digester designs for applications to feedlots.

(2) W.J. JEWELL, CORNELL UNIVERSITY, Dept. of Agricultural Engineering
"Anaerobic Fermentation of Animal Waste": Design and operation criteria. Operating a 5,000-liter plug flow digester along with many other small laboratory-size digesters to evaluate variations in temperature, residence time and flow characteristics.

(3) JOHN T. PFEFFER, University of Illinois, Dept. of Civil Engineering
"Manure Fermentation to Methane": Operating and evaluating four 200-gallon digesters for effects of temperature and rates of mixing on digestion efficiencies.

(4) PERRY L. McCARTY, Stanford University, Dept. of Civil Engineering
"Effect of Heat Treatment on Lignin." Laboratory analysis of the effect of heat on lignin to improve biodegradability of cellulose.

(5) ANDY HASHIMOTO, USDA-U.S. Animal Meat Research Center, Clay Center, NB
Operating a 1,500-gallon complete-mix thermophylic digester to evaluate gas production and effluent refeeding for beef production.

(6) WARREN COE, Hamilton-Standard Laboratories, Windsor Locks, CT
"Fuel Recovery Potential from the Montfort-Feedlots, Greeley, Colorado": Operating a 1,500-gallon complete-mix, thermophylic digester to assess the feasibility of using open, dirt feedlots for fuel gas production.

(7) KEN SMITH, Ecotope Group, Seattle, WA
"Methane Project: Monroe State Dairy Farm": Operation and evaluation of a 350-cattle unit digester constructed in Monroe, WA.

Other applications include both private companies and individuals as well as other university and government programs. The following list is representative of some of these activities:

(1) TOM ABELES, IE Associates, 5 Woodcrest Dr., Burnsville, MN 55337
"EPA Methane Digester State-of-the-Art Report": $100,000 study of design, safety and utilization of digester technology.

(2) S. PERSSON and M.D. BARTLETT, Agricultural Engineers, and R.W. REGAN, Civil Engineer, Pennsylvania State Univ., University Park, PA
"Experiences from Operating Fullsize Anaerobic Digester."

(3) JOHN M. BRINKER, President, A.O. Smith-Harvestore Products, Inc., Arlington Heights, IL
"Methane R & D" and sale of digester tanks: Harvestore built and operated a 32,000-gallon digester for two years in Wisconsin. They are now evaluating its potential, and sell tanks which can be adapted to make digesters.

(4) E.N. DALE, President, Agriculture Energy, Inc., Ludington, MI.
"Plug Flow Digesters": This company has completed two farm digester systems: (1) a 350-head feedlot near Custer, Michigan, and (2) a 250-head dairy in Rice Lake, Wisconsin. Both are plug flow systems which are being evaluated as marketable products.

There are, of course, hundreds of other projects going on throughout the country and abroad. They are certainly too numerous to mention here.

Methane digesters work, but they are like children, they must be fed and pampered and watched carefully and they need a sitter when you're away. There are no panaceas and no free lunches with renewable energy systems, and certainly not methane digesters.

METHANE REVIEWS

METHANE: FUEL OF THE FUTURE

Popular style, brief overviews of anaerobic process, digester design, etc., plus old and new ways of using gas and sludge (municipal, agriculture, *Methane in Developing Countries*, *Domestic Use* and *Transportation*). Good sections: *A Digester in Every Garden* (not practical without lifestyle changes): *Research Needs* and *References* (mostly in England). No practical designs, mostly descriptions, potentials and inspirations nicely put.

—R.M.

Methane: Fuel of the Future
Bell, Boulter, Dunlop and Keiller
1973; 86 pp

$1.60 paper (plus postage)
$4.50 cloth (plus postage)

from:
Book People
2940 Seventh St.
Berkeley, CA. 94710

or WHOLE EARTH TRUCK STORE
$3.50 paper

THE BIOMASS INSTITUTE

The Biomass Energy Institute was founded in Manitoba, Canada in 1971. The Institute's goals are to "Accumulate, develop, classify and disseminate information on biomass energy, and to act as a coordinating body between business, government and the academic community to optimize the use of biomass energy."

The Institute sponsored an *International Biomass Energy Conference* in Winnipeg in May 1973, and the Proceedings have been published in a very informative book ($10.00 from the Institute). There is also a newsletter of activities (2-3 times per year) and a pamphlet: *The Renewable Biomass Energy Guidebook* ($2.00 from the Institute).

The Institute has organized a lot of people, proposals and ideas about biomass energy, e.g., a pilot (25 hog) digester, a prototype 1000 hog waste recycler, Institute of Algology, etc.

— R.M.

For information write:
Biomass Energy Institute
310-870 Cambridge Street
Winnipeg, Manitoba R3M 3H5

METHANE: PLANNING A DIGESTER

This seems to be the definitive popular British publication on methane systems. It is an excellent summary of information presented in British research papers and books in this section, and it is also a no-nonsense perspective of the limited but valuable potential of methane technology. The book presents a much needed state-of-the-arts section on British research groups, private operations and commercial organization together with methane activities in Kenya, Australia and India. For learning about methane systems read first: The New Alchemy Newsletter, next this book and finally the Ecotope Feasibility Study.

—R.M.

Methane: Planning A Digester
Peter-John Meynell
Published by Conservation Tools and Tech. Ltd.
1976; 150 pp.

$4.95

from:
Schocken Books Inc.
200 Madison Avenue
New York, NY 10016

$5.00

from:
Prism Press
Stable Court, Chalmington
Dorchester, Dorset, England

GOBAR GAS RESEARCH STATION . . . INDIA

In India cows roam free, manure is plentiful and energy and fertilizer are scarce. Usually cow dung and local forests are burned to the detriment of the land. In the early 1940's the India Agricultural Research Institute started a program to design and develop digesters that could provide families and small farmers with *both* fuel and fertilizer from cow manure ("Gobar"). Ram Bux Singh has pulled together information from his own work and that of his predecessors (e.g., Desai, Acharya, Khadi and Village Industry Comm.) in two semi-popular manuals: (1) *Bio-Gas Plant: Generating Methane from Organic Wastes* is a general introduction and synthesis of two earlier pamphlets. It gives a general background and history of methane research throughout the world together with rough schematic drawings and descriptions of five basic "gobar" plants that have been tested to operate under a variety of conditions (small

digesters .. single and double chamber; large digesters .. single and double stage; batch digesters with animal/plant waste mixture). The author describes ways of using the gas (light, heat, cooking, incubator, refrigerator, fan, water pump, rice huller, generator), and lists results of experiments with the different kinds/sizes of digesters operating at various times of the year. (2) *Bio-Gas Plants: Designs with Specifications* contains 48 pages of roughly drawn schematics and photos of different scaled digesters and gas holders for temperate and cold climates (including a privy digester). Not quite detailed nuts and bolts, but enough for first approximations. Digesters are built from bricks, cement, mortar and metal bracings. One could argue with the basic design of these digesters. Being vertical and buried in the ground they would seem to be hard to clean out when operating on a continuous feed. Also there is no provision for the elimination of scum. But they are simple and provide good models for the individual or small group working with easily available materials.

—R.M.

Dear Primer,
By now (June 1974) more than 7,000 Bio-Gas plants are operating in India and a target for the installation of 100,000 plants has been fixed by the Government in the fifth Five Year Plan. At present some plants are under trial which could be used in cold climate areas.

—Ram Bux Singh

Bio-Gas Plants: Generating Methane From Organic Wastes
1971; 71 pp

$5.00 (including air post)

Bio-Gas Plants: Designs with Specifications
Ram Bux Singh
1973; 54 pp

$7.00 (including air post)

both books from:
Gobar Gas Research Station
Ajitmal, Etawah (U.P.)
India

or WHOLE EARTH TRUCK STORE

BIO-CONVERSION OF AGRICULTURAL WASTE

This book details the background and history of the European experiences with methane digesters. It also gives an up-to-date account of various operating digesters in the U.S. There are the usual gross estimates of energy potential from plant and animal resources, but it goes way beyond that to describe and evaluate farm energy requirements for small dairies (40-100 head). Thousand-head feed-lots are evaluated, also. All of these are done with projected cost. Finally, the book makes a full assessment of the feasibility of using anaerobic fermentation on dairies and beef feedlots. This is a detailed professional account by an engineering group that's not afraid to evaluate technology and make it understandable to the lay person or farmer. If you're serious about digesters, this book is a must for review and consultation.

— Ken Smith

Bio-Conversion of Agricultural Waste for Energy Conservation and Pollution Control
William J. Jewell
1976

$10.00

available from:
ERDA Technical Info Center
P.O. Box 62
Oak Ridge, TN 37830

RECYCLING TREATED MUNICIPAL WASTEWATER AND SLUDGE THROUGH FOREST AND CROPLAND

The Wastewater Renovation and Conservation Project was started at The Pennsylvania State University in 1962 to investigate the feasibility and environmental impact of disposal of treated municipal wastewater on land through spray irrigation. Since then the term 'Living Filter' has become more or less synonymous with the idea of spray irrigation of municipal wastewater on land.

This book is a symposium held at Penn State in August of 1972 to review current info related to the use of land for disposal of wastewater. It is the most complete book on the subject that I have seen. There are sections on: (1) properties of wastewater, (2) properties of soil to receive wastewater, (3) properties of wastewater during recycling, (4) properties of soil during recycling (especially valuable is an entire chapter on "Microbial Hazards of Disposing of Wastewater on Soil"), (5) responses of crops to wastewater, (6) systems design and economics and (7) examples of operating systems. As with most symposia there are lots of numbers, references and some esoteric descriptions. Furthermore, municipal wastewater is not your average homestead/small farm digester sludge. But if you are serious about using your garden plot or farmland as a 'living filter' for digester effluent there is good information and much food for thought here.

—R.M.

Recycling Treated Municipal Wastewater and Sludge Through Forest and Cropland
William E. Sopper & Louis T. Kardos (editors)
1973; 479 pp

$16.50

from:
The Pennsylvania State University Press
215 Wagner Building
University Park, PA 16802

CAPTURING THE SUN THROUGH BIOCONVERSION

Methane is part of a larger field of activity called Bioconversion. The energy available and waste utilization possibilities are covered in numerous "state-of-the-art" reports on: (1) urban, industry, agricultural and forestry waste biomass sources; (2) land, fresh-water and ocean farming of energy crops, such as wood and kelp; (3) processes producing gaseous, liquid and solid fuels and their further products; (4) technology assessment; (5) economic and social impacts and (6) environmental impacts. Displays the scope of bioconversion better than any other single source. Ask your library to buy it.

— Ken Smith

Capturing the Sun Through Bioconversion
the Conference Proceedings, 865 pp
March 10-12, 1976, Washington, D.C.

$18.00

from:
The Washington Center
1717 Massachusetts Ave., N.W.
Washington, D.C. 20036

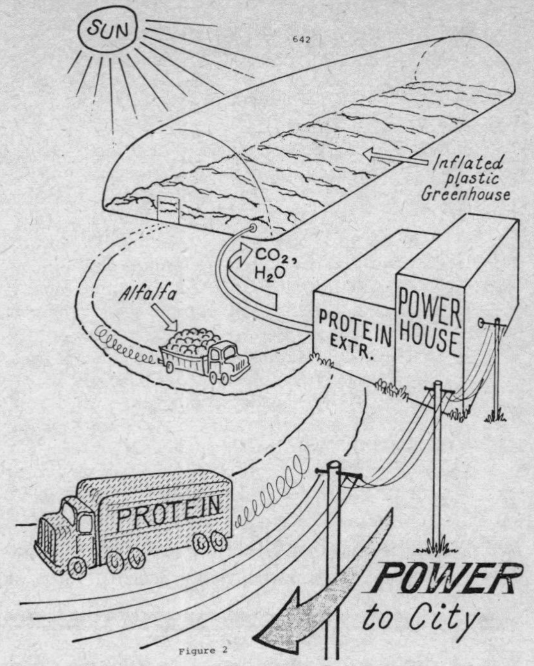

THE ANAEROBIC DIGESTION OF LIVESTOCK WASTES

Most of the current interest in methane is focused on manure as a prime resource. This is the best annotated bibliography on the subject, tracing the history of the process from its beginning, through many foreign experiences (Germany, India, France), laboratory results, and current projects. Smaller sections focus on digestion of farm-generated cellulosic materials (straw, cornstalks, etc.) and the fertilizing qualities of digester effluent. At only $2.00, you should order this right away. It is access to the popular, the technical, the obscure and the esoteric of methane fermentation.

— Ken Smith

ENERGY, AGRICULTURE, AND WASTE MANAGEMENT

Jewell has collected some of the best and most up-to-date articles on waste management and its effects on energy use in agriculture. Techniques, economics, alternatives and limitations are presented throughout this invaluable collection. Of special interest on methane processes are the following chapters:

18. "From Biodung to Biogas—Historical Review of European Experiences" — C. Tietjen.

21. "Energy Recovery and Feed Production from waste" — Hassan, Hassan, Smith,

22. "Anaerobic Digestion of Swine Wastes" — Fischer, Sievers, Fulmage,

23. "Alternative Animal Waste Fermentation" — Morris, Jewell, Casler,

and many other good chapters on animal waste related to methane production.

— Ken Smith

Energy, Agriculture and Waste Management:
Proceedings of the 1975 Cornell Agricultural Waste Management Conference
William Jewell, ed.
540 pp

$22.50

from:
Ann Arbor Science
Box 1425
Ann Arbor, MI 48106

Figure 2

The Anaerobic Digestion of Livestock Wastes to Produce Methane, 1946-June 1975: A Bibliography with Abstracts
Gregg Schadduck and James A. Moore
1975; 103 pp

$2.00

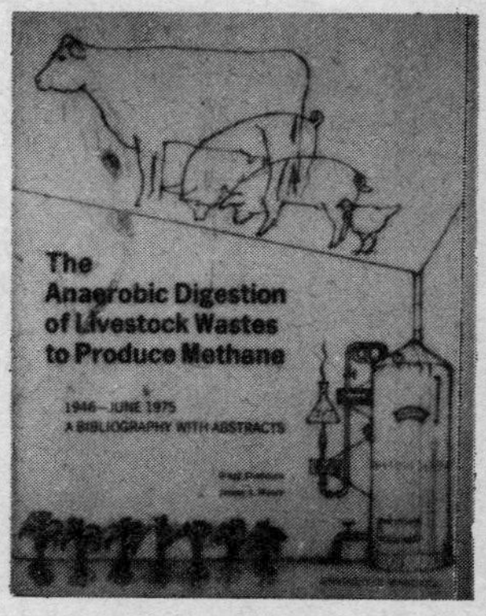

from:
J. A. Moore
Ag. Engineering Dept.
University of Minnesota
St. Paul, MN 55108

MAX'S POT, DRUM BIO-GAS PLANT

Max's Pot has evolved from a simple oil drum vertical digester to a 2-drum solar heated horizontal digester, to its current digester system. The present bio-gas plant (still 50 gallon oil drums) consists of: 1) Four *holding tanks* for receiving human feces, kitchen wastes, animal manure, and garden wastes. 2) A *solar collector* for heating water which is circulated through coils in the digester, keeping it at an optimum 95° F. 3) *Storage drums* for storing the solar-heated water. 4) *Mixing drums* for mixing the wastes from the four holding tanks. 5) A *digester* made from three 50-gallon oil drums. 6) *Gas storage tanks*. Presently Max's Pot is putting together a brochure that not only explains the design of their bio-gas plant, but also gives information about performance specifications of materials used to build bio-gas plants (hand pumps, toilets, valves, oil drums, etc.) plus directions for sizing digesters in various climates of the United States. For general inquiries send self-addressed stamped envelope to:

— R.M.

Max's Pot, Drum Bio-gas Plant
Center for Maximum Potential
Building Systems
1974

$5.00

from:
Center for Maximum Potential
Building Systems
6438 Bee Caves Road
Austin, Texas 78746

METHANE FEASIBILITY STUDY

Here's a minutely detailed paper proposing a methane-fertilizer plant for a dairy having 350 cows. The conclusion is that it will "pay" (in the commonly accepted sense of the word); about breaking even on methane alone, not counting fertilizer production. With a little luck, it should be operational by September 1975 and then we'll know. The concepts of net energy have been duly recognized; a rarity. Serious builders of biogas facilities would do well to read this all carefully. It's a model of what such a paper should be.

— J. Baldwin
CoEvolution Quarterly

A feasibility study has been completed. Manure from 350 cattle units (one cattle unit = 1000 pound animal) dairy operation will be scraped from loafing shed and deposited into the digester to produce a combustible gas and ammoniated nitrogen fertilizer. Digestion technique will utilize high rate gas recirculation mixing. Net production (after heating needs) of biogas (600 Btu/cu ft) was estimated to be 7500 to 9500 cu. ft./day in cold and warm seasons respectively. Biogas will be used to replace #2 fuel oil, Btu replacement savings estimated to be $4488.00/year. Fertilizer earnings are based on transformation of organically bound nitrogen to the ammoniated form. This was estimated to be 8 tons/year for earnings of $6344.00/year to replace ammoniated nitrogen fertilizer. Total direct cost of system was estimated to be $56,700.00 Evaluation of scale of project shows the amortized cost per cattle unit is near the break-even point for investment payback from biogas earnings alone.

Process Feasibility Study:
(The Anaerobic Digestion of Dairy Cow Manure at the State Reformatory Honor Farm, Monroe, Washington)
Ecotope Group and Parametrix, Inc.
1975; 119 pp

$8.00

from:
Ecotope Group
747 16th Ave. E.
Seattle, WA 98112

METHANE DIGESTERS FOR FUEL GAS AND FERTILIZER

"(This book) is highly recommended by this reviewer to anyone intent upon building a digester for biogas production . . . The chapter on the biology of the digestion process is written in a style both lucid and in a language intelligible to the non-specialist in biology. It is a valuable section in that it provides a rational basis for the design and operation of the digester . . . The section on design covers a wide variety of possible designs (actually only 3 . . . RM). Especially fascinating to this reviewer was the section on the 'innertube' digester."

—Clarence Golueke
Compost Science

This. . . newsletter, for all its shortcomings, must, I think, be judged the most comprehensive and available treatise on the subject. Other publications which may best it are either available only with difficulty, in a foreign language, or the cost. . . exceeds the $4.00 demanded for this primer.

—Gregg Shadduck, J. Moore
University of Minnesota

INDEX—Introduction; Background; History; Biology of Digestion; Raw Materials; The Gas; Digester; Design; Using Gas; Using Sludge; Building a Sump Digester; Building an Innertube Digester; Necessity is the Mother of Invention; References (68).

* *

Starter brew can be generated in a 1 or 5 gallon glass bottle. Care must be taken to fill the bottle only about ¼ full with either (a) active supernatant from a local sewage works or (b) the runoff from the low point on the land of any intensive stock farm in your district. Fill ¼ more with fresh dung. Leave the other ½ of the bottle for fresh manure additions at weekly intervals. Never fill too near the screw cap, since foaming could block off the opening and burst the bottle. Of course, the screw cap must be left loose to keep the bottle from exploding, except when agitating the bottle. It is a peculiarity of methane brews that a slight agitation when adding material is beneficial, but that continuous agitation has an adverse effect.

Methane Digesters for Fuel Gas and Fertilizer
L. John Fry, Richard Merrill
1973; 46 pp

$4.00

from:
New Alchemy Institute West
P.O. Box 2206
Santa Cruz, CA 95063
or WHOLE EARTH TRUCK STORE

PRACTICAL BUILDING OF METHANE POWER PLANTS

L. John Fry is a pioneer methane digester builder. This book is the history and some details of the horizontal displacement digester that Fry developed and built using low-cost, labor-intensive techniques in South Africa.

This is a passive, "plug-flow" type system. In the case of a digester this means that organic material must move passively along the digester through succeeding steps of digestion. This ensures almost total digestion of the material, but it requires long detention times (30-40 days). Conventional sewage treatment digesters, using mechanical mixers, achieve 95-98% digestion in 15-20 days. The biggest trade-off between the two systems seems to be in tank size, e.g., a 20-day digester is 1/2 the cost of a 40-day digester. The plug-flow system seems reasonable where tanks are cheap or can be made with labor-intensive methods. The rapid mixing digester is a better choice where time and labor cost are major considerations. A small tank cost will justify additional mixing equipment cost, which promotes rapid digestion and prevents scum build-ups.

Of special value in the book are 1) A generation of experiences with methane digesters *and* digester systems, from the simple to the complex. 2) Step-by-step designs for tested models built from simple materials (inner tubes, oil drums), plus a description of a 400 cubic foot concrete-building digester. 3) Plans for converting diesel motors to methane. 4) Ways of using sludge as a farmland fertilizer.

Some of the book is pulled from the New Alchemy newsletter which the author helped to write. The price is too bad, but for practical knowledge, you can't beat it.

—Ken Smith, Evan Brown

Practical Building of Methane Power Plants for Rural Energy Independence
L. John Fry
1974; 96 pp

$12.00

from:
L. John Fry
1223 N. Nopal
Santa Barbara, CA 93102

or WHOLE EARTH TRUCK STORE

A HOMESITE POWER UNIT: METHANE GENERATOR

Gives a background and, for general design purposes, shows how to calculate digester and gas holder size plus loading rate. A bit pedantic in the algebra, nonetheless it gives detailed, step-by-step design specifications for a digester built and working in California

—R.M.

An estimating procedure is presented in Table 6 where total daily gas needs are calculated at 17.3 ft^3 while special weekly needs are 89.1 ft^3. Thus, every week, 106.4 ft^3 storage space is required. This example illustrates an important design constraint: The size of the gas holder is based essentially on that use which consumes gas at the greatest rate. This is a key factor in all systems which contain storage components.

* *

A BATCH DIGESTER

The digester is a steel tank 66 inches long, having a diameter of 45 inches, located one foot above the ground on about a 5 degree incline in the direction of the 4 inch slurry exit valve (C). Since the unit was located in a moderate climate, the above ground digester used the heat from the sun for creating optimal temperatures. The incline toward the exit valve allows for easy slurry flow out of the digester. . . The loading apparatus is made of a drum (A), 22 inches in diameter and 34 inches high, to which a funnel (B), 22 inches in diameter at one end and 4 inches in diameter at the other, is attached. The loader is secured by way of a 4-inch galvanized union to a 4-inch gate valve (E) used as the continuous feed inlet to the digester. Extending 6 inches above the tank and attached to the valve (E) is a 4-inch diameter pipe (F) extending 2 feet into the tank. This pipe accepts new organic matter, empties directly into the existing slurry, and prevents the gas located in the upper part of the tank from escaping during loading operations.

A Homesite Power Unit: Methane Generator
Les Auerbach, Bill Olkowski, Ben Katz
1974; 50 pp

$5.00

from:
Les Auerbach
Alternative Energy Systems, Inc.
242 Copse Road
Madison, CT 06443

METHANE BIBLIOGRAPHY

Intormation about methane digestion is scattered widely in many places: in agricultural/municipal/industrial waste-recycling literature, in specialized microbiology texts and journals, in popular magazines like *Compost Science* and in many recent alternative publications (see Reviews). Some of the most valuable research has come out of the Sanitary Engineering Research Laboratory (SERL) in Richmond, California. Unfortunately, much of their information is hard to come by. Another good source of technical information has been the series *Sewage Works Journal* (1928-1948); *Sewage and Industrial Wastes* (1949-1958) and now the *Journal of the Water Pollution Control Federation* (1959-present). *Sewage Works Journal* is especially valuable since it represents the science and technology of waste recycling before the trend to make treatment centers bigger, and information more and more incomprehensible.

The following are but a *very* few of the key sources of information about methane systems. References are keyed in the following way: ASAE = American Society of Agricultural Engineers, St. Joseph, Michigan. ASCE = American Society of Civil Engineers. EPA = Environmental Protection Agency. GPO = Superintendent of Documents, Government Printing Office. SERL = Sanitary Engineering Research Laboratory, University of California, Berkeley, at Richmond, California. WPCF = Water Pollution Control Federation, 2900 Wisconsin Ave. N.W., Washington D.C. 20016.

Methane Digestion: Municipal/Industrial

Dept. of the Environment, 1965. SMALL SEWAGE WORKS OPERATORS' HANDBOOK. HMSO Press, P. O. Box 569, London S.E. 1 9 NH. From: Pendragon House Inc., Palo Alto, Calif. 94301.

Fair, G.M. et al. 1968. WATER AND WASTEWATER ENGINEERING, Vol. 2, Wiley, N.Y. (A)

Geyer, J. 1972. LANDFILL DECOMPOSITION GASES. EPA, NTIS, PB-213 487, $3.75. Annotated bibliography of 48 articles about anaerobic gases generated by landfills.

Greeley, S.A. and C.R. Velzy 1936. "Operation of Sludge Gas Engines," p. 57-62, SEWAGE WORKS JOURNAL, 8(1):57-62

Hawkes, H.A. 1963. THE ECOLOGY OF WASTE WATER TREATMENT. Pergamon Press, N. Y.

Imhoff, K. and G. Fair. 1956. SEWAGE TREATMENT (2nd ed.). John Wiley and Sons. N. Y. (B) Best semi-technical overview.

McCabe, B.J. and W.W. Eckenfelder. 1958. BIOLOGICAL TREATMENT OF SEWAGE AND INDUSTRIAL WASTES, Vol 2. Reinhold, N. Y.

MANUAL OF INSTRUCTION FOR WATER TREATMENT PLANT OPERATORS. New York State Dept. of Health. (From: HES, P. O. Box 7283, Albany, N.Y. 12201 Great tool!

Metcalf and Eddy, Inc. 1972. WASTEWATER ENGINEERING. McGraw-Hill, N. Y. 782 pp. (A) A technical overview; up to date.

Stander, G.J. et al. 1968. INVESTIGATION OF THE FULL-SCALE PURIFICATION OF WINE DISTILLERY WASTES BY THE ANAEROBIC DIGESTION PRO—CESS. CSIR Research Dept. No. 270, Bellville, S. Africa.

Young, J.C. and P.L. McCarty. 1969. THE ANAEROBIC FILTER FOR WASTE TREATMENT. Jour. WPCF, 41:R160-R173. Speeding up detention time with submerged 1" round stones. More surface area for bacteria. Works best with dilute soluble organic wastes (i.e., not manures).

Methane Digestion: Agricultural Wastes

Baines, S. 1970. ANAEROBIC TREATMENT OF FARM WASTES. Proceedings, Symposium Farm Wastes, Univ. of Newcastle Paper 18 (Univ. of Wisc. Extension), pp. 132-137.

Barth, C.L. and D. T. Hill 1975. "Energy and Nutrient Conservation in Swine Waste Mgmt.", Annual Mtg. 1975 Am. Soc. of Ag. Engr., St. Joseph, MO.

Boruff, C.S. and A.M. Buswell 1934. JOURNAL OF AMERICAN CHEMISTRY SOCIETY, 56:886

Boruff, C.S. and A.M. Buswell. INDUSTRIAL ENGINEERING CHEMISTRY No. 21 (1929), No. 22 (1930).

Buswell, A.M. and C.S. Boruff, U.S. PATENTS: 1,880,773 (1932); 1,990,523 (1935); 2,029,702 (1936).

Buswell, A.M. ILLINOIS STATE WATER SURVEY, Bulletins No. 30 (1930) and No. 32 (1936).

Cross, O. and A. Duran. 1970. ANAEROBIC DECOMPOSITION OF SWINE EXCREMENT. Trans. ASAE, 13(3): 320-325.

Hart, S. 1963. DIGESTION OF LIVESTOCK WASTES. Journal WPCF. 35: 748-757.

Hobson, S.A. 1973. DIGESTION OF LIVESTOCK MANURES. Jour. WPCF. 35: 748-757.

Jewell, W.J. 1974. ENERGY FROM AGRICULTURAL WASTE. METHANE GENERATION. Agric. Engin. Bull. 397, New York State College of Agric., Ithaca, N. Y. 13 pp. (C) Popular overview.

Loehr, R.C. 1974. AGRICULTURAL WASTE MANAGEMENT. Academic Press, N. Y.

MANAGEMENT OF FARM ANIMAL WASTES. 1967. Proc. Nat. Symp. on Animal Waste Management, ASAE, $5.00. 1966, $14.00.

Meenaghan, G. et al. 1970. GAS PRODUCTION FROM CATTLE WASTES. Paper no. 70-907. ASAE.

Parker, R. et al. 1974. METHANE PRODUCTION FROM SWINE WASTE WITH A SOLAR REACTOR. Presented at Southeast Region Meeting of the ASAE, Memphis, Tenn., Feb. 5, 1974.

Savery, C.W. and D. Cruzan. 1972. METHANE RECOVERY FROM CHICKEN MANURE DIGESTION. Jour. WPCF. 44(12): 2349-2354.

Taiganides, E.P. et al. 1963. ANAEROBIC DIGESTION OF HOG WASTES. Journal of Agric. Engin. Res., 8: 327-333.

Taiganides, E. and T.E. Hazen. 1966. PROPERTIES OF FARM ANIMAL EXCRETA. Trans. ASAE, 9: 374-376.

Taiganides, E.P. 1963. ANAEROBIC DIGESTION OF POULTRY MANURE. World's Poultry Science Journal, 19: 252-261.

Winters, P.C. 1957. "The Production of Methane Gas from Farm Manure and Wastes," p. 29-35, September 1957, POWER FARMING AND BETTER FARMING DIGEST, Australia and New Zealand.

Popular Overviews and Scaled Down Designs

Acharya, C.N. 1958. PREPARATION OF FUEL GAS AND MANURE BY ANAEROBIC FERMENTATION OF ORGANIC MATERIALS. I.C.A.R. Research Series, no. 15, India. 58 pp.

Boshoff, W.H. 1963. METHANE GAS PRODUCTION BY BATCH AND CONTINUOUS FERMENTATION METHODS. Tropical Science, 5(3): 155-165. (Britain). Using plant wastes in tropical (East Africa) conditions.

Diaz, L.F. et al. 1974. METHANE GAS PRODUCTION AS PART OF A REFUSE RECYCLING SYSTEM. Fourth Annual Composting and Waste Recycling Conference, Proceedings, El Paso, Texas. In: Compost Science, 15(3): 7-13. Popular description of recent research.

Fairbank, W.C. 1974. FUEL FROM FECES? The Dairyman, May, 1974. Popular putdown of digestion. "Direct combustion is the simplest way of converting feces to fuel . . . digesters are dangerous . . . on-farm digesters are not practical . only municipal or corporate industry can muster money . . ." etc., etc.

Lapp, H.M. et al. 1974. METHANE GAS PRODUCTION FROM ANIMAL WASTES. Information Division, Canada Dept. of Agriculture, Ottowa, Canada K1A OC7. 12 pp. (C)

Mother Earth News. 1972. PLOWBOY INTERVIEW WITH RAM BUX SINGH. Mother Earth News, 18. Hendersonville, N. C.

Pfeffer, J.T. 1973. RECLAMATION OF ENERGY FROM ORGANIC REFUSE. EPA - R - 800776.

Po, Chung. 1973. PRODUCTION OF METHANE GAS FROM MANURE. In: "Proceedings of Biomass Energy Conference," Winnipeg. See Reviews. 80 cubic feet, and 230 cubic feet square block digesters connected in series, built from bricks and concrete,flowing into algae pond. For 20-head pig sty and family of 12 in tropics.

PROCEEDINGS: BIOCONVERSION ENERGY RESEARCH CONFERENCE. Institute for Man and His Environment. Univ. Mass., Amherst, Mass., June 25-26, 1973. Conference about conversion of waste and feedlot materials to methane.

Singh, Ram Bux. 1972. BUILDING A BIO-GAS PLANT. Compost Science, March-April, 1972.

Singh, R.B. 1971. GOBAR GAS EXPERIMENTS IN INDIA. Mother Earth News, 12: 28-31, Hendersonville, N.C.

Taiganides, P.E. et al. 1963. SLUDGE DIGESTION OF FARM ANIMAL WASTES. Compost Science 4:26.

Water Pollution Control Federation. 1968. ANAEROBIC SLUDGE DIGESTION. WPCF, 3900 Wisconsin Ave.,Washington D.C. 20016 $2.50.

Whitehurst, Sharon and James. 1974. OUR FOUR-COW BIO-GAS PLANT. In: Producing Your Own Power, Rodale Press, Emmaus, Pa. See Reviews, Integrated Systems. Experiences with building and operating a 225 cubic foot continuous feed digester (old boiler tank) that utilizes manure from 4 cows on a Vermont dairy farm. Built with the help of Ram Bux Singh.

Biology/Chemistry

Barker, H.A. 1956, BACTERIAL FERMENTATION. John Wiley & Sons, N.Y. 95 pp. (B)

Buswell, A.M. and N.D. Hatfield. 1936. ANAEROBIC FERMENTATION. Illinois State Water Survey Bulletin, no. 32.

Chamberlin, N.S. 1930. ACTION OF ENZYMES ON SEWAGE SOLIDS. N. Jers. Agric. Exper. Sta., Bulletin 500.

Fair, G.M. and E.W. Moore. 1934. TIME AND RATE OF SLUDGE DIGESTION AND THEIR VARIATION WITH TEMPERATURE. Sewage Works Journal 6:3-13.

Fair, G.M. and E.W. Moore. 1932. HEAT AND ENERGY RELATIONS IN THE DIGESTION OF SEWAGE SOLIDS Sewage Works Jour. 4: 242.

Golueke, C.G. 1958. TEMPERATURE EFFECTS ON ANAEROBIC DIGESTION OF RAW SEWAGE SLUDGE. Sewage Works Jour. 30: 1225-1232.

Gosh, S., J.R. Conrad and Donald L. Klass 1975. "Anaerobic Acidogenesis ot Wastewater Sludge," Journal WPCF, January 1975, 47(1):30-47.

Gotaas, H.B. et al. 1954. ALGAL-BACTERIAL SYMBIOSIS IN SEWAGE OXIDATION PONDS. 5th Progress Rept., SERL.

Gould, R.F. (ed.). 1971. ANAEROBIC BIOLOGICAL TREATMENT PROCESS. ACS No. 15, Amer. Chem. Soc., 1155 16th St., Washington D.C. 20036. 196 pp. $9.00. (A)

Heukelekian, H. and M. Berger. 1953. VALUE OF CULTURE AND ENZYME ADDITIONS IN PROMOTING DIGESTION. Sewage and Indust. Wastes, 25(11): 1259.

Isaac, P.C. 1952. THE EFFECTS OF SYNTHETIC DETERGENTS ON SEWAGE TREATMENT: A REVIEW OF THE LITERATURE. New England Sewage and Industrial Wastes Assn.

Jeris, J.S. and P.L. McCarty. 1965. THE BIOCHEMISTRY OF METHANE FERMENTATION USING C^{14} TRACERS. Jour. WPCF., 37(2): 178-192. 70% of methane produced in anaerobic digestion comes from acetic acids.

Kinugasa, Y. et al. 1968. ANAEROBIC DIGESTION OF FECES WASTE BY ADDITION OF ENZYME PREPARATIONS. Suido Kenkyusho Hokoku, 5(1): 68-72, Japan.

Kotz, J.P. et al. 1969. ANAEROBIC DIGESTION. Water Research, 3: 459-493.

Lawrence, A.W. and P.L. McCarty. 1969. KINETICS OF METHANE FERMENTATION IN ANAEROBIC TREATMENT. Journal of the WPCF, 41 (2): R1-R17.

McCarty, P.L. 1964. ANAEROBIC WASTE TREATMENT FUNDAMENTALS:
I. "Chemistry and Microbiology." II. "Environmental Requirements and Control."
III. "Toxic Materials and Their Control." IV. "Process Design." Public Works 95:
(9): 107-112; (10): 123-126; (11): 91-94; (12): 95-99. Great semi-technical
introduction.

McCarthy, P.L., J. M. Gossett, J. B. Healy, W. F. Owen, D. C. Stuckey, and
Lily Young 1976. HEAT TREATMENT OF REFUSE FOR INCREASING
ANAEROBIC BIODEGRADABILITY, Report NSF/RANN/SE/AER-74-17940-
A01/PR/76/2. Dept. of Civil Eng., Stanford Univ., Stanford, CA 94305.

McKinney, R.E. 1962. MICROBIOLOGY FOR SANITARY ENGINEERS. McGraw-
Hill, New York.

Nelson, G. et al. 1939. EFFECT OF TEMPERATURE OF DIGESTION, CHEMICAL
COMPOSITION AND SIZE OF PARTICLES ON PRODUCTION OF FUEL GAS
FROM FARM WASTES. Jour. of Agric. Research (USDA). 58(4): 273-287.

Pohland, F.G. and S. Gosh, 1974. "Developments in Anaerobic Treatment
Processes," BIOTECHNOL. AND BIOENG. SYM. 2:107-129, John Wiley and
Son. NY, NY.

Sawyer, C.N. and P.L. McCarty. 1967. CHEMISTRY FOR SANITARY ENGINEERS.
2nd ed., McGraw-Hill, N. Y.

Schaezler, D.J., W.H. McHarg, and A.W. Bush, 1971. "Effect of Growth Rate on
the Transcient Responses of Batch and Continuous Microbial Cultures,"
BIOTECH. and BIOENG. Sym. No. 2, 107-129, John Wiley and Son, NY, NY

Stanwick, J.D. and M. Foulkes. 1971. INHIBITION OF ANAEROBIC DIGESTION
OF SEWAGE SLUDGE BY CHLORINATED HYDROCARBONS. Water Pollution
Control 70(1).

Sludge: Analysis

Anderson, Myron S. 1956. COMPARATIVE ANALYSES OF SEWAGE SLUDGES.
Sewage and Ind. Wastes 28: 132-135.

Bender, D.F. et al. (ed.). 1973. PHYSICAL, CHEMICAL AND MICROBIOLOGICAL
METHODS OF SOLID WASTE TESTING. EPA, NTIS. $6.75. Analysis of com-
post and sludge.

Coker, E.G. 1966. THE VALUE OF LIQUID DIGESTED SEWAGE SLUDGE. J. Agri.
Sci., Camb. 67:91-97.

Moore, E.W. et al. 1950. SIMPLIFIED METHOD FOR ANALYSIS OF B.O.D. DATA.
Sewage and Industrial Wastes, 22(10).

Vieiti, D. 1971. LABORATORY PROCEDURES: ANALYSIS FOR WASTEWATER
TREATMENT PLANT OPERATORS. EPA/NTIS, $4.85.

Vlamis, J. and D. Williams. 1961. TEST OF SEWAGE SLUDGE FOR FERTILITY
AND TOXICITY IN SOILS. Compost Science 2(1): 26-30.

WATER POLLUTION CONTROL FEDERATION, WPCF No. 18: Simplified
Laboratory Procedures, WCPF, 2626 Pennsylvania Ave., N.W., Washington D.C.
20037.

Sludge: Gardening and Farming

AGRICULTURAL BENEFITS AND ENVIRONMENTAL CHANGES RESULTING
FROM THE USE OF DIGESTED SEWAGE SLUDGE ON FIELD CROPS. 1971.
EPA, GPO. 62 pp. Results of experiments with Metropolitan Sanitary District,
Chicago.

Allen, J. 1973. SEWAGE FARMING. Environment, 15(3): 36-41. (C) Popular
overview.

Anderson, M.S. 1955. SEWAGE SLUDGE FOR SOIL IMPROVEMENT. USDA,
Circular 972. GPO.

Glathe, H., and A. Makawi. 1963. THE EFFECT OF SEWAGE SLUDGE ON SOILS
AND MICRO-ORGANISMS. Soils and Fert. 26: 273

Hinesly, T. et al. 1971. AGRICULTURAL BENEFITS AND ENVIRONMENTAL
CHANGES FROM THE USE OF DIGESTED SEWAGE SLUDGE ON FIELD
CROPS. EPA, GPO. $0.65.

INCORPORATION OF SEWAGE SLUDGE IN SOIL TO MAXIMIZE BENEFITS
AND MINIMIZE HAZARDS TO THE ENVIRONMENT. USDA. 1970. Agricultural Research Station, Beltsville, Md. Most extensive and current USDA research
on sludge farming.

Laura, R.D. and M.A. Idnani. 1972. MINERALIZATION OF NITROGEN IN
MANURES MADE FROM SPENT-SLURRY. Soil Biol. Biochem. 4:239-243.
Liquid slurry superior to sun-dried slurry and farm compost.

Law, J. 1968. AGRICULTURAL UTILIZATION OF SEWAGE EFFLUENT AND
SLUDGE. An Annotated Bibliography. WPCF, GPO. 89 pp.

Shreir, Franz. 1950. PROBLEMS IN SEWAGE FARMING. Abstracts, Sewage & Indust. Wastes, 25: 241. Spraying onto grazing areas is hygienically and biologically the best means of sewage utilization.

Van Kleeck, W. Leroy. 1958. DO'S AND DON'TS OF USING SLUDGE FOR SOIL
CONDITIONING AND FERTILIZING. Wastes Engr. 29: 256-257.

Wylie, J.C. 1955. FERTILITY FROM TOWN WASTES. Faber & Faber, London.
244 pp. (B) (OOP) Methods of sewage sludge disposal in agricultural country
in Scotland.

Sludge: Algae Systems; Growth, Harvesting, Digestion

Foree, E.G., and P.L. McCarty. 1970. ANAEROBIC DECOMPOSITION OF ALGAE.
Environmental Science and Tech., 4: 842-849.

Golueke, C. et al. 1957. ANAEROBIC DIGESTION OF ALGAE. Appl. Microbiol.
5: 47-55.

Golueke, C.G. and W.J. Oswald. 1964. HARVESTING AND PROCESSING SEWAGE-
GROWN PLANKTONIC ALGAE. SERL, Rept. 64-8. Also Journal, WPCF, April
1965. Handy description of process.

Lawlor, R. 1974. ALGAE RESEARCH IN AUROVILLE. Alternative Sources
of Energy, 16:2-9. Research in India on simple algae systems.

Oswald, W. 1960. LIGHT CONVERSION EFFICIENCY OF ALGAE GROWN IN
SEWAGE. J. Sanit. Engin. Div., ASCE. 86: 71-94.

Oswald, W.J. and H.B. Gotaas. 1957. PHOTOSYNTHESIS IN SEWAGE TREAT-
MENT. Trans. ASCE, 122: 73-105. The paper that started people rethinking the
potentials of algae production.

Po, C. 1973. PRODUCTION AND USE OF METHANE FROM ANIMAL WASTES
IN TAIWAN. In: Proc. Internat. Biomass Energy Conf., Winnipeg. See Reviews.
Simple description of a small sludge-algae operation.

TRANSACTIONS OF THE SEMINAR ON ALGAE AND METROPOLITAN WASTES.
April 27-29, 1960. U.S. Public Health Service, Robert A. Taft Sanitary Engineering
Center, Cincinnati, Ohio.

Water Pollution Control Federation. 1969. SLUDGE DEWATERING. Manual of
Practice, 20. WPCF.

Sludge: Algae Systems; For Methane Power

Golueke, C. and W. Oswald. 1963. POWER FROM SOLAR ENERGY VIA ALGAE-
POWERED METHANE. Solar Energy, 7(3): 86-92.

Golueke, C.G. and W.J. Oswald. 1959. BIOLOGICAL CONVERSION OF LIGHT
ENERGY TO THE CHEMICAL ENERGY OF METHANE. Appl. Microbiol.,
7: 219-227.

Oswald, W.J. and C.G. Golueke. 1964. SOLAR POWER VIA A BOTANICAL PRO-
CESS. Mech. Engin., Feb. pp. 40-43.

Oswald, W.J. and C.G. Golueke. 1960. BIOLOGICAL TRANSFORMATION OF
SOLAR ENERGY. Adv. Appl. Microbiol. 2: 223-262.

Sludge: Algae Systems; Algae as Livestock Feed

Cook, B. et al. 1963. THE PROTEIN QUALITY OF WASTE-GROWN GREEN
ALGAE. Jour. Nutrition, 81: 23.

Dugan, G. et al. 1970. PHOTOSYNTHETIC RECLAMATION OF AGRICULTURAL
WASTES. SERL Rept., 70-1. (A) Classic study of chicken, digester, algae, feed
systems.

Dugan, G. et al. 1972. RECYCLING SYSTEM FOR POULTRY WASTES. JWPFC,
44(3): 432-440. Popular account of 1970 study above.

Gran, C.R. and N. Klein. 1957. SEWAGE-GROWN ALGAE AS A FEEDSTUFF FOR
CHICKS. SERL.

Sludge: Hydroponics

Bridwell, R. 1972. HYDROPONIC GARDENING. Woodbridge Press Publishing Co.,
Santa Barbara, Calif. 224 pp. (C)

Douglas, J.S. 1973. BEGINNER'S GUIDE TO HYDROPONICS. Drake Publishers,
Inc., N.Y. 156 pp. (C)

Douglas, J.S. 1959. HYDROPONICS: THE BENGAL SYSTEM. Oxford Univ. Press.
(C) A simple method making use of low cost materials.

Iby, H. 1966. EVALUATING ADAPTABILITY OF PASTURE GRASSES TO HYDRO-
PONIC CULTURE AND THEIR ABILITY TO ACT AS CHEMICAL FILTERS. In:
Farm Animal Wastes. Symposium, 1966. Beltsville, Md.

Law, J.P. 1969. NUTRIENT REMOVAL FROM ENRICHED WASTE EFFLUENT
BY THE HYDROPONIC CULTURE OF COOL SEASON GRASSES. Federal
Water Quality Admin., GPO.

Saunby, T. 1970. SOILLESS CULTURE. Transatlantic Arts, Inc., Levittown, N.Y.
104 pp. (C)

Sludge/Organic Hydroponics

DeKorne, J.B. 1975. THE SURVIVAL GREENHOUSE, pp. 80-100. Walden
Foundation, P.O. Box 5, El Rito, New Mexico 87530. 165 pp, $7.50.

Institute for Local Self-Reliance 1976. HYDROPONIC PACKET. ILSR, 1717
18th St., N.W., Washington, D.C. 20009. 10 pp, $1.25.

Smith, M. 1977. ORGANIC HYDROPONICS: A SIMPLE SOLUTION. Mother
Earth News. 44:106-107.

Wolverton, B.C. and R.C. McDonald 1976. "Water Hyacinths, *Eichhornia Crassipes*
(Mart.) Solms, A Renewable Source of Energy," Proceedings: Capturing the Sun
through Bioconversion, Wash. Ctr. for Metro. Studies, 1717 Mass. Ave. N.W.,
Wash. D.C. 20036.

Digester Privies

Foster, D.H. and R.A. Engelbrecht. 1972. MICROBIAL HAZARDS IN DISPOSING
OF WASTEWATER ON SOIL. In: Recycling Treated Municipal Wastewater and
Sludge Through Forest and Cropland. See Reviews. Most complete recent
review.

Gainey, P.L. and T.H. Lord. 1952. MICROBIOLOGY OF WATER AND SEWAGE.
Prentice-Hall, N.Y. Good section on Coliform bacteria.

Gotaas, Harold. 1956. MANURE AND NIGHT SOIL DIGESTERS FOR METHANE
RECOVERY ON FARMS AND IN VILLAGES. In: Composting (H. Gotaas)
World Health Organization, Geneva.

Gram, E.B. 1943. EFFECTS OF VARIOUS TREATMENT PROCESSES ON PATH-
OGEN SURVIVAL. Sew. Wrks. Jour. 15:

Nishihara, S. 1935. DIGESTION OF HUMAN FECAL MATTER WITH pH ADJUST-
MENT OF AIR CONTROL. Sewage Wrks. Journal, 7(5): 798-809.

Wagner, E.G. and J.N. Lanoix. 1958. EXCRETA DISPOSAL FOR RURAL AREAS
AND SMALL COMMUNITIES. World Health Organization, Geneva. 187 pp.
$5.00. Many designs for privies . . . models for privy digesters.

Wiley, B. and S.C. Westerberg. 1969. SURVIVAL OF HUMAN PATHOGENS IN
COMPOSTED SEWAGE. Appl. Microbiol. 18(6): 994-1001.

ALCOHOL

ALCOHOL AS A BIOFUEL: COMING AROUND AGAIN??
by Richard Merrill and Tom Aston
PERSPECTIVE

For most people alcohol means liquor. But in its pure form alcohol is also a source of energy. It can be used for heating, cooking, lighting and as a motor fuel. We are not totally convinced that the production of alcohol is really practical for individuals or small groups seeking energy self-sufficiency. For one thing, the biggest drawback to alcohol as a fuel is that you can also drink it,[*] and the legal red tape controlling its production is almost insurmountable for small operations. For another thing it is becoming more and more difficult to justify using plants for anything except food and fodder. You can produce alcohol from organic wastes, but this requires a technology that is generally impractical for families, groups, or even cottage industries. As a final blow, sugar and materials for building stills (especially copper) are becoming harder to get . . . even the moonshiners are shutting down. In any event we only want to sum up briefly some information concerning alcohol as a biofuel. And in spite of the legal and material difficulties, alcohol is a beautiful, high energy, clean burning and totally renewable liquid fuel. Ideally it makes real sense.

		Fermentation			Synthetic		
		Sulfite			Ethyl	Direct	
Year	Molasses	Liquors	Grain	Other	Sulfate	Hydration	Total
1934	72.5	--	5.5	2.6	6.4	--	86.9
1944	113.1	--	313.6	33.6	59.9	--	519.9
1954	36.8	2.7	4.8	1.2	126.3	27.4	199.2
1964	0.7	4.1	43.9	16.3	232.7	51.7	341.1

TABLE I ETHYL ALCOHOL PRODUCTION IN THE U.S., MILLION GALLONS (190 PROOF). FROM MONICK, 1969 (REF. 1).

There are two kinds of common alcohols: ethyl alcohol (ethanol or "grain alcohol" . . . the stuff you drink), and methyl alcohol (methanol or "wood alcohol"), a toxic alcohol familiar as a heating fuel for *sterno*. At the turn of the century, before cheap crude oil became plentiful, both ethanol and methanol were seriously considered as motor fuels in the U.S. and Europe. Today almost all non-beverage alcohol is produced synthetically from natural gas or ethylene and used as industrial chemicals rather than as a fuel source. Beverage ethanol is distilled commercially from grains and fruits and its production is strictly controlled at all levels by Federal

* A general rule of thumb: IF YOU PRODUCE ETHANOL FOR POWER, DON'T DRINK IT!! Unless you are really experienced or on top of it, some of the byproducts of distillation, especially methanol, can kill you. Also alcohol is an organic solvent and will pick up most toxic organic substances that happen to be in the still or condenser.

regulations. For industry, ethanol is denatured (laced with toxic materials) and, like methanol, it is used in plastics, solvents, drugs and food processing, etc.

In short, both ethanol and methanol can be produced in a variety of ways: from the organic compounds of fossil fuels or the sugars of natural plant products (Fig. 1). Only certain processes, though, qualify alcohol as a biofuel: ethanol, from the distillation of plant sugars; and methanol, from the destructive distillation of wood and heat conversion[1,2] of refuse, or from methane gas obtained by anaerobic digestion.

Generally speaking, methanol production, by whatever means, is a high-tech process and usually impractical for small operations. This leaves us with the traditional method of making ethanol . . . fermenting and distilling certain plants. As an example, consider the making of ethanol from a starch crop. This will give us a handle on the terminology and basic processes involved.

First the plants are finely ground and heated in water to form a gelatine or *mash*. The starch in the mash is then converted to sugar by malt (sprouted grain) which contains an enzyme *diastase* designed especially for this task.[**] Yeast is then added to the mash. The yeast contains another enzyme that *ferments* the sugar produced by the malt into ethyl alcohol.

The next step is to separate or *distill* the alcohol from the mash (mostly water). Fortunately alcohol boils into vapors at a lower temperature ($77°C$; $171°F$) than water ($100°C$; $212°F$). So by heating the mash to slightly over $77°C$, alcohol vapors can escape and *condense* elsewhere on a cool surface as a more or less pure liquid alcohol. The purity of the alcohol produced depends on how efficiently the fermentation, distillation and condensation processes have converted sugar to alcohol, and separated alcohol from water. In fact the history of alcohol technology can be described in terms of fancier and fancier devices designed to make alcohol as pure as possible . . . without water or other byproducts. Many of these early devices can serve as experimental or even production models today.

ALCOHOL HISTORY

The earliest forms of alcohol were simple fermented beverages. As early as 4000-6000 B.C., the art of making crude beers and wines was flourishing in the Middle-East. By 2000 B.C., production breweries were well established in Egypt.

The Chinese were probably the first to distill alcohol directly from a fermented (rice) liquor around 800 B.C. About the earliest known apparatus used for distillation was the *Tibetan Still* (Fig. 2a), a clay pot in which various ferments were heated to drive off alcohol vapors. The vapors rose and condensed on a saucer-cover filled with cool water. Another pot (*receiver*) suspended inside the still collected the dripping alcohol which was then used for liquors. A major advance was made by the Peruvians who

** Enzymes do not enter into a chemical reaction directly. They merely speed up the reaction by virtue of their chemical structure.

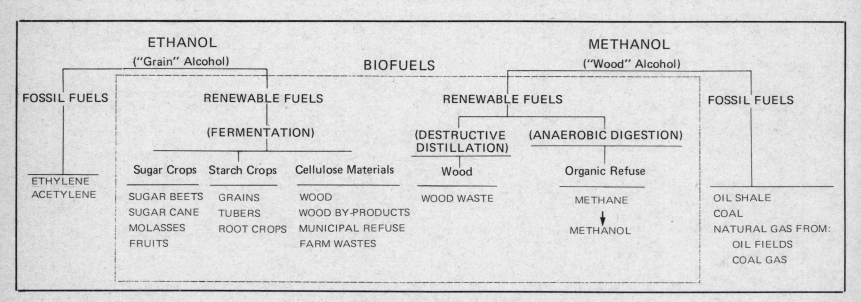

FIG. 1 PROCESSES FOR PRODUCING ALCOHOL

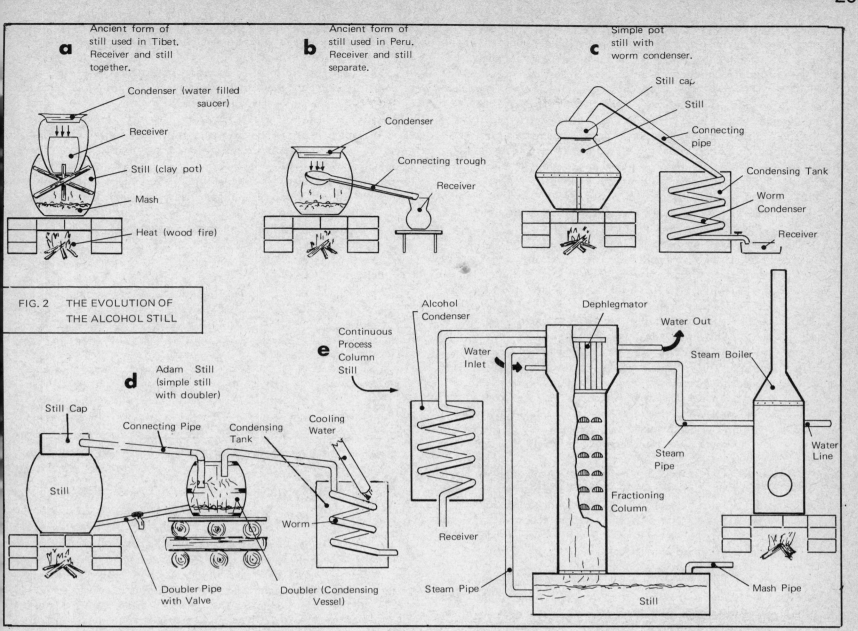

a. Ancient form of still used in Tibet. Receiver and still together.
- Condenser (water filled saucer)
- Receiver
- Still (clay pot)
- Mash
- Heat (wood fire)

b. Ancient form of still used in Peru. Receiver and still separate.
- Condenser
- Connecting trough
- Receiver

c. Simple pot still with worm condenser.
- Still cap
- Still
- Connecting pipe
- Condensing Tank
- Worm Condenser
- Receiver

FIG. 2 THE EVOLUTION OF THE ALCOHOL STILL

d. Adam Still (simple still with doubler)
- Still Cap
- Connecting Pipe
- Condensing Tank
- Cooling Water
- Still
- Worm
- Doubler Pipe with Valve
- Doubler (Condensing Vessel)

e. Continuous Process Column Still
- Alcohol Condenser
- Water Inlet
- Dephlegmator
- Water Out
- Steam Boiler
- Steam Pipe
- Water Line
- Fractioning Column
- Receiver
- Steam Pipe
- Still
- Mash Pipe

separated the receiver from the still by a connecting trough (Fig. 2b). They were thus able to reclaim more alcohol by increasing the space of the still. The *Tahitian Still*, probably also of B.C. vintage, was similar to the Peruvian still except that a pipe, rather than an open trough, was used to run alcohol from the condenser to the receiver. This prevented evaporation and increased yields.

By 500 A.D. distillation technology had advanced to the point where relatively pure forms of alcohol were being used in cosmetics, perfumes and medicines. During the "Dark" Ages, knowledge of alcohol production spread through Europe, and by 1300, Italian monasteries were distilling alcohol from wine as "aqua-vitae." *Simple Pot Stills* (Fig. 2c) were used throughout Renaissance Europe. Many of them used a new kind of condenser called a *worm*. This was a coiled metal tube that led from the still to a container holding cool water. The worm increased the area of condensation, the purity and quantity of alcohol, and separated the condenser from the receiver once and for all.

From the 18th century to the beginning of this century, major discoveries about the chemistry and technology of distillation made it possible to produce ethanol cheaply from a variety of organic materials. Although our main interest is ethanol production as the most popular means of making fuel alcohol, it is worth noting some of the important finds in general alcohol technology over the last 175 years.[2-6]

a. 1801 *Adam Still:* First still to be used on an industrial scale. Used a condensing vessel (Fig. 2d). Vessel was kept at 78°C to condense as much water as possible. Extra ferment was also kept in the vessel. Thus vapors from the still were not only better dehydrated, but mixed with fresh vapors from the vessel.

b. 1819 *Acid Hydrolysis:* Discovery that wood wastes heated with sulfuric acid yielded a product that contained sugars suitable for fermentation.

c. 1820's *Continuous Still:* Allowed mash to be loaded and alcohol received without shutting still down. The savings in time and energy was revolutionary. With the simple still the weight of fuel consumed was nearly 3 times that of the alcohol produced. The continuous still reduced this to one quarter.

d. 1823 *Methyl Alcohol:* Discovery of a new alcohol produced by condensing gases from burning wood *(Destructive Distillation).*

e. 1826 *Column Still:* Alcohol distilled by blowing steam through mash spread in layers over perforated plates in a column (rectifier). The column still was the first to produce nearly pure alcohol (94-96%) on a commercial basis.

The development of the *Continuous Process Column Still* (Fig. 2e) made possible for the first time the manufacture of ethanol in large enough quantities and at low enough prices for it to be available as a fuel by the 1830's. All subsequent advances in distillation technology were essentially variations on this theme.

f. 1828 *Synthetic Ethanol:* Synthesis of ethanol from ethylene gas dissolved in sulfuric acid.

g. 1830's *Alcohol Lamps:* Ethanol mixed with turpentine, coal tar or naptha and used as a light source. Ethanol begins replacing whale oil as a popular fuel.

h. 1900 *Commercial Wood Alcohol:* Industrial manufacturing of ethanol using acid hydrolysis.

In recent history, public interest in alcohol as a fuel has changed with the pattern of our war-prosperity-depression cycles: the development of distillation technology during wars, absence of public interest during times of prosperity and renewed interest during hard times. In this sense, alcohol is not much different than other energy sources described in the *Primer.*

Other things besides economic cycles have had an influence on the production of ethanol as an alternative fuel source. For one thing ethanol is not only a fuel but also a beverage. Since 1861 there have been a series of taxes and laws regulating its production. In general these laws have been so complex and arbitrary that they have all but prohibited the individual or small group from legally producing ethanol fuel in any practical sense. For another thing, since the 1900's alcohol has had to take a back seat to oil as a liquid fuel source. Until recently oil has been cheap, predictable and easily available, while the raw materials for ethanol production (grains, molasses, etc.) have been subject to the vagaries of weather, pest outbreaks and foreign policies. Except during wars and depressions, there has never been any real incentive to produce and use ethanol (or any alcohol) as a fuel. These trends are reflected in other historical events:

i. 1861   First *Federal Tax* on alcohol production (mostly beverage tax). Taxes plus rising prices from Civil War made alcohol too expensive to compete with kerosene.

j. 1890's   Rising taxes on ethanol limited its use as a fuel. Methyl alcohol (from destructive distillation) became a substitute for most manufacturers. Early *Heating Devices* used a Bunsen burner supplied with carborated alcohol from a pressurized reservoir. Alcohol stoves appeared.

k. 1907   *Denatured Alcohol:* U.S. tax laws changed, excluding denatured alcohol (95% ethanol, 5% toxic additives).

l. 1900-1910   Determined effort to expand use of denatured alcohol as all-purpose fuel. First major study[7] concluded that alcohol gave greater horsepower, higher compression, same mileage but lower thermal efficiency than gasoline. However, competition from petroleum continued to make alcohol non-competitive.

m. 1923   War-surplus alcohol mixed with gasoline (20/80 blend) and sold in some cities. Small corn crop of 1924 discontinued supplies.

n. 1920's-1930   European countries cut off from petroleum supplies during the war began post-war use of ethanol as a fuel. Sweden began production of ethanol from *Sulfite Liquors*, used 25/75 blend as motor fuel. Germany began alcohol economy from agricultural crops: 2/3 potatoes, 1/3 molasses, grains and sulfite liquors. Improved foreign balance of payments and reduced need for energy imports.

o. 1920's-1930's   United States industrial alcohol made mostly from molasses. Used in synthetic rubber, anti-freeze, munitions, cosmetics, etc. During 30's farm production of alcohol was slowly replaced by synthetic ethanol.

p. 1930's   Midwest research advocated use of alcohol in gasoline. Nebraska legislature passed a law that would refund $0.02/gallon to motorists who bought the blend. Petroleum industry campaign knocked out the plan.

q. 1940's   WWII put demands on alcohol industry, increased grains output of alcohol to 41% of total production by 1944. Many whiskey distilleries commandeered into war production and molasses-using plants converted to grains. Food shortages caused raw materials to change from corn to wheat. After the war, grain shortages caused potatoes to be used on a large scale for the first time.

r. 1950-present   Early 50's marked the end of farm products as a large source for alcohol. Since then, synthetic alcohol from fossil fuels has dominated the alcohol markets.

Today the specter of an "energy crisis" and fuel shortages have once again prompted interest in alcohol as a possible energy source. Special attention is being given to alcohol as an additive to gasoline in place of lead, since it would reduce pollutants and improve octane and mileage. Proponents have suggested that alcohol be produced from domestic grains,[8-11] natural gas from oil fields and coal gasification plants,[12] organic refuse[13] and other "waste" materials such as garbage, sawdust, scraps from logging, lignin discarded by paper mills, coal tailings, mine washings, etc.

Properties of Alcohol

Property	Ethanol	Methanol
Boiling point	172°F	148°F
Freezing point	-205°F	-176°F
Btu/Lb.	11,500-12,800	8200-9000
Btu/Gal.	90-100,000	65-70,000
Lb/Gal.	6.58	6.59
Specific gravity (at 20°C, water = 1.0)	.7905	.7924
Octane	99	106

TABLE II   BASIC PROPERTIES OF METHYL AND ETHYL ALCOHOL

ETHANOL PRODUCTION FROM FARM PRODUCTS

Basics

Ethanol can either be produced from gases such as ethylene, acetylene and carbon monoxide found in various fossil fuels or from carbohydrates (sugars, starches, cellulose)** found in various farm products. In the case of carbohydrates, the traditional methods of fermenting sugar to alcohol are the only ways to produce ethanol from renewable sources. Either the raw material contains sugars that can be fermented directly, or it contains the more complex carbohydrates (starches and cellulose) that must be broken down first to simple sugars before the yeast can do its work (Fig. 3).

** *Carbohydrates are compounds of carbon, hydrogen and oxygen. The hydrogen and oxygen are present in the same proportion as they are in water (H_2O), hence the name carbo-hydrate.*

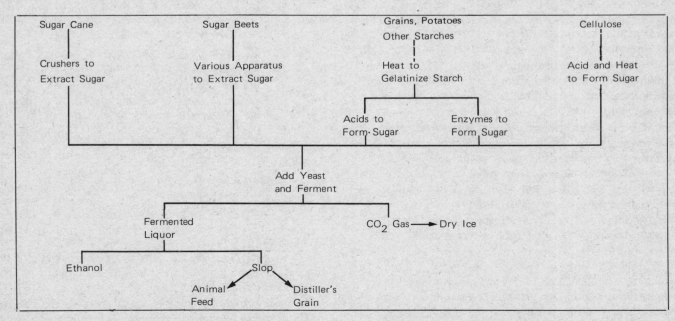

FIG. 3   BASIC STEPS IN ETHANOL PRODUCTION FROM FARM PRODUCTS.

The simplest sugars contain only one molecule (monosaccharides); these include dextrose (glucose or grape sugar) and fructose (fruit sugar). Still other sugars such as sucrose (table, beet or cane sugar), maltose (malt sugar) and lactose (milk sugar) contain two simple sugar molecules (disaccharides). The starches are even more complex carbohydrates made up of many simple sugar molecules held together by splitting out water molecules. When starch is treated with certain chemicals (acids and enzymes) the molecule takes on water and breaks apart into the original sugar molecules . . . a reaction called *hydrolysis*. Even more complex carbohydrates are the fibrous compounds called *cellulose*. Examples of cellulose are cotton, wood, plant stalks, paper and most organic refuse.

All of these carbohydrates can be converted to ethanol, but the more complex the compound, the more difficult and therefore the more costly and impractical the process becomes for small operations.

Ethanol From Sugar Crops

The simplest method of producing ethanol is fermentation and distillation of sugar beets, sugar cane, molasses and fruits. These crops contain high amounts of sugar which can be converted directly to alcohol by various yeast enzymes:

$$C_{12}H_{22}O_{11} + H_2O \xrightarrow[\text{Enzyme}]{\text{Maltase}} 2C_6H_{12}O_6$$

maltose + water   dextrose

or

$$C_{12}H_{22}O_{11} + H_2O \xrightarrow[\text{Enzyme}]{\text{Invertase}} 2C_6H_{12}O_6$$

sucrose + water   dextrose

then

$$C_6H_{12}O_6 \xrightarrow[\text{Enzyme}]{\text{Zymase}} 2C_2H_2OH + 2CO_2$$

dextrose   ethanol + carbon dioxide

There are, of course, many more considerations here than just the simple addition of yeast to a slurry of mashed fruits or sugar beets. Alcohol fermentation, like anaerobic digestion, is a biological process and requires special attention through a series of living reactions. For example:

1) Pure yeast will not produce alcohol from pure sugar, because a pure sugar solution does not contain the substances required to nourish the yeast. There are usually enough extra nutrients in the raw materials besides sugars to properly feed the yeast and set them on their way, but this may not always be the case. Special malts or other supplements may have to be added.

2) Like most biological processes, fermentation is actually a series of chemical steps . . . each one providing the basic materials for the others to follow in a *succession* of reactions. During the first period, the yeast grows in the mash at a temperature of about 82°F. The growth of the yeast becomes apparent by the gradual bubbling of carbon dioxide up through the mash. When alcohol is produced to an extent of 5%, the growth of the yeast stops. During the second period, carbon dioxide bubbles freely, and the sugar is converted to alcohol rapidly. The temperature at this point should not exceed 81°F. Water is often added here to dilute the mixture and permit further growth of the yeast. Finally there is a lessening of carbon dioxide formation and a lowering of temperature (77°-80°F).

There are really dozens of tricks to successful fermentation. Useful "how-to" descriptions of preparing mashes from molasses and sugar beets are found in books by Simmond[2] and Wright.[6] As far as fermenting fruits are concerned, probably the best sources of information are the many popular wine-making books now available on the market (e.g.[14-18]).

Ethanol From Starch Crops

Producing alcohol from potatoes, grains, etc., requires that the starch first be converted to sugar before it can be fermented. Depending on the crop, the conversion of starch to sugar is accomplished in three steps:

(1) Corn, rice and potatoes are steamed under pressure in a *converter* to gelatinize the starch and form the mash. (2) The enzyme diastase then converts the mash into a sugar solution. Diastase is found abundantly in all sprouted grains (malt), especially barley. This is why grains are often sprouted before they are fermented. (3) Yeast enzymes then convert the sugars into alcohol as described above.

An important step in the starch/ethanol processes is the preparation of the malt extract. An excellent description of do-it-yourself malting techniques is given by Wright.[6] The preparation of malts from commercial products is described in various popular books on home brewing (e.g.[19]). Good descriptions of preparing grain mashes are given by Carr,[3] Simmonds[2] and Wright.[6] Wright, Simmonds and Wente[20] deal well with methods of preparing potato mashes.

One starch crop that deserves real attention as a raw material for ethanol production is the Jerusalem artichoke or wild sunflower *(Helianthus tuberosus)*. Inulin, its major carbohydrate, is easily broken down into fermentable sugars. The plant is also easy to grow (a "weed" in many places), hardy and productive; its stalk can be burned for fuel and its flower serves as a food and habitat for many natural insect enemies of crop pests.

Ethanol From Cellulose Materials

It has been known since 1819 that when wood wastes are heated with a strong acid, a solution of fermentable sugars is obtained (Acid Hydrolysis). The idea was first put into commercial use in 1900. At about the same time it was discovered that sulfuric acid converted wood cellulose into dextrose when the mixture was heated to about 130°C, under pressure. The solution was then neutralized to allow the growth of microbes, inoculated with yeast and fermented in the usual manner.

Various processes since then have yielded from 16-40 gallons of 100% ethanol per ton of dry wood.[21, 22, 23] Other materials used in acid hydrolysis have been sulfite liquors (by-product of the wood pulp industry), sawdust, straw, cornstalks, nutshells and urban refuse.[13] However, acid hydrolysis doesn't seem to be a practical way to make fuel ethanol at home or on any small scale. The need for a system of recycling strong acids (to keep expenses and pollution down) and high pressure containers seems to make the process quite impractical. Perhaps a small acid hydrolysis plant could be operated in connection with a cottage industry mill or wood craft center . . . but it doesn't seem likely. There just are no simple ways to get alcohol from wood.

Ethanol Yields From Farm Products

It is impossible to obtain a complete conversion of starch into sugar and alcohol. Even under ideal conditions the yield of alcohol is, for various reasons, appreciably lower than the quantity theoretically possible if all starch or sugar could be converted into ethanol and carbon dioxide. The theoretical quantities are:

From simple sugars (dextrose etc.)	51.1%
From complex sugars (maltose, sucrose etc.)	53.8%
From starch	56.8%

Taking starch as the starting point, about 12-20% is generally lost in one way or another by the time the fermentation is finished. From 6-10% of the starch remains unfermented. During fermentation 2-3% glycerol is formed; a part of the sugar is used up in providing food for the yeast, and a little alcohol is lost by evaporation. In a badly conducted operation the alcohol produced may represent not more than 72% of the theoretical yield. A good yield is 6 gallons of absolute alcohol per 100 lbs. of starch, as compared with a theoretical yield of 7.2 gallons. Anything less than 5½ gallons per 100 lbs. of starch indicates something is wrong.

Table III gives some general information about the yield of ethanol from various farm products. The numbers are general and should not be taken too seriously. For one thing the crop yields are from 1970 USDA and California crop statistics. They are thus averages for large commercial farming operations. For another thing the estimates of ethanol yield per ton of raw material are in terms of the theoretical yield described above.

Raw Material (Lbs/Bushel)	Average U.S. Crop Yield (1970) (Tons/Acre)	Fermentable Content	Yield of 99.5 Ethanol			Residual Solids (Lbs/Ton)
			Gallon Per Bushel	Gallon Per Ton	Gallon Per Acre	
SUGAR CROPS						
MAIN CROPS						
Sugar Cane	41.1	11.0%	--	15	623	--
Molasses	237 gallons	51.0%	--	70	97	--
Sugar Beet	19.1	16.0%	--	22	420	100
FRUIT CROPS						
Apples (48)	14.4	11.0%	0.4	14	207	40
Apricots	5.2	10.4%	--	14	71	46
Grapes	7.9	11.5%	--	15	119	76
Peaches (48)	11.3	8.7%	0.3	12	130	34
Pears (50)	6.8	8.9%	0.3	11	78	58
Prunes, dry	2.3	55.0%	--	72	166	152
Raisins, dry	2.4	62.0%	--	81	195	166
STARCH CROPS						
GRAINS						
Barley (48)	0.9	54.3%	1.9	79	71	646
Corn (56)	2.0	57.8%	2.4	84	168	446
Sorghum (56)	1.4	54.5%	2.2	80	111	488
Oats (32)	0.8	43.6%	1.0	64	51	846
Rice (45)	2.3	54.6%	1.8	80	183	520
Rye (56)	0.7	54.0%	2.2	79	55	542
Wheat (60)	0.9	58.6%	2.6	85	77	538
TUBERS AND ROOTS						
Carrots (55)	11.8	7.5%	0.3	10	116	76
Jerusalem Artichokes (60)	9.0	15.2%	0.6	20	180	104
Potatoes (60)	11.5	15.6	0.7	23	263	76
Sweetpotatoes (55)	5.2	23.3%	0.9	34	178	92

TABLE III   ETHANOL YIELDS FROM VARIOUS FARM PRODUCTS. CROP YIELDS (EXCEPT FRUIT) ARE FROM 1970 USDA AGRICULTURAL STATISTICS. FRUIT YIELDS ARE FOR 1970, CALIFORNIA AGRICULTURAL EXTENSION SERVICE.

Methanol From Wood Wastes

Wood, sawdust, farm wastes and urban refuse can be treated with destructive distillation to produce methanol. Like acid hydrolysis, the process has little practical value for small operations since it requires such high-tech equipment. For information, it is described here only briefly.

First the raw material is dried thoroughly and placed in ovens that are connected to condensers. The dried material is then heated to about 400-500°F.

Gases from the organic matter are then condensed and run into tanks. Charcoal is left in the ovens after heating. This can be used as a fuel also. While in the tanks, the distillate separates into a tar (which settles to the bottom) and pyroligneous acid (which forms an upper layer). Pyroligneous acid consists of acetic acid, 4% methanol, acetone and allyl alcohol. Pyroligneous acid is distilled to remove all tar. Water from the settled tar is then combined with the distilled acid. Pyroligneous acid is then neutralized with lime leaving acetic acid and alcohol. Acetic acid and alcohol are separated by further distillation leaving acetic liquors and crude alcohol liquors. The latter are distilled again in column stills to produce crude methyl alcohol.

Methanol From Bio-Gas

Last year over 1 billion gallons of methanol were produced, mostly from natural gas (methane). However methane gas can also be made from wastes and other renewable resources in anaerobic digesters. Since methane is so inconvenient to compress and carry in vehicles, it has been suggested that the conversion of organic wastes to methanol instead of methane would make this fuel more practical.[12] The conversion process involves the synthesis of carbon monoxide (CO) and hydrogen (H_2) by partial oxidation of the methane with water. In the original high pressure process, pressures of 300 atmospheres at 200°C were used in the presence of a zinc-chromium oxide catalyst. In 1968 a low pressure process was developed that used 50 atmospheres at 250°C. A number of other processes have also been described,[24] but again, they are not what you would call backyard or farmstead type technology.

LEGAL PROBLEMS

There is no Federal law that prohibits the manufacture of ethanol. Yet no one can make, sell or store ethanol without a permit from the Bureau of Alcohol, Tobacco and Firearms (ATF). The ATF used to be a division of the IRS, but since July 1972 it has been a separate bureau of the Federal Government. The ATF now monitors and regulates all production of ethanol in this country. If you write the ATF office in Washington (see reviews) you can get various pamphlets that outline the mind-boggling regulations concerning ethanol production.

As far as we can tell there are no explicit regulations covering the production of ethanol *as an energy source*; it would probably be a matter of interpretation, in or out of the courts. However, at this point in time, you will almost surely be subject to the same taxes and laws regardless of whether you are using your ethanol as a beverage or a fuel. This is an area where the right pressures might force legal changes. If you want to deal with the legal hassles, the thing to do is write the ATF office in Washington and ask for the address of your nearest ATF Regional office. Then start making contacts inside. Use words like "experimental," "small-scale" and "research." If you are still forced to oblige the laws, here are a few of the things you will have to do: (a) Register with the Regional Commissioner of the ATF and apply for a permit to erect a still. Failure to do so . . . $1000 fine or 1 year in jail. (b) Buy a bond making the Government the beneficiary. This is supposed to protect the Government against tax losses. (c) Build a special accommodation for your still, which may not be located in a private dwelling, shed, backyard or boat. (d) Provide an office for the exclusive use of the government inspector including a toilet and office furniture subject to the approval of regional commissioner. (e) Keep accurate records of all production and dispensation of your ethanol. (f) Pay an alcohol tax of $10-20 per 100-proof gallon, $55 per still and $22 per condenser every year. The list of regulations goes on. At this stage it would seem technically and economically absurd to produce ethanol for energy unless the laws are either changed . . . or bent . . . or ignored.

APPLICATIONS

Ethanol Plants on the Farm

Suggestions have been made that the United States follow the example of some European countries during WWII in developing an alcohol industry on the farm, consisting of very small plants, privately owned and operated, and adapted for using local raw materials. Indeed, the industrial-alcohol law of 1906 provided specifically for such a program and the production of ethanol from farm products has been part of several programs to boost rural economies in the U.S.[8-11,25] Clearly the present method of making ethanol from ethylene will be short-lived because of impending oil shortages. Instead we might return to the old industrial process of fermenting grains. One suggestion[26] describes a rural industry of 40 acre farms . . . 1.5 million of them . . . that combines the fermentation of ethanol from corn with the raising of pigs and cattle. The corn acreage would be fertilized with urban sewage effluent, the stills fueled with corn stalks, the labor provided by dispossessed urban dwellers and the byproduct slop would be fed back to the livestock.

It is hard to imagine how such a scheme would work given *today's* conditions. For one thing, the U.S. is going to come under increasing pressure to use its domestic grain supplies for foreign barter (oil and balance of trade), and diplomacy (starving Third World Countries). In light of this, the wide spread use of food crops to produce an additive for gasoline seems absurd. For another thing, the availability of 1.5 million, forty-acre farms makes little sense in a society where land values and land monopolies are increasing.

Finally, any scheme that promotes the increased production of livestock has to contend with the gross inefficiency of our already over-meated diets. Perhaps the creation of small rural fermentation plants using farm *wastes* and within hauling distance to a central refinery might work, especially if the ethanol were used to supplement fuels used in rural areas. But speaking realistically, if alcohol is to become part of our fuel economy, it will probably be through the manufacture of alcohol from organic refuse in large still-factories near urban centers, or some other such operation.[12,13]

Alcohol Motor Fuel

The idea of using alcohol as a motor fuel has been around since the early 1900's[27-32] and has been suggested recently as an offshoot to the current "energy crisis."[8-10,33] Ethyl alcohol from farm products and blended with gasoline has been used as a motor fuel for years in several foreign countries (10-30% alcohol in gasoline). In the U.S. absolute limits to a gasoline economy are now obvious and anything that stretches the supplies will probably be used. Besides, alcohol is a good anti-knock additive for gasoline. High compression engines are more efficient than low-compression engines (more power and better mileage), but they have the tendency to "knock." Tetra-ethyl lead is now the most popular anti-knock compound, but the dangers of excessive lead in the environment are now well known.

There are real difficulties to overcome before alcohols can be used as undiluted or unblended motor fuels. They have high latent heats and thus don't vaporize easily in an ordinary carburetor. Also their low vapor pressure makes alcohol engines hard to start in the cold.

A comparison of gasoline and alcohol fuels at the same compression ratio and overall efficiency shows that the relative fuel consumption for a given power output is considerably higher for alcohol than gasoline:

Gasoline (grade 1)	100
Ethanol	161
Methanol	222

Alcohol for Lighting and Heating

Alcohol appliances (lamps, stoves and heaters) were in widespread use during the late 19th and early 20th century.[2,5,34] It is hard to find information about them these days. We located alcohol lamps in marine supply houses . . . but that was about it.

References Cited

[1]Monick, John. 1969. ALCOHOLS: THEIR CHEMISTRY, PROPERTIES, AND MANUFACTURE. Reinhold, N.Y.

[2]Simmonds, Charles. 1919. ALCOHOL. ITS PRODUCTION, PROPERTIES, CHEMISTRY AND INDUSTRIAL APPLICATIONS. Macmillan and Co., Ltd., London. 574 pp. OOP.

[3]Carr, Jess. 1972. THE SECOND OLDEST PROFESSION: AN INFORMAL HISTORY OF MOONSHINING IN AMERICA. Prentice-Hall, N.J. See reviews.

[4]Brachvogel, John. 1907. INDUSTRIAL ALCOHOL. Munn and Co., New York. OOP.

[5]Herrick, Rufus. 1907. DENATURED INDUSTRIAL ALCOHOL. John Wiley and Sons, London. OOP.

[6]Wright, F. B. 1906. DISTILLATION OF ALCOHOL. Spon and Chamberlain, New York. 194 pp. OOP. Describes processes of malting, mashing, fermenting and distilling alcohol from grain, beets, potatoes, molasses. Also alcoholometry and the de-naturing of alcohol for use in farm engines, autos, heating and lighting, etc. One of the best nitty-gritty, how-to books. Look real hard in libraries.

[7]U.S. Geological Service. 1909. COMMERCIAL DEDUCTIONS FROM COMPARISONS OF GASOLINE AND ALCOHOL ON INTERNAL COMBUSTION ENGINES. U.S. Geological Survey Bulletin, No. 392.

[8]Miller, D. L. 1970. INDUSTRIAL ALCOHOL FROM WHEAT. In: Report of 6th National Conference on Wheat Utilization Research, Oakland, California. November 1969. Agricultural Research Service, ARS 74-54.

[9]Miller, D. L. 1972. FUEL ALCOHOL FROM WHEAT. In: Proceedings of the 7th National Conference on Wheat Utilization Research, Manhattan, Kansas. November 1971. U.S. Agricultural Research Service, ARS-NC-1. Especially good bibliography on alcohol as a source of motor fuels.

[10]Scheller, W. A. 1974. AGRICULTURAL ALCOHOL IN AUTOMOTIVE FUEL . . .GASOHOL. In: Proceedings of the 8th National Conference on Wheat Utilization Research, Denver, Colorado. October 1973. U.S. Agricultural Research Service, ARS W-19.

[11]Acres Magazine. 1972. FARMER'S ALCOHOL. Acres, USA, June 1972. Box 1456, Kansas City, MO 64141.

[12]Reed, T. B. and R. M. Lerner. 1973. METHANOL: A VERSATILE FUEL FOR IMMEDIATE USE. Science, 182 (4119). Good description of methanol production schemes and overview of methanol as part of basic fuel economy. Good bibliography especially on performance of methanol in engines.

[13]Converse, A. O. et al. 1973. ACID HYDROLYSIS OF CELLULOSE IN REFUSE TO SUGAR AND ITS FERMENTATION TO ALCOHOL. EPA/NTIS PB 221-239.

[14]Mitchell, J. R. 1969. SCIENTIFIC WINEMAKING MADE EASY. British Book Center, New York. 246 pp.

[15]Mahan, Paul E. 1973. SMOKEY MOUNTAIN WINES AND HOW TO MAKE THEM. Arco Publishing Co., New York. 114 pp.

[16]Hardwick, Homer. 1954. WINEMAKING AT HOME. Simon & Schuster, Inc., New York. 218 pp.

[17]Carey, Mary. 1973. STEP BY STEP WINEMAKING. Golden Press, Racine, WI. 64 pp.

[18]Bravery, H. E. 1961. SUCCESSFUL WINE MAKING AT HOME. Arc Books, New York. 151 pp.

[19]Bravery, H. E. 1965. HOME BREWING WITHOUT FAILURE. Arc Books, New York. 159 pp.

[20]Wente, A. O. and L. Tolman. 1910. POTATO CULLS AS A SOURCE OF INDUSTRIAL ALCOHOL. U.S. Department of Agriculture Farmers' Bulletin 410, 44 pp.

[21]Kressman, F. W. 1922. THE MANUFACTURE OF ETHYL ALCOHOL FROM WOOD WASTES. U.S. Department of Agriculture Bulletin 983.

[22]Wise, L. 1949. WOOD CHEMISTRY. Reinhold, New York.

[23]Hawley, Lee F. 1923. WOOD DISTILLATION. Chemical Catalogue Department, Book Department, New York.

[24]Stanford Research Institute. 1973. CHEMICAL ECONOMICS HANDBOOK. SRI, Menlo Park, California.

[25]POWER ALCOHOL AND FARM RELIEF. 1936. "The Deserted Village," No. 3, Department of Chemistry, Iowa State College, 1936.

[26]Editorial. 1974. FEEDBACK: THE ULTIMATE PANACEA. New Scientist. May 16, 1974.

[27]Shepard, G. 1940. POWER ALCOHOL FROM FARM PRODUCTS: ITS CHEMISTRY, ENGINEERING AND ECONOMICS. Iowa Corn Research Institute, Iowa Agricultural Experimental Station, 1(3), June 1938.

[28]Jacobs, P. B. 1938. MOTOR FUEL FROM FARM PRODUCTS. USDA, Miscellaneous Publication, No. 327. Annotates early studies of alcohol as a motor fuel. Two and four cycle stationary engines, pure alcohol, mixtures and performance tests. Good overview of production techniques up to the times. Excellent bibliography.

[29]Ogston, A. R. 1937. ALCOHOL MOTOR FUELS. Journal Institute Petroleum Technology 23:506-523.

[30]Monier-Williams, G. W. 1921. POWER ALCOHOL: ITS PRODUCTION AND UTILIZATION. Henry Frowde, Hodder and Stoughton, London. OOP. Workings of internal combustion engines and applications for alcohol.

[31]Lucke, C. E. and S. M. Woodward. 1907. USE OF ALCOHOL AND GASOLINE IN FARM ENGINES. USDA, Farmers' Bulletin, 227.

[32]Davidson, J. B. and M. L. King. 1907. COMPARATIVE VALUES OF ALCOHOL AND GASOLINE FOR LIGHT AND POWER. Iowa Agriculture Experimental Station Bulletin 93.

[33]METHYL FUEL COULD PROVIDE MOTOR FUEL. 1973. Chemical and Engineering News. September 17, 1973.

[34]M'Intosh. J. 1923. INDUSTRIAL ALCOHOL. Scott, Greenwood & Son. OOP. Valuable for its discussion of the inner workings of early alcohol appliances.

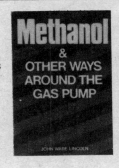

WOOD

WOOD FOR FUEL
Thomas Gage

Wood is one of the most abundant and useful natural resources on this planet. Trees are a renewable resource which today cover over 30% of the earth's land surface. Even though trees take 50 to 100 years to reach maturity, we can use this valuable resource in perpetuity if we grow and harvest trees with care and planning.

The United States also has 30% of its land area in forests (750 million acres). Of this, roughly 2/3 is "commercial" forests capable of producing at least 20 cubic feet of wood per acre per year. The remaining 250 million acres are in either natural or other public parks or are on marginal land.[1] Most of the wood produced or imported in the United States is consumed as sawlogs for lumber, plywood and pulp for paper products. In 1970, the U.S. consumed about 250 million tons of wood. In that same year the total U.S. energy consumption was 68 quads (one quad equals 10^{15} or one quadrillion BTU's). Roughly one quad or 1.5% was provided by the burning of wood (0.8 quad in the burning of tree wastes by the forest industry and 0.2 quad by the burning of fuelwood).[2] This figure can be compared with 8% of Sweden's and 15% of Finland's energy needs being supplied by wood.

Historically wood has played a much larger role in meeting U.S. energy needs. In 1860 the use of wood in the U.S. reached a peak at 4 tons per person per year. The current "energy crisis" has led to a renewed interest in the use of wood for fuel. Indications are that domestically more people are buying wood stoves and heaters and that industrially more thought is being given to the use of fuelwood in boilers for electrical generating plants.

Large scale use of wood fuel by industries is not new. Many forest product companies, including Weyerhaeuser in Tacoma, Washington, burn waste wood from their milling operations to produce some of the electricity needed to power their machinery. Recently, municipalities and small companies have begun to use waste wood from local sawmills and tree pruning operations in their boilers in place of oil or natural gas.[3] The Forest Service estimates that 30% of the waste material in urban garbage dumps is wood and paper products, which suggest that further use of this wasted renewable resource is possible.

The most ambitious plan for the use of wood fuel is the "energy plantation." These are large tracts of land devoted to the production of trees for use in a nearby electrical generating plant. The forest is cleared in a cyclical rotation of perhaps every ten to fifteen years. This rotation cropping method allows the newly planted trees to grow to a size worth harvesting for their fuel value. The primary question is how large the forest plantation would have to be.

One estimate concluded that a 400 Megawatt (400,000 kilowatts per hour) plant would only require 350 square miles of densely planted forest, with the "whole tree" being harvested.[4] By contrast, a Canadian study reported that a 400 Megawatt power plant would require more than 800 square miles of forest land, with "above-ground" harvesting methods (leaving between 20% and 35% of the wood in the forest in the form of roots, leaves, branches and tops).[5]

There is, however, a major problem with the energy plantation concept. The removal of whole trees is likely to deplete the soil nutrients by as much as eight times the amount they are now depleted with removal of the stem and larger diameter branches.[6] Some soils regain nutrients quickly and can offset these withdrawals. But others, particularly coarse sands or leachable tropical soils, can have considerably reduced yields after two or three rotations. In these cases, leaving branches and tree tops that often contain much greater nutrient content than stemwood is a means of recycling needed nutrients. Another possible alternative is to return the ash, containing many of the nutrients, to the soil. For other problems with the energy plantation concept see pg. 160.

One of the most interesting plans for the large scale use of wood for fuel involves the owners of small woodlots. A study by the Wood Energy Institute in Vermont polled small woodlot owners in a fifty-mile radius of a proposed electrical plant site.[7] Twenty-year contracts were offered for sustained yield timber management and continuous purchase of fuelwood. While the plan has not yet come to fruition, it offers a means of upgrading the quality of vast numbers of forests that are currently not being managed, while providing both fuel for energy and better trees for useful products.

Large scale use of wood can produce a lot of needed energy, but can never totally satisfy our energy needs. There are simply too many of us and we want far more energy than our parents and grandparents did. But individuals can use wood to save fuel costs. More and more people are purchasing wood heaters or stoves to supplement their use of fossil fuels for heating. Many people buy their wood from amateur woodsmen who have gone into the cordwood business. But some people are beginning to consider managing woodlots for the production of fuelwood for their own consumption or for sale to others. The next article outlines the techniques of small woodlot management for the production of fuelwood.

(See References Cited for this article on pg. 226)

WOODLOT MANAGEMENT
Thomas Gage

There are just under 2½ million small forest owners in the U.S. with an average holding of 131 acres.[8] About half of these owners are small farmers who also own woodlots. Obviously most people are not in a position to manage woodlots for their energy supplies. Nevertheless, methods for small woodlot management are outlined here for three basic reasons: (1) To provide information for those who can utilize woodlots. (2) To explain the fundamental techniques of husbanding our valuable wood resources. If we are going to use wood for fuel, we must take care of existing supplies with ecological wisdom and a sense of future generations. To extract wood from forests and woodlots without proper management techniques will only lead us to the deforested conditions that now exist in Nepal, South Korea, India, Tanzania, etc. In the developing countries, wood, not fossil fuels, is the source of the "energy crisis."[9, 10] (3) To encourage even the casual woodsman or weekend forest scrounger to understand the implications of his or her activity on the forest environment, which is not just a source of wood. It is also an important ecosystem that nurtures wildlife, builds topsoil, and contributes to the stability of the local watershed. Good woodlot management is not difficult, but it involves much more than just cutting any old tree down, or allowing them to grow willy nilly. It involves techniques that provide for the continuous regeneration of the forest in harmony with the natural environment while at the same time providing a favorable yield of wood at a reasonable cost.

Most woodlots can be managed for fuelwood _and_ for timber. Even the smallest woodlots will grow trees that can be sold commercially as saw timber. A combined management program can be used in which fuelwood is taken from thinnings, prunings and occasional harvests, _and_ in which saleable lumber is taken from some of the largest and best quality trees in a final harvest.

Determining Woodlot Production

Determining the amount of wood that a given woodlot will produce in a year is not a simple task. Very few rules of thumb exist because climates, species, stand densities and management techniques differ. Both the total amount of wood present and the annual growth rate should be calculated

The first step is taking an inventory of the existing wood. Foresters commonly refer to this as a "Timber Cruise." It involves measuring: (1) the density of the stand, that is, how many trees are present in a given area (acre), (2) the overall cubic footage of the standing timber and (3) the number and types of different species present. In all areas of the country, federal, state and county foresters will help in this task at the landowner's request. They usually offer their services and sources of information for free and can save a woodlot owner a lot of time and money. In some cases the woodlot in question will have already been surveyed from the air. A study of these photographs by someone trained in the art will give rough estimates as to the density of the stand. Even with aerial photography, however, field site inspections will be necessary to determine the species present and the quality of these trees.

In general, a timber cruise should determine the height and width of all the trees by species. Usually diameter is measured in terms of the diameter at breast height (DBH), or 4½ feet off the ground. Height is normally measured in terms of marketable timber for making into lumber. If fuelwood is the object then the limbs and the smaller diameter top growth can be calculated also. Table I illustrates the number of trees needed at various diameters to yield a cord of wood including the bark. More exact tables and descriptions of how to take a timber cruise can be found in references at the end of this article.[11-15]

Diameter breast height (inches)	Volume of tree in cords when total height is—									Percent of total height used[1]
	20 feet	30 feet	40 feet	50 feet	60 feet	70 feet	80 feet	90 feet	100 feet	
	Cords	Cords	Cords	Cords	Cords	Cords	Cords	Cords	Cords	
4	0.005	0.007	0.010	0.012	0.015					33
6	.016	.022	.030	.037	.045	.054				50
8		.043	.059	.074	.090	.106	.121	0.137		59
10		.070	.095	.120	.145	.171	.197	.222	0.248	62
12			.138	.175	.212	.250	.287	.324	.361	65
14			.192	.243	.295	.347	.40	.45	.50	68
16			.257	.325	.394	.46	.53	.60	.67	70
18			.328	.42	.50	.59	.68	.77	.86	73
20			.41	.52	.63	.74	.85	.96	1.07	75
22				.64	.77	.91	1.04	1.18	1.31	77
24				.76	.93	1.09	1.26	1.42	1.58	78
26					.90	1.09	1.27	1.47	1.65	1.85	80
28					1.04	1.27	1.49	1.71	1.93	2.15	81
30					1.21	1.47	1.72	1.98	2.24	2.49	83

TABLE I. GROSS VOLUME IN ROUGH CORDS OF THE AVAILABLE LUMBER (OR FUELWOOD) IN A TREE IN TERMS OF ITS "DIAMETER AT BREAST HEIGHT" (DBH OR 4.5 FEET) AND TOTAL HEIGHT. Tops and small branches are not included but can obviously be used for fuelwood and this will increase the net useable volume.

[1] ... includes a stump height of 1 foot. If the "percentage of total height used" is different than that listed, the estimated volume in cords will be correspondingly different. From Ref. 11.

What information can one gain from a timber cruise? Again, the evaluation will be different for each site and species but general data can give us guidelines. In New York State, average healthy forest lands have between 30 and 40 cords per acre of standing hardwoods and between 40 and 50 cords per acre of standing softwoods. Stands in the Northwest, most notably those planted by the forests industries, can have as much as 60 cords per acre in dense, well managed sites. Both these figures represent high ends of the scale and it would be reasonable to assume that densities of 10 to 15 cords per acre are the norm in many areas of the U.S., and that a stand containing 20 cords should be considered respectable.

Determining the amount of annual tree growth is the next step. Most woodlots are too small to warrant the time and expense involved in calculating the actual tree growth. However, average net annual growth per acre for all types of forest ownership and all regions of the U.S.A. is approximately 38 cubic feet or almost 1/2 of one solid 80 cubic foot cord of wood.[8] Included in this average are Forest Service Site Class I lands (3% of commercial forest lands), which represent an annual growth of at least 165 cubic feet per acre per year. Another study showed a rough average for several sites of 3 dry tons per acre per year (about one solid 80 ft³ cord), and a few measurements have shown maximum growth rates of 5 cords and more under intensive management, in good site locations, with particular species.[5]

Caution must be taken in evaluating these studies. With the exception of the national average they all represent growth in managed woodlands. They do, however, lead to the conclusion that a one cord (80 cubic feet solid) per acre per year yield can be expected in a well-managed stand. The PRESIDENT'S ADVISORY PANEL ON TIMBER AND THE ENVIRONMENT stated that with improved management techniques a yield of 80 cubic feet per acre per year nationally is well within reach.[1] Certainly greater yields are possible in some situations, but in general, growth of one cord per acre per year should be considered to be an extremely healthy forest. We can thus assume that a homeowner who needs ten cords of wood per year could harvest this wood from ten acres of land without taking any timber from the land other than an amount equivalent to the previous year's growth.

Tree Growth

The sun is the fundamental ingredient of photosynthesis which causes plants to grow. Trees and other plants absorb solar radiation and transform some of it into the stored chemical energy of their bodies (see pg. 155). Within a moderate range of light levels the rate of photosynthesis is proportional to the amount of light that a tree receives. However, in very intense light, trees may become "burned" (sunscald) and at very low levels of light trees may become stunted. The ability of a tree to grow and develop in the shade of other trees is known as its tolerance (Table II). Tolerance is an extremely important concept in forest management because it indicates the amount of light different species need for maximum growth and survival. For example, sugar maple is a very tolerant species which can withstand the effects of very little sunlight for many years.[20] If liberated from this shade, the sugar maple will usually grow to its normal full height. Aspen, on the other hand, requires good sunlight at all stages of growth and will usually die if shade is too prevalent.

Forest soils are another important factor affecting tree growth; they are also an integral part of the nutrient cycle of a tree. Soils store nutrients from decaying plants, animal droppings, weathered rock material and dead animals as well as storing water after rains. Bacteria, fungi, nematodes and earthworms help decay this matter into a rich source of natural fertilizer. Trees also play a critical role in their own nutrition. When a tree dies its complex nutrients are returned to the soil and broken down into simpler forms by soil microbes. Thus, when a tree is removed from a woodlot, nutrients vital to other trees are also removed. Though it is hard to know the exact nutrient loss caused by removing trees, presumably prolonged harvesting will cause a nutrient deficiency after two or three rotations, expecially with clearcutting. This nutrient deficiency can be lessened by leaving the new growth on the site after removal of the main stem. The tops, leaves, twigs and roots contain 20% to 35% of a tree's nutrients. The leaves and other new growth of twigs contain a higher percentage by weight of nutrients than the stem and larger branches. Hence, a woodlot owner who wishes to harvest all the branches above one inch in diameter for kindling and leave the rest behind can probably do so with little fear of appreciably depleting the soil nutrients.

Figure 1 shows the most important factors which affect tree growth. Many of these factors can be altered by man to one extent or another but the costs and ecological considerations should bear heavily on decisions to improve woodlots.

Silviculture

Silviculture is the "theory and practice of controlling forest establishment, composition and growth"[19]. It is the basis of forest management. Good silvicultural techniques can increase woodlot yield considerably, produce the type of species that is desired and provide an ecologically sound enviornment. Management takes time, costs money and must be weighted in proportion to its benefit, financial and otherwise.

The two commonly used management methods are: (1) even-age and (2) uneven-age. Much of the Northeastern U.S. is even-age forests. These are the second growth forests that grew after logging removed most of the

TABLE II. TOLERANCE AND SPECIFIC GRAVITY OF VARIOUS TREE SPECIES. Tolerance is the ability of a tree to grow and develop in the shade of other trees: VT—very tolerant, T—tolerant, M—medium, I—intolerant, VI—very intolerant. Specific gravities (S.G.) are based on oven-dry weight and a moisture content of 12%. The higher the specific gravity the more dense the wood is and the greater fuel value per unit volume it will have. Weight per cord is based on an 80 cubic foot solid cord and a weight of water of 62.4 pounds per cubic foot: to calculate lbs./cord multiply 80 ft^3 times specific gravity times 62.4 lbs/ft^3. After Ref. 15–19.

SOFTWOODS

SPECIES	TOLERANCE	S.G.	SPECIES	TOLERANCE	S.G.
Baldcypress	I	.46	Pine:		
Cedar:			Eastern white	M	.35
Alaska-	M	.44	Jack	VI	.43
Atlantic white-	—	.32	Loblolly	I	.51
Eastern red	—	.47	Lodgepole	M	.41
Incense-	T	.37	Longleaf	VI	.59
Northern white-	M	.31	Pitch	I	.52
Port-Orford-	T	.43	Pond	—	.56
Western red	VT	.32	Ponderosa	VI	.40
Douglas-fir:			Red	—	.46
Coast	M	.48	Sand	—	.48
Interior West	M	.50	Shortleaf	I	.51
Interior North	M	.48	Slash	I	.59
Interior South	M	.46	Spruce	—	.44
Fir:			Sugar	M	.36
Balsam	VT	.36	Virginia	I	.48
California red	T	.38	Western white	M	.38
Grand	—	.37	Redwood:		
Noble	T	.39	Old-growth	T	.40
Pacific silver	—	.43	Young-growth	T	.35
Subalpine	VT	.32	Spruce:		
White	T	.39	Black	—	.40
Hemlock:			Engelman	VT	.35
Eastern	VT	.40	Red	T	.41
Mountain	VT	.45	Sitka	T	.40
Western	VT	.45	White	T	.40
Larch, western	I	.52	Tamarack	VI	.53

HARDWOODS

SPECIES	TOLERANCE	S.G.	SPECIES	TOLERANCE	S.G.
Alder, red	—	.41	Magnolia:		
Ash:			Cucumbertree	M	.48
Black	M	.49	Southern	M	.50
Blue	M	.58	Maple:		
Green	M	.56	Bigleaf	—	.48
Oregon	M	.55	Black	—	.57
White	M	.60	Red	T	.54
Aspen:			Silver	M	.47
Bigtooth	VI	.39	Sugar	VT	.63
Quaking	VI	.38	Oak, red:		
Basswood, American	T	.37	Black	M	.61
Beech, American	VT	.64	Cherrybark	—	.68
Birch:			Laurel	—	.63
Paper	I	.55	Northern red	M	.63
Sweet	M	.65	Pin	M	.63
Yellow	M	.62	Scarlet	M	.67
Butternut	—	.38	Southern red	M	.59
Cherry, black	I	.50	Water	M	.63
Chestnut, American	M	.43	Willow	—	.69
Cottonwood:			Oak, white:		
Balsam poplar	—	.34	Bur	M	.64
Black	—	.35	Chestnut	M	.66
Eastern	VI	.40	Live	M	.88
Elm:			Overcup	—	.63
American	M	.50	Post	—	.67
Cedar	—	—	Swamp chestnut	—	.67
Rock	T	.63	Swampy white	I	.72
Slippery	I	.53	White	—	.68
Hackberry	M	.53	Sassafras	T	.46
Hickory, pecan:			Sweetgum	—	.52
Butternut	T	.66	Sycamore, American	M	.49
Nutmeg	—	.60	Tanoak	—	.64*
Pecan	—	.66	Tupelo:		
Water	—	.62	Black	I	.50
Hickory, true:			Swamp	—	.50
Mockernut	T	.72	Water	—	.50
Pignut	M	.75	Walnut	I	.55
Shagbark	M	.72	Willow, black	I	.39
Shellbark	—	.69	Yellow-poplar	I	.42
Honeylocust	M	.66*			
Locust, black	VI	.69			*Estimates

trees in this area a century or so ago. Even-age forests may contain trees of slightly different ages and will certainly contain trees of different diameters and heights since trees of the same age will grow at different rates. Moreover, even-age forests can be converted over time into uneven-age forests and vice versa.

Compartment managing is a specialized technique that can be applied to either uneven-age or even-age timber stands. It involves division of the wood lot into easily manageable "compartments." For example, a small woodlot owner with ten acres can divide his land into eight compartments. Division would be based on natural and man-made boundaries such as streams, roads, fences and concentrations of tree species. Each year the woodlot owner thins and prunes one of the compartments until at the end of eight years the whole woodlot would have received care and the process would start over again. Compartment size should be based on the amount of labor available each year as well as the size of the woodlot. Most compartments will not require more intensive management than one thorough thinning and pruning every eight to twelve years.

Fig. 1 FACTORS INFLUENCING TREE GROWTH

Uneven-age management is often used where the woodlot owner intends to harvest only a few trees each year. This makes it an excellent method for use in woodlots where fuelwood is the object of production. Harvesting only a few trees each year is known as "selective cutting" or "sustained yield." It allows for the annual growth increment referred to earlier to be harvested each year while still leaving the bulk of the woodlot intact.

Even-age management is often used in forest industry lands especially in the West where high yields are desired. Short rotation cropping, in which trees are only allowed to grow to their peak growth potential before harvesting is often used in conjunction with even-age management. Short rotations can be used with unven-age management but are much more difficult to employ. Both even-age management and short rotations have received a lot of attention in connection with energy plantations. They are generally considered the most efficient high yield management techniques. Clearcutting is the harvesting plan used in even-age stands. Dreaded by environmentalists for many obvious reasons, clearcutting can be used effectively with minimal environmental harm under certain conditions. More on this later.

Thinning

Thinning a woodlot is the real work of good silvicultural techniques. Many foresters refer to this as Timber Stand Improvement or TSI. It can be likened to weeding a garden and should include removal of the low-quality plants called cull, thus giving the high-quality plants room to grow.

Thinnings should remove undesirable species: insect, wind, disease or lightning damaged trees. In young stands this is referred to as a cleaning and will also include the removal of shrubs and vines which are choking the growth of young saplings. In older stands, especially uneven-age stands, liberation cuttings are sometimes necessary. Here young understory growth is set free from the oppression of larger overstory trees. Removing the overstory trees to give the younger trees more light and less competition for soil nutrients will often result in remarkable growth especially if the understory trees are tolerant species (Table II). Liberation cuttings, however, must be done with care so that the younger growth is not damaged. In some cases girdling a tree will be preferable to its actual removal. Girdling involves cutting a ring around the tree at breast height with a hatchet. Kerosene or other chemicals can be applied to the open wood, but as long as the cut reaches well into the sapwood the tree will die after a few years.

When timber stands are young, TSI will yield little slash material but as larger trees are thinned from the woodlot they can be used for fuelwood. This is the time when dual purpose management of producing fuelwood and marketable crop trees for sale as lumber must be carefully considered. Crop trees are selected on the basis of good height and diameter growth and overall healthy appearance — free of rot, insects and crooked stems. These trees are then marked with bright-colored paint and thinning goes on around them. A young timber stand may have as many as 6,000 stems or more per acre. But as the trees reach maturity of approximately 20 inches D.B.H. usually no more than 100 will remain. Thus during a 60 or more year period thinnings of cull trees and competing trees will provide fuelwood while at the same time allowing more space for growth of the crop trees. The crop trees can eventually be sold for lumber. In an uneven-age woodlot different compartments can be managed to reach maturity at different times. This will allow for a constant supply of fuelwood.

In most woodlots it is preferable to keep trees fairly closely spaced until they reach middle age, referred to as the "young timber stage." Close spacing will allow for natural pruning to occur, in which trees shed their lower limbs due to the insufficient light which reaches them. Prohibiting growth of lower limbs in this manner will focus more of the tree's growth energy into the main stem.

Harvesting

The method of harvesting used will usually have been determined when the management technique is developed. In the case of small woodlots managed for fuelwood some sort of sustained yield uneven-age management will probably be preferable and harvesting will simply involve thinning the woodlot periodically of the cull, crooked, damaged and competing trees that will not make good lumber. The so-called final harvest will take place when the crop trees are full grown and ready to be cut down.

Certain biological, ecological and economic considerations must be taken into account when a harvesting method is used. Factors such as tolerance of the species present and age of the trees as well as the amount of seed the trees are giving each year are important. Since most trees do not bear seed until they are at least 20 years old, a harvesting method that plans on natural seedings from the trees left standing will have to take this into account. Moreover, harvesting may be a wise investment when the market for wood products is high. It will then make more sense to harvest and sell the valuable timber rather than wait for the planned harvest date to come due. Conversely, when the price of lumber is low it will probably be worth waiting until it increases to at least a "normal" level before selling the timber.

There are five traditional harvest and regeneration methods: (1) clear-cutting, (2) seed-tree cutting, (3) shelterwood cutting, (4) diameter limit cutting and (5) selection cutting. Clearcutting is not a new concept but it has received a lot of attention recently. Clearcutting methods were employed on most of the virgin forests of the northeast a century ago. Little thought was given to the type of species that would reproduce after the cutting had occurred—and low yield, poor species forests in much of this area exists today as a result. Clearcutting of large areas, like what is going on now in parts of the Northwest, can be extremely damaging to the environment. It is often argued that douglas fir, the species most often replanted, is an intolerant species that requires a lot of sunshine for growth. Such justification fails to consider the long-term consequences of monoculture tree planting and the immediate environmental impact. A study commissioned by the State of Vermont reports: "The most significant problems in terms of soil erosion and the consequent degradation of stream quality by heavy sedimentation does not occur because of the cutting of trees. Rather it occurs, because of the disruption of the protective soil organic layer (humus horizon) by the logging operation itself."[6]

The type of clearcutting referred to above is large area clearcutting. Two other types which are used and can be less harmful and even beneficial are strip clearcutting and patch clearcutting. They both do less environmental damage and are relatively inexpensive harvesting methods. In strip clearcutting alternate strips of land are cleared every five to ten years, allowing new growth to get started before the adjacent strip is

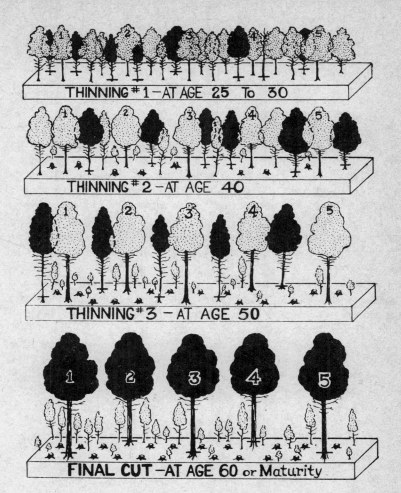

FIG. 2 STEPS IN THINNING. Before thinning, #1 crop trees (numbered) are selected on basis of good growth and health characteristics. Other trees (blackened) are then removed in periodic thinnings so that crop trees have the best possible growth environment. As the stand is thinned, reproduction begins, but growth is slow until given full sunlight. After ref. 20.

cleared. In patch clearcutting the size of the patch is the issue. Small patches can be a sensible way to harvest timber and do little damage to the soil or the surrounding standing timber. In both cases the uncut forest areas tend to hold down the erosion resulting from the exposed clearcut area. Moreover, the damage from continuous logging of a woodlot by other harvesting methods is reduced because a small area is only harvested once in a very many years.

Seed tree cutting is similar to clearcutting except that it leaves a few of the best seed-producing trees well distributed over the land area. These trees then scatter their seed to reforest the area. Seed tree cutting has many of the environmentally damaging characteristics of clearcutting. However, it does not require replanting by man and is thus favored in some areas.

Shelterwood cutting involves leaving even more trees standing than with seed tree cutting. Here the object is often the protection of the new growth from excessive sun. Shelterwood cutting leaves a relatively attractive landscape behind and also protects the soil and wildlife from overexposure to the elements.

Diameter limit cutting is often used by sawmill companies when they have contracted with a woodlot owner to remove the marketable timber. The harvest will then involve cutting every tree that is above a certain pre-specified diameter and leaving behind the smaller growth. This method lacks any silvicultural basis and is used mostly for its monetary return. It fails to remove the cull or damaged trees and may remove only the best timber thus leaving the woodlot in poor condition.

The best of the cutting methods is selection cutting. Most often used in uneven-age forests it calls for the selective cutting of undesirable trees and thus the use of good silvicultural techniques. The sustained yield concept referred to earlier goes hand in hand with selection cutting as a means of removing a portion of the trees in the woodlot according to the compartment thinning and harvesting schedule. Selective cutting is an ideal harvesting method for fuelwood production (see Fig. 2).

(See References Cited for this article on pg. 226)

FUELWOOD BASICS
Thomas Gage

Fuelwood is usually measured by the "cord." A cord is a stack of wood four feet high by four feet wide by eight feet long or 128 cubic feet (4 × 4 × 8 = 128). Because of the air spaces from stacking, however, a cord is only about 80 cubic feet of solid wood. Wood is also measured by the face cord which is a stack four feet high by eight feet long, by any length wide. Very often both standard cords and face cords will have wood cut into two foot lengths.

Most wood heaters and fireplaces are not large enough to accommodate logs greater than two feet long and many will only take an 18 inch log or shorter. Larger logs should also be split so that more surface area is open to the fire and so that wood pieces can fit inside the small openings of some wood heaters. Both cutting and splitting are time-consuming tasks that require patience and labor. Cutting fuelwood must be evaluated in terms of its economic and other rewards.

Wood is extremely heavy. This weight is a result of both the moisture content (mc) and the density of the wood, usually measured as specific gravity (sg). Moisture content is most often measured as the percent of water in the wood (see Eq. 1). A newly cut green cord of

$$\text{Moisture Content (oven-dry basis)} = \frac{\text{(original weight)} - \text{(oven-dry weight)}}{\text{oven-dry weight}} \quad \text{(Eq. 1)}$$

wood may weigh 5000 pounds but if oven-dried, to zero moisture content, it might weigh only 2000 pounds. The cord therefore has 150% moisture content. Most woods when green will have between 30% and 200% moisture content.

By knowing the specific gravity of a species of wood (from Table II pg. 216) you can calculate the weight of that wood species per cord. Specific gravity is the ratio of the weight of a given volume of wood compared with the weight of the same volume of water. Water weighs approximately 8.3 pounds per gallon and 62.4 pounds per cubic foot. Thus the weight of a cord of (say) American elm with 12% moisture content is:

80 ft³   ×   .50   ×   62.4 lbs/ft³   = 2496 lbs per cord

(solid wood)   specific gravity   weight of
volume of a   of American   water
cord   elm

The greater the specific gravity of a species of wood the greater the amount of burnable wood fibre is present. This is because the wood is denser. Dense wood, however, is heavier and more costly to transport than lighter wood species.

Fuel Value

When wood is measured by weight all species have the same fuel value of approximately 7250 Btu's per pound with zero moisture content.[21] When moisture comprises part of the weight of the log the fuel value declines. This is because water is heavy but has no fuel value itself and also because it must be "boiled off" by heat energy from the burning wood fibre. A log with 20% moisture content has approximately 5800 Btu's per pound. A green log with 50% moisture content has about 3700 Btu's per pound or 37% less energy by weight than a log with 20% moisture content.

When wood is measured by volume different species vary widely in their fuel value. Table I shows a few commonly burned species and their fuel value per cord. In general, hardwoods have more heat potential than softwoods because their specific gravity is greater. But softwoods partially make up for lower heat potential by having a greater amount of resin. Resin has a fuel value of 17,400 Btu's per pound or about twice that of wood fibre.[16] Softwoods also ignite much faster than hardwoods and are useful as kindling.

TABLE I   HEAT EQUIVALENTS OF WOOD BY SPECIES

Species (1 std. cord)[a]	Available heat of 1 cd. wood million (Btu's)[b]	Species (1 std. cord)[a]	Available heat of 1 cd. wood million (Btu's)[b]
Hickory, shagback	24.6	Ash, green	18.4
Locust, black	24.6	Pine, pitch	18.0
Ironwood (hardhack)	24.1	Sycamore, American	18.0
Apple	23.8	Ash, black	17.3
Elm, rock	23.5	Elm, American	17.2
Hickory, butternut	23.5	Maple, silver	17.0
Oak, white	22.7	Spruce, red	13.6
Beech, American	21.8	Hemlock	13.5
Birch, yellow	21.3	Willow, black	13.2
Maple, sugar	21.3	Butternut	12.8
Oak, red	21.3	Pine, red	12.8
Ash, white	20.0	Aspen (poplar)	12.5
Birch, white	18.9	Pine, white	12.1
Cherry, black	18.8	Basswood	11.7
Tamarack (eastern larch)	18.7	Fire, balsam	11.3
Maple, red	18.6		

[a] 1 standard cord = 128 cubic feet wood and air; 80 cubic feet solid wood; 20% moisture content.

[b] It is assumed that available heat of wood is oven-dry, or calorific value, minus loss due to moisture, minus loss due to water vapor formed; minus loss due to heat carried away in dry chimney gas. Stack temperature 450°F. No excess air. Efficiency of burning unit = 50 to 60%. After Ref. 22.

Moisture content does not affect the fuel value of a cord (volume) of wood as much as it does the weight of wood. This is because wood only swells a small amount when moisture is present. Wood with 20% moisture will only have 3% less energy per cord than one with zero moisture content; and wood with 50% moisture content will only have 4% less usable energy per cord than one with 20% moisture content (see Table II).

Measured by volume, therefore, dry wood will not produce much more heat than wet wood. And unless transportation costs or other factors are of concern, wood need not be dried to reduce its weight either. But dry wood does have benefits. It will burn more easily, ignite faster, stay lit longer, produce less creosote in the flue or chimney and will be easier to handle than wet wood. Therefore, if time is available, drying or "seasoning" wood is worthwhile. This means allowing the wood to lose some of its moisture while sitting outside.

A particularly good practice is to cut the cords needed for winter in the preceeding spring. The summer heat will then dry the wood. It is preferable to stack the wood so air can easily get to all the logs and to cover the stack with a tarp or place it under a shed so that it does not get wet from rain and snow. One easily built structure is a open-faced greenhouse with clear plastic sheeting. The plastic will let sunlight in and keep the rain out. However, it must be well ventilated so that moisture doesn't condense and run onto the logs. Drying time depends on the level of humidity in the air and the wetness of the wood, but seasoning will usually take no more than six months. A 15% to 20% moisture content, or more, can be expected, even after thorough air-drying.

TABLE II   COMPARISON OF FUEL VALUE BY WEIGHT AND VOLUME. (mc = MOISTURE CONTENT). 20 MILLION BTU's IS AVERAGE FUEL VALUE OF A CORD OF WOOD.

	Dry (% mc)	Seasoned (20% mc)	Green (50% mc)
fuel value per lb. (Btu)	7,250	5,800	3,700
fuel value per cord (Btu)	20.6 million	20 million	19.2 million

Heating Requirements

How much wood will you need each year? Simple rules of thumb indicate that the heating needs of an average size house can be provided by the burning of 3 to 12 cords of wood per year. Using an average of 20 million Btu's per cord from Table I this means a production of 60 million to 240 million Btu's per year. These figures must be divided by the efficiency rating of your wood heater. At an average burning efficiency of 50%, between 30 million and 120 million Btu's will be needed.

For a more specific calculation you need to know: (1) the climate where you live; (2) the size of your house and its heat retention capabilities; (3) the species of wood(s) you are burning; and (4) the efficiency of your wood heater. Climate is by far the most important consideration, and is best calculated and understood as the number of "degree days." Degree days gauge heating requirements over a year. One degree day accumulates for each degree the average temperature is below 65 degrees Fahrenheit multiplied times the time period being considered. In other words, if it is 24 degrees Fahrenheit for the entire month of January then the degree days for that month are (65 - 24) $\times$ 31 days = 1271 degree days. This is the actual number of degree days for Syracuse, New York for the month of January in an average winter. There are many published tables that list average degree days for various locations.[23] For example, in Syracuse, New York the yearly total is 6756 degree days while Bakersfield, California has only 2122 degree days and St. Louis, Missouri has 4900.

Now let's consider a specific example. Suppose we want to heat a home in Syracuse, New York using white ash. From Table I we know the heat value of white ash is 20 million Btu per cord. The reader should then refer to the SOLAR ARCHITECTURE section of this book for heat loss and retention (seasonal heat load) calculations for a building. Determining heat loss calculations for your particular house will be a helpful step in calculating the number of cords you will need for heating each year. If we assume that a seasonal heat load of 97 million Btu's has been calculated for the house, we can continue with our example in Syracuse by considering the efficiency of our wood heater. Most well-constructed heaters will have an efficiency of between 40% and 65%.[16] Using a low average of 50% will insure that our calculation is not overly optimistic. An efficiency of 50% will then double the required number of Btu's to 194 million. The required number of cords, then is:

194 million Btu's ÷ 20 million Btu/cord = 9.7 cords

To be safe we should assume about 20% error due to imprecise inputs. Furthermore, the amount of insulation in the house, the efficiency of the heater or extremes in the local climate will all affect the figures. It is logical then to spend more on insulation and chop an extra cord of wood or perhaps purchase a slightly bigger or better heater so as not to be caught short in a cold winter. The dollar, energy and comfort savings should be worth the added effort and expense. Note that if we had used Bakersfield or St. Louis as the example climate, the calculations would have resulted in a three and seven cord heating requirement respectively.

(See References Cited for this article on pg. 226)

EFFICIENT HEATING WITH WOOD
Thomas Gage and Ken Smith

In the past, most cultures have depended on wood as the primary source of fuel for cooking and heating. Even today more than one-third of the world's population uses wood as its main energy source. Unfortunately much of this wood is burned inefficiently. Even fireplaces, the traditional American heating source, have combustion-burning efficiencies of only 10% or less. Today some automatic and airtight wood heaters have efficiencies of 40% and 65% and provide much more usable heat.

Sole dependence on wood for fuel is certainly not possible for everyone. Even with more efficient heating units, limited wood resources would be rapidly depleted if we all began to use wood for fuel. But as a primary source of energy for some of us, and a secondary source for others, wood can play an important role in fulfilling our energy needs. Wood can be used as a supplement to gas, oil or electric heating, conserving non-renewable fuels and saving money. Wood can also be used as an effective back-up for solar space heating systems (burning wood when the sun doesn't shine). Wood heaters can be integrated into hot water heating systems to supplement gas, electric or solar water heating.

The following discussion is intended to introduce the reader to the efficient use of fuelwood. Those who want a more complete and comprehensive treatment of this subject will find the WOODBURNER'S ENCYCLOPEDIA (see pg. 230) an invaluable resource. It is the basis for much of the information in this article.

Heat Transfer

Radiant heat is caused by waves of infrared ("heat") radiation. Infrared is part of the light spectrum beyond the darkest red visible to the human eye. It is the only source of heat that can travel through a vacuum, and is the sole source of heat reaching us from the sun. Radiation is familiar to anyone standing close to a fireplace or stove; it's the reason one side of you gets hot while the other side remains cool. Most open fires transmit the majority of their usable heat through radiation. Wood stoves use this principle also. Black and dark colored surfaces emit more radiant heat than polished or shiny surfaces.

Heat transfer by conduction is the passage of heat from one part of a material to another part of that or some other material through the interaction of the molecular structure of the body(ies) involved. As flames touch the sides of a stove body, the molecules of the steel are warmed. This heat is transferred from the inside wall to the outside wall by the increased agitation of the molecules in the steel of the stove body. The warmth of a masonry fireplace after the fire has died out is an example of heat by conduction. Radiant heat from the fire is absorbed and stored in the stone by the conductive transfer of heat from the surface to the mass of the stone. When the heat of the fire dies, the cooler room temperature draws the heat from this mass.

The rate at which heat can be conducted through a material is known as its K-value. K-values are measured in Btu's of heat per inch thickness of material, per hour, per square foot of surface, per degree Fahrenheit (Btu-in/hr-ft^2-°F).

Each material also has a capacity for storing or retaining heat. This is referred to as "specific heat," which is given in values relative to the heat retaining capacity of water. The specific heat of water is one Btu per degree Fahrenheit per pound. Table I gives a comparison of conductivity, specific heat and density of various stove and fireplace materials.

TABLE I	THERMAL PROPERTIES OF STOVE AND FIREPLACE MATERIALS		
Material	Conductivity K = Btu-in/hr-ft^2-°F	Specific Heat BTU/lb. Water = 1.0	Density (lb./ft^2)
Cast Iron	28.0	0.12	450
Steel (1% carbon)	26.2	0.12	490
Stone (limestone)	20.0-32.0	0.22	117-175
Stone (granite)	20.0-35.0	0.20	160-190
Concrete	1.0	0.16	142-155
Cinder Block	0.4	0.18	97
Brick (building)	0.4	0.20	120
Brick (insulating)	0.6	0.20	20-50

Convection is the transfer of heat by a liquid or gas in conjunction with conductive heat transfer. Air or water, for example, heated by conduction can transfer this heat by moving to another place. In natural convection systems, the heated gas or liquid rises (being less dense) and, as it cools, falls back down. If it is heated further (say by a fire or a solar collector) it rises again and the process of natural convection occurs, resulting in a "thermosyphon" or circulation about the heat source. In forced convection systems fans or pumps are used either to speed up the

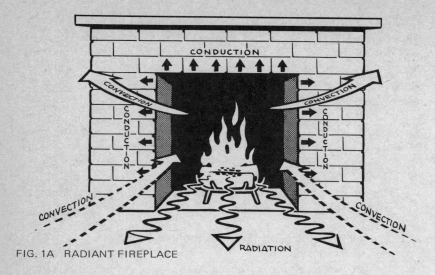

FIG. 1A RADIANT FIREPLACE

<u>RADIANT HEAT</u>: OPEN FIRES, OPEN FIREPLACES, WOOD BURNING HEATERS. PRIMARY HEAT COMES FROM RADIATION EITHER DIRECTLY FROM THE OPEN FIRE OR FROM THE HOT HEATER BODY. LESSER AMOUNTS OF HEAT COME FROM CONDUCTION THROUGH THE FIREPLACE MASONRY OR HEATER BODY AND FROM CONVECTION THROUGH HEATING OF THE AIR WHICH COMES IN CONTACT WITH THE FIRE, BRICK OR HOT STOVE BODY.

FIG. 1B RADIANT HEATER

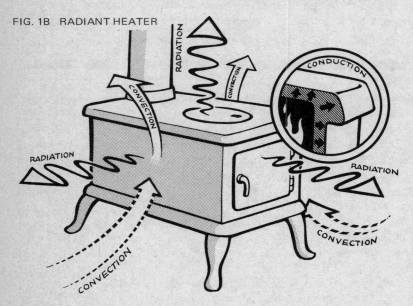

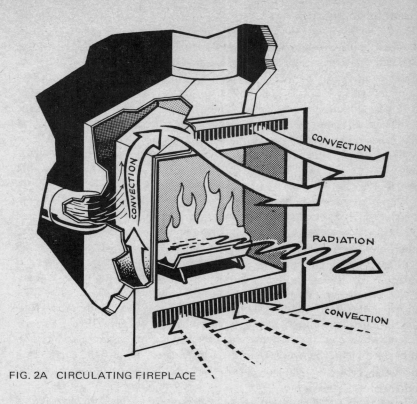

FIG. 2A CIRCULATING FIREPLACE

<u>CONVECTIVE HEAT</u>: CIRCULATING FIREPLACES AND CIRCULATING WOOD HEATERS. PRIMARY HEAT COMES FROM CONVECTION WHERE THE AIR MOVES ALONG THE HEATED SURFACE OF THE FIREPLACE OR HEATER. THE AIR CAN MOVE NATURALLY (WARM AIR RISES) OR CAN BE FORCED BY A SMALL FAN. RADIANT AND CONDUCTIVE HEAT WORK IN THE SAME WAY AS IN FIG. 1.

FIG. 2B CIRCULATING
         HEATER

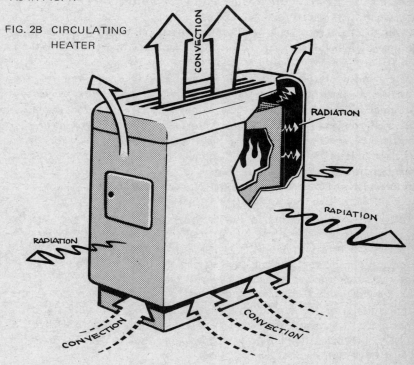

natural flow of heat or to move the heated medium counter to the natural flow (e.g., moving hot air at the top of a room down to the floor).

Fireplaces and stoves use various combinations of radiation, conduction and convection to transfer useful heat to a space. In fact, stoves and fireplaces can be put into two basic categories based on these heat transfer principles: (1) radiant heaters and (2) circulating heaters.

<u>Radiant heaters</u> rely primarily on heat from an open flame, as in a fireplace or from a radiating surface of a stove body. Radiant heat will, of course, create convective currents in a room, but for the most part, radiant heaters are limited to heating single open spaces and objects in their immediate vicinity (see Fig. 1a, b).

<u>Circulators</u> are characterized by their use of conduction-convection heat transfer. Circulating fireplaces (see Fig. 2a) usually have two metal walls with a cavity through which air is passed either by natural convection or forced air. There are accessories to convert masonry fireplaces to circulators. Circulating heaters (see Fig. 2b) have an envelope of metal around the outside of the stove body through which air is circulated by natural convection or by a forced air fan. A few circulators heat water which is then transported to radiators or to a water storage tank.

EFFICIENCY OF A WOOD BURNING UNIT

The efficiency of a wood heater depends upon its ability to convert fuelwood into heat. The complete combustion of wood is difficult because wood is a solid fuel that is chemically complex and that changes form when it burns. Also, the various densities of different wood species and the different sizes of wood pieces cause wood to burn at different rates. Moreover, wood contains carbon, hydrogen, oxygen, water, tars, and resins which all burn at different rates.

When wood burns it goes through three stages that change its chemical composition. First, it is warmed and the water (moisture content) is boiled away. In the second stage the wood fiber begins to burn, and is broken down into charcoal through the process of carbonization or pyrolysis. Many volatile gases, which are composed primarily of carbon monoxide and hydrogen are emitted. In the final stage, the charcoal and some of the volatile gases are burned. In most fires all three burning stages occur simultaneously.

Energy conversion efficiencies are complicated concepts (see pgs. 8-12). A simple type of efficiency, however, is the ratio of energy into a system to the useful energy obtained from that system. In the case of wood heaters and fireplaces, the potential energy of a cord of wood may be 20 million Btu's, but the efficiency of the heater may be only 50%. Thus the net usable energy when the cord is burned is 10 million Btu's (.5 × 20 million). The energy efficiency of a wood heater can be expressed as:

$$E = \frac{\text{usable heat output of a fireplace or heater}}{\text{total energy potential of wood}}$$

Efficiency combustion of fuelwood in a wood heater requires mixing the proper amount of oxygen with the burning fuel and the maintenance of high temperatures in the heater. This dual requirement is difficult to meet and is the main reason that efficient wood heaters are difficult to design.

Wood requires oxygen to burn. Insufficient oxygen will snuff a fire. Too much oxygen will fan a fire and cause it to burn very hotly for a short period of time, until all the fuel is consumed. When too much oxygen enters a wood heater the fire is unable to consume all the oxygen that is available to it and a great deal of wasted heat escapes up the chimney or fluepipe. This lost heat is referred to as "stack losses." Often a roaring fire is desirable at first to warm a cold house, but in general, a slow steady burn with a controlled intake of oxygen will generate more heat over a longer period of time.

On the other hand, a high temperature is necessary for efficient burning of wood because volatile gases emitted from wood only burn at high temperatures. Volatile gases produce as much as 60% of the heat potential in a piece of wood. These gases have an ignition temperature of around 1100°F, whereas wood ignites at 750°F or lower.[21] Volatiles also need air to burn, and thus sufficient oxygen must be introduced into the fire. Too much air, though, can lower the fire's temperature. So some of the better designed wood heaters introduce "primary air" to burn the wood and "secondary air" to burn the gases (Fig. 3). The use of secondary as well as primary air inlets allows sufficient oxygen in to burn the gases and wood, without lowering fire temperatures or causing "stack losses." A few manufacturers reviewed in the Wood Hardware Section utilize this double air technique.

WOOD BURNING UNITS

There are hundreds of different designs of wood burning units. Heaters, stoves, fireplaces and furnaces can all be used to heat a house with wood. But some designs are considerably more efficient than others and will require less wood during a winter heating season. In most cases these efficient heaters will also be more convenient to use and will require less attention. In general airtight heaters, automatic heaters and circulating heaters are more efficient than fireplaces and poorly fitted cast iron heaters.

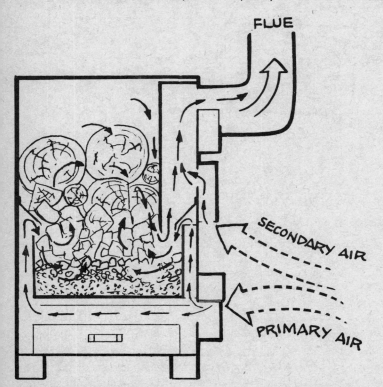

FIG. 3 RADIANT HEATER with primary and secondary air inlets for increased efficiency and burning of volatile gases. (Riteway Manufacturing Co.)

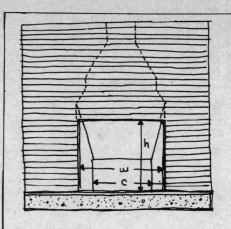 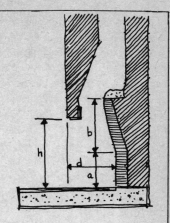

FIG. 4 FIREPLACE PROPORTIONS

1. Height of fireplace (h) = 2/3 to 3/4 of the width.
2. Depth of fireplace (d) = 1/2 to 2/3 of the height.
3. Effective area of the flue (f), the cross-section area of the pipe carrying away smoke, measured to interior dimensions:

   (f) = 1/8 x width x height of front opening (unlined flue)
   (f) = 1/10 x width x height of opening (rectangular lining)
   (f) = 1/12 x width x height of opening (round lining)

4. Throat area = 1.25 to 1.5 x flue area.
   Throat width = 3" minimum to 4—1/2" maximum
   (The throat of a fireplace is a slot-like opening above the firebox where fumes from the fire pass into the smoke chamber.)

TABLE II   RECOMMENDED DIMENSIONS FOR FIREPLACES AND SIZE OF FLUE LINING REQUIRED
(Letters at heads of columns refer to Fig. 4)

Size of fireplace opening			Minimum width of back wall	Height of vertical back wall	Height of inclined back wall	Size of flue lining req'd.	
Width w	Height h	Depth d	c	a	b	Standard rectangular (outside dimensions)	Standard round (inside diameter)
inches	inches	inches	inches	inches	inches	inches	inches
24	24	16-18	14	14	16	8½x13	10
28	24	16-18	14	14	16	8½x13	10
30	28-30	16-18	16	14	18	8½x13	10
36	28-30	16-18	22	14	18	8½x13	12
42	28-32	16-18	28	14	18	13x13	12
48	32	18-20	32	14	24	13x13	15
54	36	18-20	36	14	28	13x18	15
60	36	18-20	44	14	28	13x18	18
54	40	20-22	36	17	29	13x18	15
60	40	20-22	42	17	30	18x18	18
66	40	20-22	44	17	30	18x18	18
72	40	22-28	51	17	30	18x18	18

SOURCE: U.S. Department of Agriculture

For convenience, we have categorized wood burning units into eight types: (1) fireplaces; (2) fireplace heat reclaiming accessories; (3) cast iron heaters; (4) airtight cast iron heaters; (5) metal heaters; (6) circulating heaters; (7) cookstoves; and (8) furnaces.

Fireplaces

There are basically two types of fireplaces; masonry built-in units and metal fireplaces which can be either built-in or free standing.

Masonry fireplaces as in Fig. 1a are the traditional fireplaces in many American buildings: nevertheless they are very inefficient heaters (about 10% efficiency). By contrast a wood heater can provide up to three times the usable heat as an open fireplace from the same amount of burning wood. Because of their open hearth and aesthetic appeal, fireplaces have value other than heat output. They can also provide a focal point for household activities which can give more pleasure to the house occupants than a wood heater.

Masonry fireplace construction is complicated. (See Fig. 4 and Table II). Fireplaces may be constructed of brick, stone or other masonry materials. The reviews at the end of this article include a few books which describe construction techniques in detail.

Prefabricated metal fireplaces are generally of two types: built-in and freestanding. Built-in metal fireplaces often have optional forced air fans to make the units circulators. Circulating units (Fig. 2a) generate more usable heat and are more efficient but the added cost of the fan and its electrical bills must be considered.

In general both circulating and non-circulating units are insulated and constructed so that they can be easily fitted against wood stud walls without fire hazard. They can be adapted to use existing masonry chimneys, but they usually use double or triple walled insulated flue pipe. Prefabricated metal fireplaces are easy to install and they usually cost much less than masonry fireplaces. But metal fireplaces cool quickly once the fire has died down and do not have the heat storage capacity of masonry. Non-circulating units are simply metal fireplaces which radiate heat and do not have space for circulating convective air flow. They are no more efficient than masonry fireplaces.

Freestanding or "contemporary" fireplaces (Fig. 5) consist of a large sheet metal hood fitted over a metal base which holds the fire. They offer relatively high efficiencies because the metal hood radiates a lot of the heat from the fire outward into the room. They can also provide even heat distribution by being placed in the middle of the room.

The "Franklin stove" is a familiar and popular variation of the freestanding fireplace. The distinguishing characteristic of this cast iron fireplace is that most designs provide folding doors which can be closed to reduce the air flow through the combustion zone and up the chimney. With the doors closed the open-flame fireplace can be changed to a more efficient closed heater. The performance and efficiency are dependent on both the tight fit of the doors and the restriction of draft in the chimney with either a built-in chimney damper or one installed in a section of stove pipe. Different models of Franklin Stoves vary considerably in their quality of construction

"Combination" fireplaces are actually "airtight" heaters which can be opened and used as freestanding fireplaces. Some Franklin Stoves are considered to be "combis," as they are often called. The major distinction of a combination fireplace is that it can be closed and secured as an airtight stove. This airtight aspect allows the combustion rate to be metered by the heater air supply without the need for a chimney damper. Most combination fireplaces are of European and especially Scandinavian make.

Heat Reclaiming Accessories

The inherent inefficiency of traditional fireplaces can be improved by using various accessories and heat exchangers. These divert heat out of the fireplace and into the room, thus reducing heat loss up the chimney. The most common model is a u-shaped hollow tubular grate that convects heat from the bottom of the fireplace around the back of the fire and then directs the heated air into the room (see Heat Reclaiming Accessories, pg. 237). Other designs use glass tempered doors that can be fitted in the front of the fireplace. The doors restrict the flow of air into the fire and therefore slow the rate of burn.

Accessories that increase the efficiency of wood heaters attach to the flue. They work to capture heat that is moving up the flue before it moves out of the house. Many are actually parts of flue pipe with special blower fans or other air capturing devices that supposedly add heat to the room. Some of these have heat exchangers for heating water.

Very little research has been conducted into the efficiency of heat reclaiming accessories. Many different models purport to reclaim heat in different ways. Buyers should be sure that the materials and workmanship of the unit they purchase is of the highest quality and that the unit utilizes a good technique to capture the heat.

Cast Iron Heaters

Heaters constructed of cast iron are almost always cast in various pieces which are then fitted together, bolted and sealed with furnace cement. The quality of these heaters lies in the type of casting and fitting of the various pieces.

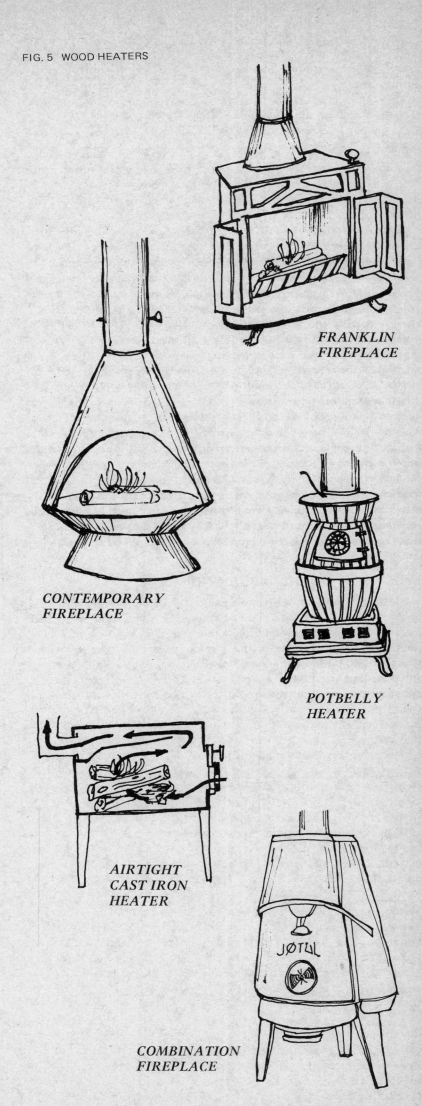

FIG. 5 WOOD HEATERS

FRANKLIN FIREPLACE

CONTEMPORARY FIREPLACE

POTBELLY HEATER

AIRTIGHT CAST IRON HEATER

COMBINATION FIREPLACE

Most older heaters and what are considered antiques were made of cast iron. The nostalgic "potbelly" heater is a good example of a cast iron heater. Other examples include the box stove, laundry stove and parlor stove.

Cast iron heaters are not "airtight" and require the addition of a stove pipe damper to control the rate of combustion. Because the doors and various parts are not designed to form an airtight seal, it is difficult to maintain a fire in a cast iron stove overnight. Burning efficiency is greatly affected by the regulation of the air damper and the type of wood burned. Unlike sheet steel stoves, cast iron heaters are able to hold heat long after the fire has gone out, but they take longer to conduct and radiate heat to the room. Cast iron heaters deliver up to 70% of their heat by radiation.

Airtight Cast Iron Heaters

Most of these heaters are imported from Scandinavian countries. They are characteristically enameled cast iron. The stove parts are cast and machined to a tight-fitting precision. In this way the air is carefully regulated and directed into the combustion chamber so that no chimney damper is required. Another characteristic of these stoves is an internal baffle which encourages an "S" flow of gases and smoke from the combustion chamber into a secondary burning chamber. In addition, the baffle maintains a very high temperature to aid in the combustion of the volatile gases. Some of the heaters provide secondary air inlets to aid in combusting the bases.

Even if they are very small, the airtight cast iron heaters can easily sustain a slow burning fire overnight. This is largely due to their airtight, high quality construction. A drawback to slow burning, overnight fires, however, is the formation of creosote in the chimney. Creosote is a dark colored, tar-like substance produced when wood is burned in the absence of oxygen. Creosote is a potentially dangerous condensate which may cause sudden chimney fires. Creosote formation is a hazard to be considered in the use of any airtight heater. It can be prevented with regular cleaning of the flue or chimney.

Metal Wood Heaters

Metal wood heaters are manufactured of sheet steel or a combination of sheet steel and cast iron. They are primarily radiant heating units. The more durable and expensive units use heavy gauge steel plate (1/8" - 5/16" thick). Welding of the sheets decreases the possibility of air leaks at the joints and adds to the sturdiness of the unit.

Unlike cast iron heaters, sheet steel heaters do not require a foundry and casting facilities for their manufacture, so they tend to be less expensive. They also have the advantage of not cracking like cast iron and of conducting heat through their walls and radiating it to the room faster than cast heaters.

The major drawback of sheet steel units is that the steel is prone to warpage under high temperature stress. This is especially true with rectangular or box construction or where very thin gauge plate is used. Very thin "tin" or "blue-steel" heaters often last only a season or so. Warpage at the door or door frame is particularly bad for maintaining an airtight combustion chamber. Many of the better made sheet steel units use cast iron doors to insure a tight fit. Some use cast iron at the top or bottom of the heater, or on the flue collar, and other areas that receive a great deal of heat stress.

Airtight metal heaters are relatively new on the market. The Fisher heaters (see pg. 235), which have gained wide acceptance are an example. They utilize a controlled draft system that allows a long slow burning fire but they require attention to the draft system to insure a constant burn. Fisher heaters are made of heavy gauge sheet steel and have cast iron doors. They have fire brick on the inside which keeps the sides from warping and adds to the heat retention capability of the heater when the fire has died out.

Automatic metal heaters are fairly common. The Ashley and Riteway heaters have been available for many years (see Fig. 3). The Ashley uses cast flue collars and a cast top but the rest of the unit is sheet steel. The Riteway is all sheet steel. Automatic units have a bimetallic thermostat to control the air flow to the combustion chamber. The thermostat has two different metal strips which are bonded together. They expand or contract at different rates, opening or closing the air inlet to the heater.

In most heaters, the bimetallic strip is normally open at room temperature, and as the room temperature rises the strip bends to close off the air supply. Thermostats are adjustable, reliable and usually a convenient means of insuring constant heat for a house. Most units can maintain a fire for several hours, and the Riteway and Ashley can easily sustain a fire for 12 hours without attendance or recharge of fuel. Many of these heaters have hot water heating coils or special shrouds that can be adapted to existing forced air ducts in a house, and in this way act as a central heating unit.

Circulating Heaters

Circulating heaters (Fig. 2b) are usually made of sheet steel and cast iron. They differ from radiant heaters because they provide most of their heat through convection. The heating unit itself is jacketed with a medium gauge sheet steel cover. Air moves between the heater and the jacket and is warmed, and then moves out the top of the heater. The air can either be "forced" with a small electric fan built into the unit, or it can be naturally convected between the heater and the jacket. Some manufacturers offer a fan as optional equipment. Forced air units are usually more efficient than natural convection units but the small cost of purchasing and running the fan should be considered.

By convecting heat, circulating heaters warm a house in the same way that most oil or gas furnaces do. However, because they are centrally located in a house, much of their heat remains in the room in which the heater is located. Circulating units are usually designed to look modernistic rather than traditional like many radiant heaters. Circulators are also often thermostatically controlled in a similar way to radiant automatics.

Cook Stoves

The wood ranges that are available today have not substantially changed in 150 years. Wood ranges were developed during the first half of the 19th century and the most common and popular design was patented in the 1830's by Thomas Woodson of New Hampshire. This basic design consists of a small fire box and an oven. The fire box is near the cooking surface and the combustion area is lined with cast iron or fire bricks placed above cast iron grates. Below the grates is an ash box to collect spent ashes from the fire. The oven occupies most of the space below the flat cast iron cooking surface.

Cookstoves can give off quite a lot of radiant heat. However, they are not very useful as heaters because their firebox is small and needs constant refueling . . . even when coal is used.

Most wood cookstoves have two openings in the flue that are located in the back of the stove along the flue pipe. By variously opening and closing the top opening (oven damper) and bottom opening in the flue, the heat of the firebox can be directed to the heating surface, to the oven or to both (Fig. 6).

For most efficient use of fuel for both cooking and heating, the stove should be operated in the oven position. This allows for the longest passage and subsequent extraction of heat from the flue gases, plus an even flow of heat across the cast iron cooking surface. When the stove is operated with the oven damper open, most of the heat is lost up the chimney. This may, however, be desirable if a lot of heat is required over a small cooking area.

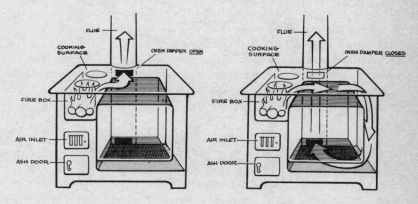

FIG. 6 WOOD COOKSTOVE showing two modes of operation.

Wood cookstoves are usually made of cast iron and sheet steel. Parts of the heater that come in contact with the fire and decorative parts are usually cast. The oven and body of the stove are usually made of sheet steel. Most local stoves are well constructed and made to withstand long-term use. Some recently manufactured styles resemble gas and electric stoves and have enameled sheet steel exteriors. Some of these are available in combination wood/electric, wood/gas or wood/kerosene models (see Wood Cookstoves, pg. 236).

Cooking with wood is a special pleasure for many people. However, it does take time, patience and experience; the art of cooking with wood is not easily learned from a book. Most wood cookstoves have their own characteristics that need to be understood before one can cook with ease and efficiency. Many people who would simply like to boil water or cook an occasional meal are better off purchasing a wood heater with a top surface, that gets hot enough to cook on.

A particular advantage of many wood/cookstoves is a built-in water jacket or water reservoir that will heat water while the stove is in operation. This constant source of hot water can be very useful. But again, a simple tea kettle or large pot on top of a wood heater will provide hot water if only occasional use is needed.

Wood Furnaces

Wood furnaces are quite different from wood heaters previously discussed because they are centrally located, and use either water or air to distribute heat to the building. There are basically four types of furnaces available: (1) forced air; (2) boilers; (3) gasifiers; (4) add-on stoves.

Forced Air Furnaces are large wood heaters and look very much like conventional furnaces. The major difference is the presence of a large fuel chamber. Controls for distribution of the hot air are identical to conventional furnaces.

Boilers: Manufacturers of wood furnaces usually offer both a forced air furnace and a boiler of the same heat output. The forced air furnace distributes heated air and the boiler distributes heated water. Boilers can be connected to either gravity or pumped circulation systems. Like their forced air counterparts, they have large fire boxes and can sustain a fire for a long period at low burning rates. Fire boxes on many of the furnaces and boilers will accept logs up to 48" long depending on the size and output of the furnace (see Fig. 7).

Gasifiers are a special type of wood furnace in which the air supply to the combustion zone is carefully metered to promote the production of volatile gases. The gaseous fuel can then be burned in almost any kind of oil or gas furnace or boiler (see Wood Gas pg. 232).

Add-on Stoves are typically sheet steel (automatic or air-tight) heaters that are adapted to existing basement or garage furnaces using oil, gas or electricity. Usually a galvanized sheet metal shroud is made to fit around the radiant heater body. This shroud is then connected to the duct work of the existing furnace. A similar integration can be accomplished by connecting a boiler or a wood stove with hot water coils in line with the existing plumbing.

When the add-on wood stove is working, the existing furnace burner is turned off. Frequently the fan or pump in the existing furnace is used to distribute the heat from the wood stove to the house and is controlled by the existing wall thermostat. This is probably the least expensive way for anyone who has a non-wood furnace to utilize wood heating as a back-up. In some cases it might be less expensive for a new home to provide a conventional furnace and an add-on wood stove rather than buying an expensive oil or gas/wood combination furnace.

With the exception of some "add-on" stoves, all the furnaces use either a bimetallic strip or an electrically operated control to modulate air supply to the furnace. Many of these furnaces feature an optional oil or gas burner that is automatically activated if the wood fuel is exhausted. Most manufacturers have been around for two or three generations, but new manufacturers are appearing all the time. Wood furnaces range in price from about $500 for add-on units, to more than $3,000 for boilers and gasifiers (see Wood Furnaces, pg. 236).

INCREASING FIREPLACE EFFICIENCY

On the average, fireplaces are only about 10% efficient at converting wood into useful heat. However several things can be done to improve this efficiency by design and construction. First, fireplaces should use cool outside or basement air to keep the fire burning. Since so much air escapes up the chimney when the fire is burning, it is sensible to fuel the fire with air other than warmed inside air. Construction of small vents immediately in front of the fireplace will suck cooler outside air or basement air to "oxygenate" the fire. Older, draftier houses may not have the problem of using cooler outside air because a lot of this air probably already sneaks into the house from window and door cracks. Using cool outside air will also create a positive air pressure in the house because heated air from the fireplace will move out into the house. This will tend to lower infiltration of outside air through cracks around windows or doors. Second, metered control of the air coming into the fireplace is essential. A glass or metal door in front of the fire will limit the air coming into the fireplace. Since this also blocks the radiant heat coming out from the unit, convection air flow tubes and vents must also be added to the unit. A number of hardware units are available to make this modification (see Heat Reclaiming Accessories pg. 237). Third, no fireplace should be constructed without a damper which can be used to regulate the fire and can also be closed. Fireplaces are sometimes accused of actually losing more heat than they give to a house. This is because when a fireplace is not in use, room air, which may have been heated by another unit tends to flow up the chimney and out of the house. The construction of a good airtight damper will prevent this heat loss. Finally, the use of a prefabricated metal circulating fireplace fitted into the existing masonry is usually a more efficient way to use the fireplace.

FIG. 8 PARTS OF A TRADITIONAL FIREPLACE

A. The flue, or the hollow center of the chimney, carries combustion gases out of the house.

B. The smoke chamber is the transition area which connects the fire chamber with the flue.

C. The smokeshelf is a horizontal surface which deflects any downdrafts coming down the chimney.

D. The damper, located in the throat of the chimney, is a movable metal plate which is used to close the flue when the fireplace is not in use.

E. The fire chamber is the space in which the fire burns.

F. Firebrick is used for lining surfaces near intense heat.

G. The fire is built on the back hearth.

H. The front hearth protects the floor from sparks and embers.

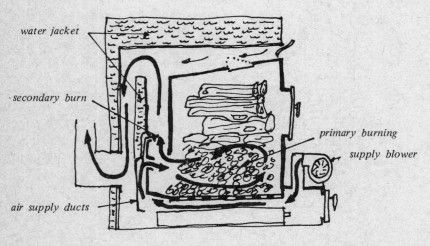

water jacket

secondary burn

air supply ducts

primary burning

supply blower

FIG. 7 RITEWAY BOILER

CHOOSING A HEATING UNIT

Oil, gas or electric furnaces or heating units usually distribute their heat about a building through a series of ducts. Wood furnaces operate in a similar fashion. But wood heaters and fireplaces are localized heat sources which can only be placed in one room. To effectively use them to heat a building, their placement and the building's design must be carefully considered.

Since warm air rises, a heater on the bottom floor of a house will also tend to heat part of the upstairs area. The extent to which the upstairs is heated will depend on the amount of conductivity through the ceiling and, more importantly, the amount of convection between floors. Stairwells, heat registers or simple floor vents will allow heat to rise naturally and warm the upstairs. Placement of the heater to take advantage of the stairwell and proper placement of heat registers and vent holes is therefore very important.

It is more difficult to heat separate rooms on the same floor when a heater is located in one room—heat does not travel well around corners. Placement of the heater near a doorway between rooms or the use of vents high in the wall and transoms (windows above doors) will allow warm air (especially at the ceiling) to move to another room. In the end, however, more than one heater will probably have to be used. With a large heater in the living room and smaller units in the bedrooms and smaller rooms, adequate heating and comfort can be provided at moderate cost. The major problem will be keeping a few units burning simultaneously while maintaining efficient heating and safety. Part of the way many wood heating units can be used effectively is through lowering heating needs and expectations. This sort of lifestyle change may mean wearing more sweaters indoors and keeping the house warmed to only 50° or 60°F.

Cost will certainly be an important consideration in choosing wood burning unit. Since prices run from less than $100 to more than $1000, planning for your needs is mandatory. Most heaters are not rated in terms of the Btu output or efficiency. We have, however, made some generalizations in Table III, using 50% efficiency as an average.

Another important consideration in choosing a unit is whether you want radiant or convective heat. Radiant heat is more efficient because it heats objects and people in the room rather than the air. Radiant heat will usually provide an uneven heat in the house because areas that are not in a direct line with the radiation from the unit will tend to stay cool. Convective heat heats the air, which in turn warms people. Convective heat is most familiar to us in the United States since most gas, oil or electric units utilize this heat transfer method. Convective heat will tend to warm the "whole" house if the air is allowed to circulate freely.

TABLE III COMPARISON OF WOOD HEATING UNITS

Heat Unit	Relative Efficiency[a]	Primary Heat Produced	Area Heated	Relative[e] Cost
Masonry	inefficient	radiant	c	H
Metal PreFab.	inefficient	radiant	c	M
Metal PreFab. Circulating	efficient	convection	large	M
Metal Freestanding	efficient	radiant	large	L
Cast Iron Combinations	efficient[b]	radiant	large	H
Cast Iron	efficient	radiant	small[d]	M
Cast Iron Airtight	efficient	radiant	c	H
Metal automatic	efficient	radiant	large	M
Metal airtight	efficient	radiant	large	M
Metal	inefficient	radiant	small[d]	L
Circulating	efficient	convection	large	H
Cook Stoves	efficient	radiant	small	H
Furnaces	efficient	convection	large	H

[a]"Efficient" means a heater efficiency of greater than 50% and "Inefficient" means a heater efficiency of less than 50%.

[b]Includes Franklin-Fireplaces of which the more airtight models are efficient and the efficient Scandinavian "Combis."

[c]Area heated will be dependent upon the size of the heating unit.

[d]Although cast iron and metal heaters may be large enough to heat large spaces because of their inefficiency they are recommended for small spaces.

[e]High (H); Medium (M); Low (L) cost.

Either circulating wood heaters or circulating built-in fireplaces will provide convective heat. The former are probably more efficient but the latter allows viewing of the fire. Convective heating units tend to be more expensive than many radiant heaters.

The bulk of heaters on the market are radiant. Masonry or non-circulating fireplaces are radiant units, and although they are very inefficient heaters, their aesthetic qualities are desirable. Radiant wood heaters come in many shapes and sizes. For large spaces, especially living rooms, efficient automatic and airtight heaters should be used. In smaller spaces, smaller cast iron heaters are probably the best available units. Even though they are inefficient, they will provide a small amount of heat, and are generally well constructed, attractive units. Kitchens that do not already have wood cookstoves may best be served with the use of a small sheet steel heater. These inexpensive units can be used to burn household paper wastes and also provide heat to the kitchen.

Finally, there is a list showing some basic principles concerning wood heating devices. The wood hardware section following this article can be a starting place to compare and evaluate wood heating units.

WOOD HEATER BASIC PRINCIPLES

1. The size of the heater should be appropriate for the heating needs of the space.
2. Cast iron retains heat longer than sheet steel and is less prone to warpage.
3. Sheet steel conducts heat faster than cast iron but is prone to warpage especially if the fire is hot and the sheet steel is thin.
4. Thin gauge sheet steel heaters will "burn out" in a short period of time, in comparison to heavier gauge (1/4" or thicker) sheet steel or cast iron.
5. Firebrick lining adds to the heat retaining capability and can add to the longevity of the wood heater.
6. Airtight and automatic airtight heaters are more efficient than non-airtight units and they require less attention when a fire is burning.
7. Non-airtight heaters require the use of a flue damper to regulate the air outflow from the heater.
8. Wood heaters burn more efficiently with their doors closed and with the air intake as controlled as possible to let in just the right amount of air. Air cracks in the unit as a result of poor construction will add to the heater's inefficiency.
9. Heaters should be designed to burn the gases from the burning wood since these gases can contain up to 60% of the heat potential in a piece of wood. Usually a heater which has a secondary air inlet will be able to burn the gases.
10. The size of the heater and the size of the fuel door should correspond to the size of the available wood supply or vice versa.
11. In general, the smaller the flue coming out of a wood heater the less heated air is likely to flow out of the heater, and out of the house, unused.
12. When furnace cement is used in a heater it should be checked annually and replaced if it has cracked and worn away.
13. Flues and chimneys should be checked periodically for creosote build-up. This is especially important with airtight heaters.

SAFETY AND THE USE OF WOOD HEATERS

It is very important to be safe with wood heaters, and fireplace safety cannot be stressed too strongly. Too many houses have burned down and too many people have been injured as a result of improper heater installation or carelessness. Most safeguards involve just common sense, careful planning and a little installation and maintenance time. Wood heaters are not as easily turned on and off as oil and natural gas heaters. They demand more attention, as do all renewable energy systems.

Most manufacturers include safety instructions with their heater. Local building codes will have wood heater safety guidelines that are usually sensibly written and provide adequate safety rules without requiring unnecessary precautions and too much additional cost. Precautions should include the purchase of a fire extinguisher and smoke detector.

Certain precautions are a must for fireplaces. First, no fireplace should be left unattended without placing a fine mesh metal screen in front of it to prevent sparks from jumping out of the fire. Second, chimneys should be checked annually for cracks or loose brick or mortar. Third, for newly constructed fireplaces, any one of the fireplace books reviewed at the end of this article should be consulted so that safety precautions, proper placement, materials and the like are taken into consideration.

For wood heaters, four safety precautions must be taken. First, adequate clearance between the heater and the nearest wall must be made. Most building codes require 18 inch distance and less if reflective or insulating materials are attached to the wall. Usually either some type of sheet steel is used or brick or asbestos composite board. Second, depending how high the legs of the heater raise it off the floor, the floor may need protection from scorching and possible ignition. Usually a foot will be adequate space but if the legs are shorter, some protection should be

placed under the heater. Either bricks or a piece of sheet metal will usually do the job. Third, chimneys and/or flues should be constructed with care and inspected regularly. Chimney fires are one of the most serious types of wood heating hazards. They occur when the build-up of creoste in the chimney or flue ignites (usually as a result of a particularly hot fire in the heater). Periodic cleaning of the chimney and connecting stovepipes will help eliminate cresote build-up. In some cases the stovepipes may need cleaning every month. This will especially be the case where slow overnight burning is a practice and cresote build-up is severest. Four, flue pipes and chimney must be insulated where they travel through combustible walls. Most codes will outline the necessary insulation requirements.

References Cited:

[1] ____, 1973. Report of the President's Advisory Panel on Timber and the Environment. 541 pp. U.S. Government Printing Office, Washington, D.C.

[2] ____, 1976. Renewable Resources for Industrial Materials. 266 pp. National Academy of Sciences, Washington, D.C.

[3] ____, 1976. Power Plants Turn to Good Old Wood. 2 pp. Business Week, 3/15/76.

[4] Kemp, C.C. and G.C. Szego, 1975. The Energy Plantation, in Energy Book 2, 125 pp. Ed. John Prenis. Running Press, Philadelphia, PA (see also Ref. 22, pg. 161 in Energy Primer.)'

[5] Evans, R.S. 1974. Energy Plantations: Should We Grow Trees For Power and Fuel. 15 pp. Department of the Environment, Canadian Forestry Service, Vancouver, B.C. Canada.

[6] J.P.R. Associates. 1975. The Feasibility of Generating Electricity in the State of Vermont Using Wood as a Fuel. 127 pp. State of Vermont, Department of Forests and Parks, Montpelier, VT.

[7] Wood Energy Institute. 1976. A Preliminary Study for Aspects of a Wood-Fired Electrical Generating Plant in Lamoille County Vermont. 15 pp. Wood Energy Institute, Waitsfield, VT.

[8] Clawson, Marion. 1975. Forests for Whom and for What? 175 pp. Resources for the Future, Washington, D.C.

[9] Earl, D.E. 1975. Forest Energy and Economic Development. 128 pp. Oxford University Press. NY.

[10] Eckholm, Erick. 1975. The Other Energy Crises: Firewood. Worldwatch paper 1. Worldwatch Institute, Washington, D.C.

[11] Gevorkiantz, S.R. and I.P. Olsen. 1955. Composite Volume Tables for Timber and Their Applications in the Lake States. 51 pp. U.S.D.A. Technical Bulletin No. 1104. U.S. Government Printing Office, Washington, D.C.

[12] Bedard, J.R. The Small Forest and the Tree Farm. 145 pp. Department of Forestry, McDonald College, Quebec, Canada.

[13] Dickson, Alex. 1973. Growing Trees for Timber in New York's Small Woodlands. 15 pp. Cornell Extension Bulletins, Cornell University, Ithaca, NY.

[14] ____, Managing the Family Forest. 1967. Farmer's Bulletin No. 2187, U.S.D.A. U.S. Government Printing Office, Washington, D.C.

[15] Forbes, R.D. 1971. Woodlands for Profit and Pleasure. 75 pp. American Forestry Association, Washington, D.C.

[16] Shelton, Jay and Andrew Shapiro. 1976. The Woodburner's Encyclopedia. 155 pp. Vermont Crossroads Press, Waitsfield, VT.

[17] ____, 1974. Wood Handbook: Wood as an Engineering Material. 400 pp. U.S. Government Printing Office, Washington, D.C.

[18] Stoddard, Charles H. 1968. Essentials of Forestry Practice. 362 pp. Ronald Press, NY.

[19] Toumey, J.W. and C.F. Korstian. 1947. Foundations of Silviculture. 468 pp. John Wiley, NY.

[20] Morrow, R.R. 1975. Sugar Bush Management. 19 pp. Cornell Extension Bulletins, Cornell University, Ithaca, NY.

[21] Reineke, L.H. 1961. Wood Fuel Combustion Practice. 9 pp. Forest Products Laboratory, Report No. 1666-18, Madison, Wisconsin.

[22] Foulds, R.T., Jr. Wood as a Home Fuel. 17 pp. Cooperative Extension Service, University of Vermont, Burlington, VT.

[23] ASHRAE. 1970. Guide and Data Book. American Society of Heating, Refrigeration and Air Conditioning Engineers. NY.

[24] Hand, A.J. October, 1976. Wood as Fuel. p. 104-107. Popular Science, New York, NY.

WOOD BOOK REVIEWS   FORESTRY AND WOODLOT MANAGEMENT

FOREST ENERGY AND ECONOMIC DEVELOPMENT

Covering over 30% of the earth's land surface, forests present a plentiful resource for useful products and energy. Earl has covered most of the pressing forestry and energy questions around the world. Although the focus of the book is on developing countries, especially in tropical areas, there is plenty of information that is applicable to the U.S. Carbonization for charcoal production is discussed in full and the many advantages of light weight and great fuel value are touted. Fuelwood may not be the way that developing countries should strive for energy self-sufficiency, but this book suggests that it is both a viable and desirable source of fuel. One that cannot be overlooked.

— T.G.

Forest Energy and Economic Development
D.E. Earl
1975; 123 pp.
$13.00

from:
Oxford University Press
200 Madison Avenue
New York, NY 10016

FORESTS FOR WHOM AND FOR WHAT?

Resources for the Future (RfF) is a non-profit corporation founded in 1952 with help from the Ford Foundation. Their main activity has been the research of forest lands and policies, energy conservation and water resources. This is one of their best books by one of their foremost authors. It details and summarizes information from two government reports and adds new data and interpretive comments. The reports are THE PRESIDENT'S ADVISORY PANEL ON TIMBER AND THE ENVIRONMENT, 1973,

and THE OUTLOOK FOR TIMBER IN THE UNITED STATES, Forest Service Report 20, 1973. Topics covered include forest statistics, biology, economics, utilization and future concerns.

We need more good analyses from people outside the forest industries. Unfortunately this book, like many RfF publications, lacks hardnosed suggestions for pressing forest policy questions. RfF has a publications list available for free.

— T.G.

Forests for Whom and For What?
Marion Clausen
1975; 171 pp
$3.65

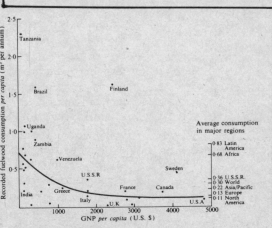

Relationship between fuelwood consumption and GNP for selected countries.

from:
Resources for the Future
1755 Massachusettes Ave., N.W.
Washington, D.C. 20036

PRESIDENT'S ADVISORY PANEL ON TIMBER AND THE ENVIRONMENT

Here is the current U.S. encyclopedia of forest statistics. Included are information, tables and graphs on ownership and use patterns, tree growth rates, forest economics, and evaluations of various silviculture techniques. A few short articles at the end of the report on timber and wildlife, recreation in forest lands, and the effects of harvesting on soil erosion and water quality are excellent. Compiled by a team knowledgeable in forest policy matters it serves as a supplement to the U.S. Department of Agriculture book, THE OUTLOOK FOR TIMBER IN THE UNITED STATES. Forest Service Report No. 20, 1973 available from the U.S. Government Printing Office.

The overwhelming problem with this report, and indeed many government forestry reports is that they attempt to develop a national forest policy based on statistical data. It is impossible to direct forest policy on a national level when it is so clearly a regional problem. Different types of forests, from the New England hard-woods to the Pacific Coast conifers need to be considered as local, micro-climate resources. Each stand of trees needs to be cared for and utilized with due consideration for its habitat as well as for the nation's and the local communities' needs.

— T.G.

President's Advisory Panel on Timber and the Environment
1973; 541 pp
$4.80
from:
Superintendent of Documents
U.S. Government Printing Office
Washington, D.C. 20402

FOREST FARMING

Forest farming is the growing and cultivation of tree crops; trees which produce fruits, nuts, berries or edible leaves for livestock or humans. Tree crops have the potential of providing part of the long term answer to worldwide food shortages. In many arid lands the planting of tree crops can hold back encroaching deserts, supply food and supply fuel. In general, forest farming is best suited for marginal lands which because of their slope, soil or other character-istics are not suitable for traditional agricultural plantings. Tree crops can, however, produce as much food per acre (acorns for example) as agriculture.

This excellent book coupled with Smith's TREE CROPS (see Forestry Bibliography) explains the many ways in which tree crops can be used to produce food. The first five chapters describe a theoretical basis for the practical application of forest farming. The next six chapters describe the design of a forest farm and the different species of trees that can be planted. Finally, the last two chapters go back to the theoretical and explore applications of forest farming techniques in temperate uplands and arid lands.

— T.G.

Forest Farming
J. Sholto Douglas and
Robert A. de J. Hart
1976; 197 pp
$7.50
from:
Watkins Publishing
45 Lower Belgrave Street
London SW1W 0LT
England
or WHOLE EARTH TRUCK STORE

FOREST PRODUCTS LABORATORY

The Forest Products Laboratory in Madison, Wisconsin studies wood. They review and re-search the properties and problems associated with timber used for industry and building as well as wood used for fuel. They provide most if not all of their information free. A few of their useful pamphlets are listed below.

— T.G.

Wood Fuel Combustion Practice
Fuel Value of Wood; Technical Note Number 98
Wood Fuel Preparation
Wood as Fuel for Heating

free

from:
U.S. Department of Agriculture
Forest Service
Forest Products Laboratory
P.O. Box 5130
Madison, Wisconsin 53705

The following table is an approximation of the number of cords of seasoned wood of various kinds needed to give the same amount of heat as a ton of coal, on the basis of 80 cubic feet of wood, with a moisture content of 15-20 percent, to the cord:

1 cord	hickory oak beech birch hard maple	ash elm locust longleaf cherry	= 1 ton coal
1½ cords	shortleaf pine western hemlock red gum	Douglas-fir sycamore soft maple	= 1 ton coal
2 cords	cedar redwood poplar catalpa	cypress basswood spruce white pine	= 1 ton coal

From FUEL VALUE OF WOOD; Technical Note Number 98

FORESTRY JOURNALS

There are a number of forestry journals pub-lished in the U.S., and many published over-seas also. Most are geared to the forest industry and government forestry departments. Two of the best magazines are the *Journal of Forestry* and *American Forests*. They have articles on the newest methods of harvesting, pruning and management as well as research reports on fertilizers, new growth techniques and forest energy resources. The *Journal of Forestry* is the better of the two, with more professional and authoritative information.

— T.G.

Journal of Forestry
Monthly
$18 per year for individuals
$24 per year for institutions
from:
The American Society of Foresters
5400 Grosvenor Lane
Washington, D.C. 20014

American Forests
Monthly
$8.50 per year
from:
The American Forestry Association
1319 Eighteenth Street, N.W.
Washington, D.C. 20036

WOODLANDS FOR PROFIT AND PLEASURE

Sometimes only the more rigorous treatments of a subject are worth reading. In the case of forestry, there is so much information, even the more shallow publications have a great deal to offer. This book is a case in point. The best data offered is on taking a "timber cruise"— measuring and determining the species of trees present in a woodlot. Many other brief sections discuss how to make a woodlot profitable.

— T.G.

Woodlands for Profit and Pleasure
R.D. Forbes
1971; 77 pp.
$5.00
from:
The American Forestry Association
1319 - 18th Street, N.W.
Washington, D.C. 20036

THE SMALL FOREST AND THE TREE FARM

This is a three booklet series that covers wood-lot management, woodlot development and protection, and logging the small forest for profit. A lot of useful tips and good illustra-tions combine to make this a succinct overview for the woodlot owner.

— T.G.

ESSENTIALS OF FORESTRY PRACTICE

There is no substitute for practical experience in forestry. But here is an excellent text used by many forestry schools, which will give the arm-chair forestry student a start. Most all of the forest basics are outlined: forest mensuration, management and finance; growth requirements; prevention of forest fires; prevention of insect damage, and more. Coupled with Toumey's two books listed in the FORESTRY BIBLIOGRAPHY this book should serve as a basis for any serious study of forestry and forestry practices.

— T.G.

Essentials of Forestry Practice
Charles H. Stoddard
1968; 362 pp
$9.00
from:
The Ronald Press Company
79 Madison Avenue
New York, NY 10016

The Small Forest and The Tree Farm
J.R. Bedard
145 pp total
FREE (as far as we can tell)
from:
Department of Forestry
McDonald College
Ste. Anne de Bellevue
Quebec, Canada

THE OWNER-BUILT HOMESTEAD

This is Ken Kern's sequel to THE OWNER-BUILT HOME. It stands by itself as a useful introduction and reference book for "how-to" information about homesteading. Kern is a pragmatist with some decentralist theorizing thrown in who puts homesteading in the perspective of being the real hard work it is. The book discusses water, soil and plant management, site selection, roads, shop and tools, animals, sanitation, human nutrition and has a good chapter on woodland management.
— T.G.

The Owner-Built Homestead
Ken Kern
1974; 207 pp

$5.00

from:
Owner-Builder Publications
P.O. Box 550
Oakhurst, California 93644
or WHOLE EARTH TRUCK STORE

PLANT A TREE

This is a conservationist's handbook with a focus on planting trees for beautification, shade and ecological enrichment. It reads a little like a Johnny Appleseed guide to regreening America. Most of the information presented here is available in other books; but many of them are now out of print.

All in all there is useful information here. A section on selecting a tree for the intended use (e.g. shade) and one which will also live in the climate selected is particularly well presented. Tree profiles for over 160 species which specify the area in the U.S. where they will grow, their expected height, and growth rate as well as data on preferred soil conditions and propagation instructions fill more than half the pages of this book. Unfortunately there is no index for ready reference.
— T.G.

Plant a Tree
Michael A. Weiner
1975; 276 pp

$6.95

from:
Collier Macmillan
866 Third Avenue
New York, NY 10022

or WHOLE EARTH TRUCK STORE

KNOW YOUR WOODS

There is a certain childlike fascination with which this book is written that rubs off on the reader. The author has been in the fine wood trade all his life in New York City. He is a collector of wood samples and a knowledgeable writer on many facets of various wood species and lumbering.

Part One is a story of woods from the forest to their intended use. Many of the species discussed are exotic hardwoods from around the world. How trees are named, how they are made into veneer; an abridged list of drugs from trees is presented. Over four-hundred different types of trees and woods are described in the second part of the book. Unfortunately this section lacks the usefulness of a field manual and lacks sufficiently precise descriptions of woods to be used by a cabinet maker in selecting lumber. Still, a lot of very useful information is to be found here and the 350 or more illustrations are first rate.
— T.G.

Shagbark hickory.

Know Your Woods
Albert Constantine, Jr.
1959; 360 pp

$10.00

from:
Charles Scribner's Sons
Vreeland Ave.
Totowa, NJ 07512

or WHOLE EARTH
TRUCK STORE

TREES OF NORTH AMERICA

TREES OF NORTH AMERICA is an attractive, tasteful, lively little book that lets you identify trees with a minimum of drudgery—and even some enjoyment. This book is a Golden Field Guide, created by people who started out making children's books. In comparison every other guide on the market looks stuffy.

TREES OF NORTH AMERICA covers all native trees, most larger shrubs, and just about every foreign exotic that has carved out its niche in our environment. The written descriptions are precise and helpful, but the strength of this book lies in its graphics—color drawings that are botanically accurate and at the same time pleasant to look at. A tree might be illustrated by drawings of its leaves, buds, twigs, bark, its general shape, perhaps a stunted form, the fruit, or the flowers—in short, by whatever features are most distinctive for that tree. There are also distribution maps to give the tree's range. Finally the cover and all the pages are glossy and water repellent—a feature that should be a statutory requirement for all field guides.

—Malcolm Margolin

Trees of North America
C. Frank Brockman
1968; 280 pp

$3.95

from:
Western Publishing
Company, Inc.
1220 Mound Ave
Racine, Wisconsin
53404

ONE MAN'S FOREST

Short on technical information but long on personal experience, this book explores, rather thoroughly, the joys, dangers, and considerations involved in "developing" timber on an elderly couple's land in Vermont. It is a down-home look at wood lot management.
— T.G.

One Man's Forest
Rockwell R. Stephens
1974; 128 pp

$3.95

from:
The Stephen Greene Press
Box 1000
Brattleboro, VT 05301

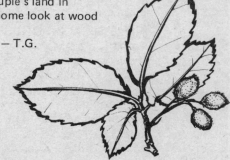

FORESTRY BIBLIOGRAPHY

Foundations of Silviculture; Upon an Ecological Basis
James W. Toumey, Clarence F. Korstian
1937; 468 pp.
John Wiley & Sons (out of print)
The plethora of information here includes discussion of the factors which affect tree growth. Most of the now outdated early forestry research information is referenced and serves to make the book a classic.

Seeding and Planting in the Practice of Forestry
James W. Toumey
1916; 455 pp
John Wiley & Sons (out of print)
Somewhat outdated but still useful information on forestry. Aspects of gathering, drying, storing, and choosing seeds as well as planting, reforestation, aforestation, planting sites, plant spacing, nurseries and more are thoroughly reviewed.

Firewood: The Other Energy Crisis
Erick P. Eckholm
1975; 10 pp; $2.00
World Watch Institute
1776 Massachusetts Avenue
Washington, D.C. 20036
This pamphlet discusses the energy crises in developing countries which depend on wood fuel as a primary fuel source. According to Eckholm, vast amounts of land are being denuded in Nepal, Pakistan, and much of Northern Africa. The damage that is being done to wildlife, watersheds and ecosystems will take many decades to correct. Ironically, these poor countries, which can ill afford to purchase fossil fuels for energy, must do so if they are to avert ecological disaster.

Tree for the Yard, Orchard, and Woodlot
Roger B. Yepsen, Jr., ed.
1976; 305 pp; $8.95
Rodale Press, Inc.
Emmaus, PA 18049
Nut trees, fruit trees and Christmas trees can be enjoyable to grow and provide an income also. A compilation of articles by different authors with different levels of expertise is presented here. It doesn't make sense to chop down fruit bearing trees for fuel but they can be a useful part of an integrated woodlot culture.

Wood as a Source of Energy
E. Bradford Walker
1976; free
Director of Forests
Agency of Environmental Conservation
Montpelier, VT 05602
A task force study which says that there is enough annual growth of cull and unmarketable wood in Vermont to supply all of the State's fuel requirements. The study states however, that only 25% of the available wood is usable. But even that would be a considerable energy savings.

Plants for Man
Robert W. Schery
1977; 657 pp; $18.95
Prentice Hall, Inc.
Englewood Cliffs, NJ
Unbelievably complete with sections on trees, fruit bearing plants and use of plants for dyes, medicines, oils, cereals, sugars and more. The section on trees presents a good overview of worldwide wood biology. This would make a fine college plant course textbook.

The Demand and Price Situation for Forest Products
U.S. Department of Agriculture
Annual
Superintendent of Documents
U.S. Government Printing Office
Washington, D.C. 20402
This is an annual study that lists forest products like sawlogs, veneer logs, and wood for paper and pulp. It is a general economic indicator of the demand for and the price of forest products in the last year.

Managing the Family Forest
U.S.D.A. Farmer's Bulletin 2187
1967; 61 pp; $.20
Superintendent of Documents
U.S. Government Printing Office
Washington, D.C. 20402
One of the best introductory woodlot management guides available. Although very short on theory, it presents most of the thinning and harvesting information necessary for good silvicultural practices. It also contains information on forest products and their markets.

Forestry Handbook
Reginald A. Forbes
1955; 1,201 pp; $19.00
Ronald Press
79 Madison Avenue
New York, NY 10016
This large text should serve as a useful resource book for anyone interested in forest management. Forest measurements, geology and soils, economics and finance, forest recreation, wildlife management and wood technology are all covered in detailed articles.

What Wood is That?
Herbert L. Edlin
1969; 160 pp;
Viking Press (out of print)
Here is a fantastic overview of how to distinguish woods from one another in the forest or in their man-transformed state of furniture, building materials and the like. Wood identification is covered in general and forty wood specimens, which are attached to the book, are discussed in particular.

The Coming Age of Wood
Egon Glesinger
1947; 279 pp
Simon & Schuster (out of print)
Somewhat outdated information on wood: how it grows, what it is used for—plastics, alcohol, sugar, food, wood alloys, lumber, chemicals, wood gas, insulation, fiber and fuel, where it grows, and where it is consumed or used. A complete book about wood from a time (1940's) when wood was more of a universally used material than it is today.

Tree Crops
J. Russell Smith
1953; 408 pp; $7.95
The Devin-Adair Co.
One Park Ave.
Old Greenwich, Connecticut 06870
A call for the planting of trees to prevent soil erosion and depletion, specifically trees which produce food—tree crops—nut trees, fruit trees and acorn bearing trees. Tree crops around the world are discussed and so is tree crop management.

The Feasibility of Generating Electricity in the State of Vermont Using Wood as a Fuel
J.P.R. Associates
1975; 127 pp; free
Director of Forests
Agency of Environmental Conservation
Montpelier, VT 05602
An incredible study of the feasibility of using wood for fuel on a large scale. A terrific introduction outlines basic forest biology and ecology concepts. Excellent references are in abundance at the end of each chapter, and sensible recommendations are made.

Growing Timber in New York's Small Woodlands
Alex Dickson; Information Bulletin 67
15 pp; $.35
N.Y. State College of Agriculture
Natural Resources Department
Cornell University
Ithaca, NY 14853
Almost every forested state publishes extension bulletins on woodland management. This is one of the better, inexpensive booklets which should provide a starting point for people interested in managing their woodlot. Mapping the woods, wild and domestic animals in the woodlot, and the value of different tree species are just a few of the areas covered.

CROWN

HEARTWOOD (INACTIVE) GIVES STRENGTH.

SAPWOOD CARRIES SAP FROM ROOT TO LEAVES.

CAMBIUM (MICROSCOPIC) BUILDS THE CELLS.

INNER BARK CARRIES PREPARED FOOD FROM LEAVES TO CAMBIUM LAYER

OUTER BARK PROTECTS TREE FROM INJURIES.

TRUNK

ENRICHED SOIL LAYER, SOURCE OF MUCH OF THE TREE'S FOOD. THE HOME OF EARTHWORMS, WHICH LEAVE MANY CHANNELS FOR WATER AND AIR TO ENTER THE SOIL.

NATURAL MULCH OF LEAVES (FOREST LITTER) PROTECTS THE SURFACE FROM DRYING AND ERODING.

DECOMPOSING LAYER (FOREST HUMUS) INHABITED BY BENEFICIAL INSECTS

ROOTS

SUBSOIL, COMPOSED OF SOIL PARTICLES AND PARENT MATERIAL. THE DEEP ROOTS FIND THEIR HOME HERE, WHERE THEY ABSORB WATER AND ANCHOR THE TREE FIRMLY IN PLACE.

INSECT PASSAGES THROUGHOUT THE SOIL

Sugar Bush Management
Robert R. Morrow; Information Bulletin 110
19 pp; $.60
N.Y. State College of Agriculture
Natural Resources Department
Cornell University
Ithaca, NY 14853
Growing sugar maple trees for syrup in conjunction with fuelwood production can be a sensible approach to woodlot management. Some good thinning and pruning methods are outlined here in theory, and for practical application.

Woodland Ecology: Environmental Forestry for Small Owners
Leon S. Minckler
1975; 229 pp; $9.95
Syracuse University Press
1011 East Water Street
Syracuse, NY 13210
Long on theory, this is forest ecology told in a narrative style by an experienced forester. Minckler has outlined the basics of integrated uses for forests, watersheds, wildlife and timber production.

HEAT FROM WOOD

THE WOODBURNERS ENCYCLOPEDIA

Without a doubt, this is the finest book to date on the subject of wood combustion as it applies to residential space heating. This book truly explains everything one would want to know about wood burning. In a very organized and detailed fashion, Shelton explains the chemistry, the physics and the thermodynamics of wood combustion and its use as a heat source.

Clear and precise definitions are given of such things as hardwoods, softwoods, moisture content, density and the heating value of wood. A full description of the combustion processes and stove efficiencies are covered in precise detail. Included is a clear analysis of the work which Shelton has been doing at Williams College on stove efficiencies and energy efficiencies for different makes and types of wood heaters.

In addition to all this, such issues as safety, chimneys, accessories, fireplaces, and various installations are discussed. A final chapter on economics is also included.

Section Two of the book, prepared by Andy Shapiro of the Wood Energy Institute, is a superb listing of manufacturers and products. If you are into wood or solar/biomass, this book is a must.

— Ken Smith

The Woodburners Encyclopedia
Jay W. Shelton and Andrew B. Shapiro, (contributing editor)
1976; 156 pp.
$6.95

from:
Vermont Crossroads Press
Box 333
Waitsfield, VT 05673
(802) 496-2469

or WHOLE EARTH TRUCK STORE

THE COMPLETE BOOK OF HEATING WITH WOOD

Larry Gay has taken an optimistic view of heating with wood. He suggests that wood, even on a large scale, could be used to fill much of America's heating needs. He discusses wood lot management, what species of wood give the most heat, and what tools to use to cut down trees. Perhaps the best part of this book is a good look at how the three most efficient wood burning heaters operate; the Ashley, Riteway and Jøtul.

— T.G.

The Complete Book of Heating with Wood
Larry Gay
1974; 128 pp
$3.00

from:
Garden Way Publishers
Charlotte, VT 05445
or WHOLE EARTH
TRUCK STORE

WOODBURNERS HANDBOOK

Efficiency and care with a limited but renewable source of energy. This book gives a general overview of burning wood for heating and cooking. It has some practical tips on repairing old stoves, building stoves from 55 gallon drums, putting in chimneys and stove pipes and keeping them clean and safe from fire. Has some history of wood stoves, and information on the hardness, splitability, and moisture content of different woods.

— T.G.

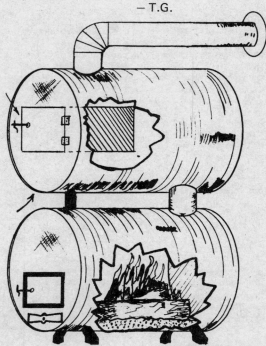

Woodburners Handbook
David Havens
1973; 107 pp
$2.50

from:
Media House
Box 1770
Portland.
Maine
or WHOLE EARTH
TRUCK STORE

COOKING WITH WOOD

Let me start off by saying that you don't learn to cook and bake on a wood range from reading a book. It takes experience, advice, patience and an agreement between you and the stove. Sondra Bidstrup's book is the best I've seen on explaining the basics of a wood range. There's a good historical discussion on the evolution of wood ranges. The information presented seems accurate and there are a few good recipes to go along.

— Ken Smith

Cooking With Wood
Sondra Bidstrup
1976; 44 pp.
$2.50

from:
Coyote
Georgetown, ME

MODERN AND CLASSIC WOODBURNING STOVES

A most interesting and refreshing addition to this otherwise typical book on woodheating is a special set of appendices which actually describe home designs. The designs specifically deal with the integration of solar and wood heating systems. Several houses are described, including the Gilsland Farm Maine Audubon building.

Another interesting and thought-provoking chapter of the book discusses "Woodburning and the Energy Crisis." This is a philosophical look at national energy policy, and how it neglects what we can do as individuals to improve our lives, environment and well-being. The book argues strongly for the efficient use of wood in spite of the apparent disregard of this fuel source by the U.S. Energy Research and Development Administration (ERDA). "At present ERDA has no such recognition of wood as a fuel."

— Ken Smith

Modern and Classic Woodburning Stoves and the Grassroots Energy Survival: A Complete Guide
Bob and Carol Ross
1976; 143 pp.
$10.00

from:
The Overlook Press
Lewis Hollow Road
Woodstock, NY 12498

or WHOLE EARTH TRUCK STORE

RAINPAPER NUMBER ONE: WOOD STOVES

This consumer guide discusses various stoves, their quality or lack of it, and prices. It has a good section on wood cookstove repair and an annotated bibliography. Well worth a buck.

— Ken Smith

RAINPAPER Number One
Bill Day's Consumer Guide to Wood Stoves
Bill Day
1976; 9 pp.
$1.00

from:
RAIN Magazine
2270 N.W. Irving
Portland, OR 97210

ANTIQUE WOOD STOVES

Extremely well illustrated, this book shows many of the designs used in the 1800's; soapstone heaters, cookstoves, box heaters, parlor stoves and coal burning heaters. The text is scant but the descriptions of each stove are interesting and the address of the owner of each stove pictured is given. Excellent history and good design ideas to boot!

— T.G.

Antique Wood Stoves
Will and Jane Curtis
1975; 63 pp.
$3.95

from:
Cobblesmith
U.S. Route 1
Ashville, ME 04607

HOW TO PLAN AND BUILD FIREPLACES

The variations in types of fireplaces and how to build them could easily fill a book. Instead this book spends two-thirds of its pages illustrating and discussing where to put a fireplace in the home—in which room, where in the room and against which wall. A good deal of the discussion refers to the different styles of fireplaces and their "designy" aspects. The assumption is made that a fireplace will be used for "supplementary heating."

The remaining third discusses how to build a masonry fireplace in an existing house, how to install a prefabricated fireplace, and gives a number of useful tables and tips on flue-sizing, fireplace sizing, dampers, etc. In all, certainly worth the money but I can't help but thinking that more could have easily been said.

— T.G.

How to Plan and Build Fireplaces
Sunset Books
1973; 96 pp
$1.95
from:
Lane Book & Magazine Co.,
Menlo Park, CA 94025
or WHOLE EARTH TRUCK STORE

BOOK OF SUCCESSFUL FIREPLACES

The BOOK OF SUCCESSFUL FIREPLACES is a complete volume. It presents tables and diagrams, information on construction do's and don'ts and explores the various different types of fireplace designs. A fascinating section on the history of the fireplace discusses the two great fireplace engineers—Ben Franklin and Count Rumford—as well as others.

— T.G.

The Book of Successful Fireplaces
R.J. Lytle & M.J. Lytle
1971; 104 pp

$9.00
from:
Structures Publishing Co.
P.O. Box 423
Farmington, MI 48024
or WHOLE EARTH TRUCK STORE

FIREPLACES AND CHIMNEYS

The basics: construction, design and modification. Good advice, well illustrated.

—T.G.

Fireplaces and Chimneys
USDA
1963; 23 pp
$0.40

from:
Superintendent of Documents
U.S. Government Printing Office
Washington, D.C. 20402

WOOD 'N ENERGY: BIMONTHLY

"Wood 'N Energy" is a newsletter from the Wood Energy Institute and The Society for the Protection of New Hampshire Forests (SPNHF). The newsletter is published six times yearly (four guaranteed) ". . . with articles, photographs, charts, and drawings on Wood to Energy topics."

The following areas are covered: harvesting techniques, equipment systems, electrical power generation, industrial processes, commercial applications and residential heating. It also includes information on recent publications, conferences and politics. If you want to stay on top of developments in the area of wood fuels, this is a publication to have.

— Ken Smith

Wood 'N Energy
Wood Energy Institute
Andrew B. Shapiro, President
Bimonthly; 8-10 pp.
$4.95 year
from:
Wood 'N Energy
SPNHF
5 South State Street
Concord, NH 03301

SOLAR/WOOD SYSTEM OF MAINE AUDUBON HEADQUARTERS

This is a publication which describes the solar/wood heating system on the new Maine Audubon Society Headquarters building at Gilsland Farm.

A unique stove design by Prof. Richard Hill of the University of Maine is installed as a back-up to the solar heating system. This stove is insulated, and all heat is dumped into the isothermic rock bin storage used for the solar system. High efficiencies are experienced since the stove is run at full combustion. Plans for the stove are part of the $8.00 package.

— Ken Smith

Gilsland Farm
Maine Audubon Headquarters
Energy Systems
1976; 50 pp. + plans
$8.00

from:
Maine Audubon Society
Gilsland Farm
118 U.S. Route One
Falmouth, ME 04105

MASONS AND BUILDERS LIBRARY

As with most Audel books these two are good complete basic sources. With the information here and a little common sense anyone ought to be able to handle the immediate technical aspects of whatever masonry he/she might want to undertake.

— Bill Duncan

Masons and Builders Library
Vol. I   Concrete·Block·Tile·Terrazo
Vol. II Bricklaying·Plastering·Rock Masonry·Clay Tile
Louis M. Dezettel
1972;
$5.65 each

from:
Theodore Audel & Company
4300 West 62nd Street
Indianapolis, Indiana 46268

WOOD BURNING QUARTERLY

If you want to keep on top of what is happening in wood burning for home and small industry, don't hesitate to order this periodical. Contributors include Larry Gay, Bob Ross, Jay Shelton and others who have written books and are working with the technology of burning wood. Paid advertising and classifieds are featured. Short articles describe improved efficiencies for stoves and fireplaces, reports on research, conferences, safety and new products. It was appropriately described by RAIN MAGAZINE as having an *Organic Gardening and Farming* format for people who heat with wood. It is an excellent place to start reading if you are just starting to think about wood heaters.

— Ken Smith

Wood Burning Quarterly
James R. Cook
(editor/publisher)
Quarterly
$4.95 year
$1.50 issue

from:
Wood Burning Quarterly Co.
Div. of Investment Rarities, Inc. ,
8009 - 34th Avenue So.
Minneapolis, MN 55420

WOOD BURNING SAFETY

The following pamphlets on wood burning safety are available from the National Fire Protection Association:

NFPA No. 89-M
Heat Producing Appliance Clearances 1976
$2.50 postpaid

NFPA No. 21
Chimneys, Fireplaces, and Vents 1972
$2.50 postpaid

Using Coal and Wood Stoves SAFELY!
A Hazard Study
NFPA No. HS-8
National Fire Protection Association
1974; 12 pp
$2.25 postpaid

from:
NFPA
470 Atlantic Avenue
Boston, MA 02210

GARDEN WAY WOOD PAMPHLETS

Garden Way Publishing has three useful pamphlets on using wood for heating.

How to Sharpen and Use an Axe and Get the Most Use of Fuel Wood, 6 pp $0.75

Curing Smoky Fireplaces, 24 pp. 1974, discusses 20 or more problems and solutions to smoky fireplaces $0.90

Wood Stove Know How, 1974, 31 pp. A basic introduction to wood stove use and maintenance. $1.50

from:
Garden Way Publishing
Charlotte, Vermont 05445

WOOD GAS

Wood gas (also known as producer gas, gasogen, gengas, portable gas and charcoal gas) is essentially gas given off as a result of the burning of wood or charcoal. Thirty-five cubic feet of wood gas is approximately equivalent to 53 gallons of gasoline or 32 gallons of diesel oil. Wood gas has been used in industry and in home furnace heating units but its primary use has been in internal combustion engines during periods of petroleum fuel shortages. Both World Wars saw the use of wood gas to power vehicles. In 1942 over 90% of the vehicles registered in Sweden were operating using wood gas. In the same year approximately 700,000 vehicles were using wood gas in Europe—primarily in England, Sweden and Germany. World War II was the height of wood gas use in the world and little interest in or use of the gas has been made since. Very limited use of wood gas was seen in the U.S. during any of these periods, though, as throughout the world, W.W. II was the time of greatest use.

In general wood gas operates in internal combustion engines in this manner: small chunks of wood or coal, about the size of a person's fist are placed in a generator fuel hopper usually made of sheet steel and sometimes lined with firebrick. The hopper is most often cylindrical, about the size of a home electric or gas water heater. The wood is ignited using a kerosene or natural gas torch. As the wood burns it produces a gas which is sucked through a series of cooling tanks and then through a steel wool, oil, and/or fabric filter. The cooling

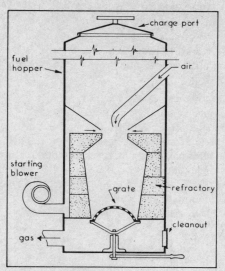

An early and heavy gasogen design.

tanks serve to cool the gas and condense it to provide a denser fuel. The filter serves to purify the gas so that tars and resins do not clog the engine. Coal gas does not give off the tars and resins that wood gas does and is thus a more desirable fuel.

For the generator to build up enough gas to power the engine it is necessary to provide a draft to increase the fire. When the engine is running the natural draw of the engine keeps the fire in the generator going. Obviously when the engine is running the hardest, the draw is

the greatest, to provide the greater amount of fuel which is necessary. There is, however, a problem in starting the engine. Since there is no draw, the generator is not operating very well. Two methods are usually used to overcome this difficulty. Either the engine is started using conventional gasoline and then switched over to wood gas or an electric fan is used to suck the gas into the engine and "fan the fire" to increase the gas production.

The problems associated with the use of wood gas are numerous and have resulted in limited use of this fuel except in times of petroleum fuel shortages. The major problem is maintenance. The generator needs to be filled with wood and then cleaned out of ash with every use. The filters need to be cleaned and the system checked for build up of tars and resins in the piping and cooling tanks. Further, because of the impurities in the wood gas, the engine needs to be overhauled more often than an engine using gasoline. One source indicated that the engine would need to be overhauled every five to eight thousand miles. The bulk and weight of the system needed to use wood gas is also a problem especially for today's compact cars. Start up time to allow the wood gas to reach sufficient peak to provide power is also a problem for those of us used to turning a key and driving off.

All in all wood gas is clearly not a solution to fuel needs for vehicles.

—T.G.

WOOD GAS BIBLIOGRAPHY

Can We Use Wood to Beat the Gasoline Shortage
Popular Science, January 1944; reprinted in Mother Earth News No. 27
   Brief general article explaining basic theory and practice of a particular truck converted to use wood gas.

Charcoal and Water are Converted into Gas That Runs Trucks
Popular Science
September 1944, page 141

Charcoal Gas for Motors: American Made Unit Available for Motor Vehicles
Scientific American
October 1944, page 174

Engine Tests with Producer Gas
A. Middleton and C. S. Bruce
February 1946; 14 pp
Journal of Research of the National Bureau of Standards; Vol. 36; Research Paper RP 1698.
   Bench tests with a four-cylinder stationary engine were made with gasoline and producer gas from charcoal as the fuels. A comparison of their performance revealed that maximum power from producer gas from charcoal is 55% of gasoline power, and that about 11.4 pounds of charcoal is equivalent to one gallon of gasoline.

Ersatz Motor Fuels
Gustav Egloff
Scientific American, July 1939
   A largely economic look at Europe's use of producer gas for vehicle fuel. Illustrating that producer gas at that time (1939) was not economical on a large scale.

Experiments on a High-Speed Producer Gas Engine
A. F. Burstall and M. W. Woods
1939, 2 pp
London Engineer No. 167

Gas Producers for Motor Vehicles
E. A. Allcut and R. H. Patten
1943; 162 pp
National Research Council of Canada
   A committee report on possibility of replacing gasoline by gas made from charcoal or wood, as a wartime emergency measure. Road and stationary tests were made on a number of engines, and tables and graphs of results are plentiful.

Gas Producers for Motor Vehicles and Their Operation with Forest Fuels
I. Kissin; 1942
Technical Communication No. 1
Imperial Forestry Bureau, England

Gasogens
Forest Products Laboratory
1944; revised 1962; 6 pp
Forest Products Lab.
Madison, Wisconsin
Report No. 1463
   General overview of fuels used, principles, design and equipment for mobile and stationary engines using gasogens.

How to Run Your Car on Wood
Travis Brock; 5 pp; The Mother Earth News No. 27
   A good overview of the various different types of producer gas generators as well as general information on the subject.

The Modern Portable Gas-Producer: Theory, Design, Fuels, Performance Utilization and Economics
B. Goldman and N. Clarke Jones
1939; Journal of the Institute of Fuels No. 12

Producer Gas for Motor Transport
E. A. Allcut; Part I 4 pp; August 1, 1943
Part II 5 pp August 15, 1943
Automotive and Aviation Industries.
Issues of the same journal published May 1 and May 15, 1941 and August 1, 1941 have articles on producer gas.
   General overview of fuels, design of plant, amount of power, cooling the gas and comments as to efficiency, problems and widespread application.

Producer Gas for Motor Vehicles
Cash and Cash; 1942
Angus and Robertson Ltd.
Sydney, Australia

Thermodynamics of Producer Gas Combustion
A. P. Oleson and Richard Wiebe
July 1945; 9 pp
Industrial and Engineering Chemistry Vol. 37, No. 7
   The composition and heats of combustion of producer gas from various raw materials including wood. The application of this gas to internal combustion engines—carburetor modification, pressures, compression and efficiency are all discussed.

CHARCOAL PRODUCTION, MARKETING, AND USE.

This technical booklet provides useful information for people interested in going into the charcoal business. While some of the report covers large-scale charcoal production, most of the material is geared to the small operator with limited capital who wants to have a supplementary income. The booklet includes useful construction diagrams for various types and sizes of charcoal kilns. Regrettably, only brief mention is made of very small backyard kilns that can supply enough charcoal for home use. Also, little mention is made of the ecological implications of charcoal production, although the report does stress the value in using waste wood for making charcoal.

—Dennis Dahlin

Charcoal Production, Marketing and Use
U.S. Forest Service Report No. 2213
1961; 137 pp

Free

from:
U.S.F.S. Forest Products Laboratory
P.O. Box 5130
Madison, Wisconsin 5705

WOOD HARDWARE

PREFABRICATED BUILT-IN FIREPLACES

Prefabricated built-in fireplaces are usually constructed of sheet steel. They are made to fit into existing masonry fireplaces or into the wall of a house. Some models are circulators and have forced air fans. The forced air units are much more efficient. Built-in fireplaces range in price from $175 to $500.

F & W Econoheat, Inc.
222 East Railroad Avenue South
P.O. Box 591
Manhattan, MT 59741
"BHV" Fireplace furnace; forced air circulating; heat can be ducted throughout house; partial outside air inlet; efficient.

Heatilator
Box 409
Mount Pleasant, IA 52641
"Heatilator®" circulating sheet steel.

Home Fireplaces
Marhan, Ontario L3R1GE
and
971 Powell Ave.
Winnipeg, Manitoba R3H0HY
"Northern Heatliner" double wall sheet. steel circulating with outside air inlet.

KNT, Inc.
PO Box 25
Hayesville, OH 44838
"Impression Mark V" outside air inlet; forced air circulating; glass doors.

Majestic Company
Huntington, IN 46750
Many different models, some circulating, some firebrick lined.

Malm Fireplaces, Inc.
368 Yolanda Ave.
Santa Rosa, CA 95404
"Woodside" circulating sheet steel.

Preway, Inc.
Wisconsin Rapids, WI 54494
Circulating sheet steel; firebrick lined; optional outside air inlet.

Ridgeway Steel Fabricators, Inc.
Box 382
Ridgeway, PA 15853
Sheet steel with water jacket, water heating grate; optional outside air inlet.

Tempco
P.O. Box 1184
Nashville, NH 37202
Sheet steel; some models firebrick lined.

CONTEMPORARY FIREPLACES

Most of these units are conical in shape but some are square or oblong. They are relatively efficient in that they radiate a lot of heat to a room through their thin sheet metal bodies. They have modern designs and often are available in enamelled colors. Cost runs between $150 and $500.

Glo-Fire
Spring and Sumner Streets
Lake Elsinore, CA 92330

Home Fireplaces
Markham, Ontario L3R1GE
or
971 Powell Ave.
Winnipeg, Manitoba R311OH4

Majestic Company
Huntington, IN 46750

Malm Fireplaces, Inc.
368 Yolanda Ave.
Santa Rosa, CA 95404

Preway, Inc.
Wisconsin Rapids, WI 54494

Temco
PO Box 1184
Nashville, NH 37202

FRANKLIN FIREPLACES

The traditional Franklin is usually made of cast iron and can operate with the doors open as a fireplace or with the doors closed as a closed heater. When the doors are closed the Franklin becomes much like a cast iron heater. These units can be efficient if the door seals and other joints are tight fitting such that the unit becomes nearly airtight. Franklins cost between $200 and $600.

Atlanta Stove Works, Inc.
P.O. Box 5254
Atlanta, GA 30307

Autocrat
New Athens, IL 62264

Double Star
c/o Whole Earth Access Co.
2466 Shattuck Ave.
Berkeley, CA 94704

Home Fireplaces
Markham, Ontario
L3R1GE, Canada

Malm Fireplaces
368 Yolanda Ave.
Santa Rosa, CA
95404

Martin Industries
P.O. Box 230
Sheffield, AL 35660

Portland Stove Foundry
57 Kennebec Street
Portland, ME 04104

Radke Imports, Ltd.
P.O. Box 545
Emmeh, ID 83617

United States Stove Co
P.O. Box 151
South Pittsburg, TN 37380

Washington Stove Works
P.O. Box 687
3402 Smith St.
Everett, WA 98201

COMBINATION FIREPLACES

Combination fireplaces are much like the Franklin. They usually, however, have a tighter fitting door that can be closed to make the fireplace an efficient airtight heater. The most efficient and attractive units are made in Scandinavia with cast iron and firebrick. They are meant to heat large areas and cost between $300 and $650.

Bow & Arrow Stove Co.
14 Arrow St.
Cambridge, MA 02138
"Fyrtöndon"

Earth Stove, Inc.
19440 S.W. Boones Ferry Rd.
Tualatin, OR 97062
Sheet Steel, thermostatically controlled.

Fireplace Distributors
5900 Empire Way South
Seattle, WA 98118
Trolla

Fisher Stoves
135 Commercial
Springfield, OR 97477
"Grandpa Bear" heavy gauge sheet steel with cast iron doors, firebrick floor.

Greenbriar Products, Inc.
Box 473
Spring Green, WI 53588

J & J Enterprises
4065 W 11th
Eugene, OR 97402
"Frontier" heavy gauge sheet steel, firebrick lined, optional screen.

Kristia Associates
P.O. Box 1118
Portland, ME 04104
"Jotul Combi-Fire No. 4" Norwegian import enamelled cast iron with firebrim lining.

Markade-Winnwood
4200 Birmingham Rd., NE
Kansas City, MO 64117

Southport Stoves
250 Tolland St.
East Hartford, CT 06108
"Morso" Danish import, enamelled cast iron with firebrick lining.

Vermont Castings, Inc.
Box 126
Prince Street
Randolph, VT 05060
Double wall cast iron, thermostatically controlled.

CAST IRON HEATERS

There are many different types of radiant cast iron heaters. The potbelly, box heater and laundry stove are examples. These heaters have relatively low efficiencies. These heaters have attractive and durable units, adequate for heating small rooms. They usually cost between $125 and $500 depending on the size and style.

Abundant Life Farm
P.O. Box 63
Lochmere, NH 03252
"Cozy" parlor stove.

Atlanta Stove Works
P.O. Box 5254 Sta. E.
Atlanta, GA 30307
box heater, potbelly, parlor heater, laundry stove.

Birmingham Stove & Range Co.
P.O. Box 2647
Birmingham, AL 35202
potbelly, laundry.

Double Star
2466 Shattuck Ave.
Berkeley, CA 94704
parlour heater, imported from Asia.

Edison Stove Works
P.O. Box 493
Edison, NJ 08817
potbelly.

Home Fireplaces
Markham, Ontario
L3R1GE
many different models.

Locke Stove Co.
114 W. 11th Street
Kansas City, MO 64105
"Warm-ever" firebrick lined box heater.

Martin Industries
Cast Iron Products
P.O. Box 128
Florence, AL 35630
box and parlor heaters.

Radke Importers, Ltd.
P.O. Box 545
Emmett, Idaho 83617
Imported potbelly; box heater; laundry heater; parlor stove; coal heater.

Vermont Iron Stove Works
The Bobbin MiW
Warren, VT 05674
old fashioned heaters.

Washington Stove Works
P.O. Box 687
Everett, WA 98201
box heater, parlor heater, pot belly, laundry stove.

CAST IRON AIRTIGHT HEATERS

Most of these radiant heating units are made in Scandinavia and have a special baffle system that forces the wood to burn like a cigar from front to back. The baffle also aids in the burning of the wood gases. These are very efficient units that can heat large areas. They are all very attractive and are solidly constructed. Many models come with colored enamel baked onto the cast iron. They usually cost between $250 and $750 depending on the size and design.

Fireplace Distributors
5960 Empire Way So.
Seattle, WA 98118
"Trolla", Norwegian Import, many different models including small cookstove and combination fireplace.

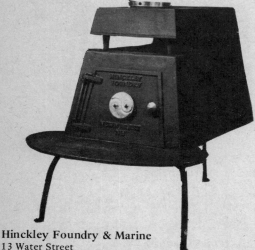

Hinckley Foundry & Marine
13 Water Street
Newmarket, NH 03857
"Hinckley" Domestic (U.S.), box heater.

METAL HEATERS

Metal heaters are similar to cast iron heaters except that their construction material is different. They are generally inefficient heaters that are only adequate for heating a small space. The quality of the unit is dependent on the degree of airtightness and the thickness of the sheet steel (the thicker the better). These units range in price from $75 to $200.

Autocrat Corp.
New Athens, IL 62264
very thin gauge sheet steel, drum heaters.

Garden Way Research
P.O. Box 26W
Charlotte, VT 05445
nearly airtight box heater, sheet steel.

Inglewood Stove Company
Rte. 4
Woodstock, VT 05091
"Tortoise" antique design sheet steel and cast iron.

Louisville Tin and Stove Co.
P.O. Box 1079
Louisville, KY 40201
very thin gauge sheet steel drum heaters.

United States Stove Co.
P.O. Box 151
South Pittsburg, TN 37380
"Boxwood" medium gauge sheet steel, fitted.

Yankee Woodstoves
Cross St.
Bennington, NH 03442
cylindrical sheet steel and cast iron.

Kristia Associates
Box 1118
Portland, ME 04104
"Jøtul" Norwegian import, many different models including ones with heat exchangers mounted on top.

S/A Distributors
730 Midtown Plaza
Syracuse, NY 13210
"Reginald," Irish Import, small box heater in kit.

Scandia Wood Stoves
Box 235
Lanaska, PA 18931
"ULEFOS," Norwegian Import, many different models of box heaters.

Scandinavian Stoves, Inc.
Box 72, Rte. 12-A
Alstead, NH 03602
"Lange," Danish import, many different models including one with heat exchangers mounted on top.

Southport Stove Works
250 Tolland St.
East Hartford, CN 06108

METAL AUTOMATIC HEATERS

Automatic heaters utilize a bimetallic "thermo-stat" to regulate airflow into the combustion chamber. Automatics are efficient, easy to use and usually will heat large spaces. Most are made of sheet steel and cast iron. Prices range from $200 to $450 depending on size and style.

Ashley Automatic Heater Co.
1604 17th Ave.
P.O. Box 730
Sheffield, AL 35660
sheet steel body and liner, cast iron top, bottom, door

Atlanta Stove Works, Inc.
P.O. Box 5254
Atlanta, GA 30307
thin oblong cylindrical sheet steel body, cast iron top, bottom and door.

Martin Industries
P.O. Box 730
Sheffield, AL 35660
"6600" and "Wood King Thiv," oblong cylindrical sheet steel cast iron top, doors and bottom.

Riteway Manufacturing Co.
P.O. Box 6
Harrisburg, VA 22801
medium gauge sheet steel, rectangular construction, cast iron fire grate.

Scot's Stove Co.
11 Ells St.
Norwalk, CT 06850
sheet steel.

Shenandoah Mfg. Co., Inc.
P.O. Box 839
Harrisonburg, VA 22801
medium gauge cylindrical sheet steel, firebrick lining.

Vermont Woodstove Co.
307 Elm Street
Bennington, VT 05201
"Down drafter" sheet steel, secondary air inlet, very efficient, can be used with forced air fan.

CIRCULATING HEATERS

These convective heating units are usually made of a combination of cast iron and sheet steel. Air is either forced (with a small fan) or allowed to naturally convect (warm air rises) around the heater. Many units are thermostatically controlled. All models have modern contemporary designs. They usually cost between $200 and $500.

Ashley Automatic Heater Co.
1604 17th Ave.
P.O. Box 730
Sheffield, AL 35660
"Imperial" and console models, cast iron and sheet steel, circulating fan, thermostatic draft

Atlanta Stove Works, Inc.
P.O. Box 5254
Atlanta, GA 30307
"Homesteader," cast iron, circulating fan, thermostatic draft control.

Autocrat Corporation
New Athens, IL 62264
Cast iron and sheet steel, thermostatic draft control, circulating fan.

Hunter Enterprises Orillia Limited
PO Box 400
Orillia, Ontario, Canada
"Valley Comfort" secondary combustion chamber for gases, thermostatic draft control.

Locke Stove Co.
114 West 11th St.
Kansas City, MO 64105
"Warm Morning" cast iron and sheet steel; thermostatic draft control, circulating fan.

Martin Industries
P.O. Box 730
Sheffield, AL 35660
"King" thermostatic draft control, circulating fan.

Metal Building Products
35 Progress St.
Nashua, NH 03060
"Nashua" forced air.

Riteway Manufacturing Co.
P.O. Box 6
Harrisonburg, VA 22801
thermostatic draft control, circulating fan.

Suburban Manufacturing Company
4700 Forest Dr.
P.O. Box 6472
Columbia, SC 29206
"Wood Master," thermostatic draft control, circulating fan.

U.S. Stove Company
PO Box 151
South Pittsburg, TN 37380
"Wonderwood" thermostatic draft control, circulating fan.

Bow & Arrow Stove Co.
14 Arrow St.
Cambridge, MA 02138
"Supra" French import; firebrick lined firebox, cast iron and sheet steel, manual draft control.

Washington Stove Works
PO Box 687
3402 Smith St.
Everett, WA 98201
"Norwester" thermostatic draft control.

METAL AIRTIGHT HEATERS

Metal airtight heaters are very similar to automatic metal heaters except that they do not have thermostats. Many units have cast iron doors and flue collars. These units are efficient and sturdy, especially if the sheet steel bodies are made of heavy gauge construction and are welded together securely. These heaters range in price from $200 to $500.

Damsite Dynamite Stove Co.
RD3
Montpelier, VT 05602
sheet steel box heater.

Fire-View Distributors
P.O. Box 370
Rogue River, OR 97537
cylindrical sheet steel with glass door for viewing the fire, firebrick lining.

Fisher Stoves
135 Commercial
Springfield, OR 97477
heavy gauge sheet steel, cast iron door, rugged construction.

Kickapoo Stove Works
Rt. 1-A
La Farge, WI 54369
sheet steel rugged construction

L.W. Gay Stove Works, Inc.
Marlboro, VT 05344
"Independence" Norwegian fire box heater, can be converted to circulating unit or used to heat water, sheet steel.

Mohawk Industries, Inc.
173 Howland Ave.
Adams, MA 01220
"Tempwood" sheet steel with cooking grate.

Southeastern Vermont Community Action, Inc.
7-9 Westminster St.
Bellows Falls, VT 05101
cylindrical steel made from recycled propane tanks, cooking surface, heat exchanger.

Sunshine Stove Works
RD 1 Box 38
Norridgewock, ME 04957
baffled "Scandinavian type" sheet steel, secondary air draft.

WOOD FURNACES

Wood furnaces come in three varieties: 1) add-ons to existing oil, gas or electric furnaces, 2) forced air wood furnaces, and 3) boilers (water heating wood furnaces). Many companies offer both a forced air unit and a boiler. Some of the available furnaces have optional oil, gas or electric conversion systems that can be used as a back-up to wood. Furnaces are expensive and range in price from $500 to $2000 or more.

Bellway Manufacturing
Grafton, VT 05146
thermostatically controlled; gravity fuel feed; firebrick lined, combination coal/oil; boilers and forced air.

Carlson Mechanical Contr's, Inc.
Box 242
Prentice, WI 54556
thermostatically controlled; boilers and forced air.

Charmaster Products, Inc.
2307 Highway 2 West
Grand Rapids, MN 55744
thermostatically controlled, gravity fuel feed.

Enwell Corp.
750 Careswell St.
Marshfield, MA 02050
"Spaulding Concept Furnace" thermostatically controlled, forced air.

Hunter Enterprises Orilla/Ltd.
P.O. Box 400
Orilla, Ont. L3Y6K1 Canada
forced air, thermostatically controlled.

Len Jay Furnace Co.
Underwood, MN
manual draft control; gravity fuel feed.

Longwood Furnace Co.
Gallatin, MO 64640
forced air, thermostatically controlled, wood/oil and wood/gas, 60%-75% efficiency.

Lynndale Manufacturing Co.
P.O. Box 1154
Harrison, AV 72601
thermostatically controlled; connects to hot water heating system.

Malleable Iron Range Co.
715 N. Spring Street
Beaverdam, WI 53916
forced air add-on.

Marathon Heater Co., Inc.
Box 165, RD 2
Marathon, NY 13803
add on forced air, optional gas burner.

Markade Winnwood
4200 Birmingham Rd., N.E.
Kansas City, MO 64117
firebrick lined steel.

Newmac Mfg. Inc.
236 Norwich Ave.
Box 545
Woodstock, Ontario N48 7W5, Canada
oil/wood combination; thermostatically controlled.

Ram and Forge
Brooks, ME 04921
oil/wood combination, thermostatically controlled.

Riteway Manufacturing Co.
Division of Sarco Corp.
P.O. Box 6
Harrisonburg, VA 22801
forced air, boilers, wood/gas, wood/oil and add-on forced air, thermostatically controlled.

Solar Wood Energy Corp.
Waldo G. Cummings
E. Lebanon, ME 04027
thermostatically controlled, forced air, many models available.

Tekton Design Corp.
Conway, MA 01341
"OT" boiler Danish import, wood/electric, wood/gas combination.

Wilson Industries
2296 Wycliff Street
St. Paul, MN 55114
forced air, wood/oil, thermostatically controlled.

WOOD COOKSTOVES

Wood cookstoves come in many different shapes and sizes but all are primarily designed for cooking. They have small fireboxes and some have ovens. Many have optional or already built-in water heating reservoirs. Most cookstoves cost between $300 and $1500. Well constructed stoves with an oven, water reservoir and large cooking surface cost approximately $800.

"Old Country" Appliances
P.O. Box 330
Vacaville, CA 95688
"Tirolia" Austrian import; five models available; cast iron enamelled; water reservoir; high quality.

Atlanta Stove Works
Atlanta, GA 30307
old fashioned design, cast iron, some with water reservoir.

Autocrat Corporation
New Athens, IL 62264
modern design enamelled; some with water reservoir.

Bow and Arrow Stove Co.
14 Arrow Street
Cambridge, MA 02138
(617) 492-1411
"Preporod" Yugoslavian import; small cast iron enamelled low cost.

Home and Harvest
2517 Glen Burnie Drive
Greensboro, NC 27406
"Victor Jr" cast iron.

Louisville Tin & Stove Co.
P.O. Box 1079
Louisville, KY 40201
block steel stovepipe oven.

Martin Industries
P.O. Box 730
Sheffield, AL 35660
"Marco Pride" and "Supreme" old-fashioned design; water reservoir with "Supreme;" cast iron.

Merry Music Box
20 McKown Street
Boothbay Harbor, ME 04538
"Styria" Austrian import; four models available; cast iron enamelled, water reservoir, high quality.

Monarch Kitchen Appliances
a division of the Malleable Iron Range Co.
Beaver Dam, WI 53916
modern design; a few different models available; most with gas or electric combinations.

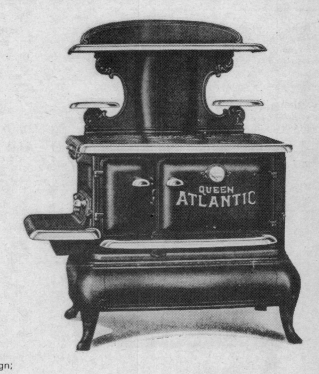

Portland Stove Foundry
57 Kennebec St.
Portland, ME 04104
"Queen Atlantic" and "Atlantic" cast iron, some models with water reservoir.

Preston Distributing Co.
Division of Preston Fuels, Inc.
Foot at Whidden Street
Lowell, MA 01852
"Chappee" French import.

KITS

Most kits have cast iron doors and steel legs that can be attached to a 15, 30 or 55 gallon oil drum so that they can be used as a heater. These are probably the best way to make an inexpensive, reasonably durable heating unit. Kits usually cost less than $75.

Country Craftsmen
PO Box 3333
Santa Rosa, CA 95402
cast iron door and fluepipe collar and 4 steel legs for 15 or 30 gallon drum.

Fisher's
Rt. 1 Box 63A
Coniter, CO 80433
Pyrex® glass door and 4 legs for 15, 30 or 55 gallon drum.

Fisk Stove
Tobey Farm
Box 935
Dennis, MA 02638
plans for airtight, thermostatically controlled bonell heater.

Whole Earth Access Company
2466 Shattuck Ave.
Berkeley, CA 94704
oil drum conversion.

S/A Import Division
730 Midtown Plaza
Syracuse, NY 13210
"Stanley," Irish import; cast iron, high quality; water reservoir.

Washington Stove Works
P.O. Box 687
3402 Smith Street
Everett, WA 98201
"Olympic" and other models; some with hot water heating coils.

Waterford Ironfounders Limited
Waterford, Ireland
"Waterford" cast iron; many models available, high quality and "Stanley" cookstove (see S/A Import Division).

BLAZING SHOWERS

The Blazing Showers Group sells two wood-water heating systems that include heating element, storage tank adapter and manual:

Stove pipe hot water heater	$59
Firebox hot water heater	$49
Storage tank adapter (alone)	$14.50
Manual	$2.50

They have also written a very practical manual on their system. It is a bit hard to read (being hand written), but it explains in a simple step-by-step fashion how to adapt your wood heater or fireplace to heat water. A variety of situations are presented and techniques for mounting, adapting and plumbing the storage tanks as well as installing the heat exchanger are described and illustrated. As a final point

the authors present instructions for building and integrating a solar collector into the wood-water heating system . . . renewable energy backing up renewable energy.

— Richard Merrill

Blazing Showers
1976; 33 pp
$2.00

from:
Blazing Showers
Box 327
Point Arena, CA 95468

COMPATIBILITY OF BLAZING SHOWERS UNIT WITH VARIOUS TYPES OF WOOD BURNING UNITS

Type of Stove	Stovepipe Heater	Firebox Heater
Box Stove	EXCELLENT available from Blazing Showers	EXCELLENT not available from Blazing Showers
Potbelly Stove	EXCELLENT available from Blazing Showers	EXCELLENT not available from Blazing Showers
Drum Stove Manual	EXCELLENT available from Blazing Showers	EXCELLENT available from Blazing Showers for larger sizes
Automatic	POOR	EXCELLENT available from Blazing Showers
Fireplace Stove	EXCELLENT WHEN USED AS HEATER WITH DOORS CLOSED available from Blazing Showers	EXCELLENT not available from Blazing Showers
Barrel Stove	EXCELLENT available from Blazing Showers	EXCELLENT available from Blazing Showers
Cook Stove	FAIR available from Blazing Showers	EXCELLENT not available from Blazing Showers
Jotul Stoves Fisher "Bean" Stoves Riteway 2,000	POOR stovepipe smoke is too cool	EXCELLENT not available from Blazing Showers

HEAT RECLAIMING ACCESSORIES

There are primarily two types of heat reclaiming accessories. For fireplaces, add-on units are manufactured which attempt to extract more convective heat from the fireplace. Many models incorporate hollow metal tubular fireplace grates which suck air in the bottom and blow it out the top of the fireplace opening. For wood heaters, flue pipe heat extractors are manufactured which attempt to blow some of the heat that usually escapes up the flue into the room. They usually hook directly to the flue. Many of these accessories use small electric fans to force the air around the heat source. Heat reclaiming accessories vary in price (usually under $200) and in quality.

American Stovalator
Rte. 7
Arlington, VT 05250
Sheet steel, glass tempered doors, fits in fireplace.

C & D Distributors, Inc.
Box 766
Old Saybrook, CT 06475
"Better n'Ben's" sheet steel wood heater attachment to a fireplace.

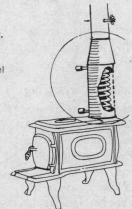

The Hubbard Creek Trading Co.
Box 9
Umpqua, OR 97486
"Heat Saver" flue pipe heat extractor with forced air fan.

Lance International
PO Box 562
1391 Blue Hills Ave.
Bloomfield, CT 06002
"Heat Reclaimer" works

Ridgeway Steel Fabricators, Inc.
Box 382, Bark St.
Ridgeway, PA 15853
"Hydroheat" water circulating in heater flue pipe.

Sturges Heat Recovery, Inc.
PO Box 397
Stone Ridge, NY 12484
Metal chimney devices; firebrick glass device for fireplaces.

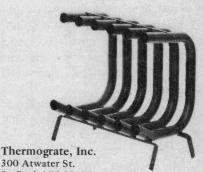

Thermograte, Inc.
300 Atwater St.
St. Paul, MN 55117
U-shaped, tubular steel, fireplace grate. The original "Thermograte."

Torrid Mfg. Co., Inc.
1248 Poplar So.
Seattle, WA 98144
forced air flue pipe heat reclaimer.

INTEGRATED SYSTEMS

So far in the *Energy Primer* we have been talking about separate energy systems . . . solar, wind, water and biofuels. Unfortunately, no *single* source of energy is likely to supply all, or even most, of the power needs of a home, group or community. The supply of each of these natural energy sources is intermittent and at any one time energy demand is likely to exceed energy supply. Furthermore, each energy source comes in a different form which makes it suitable for different uses. Solar energy comes as heat. Wind and water energy provide mechanical power. Biofuels are forms of chemical energy that are generally more portable and versatile. But integration of these sources, combining and sharing their energy loads and waste products, creates the real possibility of providing reliable and economic power supplies under a variety of conditions.

Integrated energy systems can be defined as diverse energy sources combined with one another to provide continuous energy. Picture them as inter-dependent. They are integrated during the process of production: a methane plant may receive waste heat from a solar collector and the gas produced may be compressed with the energy from a wind generator.

The trouble is that very little is known about integrated systems which combine *renewable* energy sources. The classic papers of Golding outlined some of the problems of arid regions[1] and remote communities in Third World countries.[2] A few recent papers have described integrated systems as alternatives in developed countries[3-7], and a few groups and research institutes in the U.S. and elsewhere have begun experimenting with a variety of combined energy resources.[8-16] But there is really little information about the economics, reliability, mechanics, or practicality of integrated solar, water, wind and biofuel operations.

PRACTICAL SMALL-SCALE RENEWABLE ENERGY SYSTEMS

What kind of renewable energy systems are really practical enough to deal with here and now? Table I lists what we believe to be technologies and processes that are available today and not totally out of economic reach, bogged down by laws, grossly inefficient or still on the drawing boards. Admittedly, money or clever minds can create more options. But for our purposes here we will limit our discussion to the "practical" sources listed in Table I and their relative conversion efficiencies in Table II. Table I cannot stand by itself as the final word. The discussions in the previous sections of the *Energy Primer* about these sources and their applications must be analyzed for this table to be meaningful in terms of specific situations.

BASIC ENERGY CONVERSIONS

The first thing to consider is how the five basic forms of energy interact with one another with respect to the "practical" systems outlined in Table I (see Fig. 1). When one form of energy is converted into another there is a loss (in the form of heat) of useable work that accompanies the conversion. For example, as explained in the *Biomass* section, the conversion of solar energy into chemical (plant) energy is only about 1-2% efficient, whereas the conversion of solar energy into useable heat energy is about 40-60% efficient. The important point here is that the more conversions there are between the forms of energy, the less efficient and (usually) the more costly the process will be.

There are exceptions to this rule. For example: in most cases it is more efficient to pump water with a wind generator than a windmill. A wind generator pumping water without batteries will go through two conversions: mechanical to electrical and electrical to mechanical (pump). The entire process will be about .40 x .90 = .36% efficient (Table II). On the other hand a windmill pumping water, which goes through one less conversion than a wind generator, is only about 30% efficient.

The reverse is also true; in some cases, depending on local conditions it is more practical to use less efficient systems to generate needed energy. For example, imagine a site with lots of wind and very little sun (like many

	Device	Practical	Impractical
Source	Process Source	Use	Use
SOLAR	CONVENTIONAL GREENHOUSE	Food production	
	SOLAR-HEATED GREENHOUSE	Food production Space heating	
	COLLECTOR ABSORBER (Active and Passive Systems)	Food drying Water distillation Water heating Space heating	Electricity Cooling
	CONCENTRATOR	Cooking Sterilization Process heat	Electricity* Cooling*
	PHOTOVOLTAICS	Electricity (remote)	Electricity* (large-scale)
WATER (where available)	WATER TURBINE	Electricity	Air compression Electrolytic cell
	WATER WHEEL	Mechanical power DC Electricity	Flywheel A.C. Electricity
	HYDRAULIC RAM	Pumping water	
WIND	GENERATOR	Electricity	Air compression Electrolytic cell*
	WINDMILL	Water pump Mechanical power Air compression Heat pump	
BIOFUELS (photo-synthesis)	PLANTS AND ANIMALS	Food, feed, fiber	
	ORGANIC WASTES (Agricultural and Municipal Refuse)	Compost (fertilizer, heat) Methane (light, heat electricity)	Biomass-Electricity Methane (vehicle fuel, methanol) Ethanol* (light, engines)
	WOOD AND WOOD WASTES	Water heat Space heat	Methanol* Electricity* Wood gas

TABLE I

"PRACTICAL" AND "IMPRACTICAL" CONVERSION AND STORAGE USES OF SMALL SCALE RENEWABLE ENERGY SYSTEMS. LIST IS BASED ON WHETHER THE TECHNOLOGIES ARE PRESENTLY OUT OF ECONOMIC REACH, BOGGED DOWN BY LAWS, GROSSLY INEFFICIENT, STILL ON THE DRAWING BOARDS OR SOCIALLY UNDESIRABLE. (Mid-1977).

*Technologies that are looking increasingly practical.

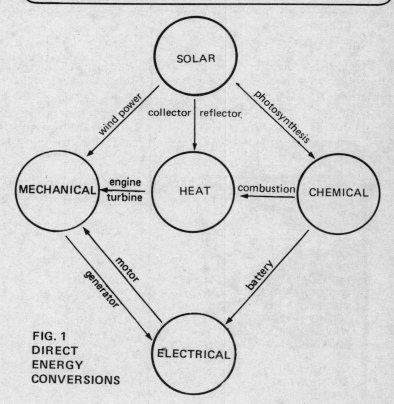

FIG. 1 DIRECT ENERGY CONVERSIONS

coastal climates). Probably the best way to get hot water under these conditions would be to go through the extra conversion and heat water with a wind generator (40% efficient), rather than rely on a slightly more efficient but less practical solar collector.

TABLE II ENERGY CONVERSION OF PRACTICAL SYSTEMS - Maximum Efficiency.

MECHANICAL	Wind Generator 40% Water Turbine 68-93% Steam Power Plant 40% Electric Motor 90%	ELECTRICAL
MECHANICAL	Windmill 20-30% Waterwheel 70-85%	MECHANICAL
SOLAR	Flat Collector 40-60% Concentrator 80-90%	THERMAL (Heat)
SOLAR	Photosynthesis 1-2%	CHEMICAL
CHEMICAL	Wood Combustion Burner 85% maximum Oil Furnace 65%	THERMAL (Heat)
CHEMICAL	Battery 80% (Storage)	ELECTRICAL
HEAT	Engine 25-36% Turbine 35-45%	MECHANICAL
CHEMICAL	Methane Digester 40-60%	CHEMICAL
ELECTRICAL	Resistance heating 99%	THERMAL

TABLE IV MEASURING LOCAL ENERGY SOURCES
see the following pages in the Energy Primer

Wind
   velocity . . . 120, 122-124, 146, 147
   generator . . . 120-126, 128, 130, 131, 139

Biofuels
   methane . . . 197-199
   wood . . . 214, 215, 218-219
   alcohol . . . 211, 212

Water
   head . . . 103
   flow . . . 101-103, 105
   power . . . 101-104

Solar
   insolation . . . 43-48, 51-52
   heaters . . . 48-49

STORAGE SYSTEMS

Inherent in any energy system with intermittent power sources is the need for various storage devices to hold the energy received until it can be used whenever needed. Again the question of practicality comes up. Table V attempts to put in perspective the practical and proven storage devices.

TABLE V RENEWABLE ENERGY STORAGE

	Storage Medium	Storage Efficiency (%)	Major Loss Characteristics	A*	B*
SOLAR (heat)	Water	75-90	Leaks thermal & physical	2	1
	Earth	varies	Leaks (thermal)	1	1
	Rock	60-80	Leaks (thermal)	2	2
	Salt hydrates	75-95	Material breakdown	3	3
WIND/WATER (Mechanical)	Pumped water	50-70	Evaporation Friction	1	2
	Compressed air (Compressor)	40-50	Leaks Friction	2	2
WIND/WATER (electric)	Battery	70-85	Internal discharge	3	3
METHANE	Tank			1	2
	Tank (compressed)	50-60		3	3
ALCOHOL	Tank		Leaks Evaporation	1	1

A⁺ **Relative Cost**
B* **Degree of Mechanical Complexity**

1. Negligible
2. Intermediate
3. Considerable

ASSESSING ENERGY NEEDS

If we are to build workable integrated energy systems we must carefully and accurately assess our energy needs (see pgs. 8-13). As we pointed out in the introduction to this book it makes little sense to gear renewable energy systems to meet current high energy demands. The first step in assessing your needs is *energy conservation*. Make the distinction between your necessities and your luxuries. This does not mean that we must live in a primitive or totally austere way. It means that in measuring our needs we must first conserve what we have and then gear our needs to our resources . . . not the other way around. We cannot apply renewable energy systems to fill present high energy demands and think that we are doing anything about the environment in which we live.

Table III illustrates the energy source which may be applied to each major type of energy need. It should be obvious from looking at the table that the use of any one energy source will not produce sufficient power to meet all energy demands, but that an integrated system has the potential to do so.

MEASURING LOCAL ENERGY SOURCES

We can get a better idea of the availability of energy supplied by actually measuring our endemic resources. Gathering this information by calculating the available solar incidence, wind velocity, water head and flow, and organic fuels makes it easy to see the possibilities for integration of these sources. Table IV gives the page numbers on which measurement data and methods may be found for each energy source discussed in previous sections.

ECONOMIES OF SCALE

Renewable energy systems cost money. . . to build or buy. In some cases (e.g., water turbines and wind generators) the cost of technology is beyond the means of the average person. In most cases it is cheaper to develop renewable energy systems when the cost is spread over a large number of people. Economists refer to this as an economy of scale.

For example, imagine a wind system supplying 1 kw/per person. A 2 kw Quirk's with inverter costs about $9,000 installed or $4,500 per person for two people (1974). A 70kw NOAH rotar wind generator costs about $45,000 (1974) or about $642 per person, for 70 people. Another example: the solar space heating costs (60% of the heat) of a 1000 ft² house (for 2 people) is about $2500. A small community of 45 houses buying similar equipment to solar-heat their individual houses would receive a standard price break of about 25% or a savings of $625 per house. These are economies of scale. Notice that the solar collectors are used in terms of the individual living units whereas the wind generator (and presumably digesters, turbines, greenhouses etc.) are more efficiently used on a cooperative basis.

TABLE III ESTIMATES OF LOCAL ENERGY REQUIREMENTS FOR A SMALL RURAL COMMUNITY
(1 FAMILY = 4 PEOPLE). from: Ref. 6

Energy Use	Annual Energy (kw-hr)*	Timing	Energy Source	Basis of Estimation
COOKING	400/family	day, precise	Or, S	1—1.5 kw stove used 1 hr/day
DOMESTIC WATER	50/family	random	W, Wa, Or	need 150 gal/day; (see *Irrigation*)
REFRIGERATION	200/family	random	W, Wa, S, Or	½ kw unit run 1 hr/day
WATER HEATING	3,500/family	random	S, Or, W	50 gal/day; 50—130°F
LIGHTING & DOMESTIC POWER	850/family	night & day, precise	W, Wa, Or	Five 100-watt lights used 4 hrs/day; 1/3 kwh/day power tools, small appliances, etc.
WATER DISTILLATION	130/family	day, random	S, Or	5/gal/day. Solar collector of 5m² giving 3/4 kw over 24 hours.
SOLAR HOUSE HEAT	4,100—15,000	day & night, precise	S, Or	7.3—10 kw average heat requirements. 70—750 kw-hr heat storage (existing solar houses)
IRRIGATION	30/acre-ft	random	W, Wa, S, Or	Pump — 500 gal/hr; 200 head
CULTIVATION SEEDING	5,600/25hp/ 75 hrs	variable	Or, W, S,	25-hp tractor used 75 hrs/yr.
FOOD/FODDER GRINDING	50/family	random	W, Wa, Or	Average of 1—2 tons/family/hr including animals
THRESHING	50/family	random	Or, W	4 tons/family/year
MISC. SMALL POWER	100/family	day & night, precise	W, Wa, Or	5 kw total power

NOTE: Or = Organic fuels; W = Wind; S = Solar; Wa = Water.   (arranged in order of preference)   *1 kw-hr = 3,413 Btu's.

However there need to be some qualifications. Most economies of scale are based on the availability of cheap fossil fuels. To use fossil fuels economically, large generation and distribution facilities are necessary. As these operations become larger, this leads to a dis-economy of scale in which sheer size and complexity begin to create hidden costs. Although there is very little information about it, the same can be said for

renewable energy systems. The difference is that fossil fuels require centralized economies and centralized politics and both seem undesirable. As our graph below suggests, small-scale renewable sources have small-scale economies. Fossil fuels can't be "used" on a small scale; renewable energy resources can.

References cited:

[1] Golding, E.W. and M.S. Thacker. 1956. THE UTILIZATION OF WIND, SOLAR RADIATION AND OTHER LOCAL ENERGY RESOURCES FOR THE DEVELOPMENT OF A COMMUNITY IN AN ARID OR SEMI-DESERT AREA. Proceedings of the New Delhi Symposium on Wind and Solar Energy, New Delhi.

[2] Golding, E.W. 1958. THE COMBINATION OF LOCAL SOURCES OF ENERGY FOR ISOLATED COMMUNITIES. Solar Energy Journal, 11(1): 7-12.

[3] Clarke, R. 1973. TECHNOLOGY FOR AN ALTERNATIVE SOCIETY. New Scientist, 11 Jan., 1973.

[4] Merriam, M. 1972. DECENTRALIZED POWER SOURCES FOR DEVELOPING COUNTRIES. International Development Society, International Development Review, 14(4): 13-18.

[5] Thring, J.B. and G.E. Smith. 1974. INTEGRATED POWER, WATER, WASTE AND NUTRIENT SYSTEM. Working Paper, Technical Research Division, Dept. of Architecture, Univ. Cambridge, England.

[6] Weintraub, R. and D. Marier. 1976. LOCAL ENERGY PRODUCTION FOR RURAL HOMESTEADS AND COMMUNITIES. In: "Radical Agriculture." R. Merrill (ed.). Harper and Row, New York.

[7] Marier, D., R. Weintraub and S. Eccli. 1974. COMBINING ALTERNATIVE ENERGY SYSTEMS. In: "Producing Your Own Power," Rodale Press, Emmaus, Pa. Revised and abridged from reference 6 above.

[8] NEW ALCHEMY INSTITUTE. 1974. Journal of the New Alchemists, No. 2. Box 432, Woods Hole, Mass. 02543. See also pages 133-135 and 181 of PRIMER.

[9] Blazej, R. et al. 1973. PLANS FOR GRASSY BROOK VILLAGE . . .A PROSPECTUS. RFD 1, Newfane, Vt. 05345.

[10] INTEGRATED LIVING SYSTEMS. See pages 181 and 189 in PRIMER.

[11] CENTER FOR MAXIMUM POTENTIAL BUILDING SYSTEMS. Austin, Texas. See pages 100, 144, 181, 192 in PRIMER.

[12] Hughes, F.P. 1973. THE ECO-HOUSE. Mother Earth News, Vol. 20, March 1973. Description of the autonomous house being built in England by Graham Caine.

[13] Vale, R.J. 1974. SERVICES FOR AN AUTONOMOUS RESEARCH COMMUNITY IN WALES. Working Paper, 5, Technical Research Division, Cambridge University, (Dept. Architecture).

[14] Golueke, C.G. and W.J. Oswald. 1973. AN ALGAL REGENERATIVE SYSTEM FOR SINGLE-FAMILY FARMS AND VILLAGES. Compost Science 14(3).

[15] THE ECOL PROJECT, Minimum Cost Housing Project, McGill University, Montreal, Canada.

[16] Davis, A.J. and R.P. Shubert. 1974. ALTERNATIVE NATURAL ENERGY SOURCES IN BUILDING DESIGN. College of Architecture, Virginia Polytechnic Institute, Blacksburg, Va. Last section of the book describes several working or proposed autonomous housing units using integrated energy systems.

FIG. 2 AN ECONOMY OF SCALE FOR SOME RENEWABLE SOURCES OF ENERGY

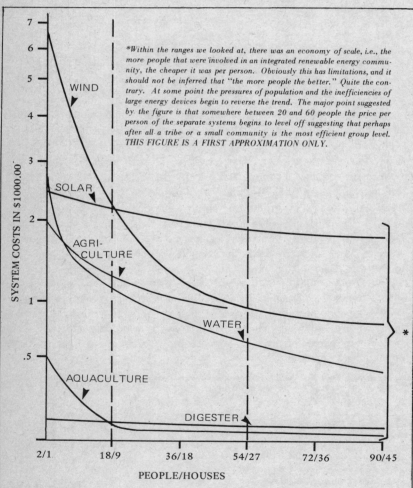

Within the ranges we looked at, there was an economy of scale, i.e., the more people that were involved in an integrated renewable energy community, the cheaper it was per person. Obviously this has limitations, and it should not be inferred that "the more people the better." Quite the contrary. At some point the pressures of population and the inefficiencies of large energy devices begin to reverse the trend. The major point suggested by the figure is that somewhere between 20 and 60 people the price per person of the separate systems begins to level off suggesting that perhaps after all a tribe or a small community is the most efficient group level. THIS FIGURE IS A FIRST APPROXIMATION ONLY.

ASSUMPTIONS

1. Wind or water (electricity): 2KW/2 people; 10 mph wind or equivalent head and flow of water; system has 20 year life. Source: T.W.
2. Solar heat: 2 people/1000 ft^2 house; 60% space heating; below 40° latitude; 20 year life. Sources: Löf and Tybout in **Solar Energy**, (vol. 14:253), Zomeworks, and Ecology Action/ Palo Alto.
3. Digester (gas): 15 ft^3/person/day; available organic matter; cooking only; 5 year life. Sources: Ram Bux Singh, Spirogester Co., Lakeside Engineering Corporation, 222 W. Adams St., Chicago 6, Illinois 90606, and Maximum Potential Building Systems.
4. Aquaculture (food): 70 grams of animal protein per day/person; 10 year life; see page 133. Sources: Solar Aquafarms and New Alchemy Institute.
5. Agriculture (food): .5 acre of land; sufficient tools, water, seeds, etc.; vegetarian diet (vegetables, grains, herbs, fruits); intensive garden farming. Sources: Ecology Action/Palo Alto and New Alchemy West.

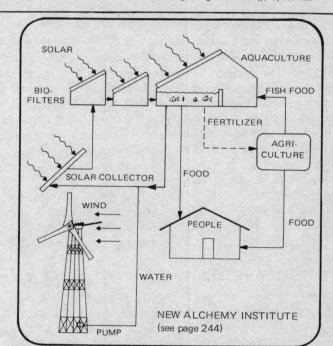

NEW ALCHEMY INSTITUTE
(see page 244)

FIG. 3 TWO PROTOTYPE INTEGRATED SYSTEMS NOW WORKING IN THIS COUNTRY

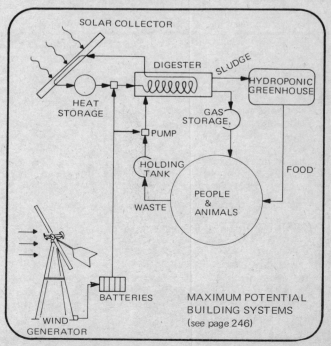

MAXIMUM POTENTIAL BUILDING SYSTEMS
(see page 246)

INTEGRATED SYSTEMS REVIEWS
RENEWABLE ENERGY ANTHOLOGIES

OTHER HOMES AND GARBAGE

This book is a joy and a bargain. Don't be put off by the title; it's not a doctrinaire tract. Though there is an underlying current of conviction in the direction of small is beautiful, independence is worthwhile, etc., the vast bulk of the printed material is facts and figures and how-to-do-it explanations.

The range of material is broad, there are chapters on *Alternative Architecture* (energy conserving design, climate factors, equipment available); *Small Scale Generation of Electricity from Renewable Sources* (wind and water); *Solar Heating* (climatology, types of systems, human comfort requirements); *Waste Handling Systems* (conventional and unconventional, aerobic and anaerobic); *Water Supply* (hygeine and engineering); and *Agriculture* and *Aquaculture* (crop requirements, composition of residues, operational procedures). The material is factual and distilled. There are a lot of tables of useful numbers needed for design, some graphs and charts, and some equations. The graphics are unusually well done.

The authors are junior faculty at Stanford University; the book was put together as a outgrowth of a year-long course for undergraduates and interested community people. For anyone working on or planning a semi-independent living style in an isolated rural area this book will be extremely useful.

However, the procedure for sizing solar collector systems contains several errors and appears to lead to performance estimates that are up to 100% higher than other more recent methods. Design calculations should be checked against other sources such as ERDA'S PACIFIC REGIONAL SOLAR HEATING HANDBOOK.
—Marshal Merriam

EARTH, WATER, WIND AND SUN

This is a fine objective overview of various renewable energy systems (geothermal, hydro, solar, wind and biofuels). Halacy discusses the history and potential of each resource and, although small independent systems are dealt with, the focus is on megascale-government projects. The wind-solar sections seem superficial and too pessimistic, and some of the references are inaccurate. Nevertheless, the graphics and description of geothermal, seathermal, hydro and tidal power plants are especially valuable.
— R.M.

Earth, Water, Wind and Sun
D.S. Halacy Jr.
1977; 186 pp.

$8.95

from:
Harper and Row
10 East 53rd St.
New York, NY 10022

Other Homes and Garbage
Jim Leckie, Gil Masters, Harry Whitehouse, Lily Young
1975; 302 pp.

$9.95 postpaid

from:
Sierra Club Books
c/o Book Warehouse Inc.
Vreeland Ave.
Totowa, NJ 07512

or WHOLE EARTH
    TRUCK STORE

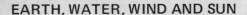

ENERGY: THE SOLAR PROSPECT

This paper is the most concise and up-to-date review of renewable energy systems . . . large and small scale. Chapters include: *Dawn of an Era, Solar Heating and Cooling, Electricity From the Sun, Catching the Wind, Falling Water, Plant Power, Storing Sunlight, Turning Toward the Sun.* It is adapted from the author's forthcoming book, RAY S OF HOPE: THE TRANSITION TO A POST—PETROLEUM WORLD (W.W. Norton, 1977; $3.95) which promises to be the up-date sequel to Wilson Clark's fine book ENERGY FOR SURVIVAL. Like Clark's book, THE SOLAR PROSPECT discusses not only the practical applications and ecological imperatives of renewable energy systems, but also their political implications.

If the (1/3 present U.S.) postulated energy demand (of the world) were met with nuclear fission, about 15,000 reactors as large as the biggest yet built would have to be constructed . . . one new reactor a day for 50 years. Sustaining these reactors would require the recycling of 20 million kilograms of plutonium annually. Every year, enough plutonium would be recycled around the world to fabricate four million Hiroshima-size bombs. Such a prospect cannot sanely be greeted with equanimity.

. . . Thus we are left with the solar options: wind, falling water, biomass, and direct sunlight. Fortunately, they are rather attractive. Solar sources add no new heat to the global environment and . . . when in equilibrium . . . they make no net contribution to the atmospheric carbon dioxide. Solar technologies fit well into a political system that emphasizes decentralization, pluralism, and local control . . . The sun's inconstancy is regional and seasonal, not arbitrary or political, and it can therefore be anticipated and planned for.
— R.M.

Energy: The Solar Prospect
Denis Hayes
1977; 79 pp

$2.00

from:
Worldwatch Institute
1776 Massachusetts Ave., N.W.
Washington, D.C. 20036

PRODUCING YOUR OWN POWER

PRODUCING YOUR OWN POWER is an anthology of articles about solar—water—wind power and methane digesters. One third of these articles are reprints from other publications: the water section is reprinted information from VITA on small water power sites; half of the methane section is a partial reprint of the New Alchemy West Newsletter; one of the wind section articles is largely a reprint of Henry Clew's own pamphlet *Electric Power From The Wind*; Eugene Eccli's *Conservation of Energy in Existing Structures* appeared previously in an issue of "Alternative Sources of Energy" Newsletter; and the fine article by Dan Marier et al on *Combining Alternative Energy Systems* is an abridged version of an article appearing in RADICAL AGRICULTURE. The best original article is *Heating and Cooking with Wood* by Ken Kern.

Both solar articles are original. They cover space heating and water heating and are elementary enough to be easily understood and meaty enough to provide some design information. The second methane article discusses a four-cow biogas plant built on a Vermont dairy farm. The second wind article by Jim DeKorne discusses using an air brake governor (Wincharger) and a surplus aircraft generator to build a wind generator modeled after a "Jacob's Model No. 15."
— T.G.

Producing Your Own Power
C.H. Stoner, Editor
1974; 322 pp

$8.95

from:
Rodale Press, Inc.
Book Division
Emmanus, PA 18049

or WHOLE EARTH
    TRUCK STORE

MORE ANTHOLOGIES

Wind/Solar Energy
Edward M. Noll
1975; 208 pp.

$7.95

from:
Howard W. Sams & Co., Inc.
4300 West 62nd St.
Indianapolis, ID 46268

This book is a beautifully illustrated introduction to the techniques of converting light energy and wind power into electricity. The book also discusses conversion and storage systems common to photovoltaics and wind generators, and lists various manufacturers. A handy solar-electric sourcebook for the home and small business.

The Handbook of Homemade Power
The Mother Earth News
1974; 374 pp.

from:
Bantam Books, Inc.
666 Fifth Ave.
New York 10019

$1.95

One of the first widely available alternative energy anthologies, this book is mostly a compilation of articles from back issues of MOTHER EARTH NEWS. It lacks technical information, but the basics are there. The most complete water power article written to date, a 1947 *Popular Science* piece, is reprinted entirely.

Survival Scrapebook #3, Energy
Stefan A. Szczelkun
1973; 112 pp.

$3.95

from:
Schocken Books
New York

A very loose and easy book that discusses solar, wind, tidal, bio-gas and animal power. Simple hand-written text and illustrations show basic principles and provide ideas for building small demonstration models.

Natural Energy Workbook
Peter Clark with Judy Landfield
1976 (rev. ed.); 128 pp.

$3.95

from:
Visual Purple
Box 996
Berkeley, CA 94701

A pleasant booklet that gives some background, technical details and lots of sketches of plans for doing-it-yourself with "locally regenerative resources (sun, wind, water and photofuels)." Poorly organized, but useful ideas can be gleaned with patience (e.g., instructions for making a parabolic mirror).

Alternative Sources of Energy
Eugene Eccli et al (eds)
1974; 278 pp

$5.00

from:
ASE Newsletter
Route 2
Milaca, MN 56353

Subtitled "Practical Technology and Philosophy for a Decentralized Society," this classic book includes the articles and letters from the first 10 issues of ALTERNATIVE SOURCES OF ENERGY Magazine. It is especially interesting for its early perspective of small-scale renewable energy systems . . . before they came out of the closet.

Energy for the Home
Peter Clegg
1975; 252 pp

$5.95

from:
Garden Way Publishing Co.
Charlotte, VT 05445

Very short on original material, here is a fine catalog of solar, wind, water, wood and methane systems . . . hardware and books. Each energy section is briefly introduced and then pages are reproduced from manufacturers' brochures.

EnergyBook 2
John Prenis (ed)
1977; 125 pp

$5.95

from:
Running Press
38 South 19th St.
Philadelphia, PA 19103

Far better than ENERGYBOOK 1, this is yet another anthology of renewable energy systems. Some information found here is not found elsewhere (e.g., methane-powered vehicles).

Environmentally Appropriate Technology
Bruce McCallum
1977; 155 pp

$3.75 (Canada)
$4.50 (elsewhere)

from:
Publishing Centre
Printing & Publishing
Supply and Services Canada
Ottawa KIA 059

Excellent review of existing renewable energy systems, their principles and applications to Canada (and other northern regions). Well-done graphics; state-of-the-art is good.

APPROPRIATE TECHNOLOGY

Modern technology is too large, too complex, too expensive and too violent. It does not solve the problems of poverty. It destroys ecosystems and it causes us to slip further and further away from our fundamental sources. Relying on it, we trade convenience for the freedom to design and manage our own modes of food and energy production, waste recycling, shelter construction, health and the other basic needs of our lives. Appropriate technology (AT), on the other hand, embraces tools and techniques that are smaller, cheaper, simpler and ecologically wise—appropriate to the human scale. Appropriate technologies are adapted to local conditions and renewable resources; by their very nature they embrace a philosophy of decentralization.

RAINBOOK: RESOURCES FOR APPROPRIATE TECHNOLOGY

RAIN (pg. 248) is one of the few periodicals I look forward to reading from cover to cover and back again. RAINBOOK is a sifting of the best from RAIN through mid-1977, plus additional comments and articles. It is a beautifully edited, well-written, annotated catalog of "resources for appropriate technology" (Place, Economics, Community Building, Communications, Transportation, Shelter, Agriculture, Health, Waste Recycling, Energy). It is definitive. In fact, RAINBOOK is the reason our "Appropriate Technology" section in the ENERGY PRIMER is so small . . . all the groups (foreign and domestic), books, and access sources are well described here . . . so why be redundant, especially when something of this quality is available at a reasonable price. Highly recommended by all of us.

— R.M.

Ironically, AT has now been accepted at every level of established government. From the United Nations—to the federal level (National Center of Appropriate Technology, Box 3838, Butte, MT 59701)—to the state level (Office of Appropriate Technology, Box 1677, Sacramento, CA 95808). More importantly, local AT groups are growing rapidly as more and more people come to understand the unifying environmental and social principles behind AT.

Because AT utilizes less energy-consuming tools and habits, much of the resource material on AT includes information on solar, wind, water and biofuels. Check the books below

(especially RAINBOOK). Also: TRANET, 7410 Vernon Square Dr., Alexandria, VA 22306, for international AT networking; Integrative Design Associates, 1740 H St., N.W., Washington D.C. 20036 for comprehensive lists of AT resources; A HANDBOOK ON APPROPRIATE TECHNOLOGY, Canadian Hunger Foundation, 75 Sparks St., Ottawa KIP 5A5 for a catalog of equipment, and, finally, the books below.

The lesson of the 70's is that change must begin at home, in the community and within the region. Technology is not the problem nor is it the solution. But certain kinds of tools will make the path to a just, stable and fulfilling society more possible.

— R.M.

RAINBOOK
1977; 251 pp.

$7.95

from:
Schocken Books
200 Madison Avenue
New York City 10016

THE LAST WHOLE EARTH CATALOG
WHOLE EARTH EPILOG

These are probably two of the finest source books in print; their format has been an inspiration to dozens of subsequent catalogs . . . including the ENERGY PRIMER. These two classic books are generally well known, not much need be said other than the list and review books and hardware on topics such as crafts, building, education, cybernetics, gardening, community, land use, communications and more.

The Soft-Technology sections of each book includes reviews of energy-related books, and hardware. Information on windmills, water turbines, solar power, composting and methane is presented. The information in the Soft-Technology section of the EPILOG is more plentiful and current than CATALOG listings.

—T.G.

The Last Whole Earth Catalog
1971; 448 pp.

$5.00

from:
Random House
457 Hahn Road
Westminster, MD 21157

or WHOLE EARTH
TRUCK STORE

Whole Earth Epilog
1974; 319 pp.

$4.00

from:
Penguin Books, Inc.
7110 Ambassador Rd.
Baltimore, MD 21207

or WHOLE EARTH TRUCK STORE

RADICAL TECHNOLOGY

Ah yes, Wind machines, poorly detailed solar collectors, goats, pyramids, and typical bourgeois schemes masquerading as environmentally OK because they are wearing the less-embarrassing pre-washed Levis which make them look like they've been working. You can find lots of books that look like this one in any "organic bookshop." But this book is different. It has sharp criticism of society and just about everything else you might think of (done in that sly British manner) and this is coupled with the best presentation of "Visions" of what may be done that I've seen. The emphasis is on changing social order by taking responsibility for your actions into your own hands. The publisher's blurb calls it "the encyclopedia of a multifaceted quest." The only book in this part of the culture that I have personally found exciting and excited. It's by the *Undercurrents* people.

— J. Baldwin

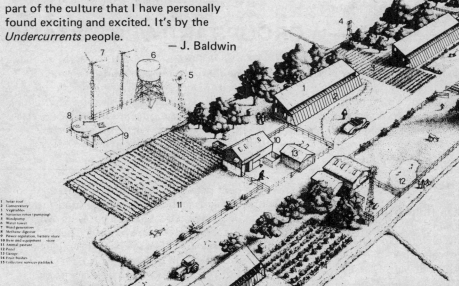

1 Solar roof
2 Conservatory
3 Vegetables
4 Saturation meter (pumping)
5 Windpump
6 Water tower
7 Wind generators
8 Methane digestor
9 Power regulation, battery store
10 Byre and equipment store
11 Animal pasture
12 Pond
13 Garage
14 Fruit bushes
15 Collective services paddock

Subjects are grouped under FOOD, ENERGY, SHELTER, AUTONOMY, MATERIALS, COMMUNICATIONS AND OTHER PERSPECTIVES. The British wit and political satire as it applies to appropriate technology are worth the price of the book alone.

— R.M.

Radical Tech'nology
Godfrey Boyle and
Peter Harper (eds)
1976; 304 pp.

$5.95

from:
Random House, Inc.
Pantheon Books
455 Hahn Road
Westminster, MD 21157

Intended for rural or semi-rural areas. These dwellings would be independent of grid services. Some services (waste treatment, some food, space and water heating) would be provided at the household level, others (electricity, water, cooking gas, some food) at the community level, where economies of scale make shared facilities cheaper. The houses are based on Brenda Vale's 'Autonomous House' design, described further in part IV of the Autonomy section.

POLITICS, RENEWABLE ENERGY AND DECENTRALIZED SOCIETY

Ever since Goldwin's POLITICAL JUSTICE, the idea of a decentralized society has rubbed like grit against the traditional centralist thinking of Socialists, Marxists and Capitalists. The political dichotomy is not between the left and the right, as we are led to believe, but rather between government and no government . . . or at least between centralized bureaucracies—industries and self-reliant local economies. As central governments decay under their own weight, there becomes a growing need for regional control of resources and institutions. Indeed, belief in a society where power rises upward from the common base rather than downward from a concentrated power runs thinly but persistently through modern history. Yet, such notions are typically suppressed as being chaos, violence and utopia.

Part of the problem is that political alternatives are usually analyzed in a political context. Few people have sought to describe the science and technology that could make a decentralized society work . . . or at least get off the ground.

Bookchin, in his now classic anthology of essays POST-SCARCITY ANARCHISM, argues that a truly alternative society must integrate an ecologically based science with a libertarian theory of human activity. Furthermore, the quality of technology is as important as the quantities that it provides, and this means

that we must scale down and diversify our energy resources and integrate them with natural laws. Dickson, in ALTERNATIVE TECHNOLOGY . . . , seems more concerned with putting down the myth that technology is politically neutral, and discussing the role of "intermediate" technology in the Third World. LIVING ON THE SUN by Boyle is a description of various renewable energy systems and their role in helping to establish an "equitable" society. Finally, Ophuls, in his book ECOLOGY AND THE POLITICS OF SCARCITY hammers away at the point that unending growth is the guiding principle of all industrial civilizations, regardless of political philosophy; that our old

social paradigms of growth and abundance are blueprints for total collapse and that some resolution might lie in the adoption of small-scale technologies and the wide acceptance of a new political "steady state."

In different ways all four books deal with an increasingly popular vision . . . a decentralized society in equilibrium with its environment, which is fueled principally by natural energy systems and technologies appropriate to a scale of local control. A far cry from reality . . . especially on the global scale . . . but fuel for the fires of necessary change.

— Richard Merrill

Post Scarcity Anarchism
Murray Bookchin
1971; 228 pp.

$2.95

from:
Ramparts Press, Inc.
P.O. Box 10128
Palo Alto, CA 94303

or WHOLE EARTH TRUCK STORE

Living on the Sun: Harnessing Renewable Energy for an Equitable Society
Godfrey Boyle
1975; 127 pp.

$4.50

from:
Calder and Boyars Ltd.
18 Brewer St.
London W1R 4AS

Alternative Technology and the Politics of Technical Change
David Dickson
1974; 224 pp.

$1.25 plus postage

from:
Collins Publishing
14 St. James Place
London SW1, England

Ecology and the Politics of Scarcity
William Ophuls
1976; 303 pp.

$6.95

from:
W. H. Freeman and Co.
660 Market Street
San Francisco, CA 94104

SMALL IS BEAUTIFUL

As a Quasi-economist, and a rather troubled one, I have searched a dozen years for a "leading" economist willing to discuss some plain, present questions that arise under the label of "people's economics."

Needless to say, the answer I usually receive has something to do with the year 2000, or after a new political system is achieved, or when the population is converted to a condition rather resembling sainthood.

E. F. Schumacher appears to be an exception. He knows how to rise above 20 year projections, statistics and theories (which he doesn't avoid, however) to deal with "intermediate technology," "Buddhist economics," and village development, for examples.

His analytical style displays his training as a Rhodes Scholar and his experience as advisor to the British Control Commission in postwar

Germany, and his twenty years of service as the top economist at the British Coal Board. But his snappy conclusions and recommendations connect plainly to his career as president of Britain's oldest organic farming organization, and as sponsor of the Fourth World Movement (for political decentralization), and as director of the renowned worker-controlled Scott Baler Company.

In his book he pops a number of dust-covered balloons. For example, he pricks the illusion that the poor will someday share the bounties of the rich, and quotes Gandhi: "Earth provides enough to satisfy every man's need, but not for every man's greed."

In this age when old-time economists have nothing to say to straighten up the world's mess, Schumacher sounds serene and sane to us who live in the cracks of this weird dinosaur establishment.

—Richard Raymond

Small is Beautiful

E. F. Schumacher, 1973; 290 pp

$3.75

from:
Harper & Row
10 East 53rd Street
New York, NY 10022

or WHOLE EARTH TRUCK STORE

GROUPS AND ORGANIZATIONS

There is an evolution to the environmental revolution . . . a symbiosis of ecological ideals and economic-political realities. Today scores of groups and organizations around the country (and abroad) are working with renewable energy systems and their application to energy production, shelter design, waste management, endemic food production/economies, and other tools for local self-reliance. Many of these groups are small non-profit organizations . . . others are associated with schools and established institutions. Some are quite scientific but with a new focus on integrating bio-technical systems for practical ends, rather than analyzing specialized fields for "objectivity." Others are demonstrating the political and social benefits inherent in a decentralized, solar-based society. But whatever their inclinations, the success of such groups around the country are making it clear that a post-industrial renaissance will require the collective wisdom and skills of diverse talents focused on a local scale. "Self-sufficiency" has evolved into "self-reliance."

— R.M.

THE NEW ALCHEMY INSTITUTE

The New Alchemy Institute (NAI) is one of the foremost groups in Northern America working with renewable food and energy systems. They have major research centers in Woods Hole and Prince Edward Island, and work groups in California (education) and Costa Rica.

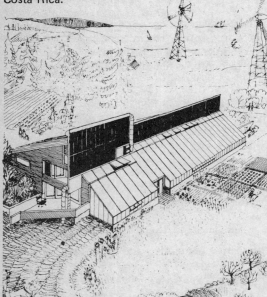

NAI is a prime mover in the area of wind power (see pg.151 in ENERGY PRIMER), aquaculture (see pg.183 in ENERGY PRIMER), and integrated food-power-waste shelters. Toward this end NAI has been creating and studying, for a number of years, small enclosed bioshelters for the culture of food. Starting with a backyard fish farm—greenhouse, enclosed in a dome, the bioshelter concept evolved into a "miniature ark," a small wind-powered and solar-heated complex for culturing fish and greenhouse foods in ecologically-linked cycles (see Journal No. 2 of the New Alchemists).

The next stage in development was the ark itself, which can be described as a food-producing bioshelter that generates its own power and recycles its own waste. The "Prince Edward Island Ark" is the latest example of this habitat design (see Journal No. 3 and No. 4 of the New Alchemists); it is a true laboratory for the study of domestic renewable energy. Space heating is accomplished by a variety of active/passive; rock/water; wood/solar systems, including a large attached greenhouse that generates heat, food, and clear air, while recycling wastes. Domestic water is solar-heated (thermosyphon). The fish system in the ark is a breakthrough in close-system aquaculture, and the perfection of a unique misting system for tree propagation will be of much value to all of us. Detailed specifications of the ark are given in AN ARK FOR PRINCE EDWARD ISLAND, available in limited supply for $200. Hopefully, requests will make it possible for NAI to publish a less expensive edition.

— R.M.

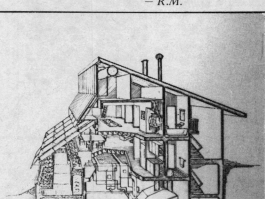

The New Alchemists also record their activities in their Journal which is available, through membership, for $25/year. For those in need there is also a $10 membership. The Journal contains information you will not find anywhere else. Write and ask for their "blurb"; it contains listings of numerous other books and articles written by New Alchemists.

Send all correspondence to:
The New Alchemy Institute
Box 432
Woods Hole, MA 02543

FARALLONES INSTITUTE

Farallones Institute is a non-profit research and educational organization working with an appropriate technology focus in the areas of alternative energy, agriculture, building design, and resource recovery. The Institute started in 1973 when a group of concerned people—architects, biologists, builders, designers, and agriculturists—formed an organization to study alternative systems: to combine the knowledge of individual experience with the scientific and educational process.

In four years Farallones Institute has grown to include urban centers in Berkeley and San Francisco (scheduled to open Spring 1977), and a rural center in Sonoma County. The purpose of the Integral Urban House in Berkeley is to provide a demonstration center for an ecologically sound and self-sustaining urban life style, and to provide convenient and accessible public education classes, workshops, and tours.

The Rural Center in Occidental is the base for the Institute's resident apprenticeship program. People come to the Rural Center to learn the elements of "whole life systems" living. Long term apprenticeships are encouraged for people interested in developing a skill in one area such as solar design, horticulture, or waste management. The Rural Center also provides a base for research and experimentation. A four unit comparison study of solar space heating systems is in progress; a small-scale greywater system for agricultural use has been installed; and a research and evaluation program to study on-site waste recycling using composting privies is in its second year. On the agricultural side, programs are being developed to study problems besetting rural areas and the small-scale family farm.

The San Francisco Center, to be located in an unused government building in the Presidio overlooking San Francisco Bay, will offer architectural design and systems design for solar space heating and hot water heating to people including homeowners, builders, and designers, who are interested in possibilities for solar energy.

THE INTEGRAL URBAN HOUSE
1516 Fifth Street
Berkeley, CA 94710   (415) 525-1150
Visiting Hours: Sat. 1—5 pm
Groups by appointment

THE RURAL CENTER
15290 Coleman Valley Road
Occidental, CA 95465   (707) 874-3060
Visiting Hours: Sat. 1—5 pm
(Closed December thru March)
Groups by appointment

Farallones Institute publishes the following pamphlets. Available by mail from either of the above addresses.

Farallones Institute information packet, including the 1977 Farallones Report describing in detail the projects and work being done by the Institute. $2.00

The Homemade Windmills of Nebraska (see page 134) $3.00

The Composting Privy, Technical Bulletin Number One $2.00

Raising Rabbits and Chickens $1.00

Backyard Composting $0.50

Plans of the Integral Urban House Solar Water Heater $3.00

AERO

The Alternative Energy Resources Organization (AERO) is a non-profit citizens' educational organization. AERO was formed in 1974 as an offshoot of the Northern Plains Resource Council, which was seeking methods for combating stripmining in the Northern Plains area. A major strategy of the Council was to promote the use of renewable energy systems. Since its inception, AERO has tried many different methods of disseminating information on renewable energy resources (solar energy, wind power, small-scale hydropower, geothermal heating, wood energy and bio-gas); they have sponsored numerous hands-on workshops and a unique traveling "New Western Energy Show" that has spawned several other such shows around the country. AERO has found out, as others have, that everybody is interested in energy, and that the benefits of renewable energies and energy conservation can be made understandable to anyone interested in finding out. AERO publishes an excellent periodical, SUN TIMES, which contains articles and news items about renewable energy, and the energy problems of the Northern Plains area. A $10.00 membership fee includes subscription to SUN TIMES.

– Michael Riordan

Alternative Energy Resources Organization
435 Stapleton Blvd.
Billings, MT 59101

CERRO GORDO

I think the most exciting new community/renewable energy project in the country is now coming together near Cottage Grove, Oregon. A rapidly growing group of families (100+) from all over the country has banded together to purchase 1200 acres of beautiful forested hillside land, fund a detailed land use study, and start planning for a new town that will eventually be home for hundreds of families. The automobile is being planned *out* of the community, and personal sharing, renewable energy sources and energy conservation are being planned in. We are looking for more people to help us plan this dream, fund its future and build its reality. Send for more free information or subscribe to the future resident's newsletter.
—C.M.

The Cerro Gordo Experiment:
A Land Planning Package
$3.00

The Community Association Newsletter
Monthly
$15/year

from:
Cerro Gordo Community Association
704 Whiteaker Street
Cottage Grove, Oregon 97424

Above all we are seeking a mode of life which promotes a deeper appreciation of our shared humanity and recognizes our interdependence. . . a mode of life which respects our ultimate dependence on natural cycles of energy and physical resources. . .an ethic of self-determination, freedom, cooperation, and personal growth. . .a decision-making process that encourages expression of individual values and reflects diverse points of view. . .a teaching-learning community with educational opportunities learner-centered and open to all. . .

CENTER FOR MAXIMUM POTENTIAL BUILDING SYSTEMS

The Center for Maximum Potential Building Systems, known as Max's Pot, is a non-profit corporation located in the hill country near Austin, Texas. Max's Pot is a group of architects, engineers, soil specialists, etc., who are focusing their efforts on the semi-arid and hot-arid Southwestern United States. The laboratory began in Fall of 1972 as a project of the School of Architecture, University of Texas. The 'initiators,' Daria and Pliny Fisk, were interested in developing appropriate technology living systems and exploring the cultural bases with which such systems could work. They did this by requesting that the early lab space should be without normal facilities such as heating, cooling, electricity and water. The intent was to make students confront the problems of making the environment habitable . . . not through the use of scarce and irreplaceable natural resources, but with natural forces and resources generally available in the environment and capable of being regenerated. Subsequent funding (mostly from private institutions and Federal grants) brought about removal of the center from University grounds and control. Early projects developed prototype systems for bio-gas units, wind generators, active and passive solar systems, greenhouses, earth heat and cooling devices and earth-building materials. Emphasis continued to be placed on using resources that were readily available or discarded in the environment.

Philosophically we are Appropriate Technologists, working holistically to integrate technology with environmental preservation. Our bias is for those people favoring self-reliance and self sustenance. Although our systems can be used commercially, their main thrust is for the Owner-Builder. . . We feel that as people develop more than one skill based on local resources, their value within the region becomes more important, and more regional knowledge with local skills can be identified, thus decentralizing more regional knowledge and more responsibility.

At present Max's Pot is especially interested in mapping local resources and developing the design of local, historic structures that utilized such natural principles as thermal chimneys, earth/air heat exchangers, caliche based adobe and sky therm roofs. Current projects include:

- An earth&caliche house for Granger, Texas, solar heated and cooled via earth/air heat exchangers, thermal chimneys and rock storage.
- An inner city, low cost, solar-heated housing project for San Antonio, Texas.
- An urban alternatives project demonstrating energy/resource conservation and cooperation in both physical, social and economic terms; at the scale of about a block, in the heart of one of the model cities neighborhoods in Austin.
- Portable earth material labs that allows people to do on site soil analysis and testing to determine suitability of local soils for building.
- Integrated system pavilion at the Center, demonstrating low water use and solar stills; solar heating and cooling via thermal chimneys and earth/air heat exchangers.
- Two solar greenhouses.

Max's Pot has several publications available describing the designs of their projects (Horizontal Drum Biogas Plant $5; 55-gallon Drum Stove $5; Wiring Diagram for a Differential Thermostat $2; Soil-Test Mini-Lab $4 and others. When you write, please send a self-addressed stamped envelope.

Center for Maximum Potential Building Systems
6438 Bee Caves Rd.
Austin, TX 78746

THE INSTITUTE OF SOCIAL ECOLOGY, GODDARD COLLEGE

Goddard College Social Ecology Institute sponsors education and research in alternative energy and agriculture.

Now in its fourth year, the Institute addresses the crucial question of how a technological society relates to ecological principles. The question is approached in three ways:

(1) through a search for the cultural, social, and economic roots of the environmental crisis; (2) through discussion of practical alternatives to existing forms of social organization, technology, and urban life and values; and (3) through "hands-on" experiments in wind and solar power, methane production, biological agriculture, aquaculture, and environmentally-sound housing.

Murray Bookchin, author of *Post-Scarcity Anarchism*, *Our Synthetic Environment*, and *Limits to the City* founded the Institute four years ago. Since that time, the Institute has evolved into a teaching and research installation that has drawn national and international attention.

A major concern of the Institute's community outreach effort is research, development, and installation of simple, low-cost winterization devices for low-income homeowners and renters in the northeast. Annual winterization workshops and conferences are held for community action workers and other interested persons.

Visiting faculty in the summer program have included Karl Hess of Community Technologies; Sam Love, co-founder of Environmental Action; John Todd, Nancy Todd, and Rich Merrill of the New Alchemy Institute; Wilson Clark, author of *Energy for Survival*; Steve Baer of Zomeworks; Eugene Eccli, editor of *Low-Cost, Energy-Efficient Shelter*; David Morris of the Institute for Local Self-Reliance; Hans Meyer and Ben Wolff, co-founders of Windworks.

During the summer of 1977, a new summer-based M.A. Program in Social Ecology was begun. Under the direction of Dan Chodorkoff,

the program combines intensive study in the summer program with a nine-month internship with one of several cooperating organizations in the northeast, including the Green Guerillas, the Urban Homestead Assistance Board, the 519 East 11th Street Project, Charas, and Consumer Action Now.

Information on any Goddard College Program can be obtained from: Summer Programs Office, Box 7, Goddard College, Plainfield, Vermont 95667; or (802) 454-8311. ext. 308.

ECOTOPE GROUP

Ecotope Group is a tax-exempt organization composed of designers, builders, researchers, educators and applied engineers working on renewable energy technologies, conservation and shelter systems. We design, build, conduct research, consult, provide information and educational workshops on the many options available for creative survival. Support for our work has come from: sale of publications; consultation; donations; research and demonstration projects funded by local, state and federal organizations.

Ongoing projects include:

(1) Solar water heating workshops—almost 1000 people have attended the dozen workshops conducted in Washington State, Canada, Idaho, Montana, Wyoming and California. The solar water heating system is either installed later by the sponsor or raffled off to workshop participants who later install the system at their home. Workshop manuals are available for $3.00 from RAIN/Ecotope (Portland office).

(2) Methane Demonstration Project—a 100,000 gallon methane digester was completed at the Monroe State Dairy Farm in January 1976. At present Ecotope is under a new contract with ERDA to test, monitor and operate the system. This activity will be a part of the ERDA— Division of Solar Energy, Fuels from Biomass Program. (In ENERGY PRIMER see Book Reviews, pg. 205).

(3) Solar Greenhouses—two solar greenhouses are presently being completed: The Parabolic/ Aquaculture Greenhouse (designed and constructed by Bear Creek Thunder, Ashland, Oregon), and the Waterwall Greenhouse (designed by Ecotope Group for the Environmental Farm Project in King County, WA). These greenhouses will be tested for their thermal and biological operations.

For more information on our activities please send a self-addressed, stamped envelope for our publications list.

Ecotope Group
747 16th East
Seattle, WA 98112

RAIN/Ecotope
2270 N.W. Irving
Portland, OR 97210

(4) An energy conservation and renewable energy plan for the State of Montana, commissioned by the Montana Energy Advisory Council. Future energy needs were determined by applying available conservation measures to the current rate of per capita energy consumption. These needs were assessed with respect to the renewable energy resources in Montana available from the sun, wind, water and biomass. The methods used in this study were similar to those developed by Amory Lovins in which the energy needs of a culture are better described by end uses than by only considering the present consumption of energy resources.

Over 80% of the U.S. population lives in urban areas . . . from the inner city to suburbia. Here is where the benefits of energy conservation and renewable energy systems will have the largest impact. However, there are only a few groups working with urban applications of renewable energy systems. The amount of land is usually small, health and sewage codes tend to discourage innovations, and wind currents and sunlight are often blocked by buildings and made unavailable . . . especially to people living in apartments and tenement buildings. So we hastily make the assumption that solar energy and other self-reliant technologies are better suited to more rural areas . . . right? Well read on!

— R.M.

INSTITUTE FOR LOCAL SELF-RELIANCE

The Institute for Local Self-Reliance (ILSR) was established to investigate the technical feasibility of community self-reliance in high density living areas and to examine the implications of such decentralization. ILSR focuses its activities on redefining urban neighborhoods and cities as self-reliance, productive systems; teaching people the tools of self-reliance; and investigating the legal powers for municipal decentralization. Currently ILSR is working on several major projects:

(1) Urban Food Systems: Experimenting with and demonstrating an integrated neighborhood food/waste/energy system. This includes rooftop hydroponic vegetable production, biodynamic/French Intensive gardening, commercial sprout production, large-scale composting, development of community gardens, and organizing for expansion of Washington's non-profit food distribution system and its diversification into food production and processing.

(2) Energy Generation and Use: Analyzing the relative costs for development of solar versus nuclear generated electricity and preparing a proposal for a city, or a consortium of cities, to assist the practical development of economic solar electricity.

(3) Neighborhood Economics: Studying the effects of credit allocation on an urban neighborhood and evaluating the feasibility and implications of community controlled credit institutions.

(4) Waste Utilization: Evaluating the economic costs of municipal sewerage and solid waste treatment in comparison with the decentralized alternatives. ILSR has studied the economics and feasibility of waterless toilets, in-house sewage treatment, community sewerage systems, decentralized solid waste collection and processing facilities, organic garbage composting and neighborhood recycling center operation including a system for household pickup of source-separated waste.

(5) Communitas: Institute members serve as resource staff for students enrolled in Communitas, an upper level college program in the University Without Walls unit of Roger Williams College, Providence, R.I. The program has affiliate status for accreditation from the Northern Central Accrediting Association and awards B.S. and B.A. degrees in fields related to urban and community development.

ILSR has a number of research reports and handy charts describing these activities. They also publish an excellent bi-monthly magazine, SELF-RELIANCE which describes on-going projects of the Institute, and reports on decentralist trends in energy, finance, agriculture, housing, manufacturing, waste, and city planning throughout the country. Subscriptions: $6.00 per year (individuals); $12.00 per year (institutions).

Institute for Local Self-Reliance
1717 18th St., N.W.
Washington, DC 20009

ENERGY TASK FORCE

In the Spring of 1976, a group of tenants and energy designers gathered at 510 East 11th St. in New York City to discuss designing and installing the first urban wind generator. It would produce electricity for their tenement building which had recently been rehabilitated and weatherproofed with funds from the Community Services Administration (CSA). Consolidated Edison had just cut power to the building over a bill dispute, and shutting off certain appliances jeopardized the tenants' welfare. A group of five designers and architects, known as the ENERGY TASK FORCE, rallied to meet the design and bureaucratic challenges of such an unusual project as a wind generator atop a 5-story tenement building in New York City.

The TASK FORCE considered decentralized wind generators as a way to assist the low and fixed income residents of the building in lifting the economic pressures of electricity price escalations. This looked attractive since Con Ed's rates are the highest in the country (10¢ per Kwh). People have only considered themselves energy consumers, but wind and solar energy technologies create the opportunity for them to become decentralized energy producers.

The wind generator project proceeded with the tenants themselves installing the 37-foot tower and the wind energy conversion system. The do-it-yourself philosophy and experience of self-help housing was carried over to an "appropriate technology" design, which the tenants themselves could carry out without expensive installation equipment and outside labor. By performing the work themselves, they have learned new skills, reduced costs, learned the system's operation and maintenance requirements, and provided themselves a way to reduce their common electric bill.

By far the most significant aspect of the 11th St. project has been the creation of an urban solar energy model for others to follow, and the establishment of revised tarriff rates with a major utility company. Excess electric power produced by the wind generator is not stored in expensive battery arrays but is put back into the transmission lines by means of a synchronous

inverter (see pg. 147). This allows a separate electric meter to register backfeed (surplus) wind energy, thus creating a credit from, rather than a debt to, the utility company. Subsequent court cases produced important legal precedents in New York that:

(1) Established the right to comingle power production, i.e. to produce local electric power without separate transmission lines.

(2) Established the right to back feed surplus electric power into established grid systems and get credited for it (at average fuel costs).

If carried to every state, the legal comingling and backfeeding of locally-produced electric power could help pave the way to a truly solar-based society.

The ENERGY TASK FORCE is willing to share their experiences with any low-income, self-help housing group (please no individual requests). Send self-addressed stamped envelope to:

Energy Task Force
156 Fifth Ave.
New York, NY 10009

The ENERGY TASK FORCE has produced two important pamphlets: *Windmill Power for City People* and *No Heat No Rent: An Urban Solar and Energy Conservation Manual*. Both of these pamphlets are published by the Community Services Administration.

PERIODICALS

With things happening so fast, it is impossible to stay on top of the latest developments in renewable energy systems and all of their cultural and technical spinoffs. The following periodicals will help you more than most. Many of the groups listed on pages 244-248 have periodicals containing valuable information (especially New Alchemy Institute, AERO and Institute for Local Self-Reliance). Also see the listing of: Solar Periodicals (pg. 66), Wind Periodicals (pg. 140) and Agriculture Periodicals (pg. 167).

— R.M.

Alternative Sources of Energy
Bi-monthly, $10.00 per year

from:
Alternative Sources of Energy
Route 2, Milaca, MN 56353

Started in 1971, this periodical has grown in size and quality with every succeeding issue. It is the best available source for continuing information on all aspects of renewable energy systems. Access information and good articles.

RAIN: Journal of Appropriate Technology
Monthly, $10.00 per year

from:
RAIN 2270 N.W. Irving
Portland, OR 97210

Without a doubt this is the finest access periodical in the country. The RAIN staff have a genius for clumping together various references related to solar energy, agriculture, community organizing etc. and binding them together with a well written technical or philosophical perspective. Their information sources and articles are up-to-date and accurate. RAIN was a constant resource for the ENERGY PRIMER. Highly recommended by all of us. The First 2½ years of RAIN (plus more) are assembled in their RAINBOOK (see pg. 242).

Coevolution Quarterly
Quarterly, $8.00 per year

from:
Coevolution Quarterly
Box 428
Sausalito, CA 94965

COEVOLUTION QUARTERLY (CQ) is a spinoff from the WHOLE EARTH CATALOG AND EPILOG. A very unique periodical, CQ has excellent feature articles on everything from Zen meditation, space colonies, and politics, to energy systems, crafts, appropriate technology and ecological (co-evolutionary) theory. CQ occasionally becomes too esoteric, and restricted to a clique of intellectuals. However the open forum of the magazine, the caliber of its contributing writers, and its ready acceptance of reader commentary, help relieve the closed circle feeling. More importantly, the "Soft Technology" section of each issue contains valuable information on energy books and hardware as well as pertinent articles.

Not Man Apart
Semi-monthly, $10.00 per year
$20.00 per year membership

from:
Friends of the Earth
529 Commercial
San Francisco, CA

This is a fine newspaper from Friends of the Earth, a world-wide environmental activists organization. NOT MAN APART offers several updates of environmental struggles (including solar energy activities) plus one or more in depth analyses each issue. Most valuable is the continuing critique ("Nuclear Blowdown") of the hazards and status of nuclear power, generally researched from the industry's own sources.

Synerjy
Semi-annual, $7.00 per year

from:
Synerjy
P.O. Box 4790
Grand Central Station, New York 10017

SYNERJY is a directory of energy alternatives. The January 1977 issue contains over 1200 worldwide entries . . . articles, books and government reports published in English; alternative energy conferences, research groups; manufacturers and facilities in fields of: Solar Energy, Geothermal Heat Transfer, Methane, Wood, Electrical Energy, Water Power, Wind Power and Energy Storage. Synerjy is a valuable resource, but it badly needs a more sophisiticated breakdown of subject matter and a better bibliographic layout for easier access.

The Elements
Monthly, $7.00 per year

from:
The Elements
1901 Q. St.
Washington, D.C. 20009

Most economists study the relationships of labor and capital and forget the study of natural resources . . . especially land. THE ELEMENTS is probably the most informative, up-to-date source of information on natural resources: uranium, fossil fuels, metals, agricultural produce etc. Here is a supplement ot the daily newspaper that will tell you the economic reasons for many political maneuvers and the importance that natural resources play in global politics.

Bulletin of Atomic Scientists
Monthly, $10.00 per year

from:
Bulletin of Atomic Scientists
1020-24 E. 58th St.
Chicago, IL 60637

Despite its title, BULLETIN OF ATOMIC SCIENTISTS often runs beautifully researched articles on Solar, Wind and Biofuel Energy systems, plus critical analyses of the nuclear power industry.

Undercurrents
Bi-monthly, $ 7.50 per year, surface

from:
Undercurrents
12 South St.
Uley, Dursley, Glos.
England

UNDERCURRENTS is a British journal of alternative energy, appropriate technology and decentralist politics. Articles include the technical and social implications of the environmental movement in England (where they live closer to the edge). Useful for its view from abroad.

REGIONAL PERIODICALS

Seriatim: Journal of Ecotopia
Quarterly, $9.00 per year

from:
Seriatim
122 Camel
El Cerrito, CA 94530

Important new journal for the Northwest US. Fine articles discuss software systems (trade networks, economic alternatives, strategies for personal and social change) and hardware systems (renewable energy applications, agri/aquaculture methods, recycling technologies). Definitely an emerging model for regional publications.

High Country News
Monthly, $10.00 per year

from:
High Country News
140 N. 7th St.
Lander, WY 82520

The RAINBOOK calls this: "A unique blend of environmental issues affecting Idaho, Montana, Wyoming, Utah and Colorado, and sensitivity to defining an area's common concerns, traditions, unique life support systems."

The Workbook
Monthly, $10.00 per year (individual)
$20.00 per year

from:
Southwest Research and Information Center
P.O. Box 4524
Albuquerque, NM 87106

A fully-indexed and annotated catalog of information about environmental, social and consumer problems. THE WORKBOOK is aimed at helping people in small towns and cities across America gain access to vital information that can help them assert control over their own lives. Much of the information seems specially applicable to the Southwest US.

Maine Times
Weekly, $12.00 per year

from:
Maine Times
41 Main St.
Topsham, ME 04086

Many articles and ads reflect the application of solar, wind, wood and agriculture technologies in the Northeast US.

Mother Earth News
Monthly, $10.00 per year

from:
Mother Earth News
P.O. Box 70
Hendersonville, NC 28739

For several years now MOTHER EARTH NEWS has enlightened city-dwellers to the potentials and advantages of more self-reliant life styles. Each issue contains articles on small-scale renewable energy devices and systems. There are no detailed designs here, only sketches of possibilities and newsbriefs of current events in the field of "alternative" energy.

MISCELLANEOUS SCIENCE PERIODICALS

It is easy (but not cheap) to stay current in the area of renewable energy and environmental issues by reading from a wide array of science periodicals. Often these include review and research articles on renewable energy systems. Check your local library.

New Scientist
Weekly, $44.30 per year
128 Long Acre
London, England WC2E 9QH

The Ecologist
Bi-monthly $14.50 per year
73 Molesworth St.
Wadebridge, Cornwall, England PL27 7DS

Nature
3 weekly editions, $108.00 per year
Macmilan Journals, Ltd.
Little Essex St.
London WC 2R 3LF, England

Science
Weekly, $50.00 per year (membership)
American Association for the Advancement of Science
1515 Massachusetts Ave., N.W.
Washington, D.C. 20005

Scientific American
Monthly, $ 18 per year
415 Madison Ave.
New York, NY 10017

Bioscience
Bimonthly through membership to AIBS
$32.00 per year, membership
American Institute of Biological Sciences
3900 Wisconsin Ave.
Washington, D.C. 20016

Environment
Monthly, $15.00 per year
Environment
438 N. Skinker Boulevard
St. Louis, MO 63130

Popular Science
Monthly, $7.94 per year
Popular Science
Boulder CO 80302

Technology Review
8 times per year, $15.00 per year
Technology Review
Room E19-430
Massachusetts Institute of Technology
Cambridge, MA 02139

CONVERSION FACTORS

Because different conventions historically have been used to measure various quantities, the following tables have been compiled to sort out the different units. The first table identifies the units typically used for describing a particular quantity. For example, speed might be measured in "miles/hour".

Most quantities can also be described in terms of the following three basic dimensions:

$$\text{length} \quad L$$
$$\text{mass} \quad M$$
$$\text{time} \quad T$$

For example, speed is given in terms of length divided by time, which can be written as "L/T". This description, called "dimensional analysis", is useful in determining whether an equation is correct. The product of the dimensions on each side of the equal sign must match. For example:

$$\text{Distance} = \text{Speed} \times \text{Time}$$
$$L = L/T \times T$$

The dimension on the left side of the equal sign is length, L. On the right side of the equal sign, the product of L/T times T is L, which matches the left side of the equation.

The second table is a Conversion Table, showing how to convert from one set of units to another. It might be necessary to take the reciprocal of the conversion factor or to make more than one conversion to get the desired results. The following handbooks might be referred to for futher information:

CRC HANDBOOK OF CHEMISTRY AND PHYSICS, 55TH ED. R.C. Weast, ed. CRC Press. Cleveland. 1974.

PHYSICS: PARTS I AND II. D. Halliday and R. Resnick. John Wiley and Sons. New York. 1966.

IES LIGHTING HANDBOOK, 4TH ED. J.E. Kaufman, ed. Illuminating Engineering Society. New York. 1966.

STANDARD HANDBOOK FOR MECHANICAL ENGINEERS, 7TH ED. Baumeister and Marks. McGraw-Hill. New York. 1967.

MEASURED QUANTITIES AND THEIR COMMON UNITS

Length(L)	Area(L^2)	Volume (L^3)
mile(mi.)	sq. mile(mi^2)	gallon(gal.)
yard(yd.)	sq. yard	quart(qt.)
foot(ft.)	sq. foot	pint(pt.)
inch(in.)	sq. inch	ounce(oz.)
fathom(fath.)	acre	cu. foot(ft^3)
kilometer(km.)	sq. kilometer	cu. yard
meter(m.)	sq. meter	cu. inch
centimeter(cm.)	sq. centimeter	liter
micron(u).		cu. centimeter
angstrom(Å.)		acre-foot
		cord
		cord-foot
		barrel(bbl.)

Mass(M)	Speed(L/T)	Flow Rate(L^3/T)
pound(lb.)	feet/minute(ft./min.)	cu. feet/min.
ton(short)	feet/sec.	cu. meter/min.
ton(long)	mile/hour	liters/sec.
ton(metric)	mile/min.	gallons/min.
gram(g.)	kilometer/hr	gallons/sec.
kilogram(kg.)	kilometer/min.	
	kilometer/sec.	

Pressure($M/L/T^2$)	Energy(ML^2/T^2)	Power(ML^2/T^3)
atmosphere(atm.)	British thermal unit(Btu.)	Btu./min.
pounds/sq. inch	calories(cal.)	Btu./hour
inches of mercury	foot-pound	watt
cm. of mercury	joule	joule/sec.
feet of water	kilowatt-hour(kw-hr.)	cal./min.
	horsepower-hour (hp.-hr.)	horsepower(hp.)

Time(T)	Energy Density(M/T^2)	Power Density(M/T^3)
year	calories/sq. cm.	cal./sq. cm./min.
month	Btu./sq. foot	Btu./sq. foot/hr
day	langley	langley/min.
hour(hr.)	watthr./sq. foot	watt/sq. cm.
minute(min.)		
second(sec.)		

TABLE OF CONVERSION FACTORS

MULTIPLY	BY	TO OBTAIN:
Acres	43560	Sq. feet
"	0.004047	Sq. kilometers
"	4047	Sq. meters
"	0.0015625	Sq. miles
"	4840	Sq. yards
Acre-feet	43560	Cu. feet
"	1233.5	Cu. meters
"	1613.3	Cu. yards
Angstroms (Å)	1x 10^{-8}	Centimeters
"	3.937x 10^{-9}	Inches
"	0.0001	Microns
Atmospheres(atm.)	76	Cm. of Hg(0^0C)
"	1033.3	Cm. of H_2O (4°C)
"	33.8995	Ft. of H_2O (39.2°F)
"	29.92	In. of Hg(32^0F)
"	14.696	Pounds/sq. inch
Barrels(petroleum, U.S.) (bbl.)	5.6146	Cu. feet
"	35	Gallons(Imperial)
"	42	Gallons(U.S.)
"	158.98	Liters
British Thermal Unit(Btu)	251.99	Calories, gm
"	777.649	Foot-pounds
"	0.00039275	Horsepower-hours
"	1054.35	Joules

MULTIPLY	BY	TO OBTAIN:
BTU	0.000292875	Kilowatt-hours
"	1054.35	Watt-seconds
Btu/hr.	4.2	Calories/min.
"	777.65	Foot-pounds/hr.
"	0.0003927	Horsepower
"	0.000292875	Kilowatts
"	0.292875	Watts (or joule/sec.)
Btu/lb.	7.25 x 10^{-4}	Cal/gram
Btu/sq. ft	0.271246	Calories/sq. cm. (or langleys)
"	0.292875	Watt-hour/sq. foot
Btu/sq. ft./hour	3.15 x 10^{-7}	Kilowatts/sq. meter
"	4.51 x 10^{-3}	Cal./sq. cm./min.(or langleys/min)
"	3.15 x 10^{-8}	Watts/sq. cm.
Calories(cal.)	0.003968	Btu.
"	3.08596	Foot-pounds
"	1.55857 x 10^{-6}	Horsepower-hours
"	4.184	Joules(or watt-seconds)
"	1.1622 x 10^{-6}	Kilowatt-hours
Calories, food unit (Cal.)	1000	Calories, gm
Calories/min.	0.003968	Btu/min.
"	0.06973	Watts
Calories/sq. cm.	3.68669	Btu/sq. ft.
"	1.0797	Watt-hr./sq. foot
Cal./sq. cm./min.	796320.	Btu/sq. foot/hr.
"	251.04	Watts/sq. cm.
Candle power (spherical)	12.566	Lumens
Centimeters(cm.)	0.032808	Feet
"	0.3937	Inches
"	0.01	Meters
"	10,000	Microns
Cm. of Hg(0^0C)	0.0131579	Atmospheres
"	0.44605	Ft. of H_2O(4^0C)
"	0.19337	Pounds/sq. inch
Cm. of H_2O(4^0C)	0.0009678	Atmospheres
"	0.01422	Pounds/sq. inch
Cm./sec.	0.032808	Feet/sec.
"	0.022369	Miles/hr.
Cords	8	Cord-feet
"	128(or 4x4x8)	Cu. feet
Cu. centimeters	3.5314667x 10^{-5}	Cu. feet
"	0.06102	Cu. inches
"	1x 10^{-6}	Cu. meters
"	0.001	Liters
"	0.0338	Ounces(U.S., fluid)
Cu. feet($ft.^3$)	0.02831685	Cu. meters
"	7.4805	Gallons(U.S., liq.)
"	28.31685	Liters
"	29.922	Quarts(U.S., liq.)
Cu. ft of H_2O (60^0F)	62.366	Pounds of H_2O
Cu. feet/min.	471.947	Cu. cm./sec.
Cu. inches(in.^{3})	16.387	Cu. cm.
"	0.0005787	Cu. feet
"	0.004329	Gallons(U.S., liq.)
"	0.5541	Ounces(U.S., fluid)
Cu. meters	1x 10^6	Cu. centimeters
"	35.314667	Cu. feet
"	264.172	Gallons(U.S., liq.)
"	1000	Liters
Cu. yard	0.76455	Cu. meters
"	201.97	Gallons(U.S., liq.)
"	27	Cu. feet
Cubits	18	Inches
Fathoms	6	Feet
"	1.8288	Meters
Feet(ft.)	30.48	Centimeters
"	12	Inches
"	0.00018939	Miles(statute)
Feet of H_2O(4^0C)	0.029499	Atmospheres
"	2.2419	Cm. of Hg(0^0C)
"	0.433515	Pounds/sq. inch

MULTIPLY	BY	TO OBTAIN:
Feet/min.	0.508	Centimeters/second
"	0.018288	Kilometers/hr.
"	0.0113636	Miles/hr.
Foot-candles	1	Lumens/sq. foot
Foot-pounds	0.001285	Btu.
"	0.324048	Calories
"	5.0505x 10^{-7}	Horsepower-hours
"	3.76616x 10^{-7}	Kilowatt-hours
Furlong	220	Yards
Gallons(U.S., dry)	1.163647	Gallons(U.S., liq.)
Gallons(U.S., liq.)	3785.4	Cu. centimeters
"	0.13368	Cu. feet
"	231	Cu. inches
"	0.0037854	Cu. meters
"	3.7854	Liters
"	8	Pints(U.S., liq.)
"	4	Quarts(U.S., liq.)
Gallons/min.	2.228x 10^{-3}	Cu. feet/sec.
"	0.06308	Liters/sec.
Grams	0.035274	Ounces(avdp.)
"	0.002205	Pounds(avdp.)
Grams-cm.	9.3011x 10^{-8}	Btu.
Grams/meter^2	4.46 x 10^{-3}	Short ton/acre
Grams/meter^2	8.93	lbs./acre
Horsepower	42.4356	Btu./min.
"	550	Foot-pounds/sec.
"	745.7	Watts
Horsepower-hrs.	2546.14	Btu.
"	641616	Calories
"	1.98x 10^6	Foot-pounds
"	0.7457	Kilowatt-hours
Inches	2.54	Centimeters
"	.083333	Feet
In. of Hg (32°F)	0.03342	Atmospheres
"	1.133	Feet of H_2O
"	0.4912	Pounds/sq. inch
In.of Water (4°C)	0.002458	Atmospheres
"	0.07355	In. of Mercury (32°F)
"	0.03613	Pounds/sq. inch
Joules	0.0009485	Btu.
"	0.73756	Foot-pounds
"	0.0002778	Watt-hours
"	1	Watt-sec.
Kilo calories/gram	1799.5	Btu/lb
Kilograms	2.2046	Pounds(avdp.)
Kilometers	1000	Meters
"	0.62137	Miles(statute)
Kilometer/hr.	54.68	Feet/min.
Kilograms/hectare	.893	lbs/acre
Kilograms/hectare	.0004465	Short ton/acre
Kilowatts	3414.43	Btu./hr.
"	737.56	Foot-pounds/sec.
"	1.34102	Horsepower
Kilowatt-hours	3414.43	Btu.
"	1.34102	Horsepower-hours
Knots	51.44	Centimeter/sec.
"	1	Mile(nautical)/hr.
Langleys	1	Calories/sq. cm.
"	1.15078	Miles(Statute)/hr.
Langleys	1	Calories/sq. cm.
Liters	1000	Cu. centimeters
"	0.0353	Cu. feet
"	0.2642	Gallons(U.S., liq.)
"	1.0567	Quarts(U.S., liq.)
Lbs./acre	.0005	Short ton/acre
Liters/min.	0.0353	Cu. feet/min.
"	0.2642	Gallons(U.S., liq.)/min.
Lumens	0.079577	Candle power(spherical)
Lumens(at 5550 Å)	0.0014706	Watts
Meters	3.2808	Feet
"	39.37	Inches
"	1.0936	Yards
Metric ton/hectare	.446	Short ton/per acre

MULTIPLY	BY	TO OBTAIN:
Micron	10000	Angstroms
"	0.0001	Centimeters
Miles(statute)	5280	Feet
"	1.6093	Kilometers
"	1760	Yards
Miles/hour	44.704	Centimeter/sec.
"	88	Feet/min.
"	1.6093	Kilometer/hr
Milliliter	1	Cu. centimeter
Millimeter	0.1	Centimeter
Ounces(avdp.)	0.0625	Pounds(avdp.)
Ounces(U.S., liq.)	29.57	Cu. centimeters
"	1.8047	Cu. inches
"	0.0625(or 1/16)	Pint(U.S., liq.)
Pints(U.S., liq.)	473.18	Cu. centimeters
"	28.875	Cu. inches
"	0.5	Quarts(U.S., liq.)
Pounds(avdp.)	0.45359	Kilograms
"	16	Ounces(avdp.)
Pounds of water	0.01602	Cu. feet of water
"	0.1198	Gallons(U.S., liq.)

MULTIPLY	BY	TO OBTAIN:
Pounds/sq. inch	0.06805	Atmospheres
"	5.1715	Cm. of mercury($0°C$)
"	27.6807	In. of water($39.2°F$)
Quarts(U.S., liq.)	0.25	Gallons(U.S., liq.)
"	0.9463	Liters
"	32	Ounces(U.S., liq.)
"	2	Pints(U.S., liq.)
Radians	57.30	Degrees
Sq. centimeters	0.0010764	Sq. feet
"	0.1550	Sq. inches
Sq. feet	2.2957×10^{-5}	Acres
"	0.09290	Sq. meters
Sq. inches	6.4516	Sq. centimeters
"	0.006944	Sq. feet
Sq. kilometers	247.1	Acres
"	1.0764×10^7	Sq. feet
"	0.3861	Sq. miles
Sq. meters	10.7639	Sq. feet
"	1.196	Sq. yards
Sq. miles	640	Acres
"	2.7878×10^7	Sq. feet
"	2.590	Sq. kilometers

MULTIPLY	BY	TO OBTAIN:
Sq. yards	9(or 3x3)	Sq. feet
"	0.83613	Sq. meters
Mile/hour	.447	Meters/second
Meters/sec.	2.24	Mile/hour
Tons,long	1016	Kilograms
"	2240	Pounds(avdp.)
Tons, metric	1000	Kilograms
"	2204.6	Pounds(avdp.)
Tons, short	907.2	Kilograms
"	2000	Pounds(avdp.)
Watts	3.4144	Btu./hr.
"	0.05691	Btu./min.
"	14.34	Calories/min.
"	0.001341	Horsepower
"	1	Joule/sec.
Watts/sq. cm.	3172	Btu./sq. foot/hr.
Watt-hours	3.4144	Btu.
"	860.4	Calories
"	0.001341	Horsepower-hours
Yards	3	Feet
"	0.9144	Meters

INDEX

This type face refers to a subject listing,
this type face refers to a book or periodical, and
this type face refers to an individual, group, or company.